ger.com/series/7412

on about this subseries at http://www.sprin

Adrien Bartoli · Andrea Fusiello (Eds.)

Computer Vision – ECCV 2020 Workshops

Glasgow, UK, August 23–28, 2020
Proceedings, Part V

Editors
Adrien Bartoli
University of Clermont Auvergne
Clermont Ferrand, France

Andrea Fusiello
Università degli Studi di Udine
Udine, Italy

ISSN 0302-9743 ISSN 1611-3349 (electronic)
Lecture Notes in Computer Science
ISBN 978-3-030-68237-8 ISBN 978-3-030-68238-5 (eBook)
https://doi.org/10.1007/978-3-030-68238-5

LNCS Sublibrary: SL6 – Image Processing, Computer Vision, Pattern Recognition, and Graphics

This Springer imprint is published by the registered company Springer Nature Switzerland AG
The registered company address is: Gewerbestrasse 11, 6330 Cham, Switzerland

Foreword

Hosting the 2020 European Conference on Computer Vision was certainly an exciting journey. From the 2016 plan to hold it at the Edinburgh International Conference Centre (hosting 1,800 delegates) to the 2018 plan to hold it at Glasgow's Scottish Exhibition Centre (up to 6,000 delegates), we finally ended with moving online because of the COVID-19 outbreak. While possibly having fewer delegates than expected because of the online format, ECCV 2020 still had over 3,100 registered participants.

Although online, the conference delivered most of the activities expected at a face-to-face conference: peer-reviewed papers, industrial exhibitors, demonstrations, and messaging between delegates. As well as the main technical sessions, the conference included a strong program of satellite events, including 16 tutorials and 44 workshops.

On the other hand, the online conference format enabled new conference features. Every paper had an associated teaser video and a longer full presentation video. Along with the papers and slides from the videos, all these materials were available the week before the conference. This allowed delegates to become familiar with the paper content and be ready for the live interaction with the authors during the conference week. The 'live' event consisted of brief presentations by the 'oral' and 'spotlight' authors and industrial sponsors. Question and Answer sessions for all papers were timed to occur twice so delegates from around the world had convenient access to the authors.

As with the 2018 ECCV, authors' draft versions of the papers appeared online with open access, now on both the Computer Vision Foundation (CVF) and the European Computer Vision Association (ECVA) websites. An archival publication arrangement was put in place with the cooperation of Springer. SpringerLink hosts the final version of the papers with further improvements, such as activating reference links and supplementary materials. These two approaches benefit all potential readers: a version available freely for all researchers, and an authoritative and citable version with additional benefits for SpringerLink subscribers. We thank Alfred Hofmann and Aliaksandr Birukou from Springer for helping to negotiate this agreement, which we expect will continue for future versions of ECCV.

August 2020

Vittorio Ferrari
Bob Fisher
Cordelia Schmid
Emanuele Trucco

Preface

Welcome to the workshops proceedings of the 16th European Conference on Computer Vision (ECCV 2020), the first edition held online. We are delighted that the main ECCV 2020 was accompanied by 45 workshops, scheduled on August 23, 2020, and August 28, 2020.

We received 101 valid workshop proposals on diverse computer vision topics and had space for 32 full-day slots, so we had to decline many valuable proposals (the workshops were supposed to be either full-day or half-day long, but the distinction faded away when the full ECCV conference went online). We endeavored to balance among topics, established series, and newcomers. Not all the workshops published their proceedings, or had proceedings at all. These volumes collect the edited papers from 28 out of 45 workshops.

We sincerely thank the ECCV general chairs for trusting us with the responsibility for the workshops, the workshop organizers for their involvement in this event of primary importance in our field, and the workshop presenters and authors.

August 2020

Adrien Bartoli
Andrea Fusiello

Preface

Organization

General Chairs

Vittorio Ferrari	Google Research, Switzerland
Bob Fisher	The University of Edinburgh, UK
Cordelia Schmid	Google and Inria, France
Emanuele Trucco	The University of Dundee, UK

Program Chairs

Andrea Vedaldi	University of Oxford, UK
Horst Bischof	Graz University of Technology, Austria
Thomas Brox	University of Freiburg, Germany
Jan-Michael Frahm	The University of North Carolina at Chapel Hill, USA

Industrial Liaison Chairs

Jim Ashe	The University of Edinburgh, UK
Helmut Grabner	Zurich University of Applied Sciences, Switzerland
Diane Larlus	NAVER LABS Europe, France
Cristian Novotny	The University of Edinburgh, UK

Local Arrangement Chairs

Yvan Petillot	Heriot-Watt University, UK
Paul Siebert	The University of Glasgow, UK

Academic Demonstration Chair

Thomas Mensink	Google Research and University of Amsterdam, The Netherlands

Poster Chair

Stephen Mckenna	The University of Dundee, UK

Technology Chair

Gerardo Aragon Camarasa	The University of Glasgow, UK

Tutorial Chairs

Carlo Colombo	University of Florence, Italy
Sotirios Tsaftaris	The University of Edinburgh, UK

Publication Chairs

Albert Ali Salah	Utrecht University, The Netherlands
Hamdi Dibeklioglu	Bilkent University, Turkey
Metehan Doyran	Utrecht University, The Netherlands
Henry Howard-Jenkins	University of Oxford, UK
Victor Adrian Prisacariu	University of Oxford, UK
Siyu Tang	ETH Zurich, Switzerland
Gul Varol	University of Oxford, UK

Website Chair

Giovanni Maria Farinella	University of Catania, Italy

Workshops Chairs

Adrien Bartoli	University Clermont Auvergne, France
Andrea Fusiello	University of Udine, Italy

Workshops Organizers

W01 - Adversarial Robustness in the Real World

Adam Kortylewski	Johns Hopkins University, USA
Cihang Xie	Johns Hopkins University, USA
Song Bai	University of Oxford, UK
Zhaowei Cai	UC San Diego, USA
Yingwei Li	Johns Hopkins University, USA
Andrei Barbu	MIT, USA
Wieland Brendel	University of Tübingen, Germany
Nuno Vasconcelos	UC San Diego, USA
Andrea Vedaldi	University of Oxford, UK
Philip H. S. Torr	University of Oxford, UK
Rama Chellappa	University of Maryland, USA
Alan Yuille	Johns Hopkins University, USA

W02 - BioImage Computation

Jan Funke	HHMI Janelia Research Campus, Germany
Dagmar Kainmueller	BIH and MDC Berlin, Germany
Florian Jug	CSBD and MPI-CBG, Germany
Anna Kreshuk	EMBL Heidelberg, Germany

Peter Bajcsy	NIST, USA
Martin Weigert	EPFL, Switzerland
Patrick Bouthemy	Inria, France
Erik Meijering	University New South Wales, Australia

W03 - Egocentric Perception, Interaction and Computing

Michael Wray	University of Bristol, UK
Dima Damen	University of Bristol, UK
Hazel Doughty	University of Bristol, UK
Walterio Mayol-Cuevas	University of Bristol, UK
David Crandall	Indiana University, USA
Kristen Grauman	UT Austin, USA
Giovanni Maria Farinella	University of Catania, Italy
Antonino Furnari	University of Catania, Italy

W04 - Embodied Vision, Actions and Language

Yonatan Bisk	Carnegie Mellon University, USA
Jesse Thomason	University of Washington, USA
Mohit Shridhar	University of Washington, USA
Chris Paxton	NVIDIA, USA
Peter Anderson	Georgia Tech, USA
Roozbeh Mottaghi	Allen Institute for AI, USA
Eric Kolve	Allen Institute for AI, USA

W05 - Eye Gaze in VR, AR, and in the Wild

Hyung Jin Chang	University of Birmingham, UK
Seonwook Park	ETH Zurich, Switzerland
Xucong Zhang	ETH Zurich, Switzerland
Otmar Hilliges	ETH Zurich, Switzerland
Aleš Leonardis	University of Birmingham, UK
Robert Cavin	Facebook Reality Labs, USA
Cristina Palmero	University of Barcelona, Spain
Jixu Chen	Facebook, USA
Alexander Fix	Facebook Reality Labs, USA
Elias Guestrin	Facebook Reality Labs, USA
Oleg Komogortsev	Texas State University, USA
Kapil Krishnakumar	Facebook, USA
Abhishek Sharma	Facebook Reality Labs, USA
Yiru Shen	Facebook Reality Labs, USA
Tarek Hefny	Facebook Reality Labs, USA
Karsten Behrendt	Facebook, USA
Sachin S. Talathi	Facebook Reality Labs, USA

W06 - Holistic Scene Structures for 3D Vision

Zihan Zhou	Penn State University, USA
Yasutaka Furukawa	Simon Fraser University, Canada
Yi Ma	UC Berkeley, USA
Shenghua Gao	ShanghaiTech University, China
Chen Liu	Facebook Reality Labs, USA
Yichao Zhou	UC Berkeley, USA
Linjie Luo	Bytedance Inc., China
Jia Zheng	ShanghaiTech University, China
Junfei Zhang	Kujiale.com, China
Rui Tang	Kujiale.com, China

W07 - Joint COCO and LVIS Recognition Challenge

Alexander Kirillov	Facebook AI Research, USA
Tsung-Yi Lin	Google Research, USA
Yin Cui	Google Research, USA
Matteo Ruggero Ronchi	California Institute of Technology, USA
Agrim Gupta	Stanford University, USA
Ross Girshick	Facebook AI Research, USA
Piotr Dollar	Facebook AI Research, USA

W08 - Object Tracking and Its Many Guises

Achal D. Dave	Carnegie Mellon University, USA
Tarasha Khurana	Carnegie Mellon University, USA
Jonathon Luiten	RWTH Aachen University, Germany
Aljosa Osep	Technical University of Munich, Germany
Pavel Tokmakov	Carnegie Mellon University, USA

W09 - Perception for Autonomous Driving

Li Erran Li	Alexa AI, Amazon, USA
Adrien Gaidon	Toyota Research Institute, USA
Wei-Lun Chao	The Ohio State University, USA
Peter Ondruska	Lyft, UK
Rowan McAllister	UC Berkeley, USA
Larry Jackel	North-C Technologies, USA
Jose M. Alvarez	NVIDIA, USA

W10 - TASK-CV Workshop and VisDA Challenge

Tatiana Tommasi	Politecnico di Torino, Italy
Antonio M. Lopez	CVC and UAB, Spain
David Vazquez	Element AI, Canada
Gabriela Csurka	NAVER LABS Europe, France
Kate Saenko	Boston University, USA
Liang Zheng	The Australian National University, Australia

Xingchao Peng	Boston University, USA
Weijian Deng	The Australian National University, Australia

W11 - Bodily Expressed Emotion Understanding

James Z. Wang	Penn State University, USA
Reginald B. Adams, Jr.	Penn State University, USA
Yelin Kim	Amazon Lab126, USA

W12 - Commands 4 Autonomous Vehicles

Thierry Deruyttere	KU Leuven, Belgium
Simon Vandenhende	KU Leuven, Belgium
Luc Van Gool	KU Leuven, Belgium, and ETH Zurich, Switzerland
Matthew Blaschko	KU Leuven, Belgium
Tinne Tuytelaars	KU Leuven, Belgium
Marie-Francine Moens	KU Leuven, Belgium
Yu Liu	KU Leuven, Belgium
Dusan Grujicic	KU Leuven, Belgium

W13 - Computer VISion for ART Analysis

Alessio Del Bue	Istituto Italiano di Tecnologia, Italy
Sebastiano Vascon	Ca' Foscari University and European Centre for Living Technology, Italy
Peter Bell	Friedrich-Alexander University Erlangen-Nürnberg, Germany
Leonardo L. Impett	EPFL, Switzerland
Stuart James	Istituto Italiano di Tecnologia, Italy

W14 - International Challenge on Compositional and Multimodal Perception

Alec Hodgkinson	Panasonic Corporation, Japan
Yusuke Urakami	Panasonic Corporation, Japan
Kazuki Kozuka	Panasonic Corporation, Japan
Ranjay Krishna	Stanford University, USA
Olga Russakovsky	Princeton University, USA
Juan Carlos Niebles	Stanford University, USA
Jingwei Ji	Stanford University, USA
Li Fei-Fei	Stanford University, USA

W15 - Sign Language Recognition, Translation and Production

Necati Cihan Camgoz	University of Surrey, UK
Richard Bowden	University of Surrey, UK
Andrew Zisserman	University of Oxford, UK
Gul Varol	University of Oxford, UK
Samuel Albanie	University of Oxford, UK

Kearsy Cormier	University College London, UK
Neil Fox	University College London, UK

W16 - Visual Inductive Priors for Data-Efficient Deep Learning

Jan van Gemert	Delft University of Technology, The Netherlands
Robert-Jan Bruintjes	Delft University of Technology, The Netherlands
Attila Lengyel	Delft University of Technology, The Netherlands
Osman Semih Kayhan	Delft University of Technology, The Netherlands
Marcos Baptista-Ríos	Alcalá University, Spain
Anton van den Hengel	The University of Adelaide, Australia

W17 - Women in Computer Vision

Hilde Kuehne	IBM, USA
Amaia Salvador	Amazon, USA
Ananya Gupta	The University of Manchester, UK
Yana Hasson	Inria, France
Anna Kukleva	Max Planck Institute, Germany
Elizabeth Vargas	Heriot-Watt University, UK
Xin Wang	UC Berkeley, USA
Irene Amerini	Sapienza University of Rome, Italy

W18 - 3D Poses in the Wild Challenge

Gerard Pons-Moll	Max Planck Institute for Informatics, Germany
Angjoo Kanazawa	UC Berkeley, USA
Michael Black	Max Planck Institute for Intelligent Systems, Germany
Aymen Mir	Max Planck Institute for Informatics, Germany

W19 - 4D Vision

Anelia Angelova	Google, USA
Vincent Casser	Waymo, USA
Jürgen Sturm	X, USA
Noah Snavely	Google, USA
Rahul Sukthankar	Google, USA

W20 - Map-Based Localization for Autonomous Driving

Patrick Wenzel	Technical University of Munich, Germany
Niclas Zeller	Artisense, Germany
Nan Yang	Technical University of Munich, Germany
Rui Wang	Technical University of Munich, Germany
Daniel Cremers	Technical University of Munich, Germany

W21 - Multimodal Video Analysis Workshop and Moments in Time Challenge

Dhiraj Joshi	IBM Research AI, USA
Rameswar Panda	IBM Research, USA
Kandan Ramakrishnan	IBM, USA
Rogerio Feris	IBM Research AI, MIT-IBM Watson AI Lab, USA
Rami Ben-Ari	IBM-Research, USA
Danny Gutfreund	IBM, USA
Mathew Monfort	MIT, USA
Hang Zhao	MIT, USA
David Harwath	MIT, USA
Aude Oliva	MIT, USA
Zhicheng Yan	Facebook AI, USA

W22 - Recovering 6D Object Pose

Tomas Hodan	Czech Technical University in Prague, Czech Republic
Martin Sundermeyer	German Aerospace Center, Germany
Rigas Kouskouridas	Scape Technologies, UK
Tae-Kyun Kim	Imperial College London, UK
Jiri Matas	Czech Technical University in Prague, Czech Republic
Carsten Rother	Heidelberg University, Germany
Vincent Lepetit	ENPC ParisTech, France
Ales Leonardis	University of Birmingham, UK
Krzysztof Walas	Poznan University of Technology, Poland
Carsten Steger	Technical University of Munich and MVTec Software GmbH, Germany
Eric Brachmann	Heidelberg University, Germany
Bertram Drost	MVTec Software GmbH, Germany
Juil Sock	Imperial College London, UK

W23 - SHApe Recovery from Partial Textured 3D Scans

Djamila Aouada	University of Luxembourg, Luxembourg
Kseniya Cherenkova	Artec3D and University of Luxembourg, Luxembourg
Alexandre Saint	University of Luxembourg, Luxembourg
David Fofi	University Bourgogne Franche-Comté, France
Gleb Gusev	Artec3D, Luxembourg
Bjorn Ottersten	University of Luxembourg, Luxembourg

W24 - Advances in Image Manipulation Workshop and Challenges

Radu Timofte	ETH Zurich, Switzerland
Andrey Ignatov	ETH Zurich, Switzerland
Kai Zhang	ETH Zurich, Switzerland
Dario Fuoli	ETH Zurich, Switzerland
Martin Danelljan	ETH Zurich, Switzerland
Zhiwu Huang	ETH Zurich, Switzerland

Hannan Lu	Harbin Institute of Technology, China
Wangmeng Zuo	Harbin Institute of Technology, China
Shuhang Gu	The University of Sydney, Australia
Ming-Hsuan Yang	UC Merced and Google, USA
Majed El Helou	EPFL, Switzerland
Ruofan Zhou	EPFL, Switzerland
Sabine Süsstrunk	EPFL, Switzerland
Sanghyun Son	Seoul National University, South Korea
Jaerin Lee	Seoul National University, South Korea
Seungjun Nah	Seoul National University, South Korea
Kyoung Mu Lee	Seoul National University, South Korea
Eli Shechtman	Adobe, USA
Evangelos Ntavelis	ETH Zurich and CSEM, Switzerland
Andres Romero	ETH Zurich, Switzerland
Yawei Li	ETH Zurich, Switzerland
Siavash Bigdeli	CSEM, Switzerland
Pengxu Wei	Sun Yat-sen University, China
Liang Lin	Sun Yat-sen University, China
Ming-Yu Liu	NVIDIA, USA
Roey Mechrez	BeyondMinds and Technion, Israel
Luc Van Gool	KU Leuven, Belgium, and ETH Zurich, Switzerland

W25 - Assistive Computer Vision and Robotics

Marco Leo	National Research Council of Italy, Italy
Giovanni Maria Farinella	University of Catania, Italy
Antonino Furnari	University of Catania, Italy
Gerard Medioni	University of Southern California, USA
Trivedi Mohan	UC San Diego, USA

W26 - Computer Vision for UAVs Workshop and Challenge

Dawei Du	Kitware Inc., USA
Heng Fan	Stony Brook University, USA
Toon Goedemé	KU Leuven, Belgium
Qinghua Hu	Tianjin University, China
Haibin Ling	Stony Brook University, USA
Davide Scaramuzza	University of Zurich, Switzerland
Mubarak Shah	University of Central Florida, USA
Tinne Tuytelaars	KU Leuven, Belgium
Kristof Van Beeck	KU Leuven, Belgium
Longyin Wen	JD Digits, USA
Pengfei Zhu	Tianjin University, China

W27 - Embedded Vision

Tse-Wei Chen	Canon Inc., Japan
Nabil Belbachir	NORCE Norwegian Research Centre AS, Norway

Stephan Weiss	University of Klagenfurt, Austria
Marius Leordeanu	Politehnica University of Bucharest, Romania

W28 - Learning 3D Representations for Shape and Appearance

Leonidas Guibas	Stanford University, USA
Or Litany	Stanford University, USA
Tanner Schmidt	Facebook Reality Labs, USA
Vincent Sitzmann	Stanford University, USA
Srinath Sridhar	Stanford University, USA
Shubham Tulsiani	Facebook AI Research, USA
Gordon Wetzstein	Stanford University, USA

W29 - Real-World Computer Vision from inputs with Limited Quality and Tiny Object Detection Challenge

Yuqian Zhou	University of Illinois, USA
Zhenjun Han	University of the Chinese Academy of Sciences, China
Yifan Jiang	The University of Texas at Austin, USA
Yunchao Wei	University of Technology Sydney, Australia
Jian Zhao	Institute of North Electronic Equipment, Singapore
Zhangyang Wang	The University of Texas at Austin, USA
Qixiang Ye	University of the Chinese Academy of Sciences, China
Jiaying Liu	Peking University, China
Xuehui Yu	University of the Chinese Academy of Sciences, China
Ding Liu	Bytedance, China
Jie Chen	Peking University, China
Humphrey Shi	University of Oregon, USA

W30 - Robust Vision Challenge 2020

Oliver Zendel	Austrian Institute of Technology, Austria
Hassan Abu Alhaija	Interdisciplinary Center for Scientific Computing Heidelberg, Germany
Rodrigo Benenson	Google Research, Switzerland
Marius Cordts	Daimler AG, Germany
Angela Dai	Technical University of Munich, Germany
Andreas Geiger	Max Planck Institute for Intelligent Systems and University of Tübingen, Germany
Niklas Hanselmann	Daimler AG, Germany
Nicolas Jourdan	Daimler AG, Germany
Vladlen Koltun	Intel Labs, USA
Peter Kontschieder	Mapillary Research, Austria
Yubin Kuang	Mapillary AB, Sweden
Alina Kuznetsova	Google Research, Switzerland
Tsung-Yi Lin	Google Brain, USA
Claudio Michaelis	University of Tübingen, Germany
Gerhard Neuhold	Mapillary Research, Austria

Matthias Niessner	Technical University of Munich, Germany
Marc Pollefeys	ETH Zurich and Microsoft, Switzerland
Francesc X. Puig Fernandez	MIT, USA
Rene Ranftl	Intel Labs, USA
Stephan R. Richter	Intel Labs, USA
Carsten Rother	Heidelberg University, Germany
Torsten Sattler	Chalmers University of Technology, Sweden and Czech Technical University in Prague, Czech Republic
Daniel Scharstein	Middlebury College, USA
Hendrik Schilling	rabbitAI, Germany
Nick Schneider	Daimler AG, Germany
Jonas Uhrig	Daimler AG, Germany
Jonas Wulff	Max Planck Institute for Intelligent Systems, Germany
Bolei Zhou	The Chinese University of Hong Kong, China

W31 - The Bright and Dark Sides of Computer Vision: Challenges and Opportunities for Privacy and Security

Mario Fritz	CISPA Helmholtz Center for Information Security, Germany
Apu Kapadia	Indiana University, USA
Jan-Michael Frahm	The University of North Carolina at Chapel Hill, USA
David Crandall	Indiana University, USA
Vitaly Shmatikov	Cornell University, USA

W32 - The Visual Object Tracking Challenge

Matej Kristan	University of Ljubljana, Slovenia
Jiri Matas	Czech Technical University in Prague, Czech Republic
Ales Leonardis	University of Birmingham, UK
Michael Felsberg	Linköping University, Sweden
Roman Pflugfelder	Austrian Institute of Technology, Austria
Joni-Kristian Kamarainen	Tampere University, Finland
Martin Danelljan	ETH Zurich, Switzerland

W33 - Video Turing Test: Toward Human-Level Video Story Understanding

Yu-Jung Heo	Seoul National University, South Korea
Seongho Choi	Seoul National University, South Korea
Kyoung-Woon On	Seoul National University, South Korea
Minsu Lee	Seoul National University, South Korea
Vicente Ordonez	University of Virginia, USA
Leonid Sigal	University of British Columbia, Canada
Chang D. Yoo	KAIST, South Korea
Gunhee Kim	Seoul National University, South Korea
Marcello Pelillo	University of Venice, Italy
Byoung-Tak Zhang	Seoul National University, South Korea

W34 - "Deep Internal Learning": Training with no prior examples

Michal Irani	Weizmann Institute of Science, Israel
Tomer Michaeli	Technion, Israel
Tali Dekel	Google, Israel
Assaf Shocher	Weizmann Institute of Science, Israel
Tamar Rott Shaham	Technion, Israel

W35 - Benchmarking Trajectory Forecasting Models

Alexandre Alahi	EPFL, Switzerland
Lamberto Ballan	University of Padova, Italy
Luigi Palmieri	Bosch, Germany
Andrey Rudenko	Örebro University, Sweden
Pasquale Coscia	University of Padova, Italy

W36 - Beyond mAP: Reassessing the Evaluation of Object Detection

David Hall	Queensland University of Technology, Australia
Niko Suenderhauf	Queensland University of Technology, Australia
Feras Dayoub	Queensland University of Technology, Australia
Gustavo Carneiro	The University of Adelaide, Australia
Chunhua Shen	The University of Adelaide, Australia

W37 - Imbalance Problems in Computer Vision

Sinan Kalkan	Middle East Technical University, Turkey
Emre Akbas	Middle East Technical University, Turkey
Nuno Vasconcelos	UC San Diego, USA
Kemal Oksuz	Middle East Technical University, Turkey
Baris Can Cam	Middle East Technical University, Turkey

W38 - Long-Term Visual Localization under Changing Conditions

Torsten Sattler	Chalmers University of Technology, Sweden, and Czech Technical University in Prague, Czech Republic
Vassileios Balntas	Facebook Reality Labs, USA
Fredrik Kahl	Chalmers University of Technology, Sweden
Krystian Mikolajczyk	Imperial College London, UK
Tomas Pajdla	Czech Technical University in Prague, Czech Republic
Marc Pollefeys	ETH Zurich and Microsoft, Switzerland
Josef Sivic	Inria, France, and Czech Technical University in Prague, Czech Republic
Akihiko Torii	Tokyo Institute of Technology, Japan
Lars Hammarstrand	Chalmers University of Technology, Sweden
Huub Heijnen	Facebook, UK
Maddern Will	Nuro, USA
Johannes L. Schönberger	Microsoft, Switzerland

Pablo Speciale	ETH Zurich, Switzerland
Carl Toft	Chalmers University of Technology, Sweden

W39 - Sensing, Understanding, and Synthesizing Humans

Ziwei Liu	The Chinese University of Hong Kong, China
Sifei Liu	NVIDIA, USA
Xiaolong Wang	UC San Diego, USA
Hang Zhou	The Chinese University of Hong Kong, China
Wayne Wu	SenseTime, China
Chen Change Loy	Nanyang Technological University, Singapore

W40 - Computer Vision Problems in Plant Phenotyping

Hanno Scharr	Forschungszentrum Jülich, Germany
Tony Pridmore	University of Nottingham, UK
Sotirios Tsaftaris	The University of Edinburgh, UK

W41 - Fair Face Recognition and Analysis

Sergio Escalera	CVC and University of Barcelona, Spain
Rama Chellappa	University of Maryland, USA
Eduard Vazquez	Anyvision, UK
Neil Robertson	Queen's University Belfast, UK
Pau Buch-Cardona	CVC, Spain
Tomas Sixta	Anyvision, UK
Julio C. S. Jacques Junior	Universitat Oberta de Catalunya and CVC, Spain

W42 - GigaVision: When Gigapixel Videography Meets Computer Vision

Lu Fang	Tsinghua University, China
Shengjin Wang	Tsinghua University, China
David J. Brady	Duke University, USA
Feng Yang	Google Research, USA

W43 - Instance-Level Recognition

Andre Araujo	Google, USA
Bingyi Cao	Google, USA
Ondrej Chum	Czech Technical University in Prague, Czech Republic
Bohyung Han	Seoul National University, South Korea
Torsten Sattler	Chalmers University of Technology, Sweden and Czech Technical University in Prague, Czech Republic
Jack Sim	Google, USA
Giorgos Tolias	Czech Technical University in Prague, Czech Republic
Tobias Weyand	Google, USA

Xu Zhang	Columbia University, USA
Cam Askew	Google, USA
Guangxing Han	Columbia University, USA

W44 - Perception Through Structured Generative Models

Adam W. Harley	Carnegie Mellon University, USA
Katerina Fragkiadaki	Carnegie Mellon University, USA
Shubham Tulsiani	Facebook AI Research, USA

W45 - Self Supervised Learning – What is Next?

Christian Rupprecht	University of Oxford, UK
Yuki M. Asano	University of Oxford, UK
Armand Joulin	Facebook AI Research, USA
Andrea Vedaldi	University of Oxford, UK

Contents – Part V

W27 - the 16th Embedded Vision Workshop

W32 - The Visual Object Tracking Challenge Workshop VOT2020

W27 - the 16th Embedded Vision Workshop

W27 - The 16th Embedded Vision Workshop

The 16th Embedded Vision Workshop brought together researchers working on vision problems that share embedded system characteristics. This year, we specifically set a focus on embedded vision for robotic applications and platforms. More precisely, this time, the focus area of the workshop was on *Embedded Vision for Augmented and Virtual Reality in Robotic Applications* and *Accelerators for Deep Learning Vision on Robotic Platforms*.

Many modern System on Chips (SoCs) processing architectures can be found in modern mobile phones, UAVs, and cars. Thus, they generate a broad range of new applications. They are developed under significant resource constraints of processing, memory, power, size, and communication bandwidth that pose significant challenges to attaining required levels of performance and speed, and frequently exploit the inherent parallelism of the specialized platforms to address these challenges. Given the heterogeneous and specialized nature of these platforms, efficient development methods are an important issue.

We had a magnificent set of invited speakers tackling the above elements in the field. Rune Storvold (Research Director, NORCE) talked about the importance of machine vision in creating safe and efficient drone-based services and gave example applications from climate research to human transport. Xianglong Liu (Associate Professor, Beihang University) discussed topics going towards low-bit network quantization and Viorica Patraucean (Research Scientist, Google DeepMind) highlighted elements of depth-parallel video models for efficient scene understanding. Davide Scaramuzza (Associate Professor, University of Zurich and ETH Zurich) talked about embedded vision for autonomous, agile micro drones and what could come next in this field. Finally, Matthias Grundmann (Research Scientist, Google AI) gave a great talk on live perception for mobile devices and web applications. These invited talks were accompanied by high-class presentations of the 11 accepted workshop papers (eight long papers and three short papers). A total of 23 papers were submitted and each submission was thoroughly reviewed by two or more reviewers. This resulted in an acceptance rate of 47.8%. The organizing committee was proud to hand over two best long papers awards of 400 euros each and one best short paper award of 200 euros. The awards were sponsored by Bosch Engineering Center Cluj, Romania.

August 2020

Stephan Weiss
Ahmed Nabil Belbachir
Marius Leordeanu
Tse-Wei Chen

Hardware Architecture of Embedded Inference Accelerator and Analysis of Algorithms for Depthwise and Large-Kernel Convolutions

Tse-Wei Chen[1(✉)], Wei Tao[2], Deyu Wang[2], Dongchao Wen[2], Kinya Osa[1], and Masami Kato[1]

[1] Canon Inc., 30-2, Shimomaruko 3-chome, Ohta-ku, Tokyo 146-8501, Japan
twchen@ieee.org

[2] Canon Information Technology (Beijing) Co., Ltd. (CIB), 12A Floor, Yingu Building, No. 9 Beisihuanxi Road, Haidian, Beijing, China

Abstract. In order to handle modern convolutional neural networks (CNNs) efficiently, a hardware architecture of CNN inference accelerator is proposed to handle depthwise convolutions and regular convolutions, which are both essential building blocks for embedded-computer-vision algorithms. Different from related works, the proposed architecture can support filter kernels with different sizes with high flexibility since it does not require extra costs for intra-kernel parallelism, and it can generate convolution results faster than the architecture of the related works.

The experimental results show the importance of supporting depthwise convolutions and dilated convolutions with the proposed hardware architecture. In addition to depthwise convolutions with large-kernels, a new structure called DDC layer, which includes the combination of depthwise convolutions and dilated convolutions, is also analyzed in this paper. For face detection, the computational costs decrease by 30%, and the model size decreases by 20% when the DDC layers are applied to the network. For image classification, the accuracy is increased by 1% by simply replacing 3×3 filters with 5×5 filters in depthwise convolutions.

Keywords: Convolutional neural networks (CNNs) · Embedded vision · Depthwise convolution · Hardware utilization

1 Introduction

Deep learning has been widely applied to image processing and computer vision applications. In the field of embedded vision and robotics, it is important to implement convolutional neural networks (CNNs) with low computational costs [1]. Many researchers propose efficient algorithms to accelerate the deep-learning-based algorithms while keeping the recognition accuracy [2,4,5,8,10,12,13].

A. Bartoli and A. Fusiello (Eds.): ECCV 2020 Workshops, LNCS 12539, pp. 3–17, 2020.
https://doi.org/10.1007/978-3-030-68238-5_1

Chollet proposes a network architecture called Xception, which includes depthwise convolution layers and point-wise convolution layers, to improve the performance [2] of image classification. Howard et al. propose a network architecture called MobileNet, which also includes depthwise convolution layers and point-wise convolution layers [5], to reduce the computational costs for embedded computing. Unlike the regular convolution layers in CNNs, the numbers of input feature maps and output feature maps in the depthwise convolution layers are the same, and the computational cost of convolution operations is proportional to the number of input feature maps or output feature maps. Sandler et al. propose MobileNetV2, in which the depthwise convolution layers are still one of the basic structures [10]. In addition to these works, there are various kinds of architectures utilizing the concept of depthwise convolutions [4,12,13].

Since depthwise convolutions have become common building blocks for compact networks for embedded vision, the hardware engineers also propose different kinds of accelerators and inference engines to implement them efficiently. Liu et al. propose an FPGA-based CNN accelerator to handle depthwise convolutions and regular convolutions with the same computation cores [6]. The same architecture is used for the depthwise convolutions and the regular convolutions, but it is difficult to increase the hardware utilization of the depthwise convolutions because the inputs of some processing elements are always set to zeros. Su et al. propose an acceleration scheme for MobileNet, utilizing modules for depthwise convolutions and regular convolutions [11]. Similar to Liu's work [6], since two separated modules are used for depthwise convolutions and regular convolutions, the hardware resource cannot be fully utilized. Yu et al. propose a hardware system, Light-OPU, which accelerates regular convolutions, depthwise convolutions, and other lightweight operations with one single uniform computation engine [18]. It efficiently utilizes the hardware by exploring intra-kernel parallelism. However, since the architecture is optimized for 3×3 filter kernels, the computational time becomes 4 times when the size of filter kernels is increased from 3×3 to 5×5. The hardware cost increases because extra line buffers are required to store the data. In modern CNN architectures, 3×3 filter kernels are commonly used, but sometimes it is necessary to increase the receptive field by using large-kernel filters [7] or dilated convolutions [15,16] for certain kinds of applications, such as face detection, image classification, and image segmentation. Therefore, a hardware system that can efficiently handle regular convolutions, depthwise convolutions, and large-kernel filters is desired.

The contribution of this paper is twofold. First, a new hardware architecture for embedded inference accelerators is proposed to handle both depthwise convolutions and regular convolutions. The proposed architecture can efficiently handle convolutions with kernels larger than 3×3, and it can achieve shorter processing time than the related works. The experimental results show that the size of the proposed hardware is 1.97M gates, while the number of parallel processing units for MAC (Multiply-Accumulate) operations is 512. Second, the features of the supported network architectures are analyzed. By replacing regular convolutions with large-kernel depthwise convolutions, we can reduce the computational costs while keeping the accuracy.

The paper is organized as follows. First, in Sect. 2, the proposed hardware architecture and supported network architectures are introduced. Then, the experimental results are discussed in Sect. 3. Finally, the conclusions are given in Sect. 4.

2 Proposed Hardware Architecture and Supported Network Architectures

The proposed hardware architecture can efficiently handle regular convolutions and depthwise convolutions with large-kernel filters. The hardware architecture and the supported network architecture are introduced in this section. The supported network architecture, the solution based on the proposed hardware, the time chart, and the algorithm are introduced in the following subsections.

2.1 Network Architecture with Two Types of Convolution Layers

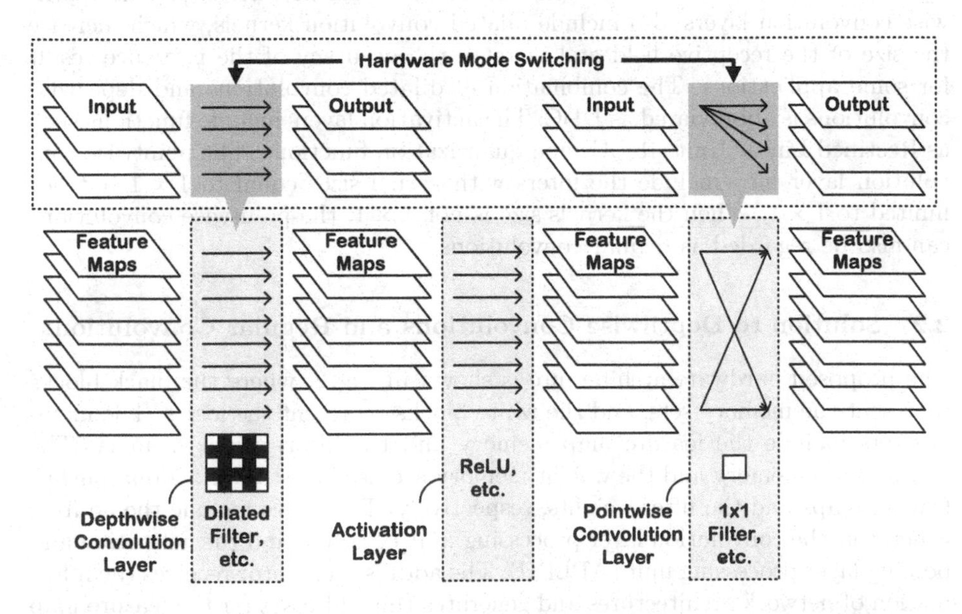

Fig. 1. Functions of the proposed hardware and an example of the supported network architectures.

The upper part of Fig. 1 shows the two modes for regular convolutions and depthwise convolutions supported by the proposed hardware architecture. The main feature of the architectures is the function of hardware mode switching. On the left side, all of the input feature maps for parallel processing are stored in the memory to generate the same numbers of output feature maps. On the right side, only 1 input feature map for parallel processing is stored in the memory to generate multiple output feature maps. When the regular convolution mode is enabled, the intermediate convolution results of multiple input feature maps are accumulated to generate the final convolution results for a single output feature map. When the depthwise convolution mode is enabled, the intermediate convolution results of multiple input feature maps become the final convolution results of multiple output feature maps. In both of the two modes, the hardware architecture can generate the same numbers of convolution results, and the multipliers in the convolution cores are fully utilized. The overhead of the hardware architecture is the buffer to store the data of multiple input feature maps in the depthwise convolution mode. However, large-kernel convolutions can be efficiently handled since the proposed hardware architecture does not require extra costs for intra-kernel parallelism [18].

The lower part of Fig. 1 shows an example of the network architecture that is focused in this paper. There are 3 layers in this example, the depthwise convolution layer, the activation layer, and the pointwise convolution layer. The depthwise convolution layer may include the filters with kernels larger than or equal to 3×3. Different from the related works, such as MobileNet [5], the depthwise convolution layers also include dilated convolution kernels, which increase the size of the receptive field and increase the accuracy of the inference result for some applications. The combination of dilated convolutions and depthwise convolutions is abbreviated as DDC. The activation layers include functions such as Rectified Linear Unit (ReLU) and quantization functions. The pointwise convolution layer may include the filters with kernel sizes equal to 1×1 but not limited to 1×1. When the kernels size is not 1×1, the pointwise convolutions can also be regarded as regular convolutions.

2.2 Solution to Depthwise Convolutions and Regular Convolutions

The proposed hardware architecture is shown in Fig. 2, where the black blocks represent the memory sets, and the white blocks represent the logics. The memory sets include the feature map memory and the filter weight memory. The feature map memory and the weight memory are used to store the input/output feature maps and the filter weights, respectively. The logics include the address generator, the convolution layer processing unit (CLPU), and the activation and pooling layer processing unit (APLPU). The address generator receives the information of network architectures and generates the addresses for the feature map memory and the weight memory. The blocks in the spatial domain of the feature maps are processed sequentially.

The CLPU includes PE_{num} convolution cores to compute the results of convolutions of the input feature maps and the input filter weights in parallel. Each convolution core contains MAC_{PE} sets of multipliers and adders to execute MAC operations.

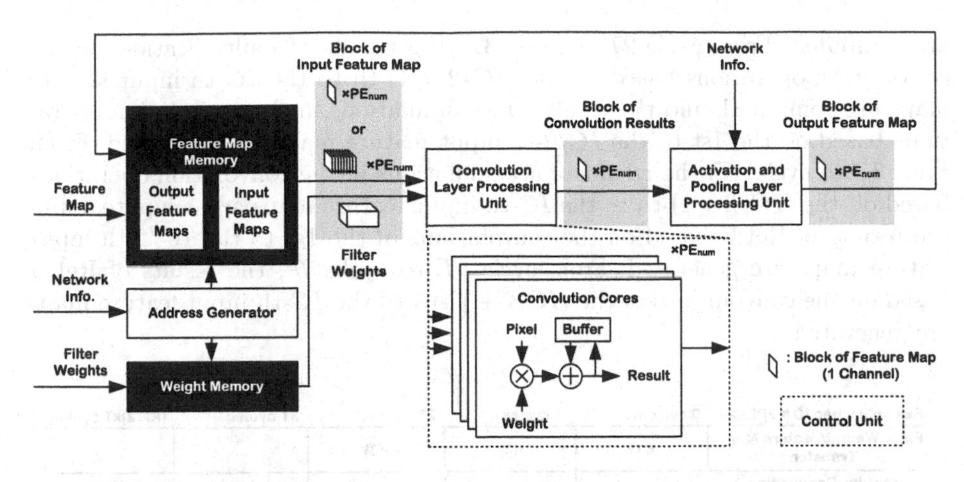

Fig. 2. Proposed hardware architecture for regular convolutions and depthwise convolutions.

2.3 Time Chart

Figure 3 shows a time chart of the hardware architecture, where the operations are executed in pipeline. In this example, we set $IC = 2PE_{num}$ for simplicity. Note that IC represents the number of input feature maps, and PE_{num} denotes the number of convolution cores. The same operations can be applied to regular convolutions and depth-wise convolutions with difference kernel sizes.

The upper part of Fig. 3 shows an example of regular convolutions. There are IC input feature maps and OC output feature maps in this layer. From cycle 0 to cycle T, the 1st input feature map and the corresponding filter weights are transferred. From cycle T to cycle $2T$, the 2nd input feature map and the corresponding filter weights are transferred, and the results of multiplications in the convolution operations based on the 1st input feature maps are computed. From cycle $2T$ to cycle $3T$, the 3rd input feature map and the corresponding filter weights are transferred. The results of multiplications in the convolution operations based on the 2nd input feature maps are computed, and the results of accumulations in the convolution operations based on the 1st input feature maps are computed. From cycle $3T$ to cycle $(IC+2)T$, the results of convolution operations based on the remaining input feature maps are computed. From cycle $(IC+2)T$ to cycle $(IC+3)T$, the results of ReLU based on the convolutions of the 1st to the IC-th input feature maps are generated.

The lower part of Fig. 3 shows an example of depthwise convolutions. There are IC input feature maps and OC output feature maps in this layer, and the values of OC and IC are the same. From cycle 0 to cycle T, the 1st to the $IC/2$-th input feature maps and the corresponding filter weights are transferred. From cycle T to cycle $2T$, the $(IC/2+1)$-th to the IC-th input feature maps and the corresponding filter weights are transferred, and the results of multiplications in the convolution operations based on the 1st to the $IC/2$-th input feature maps are computed. From cycle $2T$ to cycle $3T$, the results of multiplications in the convolution operations based on the $(IC/2+1)$-th to the IC-th input feature maps are computed, and the results of accumulations in the convolution operations based on the 1st to the $IC/2$-th input feature maps are computed. From cycle $3T$ to cycle $4T$, the results of accumulations in the convolution operations based on the $(IC/2+1)$-th to the IC-th input feature maps are computed, and the results of ReLU based on the convolutions of the 1st to the $IC/2$-th input feature maps are generated. From cycle $4T$ to cycle $5T$, the results of ReLU based on the convolutions of the $(IC/2+1)$-th to the IC-th input feature maps are generated.

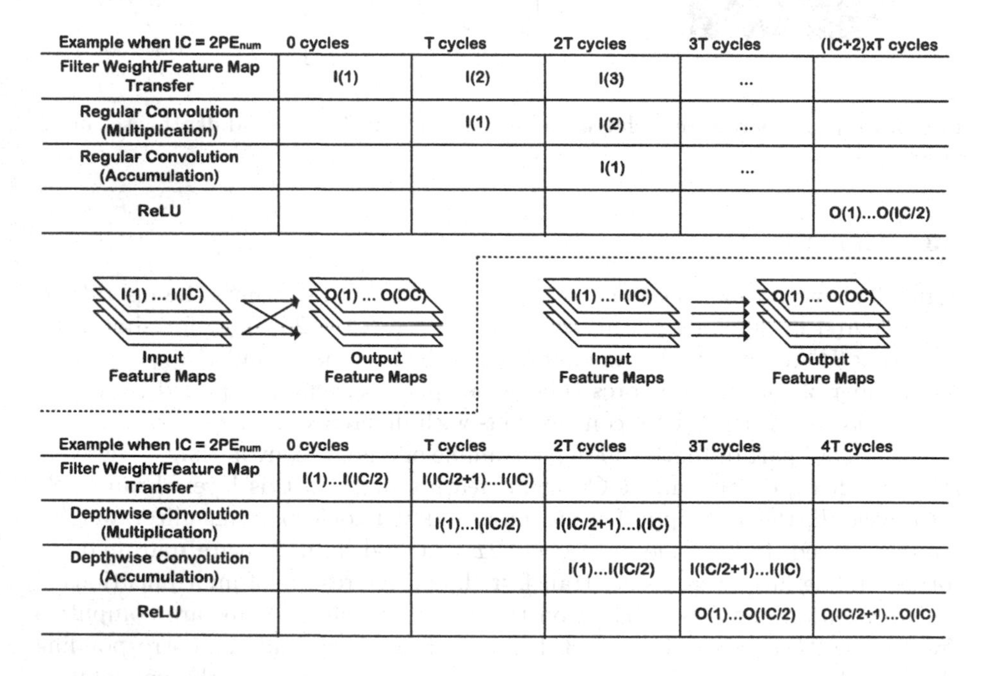

Fig. 3. Timechart for depthwise convolutions and regular convolutions.

2.4 Algorithm

Figure 4 shows the proposed algorithms. There are 4 levels in the nested loop structure. In the 1st level of loops, each convolution layer of the networks is processed sequentially, and the parameters of the corresponding layer are set. The mode for depthwise convolutions and regular convolutions is selected in this level.

When the mode for regular convolutions is selected, the operations in the 2nd level to the 4th level of loops are executed. First, in the 2nd level of loops, each output channel of the layer is processed sequentially. Then, in the 3rd level of loops, each block of the output channel is processed sequentially. Finally, in the 4th level of loops, each input channel is processed sequentially. The filter weights of the i-th convolution layer, the j-th output channel, the n-th input channel are set, and the convolution results of PE_{num} output blocks are computed in parallel.

When the mode for depthwise convolutions is selected, the operations in the 2nd level to the 3rd level of loops are executed. The number of input channels is equal to the number of output channels, and each channel is processed sequentially. The filter weights of the i-th convolution layer, the j-th output channel, the j-th input channel are set, and the convolution results of PE_{num} output blocks are computed in parallel.

```
for i = 1, 2, ..., (# of convolution layers)
  set layer parameters
  if depthwise convolution is disabled
    for j = 1, 2, ..., OC
      for m = 1, 2, ..., (# of blocks)
        for n = 1, 2, ..., IC
          extract the m-th block of the n-th channel
          set the filter weights for ( i, n, j )
          calculate the convolution result of the output block partially
        calculate the activation result of the output block
  otherwise
    for m = 1, 2, ..., (# of blocks)
      for n = 1, 2, ..., OC
        extract the m-th block of the n-th channel
        set the filter weights for ( i, n, n )
        calculate the convolution result of the output block partially
      calculate the activation result of the output block
```

Fig. 4. Algorithm of the proposed hardware.

Since the number of loops is reduced when the mode for depthwise convolutions is selected, redundant operations, in which some of filter weights are set to zeros, are not required. The common architecture for the two modes makes the hardware achieve higher utilization than the related works [6,11].

3 Experimental Results and Analysis

The section contains 3 parts. The first part is the comparison of accuracy of face detection and image classification. The second part is the analysis of the computational cost of the proposed hardware architecture. The third part is the comparison of specifications.

3.1 Comparison of Accuracy

One of the strengths of the proposed hardware architecture is the function to handle the combination of large-kernel convolutions and depthwise convolutions.

To analyze the effectiveness of large-kernel convolutions and depthwise convolutions, the experiments for two kinds of applications are performed. The first application is face detection, and the second application is image classification. Two experiments are performed for both of the applications.

In the first experiment, the accuracy of face detection on the WIDER FACE dataset [17] is analyzed. The network of RetinaFace [3] is tested with depthwise convolutions and large-kernel convolutions, and the backbone network is MobileNetV1-0.25 [5].

The WIDER FACE dataset [17] consists of 32,203 images and 393,703 face bounding boxes with a high degree of variability in scale, pose, expression, occlusion and illumination. In order to compare the proposed network architecture with RetinaFace, the WIDER FACE dataset is split into training (40%), validation (10%) and testing (50%) subsets, and three levels of difficulty (i.e. Easy, Medium, and Hard) are defined by incrementally incorporating hard samples. We use the WIDER FACE training subset as the training dataset, and the WIDER FACE validation subset as the testing dataset. The operations in the context module, which is an important structure used to increase the receptive field, include a series of 3×3 convolutions. Since the feature maps are processed with different numbers of 3×3 filters consecutively, the effects are similar to applying large-kernel convolutions (e.g. 5×5, 7×7) to the feature maps. To show the advantage of the functions of the proposed hardware architecture, some of the operations are replaced with depthwise convolutions and dilated filters. To be specific, two cascaded 3×3 convolutions are replaced with a single 3×3 convolution where the dilation rate is 1, and three cascaded 3×3 convolutions are replaced with a single 3×3 convolution where the dilation rate is 2.

Table 1. Accuracy of RetinaFace [3] and the proposed network on the WIDER FACE [17] validation subset.

Network	Easy	Medium	Hard
RetinaFace[A] [3] (with CM[a])	90.70%	87.88%	73.50%
RetinaFace[A] w/o CM[a]	89.55%	86.21%	68.83%
RetinaFace[A] with CM[a] and DDC[b]	90.28%	87.13%	73.24%
Quantized RetinaFace[A] [3] (with CM[a])	90.72%	87.45%	73.56%
Quantized RetinaFace[A] with CM[a] and DDC[b]	90.32%	87.68%	73.53%
RetinaFace[B] [3] (with CM[a])	89.98%	87.11%	72.01%
RetinaFace[B] w/o CM[a]	88.68%	84.98%	68.56%
RetinaFace[B] with CM[a] and DDC[b]	89.60%	86.13%	71.93%
Quantized RetinaFace[B] [3] (with CM[a])	89.70%	86.91%	71.89%
Quantized RetinaFace[B] with CM[a] and DDC[b]	89.56%	86.02%	71.75%

[a]CM stands for "context module."

[b]DDC stands for "depthwise and dilated convolutions."

[A]The networks are trained from a pre-trained model.

[B]The networks are trained from scratch.

The accuracy of RetinaFace [3] and its variations are shown in Table 1, which includes the proposed network architecture, the depthwise and dilated convolution (DDC) layers. Since the accuracy of the network without the context module is not available in the original paper [3], we add an ablation study to verify the effectiveness of the context module. The results show that the context module can increase the accuracy of face detection by about 1% for the Easy category and the Medium category, and by 4% for the Hard category. The proposed hardware can support the quantized RetinaFace and its variations. In the quantized networks, the feature maps and the filter weights are both quantized into 8-bit data. It is shown that the quantized versions have similar accuracy with the floating-point versions.

When the filters in the context module are replaced with the DDC layers and pointwise convolutions, the accuracy decreases by less than 1% for all the categories. The parameter settings are shown in Table 2. We trained the RetinaFace using the SGD optimizer (momentum at 0.9, weight decay at 0.0005, and batch size of 32) on the NVIDIA Titan Xp GPUs with 12 GB memory. The learning rate starts from 10^{-3}, and is divided by 10 at the 190th and at the 220th epoch. The training process terminates at the 250th epoch.

Table 2. Parameter settings of the experiments for RetinaFace [3].

Backbone	MobileNetV1-0.25
Prediction levels	8×, 16×, 32× down-sampling
Anchor settings	16 × 16, 32 × 32, 8× prediction layers 64 × 64, 128 × 128, 16× prediction layers 256 × 256, 512 × 512, 32× prediction layers
Batch size	32
No. of epochs	250

The computational costs and the model sizes of RetinaFace [3] and the proposed work are shown in Table 3. When the operations in the context module are replaced with depthwise convolutions and dilated filters, the model size of the context module decreases from 138 KB to 23 KB, and the computational cost of the context module decreases from 708 MACs per input pixel to 119 MACs per input pixel. By applying depthwise convolutions and dilated filters to the network, the total computational costs decrease by about 30%, and the total model size decreases by about 20%.

In the second experiment, to show the relation between the size of filter kernels and the inference result, the accuracy of image classification on ImageNet [9] is analyzed. Similar to the first experiment, MobileNetV1-0.25 is used for testing. Since the architecture of MobileNetV1 [5] is composed of depthwise convolution layers, it can be used to evaluate the effectiveness of large-kernel convolutions by simply increasing the kernel sizes of the filters. The accuracy

Table 3. Computational costs and the model sizes of RetinaFace [3] and the proposed network.

Network	Model size (Bytes)		Computational Cost (MACs/Input pixel)	
	CM[a]	Total	CM[a]	Total
RetinaFace[a] [3]	138K	1.12M	708	1,888
RetinaFace with CM and DDC[b]	23K	0.90M	119	1,298

[a]CM stands for "context module."
[b]DDC stands for "depthwise and dilated convolutions."

of MobileNetV1 and its variations is shown in Table 4. There are two networks with dilated convolutions, where the concept of hybrid dilated convolutions is adopted [14]. When some 3×3 filters in MobileNetV1 are replaced by 5×5 filters, the accuracy is increased by more than 1%. When some of 3×3 filters are replaced by dilated 3×3 filters, the network can also achieve higher accuracy than the original architecture.

Table 4. Accuracy of MobileNetV1-0.25 and the proposed network on ImageNet [9].

Network	Top-1 accuracy	Top-5 accuracy
MobileNetV1-0.25 [5]	68.39%	88.35%
MobileNetV1-0.25 with 5×5 filters	69.44%	89.90%
MobileNetV1-0.25 with dilated filters (A)[a]	68.52%	88.61%
MobileNetV1-0.25 with dilated filters (B)[b]	68.98%	88.85%

[a]Dilated convolutions are applied to Layer 14 to Layer 18. The kernel size of filters is 3×3, and the dilation rates are set to $2, 2, 2, 2, 2$ for the 5 layers.
[b]Dilated convolutions are applied to Layer 14 to Layer 18. The kernel size of filters is 3×3, and the dilation rate are set to $2, 3, 2, 3, 2$ for the 5 layers.

The first experiment shows that the combination of depthwise convolutions and dilated convolutions keeps the accuracy while reducing the computational costs, and the second experiment shows that the combination of depthwise convolutions and large-kernel convolutions can increase the accuracy.

In brief, these results show the importance of supporting depthwise convolutions and large-kernel convolutions, including dilated convolutions, with the proposed hardware architecture.

3.2 Computational Time

To compare the proposed hardware architecture with the related work, the processing time for regular convolutions and depthwise convolutions is expressed as equations. The processing time (number of cycles) for the regular convolutions is shown as follows.

$$T_M = \max\left(IC \times \left\lceil \frac{IN \times IM}{BW_{FM}} \right\rceil \times \left\lceil \frac{OC}{PE_{num}} \right\rceil, IC \times OC \times \left\lceil \frac{X \times Y}{BW_W} \right\rceil\right), \quad (1)$$

$$T_C = X \times Y \times IC \times \left\lceil \frac{ON \times OM}{MAC_{PE}} \right\rceil \times \left\lceil \frac{OC}{PE_{num}} \right\rceil, \quad (2)$$

where T_M and T_C are the memory access time and the computational time, respectively. The processing time can be expressed as $\max(T_M, T_C)$, and the definition of parameters in the equations are shown as follows.

IC: the number of input channels.
OC: the number of output channels.
$IN \times IM$: the size of input blocks.
$ON \times OM$: the size of output blocks.
$X \times Y$: the size of filters.
BW_{FM}: the bandwidth to transfer feature maps.
BW_W: the bandwidth to transfer filter weights.
$PE_{num} = 8$: the number of processing elements.
$MAC_{PE} = 64$: the number of parallel MAC operations implemented in 1 processing element.

The memory access time and the computational time for the depthwise convolutions are shown as follows. Similarly, the processing time can be expressed as $\max(T_M, T_C)$.

$$T_M = \max\left(OC \times \left\lceil \frac{IN \times IM}{BW_{FM}} \right\rceil, OC \times \left\lceil \frac{X \times Y}{BW_W} \right\rceil\right), \quad (3)$$

$$T_C = X \times Y \times \left\lceil \frac{ON \times OM}{MAC_{PE}} \right\rceil \times \left\lceil \frac{OC}{PE_{num}} \right\rceil \quad (4)$$

All of the processing elements are used for both regular convolutions and depthwise convolutions.

In the related work [18], the computational costs (the number of cycles) for the convolution operations are also expressed as equations. To compare the efficiency, we set $PE_{num} = 64$, and $MAC_{PE} = 8$ for the related work. For simplicity, IC and OC are set to the same value for both regular convolutions and depthwise convolutions. The results are shown in Fig. 5, where (a) and (b) refer to the cases when the size of filter kernels is 3×3, and (c) and (d) refer to the cases when the size of filter kernels is 5×5. For regular convolutions, the

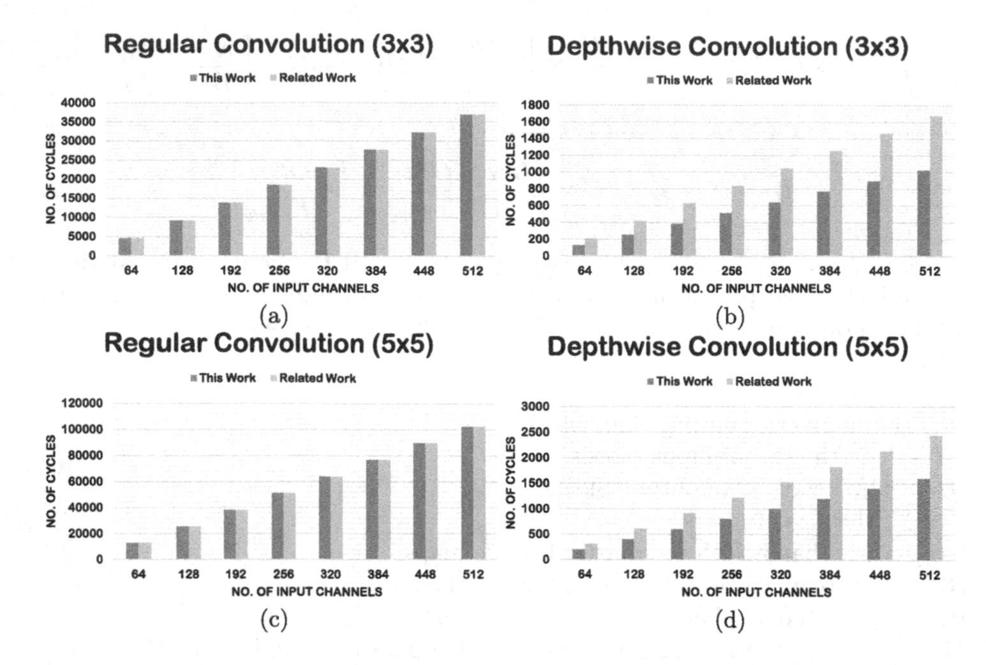

Fig. 5. Comparison of the computational time and the number of input channels with the related work [18] when the size of filter kernels is (a)(b) 3×3 and (c)(d) 5×5.

computational time of the proposed work and the related work is almost the same when the size of filter kernels is 3×3 or 5×5. For depthwise convolutions, the proposed work has shorter computational time than the related work when the size of filter kernels is 3×3 or 5×5. In the proposed architecture, since the filter weights and the feature maps can be transferred in parallel, the memory access time can be shortened so that the operations of dilated convolutions achieve faster speed than the related work. Also, since the proposed architecture can process the filter weights sequentially, the computational time is proportional to the size of filter kernels, and the computational efficiency does not decrease when the size of filter kernels becomes large.

3.3 Specifications

The specifications of the proposed embedded inference accelerator, which is designed to process modern CNNs, are shown in Table 5. The main feature of the proposed architecture is that it can compute depthwise convolutions with large kernels, including the DDC layers mentioned in the previous section.

We use the 28-nm CMOS technology library for the experiments. The resolution of input images is VGA, and the performance is 512 MAC operations per clock cycle for 8-bit feature maps and 8-bit filter weights. As shown in Fig. 2, since regular convolutions and depthwise convolutions can be selected according to the type of layers, the hardware architecture can also process RetinaFace (with MobileNetV1-0.25) efficiently. The gate counts of the CLPU and the APLPU are 1.50M and 0.47M, respectively. The processing speed for RetinaFace, which includes the post-processing algorithm, is 150fps. Compared with RetinaFace, it takes only 83% of processing time to compute the inference result of the proposed network architecture, DDC layers. Since the proposed hardware can support depthwise convolutions with large filter kernels, the target networks can be switched according to the requirement of processing time.

Table 5. Specifications of the proposed embedded inference accelerator.

Gate count (NAND-gates)	CLPU[a]: 1.50M APLPU[b]: 0.47M
Process	28-nm CMOS technology
Clock frequency	400 MHz
Memory size	Feature map memory: 4,096 KB Weight memory: 2,048 KB
Maximum input image size	VGA: 640 × 480 pixels
Processing speed	RetinaFace [3][c]: 150 fps RetinaFace with DDC[c]: 180 fps
Size of filter kernels[d]	Maximum: $N_{\mathsf{ch},i} \times 7 \times 7$
Performance	512 MACs/clock cycle

[a]CLPU stands for "Convolution Layer Processing Unit."
[b]APLPU stands for "Activation and Pooling Layer Processing Unit."
[c]The bit widths of filter weights and the feature maps are quantized to 8 bits. The post-processing time with an external CPU is included.
[d]$N_{\mathsf{ch},i}$ represents the number of output channels in the i-th convolution layer in the network.

Table 6 shows the comparison of hardware specifications with three related works. Since it is difficult to compare the performance of FPGA and ASIC, we focus on the hardware utilization of MAC units for regular convolutions and depthwise convolutions. In the first related work [6], when dealing with depthwise convolutions, the filter kernels are filled with zeros, and the advantage is that both regular convolutions and depthwise convolutions can be computed by using the same hardware architecture. Suppose that the hardware can handle regular convolutions with T_n input channels and T_m output channels in parallel, the hardware utilization becomes $\frac{100}{T_m}$% when dealing with depthwise convolutions with T_n input channels and T_m output channels. In the second related work [11], regular convolutions and depthwise convolutions are handled with different hardware architectures. The advantage is that both regular convolutions and depthwise convolutions can be handled efficiently, but some of the processing elements do not function when either depthwise convolutions or regular convolutions are handled. In the third related work [18], the processing elements can effectively handle regular convolutions and depthwise convolutions, but as discussed in Sect. 3.2, the performance decreases when the kernel size of filters is larger than or equal to 5 × 5.

The proposed architecture can handle regular convolutions and depthwise convolutions efficiently with large-kernel filters. The overhead to support the function is the buffer for input feature maps, which is mentioned in Sect. 2.1.

Table 6. Comparison with the related works.

	Liu et al. [6]	Su et al. [11]	Yu et al. [18]	This Work
Device	ZYNQ7100	XCZU9EG	XCK325T	ASIC
Frequency	100 MHz	150 MHz	200 MHz	400 MHz
Performance (GOPS)	17.11	91.2	460.8	409.6
Utilization:[a]				
Regular convolutions	100%	100%	100%	100%
Depthwise convolutions	13%[b]	50%[c]	3 × 3 filters: 100%	100%
			5 × 5 filters: 69%[d]	

[a]Utilization represents the hardware utilization of MAC processing elements for regular convolutions and depthwise convolutions in an ideal case.
[b]We suppose that $T_m = T_n = 8$ [6], and the utilization becomes $100\% \times \frac{1}{T_m} = 13\%$.
[c]We suppose that $M' = N'$ [11].
[d]Since $\alpha = 4$ [18], the utilization for 5 × 5 filter kernels is $100\% \times \left(\frac{5^2}{3^2 \cdot \alpha}\right) = 69\%$.

4 Conclusions and Future Work

In this paper, a new solution which combines the advantages of algorithms and hardware architectures, is proposed to handle modern CNNs for embedded computer vision. The contribution of this paper is twofold. First, a new hardware architecture is proposed to handle both depthwise convolutions and regular convolutions. The proposed architecture can support filters with kernels larger than 3 × 3 with high efficiency, and the processing time is shorter than the related works. The experimental results show that the gate count of the proposed hardware is 1.97M gates, and the number of parallel processing units for MAC operations is 512. Second, the features of the supported network architecture, DDC, are analyzed with two applications. By replacing regular convolutions with large-kernel depthwise convolutions, it is possible to reduce the computational costs while keeping the accuracy.

For future work, we plan to extend the functions of the proposed hardware architecture to handle other kinds of complicated network architectures. In addition, we will apply the combination of the dilated convolutions and the depthwise convolutions to other applications.

References

1. Chen, T.W., et al.: Condensation-Net: memory-efficient network architecture with cross-channel pooling layers and virtual feature maps. In: Proceedings of the IEEE/CVF Conference on Computer Vision and Pattern Recognition (CVPR) Workshops (2019)
2. Chollet, F.: Xception: deep learning with depthwise separable convolutions (2016). coRR, abs/1610.02357
3. Deng, J., Guo, J., Zhou, Y., Yu, J., Kotsia, I., Zafeiriou, S.: RetinaFace: single-stage dense face localisation in the wild (2019). coRR, abs/1905.00641
4. Han, K., Wang, Y., Tian, Q., Guo, J., Xu, C., Xu, C.: GhostNet: more features from cheap operations (2019). coRR, abs/1911.11907
5. Howard, A.G., et al.: MobileNets: efficient convolutional neural networks for mobile vision applications (2017). coRR, abs/1704.04861
6. Liu, B., Zou, D., Feng, L., Feng, S., Fu, P., Li, J.: An FPGA-based CNN accelerator integrating depthwise separable convolution. Electronics **8**(3), 281 (2019)
7. Peng, C., Zhang, X., Yu, G., Luo, G., Sun, J.: Large kernel matters - improve semantic segmentation by global convolutional network (2017). coRR, abs/1703.02719
8. Qin, H., Gong, R., Liu, X., Bai, X., Song, J., Sebe, N.: Binary neural networks: a survey (2020). coRR, abs/2004.03333
9. Russakovsky, O., et al.: ImageNet large scale visual recognition challenge. Int. J. Comput. Vis. **115**(3), 211–252 (2015). https://doi.org/10.1007/s11263-015-0816-y
10. Sandler, M., Howard, A., Zhu, M., Zhmoginov, A., Chen, L.C.: MobileNetV2: inverted residuals and linear bottlenecks (2018). coRR, abs/1801.04381
11. Su, J., et al.: Redundancy-reduced MobileNet acceleration on reconfigurable logic for ImageNet classification. In: Proceedings of International Symposium on Applied Reconfigurable Computing, pp. 16–28 (2018)
12. Tan, M., Le, Q.V.: EfficientNet: rethinking model scaling for convolutional neural networks (2019). coRR, abs/1905.11946
13. Tan, M., Le, Q.V.: MixConv: mixed depthwise convolutional kernels (2019). coRR, abs/1907.09595
14. Wang, P., et al.: Understanding convolution for semantic segmentation (2017). coRR, abs/1702.08502
15. Wei, Y., Xiao, H., Shi, H., Jie, Z., Feng, J., Huang, T.S.: Revisiting dilated convolution: a simple approach for weakly- and semi- supervised semantic segmentation. In: Proceedings of IEEE/CVF Conference on Computer Vision and Pattern Recognition (2018)
16. Wu, H., Zhang, J., Huang, K., Liang, K., Yu, Y.: FastFCN: rethinking dilated convolution in the backbone for semantic segmentation (2019). coRR, abs/1903.11816
17. Yang, S., Luo, P., Loy, C.C., Tang, X.: WIDER FACE: a face detection benchmark (2016). coRR, abs/1511.06523
18. Yu, Y., Zhao, T., Wang, K., He, L.: Light-OPU: an FPGA-based overlay processor for lightweight convolutional neural networks. In: Proceedings of the ACM/SIGDA International Symposium on Field-Programmable Gate Arrays, p. 122, February 2020

SegBlocks: Towards Block-Based Adaptive Resolution Networks for Fast Segmentation

Thomas Verelst(✉) and Tinne Tuytelaars(✉)

ESAT-PSI, KU Leuven, Leuven, Belgium
{thomas.verelst,tinne.tuytelaars}@esat.kuleuven.be

Abstract. We propose a method to reduce the computational cost and memory consumption of existing neural networks, by exploiting spatial redundancies in images. Our method dynamically splits the image into blocks and processes low-complexity regions at a lower resolution. Our novel BlockPad module, implemented in CUDA, replaces zero-padding in order to prevent the discontinuities at patch borders of which existing methods suffer, while keeping memory consumption under control. We demonstrate SegBlocks on Cityscapes semantic segmentation, where the number of floating point operations is reduced by 30% with only 0.2% loss in accuracy (mIoU), and an inference speedup of 50% is achieved with 0.7% decrease in mIoU.

1 Introduction and Related Work

Contemporary deep learning tasks use images with ever higher resolutions, e.g. 2048×1024 pixels for semantic segmentation on Cityscapes. Meanwhile, there is a growing interest for deployment on low-computation edge devices such as mobile phones. Typical neural networks are static: they apply the same operations on every image and every pixel. However, not every image region is equally important. Some dynamic execution methods [2,6,8] address this deficiency by skipping some image regions, but that makes them less suitable for dense pixel labelling tasks. Moreover, due to their implementation complexity, they do not demonstrate speed benefits [2], are only suitable for specific network architectures [8] or have low granularity due to large block sizes [6].

Some dynamic methods target segmentation tasks specifically. Patch Proposal Network [9] and Wu et al. [10] both refine predictions based on selected patches. Processing images in patches introduces a problem at patch borders, where features cannot propagate between patches when using zero-padding on each patch. Therefore, these methods use custom architectures with a global branch and local feature fusing to partly mitigate this issue. Huang et al. [3] propose to use a small segmentation network and further refine regions using a large and deeper network. However, since features of the first network are not re-used, the second network has to perform redundant feature extraction operations. Moreover, patches are 256×256 pixels or larger to minimize problems at patch borders which limits flexibility.

A. Bartoli and A. Fusiello (Eds.): ECCV 2020 Workshops, LNCS 12539, pp. 18–22, 2020.
https://doi.org/10.1007/978-3-030-68238-5_2

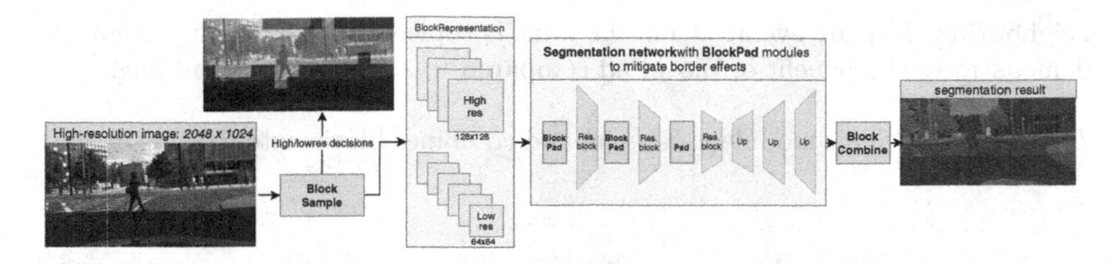

Fig. 1. Overview: The BlockSample module samples image regions in high or low resolution based on a complexity criterion. Image regions are efficiently stored in blocks. The BlockPad module enables feature propagation at block borders.

Our method differs from existing methods in two ways. First, there are no discontinuities at patch borders due to our novel BlockPad module, efficiently implemented in CUDA. BlockPad replaces zero-padding and propagates features as if the network was never processed in blocks, making it possible to adapt existing architectures without retraining. Also, we can obtain networks with different levels of complexity simply by changing a single threshold. Patches can be as small as 8×8 pixels, without major overhead. Secondly, instead of just skipping computations at low-complexity regions, we process a downsampled version of the region. This reduces the computational cost and memory consumption, while still providing dense pixel-wise labels. Our PyTorch implementation is available at https://github.com/thomasverelst/segblocks-segmentation-pytorch.

2 Method

Our method introduces several modules for block-based processing (Fig. 1):

BlockSample. The BlockSample module splits the image into blocks and low-complexity regions are downsampled by a factor 2 to reduce their computational cost. We use blocks of 128×128 pixels. As the network pools down feature maps, the block size is reduced accordingly up to just 4×4 pixels in the deepest layers, offering fine granularity. A simple heuristic determines the block complexity C_b. High-resolution processing is used when C_b exceeds a predefined threshold τ. For each image block b, we compare the $L1$ difference between the original image content of block I_b and a down- and re-upsampled version:

$$C_b = L_1(I_b - \texttt{nn\_upsampling}(\texttt{avg\_pooling}(I_b))) \tag{1}$$

BlockPad. The BlockPad module replaces zero-padding in order to avoid discontinuities at block borders, as shown in Fig. 2. For adjacent high-resolution blocks, BlockPad copies corresponding pixel values from neighboring blocks into the padding. When padding a low-resolution block, the two nearest pixels of the neighboring block are averaged and the value is copied into the padding. Figure 3 demonstrates the benefit of BlockPad compared to standard zero-padding.

BlockCombine. This module upsamples and combines blocks into a single image.

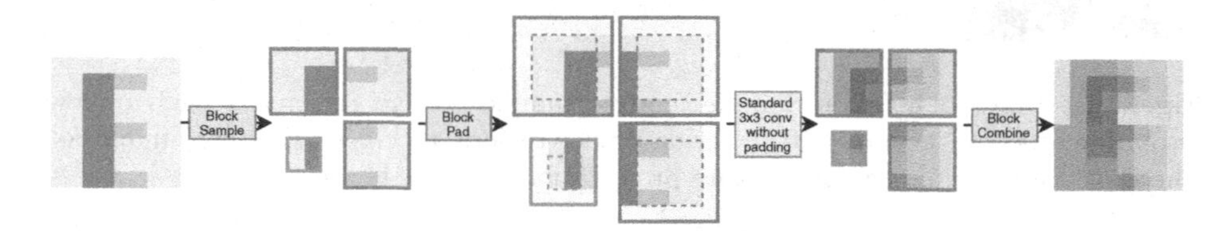

Fig. 2. Illustration of our modules. The BlockPad module replaces zero-padding for 3×3 convolutions and avoids discontinuities between patches.

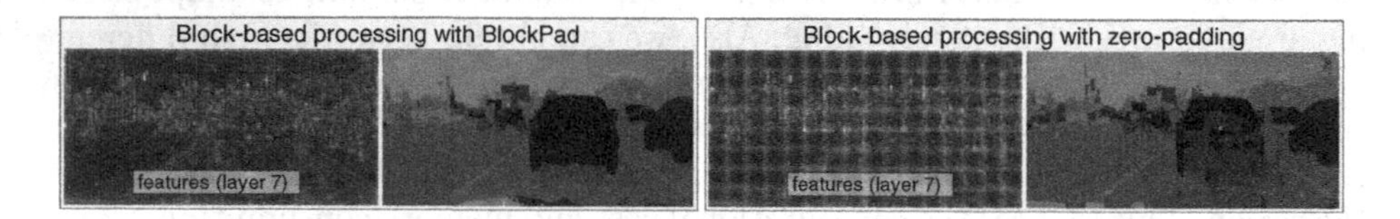

Fig. 3. Impact of the BlockPad module, by visualizing the intermediate feature maps and outputs. Zero-padding of individual blocks introduces border artifacts in the corresponding feature maps, resulting in poor segmentation results. The BlockPad module adds padding as if the image was never split in blocks.

3 Experiments and Results

We test our method on the Cityscapes dataset [1] for semantic segmentation, consisting of 2975 training and 500 validation images of 2048×1024 pixels. Our method is applied on the Swiftnet-RN18 [5] network. Table 1 shows that our SegBlocks models achieve state of the art results for real-time semantic segmentation, having 75.4% mIoU accuracy with 45 FPS and 61.4 GMACs (=GFLOPS/2) on a Nvidia 1080 Ti 11GB GPU. This outperforms other dynamic methods [3,4,9,10] as well as other static networks. We report FPS of other methods and their normalized equivalents based on a 1080 Ti GPU, similar to [5]. We did not use TensorRT optimizations. Memory usage is reported for batch size 1, as the total reserved memory by PyTorch. We also experiment with a larger model based on ResNet-50, where our method shows more relative improvement as 1×1 convolutions do not require padding. SegBlocks makes it possible to run the SwiftNet-RN50 model at real-time speeds of 21 FPS.

Table 1. Results on Cityscapes semantic segmentation

Method	Set	mIoU	GMAC	FPS	norm FPS	mem
SwiftNet-RN50 (our impl.)	val	77.5	206.0	14 @ 1080 Ti	14	2575 MB
SegBlocks-RN50 ($\tau = 0.2$)	val	77.3	146.6	17 @ 1080 Ti	17	1570 MB
SegBlocks-RN50 ($\tau = 0.3$)	val	76.8	121.7	21 @ 1080 Ti	21	1268 MB
SwiftNet-RN18 (our impl.)	val	76.3	104.1	38 @ 1080 Ti	38	2052 MB
SegBlocks-RN18 ($\tau = 0.3$)	val	75.5	61.5	45 @ 1080 Ti	45	1182 MB
SegBlocks-RN18 ($\tau = 0.5$)	val	74.1	46.2	60 @ 1080 Ti	60	1111 MB
Patch Proposal Network [9]	val	75.2	–	24 @ 1080 Ti	24	1137 MB
Huang et al. [3]	val	76.4	–	1.8 @ 1080 Ti	1.8	–
Wu et al. [10]	val	72.9	–	15 @ 980 Ti	33	–
Learning Downsampling [4]	val	65.0	34	–	–	–
SwiftNet-RN18 [5]	val	75.4	104.0	40 @ 1080 Ti	40	–
BiSeNet-RN18 [11]	val	74.8	–	66 @TitanXP	64	–
ERFNet [7]	test	69.7	27.7	11 @ Titan X	18	–

4 Conclusion

We proposed a method to adapt existing segmentation networks for adaptive resolution processing. Our method achieves state-of-the-art results on Cityscapes, and can be applied on other network architectures and tasks.

Acknowledgement. The work was funded by the HAPPY and CELSA-project.

References

1. Cordts, M., et al.: The cityscapes dataset for semantic urban scene understanding. In: IEEE CVPR 2016 Proceedings, pp. 3213–3223 (2016)
2. Figurnov, M., et al.: Spatially adaptive computation time for residual networks. In: IEEE CVPR 2017 Proceedings, pp. 1039–1048 (2017)
3. Huang, Y.H., Proesmans, M., Georgoulis, S., Van Gool, L.: Uncertainty based model selection for fast semantic segmentation. In: MVA 2019 Proceedings, pp. 1–6 (2019)
4. Marin, D., et al.: Efficient segmentation: learning downsampling near semantic boundaries. In: IEEE CVPR 2019 Proceedings, pp. 2131–2141 (2019)
5. Orsic, M., Kreso, I., Bevandic, P., Segvic, S.: In defense of pre-trained imagenet architectures for real-time semantic segmentation of road-driving images. In: IEEE CVPR 2019 Proceedings, pp. 12607–12616 (2019)
6. Ren, M., Pokrovsky, A., Yang, B., Urtasun, R.: SBNet: sparse blocks network for fast inference. In: IEEE CVPR 2018 Proceedings, pp. 8711–8720 (2018)
7. Romera, E., Alvarez, J.M., Bergasa, L.M., Arroyo, R.: ERFNet: efficient residual factorized convnet for real-time semantic segmentation. IEEE Trans. Intell. Transp. Syst. **19**(1), 263–272 (2017)
8. Verelst, T., Tuytelaars, T.: Dynamic convolutions: exploiting spatial sparsity for faster inference. In: IEEE CVPR 2020 Proceedings, pp. 2320–2329 (2020)

9. Wu, T., Lei, Z., Lin, B., Li, C., Qu, Y., Xie, Y.: Patch proposal network for fast semantic segmentation of high-resolution images. In: AAAI 2020 Proceedings, pp. 12402–12409 (2020)
10. Wu, Z., Shen, C., Hengel, A.V.D.: Real-time semantic image segmentation via spatial sparsity. arXiv preprint arXiv:1712.00213 (2017)
11. Yu, C., Wang, J., Peng, C., Gao, C., Yu, G., Sang, N.: BiSeNet: bilateral segmentation network for real-time semantic segmentation. In: Ferrari, V., Hebert, M., Sminchisescu, C., Weiss, Y. (eds.) ECCV 2018. LNCS, vol. 11217, pp. 334–349. Springer, Cham (2018). https://doi.org/10.1007/978-3-030-01261-8_20

Weight-Dependent Gates for Differentiable Neural Network Pruning

Yun Li[1(✉)], Weiqun Wu[2], Zechun Liu[3], Chi Zhang[4], Xiangyu Zhang[4], Haotian Yao[4], and Baoqun Yin[1]

[1] University of Science and Technology of China, Hefei, China
yli001@mail.ustc.edu.cn, bqyin@ustc.edu.cn
[2] Chongqing University, Chongqing, China
wuwq@cqu.edu.cn
[3] Hong Kong University of Science and Technology, Kowloon, China
zliubq@connect.ust.hk
[4] Megvii Inc. (Face++), Beijing, China
{zhangchi,zhangxiangyu,yaohaotian}@megvii.com

Abstract. In this paper, we propose a simple and effective network pruning framework, which introduces novel weight-dependent gates to prune filter adaptively. We argue that the pruning decision should depend on the convolutional weights, in other words, it should be a learnable function of filter weights. We thus construct the weight-dependent gates (W-Gates) to learn the information from filter weights and obtain binary filter gates to prune or keep the filters automatically. To prune the network under hardware constraint, we train a Latency Predict Net (LPNet) to estimate the hardware latency of candidate pruned networks. Based on the proposed LPNet, we can optimize W-Gates and the pruning ratio of each layer under latency constraint. The whole framework is differentiable and can be optimized by gradient-based method to achieve a compact network with better trade-off between accuracy and efficiency. We have demonstrated the effectiveness of our method on Resnet34 and Resnet50, achieving up to 1.33/1.28 higher Top-1 accuracy with lower hardware latency on ImageNet. Compared with state-of-the-art pruning methods, our method achieves superior performance(This work is done when Yun Li, Weiqun Wu and Zechun Liu are interns at Megvii Inc (Face++)).

Keywords: Weight-dependent gates · Latency predict net · Accuracy-latency trade-off · Network pruning

1 Introduction

In recent years, convolutional neural networks (CNNs) have achieved state-of-the-art performance in many tasks, such as image classification [12], semantic

A. Bartoli and A. Fusiello (Eds.): ECCV 2020 Workshops, LNCS 12539, pp. 23–37, 2020.
https://doi.org/10.1007/978-3-030-68238-5_3

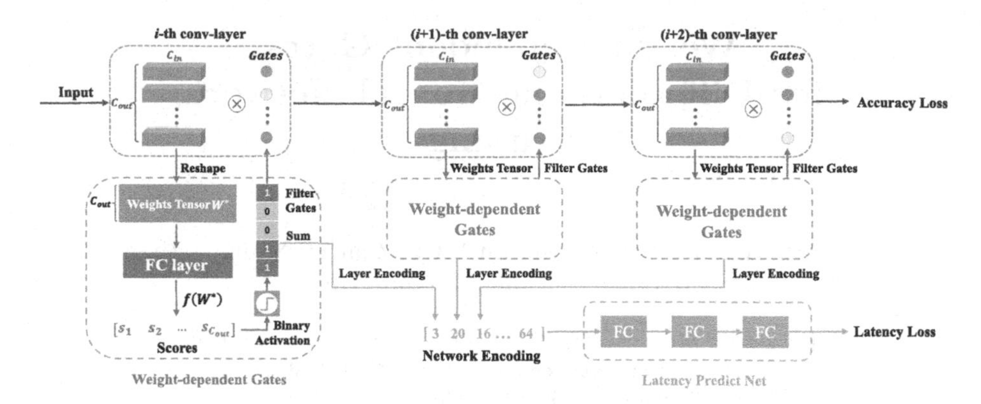

Fig. 1. Pruning framework overview. There are two main parts in our framework, Weight-dependent Gates (W-Gates) and Latency Predict Net (LPNet). The W-Gates learns the information from the filter weights of each layer and generates binary filter gates to open or close corresponding filters automatically (0: close, 1: open). The filter gates of each layer are then summed to constitute a network encoding. Next, the network encoding of the candidate pruned network is fed into LPNet to get the predicted latency. Afterwards, under a given latency constraint, accuracy loss and latency loss compete against each other during training to obtain a compact model with better accuracy-efficiency trade-off.

segmentation [30] and object detection [10]. Despite their great success, billions of float-point-operations (FLOPs) and long inference latency are still prohibitive for CNNs to deploy on many resource-constraint hardware. As a result, a significant amount of effort has been invested in CNNs compression and acceleration.

Filter Pruning [21,26,32] is seen as an intuitive and effective network compression method. However, it is constrained by the three following challenges: 1) **Pruning indicator**: CNNs are usually seen as a black box, and individual filters may play different roles within and across different layers in the network. Thus, it is difficult to fully quantify the importance of their internal convolutional filters. 2) **Pruning ratio**: The redundancy varies from different layers, making it a challenging problem to determine appropriate pruning ratios for different layers. 3) **Hardware constraint**: Most previous works adopt hardware-agnostic metrics such as FLOPs to evaluate the efficiency of a CNN. But the inconsistency between hardware agnostic metrics and actual efficiency [38] leads an increasing industrial demand on directly optimizing the hardware latency.

Previous works have tried to address these issues from different perspectives. These works mainly rely on manual-designed indicators [15,21–23] or data-driven pruning indicators [17,29,31], which have demonstrated the effectiveness of such methods in compressing the CNNs. However, manual-designed indicators usually involve human participation, and data-driven pruning indicators may be affected by the input data. Besides, the pruning ratio of each layer or a global pruning threshold is usually human-specified, making the results prone to be trapped in sub-optimal solutions [27]. Furthermore, conventional filter pruning methods only address one or two challenges above, and especially the hardware constraint is rarely addressed.

In this paper, we propose a simple yet effective filter pruning framework, which addresses all the aforementioned challenges. We argue that the pruning decisions should depend on the convolutional filters, in other words, it should be a learnable function of filter weights. Thus, instead of designing a manual indicator or directly learn scaling factors [17,29], we propose a novel weight-dependent gates (W-Gates) to directly learn a mapping from filter weights to pruning gates. W-Gates takes the filter weights of a pre-trained CNN as input to learn information from convolutional weights during training and obtain binary filter gates to open or close the corresponding filters automatically. The filter gates here are data-independent and involve no human participation. Meanwhile, the filter gates of each layer can also be summed as a layer encoding to determine the pruning ratio, and all the layer encodings constitute the candidate network encoding.

To prune the network under hardware constraint, we train a Latency Predict Net (LPNet) to estimate the latency of candidate pruned networks. After training, LPNet is added to the framework and provides latency guidance for W-Gates. As shown in Fig. 1, LPNet takes the candidate network encoding as input and output a predict latency. It facilitates us to carry out filter pruning with consideration of the overall pruned network and is beneficial for finding optimal pruning ratios. Specially, the whole framework is differentiable, allowing us to simultaneously impose the gradients of accuracy loss and latency loss to optimize the pruning ratio and W-Gates of each layer. Under a given latency constraint, the accuracy loss and the latency loss compete against each other to obtain a compact network with better accuracy-efficiency trade-off.

We evaluate our method on Resnet34/50 [12] on ImageNet dataset. Comparing with uniform baselines, we consistently delivers much higher accuracy and lower latency. With the same FLOPs and lower latency, we achieve 1.09%–1.33% higher accuracy than Resnet34, 0.67%–1.28% higher accuracy than Resnet50. Compared to other state-of-the-art pruning methods [8,13,35,47,48], our method also produces superior results.

2 Related Work

A significant amount of effort has been devoted to deep model compression, such as matrix decomposition [20,45], quantization [2,3,5,28], compact architecture learning [16,33,38,41,46] and network pruning [9,11,15,24,25,27]. In this section, we discuss the works which are most related to our work.

Network Pruning. Pruning is an intuitive and effective network compression method. Prior works devote to weights pruning. [11] proposes to prune unimportant connections whose absolute weights are smaller than a given threshold. This method is not implementation friendly and can not obtain faster inference without dedicated sparse matrix operation hardware. To tackle this problem, some filter pruning methods [15,21,29,32] have been explored recently. [15]

proposes a LASSO based method, which prunes filters with least square reconstruction. [32] prunes the filters based on statistics information computed from its next layer. The above two methods all prune filters based on feature maps. [17,29] choose to directly learn a scale factor to measure the importance of the corresponding filters. Different from the above data-driven methods, [22] propose a feature-agnostic method, which prunes filter based on a kernel sparsity and entropy indicator. A common problem of the above-described filter pruning methods is that the compression ratio for each layer need to be manually set based on human experts or heuristics, which is too time-consuming and prone to be trapped in sub-optimal solutions. To tackle this issue, [44] train a network with switchable batch normalization, which adjust its width of the network on the fly during training. The switchable batch normalization in [44] and the scale factor in [17,29] can all be seen as filter gates, which are weight-independent. However, we argue that the filter pruning decision should depend on the information in filter weights. Therefore, we propose weight-dependent gates which directly learn a mapping from filter weights to pruning decision.

Quantization and Binary Activation. [5,37] propose to quantize the real value weights into binary/ternary weights to yield a large amount of model size saving. [18] proposes to use the derivative of clip function to approximate the derivative of the sign function in the binarized networks. [40] relaxes the discrete mask variables to be continuous random variables computed by the Gumbel Softmax [19,34] to make them differentiable. [28] proposes a tight approximation to the derivative of the non-differentiable binary function with respect to real activation. These works above inspired our binary activation and gradient estimation in W-Gates.

Resource-Constrained Compression. Recently, the real hardware performance has attracted more attention compared to FLOPs. AutoML methods [14,43] propose to prune filters iteratively in different layers of a CNN via reinforcement learning or an automatic feedback loop, which take real time latency as a constraint. Some recent works [6,27,40] introduce a look-up table to record the delay of each operation or each layer and sum them to obtain the latency of the whole network. This method is valid for many CPUs and DSPs, but may not for parallel computing devices such as GPUs. [42] treats the hardware platform as a black box and creates an energy estimation model to predict the latency of specific hardware as an optimization constraint. [27] proposes PruningNet, which takes the network encoding vector as input and output weight parameters of pruned network. The above two works inspired our LPNet training. We train a LPNet to predict the latency of target hardware platform, which takes the network encoding vector as input and output the predicted latency.

3 Method

In this paper, we adopt a latency-aware approach to prune the network architecture automatically. Following the works [27,39], we formulate the network pruning as a constrained optimization problem:

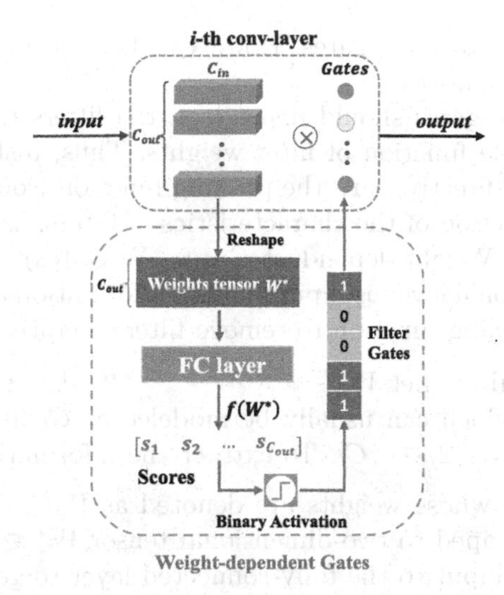

Fig. 2. The proposed W-Gates. It introduces a fully connected layer to learn the information from the reshaped filter weights W^* and generates a score for each filter. After binary activation, we obtain the weight-dependent gates (0 or 1) to open or close the corresponding filters automatically (0: close, 1: open). The gates are placed after the BN transform and ReLU.

$$\begin{aligned} &\arg\min_{w_c} \ell\left(c, w_c\right) \\ &s.t. \quad lat\left(c\right) \leqslant Lat_{const} \end{aligned} \tag{1}$$

where ℓ is the loss function specific to a given learning task, and c is the network encoding vector, which is a set of the pruned network channel width $(c_1, c_2, \cdots, c_l)$. $lat\left(c\right)$ denotes the real latency of the pruned network, which depends on the network channel width set c. w_c means the weights of the remained channels and Lat_{const} is the given latency constraint (Fig. 2).

To solve this problem, we propose a latency-aware network pruning framework, which mainly based on weight-dependent filter gates. We first construct a Weight-dependent Gates (W-Gates), which is used to learn the information from filter weights and generate binary gates to determine which filters to prune automatically. Then, a Latency Predict Net (LPNet) is trained to predict the real latency of a given architecture in specific hardware and decide how many filters to prune in each layer. These two parts complement each other to generate the pruning strategy and obtain the best model under latency constraints.

3.1 Weight-Dependent Gates

The convolutional layer has always been adopted as a black box, and we could only judge from the output what it has done. Conventional filter pruning methods mainly rely on hand-craft indicators or optimization based indicators. They share one common motivation: they try to find a pruning function which can map the filter weights to filter gates. However, their pruning function usually involves human participation.

We argue that the gates should depend on the filters themselves, in other words, it is a learnable function of filter weights. Thus, instead of designing a manual indicator, we directly learn the pruning function from the filter weights, which is a direct reflection of the characteristics of filters. To achieve the above goal, we propose the Weight-dependent Gates (W-Gates). W-Gates takes the weights of a convolutional layer as input to learn information and generate binary filter gates as the pruning function to remove filters adaptively.

Filter Gates Learning. Let $W \in \mathbb{R}^{C_l \times C_{l-1} \times K_l \times K_l}$ denotes the weights of a convolutional layer, which can usually be modeled as C_l filters and each filter $W_i \in \mathbb{R}^{C_{i-1} \times K \times K}, i = 1, 2, \cdots, C_l$. To extract the information in each filter, a fully-connected layer, whose weights are denoted as $\widehat{W} \in \mathbb{R}^{(C_{l-1} \times K_l \times K_l) \times 1}$, is introduced here. Reshaped to two-dimensional tensor $W^* \in \mathbb{R}^{C_l \times (C_{l-1} \times K_l \times K_l)}$, the filter weights are input to the fully-connected layer to generate the score of each filter:

$$s^r = f\left(W^*\right) = W^* \widehat{W}, \tag{2}$$

where $s^r = \left[s_1^r, s_2^r, \ldots, s_{C_l}^r\right]$ denotes the score set of the filters in this convolutional layer.

To suppress the expression of filters with lower scores and obtain binary filter gates, we introduce the following activation function:

$$\sigma\left(x\right) = \frac{\operatorname{Sign}\left(x\right) + 1}{2}. \tag{3}$$

The curve of Eq. (3) is shown in Fig. 3(a). We can see that after processing by the activation function, the negative scores will be converted to 0, and positive scores will be converted to 1. Then, we get the binary filter gates of the filters in this layer:

$$gates^b = \sigma\left(s^r\right) = \sigma\left(f\left(W^*\right)\right). \tag{4}$$

Different from the path binarization [1] in neural architecture search, in filter pruning tasks, the pruning decision should depend on the filter weights, in other words, our proposed binary filter gates are weights-dependent, as shown in Eq. (4).

Next, we sum the binary filter gates of each layer to obtain the layer encoding, and all the layer encodings form the network encoding vector c. The layer encoding here denotes the number of filters kept and can also determine the pruning ratio of each layer.

Gradient Estimation. As can be seen from Fig. 3, the derivative of function $\sigma\left(\cdot\right)$ is an impulse function, which cannot be used directly during the training process. Inspired by the recent Quantized Model works [4,18,28], specifically Bi-Real Net [28], we introduce a differentiable approximation of the non-differentiable function $\sigma\left(x\right)$. The gradient estimation process is as follows:

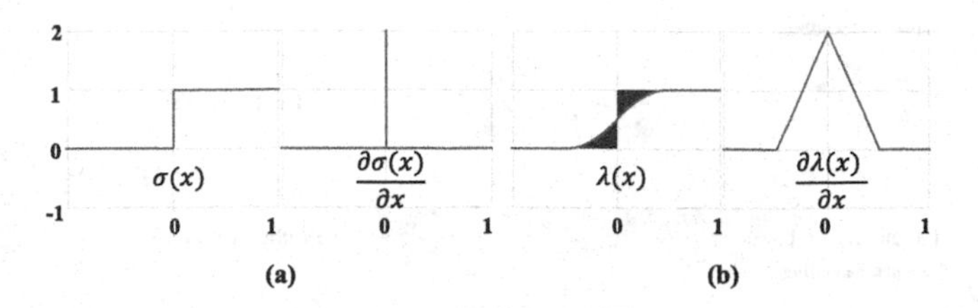

Fig. 3. (a) The proposed binary activation function and its derivative. (b) The designed differentiable piecewise polynomial function and its derivative, and this derivative is used to approximate the derivative of binary activation function in gradients computation.

$$\frac{\partial L}{\partial X_r} = \frac{\partial L}{\partial \sigma(X_r)} \frac{\partial \sigma(X_r)}{\partial X_r} \approx \frac{\partial L}{\partial X_b} \frac{\partial \lambda(X_r)}{\partial X_r}, \tag{5}$$

where X_r denotes the real value output s_i^r, X_b means the binary output. $\lambda(X_r)$ is the approximation function we designed, which is a piecewise polynomial function:

$$\lambda(X_r) = \begin{cases} 0, & if\ X_r < -\frac{1}{2} \\ 2X_r + 2X_r^2 + \frac{1}{2}, & if\ -\frac{1}{2} \leqslant X_r < 0 \\ 2X_r - 2X_r^2 + \frac{1}{2}, & if\ 0 \leqslant X_r < \frac{1}{2} \\ 1, & otherwise \end{cases}, \tag{6}$$

and the gradient of above approximation function is:

$$\frac{\partial \lambda(X_r)}{\partial X_r} = \begin{cases} 2 + 4X_r, & if\ -\frac{1}{2} \leqslant X_r < 0 \\ 2 - 4X_r, & if\ 0 \leqslant X_r < \frac{1}{2} \\ 0, & otherwise \end{cases}. \tag{7}$$

As discussed above, we can adopt the binary activation function Eq. (3) to obtain the binary filter gates in the forward propagation, and then update the weights of fully-connected layer with an approximate gradient Eq. (7) in the backward propagation.

3.2 Latency Predict Net

Previous works on model compression aim primarily to reduce FLOPs, but it does not always reflect the actual latency on hardware. Therefore, some recent NAS-based methods [27,40] pay more attention to adopt the hardware latency as a direct evaluation indicator than FLOPs. [40] proposes to build a latency look-up table to estimate the overall latency of a network, as they assume that the runtime of each operator is independent of other operators, which works well on many mobile serial devices, such as CPUs and DSPs. Previous works [6,27, 40] have demonstrated the effectiveness of such method. However, the latency generated by look-up table is not differentiable with respect to the filter selection within layer and the pruning ratio of each layer.

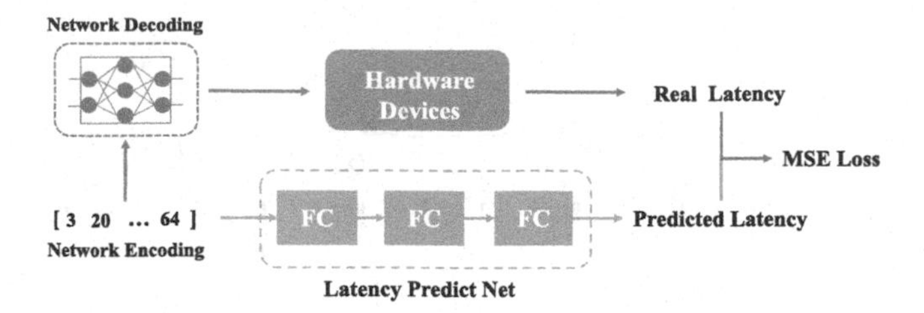

Fig. 4. The offline training process of LPNet. ReLU is placed after each FC layer. LPNet takes network encoding vectors as the input data and the measured hardware latency as the ground truth. Mean Squared Error (MSE) loss function is adopted here.

To address the above problem, we construct a LPNet to predict the real latency of the whole network or building blocks. The proposed LPNet is fully differentiable with respect to filter gates and pruning ratio of each layer. As shown in Fig. 4, the LPNet consists of three fully-connected layers, which takes a network encoding vector $c = (c_1, c_2, \ldots, c_n)$ as input and output the latency for specified hardware platform:

$$lat\,(c) = \text{LPNet}\,(c_1, c_2, \ldots, c_n)\,. \tag{8}$$

where $c_i = \text{sum}\left(gates_i^b\right) = \text{sum}\left(\sigma\left(f\left(W^*\right)\right)\right)$ in the pruning framework.

To pre-train the LPNet, we generate network encoding vectors as input and test their real latency on specific hardware platform as labels. As there is no need to train the decoding network, the training of LPNet is very efficient. As a result, training such a LPNet makes the latency constraint differentiable with respect to the network encoding and binary filter gates shown in Fig. 1. Thus we can use gradient-based optimization to adjust the filter pruning ratio of each convolutional layer and obtain the best pruning ratio automatically.

3.3 Latency-Aware Filter Pruning

The proposed method consists three main stages. First, training the LPNet offline, as described in Sect. 3.2. After LPNet is trained, we can obtain the latency by inputting the encoding vector of a candidate pruned network. Second, pruning the network under latency constraint. We add W-Gates and the LPNet to a pretrained network to do filter pruning, in which the weights of LPNet are fixed. As shown in Fig. 1, W-Gates learns the information from convolutional weights and generate binary filter gates to determine which filters to prune. Next, LPNet takes the network encoding of candidate pruned net as input and output a predicted latency to optimize the pruning ratio and filter gates of each layer. Then, the accuracy loss and the latency loss compete against each other during training and finally obtain a compact network with best accuracy while meeting the latency constraint. Third, fine-tuning the network. After getting the pruned network, a fine-tuning process with only few epochs follows to regain accuracy and obtain a better performance, which is less time-consuming than training from scratch.

Furthermore, to make a better accuracy-latency trade-off, we define the following latency-aware loss function:

$$\ell\left(c, w_c\right) = \mathrm{CE}\left(c, w_c\right) + \alpha \log\left(1 + lat\left(c\right)\right), \tag{9}$$

where $\mathrm{CE}\left(c, w_c\right)$ denotes the cross-entropy loss of an architecture with a network encoding c and parameters w_c. $lat\left(c\right)$ is the latency of the architecture with network encoding vector c. The coefficient α can modulate the magnitude of latency term. Such a loss function can carry out filter pruning task with consideration of the overall network structure, which is beneficial for finding optimal solutions for network pruning. In addition, this function is differentiable with respect to layer-wise filter choices c and the number of filters, which allows us to use gradient-based method to optimize them and obtain a best trade-off between accuracy and efficiency.

4 Experiments

In this section, we demonstrate the effectiveness of our method. First, we give a detail description of our experiment settings. Next, we carry out four ablation studies on ImageNet dataset to illustrate the effect of the key part W-Gates in our method. Then, we prune resnet34 and resnet50 under latency constraint. Finally, we compare our method with several state-of-the-art filter pruning methods.

4.1 Experiment Settings

We carry out all experiments on the ImageNet ILSVRC 2012 datasets [7]. ImageNet contains 1.28 million training images and 50000 validation images, which are categorized into 1000 classes. The resolution of the input images are set to 224×224. All the experiments are implemented with Pytorch [36] framework and networks are trained using stochastic gradient descent (SGD).

To train the LPNet offline, we sample network encoding vectors c from the search space and decode the corresponding network to test their real latency on one NVIDIA RTX 2080 Ti GPU as latency labels. As there is no need to train the network, it takes only few milliseconds to get a latency label. For deeper building block architecture, such as resnet50, the network encoding sampling space is huge. We choose to predict the latency of building block and then sum them up to get the predict latency of the overall network, and this will greatly reduce the encoding sampling space. Besides, the LPNet of building block can also be reused cross models of different depths and different tasks on the same type of hardware. For the bottleneck building block of resnet, we collect 170000 (c, latency) pairs to train the LPNet of bottleneck building block and 5000 (c, latency) pairs to train the LPNet of basic building block. We randomly choose 80% of the dataset as training data and leave the rest as test data. We use Adam to train the LPNets and find the average test errors can quickly converge to lower than 2%.

4.2 Pruning Results Under Latency Constraint

The inconsistency between hardware agnostic metrics and actual efficiency leads an increasing attention in directly optimizing the latency on the target devices. Taking the CNN as a black box, we train a LPNet to predict the real latency in the target device. For Resnet34 and Resnet50, we train two LPNets offline to predict the latency of basic building blocks and bottleneck building blocks, respectively. To fully consider all the factors and decode the sparse architecture, we add the factors of feature map size and downsampling to the network encoding vector. For these architectures with shortcut, we do not prune the output channels of the last layer in each building block to avoid mismatching with the shortcut channels.

Table 1. This table compares the Top-1 accuracy and latency on ImageNet about Resnet34 and Resnet50. We set the same compression ratio for each layer as uniform baseline. The input batch size is set to 100 and the latency is measured using Pytorch on NVIDIA RTX 2080 Ti GPU. The results show that, with the same FLOPs, our method outperforms the uniform baselines by a large margin in terms of accuracy and latency.

Model	Uniform Baselines			W-Gates		
	FLOPs	Top1-Acc	Latency	FLOPs	Top1-Acc	Latency
Resnet34	3.7G (1X)	73.88%	54.04ms	–	–	–
	2.9G	72.56%	49.23 ms	2.8G	**73.76%**	**46.67 ms**
	2.4G	72.05%	44.27 ms	2.3G	**73.35%**	**40.75 ms**
	2.1G	71.32%	43.47 ms	2.0G	**72.65%**	**37.30 ms**
Resnet50	4.1G (1X)	76.15%	105.75 ms	–	–	–
	3.1G	75.59%	97.87 ms	3.0G	**76.26%**	**95.15 ms**
	2.6G	74.77%	91.53 ms	2.6G	**76.05%**	**88.00 ms**
	2.1G	74.42%	85.20 ms	2.1G	**75.14%**	**80.17 ms**

Pruning Results on Resnet34. We first employ the pruning experiments on a medium depth network Resnet34. Resnet34 consists of basic building blocks, each basic building block contains two 3×3 convolutional layers. We add the designed W-Gates to the first layer of each basic building block to learn the information and do filter selection automatically. LPNets are added to the framework to predict the latency of each building block. The pruning results are shown in Table 1. It can be observed that our method can save 25% hardware latency with only 0.5% accuracy loss on ImageNet dataset. With the same FLOPs, W-Gates achieves 1.1% to 1.3% higher Top-1 accuracy than uniform baseline, and the hardware latency is much lower, which shows that our W-Gates can automatically generate efficient architectures.

Pruning Results on Resnet50. For the deeper network resnet50, we adopt the same setting with resnet34. Resnet50 consists of bottleneck building blocks, each of which contains a 3×3 layer and two 1×1 layers. We employ W-Gates to prune the filters of the first two layers in each bottleneck module during training. The Top-1 accuracy and hardware latency of pruned models are shown in Table 1. When pruning 37% FLOPs, we can save 17% hardware latency without notable accuracy loss.

Table 2. This table compares the Top-1 ImageNet accuracy of our W-Gates method and state-of-the-art pruning methods on ResNet34 and ResNet50.

Model	Methods	FLOPs	Top1 Acc
Resnet34	IENNP [35] (CVPR-19)	2.8G	72.83%
	FPGM [13] (CVPR-19)	2.2G	72.63%
	W-Gates (ours)	2.8G	**73.76%**
	W-Gates (ours)	2.3G	**73.35%**
Resnet50	VCNNP [47] (CVPR-19)	2.4G	75.20%
	FPGM [13] (CVPR-19)	2.4G	75.59%
	FPGM [13] (CVPR-19)	1.9G	74.83%
	IENNP [35] (CVPR-19)	2.2G	74.50%
	RRBP [48] (ICCV-19)	1.9G	73.00%
	C-SGD-70 [8] (CVPR-19)	2.6G	75.27%
	C-SGD-60 [8] (CVPR-19)	2.2G	74.93%
	C-SGD-50 [8] (CVPR-19)	1.8G	**74.54%**
	W-Gates (ours)	3.0G	**76.26%**
	W-Gates (ours)	2.4G	**75.96%**
	W-Gates (ours)	2.1G	**75.14%**
	W-Gates (ours)	1.9G	**74.32%**

4.3 Comparisons with State-of-the-Arts

We compare our method with several state-of-the-art filter pruning methods [8,13,35,47,48] on ResNet34 and ResNet50, shown in Table 2. As there is no latency data provided in these works, we only compare the Top1 accuracy with the same FLOPs. It can be observed that, compared with state-of-the-art filter pruning methods, W-Gates achieves higher accuracy with the same FLOPs.

5 Conclusion

In this paper, we propose a novel filter pruning method to address the problems on pruning indicator, pruning ratio and platform constraint on the same time. We first propose a W-Gates to learn the information from convolutional weights and generate binary filter gates. Then, we pretrain a LPNet to predict the hardware latency of candidate pruned networks. The entire framework is differentiable with respect to filter choices and pruning ratios, which can be optimized by gradient-based method to achieve better pruning results.

Acknowledgement. This work is supported the Equipment Pre-Research Foundation of China under grant No. 61403120201.

References

1. Cai, H., Zhu, L., Han, S.: ProxylessNAS: direct neural architecture search on target task and hardware. arXiv preprint arXiv:1812.00332 (2018)
2. Cao, S., et al.: SeerNet: predicting convolutional neural network feature-map sparsity through low-bit quantization. In: Proceedings of the IEEE Conference on Computer Vision and Pattern Recognition (CVPR), pp. 11216–11225 (2019)
3. Chen, W., Wilson, J., Tyree, S., Weinberger, K., Chen, Y.: Compressing neural networks with the hashing trick. In: International Conference on Machine Learning (ICML), pp. 2285–2294 (2015)
4. Courbariaux, M., Bengio, Y., David, J.P.: BinaryConnect: training deep neural networks with binary weights during propagations. In: Advances in Neural Information Processing Systems (NeurIPS), pp. 3123–3131 (2015)
5. Courbariaux, M., Hubara, I., Soudry, D., El-Yaniv, R., Bengio, Y.: Binarized neural networks: training deep neural networks with weights and activations constrained to +1 or -1. arXiv preprint arXiv:1602.02830 (2016)
6. Dai, X., Y., et al.: ChamNet: towards efficient network design through platform-aware model adaptation. In: Proceedings of the IEEE Conference on Computer Vision and Pattern Recognition (CVPR), pp. 11398–11407 (2019)
7. Deng, J., Dong, W., Socher, R., Li, L.J., Li, K., Fei-Fei, L.: ImageNet: a large-scale hierarchical image database. In: IEEE Conference on Computer Vision and Pattern Recognition (CVPR), pp. 248–255 (2009)
8. Ding, X., Ding, G., Guo, Y., Han, J.: Centripetal SGD for pruning very deep convolutional networks with complicated structure. In: Proceedings of the IEEE Conference on Computer Vision and Pattern Recognition (CVPR), pp. 4943–4953 (2019)
9. Ding, X., Ding, G., Guo, Y., Han, J., Yan, C.: Approximated oracle filter pruning for destructive CNN width optimization. arXiv preprint arXiv:1905.04748 (2019)
10. Girshick, R., Donahue, J., Darrell, T., Malik, J.: Rich feature hierarchies for accurate object detection and semantic segmentation. In: Proceedings of the IEEE Conference on Computer Vision and Pattern Recognition (CVPR), pp. 580–587 (2014)
11. Han, S., Mao, H., Dally, W.J.: Deep compression: compressing deep neural networks with pruning, trained quantization and Huffman coding. In: Proceedings of International Conference on Learning Representations (ICLR) (2016)
12. He, K., Zhang, X., Ren, S., Sun, J.: Deep residual learning for image recognition. In: Proceedings of the IEEE Conference on Computer Vision and Pattern Recognition (CVPR), pp. 770–778 (2016)

13. He, Y., Liu, P., Wang, Z., Hu, Z., Yang, Y.: Filter pruning via geometric median for deep convolutional neural networks acceleration. In: Proceedings of the IEEE Conference on Computer Vision and Pattern Recognition (CVPR), pp. 4340–4349 (2019)
14. He, Y., Lin, J., Liu, Z., Wang, H., Li, L.-J., Han, S.: AMC: AutoML for model compression and acceleration on mobile devices. In: Ferrari, V., Hebert, M., Sminchisescu, C., Weiss, Y. (eds.) ECCV 2018. LNCS, vol. 11211, pp. 815–832. Springer, Cham (2018). https://doi.org/10.1007/978-3-030-01234-2_48
15. He, Y., Zhang, X., Sun, J.: Channel pruning for accelerating very deep neural networks. In: Proceedings of the IEEE International Conference on Computer Vision (ICCV) (2017)
16. Howard, A.G., et al.: MobileNets: efficient convolutional neural networks for mobile vision applications. arXiv preprint arXiv:1704.04861 (2017)
17. Huang, Z., Wang, N.: Data-driven sparse structure selection for deep neural networks. In: Ferrari, V., Hebert, M., Sminchisescu, C., Weiss, Y. (eds.) ECCV 2018. LNCS, vol. 11220, pp. 317–334. Springer, Cham (2018). https://doi.org/10.1007/978-3-030-01270-0_19
18. Hubara, I., Courbariaux, M., Soudry, D., El-Yaniv, R., Bengio, Y.: Binarized neural networks. In: Advances in Neural Information Processing Systems (NeurIPS), pp. 4107–4115 (2016)
19. Jang, E., Gu, S., Poole, B.: Categorical reparameterization with gumbel-softmax. arXiv preprint arXiv:1611.01144 (2016)
20. Jia, K., Tao, D., Gao, S., Xu, X.: Improving training of deep neural networks via singular value bounding. In: Proceedings of the IEEE Conference on Computer Vision and Pattern Recognition (CVPR), vol. 2017, pp. 3994–4002 (2017)
21. Li, H., Kadav, A., Durdanovic, I., Samet, H., Graf, H.P.: Pruning filters for efficient convnets. In: Proceedings of International Conference on Learning Representations (ICLR) (2017)
22. Li, Y., et al.: Exploiting kernel sparsity and entropy for interpretable CNN compression. In: Proceedings of the IEEE Conference on Computer Vision and Pattern Recognition (CVPR), pp. 2800–2809 (2019)
23. Li, Y., Wang, L., Peng, S., Kumar, A., Yin, B.: Using feature entropy to guide filter pruning for efficient convolutional networks. In: Tetko, I.V., Kůrková, V., Karpov, P., Theis, F. (eds.) ICANN 2019. LNCS, vol. 11728, pp. 263–274. Springer, Cham (2019). https://doi.org/10.1007/978-3-030-30484-3_22
24. Lin, S., Ji, R., Li, Y., Wu, Y., Huang, F., Zhang, B.: Accelerating convolutional networks via global & dynamic filter pruning. In: Proceedings of the 27th International Joint Conference on Artificial Intelligence (IJCAI), pp. 2425–2432 (2018)
25. Lin, S., et al.: Towards optimal structured CNN pruning via generative adversarial learning. In: Proceedings of the IEEE Conference on Computer Vision and Pattern Recognition (CVPR), pp. 2790–2799 (2019)
26. Liu, B., Wang, M., Foroosh, H., Tappen, M., Pensky, M.: Sparse convolutional neural networks. In: Proceedings of the IEEE Conference on Computer Vision and Pattern Recognition (CVPR), pp. 806–814 (2015)
27. Liu, Z., Mu, H., Zhang, X., Guo, Z., Yang, X., Cheng, K.T., Sun, J.: MetaPruning: meta learning for automatic neural network channel pruning. In: Proceedings of the IEEE International Conference on Computer Vision (ICCV), pp. 3296–3305 (2019)

28. Liu, Z., Wu, B., Luo, W., Yang, X., Liu, W., Cheng, K.-T.: Bi-real net: enhancing the performance of 1-bit CNNs with improved representational capability and advanced training algorithm. In: Ferrari, V., Hebert, M., Sminchisescu, C., Weiss, Y. (eds.) ECCV 2018. LNCS, vol. 11219, pp. 747–763. Springer, Cham (2018). https://doi.org/10.1007/978-3-030-01267-0_44
29. Liu, Z., Li, J., Shen, Z., Huang, G., Yan, S., Zhang, C.: Learning efficient convolutional networks through network slimming. In: Proceedings of the IEEE International Conference on Computer Vision (ICCV), pp. 2736–2744 (2017)
30. Long, J., Shelhamer, E., Darrell, T.: Fully convolutional networks for semantic segmentation. In: Proceedings of the IEEE Conference on Computer Vision and Pattern Recognition (CVPR), pp. 3431–3440 (2015)
31. Luo, J.H., Wu, J.: Autopruner: an end-to-end trainable filter pruning method for efficient deep model inference. Pattern Recognit. (PR) 107461 (2020)
32. Luo, J.H., Wu, J., Lin, W.: ThiNet: a filter level pruning method for deep neural network compression. In: Proceedings of the IEEE International Conference on Computer Vision (ICCV), pp. 5058–5066 (2017)
33. Ma, N., Zhang, X., Zheng, H.-T., Sun, J.: ShuffleNet V2: practical guidelines for efficient CNN architecture design. In: Ferrari, V., Hebert, M., Sminchisescu, C., Weiss, Y. (eds.) Computer Vision – ECCV 2018. LNCS, vol. 11218, pp. 122–138. Springer, Cham (2018). https://doi.org/10.1007/978-3-030-01264-9_8
34. Maddison, C.J., Mnih, A., Teh, Y.W.: The concrete distribution: a continuous relaxation of discrete random variables. arXiv preprint arXiv:1611.00712 (2016)
35. Molchanov, P., Mallya, A., Tyree, S., Frosio, I., Kautz, J.: Importance estimation for neural network pruning. In: The IEEE Conference on Computer Vision and Pattern Recognition (CVPR) (2019)
36. Paszke, A., Gross, S., Chintala, S., Chanan, G.: Pytorch: tensors and dynamic neural networks in python with strong GPU acceleration (2017)
37. Rastegari, M., Ordonez, V., Redmon, J., Farhadi, A.: XNOR-Net: ImageNet classification using binary convolutional neural networks. In: Leibe, B., Matas, J., Sebe, N., Welling, M. (eds.) ECCV 2016. LNCS, vol. 9908, pp. 525–542. Springer, Cham (2016). https://doi.org/10.1007/978-3-319-46493-0_32
38. Sandler, M., Howard, A., Zhu, M., Zhmoginov, A., Chen, L.C.: MobileNetV2: inverted residuals and linear bottlenecks. In: The IEEE Conference on Computer Vision and Pattern Recognition (CVPR) (2018)
39. Tan, M., et al.: MnasNet: platform-aware neural architecture search for mobile. In: The IEEE Conference on Computer Vision and Pattern Recognition (CVPR) (2019)
40. Wu, B., et al.: FBNet: hardware-aware efficient convnet design via differentiable neural architecture search. In: Proceedings of the IEEE Conference on Computer Vision and Pattern Recognition (CVPR), pp. 10734–10742 (2019)
41. Xu, Y., et al.: Latency-aware differentiable neural architecture search. arXiv preprint arXiv:2001.06392 (2020)
42. Yang, H., Zhu, Y., Liu, J.: ECC: platform-independent energy-constrained deep neural network compression via a bilinear regression model. In: Proceedings of the IEEE Conference on Computer Vision and Pattern Recognition (CVPR), pp. 11206–11215 (2019)
43. Yang, T.-J., et al.: NetAdapt: platform-aware neural network adaptation for mobile applications. In: Ferrari, V., Hebert, M., Sminchisescu, C., Weiss, Y. (eds.) ECCV 2018. LNCS, vol. 11214, pp. 289–304. Springer, Cham (2018). https://doi.org/10.1007/978-3-030-01249-6_18

44. Yu, J., Yang, L., Xu, N., Yang, J., Huang, T.: Slimmable neural networks. In: International Conference on Learning Representations (ICLR) (2019)
45. Yu, X., Liu, T., Wang, X., Tao, D.: On compressing deep models by low rank and sparse decomposition. In: Proceedings of the IEEE Conference on Computer Vision and Pattern Recognition (CVPR), pp. 7370–7379 (2017)
46. Zhang, X., Zhou, X., Lin, M., Sun, J.: ShuffleNet: an extremely efficient convolutional neural network for mobile devices. In: Proceedings of the IEEE Conference on Computer Vision and Pattern Recognition (CVPR), pp. 6848–6856 (2018)
47. Zhao, C., Ni, B., Zhang, J., Zhao, Q., Zhang, W., Tian, Q.: Variational convolutional neural network pruning. In: Proceedings of the IEEE Conference on Computer Vision and Pattern Recognition (CVPR), pp. 2780–2789 (2019)
48. Zhou, Y., Zhang, Y., Wang, Y., Tian, Q.: Accelerate CNN via recursive Bayesian pruning. In: Proceedings of the IEEE International Conference on Computer Vision (ICCV), pp. 3306–3315 (2019)

QuantNet: Learning to Quantize by Learning Within Fully Differentiable Framework

Junjie Liu[1], Dongchao Wen[1](✉), Deyu Wang[1], Wei Tao[1], Tse-Wei Chen[2], Kinya Osa[2], and Masami Kato[2]

[1] Canon Information Technology (Beijing) Co., Ltd., Beijing, China
{liujunjie,wendongchao,wangdeyu,taowei}@canon-ib.com.cn
[2] Device Technology Development Headquarters, Canon Inc., Ota City, Japan
twchen@ieee.org

Abstract. Despite the achievements of recent binarization methods on reducing the performance degradation of Binary Neural Networks (BNNs), gradient mismatching caused by the Straight-Through-Estimator (STE) still dominates quantized networks. This paper proposes a meta-based quantizer named QuantNet, which utilizes a differentiable sub-network to directly binarize the full-precision weights without resorting to STE and any learnable gradient estimators. Our method not only solves the problem of gradient mismatching, but also reduces the impact of discretization errors, caused by the binarizing operation in the deployment, on performance. Generally, the proposed algorithm is implemented within a fully differentiable framework, and is easily extended to the general network quantization with any bits. The quantitative experiments on CIFAR-100 and ImageNet demonstrate that QuantNet achieves the significant improvements comparing with previous binarization methods, and even bridges gaps of accuracies between binarized models and full-precision models.

Keywords: Deep neural networks · Quantization · Compression

1 Introduction

Deep neural networks (DNNs) have achieved remarkable success in several fields in recent years. In particular, convolutional neural networks (CNNs) have shown state-of-the-art performance in various computer vision tasks such as image classification, object detection, trajectory tracking, etc. However, an increasing number of parameters in these networks also lead to the larger model size and higher computation cost, which gradually becomes great hurdles for many applications, especially on some resource-constrained devices with limited memory space and low computation ability.

To reduce the model size of DNNs, representative techniques such as network quantization [7,18,20,25,42], filters pruning [14,30,31], knowledge distillation [16,17,27] and deliberate architecture design [6,26] are proposed. As one of

A. Bartoli and A. Fusiello (Eds.): ECCV 2020 Workshops, LNCS 12539, pp. 38–53, 2020.
https://doi.org/10.1007/978-3-030-68238-5_4

typical solutions, the quantization based method quantizes floating-point values into discrete values in order to generate the quantized neural networks (QNNs) as compact as possible. In the most extreme case, if both network weights and network activations are binarized (BNNs) [8], the computation can be efficiently implemented via bitwise operations, which enables about 32× memory saving and 58× speeding up [33] on CPUs in inference.

Despite the advantages we mentioned above, how to alleviate performance degradation of quantized networks is still under research, especially for binarized networks. In general, BNNs involve a *sign* function to obtain signs of parameters. The non-differentiable sign function leads to gradient vanishing almost anywhere. To address this issue, some works [28,44] propose low-bit training algorithms to relieve the impact of gradients quantization errors, and another works focus on estimating the vanishing gradients. The Straight-Through Estimator (STE) [3] is commonly used to estimate the vanishing gradients during the back-propagation, while the well-known gradient mismatching problem [15,19,39] is introduced.

As the number of quantized bits decrease, the gradients estimated by STE depart further from the real gradients. Thus, the gradient mismatching is considered as the main bottleneck of performance improvements of binarized models. As one of promising solutions, estimating more accurate gradients is suggested by recent methods. Some of these methods [29,37,38,40] try to refine the gradients estimated by STE with extra parameters, and others [2,5] address the problem by replacing STE with learnable gradient estimators. Different from these efforts on estimating more accurate gradients, the individual method [25] employs a differentiable function *tanh* as a soft binarizing operation, in order to replace the non-differentiable function *sign*. *Thus, it will no longer require STE to estimate gradients.*

In this paper, we follow the idea of soft binarization, but we focus on solving two important issues that are left out. Firstly, although the soft binarization solves the problem of gradient mismatching, another issue of gradient vanishing from the function *tanh* arises. It not only causes the less ideal convergence behavior, but makes the solution highly suboptimal. Moreover, as the soft binarization involves a post-processing step, how to reduce the impact of the discretization errors on performance is very important. With these motivations, *we propose a meta-based quantizer named QuantNet for directly generating binarized weights with an additional neural network. The said network is referred to as a meta-based quantizer and optimized with the binarized model jointly.* In details, it not only generates the higher dimensional manifolds of weights for easily finding the global optimal solution, but also penalizes the binarized weights into sparse values with a task-driven priority for minimizing the discretization error. For demonstrating the effectiveness of our claims, we present the mathematical definition of two basic hypotheses in our binarization method, and design a joint optimization scheme for the QuantNet.

We evaluate the performance of our proposed QuantNet by comparing it with the existing binarization methods on the standard benchmarks of classification task with CIFAR-100 [23] and ImageNet [9]. As for the baseline with different network architectures, AlexNet [24], ResNet [13], MobileNet [36] and DenseNet [11] are validated. The extensive experiments demonstrate that our method achieves remarkable improvements than state-of-the-arts across various datasets and network architectures.

In the following, we briefly review previous works related to network quantization in Sect. 2. In Sect. 3, we define the notations and present the mathematical definition of existing binarization methods. For Sect. 4, we present two basic hypotheses of our binarization method and exhibit the implementation details. Finally, we demonstrate the effectiveness and efficiency of our method in Sect. 5, and make the conclusions in Sect. 6.

2 Related Work

In this section, we briefly review existing methods on neural network quantization. As the most typical strategy to achieve the purpose of network compression, the network quantization has two major benefits - reducing the model size while improving the inference efficiency. Comparing with the strategies of network filters pruning [14,30,31] and compact architecture design [6,26], how to alleviate the performance degradation in quantized model [19] is still unsolved, especially for the binarized model [1,15,39].

Deterministic Weight Quantization. Through introducing a deterministic function, traditional methods quantize network weights (or activations) by minimizing quantization errors. For examples, BinaryConnect [8] uses a stochastic function for binarizing weights to the binary set $\{+1, -1\}$, which achieves better performance than two-step approachs [12,21] on several tasks. Besides, XNOR-net [33] scales the binarized weights with extra scaling factors and obtains better results. Furthermore, Half-Wave-Gaussian-Quantization (HWGQ) [4] observes the distribution of activations, and suggests some non-uniform quantization functions for constraining unbounded values of activations. Instead of binarizing the model, the ternary-connect network [43] and DoReFa-Net [42] perform the quantization with multiple-bits via various functions to bound the range of parameters.

These methods purely focus on minimizing quantization errors between full-precision weights and quantized weights, however less quantization errors do not necessarily mean better performance of a quantized model.

Loss-Aware Weight Quantization. As less quantization errors do not necessarily mean better performance of a quantized model, several recent works propose the loss-aware weight quantization in terms of minimizing the task loss rather than quantization errors. The loss-aware binarization (LAB) [18] proposes a proximal Newton algorithm with diagonal Hessian approximation that minimizes the loss with respect to the binarized weights during optimization. Similar to LAB, LQ-Net [41] allows floating-point values to represent the basis of quantized values during the quantization. Besides, PACT [7] and SYQ [10] suggest parameterized functions to clip the weights or activation value during training.

In the latest works, QIL [20] parameterizes the non-linear quantization intervals and obtains the optimal solution by minimizing with the constraint from task loss. And self binarization [25] employs a soft-binarization function to evolve weights and activations during training to become binary.

In brief, through introducing learnable constraints or scaling factors, these methods alleviate the performance degradation in their quantized models, but the gradient mismatching problem caused by the sign function and STE [1] *is still unconsidered.*

Meta-based Weight Quantization. As the quantization operator in training process is non-differentiable, which leads to either infinite gradients or zero gradients, MixedQuant [37] addresses a gradient refiner by introducing the assistant variable for approximating the more accurate gradients. Similar to MixedQuant, ProxQuant [2] proposes an alternative approach that formulates quantized network training as a regularized learning problem and optimizes it by the proximal gradients. Furthermore, Meta-Quant [5] proposes a gradient estimator to directly learn the gradients of quantized weights by a neural network, in order to remove STE commonly used in back-propagation.

Although such methods have noticed that refining the gradients computed by STE or directly estimating the gradients by meta-learner is helpful to alleviate the problem of gradient mismatching, the increasing complexity of learning gradients introduces a new bottleneck.

3 Preliminaries

Notations. For a vector x, where $x_i; i \leq n$ is the element of x, we use $\sqrt{x}$ to denote the element-wise square root, $|x|$ denotes the element-wise absolute value, and $\|x\|_p$ is the p-norm of x. $sign(x)$ is an element-wise function denoting that $sign(x_i) = 1; \forall i \leq n$ if $x_i \geq 0$ and -1 otherwise. We use $diag(X)$ to return the diagonal elements of matrix X, and $Diag(x)$ to generate a diagonal matrix with vector x. For two vectors x and y, $x \odot y$ denotes the element-wise multiplication and $x \oslash y$ denotes the element-wise division. For a matrix X, $vec(X)$ denotes to return a vector by stacking all the columns of X. In general, ℓ is used to denote the objective loss, and both $\partial \ell / \partial x$ and $\nabla \ell(x)$ denote the derivative of ℓ with respect to x.

Background of Network Binarization. The main operation in network binarization is the linear (or non-linear) discretization. Taking a multilayer perception (MLP) neural network as an example, one of its hidden layers can be expressed as

$$\mathbf{w}_q = f(\mathbf{w})^r \odot binarize(\mathbf{w}) \tag{1}$$

where $\mathbf{w} \in \mathbb{R}^{m \cdot n}$ is the full-precision weights, and m, n are the number of input filter channels, the number of output filter channels[1], respectively. Based

[1] In this paper, the kernels on full connected layer are regarded as a special type of the convolutional kernels.

on the full-precision (floating-point) weights, the corresponding binarized weights $\mathbf{w}_q$ is computed by two separate functions $f(\mathbf{w})^r$ and $binarize(\mathbf{w})$, and the goal is to represent the floating-point elements in $\mathbf{w}$ with one bit.

In BinaryConnect [8], $f(\mathbf{w})^r$ is defined as the constant 1, and $binarize(\mathbf{w})$ is defined by $sign(\mathbf{w})$, which means each element of $\mathbf{w}$ will be binarized to {-1,+1}. For XNOR-Net [33], it follows the definition of BinaryConnect on $binarize(\mathbf{w})$, but further defines $f(\mathbf{w}_t)^r = \|\mathbf{w}_t\|_1/(m \times n)$ and r is defined as the constant 1, where t is the current number of training iterations. Different from the determining function on $f(\mathbf{w})$, the Loss-Aware Binarization (LAB) [18] suggests a task-driven $f(\mathbf{w}_t)$ with the definition of $\|d_{t-1} \odot \mathbf{w}_t\|_1/\|d_{t-1}\|_1$, where d_{t-1} is a vector containing the diagonal $diag(D_{t-1})$ of an approximate Hessian D_{t-1} of the task loss. Furthermore, QIL [20] extends $f(\mathbf{w}_t)^{r_t}$ into a nonlinear projection by setting r_t to be learnable and $r_t > 1$ for all t. Considering the back-propagation, as STE [39] with *sign* function introduces the major performance bottleneck for BNNs, Self-Binarization [25] defines $binarize(\mathbf{w})$ as $tanh(\mathbf{w})$. In the training of BNNs, the *tanh* function transforms the full-precision weights $\mathbf{w}$ to obtain weights $\mathbf{w}_q$ that are bounded in the range $[-1, +1]$, and these weights are closer to binary values as the training converges. After the training, $\mathbf{w}_q$ are very close to the exact set of $\{+1, -1\}$, and the fixed point values will be obtained by taking the sign of the $\mathbf{w}_q$.

4 Methodology

In this section, we firstly present the mathematical definition of two basic hypotheses in our binarization method. Then we propose a meta-based quantizer named QuantNet for directly generating binarized weights within a fully differentiable framework. Moreover, a joint optimization scheme implemented in a standard neural network training is designed to solve the proposal.

4.1 Assumptions

As can be seen, the work [25] replaces the hard constraint *sign* with the soft penalization *tanh*, and penalizes the output of *tanh* to be the closest binary values. However, there are two important issues which are ignored.

In the case of binarizing weights with *tanh*, as most of the elements in binarized weights are close to $\{+1, -1\}$ at the early stage of training, these elements will reach saturation simultaneously, and then cause the phenomenon of gradients vanishing. In brief, if the element is saturated on $+1$, it will not be able to get close to -1 again. On the contrary, the case of the element saturated on -1 is the same. It means that flipping values of these elements is impossible. As a result, only a few unsaturated elements will oscillate around zero, which causes the less ideal convergence behavior and makes the solution highly suboptimal.

Moreover, different from the hard constraint methods, the soft penalization method contains a post-processing step with rounding functionality, and it rounds the binarized weights for further obtaining the fixed point (discrete)

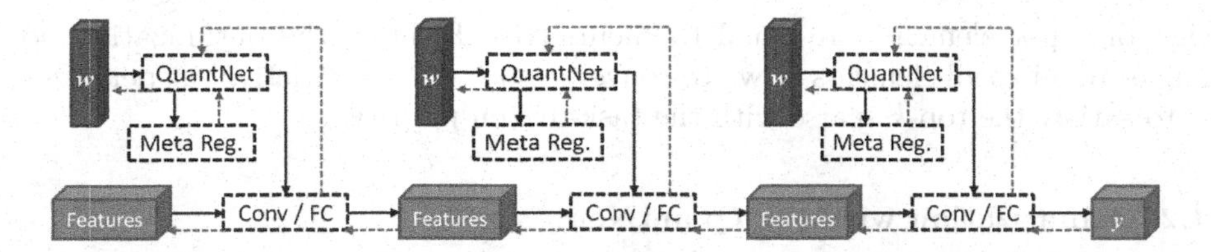

Fig. 1. The architecture of our binarization method. The solid and dashed (blue color) lines represent the feed-forward process and the gradient flow of back-propagation separately, and the dashed line with red color means the meta-gradient flow from the meta regularization (Best viewed in color). (Color figure online)

values. With the increasing number of the network parameters, the discretization error caused by the rounding function will be the major factor to limit performances. To propose our method for solving above issues, we make two fundamental hypotheses in the following.

Assumption 1: We assume that there exists the functional $\mathcal{F}$ to form $tanh(\mathcal{F}(\mathbf{w}))$, and $\lim_{\mathbf{w}\to\infty} \nabla\mathcal{F}(\mathbf{w}) \cdot (1 - tanh^2(\mathcal{F}(\mathbf{w})) \neq 0$, then the derivative of $tanh()$ with respect to $\mathbf{w}$ is expressed as

$$\begin{aligned} &\lim_{\mathbf{w}\to\infty} \frac{\partial tanh(\mathcal{F}(\mathbf{w}))}{\partial \mathbf{w}} \neq 0 \\ w.r.t. \quad &\nabla tanh(\mathcal{F}(\mathbf{w})) = \frac{\partial \mathcal{F}(\mathbf{w})}{\partial \mathbf{w}} \Big(1 - tanh^2(\mathcal{F}(\mathbf{w}))\Big) \end{aligned} \tag{2}$$

Assumption 1 derives a corollary that if $\mathbf{w}$ is out of a small range like $[-1, +1]$, the gradient of $tanh(F())$ for $\mathbf{w}$ will not severely vanish. Through generating the higher dimensional manifolds of full-precision weights $\mathbf{w}$, the gradient vanishing during optimization is relieved, which allows optimizers to solve the globally optimal.

Assumption 2: We assume for a vector $v \in \mathbb{R}^n$ that is k-sparse, there exists an extremely small $\epsilon \in (0, 1)$ with optimal $\mathbf{w}_q^*$, in the optimization of $\ell(\mathbf{w}_q)$ with the objective function ℓ, it has the property that

$$\begin{aligned} &\lim_{\mathbf{w}_q\to\mathbf{w}_q^*} \|\ell(\mathbf{w}_q) - \ell(sign(\mathbf{w}_q))\|_2^2 = 0 \\ s.t. \quad &(1-\epsilon) \leq \frac{\ell(\mathbf{w}_q^* v)}{\ell(\mathbf{w}_q^*)} \leq (1+\epsilon) \end{aligned} \tag{3}$$

Assumption 2 derives a conclusion that if the said constraint of $\frac{\ell(\mathbf{w}_q^* v)}{\ell(\mathbf{w}_q^*)}$ with ϵ is satisfied, it represents the top-k elements in $\mathbf{w}_q^*$ dominate the objective function ℓ, while the remaining elements do not affect the output seriously. In this case, the discretization error caused by the post-processing step is minimized, as the sign of top-k elements are equal to themselves. In brief, the optimization no longer requires all elements in $\mathbf{w}_q$ to converge to $\{+1, -1\}$ strictly, but penalizes it to satisfy the top-k sparse with the task-driven priority.

4.2 Binarization with the QuantNet

Based on above two fundamental hypotheses, we propose a meta-based quantizer named QuantNet for directly generating the binarized weights. Our proposal is to form the functional $\mathcal{F}$ for transforming $\mathbf{w}$ into higher dimensional mainfold $\mathcal{F}(\mathbf{w})$, and optimizing the dominant elements $\mathbf{w}_q$ to satisfy the sparse constraint.

As for implementation details of QuantNet, we design an encoding module accompanied by a decoding module, and further construct an extra compressing module. Specially, suppose full-precision weights come from a convolution layer with 4D shape $\mathbb{R}^{k\times k\times m\times n}$, where k, m and n denote the kernel size, the number of input channels and the number of output channels, respectively.

The input weights will be firstly reshaped into the 2D shape $\mathbb{R}^{m\cdot n\times k^2}$. It means that QuantNet is a kernel-wise quantizer and process each kernel of weights independently, where the batch size is the number of total filters of full-precision weights. In the encoding and decoding process, it firstly expands the reshaped weights into higher dimensional mainfold, which is achieved with the dimensional guarantee that makes the output shape of encoding module to satisfy $\mathbb{R}^{m\cdot n\times d^2}$, $s.t.\ d \gg k$. And then, the compressing module is to transform the higher dimensional manifolds into low-dimensional spaces. If the manifold of interest remains non-zero volume after the compressing process, it corresponds to a higher priority to improve the performance of binarized model on the specific task. Finally, the decoding module generates the binarized weights with the output of the compressing module and the soft binarization function, while restoring the original shape of full-precision weights for main network optimization.

The Fig. 1 provides visualization of QuantNet in the architecture.

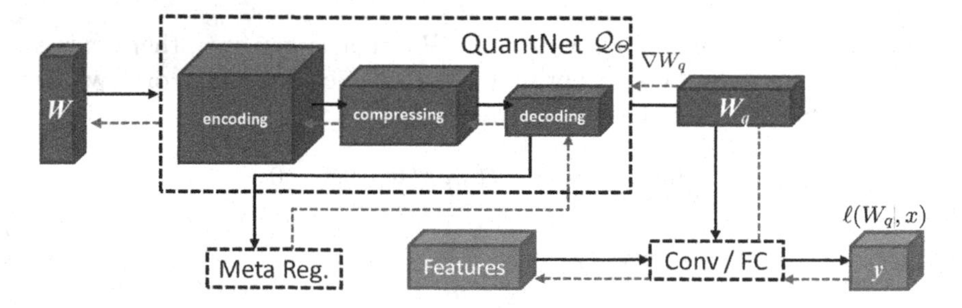

Fig. 2. The architecture of QuantNet. QuantNet $\mathcal{Q}_\Theta$ takes the full-precision weights W as the input, and directly outputs the binarized weights W_q for the network. With the loss of $\ell(W_q, x)$, W_q will be directly updated in the back-propagation, and ∇W_q will be used to update Θ in QuantNet $\mathcal{Q}$. Finally, a new W is computed during this training step.

Feed-Forward Step. Given a full-precision weights[2] W, the proposed QuantNet $\mathcal{Q}_\Theta$ incorporates the parameters Θ to generate the binarized weights W_q with $tanh(\mathcal{Q}_\Theta(W))$. After W is quantized as W_q, the loss ℓ is generated by $\ell(W_q, \{x, y\})$ with the training set {x, y} (Fig. 2).

Back-Propagation Step. The gradient of ℓ with regard to W_q in each layer is computed by the back-propagation. For example, the gradient of weights g_{W_q} in last layer is computed by $\nabla\ell(W_q)$. Then, the QuantNet $\mathcal{Q}_\Theta$ receives the g_{W_q} from the corresponding W_q, and updates its parameters Θ by

$$g_\Theta = \frac{\partial \ell}{\partial W_q}\frac{\partial W_q}{\partial \Theta} = g_{W_q}\frac{\partial W_q}{\partial \Theta} \tag{4}$$

and the gradients of g_Θ is further used to update the full-precision weights W by,

$$W^{t+1} = W^t - \eta \cdot g_\Theta \tag{5}$$

where t denotes the t-th training iteration and η is the learning rates defined for QuantNet.

In practice, QuantNet is applied layer-wise. However, as the number of extra parameters introduced by QuanNet is much less than the network weights, so the computation cost caused by our proposal is acceptable during the network training as shown in the Table 5.

4.3 Optimization

As for the optimization of QuantNet, it is included in the target network that will be binarized. Given the full-precision weights, QuantNet generates the binarized weights to apply on objective tasks, and it is optimized with the back-propagation algorithm to update all variables. In brief, our binarization framework is fully differentiable without any gradient estimators, so there is no information loss during the binarization. We now present the optimization details.

QuantNet Optimization. For satisfying the constraint in Assumption 2, we propose an objective function inspired by the idea of sparse coding. It constrains the compressing process in QuantNet during the binarization. Let $\mathbf{w}_q$ be the binarized weights and introduce a reference tensor $\mathbf{b}$, we aim to find an optimal $\mathbf{w}_q^*$ that satisfies

$$\mathbf{w}_q^* = \arg\min_{\mathbf{w}_q} \|\mathbf{b} - \sqrt{\mathbf{w}_q^2}\|_2 + \|\mathbf{w}_q\|_1, \; s.t. \; \mathbf{b} \in \{1\}^{m \cdot n} \tag{6}$$

where, tensor $\mathbf{b}$ is chosen to be all ones to make elements in $\mathbf{w}_q$ to get close to -1 or +1. As Eq. 6 is independent of the task optimization of binarized models, we alternately solve it with an extra optimizer during the standard network training. At each iteration of optimization, there is adversarial relationship between Eq. 4 and Eq. 6, and the optimization tries to find the balance between minimizing

[2] We omit the notation of layers l in W_l for simplification.

Algorithm 1: Generating the binarized weights with QuantNet

Input: the full-precision weights W, the QuantNet $\mathcal{Q}$ with parameters Θ, and the training set {X, Y}, training iteration t, $\epsilon = 1e - 5$
Output: the optimally binarized weights W_q^*
Training for each layer
for $t = 0; t \leq T$ **do**
  Feed-Forward;
    Compute W_q^t with $tanh(\mathcal{Q}_{\Theta^t}(W^t))$;
    Compute $\ell(W_q^t, \{x^t, y^t\})$ with W_q^t and $\{x^t, y^t\}$;
  Back-Propagation;
    Compute ∇W_q^t with $\ell(W_q^t, \{x^t, y^t\})$;
    Compute $\nabla \Theta^t$ with Eq. 4 and Eq. 6;
    Update the W^t with Eq. 5;
end
Discretization Step
$W_q^* = sign(W_q^T)$;

the binarization error while penalizing the sparsity of binarized weights based on the task priority.

Binarized Model Optimization. As for the optimization of binarized model, we use the standard mini-batch based gradient descent method. After the QuantNet $\mathcal{Q}_\Theta$ is constructed and initialized, the QuantNet optimization is accompanied with the training process of the binarized model, and the objective function of binarized model optimization depends on the specific task. With the specific objective function, the task-driven optimizer is employed to compute the gradients for each binarized layer. The operations for whole optimization are summarized in Algorithm 1.

5 Experiments

In this section, we firstly show the implementation details of our method and experiment settings. Secondly, the performance comparison between our method and STE-based or Non STE-based methods is generated, which further includes the analysis of convergence behaviour.

5.1 Implementation Details

As for the implementation details in experiments, we run each experiment five times with the same initialization function from different starting points. Besides, we fix the number of epoches for training and use the same decay strategy of learning rate in all control groups. At the end, we exhibit the average case of training loss and corresponding prediction accuracy. We show the implementation details as follows.

Network Architecture. We apply the unit with structure of "FC-BN-Leaky ReLU" to construct QuantNet, and each processing module in QuantNet contains at least one unit. For reducing the complexity of network training, QuantNet used in experiments contains only one unit for each processing module, and we still observe a satisfied performance during evaluation. Similar to the existing methods [33,37,41], we leave the first and last layers and then binarizing the remaining layers. For comparison, a fully binarized model by binarizing all layers is also generated. Considering that the bitwise operations can speedup the inference of network significantly, we analyze the balance between the computation cost saving and model performance boosting by these two models. The experiment result exhibits only 0.6% accuracy drop (more than 1% in previous methods) in CIFAR-10 [23] with ResNet-20, in the case that all layers are binairzed.

Initialization. In experiments, all compared methods including our method use the truncated Gaussian initialization if there is not specified in their papers, and all binarized model from experiments are trained from scratch without leveraging any pre-trained model. As for the initialization of QuantNet, we employ the normal Gaussian initialization for each layer. Furthermore, we also evaluate the random initialization, which initialize the variable with the different settings of the mean and variance, but there is not significant difference on the results.

Hyper-parameters and Tuning. We follow the hyper-parameter settings such as the learning rate, batch size, training epoch and weight decay of their original paper. For fair comparison, we use the default hyper-parameters in Meta-Quant [5] and Self-Binarizing [25] to generate the fully binarized network. As for the hyper-parameters of our QuantNet, we set the learning rate as $1e-3$ and the moving decay factor as 0.9. We also evaluate different optimization methods including SGD(M) [35], Adam [22] and AMSGrad [34]. Although we observe that the soft binarization in AMSGrad has a faster convergence behaviour than the others, we still use the SGD(M) with average performance for all methods to implement the final comparison. In future, we plan to analyze the relationship between the soft binarization and different optimizers (Table 1).

Table 1. Comparison with different optimizer on ResNet-20 for CIFAR10.

Optimizer	Accuracy (%)	Training time
SGD(M)	90.04	1.0×
Adam	89.98	~1.2×
AMSGrad	90.12	~0.9×

5.2 Performance Comparison

QuantNet aims at generating the binarized weights without STE and other estimators. Hence, we compare it with both STE-based binarization methods

[8,18,20,41,42] and Non STE-based binarization methods [2,5,25,37] with the idea of avoiding the discrete quantization. In details, the evaluation is based on the standard benchmark of classification task with CIFAR-100 [23] and ImageNet [9], and the base network architectures are based on the AlexNet [24], ResNet [13] and MobileNet [36].

Evaluation of the Discretization Error. As the soft binarization method [25] always involves a post-processing step, which aims at transforming the float-point weights into the fixed-point weights, we name this step the discretization step which is shown in Algorithm 1. For comparing the discretization error caused by the step between the self-binarizing [25] and the proposed QuantNet, we generate both two binarized models for self-binarizing and QuantNet, and use the notation (D) to denote the prediction accuracy of binarized model after the discretization step, which means the weights in the binarized model is transformed into the integer exactly. As shown in the Table 2, QuantNet achieves the best performance even better than the FP, the major reason is that the sparse constraint encourages a better generalization ability. Moreover, since the discretization error is considered in our algorithm during the binarizing process, comparing to the accuracy drop 1.85% in self-binarizing [25], QuantNet only reduces 0.59%.

Table 2. Prediction accuracy of binarized AlexNet on CIFAR-10. The FP represents the full-precision model with 32-bits for both weights and activations.

Methods	Bit-width (W/A)	Acc.(%)	Acc.(%) (after discretization)
FP	32/32	86.55	–
Self-Binar. [25]	1/1	86.91%	84.31%
Ours	1/1	**87.08%**	**86.49%**

Comparison with STE-Based Binarization. We evaluate our QuantNet with the STE-based binarization methods, and report the top-1 accuracy in Table 3. Besides, we use PACT [7] to quantize the activation into 2 bits if the compared method does not support the activation quantization. For the compared prediction accuracy used in this table, we use the results from the original paper if it is specified. Overall, QuantNet achieves the best performance compared to existing STE-based methods, which surpasses QIL [20] more than 2% before the discretization step, and even obtain a comparable performance with the full-precision model. It demonstrates the advantage of directly binarizing the weights within a fully differentiable framework. Although the discretization (rounding operation) introduces a post-processing step, the experiment results still prove the effectiveness of our binarization method, and the degradation of prediction accuracy caused by rounding is negligible.

Comparison with Non STE-Based Binarization. As for the Non STE-based binarization methods, it mainly includes two categories: *learning better*

Table 3. Top-1 accuracy (%) on CIFAR-100. Comparison with the existing methods on ResNet-56 and ResNet-110.

Method	Bit-width	ResNet-56	ResNet-110
FP	32W32A	71.22%	72.54%
BWN [8]	1W2A	64.29%	66.26%
DoReFa-Net [42]	1W2A	66.42%	66.83%
LAB [18]	1W2A	66.73%	67.01%
LQ-Nets [41]	1W2A	66.55%	67.09%
QIL [20]	1W2A	67.23%	68.35%
Ours	1W2A	**69.38%**	**70.17%**
Ours(D)	1W2A	**68.79%**	**69.48%**

gradients for non-differentiable binarization function, and *replacing the non-differentiable function with the differentiable one.* We compare the QuantNet with the representative works in these two categories - ProxQuant [2] and Meta-Quant [5] in the first and Self-Binarizing [25] in the second. With the increasing number of parameters in larger architecture [13], although the methods [2,5] related gradient refinement have improved the performance effectively, the bottleneck caused by the gradient estimation appears obviously, and our method have achieved the significant improvement than these methods (Table 4). Moreover, as the discretization error caused by the rounding operation is well considered by our method, QuantNet is affected less than Self-Binarizing [25].

Table 4. Top-1 accuracy (%) on ImageNet. Comparison with the existing methods on AlexNet (left), ResNet-34 (right).

Method	Bit-width	AlexNet	Method	Bit-width	ResNet-34
FP	32W32A	55.07%	FP [32]	32W32A	73.30%
Self-Binar. [25]	1W32A	52.89%	ProxQuant [2]	1W32A	70.42%
Self-Binar.(D) [25]	1W32A	50.51%	Meta-Quant [5]	1W32A	70.84%
Ours	1W32A	**54.06%**	Ours	1W32A	**71.97%**
Ours(D)	1W32A	**53.59%**	Ours(D)	1W32A	**71.35%**

Convergence Analysis. We analyze the convergence behaviour of our QuantNet and other binarization methods during the training process. In details, we use ResNet-34 as the base architecture, and compare with the STE-based methods and non-STE based methods separately. For the first case, QuantNet exhibits a significantly smooth loss curve over STE, including much faster convergence speed and lower loss values, and it also achieves the best prediction accuracy in the test reported in Table 3. The main reason of the better convergence of our method is that QuantNet is totally differentiable during the optimization. Furthermore, we analyze the proposed method in the second case, and we observe that all the non-STE based methods can smooth the loss curve effectively, but our method achieve the lowest loss value as there is not estimation of gradients.

Complexity of Models. As our QuantNet involves the extra computation cost and parameters during the optimization, we analyze its efficiency comparing to the traditional STE-based methods and other Meta-based methods. For QuantNet, it is independent of the scale of input resource, and its time complexity is related to the amount of its parameters. In the Table 5, the total training time cost is exhibited, and we leave the inference step since the QuantNet is removed in this step. For the setting of experiment in this table, the base architecture ResNet-34 is used, and the bitwise operation is not implemented for all cases.

Table 5. Training time on ResNet-34

Bit-width (1W/2A)	Training time(iter./s)
FP(32W/32A)	1.0×
MixedQuant [37]	1.5×
Meta-Quant [5]	2.8×
Self-Binar. [25]	1.2×
Ours	1.7×

6 Conclusions

In the paper, we present a meta-based quantizer QuantNet to binarize the neural network, which directly binarize the full-precision weights without STE and any learnable gradient estimators. In contrast to the previous soft binarizing method, the proposed QuantNet not only solves the problem of gradient vanishing during the optimization, but also alleviates the discretization errors caused by the post-processing step for obtaining the fixed-point weights. The core idea of our algorithm is to transform the high dimensional manifolds of weights, while penalize the dominant elements in weights into sparse according to the task-driven priority. In conclusion, the QuantNet outperforms the existing binarization methods on the standard benchmarks, which not only can be applied on weights, but also can be extended to activations (or quantization with other bits) easily.

References

1. Alizadeh, M., Fernández-Marqués, J., Lane, Nicholas, D., Gal, Y.: An empirical study of binary neural networks' optimisation. In: International Conference on Learning Representations (ICLR) (2019)
2. Bai, Y., Wang, Y.X., Liberty, E.: Quantized neural networks via proximal operators. In: International Conference on Learning Representations (ICLR) (2019)
3. Bengio, Y., Leonard, N., Courville, A.: Estimating or propagating gradients through stochastic neurons for conditional computation. arXiv preprint arXiv:1308.3432 (2013)

4. Cai, Z., He, X., Jian, S., Vasconcelos, N.: Deep learning with low precision by half-wave gaussian quantization. In: The IEEE Conference on Computer Vision and Pattern Recognition (CVPR) (2017)
5. Chen, S., Wang, W., Pan, S.J.: MetaQuant: learning to quantize by learning to penetrate non-differentiable quantization. In: Annual Conference on Neural Information Processing Systems (NIPS) (2019)
6. Cheng, Y., Wang, D., Zhou, P., Zhang, T.: A survey of model compression and acceleration for deep neural networks. arXiv preprint arXiv:1710.09282 (2017)
7. Choi, J., Wang, Z., Venkataramani, S., Chuang, P.I.J., Srinivasan, V., Gopalakrishnan, K.: Pact: Parameterized clipping activation for quantized neural networks. In: The IEEE Conference on Computer Vision and Pattern Recognition (CVPR) (2018)
8. Courbariaux, M., Bengio, Y., David, J.P.: BinaryConnect: training deep neural networks with binary weights during propagations. In: Annual Conference on Neural Information Processing Systems (NIPS), pp. 3123–3131 (2015)
9. Deng, J., Dong, W., Socher, R., Li, L.J., Li, K., Li, F.F.: ImageNet: a large scale hierarchical image database. In: The IEEE Conference on Computer Vision and Pattern Recognition (CVPR) (2009)
10. Faraone, J., Fraser, N., Blott, M., Leong, P.H.: SYQ: learning symmetric quantization for efficient deep neural networks. In: The IEEE Conference on Computer Vision and Pattern Recognition (CVPR) (2018)
11. Gao, H., Zhuang, L., Laurens, V.D.M.: Densely connected convolutional networks. In: The IEEE Conference on Computer Vision and Pattern Recognition (CVPR) (2017)
12. Han, S., Mao, H., Dally, W.J.: Deep compression: compressing deep neural networks with pruning, trained quantization and Huffman coding. In: International Conference on Learning Representations (ICLR) (2016)
13. He, K., Zhang, X., Ren, S., Sun, J.: Deep residual learning for image recognition. In: The IEEE Conference on Computer Vision and Pattern Recognition (CVPR), pp. 770–778 (2016)
14. He, Y., Zhang, X., Sun, J.: Channel pruning for accelerating very deep neural networks. In: The IEEE International Conference on Computer Vision (ICCV), pp. 1389–1397 (2017)
15. Helwegen, K., Widdicombe, J., Geiger, L., Liu, Z., Kwang-Ting, C., Nusselder, R.: Latent weights do not exist: rethinking binarized neural network optimization. In: Advances in Neural Information Processing Systems (NIPS) (2019)
16. Heo, B., Lee, M., Yun, S.: Knowledge distillation with adversarial samples supporting decision boundary. In: Association for the Advance of Artificial Intelligence (AAAI) (2019)
17. Hinton, G., Vinyals, O., Dean, J.: Distilling the knowledge in a neural network. Comput. Sci. **14**(7), 38–39 (2015)
18. Hou, L., Yao, Q., Kwok, J.T.: Loss-aware binarization of deep networks. In: International Conference on Learning Representations (ICLR) (2017)
19. Hou, L., Zhang, R., Kwok, J.T.: Analysis of quantized models. In: International Conference on Learning Representations (ICLR) (2019)
20. Jung, S., et al.: Learning to quantize deep networks by optimizing quantization intervals with task loss. In: The IEEE Conference on Computer Vision and Pattern Recognition (CVPR), pp. 4350–4359 (2019)
21. Kin, Y.D., Park, E., Yoo, S., Choi, T., Yang, L., Shin, D.: Compression of deep convolutional neural networks for fast and low power mobile applications. In: International Conference on Learning Representations (ICLR) (2016)

22. Kingma, D., Ba, J.: Adam: a method for stochastic optimization. In: International Conference on Learning Representations (ICLR) (2015)
23. Krizhevsky, A., Hinton, G.: Learning multiple layers of features from tiny images. Master's thesis (2009)
24. Krizhevsky, A., Sutskever, I., Hinton, G.: ImageNet classification with deep convolutional neural networks. In: Advances in Neural Information Processing Systems (NIPS) (2012)
25. Lahoud, F., Achanta, R., Márquez-Neila, P., Süsstrunk, S.: Self-binarizing networks. In: arXiv preprint arXiv:1902.00730 (2019)
26. Li, D., Wang, X., Kong, D.: DeepRebirth: accelarating deep neural network execution on mobile device. In: International Conference on Learning Representations (ICLR) (2017)
27. Liu, J., et al.: Knowledge representing: Efficient, sparse representation of prior knowledge for knowledge distillation. In: The IEEE Conference on Computer Vision and Pattern Recognition (CVPR) Workshops (2019)
28. Liu, J., et al.: BAMSProd: a step towards generalizing the adaptive optimization methods to deep binary model. In: The IEEE Conference on Computer Vision and Pattern Recognition (CVPR) Workshops (2020)
29. Liu, Z., Wu, B., Luo, W., Yang, X., Liu, W., Cheng, K.-T.: Bi-real net: enhancing the performance of 1-bit CNNs with improved representational capability and advanced training algorithm. In: Ferrari, V., Hebert, M., Sminchisescu, C., Weiss, Y. (eds.) ECCV 2018. LNCS, vol. 11219, pp. 747–763. Springer, Cham (2018). https://doi.org/10.1007/978-3-030-01267-0_44
30. Liu, Z., Li, J., Shen, Z., Huang, G., Yan, S., Zhang, C.: Learning efficient convolutional networks through network slimming. In: The IEEE International Conference on Computer Vision (ICCV), pp. 2736–2744 (2017)
31. Luo, J.H., Wu, J., Lin, W.: ThiNet: a filter level pruning method for deep neural network compression. In: Advances in Neural Information Processing Systems (NIPS), pp. 5068–5076 (2017)
32. Qin, H., et al.: Forward and backward information retention for accurate binary neural networks. In: IEEE/CVF Conference on Computer Vision and Pattern Recognition (CVPR) (2020)
33. Rastegari, M., Ordonez, V., Redmon, J., Farhadi, A.: XNOR-Net: ImageNet classification using binary convolutional neural networks. In: Leibe, B., Matas, J., Sebe, N., Welling, M. (eds.) ECCV 2016. LNCS, vol. 9908, pp. 525–542. Springer, Cham (2016). https://doi.org/10.1007/978-3-319-46493-0_32
34. Reddi, S.J., Kale, S., Kumar, S.: On the convergence of Adam and beyond. In: International Conference on Learning Representations (ICLR) (2018)
35. Robbins, H., Monro, S.: A stochasitc approximation method. Ann. Math. Stat. **22**, 400–407 (1951)
36. Sandler, M., Howard, A., Zhu, M., Zhmoginov, A., Chen, L.C.: MobileNetV2: inverted residuals and linear bottlenecks. In: The IEEE Conference on Computer Vision and Pattern Recognition (CVPR), pp. 4510–4520 (2018)
37. Uhlich, S., et al.: Mixed precision DNNs: all you need is a good parametrization. In: International Conference on Learning Representations (ICLR) (2020)
38. Wang, P., Hu, Q., Zhang, Y., Zhang, C., Liu, Y., Cheng, J.: Two-step quantization for low-bit neural networks. In: The IEEE Conference on Computer Vision and Pattern Recognition (CVPR) (2018)
39. Yin, P., Lyu, J., Zhang, S., Osher, S., Qi, Y., Xin, J.: Understanding straight-through estimator in training activation quantized neural nets. arXiv preprint arXiv:1903.05662 (2019)

40. Yin, P., Zhang, S., Lyu, J., Osher, S., Qi, Y., Xin, J.: Blended coarse gradient descent for full quantization of deep neural networks. arXiv preprint arXiv:1808.05240 (2018)
41. Zhang, D., Yang, J., Ye, D., Hua, G.: LQ-Nets: learned quantization for highly accurate and compact deep neural networks. In: Ferrari, V., Hebert, M., Sminchisescu, C., Weiss, Y. (eds.) ECCV 2018. LNCS, vol. 11212, pp. 373–390. Springer, Cham (2018). https://doi.org/10.1007/978-3-030-01237-3_23
42. Zhou, S., Ni, Z., Zhou, X., He, W., Zou, Y.: DoReFa-Net: training low bitwidth convolutional neural networks with low bitwidth gradients. arXiv preprint arXiv:1606.06160 (2016)
43. Zhu, C., Han, S., Mao, H., Dally, W.J.: Trained ternary quantization. In: International Conference on Learning Representations (ICLR) (2016)
44. Zhu, F., et al.: Towards unified INT8 training for convolutional neural network. In: The IEEE Conference on Computer Vision and Pattern Recognition (CVPR) (2020)

An Efficient Method for Face Quality Assessment on the Edge

Sefa Burak Okcu(✉), Burak Oğuz Özkalaycı, and Cevahir Çığla

Aselsan Inc., Ankara, Turkey
{burak.okcu,bozkalayci,ccigla}@aselsan.com.tr

Abstract. Face recognition applications in practice are composed of two main steps; face detection and feature extraction. In a sole vision-based solution, the first step generates multiple detections for a single identity by ingesting a camera stream. A practical approach on edge devices should prioritize these detections of identities according to their conformity to recognition. In this perspective, we propose a face quality score regression by just appending a single layer to a face landmark detection network. With almost no additional cost, face quality scores are obtained by training this single layer to regress recognition scores with surveillance like augmentations. We implemented the proposed approach on edge GPUs with all face detection pipeline steps, including detection, tracking, and alignment. Comprehensive experiments show the proposed approach's efficiency through comparison with state-of-the-art face quality regression models on different data sets and real-life scenarios.

Keywords: Face quality assessment · Key frame extraction · Face detection · Edge processing

1 Introduction

In the COVID-19 pandemic, the need and importance of remote capabilities during access control and thermography for human screening have increased. Face detection and recognition systems are the most popular and endeavored tools for remote and isolated capabilities in such scenarios. Moreover, hardware improvements and publicly available face data sets lead to increased demand for face recognition (FR) research. The FR systems can be categorized into two mainstreams according to the application area: access control and payment involving volunteered faces [1–4] and surveillance for public safety [5–8]. People try to provide recognizable faces in the systems for the first scenario where the environment is controlled in terms of lighting, head pose and motion. FR problem is mostly solved for this scenario such that access control devices are available in many facilities, and millions of people use smartphones with face recognition technology. On the other hand, new problems form for these scenarios that need special attention such as liveness detection to handle spoofing attacks [9–11] or masked faced recognition under severe conditions.

A. Bartoli and A. Fusiello (Eds.): ECCV 2020 Workshops, LNCS 12539, pp. 54–70, 2020.
https://doi.org/10.1007/978-3-030-68238-5_5

FR for surveillance involves a variety of open problems compared to the first group. In this type, cameras observe people that walk-run through streets, terminals in their natural poses indoors and outdoors. People occlude each other, lighting is not controlled and people do not look at the cameras intentionally. Face detection, mostly solved for specifically large and clear faces, encounters these uncontrolled problems as the first step. The cameras should be located appropriately to catch the faces with sufficient resolution (e.g. 60 pixels between two eyes) and as frontal as possible. During the face detection step, many unrecognizable face photos of an individual can be captured that requires additional filtering to boost FR rates. Apart from that, face databases are in the order of millions for public security that makes FR a harder problem.

Many researchers put their attention on developing new cost functions and convolutional neural networks for high performance FR algorithms [12–16]. At the same time, lightweight face detection gains attention of researchers [17–20] that enables algorithms to run on edge devices with high performance in real-time. Following face detection on the edge, FR is generally performed on servers to cope with large databases and the need for handling complex scenes. This solution requires the transfer of detected faces from edge devices to servers. At that point, face quality assessment (FQA) methods [21–25] is the key for selecting the best faces to reduce the amount of data transfer and decrease the false alarm rate in FR. In this approach, multiple faces of an individual, obtained during face detection and tracking, are analyzed and the most appropriate faces are selected for the FR step. FQA is an important feature that is applicable on both the edge and server side and removes redundant, irrelevant data in an FR system.

In this study, we focus on efficient FQA that is achieved in conjunction with face landmark detection on the edge devices. Giving the related work for FR and FQA in the next section, we provide the motivation and contribution of this study that is followed by the details of the proposed face quality measure framework in the fourth section. After the comprehensive experiments on the reliability of the proposed approach and correlation with the FR rates, we conclude in the final section.

2 Related Work

In the last decade, various approaches have been proposed to address FQA which can mainly be categorized into classical and machine learning techniques. The classical techniques rely on several image statistics while machine learning methods exploit convolutional neural networks to train FQA models. [26] defines an 'ideal' face using patch-based local analysis by taking alignment error, image shadowing and pose variation into consideration simultaneously. [21] combines the classical image quality metrics of contrast, brightness, focus, An Efficient Method for Face Quality Assessment on the Edge 3 sharpness and illumination for FQA. [27] propounds that foreground-background segmentation improves image quality assessment performance, especially for frontal facial portraits. In

[24], support vector machine is utilized to model the quality of frontal faces based on 15 features extracted for different image regions (entire image, face region, eyes region, mouth region) according to sharpness, illumination, contrast and color information. In [28], a learning to rank based framework is proposed to evaluate face image quality. Firstly, weights are trained to calculate scores from five different face feature vectors. These scores are combined using a second-order polynomial in order to obtain the ultimate face quality score. [22] expresses a convolutional neural network (CNN) based FQA algorithm which is trained to imitate face recognition ability of local binary patterns or histogram of oriented gradients based face recognition algorithms and predict a face quality score. [29] utilizes a technique for high quality frame selection along with video by fusing different normalized face feature scores which are face symmetry, sharpness, contrast, brightness, expression neutrality and openness of the eyes via a two layers neural network.

In [30], a deep feature-based face quality assessment is proposed. After deep-features are extracted in face images, the best representative features are selected using sparse coding and face quality score obtained from support vector regression (SVR). [31] a deep feature-based face quality assessment is proposed. After deep features are extracted in face images, the best representative features are selected using sparse coding and face quality score obtained from support vector regression (SVR). [23] proposed a CNN based key frame engine which is trained according to recognition scores of a deep FR system. In a similar fashion, [32] proposed a CNN based FQA module trained according to recognition scores obtained by cosine similarity of extracted deep face features in a database. Nevertheless, it is not obvious whether the best scores are attained with the same identities or not. [33] compares two face image quality assessment methods which are designed by SVR model on features extracted by a CNN with two different labelling methods: human quality values (HQV) assigned by humans via Amazon Turk and matcher quality values obtained by similarity scores between faces. The models are tested on two unconstrained face image databases, LFW and IJB-A, and it is concluded that HQV is a better method for quality assessment. In [34], Yu et al. designed a shallow CNN in order to classify images into degradation classes as low-resolution, blurring, additive Gaussian white noise, salt and pepper noise and Poisson noise. The model was trained on the augmented CASIA database. They calculated a biometric quality assessment score by weighted summation of degradation class probabilities with the weights of each class's recognition accuracy. [35] indicates an approach in order to obtain iconicity score based on a Siamese multi-layer perceptron network which is trained via features extracted from face recognition networks.

In [25], a ResNet-based CNN model, called FaceQnet, is proposed. The quality values are obtained by the similarity scores between gallery faces selected by ICAO compliance software and other faces in the database. They observed that face quality measurement extracted from FaceQnet improved performance of face recognition task on VGGFace2 and BioSecure databases. In a recent study [36], a deep CNN is utilized to calculate facial quality metric, and it is

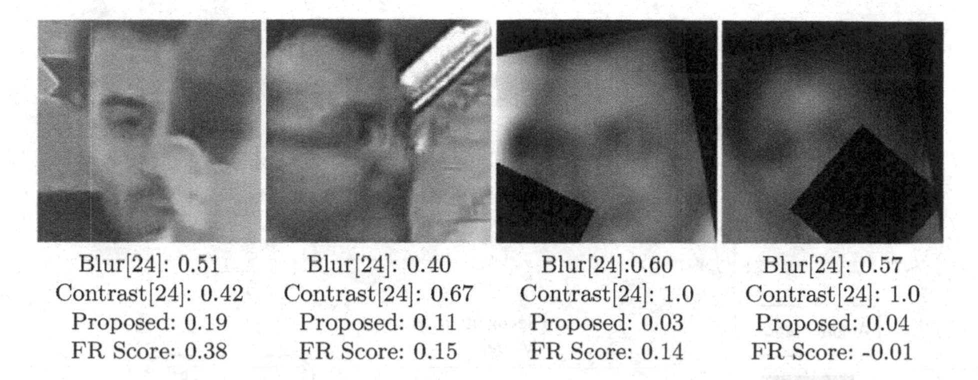

Fig. 1. The face quality assessment should be highly correlated with the face recognition scores for surveillance scenarios.

trained by using faces augmented with brightness, contrast, blur, occlusion, pose change and face matching score obtained from the distance between input image and reference image via PLDA. This work introduces use of occlusion during FQA training on the other hand just the eyes-mouth-nose landmark regions are covered in an unrealistic way that do not fit with surveillance scenarios.

3 Motivation

State-of-the-art has focused on developing algorithms that tackle FQA without much concern on computational analysis that results in complex networks with high execution times. On the other hand, FQA is a tool that is mostly utilized on edge devices incorporated with face detection, tracking and face image post-processing. Therefore, it should consume moderate computational load that is almost seamless to achieve real-time performance on the cameras.

In a complete face recognition system, the main target is to achieve high face matching accuracy with low false alarm rates. In that manner, FQA should associate with face recognition in such a way that high face matching scores are accompanied with high face quality measures. This helps to eliminate low quality faces and matching is only performed on limited number of faces that are more likely to be matched.

Classical image aesthetics measurements can easily be deployed on the edge devices for FQA however they do not provide sufficient accuracy in terms of face recognition especially for the surveillance scenarios including severe occlusions and low local contrast, as in Fig. 1. Focusing on the face landmark points and calculating image quality on those regions improves FQA for motion blur and low contrast faces but still lacks occlusion cases which are very common in surveillance.

Most of the FQA methods do not consider face distortion scenarios observed in surveillance during the augmentation steps of model training. In that manner, artificial blur and occlusion are important tools for simulating scenarios that

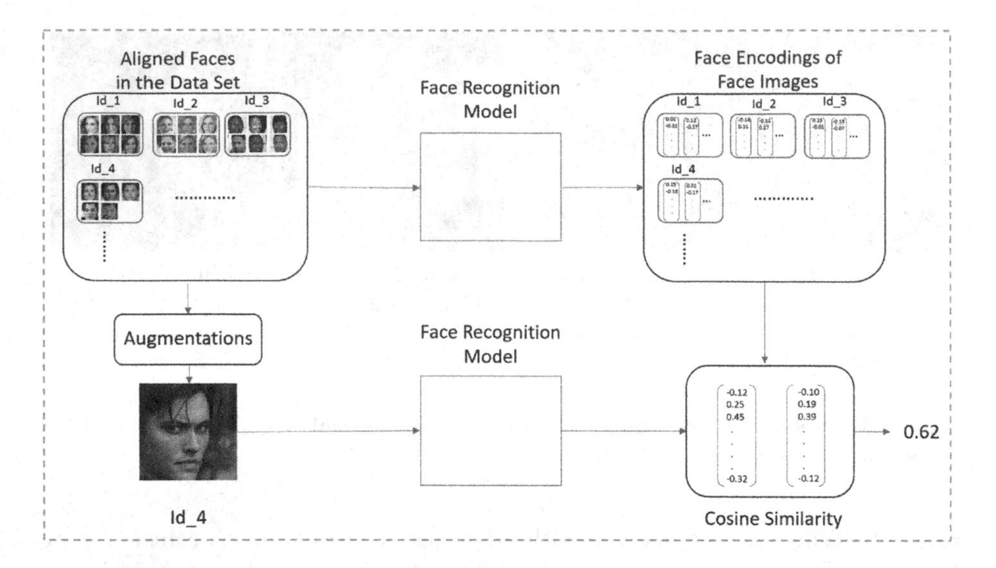

Fig. 2. Face recognition system used for extracting face similarity scores

can be faced in surveillance applications as shown in Fig. 1. Recently, [37] has introduced occlusion as a tool during augmentation of their training data, while their aim is to extract occluding regions such as mask or sun-glasses in a face image at the detection step rather than FQA.

In this study, we tackle FQA by proposing a technique following a recognition guided motivation. We utilize an extra output layer in a well known landmark extraction network without additional complexity. Besides, we exploit random rotation, Gaussian blur and occlusion as augmentation steps in the training phase, that helps to model the distortion scenarios in realistic surveillance FR. Finally, through extensive experiments, we illustrate that a simple layer extension is quite sufficient to measure face quality in surveillance scenarios that has correlation with recognition as illustrated in Fig. 1.

4 Recognition Guided Face Quality Assessment

4.1 FR System Overlook

FQA is an intermediate step along with face detection, tracking and alignment which is embedded in the landmark extraction network as an extra output for the sake of computational efficiency. The landmark extraction network's weights are frozen during the training stage and the extra added layer is trained by face recognition similarity scores that are considered as face quality score. In order to obtain these scores, we employed ArcFace [14] that is one of the state-of-the-art face recognition methods. It is also important to note that the proposed approach does not rely on ArcFace [14] and the recognition model can be replaced by another one as will be illustrated in the experimental results. As a CNN

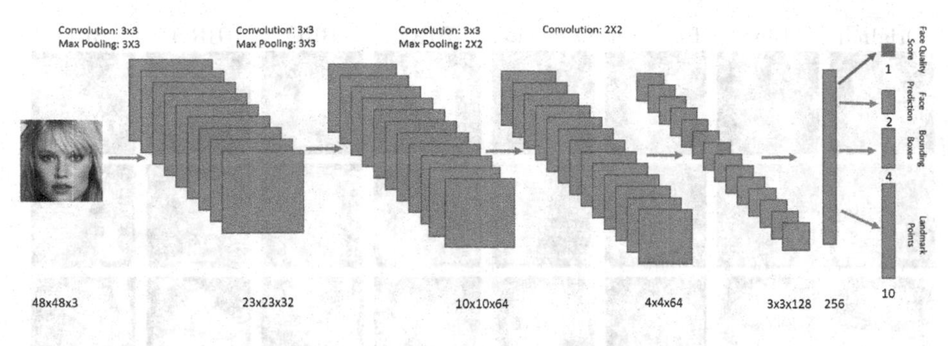

Fig. 3. Modified O-Net architecture

architecture, ResNet50 [38] is wielded with output embedding size of 512. Next, CelebA dataset [39] is utilized to train FQA layer along with cosine similarity metrics applied on the Arcface embeddings. The overall face recognition system that is exploited in order to obtain face similarity scores is shown in Fig. 2.

4.2 Network Structure

Multi-task Cascaded Convolutional Neural Network (MTCNN) [17], is one of the well known face detection and facial landmark points extraction networks. It consists of 3 cascaded stages Proposal Network (P-Net), Refine Network (R-Net), Output Network (O-Net) in order to boost both performance of face detection and face alignment. P-Net produces proposal bounding box windows and performs bounding box regression, R-net rejects false proposals and refine bounding boxes, and O-net also further refines bounding boxes and produces 5 facial landmark coordinates. In designing FQA network, our first aim is to avoid increase in computational load; hence we add only one extra node to the output layer of O-Net instead of creating a new CNN. In addition, we remove the fully connected layer which is used for predicting face class since we do not use the face detector output of this model for deployment and we input only face images to the modified o-net, aka *monet.* Adding an extra node into the overall network architecture as shown in Fig. 3 enables the estimation of face quality score with a negligible computational cost.

4.3 Augmentation

There are different types of distortions in surveillance scenarios affecting quality of captured faces as well as face recognition performance. In that manner, change in head orientation, motion blur and occlusion are the most common cases in such scenarios where the subjects are not aware of face recognition and behave naturally. Thus, we apply random rotation, random Gaussian blur and random occlusion to face images in order to duplicate these realistic distortions.

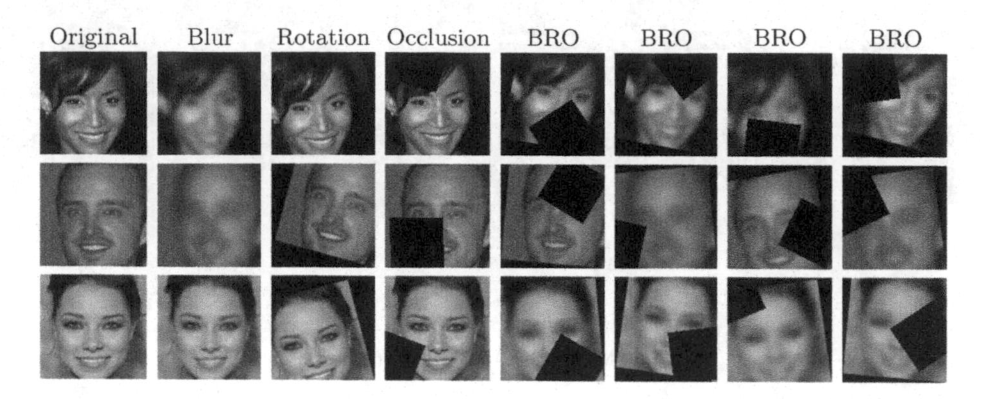

Fig. 4. Blur, rotation and occlusion augmentations are utilized in the training to simulate real life surveillance data, images are taken from CelebA dataset [39]

Rotation: In real-life scenes, head orientation can change naturally among consecutive frames. Even though faces are aligned to compensate rolling of the faces and enforced to have horizontally aligned eye locations, this process depends on detection of landmarks' points, which might cause some errors. In order to take its effect into account, we rotate faces between −15 and 15° randomly.

Blurring: Surveillance cameras are generally zoomed for capturing faces with more detail (ideally eyes are separated by 60 pixels). Besides, people move naturally without any attempt to make their faces recognizable in front of a device. These two facts result in blur which definitely drops face recognition performance. Thus, we add some random Gaussian blur to the cropped faces of 112×112 resolution with changing kernel sizes between 3 and 21.

Occlusion: Occlusion in faces is another cause that decreases face recognition performance in surveillance scenes. In the augmentation, we add random occlusions to face images by creating a rectangle with a maximum size of one-fourth of the face image size and roll it randomly. Then, the rectangle is placed to a random location in the image, which will occupy at most 25% of the image.

Some examples of augmentations on CelebA dataset are shown in Fig. 4 for three different faces. In the first three columns, single attacks of blurring, rotation and occlusion are given, while the rest four columns correspond to random combinations of these three.

4.4 Training Dataset

CelebA Faces: Large-scale CelebFaces Attributes (CelebA) dataset [39] is one of the most suitable data sets for training FQA networks due to several reasons. First of all, it has identity labels in order to extract face similarity scores required for training our FQA network. Secondly, it contains more than 200k face images with different resolutions, pose variations and background. Finally, it has also face bounding boxes, 5 facial landmark points and many attribute labels (around 40). [39] consists of 202,599 face images from 10,177 different identities and we have used training part of CelebA which is composed of 162770 face images for training our FQA layer.

4.5 Network Training

Figure 8. MTCNN Pipeline The modified version of O-Net architecture is trained on CelebA dataset with various alternatives of augmentation methods as rotation, blur, occlusion and rotation+blur+occlusion. This enables the analyses of the effect of augmentations and their representation capability. Training batch size is set to 128 and training started with learning rate of 0.01. CNN training is performed on a workstation with Nvidia GTX 1080Ti GPU by using MXNet framework [40].

5 Experimental Results

The experiments are conducted in three main sessions, where in the first section comprehensive tests are performed for the proposed approach with various alternatives on the training data. In the second section, proposed approach is compared with state of the art face quality assessment as well as classical image quality metrics. In the final section we run the proposed approach within a complete pipeline including face detection tracking and alignment on a well known mobile GPUs (Nvidia Jetson TX2). Before getting into the details of the experiments, we would like to introduce the datasets utilized throughout the experiments.

5.1 Dataset Analyses

LFW Dataset: Labeled Faces in the Wild (LFW) [41,42] is a well-known benchmark dataset for face verification. It contains 13,233 images from 5749 people and they are suitable for research on unconstrained face recognition. In this work, we utilize LFW for comparing average face quality score of original pairs with those of Gaussian blurred and/or occluded faces. The face pairs are taken from [14] Github repository, which has 3,000 true matched pairs and 3,000 false matched pairs. In order to compare the change in average face quality scores with distortions, we utilize the true matched pairs.

Color FERET Database: The Face Recognition Technology (FERET) [43,44] database consists of 14,126 images from 1199 individuals. The database has been collected by different organizations in highly controlled environment. It has been created for developing face recognition technology via new algorithms and techniques. We take advantage of FERET database for evaluating our FQA method as in LFW. For each image in the database, a match pair is assigned through face matching via *Arcface* and *facenet* under different scenarios.

Georgia Tech Face Database: Georgia Tech Face Database [45] is comprised of 750 images taken from 50 individuals. Face images have been collected under different illumination conditions with different facial expressions and orientations. The best matches are assigned for each image which could alter according to the simulated attacks during experiments.

Realistic Surveillance Data: Apart from the previously described three datasets, we also exploit a realistic surveillance scenario where 15 different people walk in a crowd in various scenes. We have recorded videos on three different sites, *lab*, *indoor_low* and *indoor_high*. The last two differentiate with respect to the location of camera, low indicates camera location at 2 m while high indicates camera located in 4 m with high zoom level. This data provides observation of the FQA performance under real scenarios where natural motion blur and occlusion as well as different head orientations and natural cases are observed including yawning, phone call, drinking, looking down.

5.2 Algorithm Tests

The proposed faceness measure approach has been tested on four different sets as referenced recently. We apply artificial attacks on the original data (the first three) in order to simulate the real surveillance scenarios such as blur and occlusion that are very common and measure the face matching scores via *Arcface* [14]. It is important to note that, we do not apply any attacks on the realistic surveillance data which has natural distortions. In that manner, we analyze three scenarios with attacks including only Gaussian blur, only partial occlusion and both. Gaussian blur is applied with random standard deviation, partial occlusion is applied by choosing four random points within the image and painting the covered area. Before getting into the results of the proposed method, we would like to share the average similarity values as well as the effects of simulated attacks on the *Georgia*, *LFW* and *Feret* datasets in Table 1. It is clear that, the face matching similarity is dropped severely (to $\frac{1}{3}$ of the average similarity in case of no attack) when both blur and occlusion attacks are applied to the test images. On the other hand, adding occlusion has less impact in *Georgia* and *LFW* compared to blurring that is vice-versa for *Feret* dataset. This behaviour is valid for both face similarity metrics.

The performance of FQA is measured through the Pearson correlation coefficient that calculates the linearity between the measurements. In that manner, after each attack, the estimated face quality and the actual matching score are measured that are subject to the Pearson correlation. This correlation approaches to "1" as the measurements are linearly related to each other.

We come up with four FQA models that are trained under various augmentation scenarios. *Blur* model is trained on the data distorted by only Gaussian blur, *Rot* model is trained by data distorted through only rotation, *Occ* model is trained by only occlusion augmentation and the final model *BRO* is trained

Table 1. The effect of attacks on different average face matching scores for different datasets

DataSet	Similarity	No Attack	Blur	Occ	Blur+Occ
Georgia	Arcface	0.85	0.61	0.65	0.24
	FaceNet	0.48	0.69	0.78	1.12
LFW	Arcface	0.71	0.34	0.53	0.21
	FaceNet	0.33	0.62	0.53	0.74
Feret	Arcface	0.59	0.44	0.39	0.19
	FaceNet	0.68	0.99	0.92	1.19

by all augmentations. These four models help us to understand if the models are trained properly and behave as expected.

These four models estimate the faceness quality of each image under three different distortions (Gaussian blur, Occlusion and both). Then for each distorted image two scores are calculated via *Arcface*, the best matching among the other images in the database and self similarity with the un-distorted original version of the image. The Pearson coefficients are calculated between the matching scores and the estimated faceness quality measures. The results are given for GeorgiaTech, LFW and Feret datasets in Table 2 correspondingly. The correlation coefficients show same tendency for each dataset, where each attack is best represented by the models that are trained over the same augmentation set. The unified *BRO* model has the best correlation with real matching scores for the attacks that include both blur and occlusion. Besides, it always has the second best Pearson correlations for the other attacks that are very close to the best correlations in each row.

Among these various alternatives on attacks and datasets, 78% of the achieved best Pearson correlations are above 0.75 which indicates very good to excellent correlation between variables. This means that the proposed framework enables high quality FQA as long as proper augmentations are applied in the training.

For the sake of completeness we also exploited *Facenet* [46] to train and evaluate the FQA models apart from *Arcface*. We calculated the Pearson correlations for two BRO models trained according to face recognition scores of different methods, namely *Arcface* and *FaceNet*. We utilized two face similarity measures and cross-checked to observe if the proposed approach is sensitive to the similarity metric choices in training phase. Since two metrics yield opposite behaviors for the best matches, where *Arcface* tends to approach *1.0* and *FaceNet* tends to approach *0.0* for a perfect match, the Pearson correlations are negated in the cross scenarios for the sake of simplicity. The results are given in Table 3, where *Arcface* based model training achieves superior correlation compared to *FaceNet*. On the other hand, the difference between two models under the same similarity during test phase (two measures on the same row) does not exceed 2% indicating that the proposed approach is almost independent of the face similarity metrics exploited during the train phase. It is also clear that, *Arcface* provides higher Pearson correlation when it is applied in matching during tests as the results in the same columns are checked.

Table 2. Face quality prediction correlation with actual scores on GeorgiaTech, LWF and Feret face datasets

		Georgia				LFW				Feret			
Attack	Similarity	Blur	Rot	Occ	BRO	Blur	Rot	Occ	BRO	Blur	Rot	Occ	BRO
Blur	Match	**0.84**	0.67	0.56	**0.84**	**0.88**	0.73	0.79	0.87	**0.71**	0.46	0.64	0.69
	Self	**0.86**	0.65	0.53	0.85	**0.62**	0.29	0.47	0.57	**0.62**	0.29	0.47	0.57
Occlusion	Match	0.73	0.80	**0.82**	**0.82**	0.67	0.71	**0.74**	0.73	0.62	0.62	**0.78**	0.72
	Self	0.68	0.76	**0.79**	0.78	0.56	0.55	**0.75**	0.66	0.56	0.55	**0.75**	0.66
Blur+Occ	Match	0.71	0.73	0.68	**0.79**	0.79	0.64	0.66	**0.82**	0.58	0.52	0.64	**0.69**
	Self	0.67	0.70	0.66	**0.77**	0.83	0.67	0.69	**0.86**	0.56	0.48	0.62	**0.66**

Table 3. Pearson correlation scores of face quality scores from BRO models trained by different FR models with respect to face similarity scores obtained from different FR models

Pearson Corr.		BRO Models	
Dataset	Face Recognition Models	Arcface	FaceNet
GeorgiaTech	Arcface	**0.787**	0.768
	FaceNet	0.732	0.730
LFW	Arcface	**0.819**	0.807
	FaceNet	0.759	0.749
Feret	Arcface	**0.686**	0.676
	FaceNet	0.609	0.583

5.3 Comparison with State-of-the-Art

In this section, we compare the proposed FQA method trained under *BRO* augmentation with a recent technique *FaceQnet* [25] developed for FQA as well as two classical measures blur and contrast proposed in [24]. We utilize both face similarity metrics and two BRO models trained through these metrics. As in the previous section, we measure correlation of the actual similarity and the estimated FQA by the Pearson correlation.

The Pearson correlations are given in Table 4 for each FQA method under three distortion attacks on three different data sets calculated for two different face matching scores. It is obvious that proposed approach provides the highest correlation with the actual face matching scores for 11 cases out of 18, where the rest is provided by *FaceQnet*. The classical measures fail to predict the FQA in terms of face matching. However, under the Gaussian blur and joint attack scenario the *blur* metric has significant correlation due to severe decrease in sharpness of the face images. *Contrast* can not provide any correlation with the face recognition score especially under the occlusion attacks. These observation validate the motivation of this study that classical FQA are not sufficient to model face matching similarity.

We extend this comparison for the realistic data that we have collected as mentioned in Sect. 5.1. In this scenario, the face images are gathered from surveillance videos to observe the real cases including occlusion and motion blur without any artificial attacks. In Table 5, the Pearson correlation of each are given for three different sites. It is clear that proposed approach has competitive results with *FaceQnet*, where it has higher correlation when *Arcface* is utilized for face similarity. The correlations for both methods are very close to each other. The sorted face crops of two individuals are given in Fig. 5 that are gathered by

Table 4. Pearson correlation of proposed, FaceQnet, blur and occlusion metrics with respect to ArcFace/FaceNet matching scores

Attack	Dataset	FQA	FaceQnet	Blur	Contrast
Blur	Georgia	**0.84/0.73**	0.78/0.52	0.58/0.59	0.21/0.29
	LFW	**0.87/0.80**	0.69/0.69	0.79/0.68	0.22/0.19
	Feret	0.69/0.65	**0.71/0.73**	0.42/0.37	0/0.08
Occlusion	Georgia	**0.82**/0.73	0.78/**0.74**	0.03/0.02	0.07/0.06
	LFW	0.72/0.68	**0.73/0.72**	0.03/0.02	0/0
	Feret	**0.72**/0.71	0.71/**0.79**	0.06/0.02	0.01/0.12
Blur+Occ	Georgia	**0.79/0.73**	0.71/0.70	0.57/0.55	0.05/0
	LFW	**0.82/0.75**	0.62/0.65	0.79/0.69	0.06/0.04
	Feret	**0.69**/0.58	0.66/**0.70**	0.49/0.03	0.03/0.06

Table 5. Pearson correlations of FQA methods with respect to FR models on surveillance videos

		FQA Models			
Dataset	Similarity	BRO	FaceQnet	Blur	Contrast
Lab	Arcface	**0.76**	0.73	0.22	0.05
	FaceNet	0.49	**0.62**	0.22	0.07
Indoor_low	Arcface	**0.79**	0.77	0.22	0.12
	FaceNet	0.64	**0.67**	0.24	0.27
Indoor_high	Arcface	**0.59**	0.54	0.11	0.04
	FaceNet	**0.69**	0.65	0.08	0.06

consecutive face detection and tracking that yield same track ID for each person. The FQ estimations of the worst (red rectangle) and the best (green rectangle) faces are given below the Fig. 5, where proposed approach has the best discrimination between the best and the worst images compared to the other techniques. Face quality estimates are significantly increased by the proposed method when faces are observed crisp and clear, while these values drop severely under motion blur and occlusion. Further results are given in Fig. 6 for surveillance scenario, where the estimated face quality and the actual recognition scores are given for each crop. It is obvious that the proposed approach is consistent with the actual face similarity metrics and provides valid prejudgement of faces before recognition.

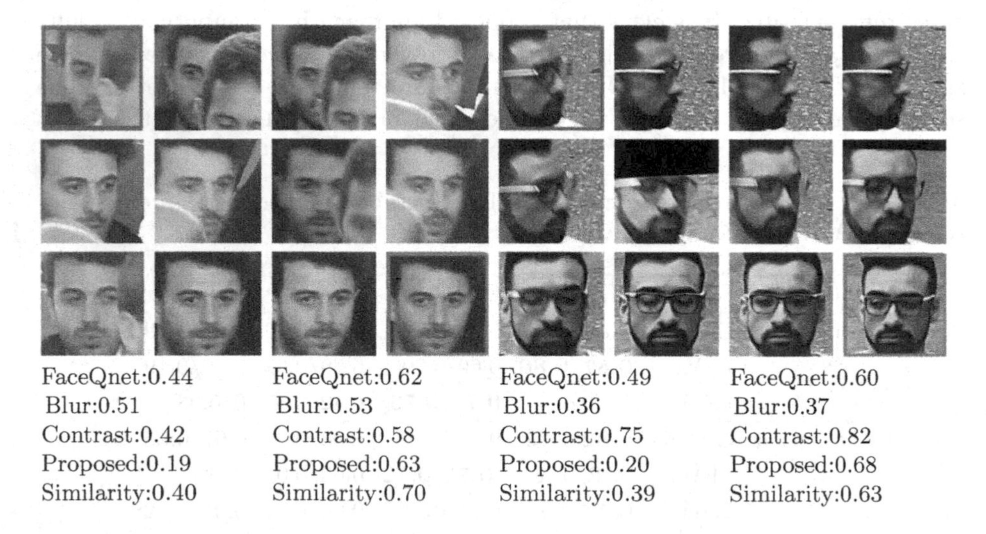

Fig. 5. The sorted faces of two individuals from our surveillance videos, red indicates the lowest quality face while green indicates the highest quality face of the tracks. The estimated face quality measures for the best and worst face crops are given per FQA model. (Color figure online)

5.4 Edge Deployment

The proposed FQA approach is incorporated with face detection, tracking and alignment to observe if it can eliminate the low quality faces and highlight ideal faces for recognition. The implementation is conducted on well known Nvidia Jetson TX2 board as efficient, small form factor platform for edge deployment. We exploit Pelee [47] that is a recent and efficient single shot detector for the initial detection of faces with input size of 304×304. The faces are tracked by bi-directional bounding box association, so each individual is assigned to a unique ID. The proposed FQA method is applied in conjunction with facial landmark extraction to each face in order to provide landmark points and the face quality estimate simultaneously. The faces are aligned for recognition and the face crops are sorted for each track ID with respect to estimated face qualities. This optimized pipeline enables real-time processing of surveillance videos. The average execution times of this application are given as 15 ms, 1.0 ms, 1.8 ms for initial face detection, bi-directional tracking and landmark extraction per face consecutively. We have also included *FaceQnet* as an alternative FQA within the pipeline, where the average execution time is around 19.5 ms. It is important to note that proposed FQA technique does not require additional complexity since it is integrated into landmark extraction framework. On the other hand, *FaceQnet* requires as higher computational burden as the initial face detection that introduces an overhead to the full system. Though, the proposed method is an efficient alternative for the state-of-the-art enabling consistent FQA in no time.

Fig. 6. The FQA of the proposed method and the face similarity (*Arcface*) values are given for ordered faces of seven individuals.

6 Conclusions

In this paper, we propose an efficient FQA method that is incorporated with landmark extraction framework by an additional single fully connected layer. The features extracted for landmark detection are reused to regress the matching scores. In order to simulate real life scenarios of surveillance cameras, that make recognition harder, we apply rotation, blur, and occlusion attacks as data augmentation in the training of FQA layer. In depth analyses indicate that the proposed single layer extension as well as surveillance guided augmentation yield high correlation between FQA and face recognition scores. This enables quite efficient selection of best representative face images on the edge for each face track and decrease the number of face images transferred to the recognition servers.

This does not only ease the network load of the face detection cameras but also potentially increases the accuracy and rate of the recognition server. Although for fixed detection and recognition networks, the introduced FQA layer provides state-of-the-art quality evaluation, further fine tuning and/or end-to-end training of detection and recognition networks with realistic augmentation attacks is the subject of future work.

References

1. Konen, W., Schulze-kruger, E.: ZN-Face: a system for access control using automated face recognition. In: International Workshop on Automated Face- and Gesture-Recognition, pp. 18–23 (1995)
2. Yugashini, I., Vidhyasri, S., Gayathri Devi, K.: Design and implementation of automated door accessing system with face recognition. Int. J. Sci. Mod. Eng. (IJISME) **1**(12) (2013)
3. Ibrahim, R., Zin, Z.M.: Study of automated face recognition system for office door access control application. In: IEEE 3rd International Conference on Communication Software and Networks, pp. 132–136 (2011)
4. Feng, W., Zhou, J., Dan, C., Peiyan, Z., Li, Z.: Research on mobile commerce payment management based on the face biometric authentication. Int. J. Mob. Commun. **15**, 278–305 (2017)
5. Wheeler, F.W., Weiss, R.L., Tu, P.H.: Face recognition at a distance system for surveillance applications. In: 4th IEEE International Conference on Biometrics: Theory, Applications and Systems (BTAS), pp. 1–8 (2010)
6. Lei, Z., Wang, C.C., Wang, Q., Huang, Y.: Real-time face detection and recognition for video surveillance applications. In: World Congress on Computer Science and Information Engineering (WRI), vol. 5, pp. 168–172 (2009)
7. Xu, X., Liu, W., Li, L.: Low resolution face recognition in surveillance systems. J. Comput. Commun. **2**, 70–77 (2014)
8. Haghighat, M., Abdel-Mottaleb, M.: Lower resolution face recognition in surveillance systems using discriminant correlation analysis. In: 12th IEEE International Conference on Automatic Face & Gesture Recognition (FG 2017), pp. 912–917 (2017)
9. Chingovska, I., Erdogmus, N., Anjos, A., Marcel, S.: Face recognition systems under spoofing attacks. In: Bourlai, T. (ed.) Face Recognition Across the Imaging Spectrum, pp. 165–194. Springer, Cham (2016). https://doi.org/10.1007/978-3-319-28501-6_8
10. Schulze-kruger, E., Konen, W.: Spoofing face recognition with 3D masks. IEEE Trans. Inf. Forensics Secur. **9**, 1084–1097 (2014)
11. Akbulut, Y., Sengür, A., Budak, Ü., Ekici, S.: Deep learning based face liveness detection in videos. In: International Artificial Intelligence and Data Processing Symposium (IDAP), pp. 1–4 (2017)
12. Liu, W., Wen, Y., Yu, Z., Li, M., Raj, B., Song, L.: SphereFace: deep hypersphere embedding for face recognition. In: IEEE Conference on Computer Vision and Pattern Recognition (CVPR), pp. 6738–6746 (2017)
13. Wang, H.J., et al.: CosFace: large margin cosine loss for deep face recognition. In: IEEE/CVF Conference on Computer Vision and Pattern Recognition, pp. 5265–5274 (2018)

14. Deng, J., Guo, J., Zafeiriou, S.: ArcFace: additive angular margin loss for deep face recognition. In: IEEE/CVF Conference on Computer Vision and Pattern Recognition (CVPR), pp. 4685–4694 (2018)
15. Chen, S., Liu, Y.P., Gao, X., Han, Z.: MobileFaceNets: efficient CNNs for accurate real-time face verification on mobile devices. In: Chinese Conference on Biometric Recognition, pp. 428–438 (2018)
16. Yan, M., Zhao, M., Xu, Z., Zhang, Q., Wang, G., Su, Z.: VargFaceNet: an efficient variable group convolutional neural network for lightweight face recognition. In: IEEE/CVF International Conference on Computer Vision Workshop (ICCVW), pp. 2647–2654 (2019)
17. Zhanga, K., Zhang, Z., Li, Z., Qiao, Y.: Joint face detection and alignment using multitask cascaded convolutional networks. IEEE Signal Process. Lett. **23**, 1499–1503 (2016)
18. Najibi, M., Samangouei, P., Chellappa, R., Davis, L.S.: SSH: single stage headless face detector. In: IEEE International Conference on Computer Vision (ICCV), pp. 4885–4894 (2017)
19. Zhang, S., Zhu, X., Lei, Z., Shi, H., Wang, X., Li, S.Z.: S^3FD: single shot scale-invariant face detector. In: IEEE International Conference on Computer Vision (ICCV), pp. 192–201 (2017)
20. Deng, J., Guo, J., Zhou, Y., Yu, J., Kotsia, I., Zafeiriou, S.: RetinaFace: single-stage dense face localisation in the wild (2019). arXiv:1905.00641
21. Abaza, A., Harrison, M.A., Bourlai, T.: Quality metrics for practical face recognition. In: Proceedings of the 21st International Conference on Pattern Recognition (2012)
22. Vignesh, S., Priya, K.M., Channappayya, S.S.: Face image quality assessment for face selection in surveillance video using convolutional neural networks. In: CIEEE Global Conference on Signal and Information Processing (GlobalSIP) (2015)
23. Qi, X., Liu, C., Schuckers, S.: CNN based key frame extraction for face in video recognition. In: 2018 IEEE 4th International Conference on Identity, Security, and Behavior Analysis (ISBA) (2018)
24. Lienhard, A., Ladret, P., Caplier, A.: Low level features for quality assessment of facial images. In: 10th International Conference on Computer Vision Theory and Applications (VISAPP 2015), pp. 545–552 (2015)
25. Hernandez-Ortega, J., Galbally, J., Fiérrez, J., Haraksim, R., Beslay, L.: FaceQnet: quality assessment for face recognition based on deep learning. In: International Conference on Biometrics (ICB) (2019)
26. Wong, Y., Chen, S., Mau, S., Anderson, S., Lovell, B.: Patch-based probabilistic image quality assessment for face selection and improved video-based face recognition. In: Computer Vision and Pattern Recognition Workshops (CVPRW) (2011)
27. Lienhard, A., Reinhard, M., Caplier, A., Ladret, P.: Low level features for quality assessment of facial images. In: 2014 International Conference on Computer Vision Theory and Applications (VISAPP) vol. 2 (2014)
28. Bai, G., Chen, J., Deng, Y., Su., G.: Face image quality assessment based on learning to rank. IEEE Signal Process. Lett. **22**, 90–94 (2015)
29. Anantharajah, K., Denman, S., Sridharan, S., Fookes, C., Tjondronegor, D.: Quality based frame selection for video face recognition. In: 6th International Conference on Signal Processing and Communication System (2012)
30. Li, Y., Po, L., Feng, L., Yuan, F.: No-reference image quality assessment with deep convolutional neural networks. In: IEEE International Conference on Digital Signal Processing (DSP), pp. 685–689 (2019)

31. Khastavaneh, H., Ebrahimpour-Komleh, H., Joudaki, M.: Face image quality assessment based on photometric features and classification techniques. In: IEEE 4th International Conference on Knowledge-Based Engineering and Innovation (KBEI), pp. 289–293 (2017)
32. Qi, X., Liu, C., Schuckers, S.: Boosting face in video recognition via CNN based key frame extraction. In: 2018 International Conference on Biometrics (ICB) (2018)
33. Best-Rowden, L., Jain, A.K.: Learning face image quality from human assessments. IEEE Trans. Inf. Forensics Secur. **13**(12), 3064–3077 (2018)
34. Yu, J., Sun, K., Gao, F., Zhu, S.: Face biometric quality assessment via light CNN. Pattern Recogn. Lett. **107**, 25–32 (2018)
35. Dhar, P., Castillo, C.D., Chellappa, R.: On measuring the iconicity of a face. In: IEEE Winter Conference on Applications of Computer Vision (WACV) (2019)
36. Zhuang, N., et al.: Recognition oriented facial image quality assessment via deep convolutional neural network. J. Neurocomput. **358**, 109–118 (2019)
37. Chen, Y., Song, L., He, R.: Adversarial-occlussion-aware-face-detection. In: IEEE International Conference on Biometrics, Theory, Applications and Systems (2018)
38. He, K., Zhang, X., Ren, S., Sun, J.: Deep residual learning for image recognition. In: IEEE Conference on Computer Vision and Pattern Recognition (CVPR), pp. 770–778 (2016)
39. Liu, Z., Luo, P., Wang, X., Tang, X.: Deep learning face attributes in the wild. In: International Conference on Computer Vision (ICCV) (2015)
40. Chen, T., et al.: MXNet: a flexible and efficient machine learning library for heterogeneous distributed system. In: Neural Information Processing Systems, Workshop on Machine Learning Systems (2015)
41. Huang, G.B., Ramesh, M., Berg, T., Learned-Miller, E.: Labeled faces in the wild: a database for studying face recognition in unconstrained environments. Technical report 07–49, University of Massachusetts (2007)
42. Learned-Miller, E., Huang, G.B.: Labeled faces in the wild: updates and new reporting procedures. University of Massachusetts, Amherst, UM-CS-2014-003 (2014)
43. Phillips, P.J., Wechsler, H., Huang, J., Rauss, P.J.: The FERET database and evaluation procedure for face-recognition algorithms. Image Vis. Comput. **16**, 295–306 (1998)
44. Phillips, P.J., Moon, H., Rizvi, S.A., Rauss, P.J.: The FERET evaluation methodology for face-recognition algorithms. IEEE Trans. Pattern Anal. Mach. Intell. **22**, 1090–1104 (2000)
45. Georgia tech face database. ftp://ftp.ee.gatech. edu/pub/users/hayes/facedb/
46. Schroff, F., Kalenichenko, D., Philbin, J.: FaceNet: a unified embedding for face recognition and clustering. In: Proceedings of the IEEE Conference on Computer Vision and Pattern Recognition, pp. 815–823 (2015)
47. Wang, R.J., Li, X., Ling, C.X.: Pelee: a real-time object detection system on mobile devices (2018)

Efficient Approximation of Filters for High-Accuracy Binary Convolutional Neural Networks

Junyong Park, Yong-Hyuk Moon, and Yong-Ju Lee(✉)

Electronics and Telecommunications Research Institute, Daejeon, South Korea
{junyong.park,yhmoon,yongju}@etri.re.kr

Abstract. In this paper, we propose an efficient design to convert full-precision convolutional networks into binary neural networks. Our method approximates a full-precision convolutional filter by sum of binary filters with multiplicative and additive scaling factors. We present closed form solutions to the proposed methods. We perform experiments on binary neural networks with binary activations and pre-trained neural networks with full-precision activations. The results show an increase in accuracy compared to previous binary neural networks. Furthermore, to reduce the complexity, we prune scaling factors considering the accuracy. We show that up to a certain degree of threshold, we can prune scaling factors while maintaining accuracy comparable to full-precision convolutional neural networks.

Keywords: Binary neural network · Convolutional neural networks · Filter approximation · Scaling factors

1 Introduction

The demand of solutions for complex problems leads convolutional neural networks (CNNs) to become deeper and wider, increasing up to billions of parameters. This makes CNNs harder to operate on many real-time applications or resource constrained environments such as embedded devices. Binarization is introduced to reduce the complexity of neural networks. Binary CNNs use binary values for filters and/or inputs. Binary operations can achieve an immense amount of computation reduction compared to full-precision operations [17]. However, binary filters and activations are difficult to converge when trained from scratch, and binarizing the full precise values reduces the accuracy of the output due to the loss of information. We propose a novel, direct approach to convert full-precision filters into binary filters with minimum accuracy loss, without extra training for the binary filters.

The contributions of this paper are as follows:

- We propose a method approximating the full precision convolutional filter by using the K binary filters with multiplicative and additional scaling factors. We provide closed form solutions for finding scaling factors. We provide

A. Bartoli and A. Fusiello (Eds.): ECCV 2020 Workshops, LNCS 12539, pp. 71–84, 2020.
https://doi.org/10.1007/978-3-030-68238-5_6

several orthogonal or quasi-orthogonal vectors of different lengths to support convolutional filters of different sizes.
- When $K \geq 2$, called *Improved Binary Filters (IBF)* is proposed on networks with two scenarios. On pretrained models with full-precision inputs, IBF can simply replace full-precision filters and achieve similar accuracy. On binary inputs, IBF can reduce the complexity using binary operations but achieve much higher accuracy than well-known XNOR-Net [17] with only $K+1$ times higher complexity.
- To reduce the complexity of IBF, we prune scaling factors considering the accuracy. We show that up to a certain degree of threshold, we can prune scaling factors but also maintain accuracy comparable to full-precision convolutional filter.

1.1 Related Work

Binary Neural Networks. Binary neural network uses binary values for its calculation. Given a neural network with real values, binarization replace weight values to $+1$ when the value is positive, and -1 when the value is negative, which is equivalent to a sign function [2]. This approach leads to massive improvement in model size compression (about 32 times reduction) and computational load ([2] shows results up 7 times faster computation time in GPU) on MNIST [12] and CIFAR-10 [10] datasets.

XNOR-Net [17] binarize both filters and input values using the sign function and a scaling factor. XNOR-Net is one of the first binary neural networks to compete with state-of-the-art models on large datasets such as ImageNet [5]. By binarizing both input and filters, the convolution operation can be replaced with binary operation resulting 58 times faster and 32 times reduction in memory size. However, since binary gradients have issues converging binary filters, they use full-precision gradients to update full-precision values, then binarize filters on every epoch in their training process. However, XNOR-Net method of binarization cannot be used on full-precision weights after training, the degradation of representation bits is too much for the model to provide a reasonable output. Our method can overcome this problem by providing a binarization method that approximates the original full-trained method with very little accuracy loss.

Linear Combinations of Binary Filters. Ever since XNOR-Net has been introduced, several methods have been proposed to compensate for the relative accuracy loss. Since binary filters themselves are memory efficient and computationally lighter in floating point multiplication and addition for convolutions, several proposals have been made to linearly combine a number of binary filters to approximate the original full precision filters.

In [13], the authors introduce linear combinations of multiple binary weight bases and employed multiple binary activation to alleviate information loss. The linear combinations of multiple binary weight bases provides a means for binary weight vectors to represent more information per layer than the sign. Also, [23] proposed a different strategy of structure approximation for binarization; the authors propose Group-Net which uses decomposition of filters into groups of binary filters, which increase the complexity layer-wise and group-wise. The high number of bases and activation made it possible for [23] to achieve exceptionally high results. Compared to these papers, our method provides a less complex algorithm, a directly calculable solution for binary filters which in most cases much faster than training/converging binary filters. This can be intuitively understood, as in practice, the size of the model in use is much smaller than the size of the data to be trained.

The adaptation of binary matrices has been attempted in deterministic binary filter network (DBFNet) [21]. They use orthogonal binary basis filters to generate a linear combination of binary operation with multiplicative factors. The binary filters are generated from a orthogonal tree generator. However, the multiplicative scaling factors are not determined. Thus, DBFNet has to re-train the entire network with the same method as previous traditional binary neural networks.

Compared to DBFNet, our proposed scheme can fully provide an accurate approximation for any type of full-precision filters. Our scheme can provide direct results without additional training.

2 Filter Approximation

2.1 Filters Approximation Using Binary Orthogonal Vectors

Consider a convolutional filter of size $n_w \times n_h \times n_c$, where n_w, n_h, n_c are size of width, height, channel of the filter, respectively. At each channel, we approximate convolutional filter $\mathbf{W} \in \mathbb{R}^{n_w \times n_h}$ by a scaled binary matrix with additive factor:

$$\mathbf{W} \approx \alpha \mathbf{B} + \beta \mathbf{1}_{N_W}, \tag{1}$$

where $\mathbf{B} \in \{+1, -1\}^{n_w \times n_h}$, $\alpha, \beta \in \mathbb{R}$, $N_W = n_w n_h$, and $\mathbf{1}_{N_W}$ is an $n_w \times n_h$ all-one matrix.

If $\beta = 0$, then the optimum $\mathbf{B}$ and α are

$$\mathbf{B} = sign\left[\mathbf{W}\right], \alpha = \frac{\|\mathbf{W}\|_1}{N_W}, \tag{2}$$

where $sign(x) = \begin{cases} +1 & x \geq 0 \\ -1 & x < 0 \end{cases}$, and $\|\|_1$ represents L1 norm. (2) are well-known solution for XNOR network [17]. When $\beta \neq 0$, we transform (1) into vector form:

$$\mathbf{w} \approx \alpha \mathbf{b} + \beta \mathbf{1}, \tag{3}$$

where $\mathbf{w} = vec\left[\mathbf{W}\right]$, $\mathbf{b} = vec\left[\mathbf{B}\right]$, and $\mathbf{1}$ is an $N_W \times 1$ all-one vector, and $vec\left[\cdot\right]$ is a vectorization. We choose $\mathbf{B} = sign\left[\mathbf{W}\right]$. To find values of α and β which minimize the error:

$$\min_{\alpha,\beta} \|\mathbf{w} - \alpha \mathbf{b} - \beta \mathbf{1}\|^2, \tag{4}$$

differentiate (4) with respect to α yields

$$\alpha = \frac{1}{N_W}\mathbf{w}^T\mathbf{b} - \frac{1}{N_W}\beta\mathbf{1}^T\mathbf{b}. \tag{5}$$

Similarly differentiating with respect to β yields

$$\beta = \frac{1}{N_W}\mathbf{w}^T\mathbf{1} - \frac{1}{N_W}\alpha\mathbf{b}^T\mathbf{1}. \tag{6}$$

From (5) and (6),

$$\alpha = \frac{\mathbf{w}^T\mathbf{b} - \frac{1}{N_W}\mathbf{w}^T\mathbf{1}\mathbf{1}^T\mathbf{b}}{N_W - \frac{1}{N_W}\mathbf{b}^T\mathbf{1}\mathbf{1}^T\mathbf{b}}. \tag{7}$$

Note that when $\beta = 0$, (5) reduces to α in (2).
Let us generalize the problem of approximating convolutional filter by K binary matrices, multiplicative and additional factors.

$$\mathbf{W} \approx \sum_{k=1}^{K} \alpha_k\mathbf{B}_k + \beta\mathbf{1}_{N_W}, \tag{8}$$

where $\mathbf{B}_k \in \{+1, -1\}^{n_w \times n_h}$, $\alpha_k, \beta \in \mathbb{R}$, and $\mathbf{1}_{N_W}$ is an $n_w \times n_h$ all-one matrix. The equivalent problem of (8) is to approximate convolutional filter vector $\mathbf{w}$ by K binary orthogonal vectors, and multiplicative and additional factors.

$$\mathbf{w} \approx \sum_{k=1}^{K} \alpha_k\mathbf{b}_k + \beta\mathbf{1}, \tag{9}$$

where $\mathbf{w} = vec\left[\mathbf{W}\right]$, and $\{\mathbf{b}_k\}$ are assumed as binary orthogonal vector of length N_W,

$$\frac{1}{N_W}\mathbf{b}_i^T\mathbf{b}_j = \begin{cases} 1 & i = j \\ 0 & i \neq j, \end{cases} \tag{10}$$

and K is the numbers of orthogonal vectors used for approximation. We can see (3) is a special case of (9) when $K = 1$. The least square (LS) problem becomes

$$\min_{\bar{\alpha},\beta} \left\| \mathbf{w} - \sum_{k=1}^{K} \alpha_k\mathbf{b}_k - \beta\mathbf{1} \right\|^2, \tag{11}$$

where $\bar{\alpha} = \left[\alpha_1, \alpha_2, \cdots, \alpha_K\right]^T$. The solution for (11) are

$$\alpha_k = \frac{\mathbf{w}^T\mathbf{b}_k - \frac{1}{N_W}\mathbf{w}^T\mathbf{1}\mathbf{1}^T\mathbf{b}_k}{N_W - \frac{1}{N_W}\mathbf{b}_k^T\mathbf{1}\mathbf{1}^T\mathbf{b}_k}, \tag{12}$$

$$\beta = \frac{1}{N_W}\mathbf{w}^T\mathbf{1} - \frac{1}{N_W}\sum_{k=1}^{K} \alpha_k\mathbf{b}_k^T\mathbf{1}. \tag{13}$$

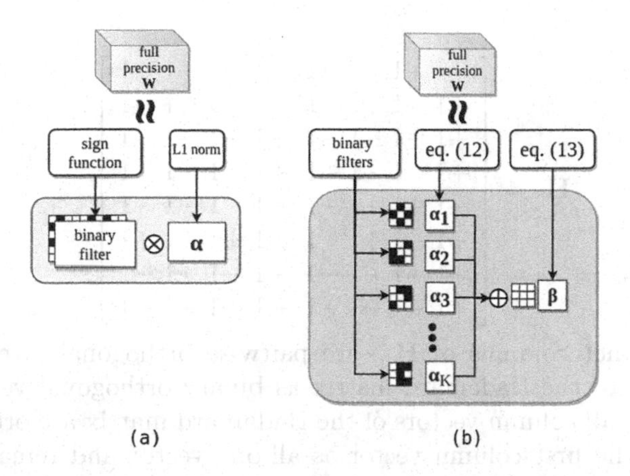

Fig. 1. Illustration of binarization methods. (a) XNOR-Net method. (b) Proposed IBF method.

Note that α_k in (12) depends only on corresponding vector $\mathbf{b}_k$ not from all other vectors due to orthogonality of $\{\mathbf{b}_k\}$. Also note that when $K = 1$, (12) and (13) reduce to (7) and (6), respectively. Once $\alpha_1, \alpha_2, \cdots, \alpha_K, \beta$ are determined, we reconstruct (8) by using $\mathbf{W} = vec^{-1}[\mathbf{w}]$ and $\mathbf{B_k} = vec^{-1}[\mathbf{b}_k]$. We call this scheme as *Improved Binary Filters* (IBF).

Figure 1 illustrates the two binarization methods. Figure 1(a) represents previous XNOR-Net binarization. The binary filter is created as the sign of the original filter, and it is scaled by the multiplication of the L1 norm. Figure 1(b) represents proposed IBF method. The original full-precision (FP) filter is approximated by a set of binary filters generated from a binary orthogonal vectors. Each binary orthogonal vectors are scaled with α_k calculated from (12) and added by the additive factor β calculated from (13).

2.2 Generation of Binary Orthogonal Vectors

Binary Orthogonal Vectors. In the previous section, we provide a closed solution for formula (9), given the binary vectors are mutually orthogonal. Binary orthogonal vectors can be extracted from a orthogonal matrix. Hadamard matrices of order N is an $N \times N$ matrix $\mathbf{H}_N$ of $\{-1, +1\}$ such that the (i, j)-th component of $\frac{1}{N}\mathbf{H}_N^T\mathbf{H}_N$ is given [16]

$$\frac{1}{N}\left\{\mathbf{H}_N^T\mathbf{H}_N\right\}_{i,j} = \begin{cases} 1 & i = j \\ 0 & i \neq j. \end{cases} \tag{14}$$

For example, Hadamard matrices of order 8 is given:

$$\mathbf{H}_8 = \begin{bmatrix} +1 & +1 & +1 & +1 & +1 & +1 & +1 & +1 \\ +1 & -1 & +1 & -1 & +1 & -1 & +1 & -1 \\ +1 & +1 & -1 & -1 & +1 & +1 & -1 & -1 \\ +1 & -1 & -1 & +1 & +1 & -1 & -1 & +1 \\ +1 & +1 & +1 & +1 & -1 & -1 & -1 & -1 \\ +1 & -1 & +1 & -1 & -1 & +1 & -1 & +1 \\ +1 & +1 & -1 & -1 & -1 & -1 & +1 & +1 \\ +1 & -1 & -1 & +1 & -1 & +1 & +1 & -1 \end{bmatrix}. \tag{15}$$

Note that distinct columns of $\mathbf{H}_N$ are pairwise orthogonal. We can use the column vectors of the Hadamard matrix as binary orthogonal vectors $\{\mathbf{b}_k\}$ in (10). Although, all column vectors of the Hadamard matrix are orthogonal each other, we use the first column vector as all-one vector and remaining column vectors as binary orthogonal vectors in (10). For Hadamard matrix of order N, the maximum number of orthogonal vectors is $N-1$. For our applications, the number of binary orthogonal vectors should be larger than N_W. For example, for convolutional filter of size 3×3, the length of vector N_W is 9, and $\mathbf{H}_{16}$ is chosen for approximation. And all columns except the first column have equal number of ones and zeros: $\mathbf{b}_k^T \mathbf{1} = 0$ in (11), (12). The order of Hadamard matrix exists only for 1, 2, or multiple of 4. There are several ways of constructing the Hadamard matrix of order N. The most popular method is referred to as the Sylvester construction:

$$\mathbf{H}_{2N} = \begin{bmatrix} \mathbf{H}_N & \mathbf{H}_N \\ \mathbf{H}_N & -\mathbf{H}_N \end{bmatrix}. \tag{16}$$

The codeword of the first-order Reed Muller code is a linear combination of column vectors of the Hadamard matrix: $\mathbf{c} = a_1\mathbf{h}_2 + \cdots + a_{M-1}\mathbf{h}_M + \beta\mathbf{1}$, where $\alpha_1, \cdots, a_M, \beta \in \{0, 1\}$ and $\{\mathbf{h}_k\}$ is the column vector of the Hadamard matrix $\mathbf{H}$. And this representation is special form of (9).

3 Binary Convolutional Neural Network

3.1 Binary Neural Network

Figure 2(a) illustrates a typical convolutional block in a CNN. Input is processed through full-precision convolution, batch normalization, non-linear activation in order. Figure 2(b) is used by previous binary neural networks such as XNOR-Net. In this block, batch normalization is performed before the binary activation to prevent binarization loss. Binary activation makes the full-precision input to be transferred into binary input. Last, binary input is computed with binary convolution. These are works of previous methods of full-precision or binary neural networks. In the next section we will explain Fig. 2(c) and (d).

3.2 Binary Input

We propose training method of full-precision filters with binary inputs. In forward pass, inputs are binarized through activations using batch normalization

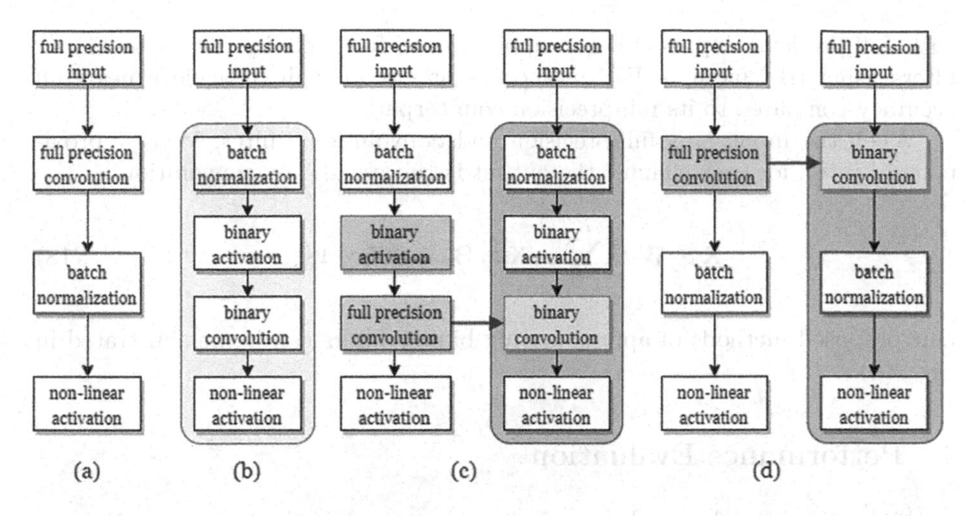

Fig. 2. Illustration of various blocks in a convolutional layer. (a) Original full-precision (FP) convolutional layer. (b) XNOR-Net binary convolutional layer. (c) our proposed BIFF method with IBF method. (d) our proposed IBF method with FP input.

and binary activation using sign function. Since sign function is not differentiable for back propagation, so we use straight-through-estimator (STE) [1] to compute the gradient for the non-differentiable function. To prevent radical changes in the gradient, clipping is used to stop gradient flow if magnitude of the input $\mathbf{X} \in \mathbb{R}^{i_w \times i_h}$ is too large [3].

We call this method as binary input, full-precision filters (BIFF) as shown in Fig. 2(c) left. BIFF is an interim reference to measure the performance of IBF. Training is performed using a small learning rate (e.g 0.0001) to maintain stability while training. After applying IBF method to FP filter in BIFF, the block becomes Fig. 2(c) right, where the block becomes fully binary neural network. As a result, we can utilize the full advantage of binary convolution operations of XNOR and pop-count operations. For each channel, the output is represented by convolution,

$$\mathbf{X} \otimes \mathbf{W} \approx \sum_{k=1}^{K} \alpha_k \mathbf{X}_b \odot \mathbf{B}_k + \beta \mathbf{X}_b \odot \mathbf{1}_{N_W}, \tag{17}$$

where $\odot$ represents binary convolution implemented by XNOR and pop-count operations. We can implement the convolution between input and filter mostly using binary operations, and its complexity is $K + 1$ times higher than XNOR net approximately.

3.3 Full-Precision Input

To obtain binary filters with full-precision inputs, we start with a full-precision neural network. Since the majority of convolutional neural networks are trained with full-precision inputs, we can approximate full-precision filters to binary filters using IBF method. IBF method is precise enough to achieve near full accuracy compared to its full-precision counterparts.

When the input $\mathbf{X}$ is full-precision and convolutional filter, $\mathbf{W}$ is approximated by (8), for each channel the output is represented by convolution,

$$\mathbf{X} \otimes \mathbf{W} \approx \sum_{k=1}^{K} \alpha_k \mathbf{X} \otimes \mathbf{B}_k + \beta \mathbf{X} \otimes \mathbf{1}_{N_W}. \tag{18}$$

Our proposed methods of approximating binary filters can be demonstrated in Fig. 2(d).

4 Performance Evaluation

In this section we validate the usage of our proposed methods for convolutional neural networks for the task of classification. We compare the performance of three popular CNN architectures; AlexNet [11], VGG [18] and ResNet [6] using four types of methods: full-precision convolutional filter with full precision input (reference), XNOR-Net, BIFF, and proposed IBF using (8). Our methods are tested on classification on famous datasets such as Fashion MNIST [22], CIFAR-10 [10], and ImageNet [5].

4.1 Dataset and Implementation

Fashion MNIST [22] is a dataset composed of Zalando's article images. Similar to MNIST [12], Fashion MNIST has a training set of 60,000 and testing set of 10,000. All images are 28-by-28 gray scale images with ten different pieces of clothing (10 classes)

CIFAR-10 [10] is a dataset composed of 60,000 and a testing set of 10,000. All images are 32-by-32 RGB color images with 10 classes: airplanes, automobiles, birds, cats, deers, dogs, frogs, horses, ships and trucks. For training, we use random cropping, random horizontal flips for data variety.

The ImageNet Large Scale Visual Recognition Challenge (ILSVRC) [5] is a large image classification dataset composed of 1.2 million high-resolution images for training of 1000 categories of objects. The validation set has 50,000 images. We test our performance on validation set using top-1 and top-5 accuracy.

For reference neural network models, we use AlexNet [11], VGG with channels adjusted [18], SqueezeNet [8], and ResNet-18 [6].

For training, we use Adam optimizer [9]. We set the learning rate to 0.0001 to maintain stability; for binary neural networks if the learning rate is too high, too many weights might change signs, causing instability when training networks [20]. For reference, networks with full-precision weights and activation is also trained for comparison. We compare the performance of XNOR-Net and our BIFF and IBF based on the same network structure of the network structure.

4.2 Performance

Binary Neural Network; Binary Input. We conduct experiments of image classification by training various neural network models using different binary methods from scratch. We show results of accuracy for image classification.

Table 1. Performance comparison on binary neural networks tested on the Fashion MNIST dataset.

Accuracy (%)	Full-precision	XNOR-Net	BIFF	IBF
AlexNet	90.91	87.34	90.40	90.34
VGG11	92.24	88.16	91.88	91.71
ResNet18	93.15	90.32	91.62	91.60
SqueezeNet	92.31	90.31	91.91	90.42

Table 1 is a comparison of binary neural network methods tested on the Fashion MNIST dataset. The reference network use original full-precision filters with full-precision inputs as in Figure 2(a). The convolutional layer in XNOR-Net as shown in Fig. 2(b) uses binary filters with binary inputs. BIFF as shown in Fig. 2(c) left trains using binary inputs with full-precision filters. IBF approximates FP filters in BIFF by binary filters as shown in Fig. 2(c) right. Both of BIFF and IBF are superior to XNOR-Net method, BIFF is almost near full-precision accuracy to reference neural networks (less than 3% accuracy drop), and proposed IBF method with K_{max} values has little to none accuracy loss when converted from BIFF (less than 1% accuracy drop).

Table 2. Performance comparison on binary neural network tested on the CIFAR10 dataset.

Accuracy (%)	Full-precision	XNOR-Net	BIFF	IBF
AlexNet	89.9	84.3	85.1	85.0
VGG11	86.14	64.87	81.48	81.42
ResNet18	92.5	83.1	91.1	91.05

Table 2 is a comparison of binary neural networks tested on the CIFAR-10 dataset. Because of the data complexity, the accuracy drop for BIFF and IBF (about 5% accuracy drop) is larger than results from Table 1, but the accuracy is higher than the results tested on XNOR-Net. Results on the ResNet-18 model are shown to be highest on both of the datasets.

Table 3 shows the maximum K value, K_{max} on AlexNet model to show how much binary filters can be used for filter approximation in IBF. The size of original full-precision layer is $n_w \times n_h \times n_c$. For each channel, the filter vector length

Table 3. The maximum number of orthogonal vectors, K_{max}, to approximate the convolutional filters on the AlexNet.

	Size of filter ($n_w \times n_h \times n_c$)	N_W	K_{max} per channel
1st layer	11 × 11 × 64	121	0 (full-precision)
2nd layer	5 × 5 × 192	25	31
3rd layer	3 × 3 × 384	9	15
4th layer	3 × 3 × 256	9	15
5th layer	3 × 3 × 256	9	15

$N_W = n_w n_h$ determines the minimum K_{max} value to set for filter approximation. When we use the Hadamard matrix of order N, the $K_{max} = N - 1$ which is the minimum value satisfying $N - 1 \geq N_W$. We leave out binarization method for first layer. It is important to note that we do not need to use full K_max value per channel to achieve full accuracy.

Table 4. Top-1/Top-5 accuracy (%) of binary neural networks on the ImageNet dataset. Activation is 1-bit. Model is based on ResNet-18.

Method Name	Top-1/Top-5 (%)
Full-precision	69.76/89.08
XNOR-Net [17]	51.20/73.20
BNN+ [4]	52.64/72.98
ABC-Net [14]	54.10/78.10
DBF-Net [21]	55.10/78.50
Bi-Real Net [15]	56.40/79.50
BIFF	60.50/82.33
IBF	60.28/82.15

Table 4 shows the state-of-the-art results on larger dataset, ImageNet. Images are resized to 256 pixels, then random crop of 224 × 224 for training. We compare BIFF and IBF with other binary neural network methods. To compare with same conditions, only models with binary activations (1-bit), with base model ResNet-18, is used. We compare BIFF and IBF to other methods such as XNOR-Net [17], BNN+ [4], ABC-Net (highest 5-weight base with 1-base activation) [14], DBF-Net [21], and Bi-Real Net [15]. Except FP filters, BIFF method shows highest accuracy compared to other methods of training binary filters from scratch, and IBF method shows very little accuracy loss compared to BIFF.

Pre-trained Model; Full-Precision Input. We compare the accuracy of state-of-the-art neural networks. We directly use filter approximation using IBF method from full-precision pre-trained filter to binary filter, as shown in Fig. 2(d).

Table 5. Top-1/Top-5 accuracy (%) of pre-trained state-of-the-art networks. Activation is 32-bit. Orthogonal vectors are used for IBF.

Accuracy (%)	Full-precision	IBF
ResNet-18	69.76/89.08	69.41/88.92
ResNet-50	76.15/92.87	75.99/92.85
VGG-11	69.02/88.63	68.92/88.59
VGG-19	72.38/90.88	71.91/90.56
SqueezeNet 1.1	58.19/80.62	58.18/80.47
Densenet-161	77.65/83.80	77.38/83.64
MNASNet 1.0	73.51/91.54	73.35/91.38

Table 5 shows the result for IBF method to apply on pre-trained full-precision filters with full-precision inputs. IBF method directly converts the convolutional layers from pre-trained models. The second column represents accuracy for the original pre-trained FP models: ResNet-18, ResNet-50 [6], VGG-11, VGG-19 [18], SqueezeNet 1.1 [8], DenseNet 161 [7], and MNASNet 1.0 [19]. The third column represents the accuracy after we perform IBF method to only the convolutional filters. We use K_{max} value for maximum accuracy. Our methods report accuracy loss less than 0.8%. This accuracy can be achieved with no additional fine-tuning.

4.3 Scaling Factors Pruning

For maximum accuracy, we assume maximum value K_{max} for all our binary filter approximation. However, to reduce computational complexity and memory size, we can modify the K value to arrange the amount of binary orthogonal filters used in proposed filter approximation. We modify (9) as $\mathbf{W} \approx \sum_{S} \alpha_k \mathbf{b}_k + \beta \mathbf{1}$, where $S = \{\alpha_k : |\alpha_k| > \alpha_{th}\}$ and $|S| < K$. We discard $\{\alpha_k\}$ whose magnitudes are smaller than pre-determined threshold α_{th} and thus reduce the complexity sacrificing accuracy. Figure 3 is a sample histogram of α_k in the 2nd block before the pruning. It shows the distribution of scaling factor α.

Since most of values are located around zero, we can choose the threshold α_{th} by considering accuracy and complexity:

$$\left\| \sum_{S^c} \alpha_k \mathbf{b}_k \right\|^2 \leq \varepsilon, \tag{19}$$

where $S^C = \{\alpha_k : |\alpha_k| < \alpha_{th}\}$, and ε is an arbitrary small number.

Table 6 shows an experiment of scaling factor pruning. We remove the multiplicative scaling factors α based on the preset threshold. The top row indicates the threshold we set for the pruning method. The columns represent the average percentage of the α remained after pruning in each blocks. The second column

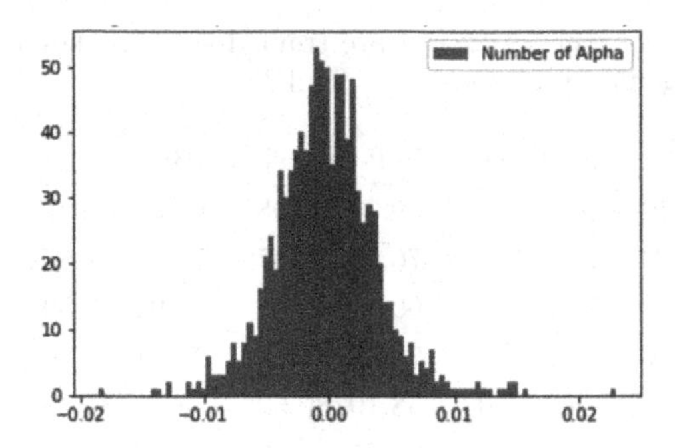

Fig. 3. Histogram of α scaling factors in BOF

Table 6. Percentage of scaling factors remained after applying threshold parameter α_{th} for pruning on pre-trained ResNet-18 network. Top-1 and top-5 accuracy are measured below.

Threshold α_{th}	(None)	1×10^{-5}	1×10^{-4}	1.5×10^{-4}	2×10^{-4}
1st block (%)	100	99.11	91.48	87.29	83.07
2nd block (%)	100	98.39	84.13	76.48	69.20
3rd block (%)	100	96.53	66.87	52.30	41.44
4th block (%)	100	91.96	37.88	23.11	14.27
Top-1/Top-5 (%)	69.76/89.08	69.74/89.07	68.94/88.68	66.27/87.05	58.64/81.63

is when the threshold is 0; no values are pruned as a result. The third column is when we set the threshold value to 1×10^{-5}, every $|\alpha|$ value that is lower than the threshold is removed. In this column we gain an average of 4% of pruning values while only losing 0.02% accuracy. When we set the threshold to 1×10^{-4}, we can prune about 30% of the scaling factors while only losing 0.98% accuracy. The fourth residual block particularly has over 60% of its values pruned. Threshold 1.5×10^{-4} and 2×10^{-4} are very similar in value range, however they contain very different results compared to previous thresholds. 1.5×10^{-4} on average pruned 67% of the scaling factors but only had 3.19% loss on accuracy. Compared to this 2×10^{-4} pruned 60% of the values on average, but the performance is much worse, 11.1 2% accuracy loss on total neural network performance.

4.4 Complexity Analysis

In a standard convolution with full-precision input, the total number of operation is $n_c N_W N_I$ where $N_W = n_w n_h$ and $N_I = i_w i_h$, where i_w and i_h are size of input. When input is binary-valued, and the convolutional filter $\mathbf{W} \in \mathbb{R}^{n_w \times n_h}$ is approximated by (2) in XNOR-Net, the binary convolution operation becomes XNOR and popcount operations. With the current generation of CPUs, up to 64 binary operations can be computed in one clock of CPU. So the total number of operations is reduced as $\frac{1}{64} n_c N_W N_I + N_I$. If IBF is used, then the total number of operations is increased by $K+1$ times over XNOR-Net; $\frac{1}{64}(K+1) n_c N_W N_I + (K+$

$1)N_I$. For example, when $n_c = 256$, $N_I = 256$, $N_W = 9$ and $K = K_{max} = 15$, the number of operations for standard convolution, XNOR-Net, and IBF are 589,864, 9,472, and 151,552, respectively. Although IBF requires 16 times higher complexity than XNOR-Net, it only requires 25.7% of standard convolution with similar accuracy. If we prune the scaling factors, then K value becomes less than K_{max}, which further reduces the complexity.

5 Conclusions

In this paper, we present an optimal design for binary filters for convolutional neural network. We propose IBF, an accurate approximation method of convolutional filters using binary orthogonal vectors with multiplicative and additive scaling factors. Using the least square method, we provide closed form for the scaling factors. Since there are a trade-off between accuracy and complexity, we prune scaling factors by predefined threshold. We propose also high-accuracy training method for binary neural networks. We perform numerical experiments on various datasets and neural network architectures. IBF shows higher accuracy than state-of-the-art binary neural networks on datasets such as ImageNet within a reasonable increase of complexity. While we have conducted our experiments using accuracy as main criterion, we plan to prove our method using more experiments based on other forms of performance, such as model computing power, inference latency or model flops. Our method can also be further optimized if the K value optimized for both complexity and accuracy can be found for each layer.

Acknowledgements. This work was supported by Institute for Information & communications Technology Promotion (IITP) grant funded by the Korea government (MSIT), (No. 2018-0-00278-0031001, Development of Big Data Edge Analytics SW Technology for Load Balancing and Active Timely Response).

References

1. Bengio, Y., Léonard, N., Courville, A.C.: Estimating or propagating gradients through stochastic neurons for conditional computation. CoRR (2013)
2. Courbariaux, M., Bengio, Y., David, J.P.: BinaryConnect: training deep neural networks with binary weights during propagations. In: Advances in Neural Information Processing Systems (2015)
3. Courbariaux, M., Hubara, I., Soudry, D., El-Yaniv, R., Bengio, Y.: Binarized neural networks: training deep neural networks with weights and activations constrained to +1 or −1. arXiv:1602.02830 (2016)
4. Darabi, S., Belbahri, M., Courbariaux, M., Nia, V.P.: BNN+: improved binary network training. CoRR (2018)
5. Deng, J., Dong, W., Socher, R., Li, L.J., Li, K., Fei-Fei, L.: ImageNet: a large-scale hierarchical image database. In: IEEE Conference on Computer Vision and Pattern Recognition (2009)
6. He, K., Zhang, X., Ren, S., Sun, J.: Deep residual learning for image recognition. In: IEEE Conference on Computer Vision and Pattern Recognition (2016)

7. Huang, G., Liu, Z., Van Der Maaten, L., Weinberger, K.Q.: Densely connected convolutional networks. In: IEEE Conference on Computer Vision and Pattern Recognition (2017)
8. Iandola, F.N., Han, S., Moskewicz, M.W., Ashraf, K., Dally, W.J., Keutzer, K.: SqueezeNet: AlexNet-level accuracy with 50x fewer parameters and <0.5 MB Model Size. arXiv:1602.07360 (2016)
9. Kingma, D.P., Ba, J.: Adam: a method for stochastic optimization. In: International Conference on Learning Representations (2015)
10. Krizhevsky, A., Hinton, G., et al.: Learning Multiple Layers of Features from Tiny Images. University of Toronto, Toronto (2009)
11. Krizhevsky, A., Sutskever, I., Hinton, G.E.: ImageNet classification with deep convolutional neural networks. In: Advances in Neural Information Processing Systems (2012)
12. LeCun, Y., Bottou, L., Bengio, Y., Haffner, P.: Gradient-based learning applied to document recognition. Proc. IEEE **86**(11), 2278–2324 (1998)
13. Lin, X., Zhao, C., Pan, W.: Towards accurate binary convolutional neural network (2017)
14. Lin, X., Zhao, C., Pan, W.: Towards accurate binary convolutional neural network. In: Advances in Neural Information Processing Systems (2017)
15. Liu, Z., Wu, B., Luo, W., Yang, X., Liu, W., Cheng, K.T.: Bi-Real Net: enhancing the performance of 1-bit CNNS with improved representational capability and advanced training algorithm (2018)
16. Moon, T.K.: Error Correction Coding: Mathematical Methods and Algorithms. Wiley, Hoboken (2005)
17. Rastegari, M., Ordonez, V., Redmon, J., Farhadi, A.: XNOR-Net: ImageNet classification using binary convolutional neural networks. In: Leibe, B., Matas, J., Sebe, N., Welling, M. (eds.) ECCV 2016. LNCS, vol. 9908, pp. 525–542. Springer, Cham (2016). https://doi.org/10.1007/978-3-319-46493-0_32
18. Simonyan, K., Zisserman, A.: Very deep convolutional networks for large-scale image recognition. In: International Conference on Learning Representations (2015)
19. Tan, M., et al.: MNASNet: platform-aware neural architecture search for mobile. In: IEEE Conference on Computer Vision and Pattern Recognition (2019)
20. Tang, W., Hua, G., Wang, L.: How to train a compact binary neural network with high accuracy? In: AAAI Conference on Artificial Intelligence (2017)
21. Tseng, V.W.S., Bhattacharya, S., Fernández-Marqués, J., Alizadeh, M., Tong, C., Lane, N.D.: Deterministic binary filters for convolutional neural networks. In: International Joint Conference on Artificial Intelligence (2018)
22. Xiao, H., Rasul, K., Vollgraf, R.: Fashion-MNIST: a Novel Image Dataset for Benchmarking Machine Learning Algorithms. CoRR (2017)
23. Zhuang, B., Shen, C., Tan, M., Liu, L., Reid, I.: Structured binary neural networks for accurate image classification and semantic segmentation (2019)

One Weight Bitwidth to Rule Them All

Ting-Wu Chin[1(✉)], Pierce I-Jen Chuang[2], Vikas Chandra[2], and Diana Marculescu[1,3]

[1] Eletrical and Computer Engineering, Carnegie Mellon University, Pittsburgh, USA
{tingwuc,dianam}@cmu.edu
[2] Facebook Inc., Menlo Park, USA
{pichuang,vchandra}@fb.com
[3] Eletrical and Computer Engineering, The University of Texas at Austin, Austin, USA
dianam@utexas.edu

Abstract. Weight quantization for deep ConvNets has shown promising results for applications such as image classification and semantic segmentation and is especially important for applications where memory storage is limited. However, when aiming for quantization without accuracy degradation, different tasks may end up with different bitwidths. This creates complexity for software and hardware support and the complexity accumulates when one considers mixed-precision quantization, in which case each layer's weights use a different bitwidth. Our key insight is that optimizing for the least bitwidth subject to no accuracy degradation is not necessarily an optimal strategy. This is because one cannot decide optimality between two bitwidths if one has smaller model size while the other has better accuracy. In this work, we take the first step to understand if some weight bitwidth is better than others by aligning all to the same model size using a width-multiplier. Under this setting, somewhat surprisingly, we show that using a single bitwidth for the whole network can achieve better accuracy compared to mixed-precision quantization targeting zero accuracy degradation when both have the same model size. In particular, our results suggest that when the number of channels becomes a target hyperparameter, a single weight bitwidth throughout the network shows superior results for model compression.

Keywords: Model compression · Deep learning architectures · Quantization · ConvNets · Image classification

1 Introduction

Recent success of ConvNets in computer vision applications such as image classification and semantic segmentation has fueled many important applications

Electronic supplementary material The online version of this chapter (https://doi.org/10.1007/978-3-030-68238-5_7) contains supplementary material, which is available to authorized users.

A. Bartoli and A. Fusiello (Eds.): ECCV 2020 Workshops, LNCS 12539, pp. 85–103, 2020.
https://doi.org/10.1007/978-3-030-68238-5_7

in storage-constrained devices, *e.g.*, virtual reality headsets, drones, and IoT devices. As a result, improving the parameter-efficiency (the top-1 accuracy to the parameter counts ratio) of ConvNets while maintaining their attractive features (*e.g.*, accuracy for a task) has gained tremendous research momentum recently.

Among the efforts of improving ConvNets' efficiency, weight quantization was shown to be an effective technique [5,10,38,39]. The majority of research efforts in quantization has targeted quantization algorithms for finding the lowest possible weight bitwidth without compromising the figure-of-merit (*i.e.*, accuracy). Mixed-precision quantization methods, which allow different bitwidths to be selected for different layers in the network, have recently been proposed to further compress deep ConvNets [6,29,31]. Nevertheless, having different bitwidths for different layers greatly increases the neural network implementation complexity from both hardware and software perspectives. For example, hardware and software implementations optimized for executing an 8 bits convolution are sub-optimal for executing a 4 bits convolution, and vice versa.

To minimize the efforts of hardware and software support, it is natural to wonder: "Is some weight bitwidth better than others?" However, this is an ill-posed problem as one cannot decide optimality between two bitwidths if one has smaller model size while the other has better accuracy. This work takes a first step towards understanding if some bitwidth is better than other bitwidths under a given model size constraint. Given the multi-objective nature of the problem, we need to align different bitwidths to the same model size to further decide the optimality for the bitwidth selection. To realize model size alignment for different bitwidths, we use the width-multiplier[1] [11] as a tool to compare the performance of different weight bitwidths under *the same model size*.

With this setting, we find that there exists some weight bitwidth that consistently outperforms others across different model sizes when both are considered under a given model size constraint. This suggests that one can decide the optimal bitwidth for small model sizes to save computing cost and the result generalizes to large model sizes[2]. Additionally, we show that the optimal bitwidth of a convolutional layer negatively correlates to the convolutional kernel fan-in. As an example, depth-wise convolutional layers turn to have optimal bitwidth values that are higher than that of all-to-all convolutions. We further provide a theoretical reasoning for this phenomenon. These findings suggest that architectures such as VGG and ResNets are more parameter-efficient when they are wide and use binarized weights. On the other hand, networks such as MobileNets [11] might require different weight bitwidths for all-to-all convolutions and depth-wise convolutions. Somewhat surprisingly, we find that on ImageNet, under a given model size constraint, a single bitwidth for both ResNet-50 and MobileNetV2 can outperform mixed-precision quantization using reinforcement learning [29]

[1] Width-multiplier grows or shrinks the number of channels across the layers with identical proportion for a certain network, *e.g.*, grow the number of channels for all the layers by 2×.

[2] Note that we use width-multiplier to scale model across different sizes.

that targets minimum total bitwidth without accuracy degradation. This suggests that searching for the minimum bitwidth configuration that does not introduce accuracy degradation without considering other hyperparameters affecting model size is a sub-optimal strategy. Our results suggest that when the number of channels becomes one of the hyperparameters under consideration, a single weight bitwidth throughout the network shows great potential for model compression.

In summary, we systematically analyze the model size and accuracy tradeoff considering both weight bitwidths and the number of channels for various modern networks architectures (variants of ResNet, VGG, and MobileNet) and datasets (CIFAR and ImageNet) and have the following contributions:

- We empirically show that when allowing the network width to vary, lower weight bitwidths outperform higher ones in a Pareto sense (accuracy *vs.* model size) for networks with standard convolutions. This suggests that for such ConvNets, further research on wide binary weight networks is likely to identify better network configurations which will require further hardware/software platform support.
- We empirically show that the optimal bitwidth of a convolutional layer negatively correlates to the convolutional kernel fan-in and provide theoretical reasoning for such a phenomenon. This suggests that one could potential categorize ConvNets based on the convolutional kernel fan-in when designing the corresponding bitwidth support from both software and hardware.
- We empirically show that one can achieve a more accurate model (under a given model size) by using a single bitwidth when compared to mixed-precision quantization that uses deep reinforcement learning to search for layer-wise weight precision values. Moreover, the results are validated on a large-scale dataset, *i.e.*, ImageNet.

The remainder of the paper is organized as follows. Section 2 discusses related work. Section 3 discusses the methodology used to discover our findings. Section 4 discusses our experiments for all our findings. In particular, Sect. 4.2 shows that some bitwidth can outperform others consistently across model sizes when both are compared under the same model size constraint using width-multipliers. Section 4.3 discusses how fan-in channel count per convolutional kernel affects the resilience of quantization for convolution layers, which further affects the optimal bitwidth for a convolution layer. Section 4.4 scales up our experiments to ImageNet and demonstrates that a single weight bitwidth manages to outperform mixed-precision quantization given the same model size. Section 5 concludes the paper.

2 Related Work

Several techniques for improving the efficiency of ConvNets have been recently proposed. For instance, pruning removes the redundant connections of a trained neural network [3,7,16,28,33,34,41], neural architecture search (NAS) tunes the number of channels, size of kernels, and depth of a network [2,25–27], and convolution operations can be made more efficient via depth-wise convolutions [11], group convolutions [12,37], and shift-based convolutions [9,30]. In addition to the aforementioned techniques, network quantization introduces an opportunity for hardware-software co-design to achieve better efficiency for ConvNets.

There are in general two directions for weight quantization in prior literature, post-training quantization [18,20,23,36] and quantization-aware training [4,10,13,14,21,35,40]. The former assumes training data is not available when quantization is applied. While being fast and training-data-free, its performance is worse compared to quantization-aware training. In contrast, our work falls under the category of quantization-aware training.

In quantization-aware training, [21] introduces binary neural networks, which lead to significant efficiency gain by replacing multiplications with XNOR operations at the expense of significant accuracy degradation. Later, [40] propose ternary quantization and [13,39] bridge the gap between floating-point and binarized neural networks by introducing fixed-point quantization. Building upon prior art, the vast majority of existing work focuses on reducing the accuracy degradation by improving the training strategy [5,17,32,38] and developing better quantization schemes [14,29,35]. However, prior art has studied quantization by fixing the network architecture, which may lead to a sub-optimal bitwidth selection in terms of parameter-efficiency (the top-1 accuracy to the parameter counts ratio).

Related to our work, [19] have also considered the impact of channel count in quantization. In contrast, our work has the following novel features. First, we find that in ConvNets with standard convolutions, *a lower bitwidth outperforms higher ones under a given model size constraint.* Second, we find that the Pareto optimal bitwidth negatively correlates to the convolutional kernel fan-in and we provide theoretical insights for it. Last, we show that a single weight bitwidth can outperform *mixed-precision* quantization on ImageNet for ResNet50 and MobileNetV2.

3 Methodology

In this work, we are interested in comparing different bitwidths under a given model size. To do so, we make use of the width-multiplier to scale the models. To be precise in the following discussion, we define an ordering relation across bitwidths as follows:

Definition 1 (bitwidth ordering). *We say bitwidth A is better than bitwidth B for a network family $\mathcal{F}$, if,*

$$Acc(N(A,s)) > Acc(N(B,s)) \quad \forall s,$$

where $Acc(\cdot)$ evaluates the validation accuracy of a network, $N(A,s)$ produces a network in $\mathcal{F}$ that has bitwidth A and model size of s by using width-multiplier.

With Definition 1, we can now compare weight bitwidths for their parameter-efficiency.

3.1 Quantization

This work focuses on weight quantization and we use a straight-through estimator [1] to conduct quantization-aware training. Specifically, for bitwidth values larger than 2 bit ($b > 2$), we use the following quantization function for weights during the forward pass:

$$Q(\boldsymbol{W}_{i,:}) = \lfloor \frac{clamp(\boldsymbol{W}_{i,:}, -a_i, a_i)}{r_i} \rceil \times r_i, \quad r_i = \frac{a_i}{2^{b-1}-1} \tag{1}$$

where

$$clamp(w, min, max) = \begin{cases} w, & \text{if } min \leq w \leq max \\ min, & \text{if } w < min \\ max & \text{if } w > max \end{cases}$$

and $\lfloor \cdot \rceil$ denotes the round-to-nearest-neighbor function, $\boldsymbol{W} \in \mathbb{R}^{C_{out} \times d}$, $d = C_{in}K_wK_h$ denotes the real-value weights for the ith output filter of a convolutional layer that has C_{in} channels and $K_w \times K_h$ kernel size. $\boldsymbol{a} \in \mathbb{R}^{C_{out}}$ denotes the vector of clipping factors which are selected to minimize $\|Q(\boldsymbol{W}_{i,:}) - \boldsymbol{W}_{i,:}\|_2^2$ by assuming $\boldsymbol{W}_{i,:} \sim \mathcal{N}(0, \sigma^2 \boldsymbol{I})$. More details about the determination of a_i is in Appendix A.

For special cases such as 2 bits and 1 bit, we use schemes proposed in prior literature. Specifically, let us first define:

$$|\bar{\boldsymbol{W}_{i,:}}| = \frac{1}{d}\sum_{j=1}^{d} |\boldsymbol{W}_{i,j}|. \tag{2}$$

For 2 bit, we follow trained ternary networks [40] and define the quantization function as follows:

$$Q(\boldsymbol{W}_{i,:}) = (sign(\boldsymbol{W}_{i,:}) \odot \boldsymbol{M}_{i,j}) \times (|\bar{\boldsymbol{W}_{i,:}}|)$$
$$\boldsymbol{M}_{i,j} = \begin{cases} 0, & \boldsymbol{W}_{i,j} < 0.7|\bar{\boldsymbol{W}_{i,:}}|. \\ 1, & otherwise. \end{cases} \tag{3}$$

For 1 bit, we follow DoReFaNets [39] and define the quantization function as follows:

$$Q(\boldsymbol{W}_{i,:}) = sign(\boldsymbol{W}_{i,:}) \times (|\bar{\boldsymbol{W}_{i,:}}|). \tag{4}$$

For the backward pass for all the bitwidths, we use a straight-through estimator as in prior literature to make the training differentiable. That is,

$$\frac{\partial Q(\boldsymbol{W}_{i,:})}{\partial \boldsymbol{W}_{i,:}} = \boldsymbol{I}. \tag{5}$$

In the sequel, we quantize the *first and last layers to 8 bits*. They are fixed throughout the experiments. We note that it is a common practice to leave the first and the last layer *un-quantized* [39], however, we find that using 8 bits can achieve comparable results to the floating-point baselines.

As for activation, we use the technique proposed in [13] and use 4 bits for CIFAR-100 and 8 bits for ImageNet experiments. The activation bitwidths are chosen such that the quantized network has comparable accuracy to the floating-point baselines.

3.2 Model Size

The size of the model (C_{size}) is defined as:

$$\mathcal{C}_{size} = \sum_{i=1}^{O} b(i) C_{in}(i) K_w(i) K_h(i) \tag{6}$$

where O denotes the total number of filters, $b(i)$ is the bitwidth for filter i, $C_{in}(i)$ denotes the number of channels for filter i, and $K_w(i)$ and $K_h(i)$ are the kernel height and width for filter i.

4 Experiments

We conduct all our experiments on image classification datasets including CIFAR-100 [15] and ImageNet. All experiments are trained from scratch to ensure different weight bitwidths are trained equally long. While we do not start from a pre-trained model, we note that our baseline fixed-point models (*i.e.*, 4 bits for CIFAR and 8 bits for ImageNet) have accuracy comparable to their floating-point counterparts. For all the experiments on CIFAR, we run the experiments three times and report the mean and standard deviation.

4.1 Training Hyper-parameters

For CIFAR, we use a learning rate of 0.05, cosine learning rate decay, linear learning rate warmup (from 0 to 0.05) with 5 epochs, batch size of 128, total training epoch of 300, weight decay of $5e^{-4}$, SGD optimizer with Nesterov acceleration and 0.9 momentum.

For ImageNet, we have identical hyper-parameters as CIFAR except for the following hyper-parameters batch size of 256, 120 total epochs for MobileNetV2 and 90 for ResNets, weight decay $4e^{-5}$, and 0.1 label smoothing.

4.2 Bitwidth Comparisons

In this subsection, we are primarily interested in the following question:

When taking network width into account, does one bitwidth consistently outperform others across model sizes?

To our best knowledge, this is an open question and we take a first step to answer this question empirically. If the answer is affirmative, it may be helpful to focus the software/hardware support on the better bitwidth when it comes to parameter-efficiency. We consider three kinds of commonly adopted ConvNets, namely, ResNets with basic block [8], VGG [24], and MobileNetV2 [22]. These networks differ in the convolution operations, connections, and filter counts. For ResNets, we explored networks from 20 to 56 layers in six layer increments. For VGG, we investigate the case of eleven layers. Additionally, we also study MobileNetV2, which is a mobile-friendly network. We note that we modify the stride count in of the original MobileNetV2 to match the number of strides of ResNet for CIFAR. The architectures that we introduce for the controlled experiments are discussed in detail in Appendix B.

For CIFAR-100, we only study weight bitwidths below 4 since it achieves performance comparable to its floating-point counterpart. Specifically, we consider 4 bits, 2 bits, and 1 bit weights. To compare different weight bitwidths using Definition 1, we use the width-multiplier to align the model size among them. For example, one can make a 1-bit ConvNet twice as wide to match the model size of a 4-bit ConvNet[3]. For each of the networks we study, we sweep the width-multiplier to consider points at multiple model sizes. Specifically, for ResNets, we investigate seven depths, four model sizes for each depth, and three bitwidths, which results in $7 \times 4 \times 3 \times 3$ experiments. For both VGG11 and MobileNetV2, we consider eight model sizes and three bitwidths, which results in $2 \times 8 \times 3 \times 3$ experiments.

As shown in Fig. 1, across the three types of networks we study, there exists some bitwidth that is better than others. That is, the answer to the question we raised earlier in this subsection is affirmative. For ResNets and VGG, this value is 1 bit. In contrast, for MobileNetV2, it is 4 bits. The results for ResNets and VGG are particularly interesting since lower weight bitwidths are better than higher ones. In other words, binary weights in these cases can achieve the best accuracy and model size trade-off. On the other hand, MobileNetV2 exhibits a different trend where higher bitwidths are better than lower bitwidths up to 4 bits[4].

4.3 ConvNet Architectures and Quantization

While there exists an ordering among different bitwidths as shown in Fig. 1, it is not clear what determines the optimal weight bitwidth. To further uncover the relationship between ConvNet's architectural parameters and its optimal weight bitwidth, we ask the following questions.

What architectural components determine the MobileNetV2 optimal weight bitwidth of 4 bits as opposed to 1 bit?

As it can be observed in Fig. 1, MobileNetV2 is a special case where the higher bitwidth is better than lower ones. When comparing MobileNetV2 to the

[3] Increase the width of a layer increases the number of output filters for that layer as well as the number of channels for the subsequent layer. Thus, number of parameters and number of operations grow approximately quadratically with the width-multiplier.

[4] However, not higher than 4 bits since the 4-bit model has accuracy comparable to the floating-point model.

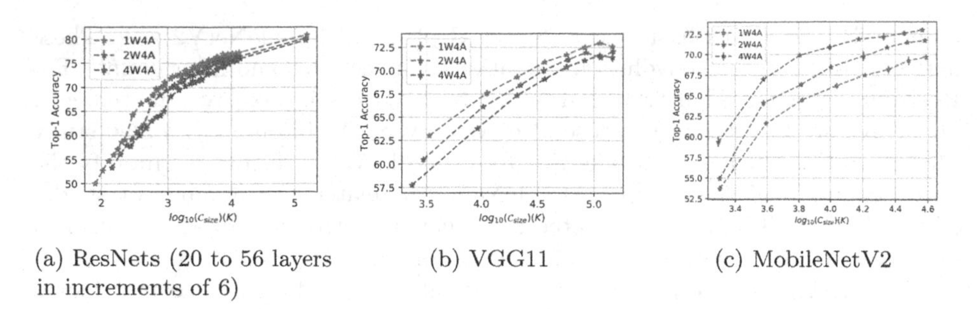

(a) ResNets (20 to 56 layers in increments of 6) (b) VGG11 (c) MobileNetV2

Fig. 1. Some bitwidth is consistently better than other bitwidths across model sizes. $\mathcal{C}_{size}$ denotes model size. xWyA denotes x-bit weight quantization and y-bit activation quantization. The experiments are done on the CIFAR-100 dataset. For each network, we sweep the width-multiplier to cover points at multiple model sizes. For each dot, we plot the mean and standard deviation of three random seeds. The standard deviation might not be visible due to little variances.

other two networks, there are many differences, including how convolutions are connected, how many convolutional layers are there, how many filters in each of them, and how many channels for each convolution. To narrow down which of these aspects result in the reversed trend compared to the trend exhibits in ResNets and VGG, we first consider the inverted residual blocks, *i.e.*, the basic component in MobileNetV2. To do so, we replace all basic blocks (two consecutive convolutions) of ResNet26 with the inverted residual blocks as shown in Fig. 2c and 2d. We refer to this new network as Inv-ResNet26. As shown in Fig. 2a and 2b, the optimal bitwidth shifts from 1 bit to 4 bit once the basic blocks are replaced with inverted residual blocks. Thus, we can infer that the inverted residual block itself or its components are responsible for such a reversed trend.

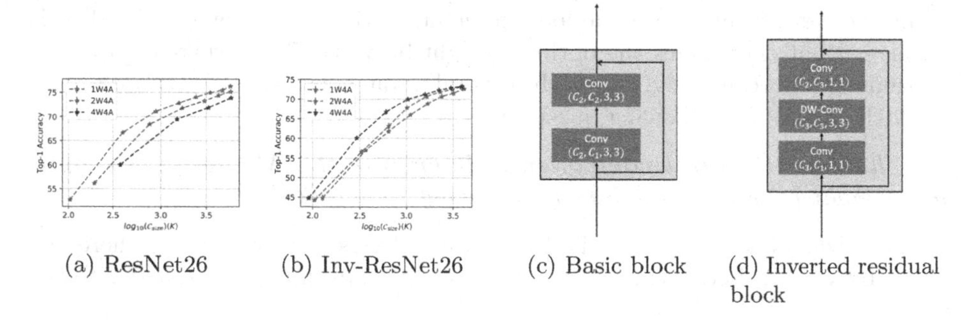

(a) ResNet26 (b) Inv-ResNet26 (c) Basic block (d) Inverted residual block

Fig. 2. The optimal bitwidth for ResNet26 changes from 1 bit (a) to 4 bit (b) when the building blocks change from basic blocks (c) to inverted residual blocks (d). $\mathcal{C}_{size}$ in (a) and (b) denotes model size. (C_{out}, C_{in}, K, K) in (c) and (d) indicate output channel count, input channel count, kernel width, and kernel height of a convolution.

Since an inverted residual block is composed of a point-wise convolution and a depth-wise separable convolution, we further consider the case of depth-wise separable convolution (DWSConv). To identify whether DWSConv can cause the inverted trend, we use VGG11 as a starting point and gradually replace each of the convolutions with DWSConv. We note that doing so results in architectures that gradually resemble MobileNetV1 [11]. Specifically, we introduce three variants of VGG11 that have an increasing number of convolutions replaced by DWSConvs. Starting with the second layer, *variant A* has one layer replaced by DWSConv, *variant B* has four layers replaced by DWSConvs, and *variant C* has all of the layers except for the first layer replaced by DWSConvs (the architectures are detailed in Appendix B).

As shown in Fig. 4, as the number of DWSConv increases (from variant A to variant C), the optimal bitwidth shifts from 1 bit to 4 bits, which implies that depth-wise separable convolutions or the layers within it are affecting the optimal bitwidth. To identify which of the layers of the DWSConv (*i.e.*, the depth-wise convolution or the point-wise convolution) has more impact on the optimal bitwidth, we keep the bitwidth of depth-wise convolutions fixed at 4 bits and quantize other layers. As shown in Fig. 4d, the optimal curve shifts from 4 bits being the best back to 1 bit, with a similarly performing 2 bits. Thus, depth-wise convolutions appear to directly affect the optimal bitwidth trends.

Is depth-wise convolution less resilient to quantization or less sensitive to channel increase?

After identifying that depth-wise convolutions have a different characteristic in optimal bitwidth compared to standard all-to-all convolutions, we are interested in understanding the reason behind this. In our setup, the process to obtain a lower bitwidth network that has the same model size as a higher bitwidth network can be broken down into two steps: (1) quantize a network to lower bitwidth and (2) grow the network with width-multiplier to compensate for the reduced model size. As a result, the fact that depth-wise convolution has higher weight bitwidth better than lower weight bitwidth might potentially be due to the large accuracy degradation introduced by quantization or the small accuracy improvements from the use of more channels.

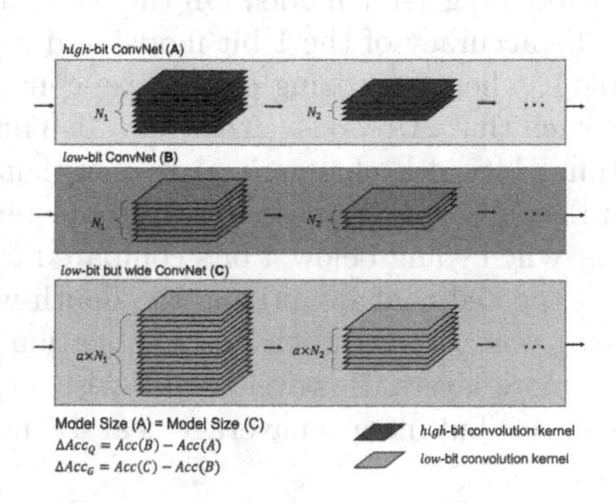

Fig. 3. Visualization of our accuracy decomposition, which is used for analyzing depth-wise convolutions.

To further diagnose the cause, we decompose the accuracy difference between a lower bitwidth but wider network and a higher bitwidth but narrower network into accuracy differences incurred in the aforementioned two steps as shown in Fig. 3. Specifically, let ΔAcc_Q denote the accuracy difference incurred by quantizing a network and let ΔAcc_G denote the accuracy difference incurred by increasing the channel count of the quantized network.

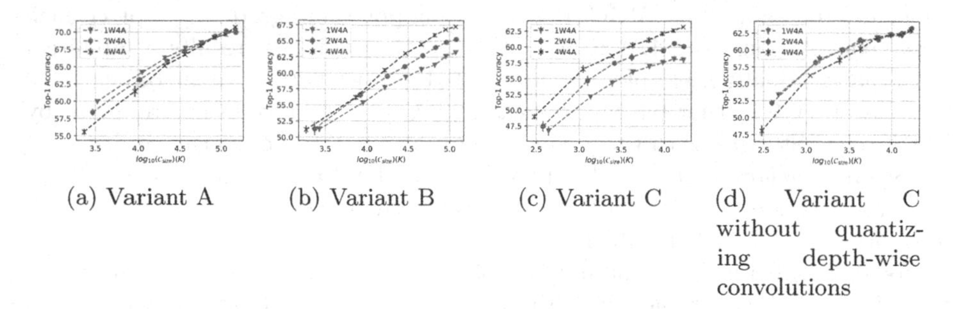

(a) Variant A (b) Variant B (c) Variant C (d) Variant C without quantizing depth-wise convolutions

Fig. 4. The optimal bitwidth for VGG shifts from 1 bit to 4 bit as more convolutions are replaced with depth-wise separable convolutions (DWSConv), *i.e.*, from (a) to (c). Variant A, B, and C have 30%, 60%, and 90% of the convolution layers replaced with DWSConv, respectively. As shown in (d), the optimal bitwidth changes back to 1 bit if we only quantize point-wise convolution but not depth-wise convolutions.

We analyze ΔAcc_G and ΔAcc_Q for networks with and without quantizing depth-wise convolutions, *i.e.*, Fig. 4c and Fig. 4d. In other words, we would like to understand how depth-wise convolutions affect ΔAcc_G and ΔAcc_Q. On one hand, ΔAcc_Q is evaluated by comparing the accuracy of the 4-bit model and the corresponding 1-bit model. On the other hand, ΔAcc_G is measured by comparing the accuracy of the 1-bit model and its 2× grown counterpart. As shown in Table 1, when quantizing depth-wise convolutions, ΔAcc_Q becomes more negative such that $\Delta Acc_Q + \Delta Acc_G < 0$. This implies that the main reason for the optimal bitwidth change is that quantizing depth-wise convolutions introduce more accuracy degradation than it can be recovered by increasing the channel count when going below 4 bits compared to all-to-all convolutions. We note that it is expected that quantizing the depth-wise convolutions would incur smaller ΔAcc_Q compared to their no-quantization baseline because we essentially quantized more layers. However, depth-wise convolutions only account for 2% of the model size but incur on average near 4× more accuracy degradation when quantized.

We would like to point out that Sheng *et al.* [23] also find that quantizing depth-wise separable convolutions incurs large accuracy degradation. However, their results are based on post-training layer-wise quantization. As mentioned in their work [23], the quantization challenges in their setting could be resolved by quantization-aware training, which is the scheme considered in this work. Hence, our observation is novel and interesting.

Why is depth-wise convolution less resilient to quantization?

Having uncovered that depth-wise convolutions introduce large accuracy degradation when weights are quantized below 4 bits, in this section, we investi-

Table 1. Quantizing depth-wise convolution introduces large accuracy degradation across model sizes. $\Delta Acc_Q = Acc_{1bit} - Acc_{4bit}$ denotes the accuracy introduced by quantization and $\Delta Acc_G = Acc_{1bit,2\times} - Acc_{1bit}$ denotes the accuracy improvement by increasing channel counts. The ConvNet is VGG variant C with and without quantizing the depth-wise convolutions from 4 bits to 1 bit.

Width-multiplier	1.00×		1.25×		1.50×		1.75×		2.00×		Average	
Variant C	ΔAcc_Q	ΔAcc_G	ΔAcc_Q	ΔAcc_G	ΔAcc_Q	ΔAcc_G	ΔAcc_Q	ΔAcc_G	ΔAcc_Q	ΔAcc_G	ΔAcc_Q	ΔAcc_G
w/o Quantizing DWConv	-1.54	+2.61	-2.76	+2.80	-1.77	+1.74	-1.82	+1.64	-1.58	+1.55	-1.89	+2.07
Quantizing DWConv	-8.60	+4.39	-7.60	+3.41	-7.74	+3.19	-8.61	+4.09	-7.49	+2.25	-8.01	+3.47

gate depth-wise convolutions from a quantization perspective. When comparing depth-wise convolutions and all-to-all convolutions in the context of quantization, they differ in the number of elements to be quantized, *i.e.*, $C_{in} = 1$ for depth-wise convolutions and $C_{in} >> 1$ for all-to-all convolutions.

Why does the number of elements matter? In quantization-aware training, one needs to estimate some statistics of the vector to be quantized (*i.e.*, $\boldsymbol{a}$ in Eq. 1 and $\overline{|\boldsymbol{w}|}$ in Eqs. 3, 4) based on the elements in the vector. The number of elements affect the robustness of the estimate that further decides the quantized weights. More formally, we provide the following proposition.

Proposition 1. *Let $\boldsymbol{w} \in \mathbb{R}^d$ be the weight vector to be quantized where $\boldsymbol{w}_i$ is characterized by normal distribution $\mathcal{N}(0, \sigma^2)$ $\forall i$ without assuming samples are drawn independently and $d = C_{in}K_wK_h$. If the average correlation of the weights is denoted by ρ, the variance of $\overline{|\boldsymbol{w}|}$ can be written as follows:*

$$\mathrm{Var}(\overline{|\boldsymbol{w}|}) = \frac{\sigma^2}{d} + \frac{(d-1)\rho\sigma^2}{d} - \frac{2\sigma^2}{\pi}. \tag{7}$$

The proof is in Appendix C. This proposition states that, as the number of elements (d) increases, the variance of the estimate can be reduced (due to the first term in Eq. (7)). The second term depends on the correlation between weights. Since the weights might not be independent during training, the variance is also affected by their correlations.

We empirically validate Proposition 1 by looking into the sample variance of $\overline{|\boldsymbol{w}|}$ across the course of training[5] for different d values by increasing (K_w, K_h) or C_{in}. Specifically, we consider the 0.5× VGG variant C and change the number of elements of the depth-wise convolutions. Let $d = (C_{in} \times K_w \times K_h)$ for a convolutional layer, we consider the original depth-wise convolution, *i.e.*, $d = 1 \times 3 \times 3$

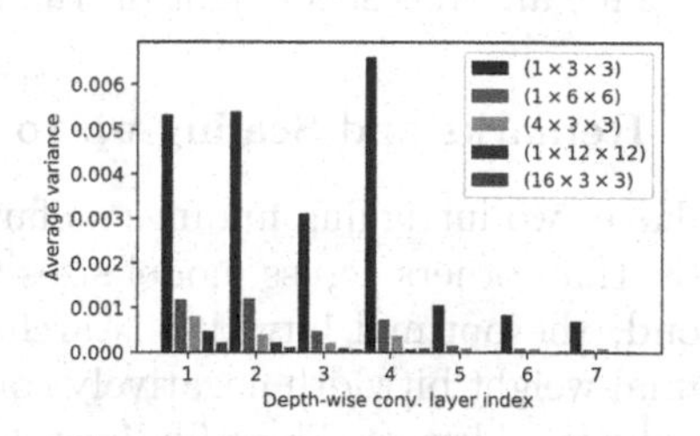

Fig. 5. The average estimate $\mathrm{Var}(\overline{|\boldsymbol{w}|})$ for each depth-wise convolution under different $d = (C_{in} \times K_w \times K_h)$ values.

[5] We treat the calculated $\overline{|\boldsymbol{w}|}$ at each training step as a sample and calculate the sample variance across training steps.

and increased channels with $d = 4 \times 3 \times 3$ and $d = 16 \times 3 \times 3$, and increased kernel size with $d = 1 \times 6 \times 6$ and $d = 1 \times 12 \times 12$. The numbers are selected such that increasing the channel count results in the same d compared to increasing the kernel sizes. We note that when the channel count (C_{in}) is increased, it is no longer a depth-wise convolution, but rather a group convolution.

In Fig. 5, we analyze layer-level sample variance by averaging the kernel-level sample variance in the same layer. First, we observe that results align with Proposition 1. That is, one can reduce the variance of the estimate by increasing the number of elements along both the channel (C_{in}) and kernel size dimensions (K_w, K_h). Second, we find that increasing the number of channels (C_{in}) is more effective in reducing the variance than increasing kernel size (K_w, K_h), which could be due to the weight correlation, *i.e.*, intra-channel weights have larger correlation than inter-channel weights.

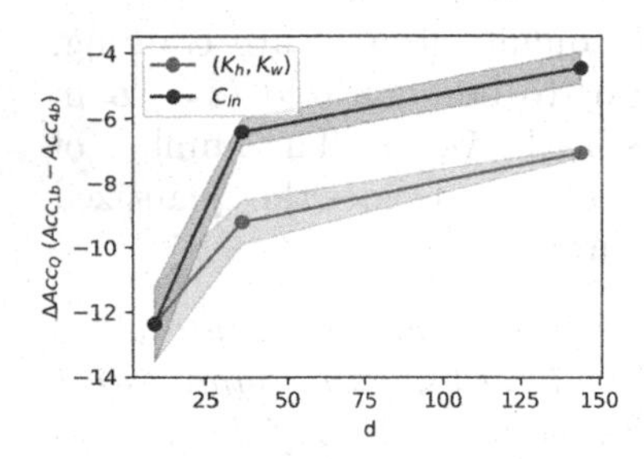

Fig. 6. d negatively correlates with the variance and positively correlates with the accuracy difference induced by quantization $\Delta Acc_Q = Acc_{1bit} - Acc_{4bit}$.

Nonetheless, while lower variance suggests a more stable value during training, it might not necessarily imply lower quantization error for the quantized models. Thus, we conduct an accuracy sensitivity analysis with respect to quantization for different d values. More specifically, we want to understand how d affects the accuracy difference between lower bitwidth (1 bit) and higher bitwidth (4 bits) models (ΔAcc_Q). As shown in Fig. 6, we empirically find that d positively correlates with ΔAcc_Q, *i.e.*, the larger the d, the smaller the accuracy degradation is. On the other hand, when comparing channel counts and kernel sizes, we observe that increasing the number of channels is more effective than increasing the kernel size in reducing accuracy degradation caused by quantization. This analysis sheds light on the two different trends observed in Fig. 1.

4.4 Remarks and Scaling up to ImageNet

We have two intriguing findings so far. First, there exists some bitwidth that is better than others across model sizes when compared under a given model size. Second, the optimal bitwidth is architecture-dependent. More specifically, the optimal weight bitwidth negatively correlates with the fan-in channel counts per convolutional kernel. These findings show promising results for the hardware and software researchers to support only a certain set of bitwidths when it comes to parameter-efficiency. For example, use binary weights for networks with all-to-all convolutions.

Next, we scale up our analysis to the ImageNet dataset. Specifically, we study ResNet50 and MobileNetV2 on the ImageNet dataset. Since we keep the bitwidth of the first and last layer quantized at 8 bits, scaling them in terms of width will grow the number of parameters much more quickly than other layers. As a result, we keep the number of channels for the first and last channel fixed for the ImageNet experiments. As demonstrated in Sect. 4.2, the bit ordering is consistent across model sizes, we conduct our analysis for ResNet50 and MobileNetV2 by scaling them down with a width-multiplier of 0.25× for computational considerations. The choices of bitwidths are limited to $\{1, 2, 4, 8\}$.

Table 2. bitwidth ordering for MobileNetV2 and ResNet50 with *the model size aligned to the* 0.25× *8 bits models* on ImageNet. Each cell reports the top-1 accuracy of the corresponding model. The trend for the optimal bitwidth is similar to that of CIFAR-100 (4 bit for MobileNetV2 and 1 bit for ResNet).

Weight bitwidth for Convs\DWConvs	MobileNetV2				ResNet50 None
	8 bits	4 bits	2 bits	1 bit	
8 bits	52.17	53.89	50.51	48.78	71.11
4 bits	56.84	**59.51**	57.37	55.91	74.65
2 bits	53.89	57.10	55.26	54.04	75.12
1 bit	54.82	58.16	56.90	55.82	**75.44**

As shown in Table 2, we can observe a trend similar to the CIFAR-100 experiments, *i.e.*, for networks without depth-wise convolutions, the lower weight bitwidths the better, and for networks with depth-wise convolutions, there are sweet spots for depth-wise and other convolutions. Specifically, the final weight bitwidth selected for MobileNetV2 is 4 bits for both depth-wise and standard convolutions. On the other hand, the selected weight bitwidth for ResNet50 is 1 bit. If bit ordering is indeed consistent across model sizes, these results suggest that the optimal bitwidth for MobileNetV2 is 4 bit and it is 1 bit for ResNet50. However, throughout our analysis, we have not considered mixed-precision, which makes it unclear if the so-called optimal bitwidth (4 bit for MobileNetV2 and 1 bit for ResNet-50) is still optimal when compared to mixed-precision quantization.

As a result, we further compare with mixed-precision quantization that uses reinforcement learning to find the layer-wise bitwidth [29]. Specifically, we follow [29] and use a reinforcement learning approach to search for the lowest bitwidths without accuracy degradation (compared to the 8 bits fixed point models). To compare the searched model with other alternatives, we use width-multipliers on top of the searched network match the model size of the 8 bit quantized model. We consider networks of three sizes, *i.e.*, the size of $1\times, 0.5\times$ and 0.25× 8-bit fixed point models. As shown in Table 3, we find that a single bitwidth (selected via Table 2) outperforms both 8 bit quantization and mixed-precision quantization by a significant margin for both networks considered. This results suggest that searching for the bitwidth without accuracy degradation is indeed a sub-optimal strategy and can be improved by incorporating channel counts into the search space and reformulate the optimization problem as maximizing accuracy under storage constraints. Moreover, our results also imply that when the number of channels are allowed to be altered, a single weight bitwidth throughout the network shows great potential for model compression, which has the potential of greatly reducing the software and hardware optimization costs for quantized ConvNets.

Table 3. The optimal bitwidth selected in Table 2 is indeed better than 8 bit when scaled to larger model sizes and more surprisingly, it is better than mixed-precision quantization. All the activations are quantized to 8 bits.

Width-multiplier for 8-bit model		1×		0.5×		0.25×	
Networks	Methods	Top-1 (%)	C_{size} (10^6)	Top-1 (%)	C_{size} (10^6)	Top-1 (%)	C_{size} (10^6)
ResNet50	Floating-point	76.71	816.72	74.71	411.48	71.27	255.4
	8 bits	76.70	204.18	74.86	102.87	71.11	63.85
	Flexible [29]	77.23	204.18	76.04	102.90	74.30	63.60
	Optimal (1 bit)	**77.58**	204.08	**76.70**	102.83	**75.44**	63.13
MobileNetV2	Floating-point	71.78	110.00	63.96	61.76	52.79	47.96
	8 bits	71.73	27.50	64.39	15.44	52.17	11.99
	Flexible [29]	72.13	27.71	65.00	15.54	55.20	12.10
	Optimal (4 bit)	**73.91**	27.56	**68.01**	15.53	**59.51**	12.15

5 Conclusion

In this work, we provide the first attempt to understand the ordering between different weight bitwidths by allowing the channel counts of the considered networks to vary using the width-multiplier. If there exists such an ordering, it may be helpful to focus on software/hardware support for higher-ranked bitwidth when it comes to parameter-efficiency, which in turn reduces software/hardware optimization costs. To this end, we have three surprising findings: (1) there exists a weight bitwidth that is better than others across model sizes under a given model size constraint, (2) the optimal weight bitwidth of a convolutional layer negatively correlates to the fan-in channel counts per convolutional kernel, and (3) with a single weight bitwidth for the whole network, one can find configurations that outperform layer-wise mixed-precision quantization using reinforcement learning when compared under a given same model size constraint. Our results suggest that when the number of channels are allowed to be altered, a single weight bitwidth throughout the network shows great potential for model compression.

Acknowledgement. This research was supported in part by NSF CCF Grant No. 1815899, NSF CSR Grant No. 1815780, and NSF ACI Grant No. 1445606 at the Pittsburgh Supercomputing Center (PSC).

A Clipping Point for Quantization-aware Training

As mentioned earlier, $\boldsymbol{a} \in \mathbb{R}^{C_{out}}$ denotes the vector of clipping factors which is selected to minimize $\|Q(\boldsymbol{W}_{i,:}) - \boldsymbol{W}_{i,:}\|_2^2$ by assuming $\boldsymbol{W}_{i,:} \sim \mathcal{N}(0, \sigma^2 \boldsymbol{I})$. More specifically, we run simulations for weights drawn from a zero-mean Gaussian distribution with several variances and identify the best $a_i^* = \arg\min_{a_i} \|Q_{a_i}(\boldsymbol{W}_{i,:}) - \boldsymbol{W}_{i,:}\|_2^2$ empirically. According to our simulation, we find that one can infer a_i from the sample mean $|\bar{\boldsymbol{W}_{i,:}}|$, which is shown in Fig. 7. As a result, for the different precision values considered, we find $c = \frac{|\bar{\boldsymbol{W}_{i,:}}|}{a_i^*}$ via simulation and use the obtained c to calculate a_i on-the-fly throughout training.

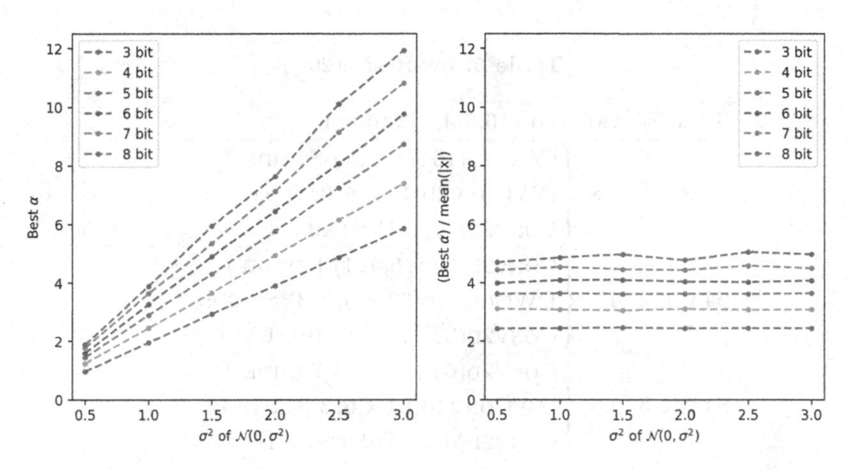

Fig. 7. Finding best a_i for different precision values empirically through simulation using Gaussian with various σ^2.

B Network Architectures

For the experiments in Sect. 4.2, the ResNets used are detailed in Table 4. Specifically, for the points in Fig. 1a, we consider ResNet20 to ResNet56 with width-multipliers of $0.5\times, 1\times, 1.5\times$, and $2\times$ for the 4-bit case. Based on these values, we consider additional width-multipliers $2.4\times$ and $2.8\times$ for the 2-bit case and $2.5\times, 3\times, 3.5\times$, and $3.9\times$ for the 1-bit case. We note that the right-most points in Fig. 1a is a $10\times$ ResNet26 for the 4 bits case. On the other hand, VGG11 is detailed in Table 6 for which we consider width-multipliers from $0.25\times$ to $2\times$ with a step of 0.25 for the 4 bits case (blue dots in Fig. 1c). The architecture details for Inv-ResNet26 used in Fig. 2b is shown in Table 5. The architecture of MobileNetV2 used in the CIFAR-100 experiments follows the original MobileNetV2 (Table 2 in [22]) but we change the stride of all the bottleneck blocks to 1 except for the fifth bottleneck block, which has a stride of 2. As a result, we down-sample the image twice in total, which resembles the ResNet design for the CIFAR experiments [8]. Similar to VGG11, we consider width-multipliers from $0.25\times$ to $2\times$ with a step of 0.25 for MobileNetV2 for the 4 bits case (blue dots in Fig. 1b).

Table 4. ResNet20 to ResNet56

Layers	20	26	32	38	44	50	56
Stem	Conv2D (16,3,3) Stride 1						
Stage 1	3 × { Conv2D(16, 3, 3) Stride 1 Conv2D(16, 3, 3) Stride 1 }	4×	5×	6×	7×	8×	9×
Stage 2	3 × { Conv2D(32, 3, 3) Stride 2 Conv2D(32, 3, 3) Stride 1 }	4×	5×	6×	7×	8×	9×
Stage 3	3 × { Conv2D(64, 3, 3) Stride 2 Conv2D(64, 3, 3) Stride 1 }	4×	5×	6×	7×	8×	9×

Table 5. Inv-ResNet26

Stem	Conv2D (16,3,3) Stride 1
Stage 1	4 × { Conv2D(16 × 6, 1, 1) Stride 1 DWConv2D(16 × 6, 3, 3) Stride 1 Conv2D(16, 1, 1) Stride 1 }
Stage 2	4 × { Conv2D(32 × 6, 1, 1) Stride 1 DWConv2D(32 × 6, 3, 3) Stride 2 Conv2D(32, 1, 1) Stride 1 }
Stage 3	4 × { Conv2D(64 × 6, 1, 1) Stride 1 DWConv2D(64 × 6, 3, 3) Stride 2 Conv2D(64, 1, 1) Stride 1 }

Table 6. VGGs

VGG11	Variant A	Variant B	Variant C
Conv2D (64,3,3)			
MaxPooling			
Conv2D (128,3,3)	{ Conv2D(128, 1, 1) DWConv2D(128, 3, 3) }	{ Conv2D(128, 1, 1) DWConv2D(128, 3, 3) }	{ Conv2D(128, 1, 1) DWConv2D(128, 3, 3) }
MaxPooling			
Conv2D (256,3,3)	Conv2D (256,3,3)	{ Conv2D(256, 1, 1) DWConv2D(256, 3, 3) }	{ Conv2D(256, 1, 1) DWConv2D(256, 3, 3) }
Conv2D (256,3,3)	Conv2D (256,3,3)	{ Conv2D(256, 1, 1) DWConv2D(256, 3, 3) }	{ Conv2D(256, 1, 1) DWConv2D(256, 3, 3) }
MaxPooling			
Conv2D (512,3,3)	Conv2D (512,3,3)	{ Conv2D(512, 1, 1) DWConv2D(512, 3, 3) }	{ Conv2D(512, 1, 1) DWConv2D(512, 3, 3) }
Conv2D (512,3,3)	Conv2D (512,3,3)	Conv2D (512,3,3)	{ Conv2D(512, 1, 1) DWConv2D(512, 3, 3) }
MaxPooling			
Conv2D (512,3,3)	Conv2D (512,3,3)	Conv2D (512,3,3)	{ Conv2D(512, 1, 1) DWConv2D(512, 3, 3) }
Conv2D (512,3,3)	Conv2D (512,3,3)	Conv2D (512,3,3)	{ Conv2D(512, 1, 1) DWConv2D(512, 3, 3) }
MaxPooling			

C Proof for Proposition 5.1

Based on the definition of variance, we have:

$$\begin{aligned}
\mathrm{Var}(\frac{1}{d}\sum_{i=1}^{d}|\boldsymbol{w}_i|) &:= \mathbb{E}\left[\left(\frac{1}{d}\sum_{i=1}^{d}|\boldsymbol{w}_i|\right)^2 - \left(\mathbb{E}\frac{1}{d}\sum_{i=1}^{d}|\boldsymbol{w}_i|\right)^2\right] \\
&= \mathbb{E}\left[\left(\frac{1}{d}\sum_{i=1}^{d}|\boldsymbol{w}_i|\right)^2 - \frac{2\sigma^2}{\pi}\right] \\
&= \frac{1}{d^2}\mathbb{E}\left(\sum_{i=1}^{d}|\boldsymbol{w}_i|\right)^2 - \frac{2\sigma^2}{\pi} \\
&= \frac{\sigma^2}{d} + \frac{d-1}{d}\rho\sigma^2 - \frac{2\sigma^2}{\pi}.
\end{aligned}$$

References

1. Bengio, Y., Léonard, N., Courville, A.: Estimating or propagating gradients through stochastic neurons for conditional computation. arXiv preprint arXiv:1308.3432 (2013)
2. Cai, H., Zhu, L., Han, S.: ProxylessNAS: direct neural architecture search on target task and hardware. arXiv preprint arXiv:1812.00332 (2018)
3. Chin, T.W., Ding, R., Zhang, C., Marculescu, D.: Towards efficient model compression via learned global ranking. In: Proceedings of the IEEE/CVF Conference on Computer Vision and Pattern Recognition (CVPR), June 2020
4. Choi, J., Chuang, P.I.J., Wang, Z., Venkataramani, S., Srinivasan, V., Gopalakrishnan, K.: Bridging the accuracy gap for 2-bit quantized neural networks (QNN). arXiv preprint arXiv:1807.06964 (2018)
5. Ding, R., Chin, T.W., Liu, Z., Marculescu, D.: Regularizing activation distribution for training binarized deep networks. In: The IEEE Conference on Computer Vision and Pattern Recognition (CVPR), June 2019
6. Dong, Z., Yao, Z., Gholami, A., Mahoney, M., Keutzer, K.: HAWQ: Hessian aWare quantization of neural networks with mixed-precision. arXiv preprint arXiv:1905.03696 (2019)
7. Frankle, J., Carbin, M.: The lottery ticket hypothesis: finding sparse, trainable neural networks. In: International Conference on Learning Representations (2019). https://openreview.net/forum?id=rJl-b3RcF7
8. He, K., Zhang, X., Ren, S., Sun, J.: Deep residual learning for image recognition. In: Proceedings of the IEEE Conference on Computer Vision and Pattern Recognition, pp. 770–778 (2016)
9. He, Y., Liu, X., Zhong, H., Ma, Y.: AddressNet: shift-based primitives for efficient convolutional neural networks. In: 2019 IEEE Winter Conference on Applications of Computer Vision (WACV), pp. 1213–1222. IEEE (2019)
10. Hou, L., Kwok, J.T.: Loss-aware weight quantization of deep networks. In: International Conference on Learning Representations (2018). https://openreview.net/forum?id=BkrSv0lA-

11. Howard, A.G., et al.: MobileNets: efficient convolutional neural networks for mobile vision applications. arXiv preprint arXiv:1704.04861 (2017)
12. Huang, G., Liu, S., van der Maaten, L., Weinberger, K.Q.: CondenseNet: an efficient DenseNet using learned group convolutions. Group **3**(12), 11 (2017)
13. Jacob, B., et al.: Quantization and training of neural networks for efficient integer-arithmetic-only inference. In: The IEEE Conference on Computer Vision and Pattern Recognition (CVPR), June 2018
14. Jung, S., et al.: Learning to quantize deep networks by optimizing quantization intervals with task loss. In: The IEEE Conference on Computer Vision and Pattern Recognition (CVPR), June 2019
15. Krizhevsky, A., Hinton, G., et al.: Learning multiple layers of features from tiny images. Technical report, Citeseer (2009)
16. Li, H., Kadav, A., Durdanovic, I., Samet, H., Graf, H.P.: Pruning filters for efficient convnets (2017)
17. Louizos, C., Reisser, M., Blankevoort, T., Gavves, E., Welling, M.: Relaxed quantization for discretized neural networks. In: International Conference on Learning Representations (2019). https://openreview.net/forum?id=HkxjYoCqKX
18. Meller, E., Finkelstein, A., Almog, U., Grobman, M.: Same, same but different: recovering neural network quantization error through weight factorization. In: Chaudhuri, K., Salakhutdinov, R. (eds.) Proceedings of the 36th International Conference on Machine Learning. Proceedings of Machine Learning Research, vol. 97, pp. 4486–4495, Long Beach. PMLR, 09–15 June 2019. http://proceedings.mlr.press/v97/meller19a.html
19. Mishra, A., Nurvitadhi, E., Cook, J.J., Marr, D.: WRPN: wide reduced-precision networks. In: International Conference on Learning Representations (2018). https://openreview.net/forum?id=B1ZvaaeAZ
20. Nagel, M., van Baalen, M., Blankevoort, T., Welling, M.: Data-free quantization through weight equalization and bias correction. arXiv preprint arXiv:1906.04721 (2019)
21. Rastegari, M., Ordonez, V., Redmon, J., Farhadi, A.: XNOR-net: ImageNet classification using binary convolutional neural networks. In: Leibe, B., Matas, J., Sebe, N., Welling, M. (eds.) ECCV 2016. LNCS, vol. 9908, pp. 525–542. Springer, Cham (2016). https://doi.org/10.1007/978-3-319-46493-0_32
22. Sandler, M., Howard, A., Zhu, M., Zhmoginov, A., Chen, L.C.: MobileNetV2: inverted residuals and linear bottlenecks. In: Proceedings of the IEEE Conference on Computer Vision and Pattern Recognition, pp. 4510–4520 (2018)
23. Sheng, T., Feng, C., Zhuo, S., Zhang, X., Shen, L., Aleksic, M.: A quantization-friendly separable convolution for MobileNets. In: 2018 1st Workshop on Energy Efficient Machine Learning and Cognitive Computing for Embedded Applications (EMC2), pp. 14–18. IEEE (2018)
24. Simonyan, K., Zisserman, A.: Very deep convolutional networks for large-scale image recognition. arXiv preprint arXiv:1409.1556 (2014)
25. Stamoulis, D., et al.: Designing adaptive neural networks for energy-constrained image classification, In: Proceedings of the International Conference on Computer-Aided Design, p. 23. ACM (2018)
26. Stamoulis, D., et al.: Single-path NAS: designing hardware-efficient convnets in less than 4 hours. arXiv preprint arXiv:1904.02877 (2019)
27. Tan, M., Chen, B., Pang, R., Vasudevan, V., Le, Q.V.: MnasNet: platform-aware neural architecture search for mobile. arXiv preprint arXiv:1807.11626 (2018)
28. Theis, L., Korshunova, I., Tejani, A., Huszár, F.: Faster gaze prediction with dense networks and fisher pruning. arXiv preprint arXiv:1801.05787 (2018)

29. Wang, K., Liu, Z., Lin, Y., Lin, J., Han, S.: HAQ: hardware-aware automated quantization with mixed precision. In: Proceedings of the IEEE Conference on Computer Vision and Pattern Recognition, pp. 8612–8620 (2019)
30. Wu, B., et al.: Shift: a zero flop, zero parameter alternative to spatial convolutions. In: Proceedings of the IEEE Conference on Computer Vision and Pattern Recognition, pp. 9127–9135 (2018)
31. Wu, B., Wang, Y., Zhang, P., Tian, Y., Vajda, P., Keutzer, K.: Mixed precision quantization of convnets via differentiable neural architecture search. arXiv preprint arXiv:1812.00090 (2018)
32. Yang, J., et al.: Quantization networks. In: The IEEE Conference on Computer Vision and Pattern Recognition (CVPR), June 2019
33. Ye, J., Lu, X., Lin, Z., Wang, J.Z.: Rethinking the smaller-norm-less-informative assumption in channel pruning of convolution layers (2018)
34. Yu, R., et al.: NISP: pruning networks using neuron importance score propagation. In: The IEEE Conference on Computer Vision and Pattern Recognition (CVPR), June 2018
35. Yuan, X., Ren, L., Lu, J., Zhou, J.: Enhanced Bayesian compression via deep reinforcement learning. In: The IEEE Conference on Computer Vision and Pattern Recognition (CVPR), June 2019
36. Zhao, R., Hu, Y., Dotzel, J., De Sa, C., Zhang, Z.: Improving neural network quantization without retraining using outlier channel splitting. In: Chaudhuri, K., Salakhutdinov, R. (eds.) Proceedings of the 36th International Conference on Machine Learning, Proceedings of Machine Learning Research, vol. 97, pp. 7543–7552, Long Beach. PMLR, 09–15 June 2019. http://proceedings.mlr.press/v97/zhao19c.html
37. Zhao, R., Hu, Y., Dotzel, J., Sa, C.D., Zhang, Z.: Building efficient deep neural networks with unitary group convolutions. In: Proceedings of the IEEE Conference on Computer Vision and Pattern Recognition, pp. 11303–11312 (2019)
38. Zhou, A., Yao, A., Guo, Y., Xu, L., Chen, Y.: Incremental network quantization: towards lossless CNNS with low-precision weights. In: International Conference on Learning Representations (2017). https://openreview.net/forum?id=HyQJ-mclg
39. Zhou, S., Wu, Y., Ni, Z., Zhou, X., Wen, H., Zou, Y.: DoReFa-Net: training low bitwidth convolutional neural networks with low bitwidth gradients. arXiv preprint arXiv:1606.06160 (2016)
40. Zhu, C., Han, S., Mao, H., Dally, W.J.: Trained ternary quantization. In: International Conference on Learning Representations (2017). https://openreview.net/forum?id=S1_pAu9xl
41. Zhuang, Z., et al.: Discrimination-aware channel pruning for deep neural networks. In: Advances in Neural Information Processing Systems, pp. 883–894 (2018)

Real-Time Detection of Multiple Targets from a Moving 360° Panoramic Imager in the Wild

Boyan Yuan(✉) and Nabil Belbachir

NORCE Norwegian Research Centre AS,
Jon Lilletuns vei 9 H, 4879 Grimstad, Norway
{boyu,nabe}@norceresearch.no

Abstract. Our goal is to develop embedded and mobile vision applications leveraging state-of-the-art visual sensors and efficient neural network architectures deployed on emerging neural computing engines for smart monitoring and inspection purposes. In this paper, we present 360° vision system onboard an automobile or UAV platform for large field-of-view and real-time detection of multiple challenging objects. The targeted objects include flag as a deformable object; UAV as a tiny, flying object which changes its scales and positions rapidly; and grouped objects containing piled sandbags as deformable objects in a group themselves, flag and stop sign to form a scene representing an artificial fake checkpoint. Barrel distortions owing to the 360° optics make the detection task even more challenging. A light-weight neural network model based on MobileNets architecture is transfer learned for detection of the custom objects with very limited training data. In method 1, we generated a dataset of perspective planar images via a virtual camera model which projects a patch on the hemisphere to a 2D plane. In method 2, the panomorph images are directly used without projection. Real-time detection of the objects in 360° video is realized by feeding live streamed frames captured by the full hemispheric (180° × 360°) field-of-view ImmerVision Enables panomorph lens to the trained MobileNets model. We found that with only few training data which is far less than 10 times of Vapnik–Chervonenkis dimension of the model, the MobileNets model achieves a detection rate of 80–90% for test data having a similar distribution as the training data. However, the model performance dropped drastically when it was put in action in the wild for unknown data in which both weather and lighting conditions were different. The generalization capability of the model can be improved by training with more data. The contribution of this work is a 360° vision hardware and software system for real-time detection of challenging objects. This system could be configured for very low-power embedded applications by running inferences via a neural computing engine such as Intel Movidius NSC2 or HiSilicon Kirin 970.

The original version of this chapter was revised: An acknowledgement text has been added. The correction to this chapter is available at https://doi.org/10.1007/978-3-030-68238-5_49

A. Bartoli and A. Fusiello (Eds.): ECCV 2020 Workshops, LNCS 12539, pp. 104–120, 2020.
https://doi.org/10.1007/978-3-030-68238-5_8

Keywords: 360° vision · Object recognition · MobileNets · Convolution neural networks · Geometrical model

1 Introduction

Embedded and mobile vision with 360° cameras finds a number of important applications such as virtual and augmented reality, autonomous driving, robotics, visual SLAN and surveillance [8,11,14,16,19–21,32,41,51]. Hemispherical and spherical imagery generation involves complicated optical design and image mapping techniques in order to produce high-resolution images with minimum distortions. In the case of ImmerVision Enables panomorph lens, optimization algorithms are utilized in lens optical design and image formation such that the spatial distribution of pixels are optimized for the purpose of the lens. This means that the lens designed for surveillance applications are different from those designed for autonomous driving or for augmented reality [3]. The customization of optical sampling of light, image mapping and projection, and distortion control algorithms generates high-quality hemispherical images, however, it may impose new challenges for object detection and recognition tasks in these imagery.

In this research, three categories of challenging objects are included. Flag is a non-rigid object which easily deforms upon an external input such as wind or rain makes its geometry varies largely [35]. Unmanned aerial vehicle (UAV, also called drone) is a flying object which changes its scales and positions rapidly without having a predictable trajectory [12]. Sandbags are both deformable and are grouped objects [5]. Furthermore, the detection must be carried out onboard a moving platform such as an automobile or a UAV where the camera is subject to motion and random vibrations. It is also required to take into consideration the power constraints in both hardware and software design and integration.

Background subtraction is an effective method for detection of moving objects, but does not work well when the camera itself is moving [6,12]. Object tracking algorithms such as based on probabilistic Kalman filter [46] and kernelized correlation filters [7] with a given initial proposal was found to quickly lose tracking because of the scale of the objects changes considerably when driving at a normal speed of e.g., 60 km/h. We then consider to take advantage of the powerful convolutional neural network (CNN) models for the simultaneous detection of multiple objects in 360° video task. As the objective of this research is to develop hardware aware embedded and mobile applications, we established a 360° dataset with very limited size which are collected in Norway and in the Netherlands, and leveraged transfer learning scheme for the simultaneous detection of multiple objects.

2 Previous Work

CNNs for Spherical Data. Coors et al. [24] presented a novel convolutional kernel which is invariant to latitudinal distortions via projection of a regular sampling pattern to its tangent plane. Similarly, Su et al. [20] proposed an adaptive CNN kernels applied to the equirectangular projection of sphere, i.e., the CNN kernels are alike the ones for the locally projected patches.

Cohen et al. [23] presented a novel way treating convolution on spherical data as a three dimensional manifold, i.e., special orthogonal group (SO(3)) and using generalized Fourier transform for fast group correlation. Subsequently, Esteves et al. [25] modeled 3D data with multi-valued spherical functions and proposed a novel spherical convolutional network that implements exact convolutions on the sphere by realizing them in the spherical harmonic domain. Yu et al. [42] pointed out that spherical CNN loses the object's location and overlarge bandwidth is required to preserve a small object's information on a sphere. And they proposed a novel grid-based spherical CNN (G-SCNN) which transforms a spherical image to a conformal grid map to be the input to the S2/SO3 convolution. Defferrard et al. and Perraudin et al. [33,38] presented a graph based spherical CNN named DeepSphere. Their idea is to model the sampled sphere as a graph of connected pixels and using the length of the shortest path between two pixels as an approximation of the geodesic distance between them. Yang et al. [51] generalized the grid-based CNNs to a non-Euclidean space by taking into account the geometry of spherical surfaces and propose a Spherical Graph Convolutional Network (SGCN) to encode rotation equivariant representations.

Tens of new models are proposed for visual understanding in spherical data recently. Zhao et al. [52] proposed a 360° detector named Reprojection R-CNN by combining the advantages of both ERP and PSP, which generates spherical bounding boxes. The method was evaluated on a synthetic dataset for detection of salient objects like person and train. Wang et al. [40] used a modified RCNN model for object detection in synthesized dataset 360GoogleStreetView. Chou et al. [45] a real-world 360° panoramic dataset containing common objects of 37 categories. Lee et al. [49] proposed to project spherical images onto an icosahedral spherical polyhedron and apply convolution on transformed images.

Transfer Learning for 360° Vision and MobileNets. Transfer learning is a proven technique to train deep neural network models especially when few custom training data are available [4,50]. It has also been used in successfully for 360° vision [26,43]. Among the many famous deep neural network architectures, MobileNets [a]nd its improved versions Dynamic CNN [44], MobileNet V2 [30], quantization-friendly seperable MobileNets [31], MobileNets V3 [36] are high-performance neural networks striking a balance between speed and accuracy.

As the objective of this work is to develop a reliable hardware and software system for real-time multiple objects detection, our focus is to optimize the visual sensor specifications dedicated to the application, and to leverage main stream neural network architectures and the transfer learning scheme. For the visual sensor, we used the ImmerVision Enables panomorph lens which is customized for surveillance applications. For dealing with the SO(3) features of the hemispherical data, we carry out both the local perspective projection to 2D plane, as well as directly labeling, training and inferencing in the hemispherical data. We believe that as our target objects is very small, the local projection onto its tangent plane would be very close to the patch on the sphere. As pointed out by Furnari et al. [9] that the inherently non-linear radial distortion can be locally approximated by linear functions with a reasonably small error. They showed that popular affine region detectors (i.e., maximally stable extremal regions, and

Harris and Hessian affine region detectors), although not convolutional operators, can be effectively employed directly on fisheye images. Zhu et al. [43] showed empirical evidence that directly detection of objects from raw fisheye images without restoration by using a deep neural network architecture based on the MobileNets and feature pyramid structure actually worked.

Furthermore, as we understand that many sophisticated and high-performance deep neural network architectures have been designed, trained and evaluated on large-size, well-annotated, high-quality datasets. It is an important and necessary task to develop real-world applications such as embedded and mobile applications in robotics and autonomous vehicles. It is also important to put these deep architectures in action for challenging real-world detection scenarios. As the detection error on the standard datasets keep going down, we need to keep in mind that the statistical learning theories on generalization capabilities and error bounds conditions their performance for real-world applications [1,17,18,22,27,28]. Our contribution is a 360° vision hardware and software system for real-time detection of challenging objects. This system is readily configured for very low-power embedded applications by running inferences via a neural computing engine such as Intel Movidius NSC2 or HiSilicon Kirin 970 [34,37,39].

3 Custom Dataset Collection

ImmerVision 360° Panoramic Imager. A critical hardware in the real-time panoramic detection system is the ImmerVision hemispherical panoramic imager (FOV: $180^\circ \times 360^\circ$) [2,3,29]. Their panomorph lenses and SDK enables live streaming of original panoramic image with minimum distortions, optimized pixel distributions and resolutions which is important for real-time embedded detection. In contrast, common panoramic imagers usually produce a single full FOV image which is projected and stitched together. Projected (mostly equirectangular projection) images are the common panoramic data format used by different backend application frameworks. The newly developed deep architectures such as spherical CNNs use equirectangular projected images as inputs. However, the projection pipelines run algorithms for pixel manipulations (stretching or compressing), interpolation, distortion control and projection which generate artifacts on the original data. Furthermore, the projection process is costly and in most circumstances cannot be done in real-time due to both battery and computing capacities. In this research, we utilize the ImmerVision panoramic imager to facilitate live streaming of originally captured 360° content. Table 1 shows the specifications of the panomorph lens used for real-time streaming and inferences with limited battery and computing capacity in an automobile while driving at normal speed. We point out that the high compression ratio of the system could have been customized to a lower compression ratio while not sacrificing live streaming capability to provide more data for feature aggregation and extraction in deep models. Figure 1 shows the ImmerVision lens geometrical projection model as stereographic projection.

Visual Targets and Setup. The targeted objects include a UAV as flying object which changes its position thus distance and angular FOV with reference to the principle point of the lens rapidly. Its trajectory, speeds and accelerations could change rapidly and randomly. The other objects, i.e., piled sandbags, isis flag, and Arabic stop sign grouped together and spatially arranged such that they represent a scene as artificial fake checkpoints. Details about the targets and the conditions for collecting the panoramic video data are presented in Table 2. We point out that limited by the scope and purpose of the project as a research activity, rather than for the development of commercial services, the scope of the dataset is very small and its statistical setup is not adequate. Figure 2 shows the configuration for data collection using the panorama camera while driving.

Table 1. The specifications of the 360° panomorph lens for live streaming of original panoramic data.

Image sensor	1/1.8 in. (9 mm diagonal) Sony progressive scan CMOS
Projection function	Stereographic projection
Focal length	1.05 mm
F-number	2.08
Total pixels	3072 × 2048
Registered panomorph lens	Immervision Enables RPL: BS55T
PTZ control	Digital pan, tilt, zoom
Shutter speed	1/10 000 s
Video compression	H.264

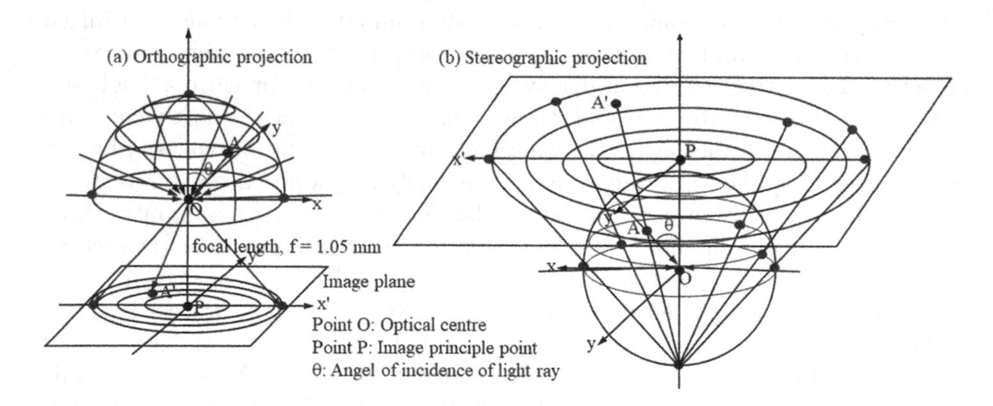

Fig. 1. Geometrical projection models orthographic project vs. stereographic project.

4 Transfer Learning in Panoramic Data Based on MobileNets Architecture

At the cost of full 360° angle of view, machine learning in panoramic images and videos is more challenging than in perspective contents. The 360° data is essen-

Table 2. The multiple challenging targets and conditions for panoramic dataset collection.

Objects	Description	Variability	Background scenes
UAV	DJI Mavic, only 1	Distance, viewing angels, speed	1. Outdoor open space in cloudy and sunny day in Norway
			2. Outdoor built area (campus) in sunny day in the Netherlands
Arabic stop sign	Printed image on A3 paper, only 1	Distance, viewing angels, speed	1. Open parking slot with snow in Norway
			2. Outdoor built area in cloudy winter day in Norway
			3. Outdoor built area (campus) in sunny day in the Netherlands
Fake isis flag	Printed image on A3 paper, only 1	Distance, viewing angels, speed	1. Open parking slot with snow in Norway
			2. Outdoor built area in cloudy winter day in Norway
			3. Outdoor built area (campus) in sunny day in the Netherlands
Sandbags	Real and piled up, 100, 60 cm × 90 cm	Distance, viewing angels, speed	Outdoor built area (campus) in sunny day in the Netherlands
Concrete jersey barrier	Both real (10) and fake (only 2)	Distance, viewing angels, speed	1. Open parking slot with snow in Norway
			2. Outdoor built area in cloudy winter day in Norway
Birds	Seagull	As caught by the camera	Outdoor open space in sunny day in Norway

tially 3D: the pixel values are defined on a 3D coordinates as (x, y, z) in Cartesian coordinates or (r, θ, ϕ) in spherical coordinates rather than (x, y) or (r, θ) [48]. Generally, panoramic data contain considerable radial distortions. In addition, we point out that today panoramic lens optical design and formation of projected images seem to be specific to different lens producers, which eventually impose inconsistency and noisy manifolds in data. Consequently, transferring of knowledge learned using perspective data such as COCO dataset to the 360° vision domain is indeed a naive approach. However, for the ImmerVision panomorph lens customized for surveillance, sophisticated distortion control and spatial distribution of pixels technology renders a hemispherical content with small distortions, especially for the zone where the angular FOV is small. Furthermore, if we focus on a local content, e.g., a UAV, the local distortions are not large. In other words, the projection of a small patch on a sphere to its tangent plane will be very close to the sphere itself.

In this work, we experimented two methods for learning the deep neural network architectures for the targets detection in panoramic images and videos. In method 1, the panoramic data is transformed to a perspective image locally by using a virtual camera model with parameters as pan, tilt and zoom. The PTZ parameters define a local patch on the hemisphere and it is projected to

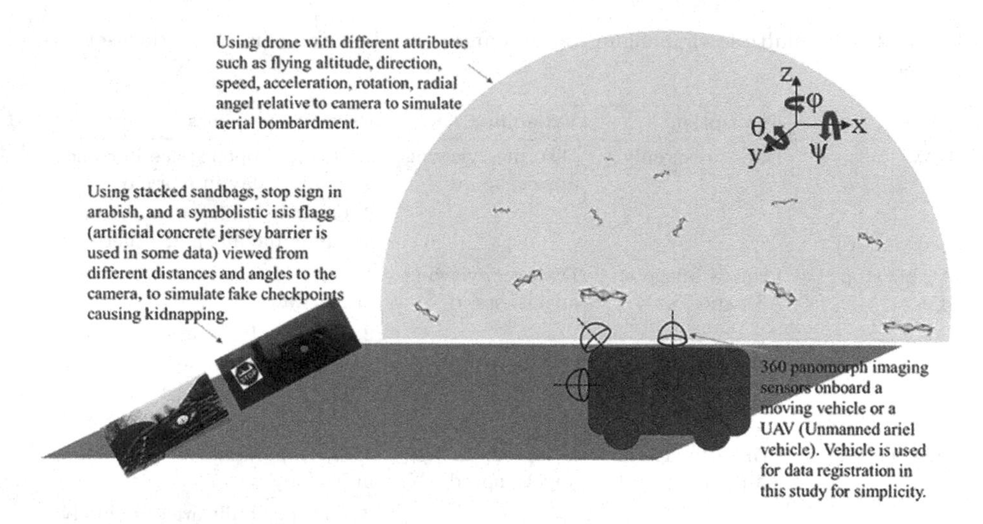

Fig. 2. The configuration for panoramic data collection.

a 2D plane to generate a perspective alike image. Those perspective images are used as inputs for transfer train the MobileNets. In method 2, the panoramic domain is treated as a 2D ellipse locally without projection. The original ellipse is used for annotation, train, validation and test. Table 3 shows the two methods with method 2 better suited for real-time detection in live streamed panoramic videos. For both methods, the whole original MobileNets model are retrained, no modifications on the architecture are done.

We tested several pre-trained networks such as ResNet [10], Mask-RCNN [15], Inception [13] and decided to use MobileNets for its optimized speed and accuracy while suited for embedded and mobile applications. Although it is not the innovation of this work, we think it necessary to present the MobileNets-SSD architecture (Fig. 3) for learning in the projected perspective images (residual distortions are obvious) and the panoramic ellipses using initial weights obtained from large scale training on the COCO dataset.

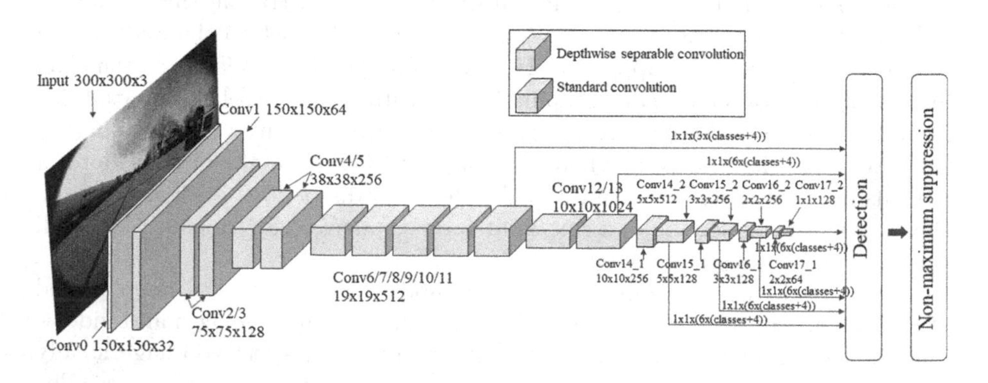

Fig. 3. Mobilenet-SSD architecture for projected rectilinear and 360 imagery.

Table 3. Detection via geometric modeling and learning in rectilinear imagery vs. directly in 360 imagery.

Method 1	Project and learn in rectilinear imagery
1	Require Pan, tilt and zoom parameters to define the region on the sphere to be projected
2	Apply stereographic project function
3	Re-train pretrained MobileNet on Coco dataset using annotated projected imagery (Gradient calculation using backpropagation and update weights using SGD with learning rate of 0.001)
4	Evaluate trained Mobilenet with test data
5	Evaluate trained Mobilenet in the wild to unseen data
Method 2	Directly learn in 360 imagery
1	Re-train pretrained MobileNet on Coco dataset using annotated 360 imagery (Gradient calculation using back-propagation and update weights using SGD with learning rate of 0.001)
2	Evaluate trained Mobilenet with test data
3	Evaluate trained Mobilenet in the wild to unseen data

5 Experiments

The dataset both projected images and panoramic images are human annotated using a open-source graphical labeling tool LabelImg. The annotated data are split into train data and validation data and is tranformed to a tensorflow record files. The MobileNets pre-trained on COCO dataset is used as the initial model with hyperparameters as 0.004 for initial learning rate and stochastic gradient descent for optimization algorithm. Total annotated training images are about 200 images, and annotated validation image of 20 images. The training was carried out on a NVIDIA DGX machine using multiple GPUs for 20,000 steps or more until the loss is reduced to less than 0.5. We show some examples for training and validation images to present an evaluation on the data characteristics. Figure 4 shows examples of training data for UAV. The scale of the UAV changes from as small as a point without texture or pattern information to find to a larger patch with geometry, symmetry and color features. It is hard even for human perception to recognize that the UAV data collected in a Norwegian landscap and weather condition may garanti extraction of common features. Similar challenges are encountered in the fake checkpoint as grouped objects (Flag, stop sign and piled sandbags). Figure 5 shows examples of train and validation data for fake checkpoint in projected images. Figure 6 shows example train and validation data as raw panoramic images. We can observe that local distortions at the region of the targets are small. Although the treatment of the data as 2D ellipse essentially loses or misuse the location information of the objects, its other features such as shape and color information are not reduced or much distorted.

Fig. 4. Example training and validation data for UAV in projected images.

Fig. 5. Example training and validation data for fake checkpoint in projected images.

6 Results and Discussion

The detection results for the test data which having the similar distribution as the training data are very promising. With only small train data, the deep neural network model are learned to achieve average precision score of 0.82 for fake checkpoint and 0.90 for UAV. Table 4 shows the average precision and F-score for UAVs and fake checkpoints as grouped objects containing piled sandbags, flag and stop sign.

Figure 7 shows UAVs which are detected successfully with probability above 0.5 and non-detections, no false positive detections were obtained. The original size of the UAV is 83 mm (height) × 83 mm (width) x 198 mm (length) when

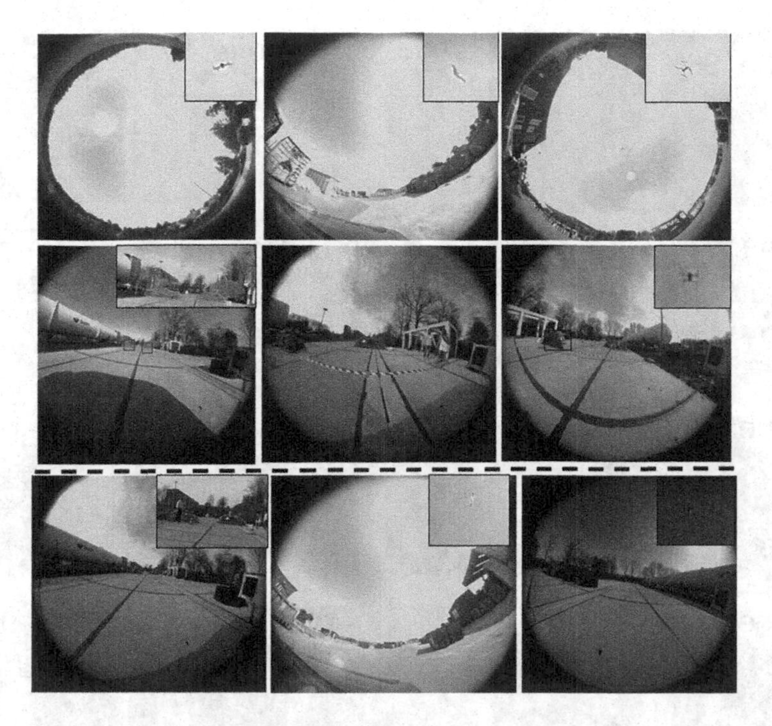

Fig. 6. Example training and validation data for UAV and fake checkpoint in panoramic images

folded down and 335 mm wide exluding propellers when unfolded. It is not straightforward to justify the reason that the deep model has failed to detect UAVs in some images. The adversarial robustness of deep models and interpretable AI have been an important research area, we think we could utilize these concrete examples to inspect the underlying mechanism of the deep neural network models.

Table 4. Average precision and f-score for UAV and fake checkpoint detection.

Objects	Average precision	F-score
UAV	0.90	0.94
Fake checkpoint (Grouped objects as sandbags, flag and stop sign)	0.82	0.90

Figure 8 shows examples of detection results for fake checkpoint in projected images. Non-detection result of sandbags and fake checkpoint in the upper left image is obtained and we did not obtain false positive detections.

Figure 9 shows detection results for raw panoramic images without projection. Good detection results are obtained for UAVs and for fake checkpoints. And the deep model was able to distinguish a bird as seagull from a UAV. No

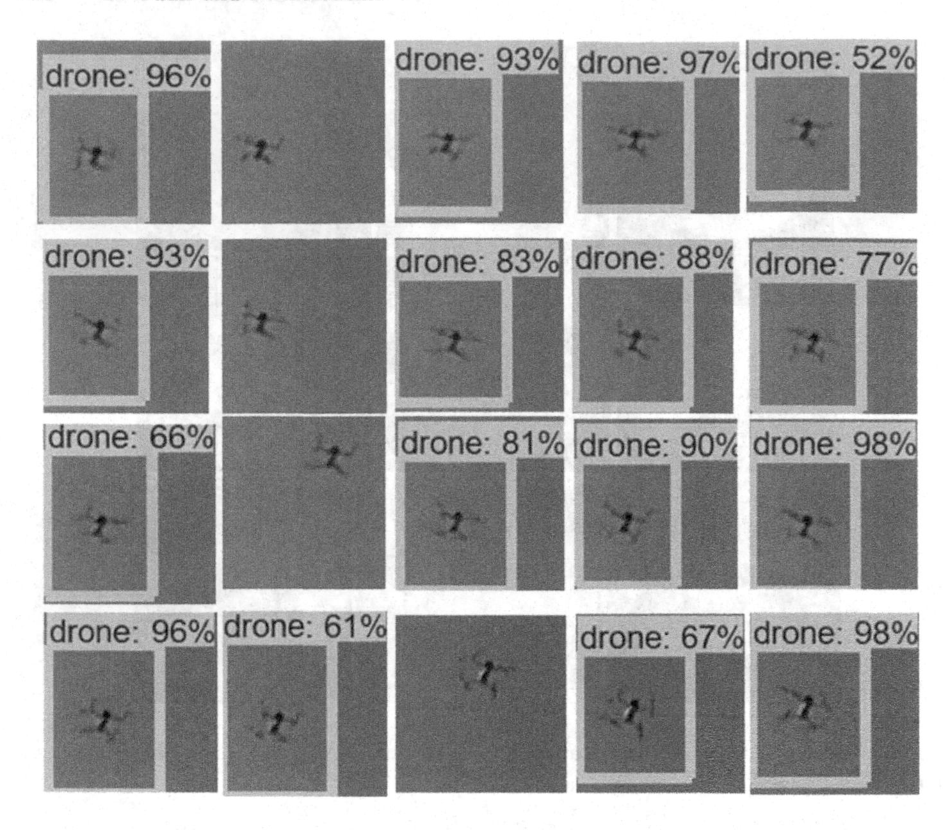

Fig. 7. Detection result for UAV in projected test dataset.

Fig. 8. Detection results for fake checkpoint in projected test dataset.

false positive detection of light poles as UAV or bird. These result shows that our method 2 without projection has shown promising results. This is important for real-time detection in live streamed panoramic videos.

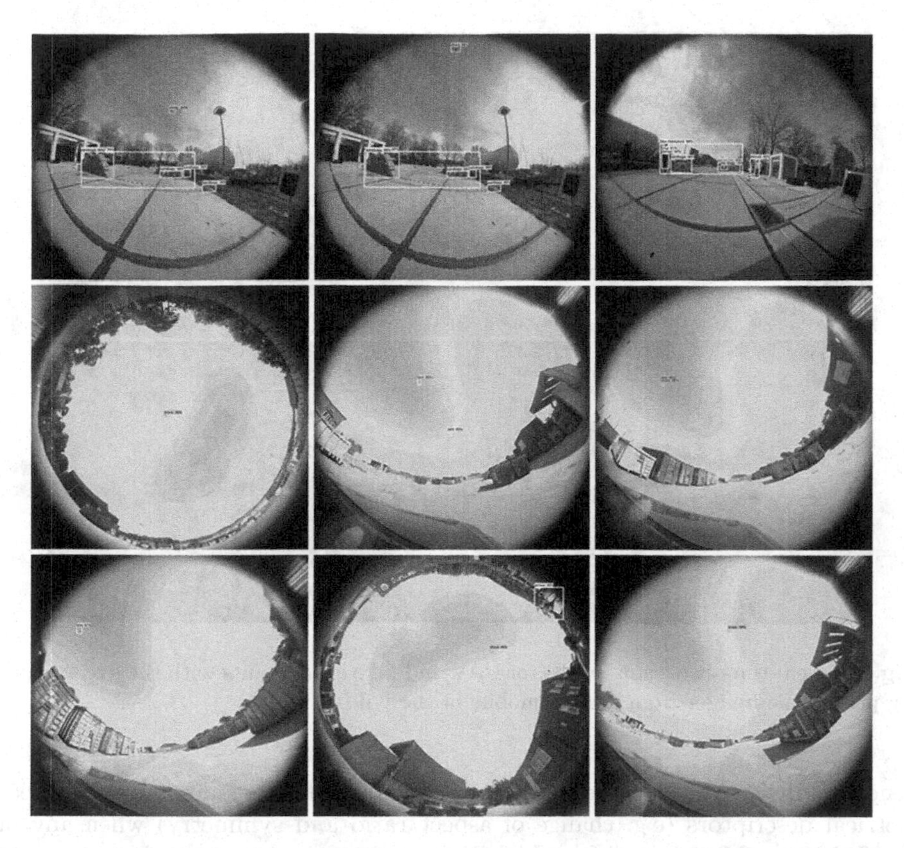

Fig. 9. Detection results for UAV and fake checkpoint directly in raw panoramic test dataset without projection.

Figure 10 shows the result obtained when the trained deep MobileNets model are put into action in the wild. Successful detection of UAV result was obtained. Although the UAV is the same DJI Mavic, and similar sandbags, flag and stop sign are used for setting up the targets, many non-detection were resulted. An explaination could be that the distribution of train data and test data. For example, the background scene appeared very different: the sky for the UAV is cloudy as opposed to clear blue sky, sandbags are piled and positioned in different manners. And that the differences in overall lighting conditions. False positive results of sandbags were also obtained.

We believe the deep MobileNets model could achieve much better performance when a larger panoramic training dataset is used to transfer learn the model. Although the data complexity in the panoramic domain is much more complex than perspective images, transfer learning scheme adapting from one domain to a new domain have been proven to be effective. For fine tuning model performance, it is necessary to take into account the complexity of the panoramic data. UAV has 6 degree of freedom, i.e, $(x, y, z, \psi, \theta, \phi)$, where ψ, θ, ϕ

Fig. 10. Real-time detection results for UAV and fake checkpoints with the live streaming panoramic imager from an automobile in the wild.

denotes pitch, roll, yaw, its rotational transformation renders different geometrical descriptors (e.g, change of aspect ratio and symmetry) when any of the 6 degree of freedom differs. Lighting condition is important for real-world visual perception and lighting estimation in panoramic images [47] could enhance robustness owing to lighting variations. We did not compare the detection accuracy and speed of the MobileNets model against the newer models proposed in previous work, this is because we leverage transfer learning scheme. This means that the model shall be trained on high-quality samples with an optimal selection of hyper-parameters. Most of the new models are trained and tested on synthetic dataset for detection of salient and common objects.

After all, we recall that statistical theory of learning [1] emphasizes that generalization is guaranteed only when the test dataset has the same distribution as the train dataset and only when the number of samples N is large, normally 10 time of Vapnik-Chervonenkis (VC) dimension of a learning model. Violating the assumptions gives rise to artifacts such as sample bias, covariate shift, and overfitting in learning system resulting in poor performance. Transfer learning is an interesting paradigm based on neurobiological analogy that knowledge learned in source domain is transferrable to a target domain thus speed up learning, i.e., using fewer training examples. There is limited theoretical derivations and empirical evidences to adequately qualify or disqualify that transfer learning could possibly bypass the data hunger and covariate shift pitfall and provide-world a robust learning system cheaply.

7 Conclusion and Future Work

We developed a real-time vision application leveraging state-of-the-art visual sensors and efficient neural network architectures. The specifications of the system include the ImmerVision 360° panoramic vision system on an automobile or UAV platform for live streaming of original panoramic data content rather than projected images. Our result showed that the light-weight neural network model based on MobileNets architecture directly learned in 360° spherical data achieved a satisfactory performance when the distribution of test data and train data are consistent. However, owing to the fewness of the training data and increased complexity of the panoramic images and videos, the performance of the trained network in the wild shall be improved by increasing the size of the training data. The contribution of this work is a large FOV vision hardware and software system for real-time detection of challenging targets in 360° videos in the wild. This system is readily configured for very low-power embedded applications in robots and UAVs with machine learning accelerators such as Movidius NCS2 or HiSilicon Kirin 970. Furthermore, this work provide empirical performance data of deep neural nets models which are pre-trained using perspective image dataset and are only few-shot transfer learned with data in a new domain. It would be interesting to implement the spherical CNNs and graph spherical CNNs, which are experimented on equirectangular projected panoramic images, to compare their performance with directly learning in raw panoramic images.

Acknowledgement. This work is supported by the project Spacetime Vision – Towards Unsupervised Learning in the 4D World funded under the EEA grant number EEA-RO-NO-2018-0496.

References

1. Bousquet, O., Boucheron, S., Lugosi, G.: Introduction to statistical learning theory. In: Bousquet, O., von Luxburg, U., Rätsch, G. (eds.) ML -2003. LNCS (LNAI), vol. 3176, pp. 169–207. Springer, Heidelberg (2004). https://doi.org/10.1007/978-3-540-28650-9_8
2. Thibault, S.: Enhanced surveillance system based on panomorph panoramic lenses. In: Optics and Photonics in Global Homeland Security III, vol. 6540, p. 65400E. International Society for Optics and Photonics (2007)
3. Thibault, S., Konen, P., Roulet, P., Villegas, M.: Novel hemispheric image formation: concepts and applications. In: Photon Management III, vol. 6994, p. 699406. International Society for Optics and Photonics (2008)
4. Pan, S.J., Yang, Q.: A survey on transfer learning. IEEE Trans. Knowl. Data Eng. **22**(10), 1345–1359 (2009)
5. Li, C., Parikh, D., Chen, T.: Automatic discovery of groups of objects for scene understanding. In: 2012 IEEE Conference on Computer Vision and Pattern Recognition, pp. 2735–2742. IEEE (2012)
6. Moo Yi, K., Yun, K., Wan Kim, S., Jin Chang, H., Young Choi, J.: Detection of moving objects with non-stationary cameras in 5.8 ms: bringing motion detection to your mobile device. In: Proceedings of the IEEE Conference on Computer Vision and Pattern Recognition Workshops, pp. 27–34 (2013)

7. Henriques, J.F., Caseiro, R., Martins, P., Batista, J.: High-speed tracking with kernelized correlation filters. IEEE Trans. Pattern Anal. Mach. Intell. **37**(3), 583–596 (2014)
8. Caruso, D., Engel, J., Cremers, D.: Large-scale direct slam for omnidirectional cameras. In: 2015 IEEE/RSJ International Conference on Intelligent Robots and Systems (IROS), pp. 141–148. IEEE (2015)
9. Furnari, A., Farinella, G.M., Bruna, A.R., Battiato, S.: Affine covariant features for fisheye distortion local modeling. IEEE Trans. Image Process. **26**(2), 696–710 (2016)
10. He, K., Zhang, X., Ren, S., Sun, J.: Deep residual learning for image recognition. In: Proceedings of the IEEE Conference on Computer Vision and Pattern Recognition, pp. 770–778 (2016)
11. Nguyen, T.B., Chung, S.T., et al.: ConvNets and AGMM based real-time human detection under fisheye camera for embedded surveillance. In: 2016 International Conference on Information and Communication Technology Convergence (ICTC), pp. 840–845. IEEE (2016)
12. Rozantsev, A., Lepetit, V., Fua, P.: Detecting flying objects using a single moving camera. IEEE Trans. Pattern Anal. Mach. Intell. **39**(5), 879–892 (2016)
13. Szegedy, C., Vanhoucke, V., Ioffe, S., Shlens, J., Wojna, Z.: Rethinking the inception architecture for computer vision. In: Proceedings of the IEEE Conference on Computer Vision and Pattern Recognition, pp. 2818–2826 (2016)
14. Zhang, Z., Rebecq, H., Forster, C., Scaramuzza, D.: Benefit of large field-of-view cameras for visual odometry. In: 2016 IEEE International Conference on Robotics and Automation (ICRA), pp. 801–808. IEEE (2016)
15. He, K., Gkioxari, G., Dollár, P., Girshick, R.: Mask R-CNN. In: Proceedings of the IEEE International Conference on Computer Vision, pp. 2961–2969 (2017)
16. Hu, H.-N., Lin, Y.C., Liu, M.Y., Cheng, H.T., Chang, Y.J., Sun, M.: Deep 360 pilot: learning a deep agent for piloting through 360deg sports videos. In: Proceedings of the IEEE Conference on Computer Vision and Pattern Recognition, pp. 3451–3460 (2017)
17. Kawaguchi, K., Kaelbling, L.P., Bengio, Y.: Generalization in deep learning. arXiv preprint arXiv:1710.05468 (2017)
18. Neyshabur, B., Bhojanapalli, S., McAllester, D., Srebro, N.: Exploring generalization in deep learning. In: Advances in Neural Information Processing Systems, pp. 5947–5956 (2017)
19. Ran, L., Zhang, Y., Zhang, Q., Yang, T.: Convolutional neural network-based robot navigation using uncalibrated spherical images. Sensors **17**(6), 1341 (2017)
20. Su, Y.-C., Grauman, K.: Learning spherical convolution for fast features from 360 imagery. In: Advances in Neural Information Processing Systems, pp. 529–539 (2017)
21. Baek, I., Davies, A., Yan, G., Rajkumar, R.R.: Real-time detection, tracking, and classification of moving and stationary objects using multiple fisheye images. In: 2018 IEEE Intelligent Vehicles Symposium (IV), pp. 447–452. IEEE (2018)
22. Chung, Y., Haas, P.J., Upfal, E., Kraska, T.: Unknown examples & machine learning model generalization. arXiv preprint arXiv:1808.08294 (2018)
23. Cohen, T.S., Geiger, M., Köhler, J., Welling, M.: Spherical CNNs. arXiv preprint arXiv:1801.10130 (2018)
24. Coors, B., Paul Condurache, A., Geiger, A.: SphereNet: learning spherical representations for detection and classification in omnidirectional images. In: Proceedings of the European Conference on Computer Vision (ECCV), pp. 518–533 (2018)

25. Esteves, C., Allen-Blanchette, C., Makadia, A., Daniilidis, K.: Learning so (3) equivariant representations with spherical CNNs. In: Proceedings of the European Conference on Computer Vision (ECCV), pp. 52–68 (2018)
26. Georgakopoulos, S.V., Kottari, K., Delibasis, K., Plagianakos, V.P., Maglogiannis, I.: Pose recognition using convolutional neural networks on omni-directional images. Neurocomputing **280**, 23–31 (2018)
27. Kolter, Z., Madry, A.: Adversarial robustness: theory and practice. Tutorial at NeurIPS (2018)
28. Novak, R., Bahri, Y., Abolafia, D.A., Pennington, J., Sohl-Dickstein, J.: Sensitivity and generalization in neural networks: an empirical study. arXiv preprint arXiv:1802.08760 (2018)
29. Roulet, P., et al.: Method to capture, store, distribute, share, stream and display panoramic image or video, US Patent App. 15/656,707, 24 May 2018
30. Sandler, M., Howard, A., Zhu, M., Zhmoginov, A., Chen, L.C.: MobileNetV2: inverted residuals and linear bottlenecks. In: Proceedings of the IEEE Conference on Computer Vision and Pattern Recognition, pp. 4510–4520 (2018)
31. Sheng, T., Feng, C., Zhuo, S., Zhang, X., Shen, L., Aleksic, M.: A quantization-friendly separable convolution for MobileNets. In: 2018 1st Workshop on Energy Efficient Machine Learning and Cognitive Computing for Embedded Applications (EMC2), pp. 14–18. IEEE (2018)
32. Xu, Y., et al.: Gaze prediction in dynamic 360 immersive videos. In: proceedings of the IEEE Conference on Computer Vision and Pattern Recognition, pp. 5333–5342 (2018)
33. Defferrard, M., Milani, M., Gusset, F., Perraudin, N.: DeepSphere: a graph-based spherical CNN. In: International Conference on Learning Representations (2019)
34. Hossain, S., Lee, D.j.: Deep learning-based real-time multiple-object detection and tracking from aerial imagery via a flying robot with GPU-based embedded devices. Sensors **19**(15), 3371 (2019)
35. Hou, Y.C., Sahari, K.S.M., How, D.N.T.: A review on modeling of flexible deformable object for dexterous robotic manipulation. Int. J. Adv. Rob. Syst. **16**(3), 1729881419848894 (2019)
36. Howard, A., et al.: Searching for MobileNetV3. In: Proceedings of the IEEE International Conference on Computer Vision, pp. 1314–1324 (2019)
37. Ignatov, A., et al.: AI benchmark: all about deep learning on smartphones in 2019. In: 2019 IEEE/CVF International Conference on Computer Vision Workshop (ICCVW), pp. 3617–3635. IEEE (2019)
38. Perraudin, N., Defferrard, M., Kacprzak, T., Sgier, R.: DeepSphere: efficient spherical convolutional neural network with HEALPix sampling for cosmological applications. Astron. Comput. **27**, 130–146 (2019)
39. Reuther, A., Michaleas, P., Jones, M., Gadepally, V., Samsi, S., Kepner, J.: Survey and benchmarking of machine learning accelerators. arXiv preprint arXiv:1908.11348 (2019)
40. Wang, K.-H., Lai, S.-H.: Object detection in curved space for 360-degree camera. In: ICASSP 2019–2019 IEEE International Conference on Acoustics, Speech and Signal Processing (ICASSP), pp. 3642–3646. IEEE (2019)
41. Wang, T., Hsieh, Y.Y., Wong, F.W., Chen, Y.F.: Mask-RCNN based people detection using a top-view fisheye camera. In: 2019 International Conference on Technologies and Applications of Artificial Intelligence (TAAI), pp. 1–4. IEEE (2019)
42. Yu, D., Ji, S.: Grid based spherical CNN for object detection from panoramic images. Sensors **19**(11), 2622 (2019)

43. Zhu, J., Zhu, J., Wan, X., Wu, C., Xu, C.: Object detection and localization in 3D environment by fusing raw fisheye image and attitude data. J. Vis. Commun. Image Represent. **59**, 128–139 (2019)
44. Chen, Y., Dai, X., Liu, M., Chen, D., Yuan, L., Liu, Z.: Dynamic convolution: attention over convolution kernels. In: Proceedings of the IEEE/CVF Conference on Computer Vision and Pattern Recognition, pp. 11030–11039 (2020)
45. Chou, S.-H., Sun, C., Chang, W.Y., Hsu, W.T., Sun, M., Fu, J.: 360-indoor: towards learning real-world objects in 360deg indoor equirectangular images. In: The IEEE Winter Conference on Applications of Computer Vision, pp. 845–853 (2020)
46. Farahi, F., Yazdi, H.S.: Probabilistic Kalman filter for moving object tracking. Sig. Process. Image Commun. **82**, 115751 (2020)
47. Gkitsas, V., Zioulis, N., Alvarez, F., Zarpalas, D., Daras, P.: Deep lighting environment map estimation from spherical panoramas. In: Proceedings of the IEEE/CVF Conference on Computer Vision and Pattern Recognition Workshops, pp. 640–641 (2020)
48. Jin, L., et al.: Geometric structure based and regularized depth estimation from 360 indoor imagery. In: Proceedings of the IEEE/CVF Conference on Computer Vision and Pattern Recognition, pp. 889–898 (2020)
49. Lee, Y., Jeong, J., Yun, J., Cho, W., Yoon, K.-J.: SpherePHD: applying CNNs on 360° images with non-Euclidean spherical PolyHeDron representation. IEEE Trans. Pattern Anal. Mach. Intell., 1 (2020). https://doi.org/10.1109/TPAMI.2020.2997045
50. Yan, X., Acuna, D., Fidler, S.: Neural data server: a large-scale search engine for transfer learning data. In: Proceedings of the IEEE/CVF Conference on Computer Vision and Pattern Recognition, pp. 3893–3902 (2020)
51. Yang, Q., Li, C., Dai, W., Zou, J., Qi, G.J., Xiong, H.: Rotation equivariant graph convolutional network for spherical image classification. In: Proceedings of the IEEE/CVF Conference on Computer Vision and Pattern Recognition, pp. 4303–4312 (2020)
52. Zhao, P., You, A., Zhang, Y., Liu, J., Bian, K., Tong, Y.: Spherical criteria for fast and accurate 360 object detection. In: AAAI, pp. 12959–12966 (2020)

Post Training Mixed-Precision Quantization Based on Key Layers Selection

Lingyan Liang(✉)

State Key Laboratory of High-End Server and Storage Technology, Inspur Group Company Limited, Beijing, China
lianglingyan@inspur.com

Abstract. Model quantization has been extensively used to compress and accelerate deep neural network inference. Because post-training quantization methods are simple to use, they have gained considerable attention. However, when the model is quantized below 8-bits, significant accuracy degradation will be involved. This paper seeks to address this problem by building mixed-precision inference networks based on key activation layers selection. In post training quantization process, key activation layers are quantized by 8-bit precision, and non-key activation layers are quantized by 4-bit precision. The experimental results indicate an impressive promotion with our method. Relative to ResNet-50(W8A8) and VGG-16(W8A8), our proposed method can accelerate inference with lower power consumption and a little accuracy loss.

Keywords: CNN · Quantization · Mixed-precision · Key layers selection

1 Introduction

The deep convolutional neural networks (CNNs) have achieved state-of-the-art performance in computer vision tasks, such as image classification, object detection and semantic segmentation. In order to improve network efficiency, model quantization is one of the most popular strategies for compressing and accelerating deep neural networks, which represents weights and/or activations by using a low bit-width precision without significant loss in the network accuracy. Generally, DNN quantization techniques include post-training and quantization-aware training techniques. Although quantization-aware techniques generally achieve better results, it requires the full dataset for re-training. However, post-training quantization methods are more easily applied in practice, without requiring additional information from the user except a small unlabeled calibration set. Unfortunately, post-training quantization below 8 bits usually incurs significant accuracy degradation. This paper seeks to address this problem by building the mixed-precision inference network. We proposed a mixed-precision quantization method based on key activation layers selection. And different layers will be

A. Bartoli and A. Fusiello (Eds.): ECCV 2020 Workshops, LNCS 12539, pp. 121–125, 2020.
https://doi.org/10.1007/978-3-030-68238-5_9

quantized based on different layers' characteristics. We use information similarity between adjacent layers to evaluate significance of layers. The proposed method is evaluated on image classification tasks (ResNet-50 and VGG-16). The experimental results illustrate the effectiveness of the proposed method.

2 Related Work

Different from quantization-aware training methods, the post-training quantization methods can compress weights and/or quantize both weights and activations to realize a faster inference with limited data or even data-free, without requiring to re-train models. Since activations need to be quantized, one needs calibration data to calculate the dynamic ranges of activations, which should minimize the information loss from quantization. The common methods include Minimal mean squared error (MMSE), Minimal KL Divergence (KL) [4], ACIQ [2]. In Nagel's work [5], they proposed a data-free quantization method for deep neural networks that does not require fine-tuning or hyper-parameter selection. At present, some 8-bit post-training quantization methods/toolboxes have been open-sourced, like Nivida's TensorRT, Xilinx's ml-suite and Tensorflow's TFLite. But post-training quantization below 8 bits still incurs significant accuracy degradation, so this paper proposed a mixed-precision post training quantization method.

3 Approach

Because each network layer contributes diversely to the overall performance and has different sensitivity to quantization. More attention should be payed to the significant layers. So we propose a mixed-precision network quantization method based on key layers selection. In this section, we will describe our methodology in detail. Figure 1 shows the framework of mixed-precision quantization.

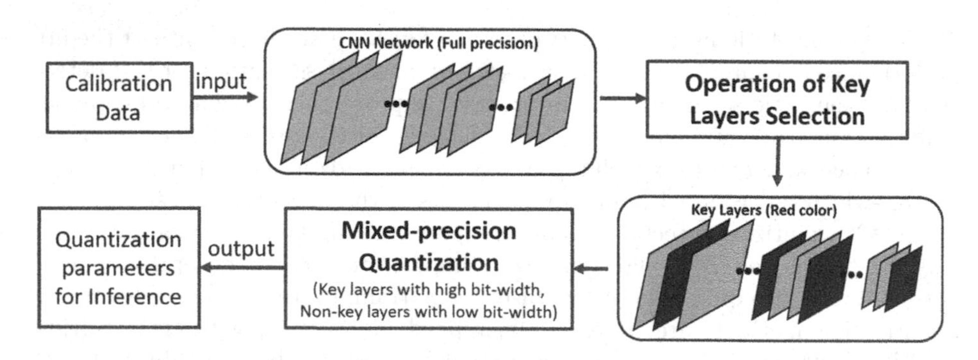

Fig. 1. The framework of mixed-precision quantization of convolutinal networks (Color figure online)

In this paper, mixed-precision quantization includes two parts: "key activation layers selection" and "low mixed-precision quantization". The main workflow is as following: 1) input a group of unlabeled images as calibration data into the pre-trained full-precision CNN network, and calculate every convolutional layer's feature maps. 2) based on feature maps' information, select the networks' important layers as key layers, and the selected key layers are marked with red color in Fig. 1) quantize pre-trained full-precision networks, and key activation layers are quantized by using high bit-width precision to maintain more information details of the network, and non-key activation layers are quantized by using low bit-width precision to maintain the semantic information of the network.

For the selection of key activation layers, we use entropy as the description of activation layer. If the information distribution on the current activation layer is not similar to its previous layer, we think there are some new features appearing. In order to preserve more information in feature maps, this layer is identified as a key layer, which will be quantized with high bit-width precision. Otherwise, it is identified as a non-key layer with low bit-width precision.

In key activation layers selection, we take every layer as a unit to count their information distribution. And the similarity calculation is based on (1), where *H(P)* is the feature maps' information entropy, which can be calculated by (3). V and M represent feature maps' variance value and mean value respectively. k and $k-1$ represent adjacent layers. α is the parameter which can balance the combined information between data distribution and information entropy. In our experiments, $\alpha = 0.5$. In (1), if *Diff* value is larger, the similarity is lower. The difference of information entropy can be calculated based on (2). In (3), b represents "bins", and N is the number of bins, set as 2048. *P(b)* is the probability value of b. In the key layers selecting process, the important of every activation layer is sorted based on *Diff* value. And based on hardware performance and accuracy requirement, we can choose suitable numbers of key layers.

$$Diff = \alpha * (|V_{k-1} - V_k| + |M_{k-1} - M_k|) + (1-\alpha) * (H(P_{k-1}) - H(P_k)) \quad (1)$$

$$H(P_{k-1}) - H(P_k) = \sum_{b=0}^{N} P_{k-1}(b) * log_2\left(\frac{P_{k-1}(b)}{P_k(b)}\right) \quad (2)$$

$$H(P) = -\sum_{b=0}^{N} P(b) * log_2(P(b)) \quad (3)$$

Based on key activation layers selection results, all activation layers are classified into key layers and non-key layers. The key activation layers are quantized using 8-bit, and the non-key activation layers are quantized using 4-bit.

4 Experiments

To evaluate the performance of our proposed method, an extensive performance analysis is illustrated in this part. In the experiments, all the weights are quantized to 8-bit precision ("W8"), and all the activations are quantized to 4-bit or/and 8-bit precision based on the key layers selection results ("A8/A4/Mixed").

We evaluate ResNet-50 and VGG-16 models on ImageNet2012's 50000 validation images. We randomly choose 500 images as calibration data for key layers selection and calculating quantizer parameters by using Minimal KL Divergence (KL) [4]. Compared with 32-bit float precision model, Table 1 shows the results with three different quantization settings: W8A8, W8A4, W8-Mixed.

Table 1. Classification Results on ImageNet dataset

Models	Key layer	Accuracy	FP32	W8A8	W8A4	W8-Mixed
ResNet-50	11 (53)	Top1 (%)	73.75	73.5 (0.25↓)	67.55 (6.2↓)	**72.35** (1.38↓)
		Top5 (%)	91.65	91.3 (0.35↓)	88.4 (3.25↓)	**90.5** (1.15↓)
VGG-16	3 (13)	Top1 (%)	70.5	70.3 (0.2↓)	67.05 (3.45↓)	68.3 (2.2↓)
		Top5 (%)	89.75	89.4 (0.35↓)	87.7 (2.05↓)	**88.9** (0.85↓)

In Table 1, ResNet-50 (including shortcut convolution layers) has 53 convolution layers, and we selected 11 activation layers as key layers, which are quantized by 8-bit precision, and other activation layers are quantized by 4-bit precision. VGG-16 model has 13 convolution layers, and only quantized 3 key activation layers by using 8-bit precision. The accuracy result is shown in Table 1. We can find that, *W8-Mixed* is little lower than *W8A8*, but it is much better than *W8A4*. So our proposed method can effectively keep models' accuracy by using 8-bit and 4-bit mixed-precision. In addition, compared with ResNet-50 (FP32) (about 41M memory footprint for all activations), ResNet-50 (W8-Mixed) only needs about 7M, which is lower than ResNet-50 (W8A8)'s 10M. For memory with fixed bandwidth, the number and time of data transmission are reduced, and the power consumption of data handling is also reduced.

5 Conclusions

In this paper, we proposed a mixed-precision network quantization method based on key activation layers selection. The effectiveness of the proposed method is evaluated based on ResNet-50 and VGG-16 models, and the experimental results proved its effectiveness. In future, we will add some other strategies to improve quantization accuracy, like per-channel quantization [3], bias correction [1,5].

Acknowledgements. The work was funded by the Key R&D Plan of Shandong Province (No. 2019JZZY011101).

References

1. Banner, R., Nahshan, Y., Hoffer, E., Soudry, D.: Post-training 4-bit quantization of convolution networks for rapid-deployment. arXiv Computer Vision and Pattern Recognition (2018)
2. Banner, R., Nahshan, Y., Hoffer, E., Soudry, D.: ACIQ: analytical clipping for integer quantization of neural networks. arXiv (2019)
3. Krishnamoorthi, R.: Quantizing deep convolutional networks for efficient inference: a whitepaper. arXiv Machine Learning (2018)
4. Migacz, S.: 8-bit inference with TensorRT. In GPU Technology Conference (2017)
5. Nagel, M., Van Baalen, M., Blankevoort, T., Welling, M.: Data-free quantization through weight equalization and bias correction. arXiv International Conference on Computer Vision (2019)

Subtensor Quantization for Mobilenets

Thu Dinh, Andrey Melnikov, Vasilios Daskalopoulos, and Sek Chai(✉)

Latent AI, Princeton, NJ 08558, USA
sek@latentai.com

Abstract. Quantization for deep neural networks (DNN) have enabled developers to deploy models with less memory and more efficient low-power inference. However, not all DNN designs are friendly to quantization. For example, the popular Mobilenet architecture has been tuned to reduce parameter size and computational latency with separable depthwise convolutions, but not all quantization algorithms work well and the accuracy can suffer against its float point versions. In this paper, we analyzed several root causes of quantization loss and proposed alternatives that do not rely on per-channel or training-aware approaches. We evaluate the image classification task on ImageNet dataset, and our post-training quantized 8-bit inference top-1 accuracy in within 0.7% of the floating point version.

1 Introduction

Quantization is a crucial optimization for DNN inference, especially for embedded platforms with very limited budget for power and memory. Edge devices on mobile and IoT platforms often rely fixed-point hardware rather than more power-intensive floating point processors in a GPU. Through quantization, parameters for DNN can be converted from 32-bit floating point (FP32) towards 16-bit or 8-bit models, with minimal loss of accuracy. While significant progress has been made recently on quantization algorithms, there is still active research to understand the relationships among bit-precision, data representation, and neural model architecture.

One of the main challenges for quantization is the degrees of non-linearity and inter-dependencies among model composition, dataset complexity and bit-precision targets. Quantization algorithms assign bits for the DNN parameters, but the best distribution depends on range of values required by each model layer to represent the learnt representation. As the model grows deeper, the complexity of analysis increases exponentially as well, making it challenging to reach optimal quantization levels.

In this short paper, we present a case-study of a state-of-the-art Mobilenetv2 [9] architecture from a quantization perspective. We propose a new algorithmic approach to enhance asymmetric quantization to support depthwise convolutions. In comparison, current research has steered towards more complicated quantization schemes that require more memory and hardware support.

A. Bartoli and A. Fusiello (Eds.): ECCV 2020 Workshops, LNCS 12539, pp. 126–130, 2020.
https://doi.org/10.1007/978-3-030-68238-5_10

Using our approach instead, the overall DNN inference uses a simpler quantization scheme and still maintain algorithm accuracy; all achieved without retraining the model. To the best of our knowledge, we offer the following contributions in this paper: (1) new algorithmic approach for enhanced asymmetric quantization, and (2) comparison of accuracy, memory savings, and computational latency.

2 Related Work

Most post-training quantization (PTQ) algorithms are based on finding the mapping between floating point and an evenly spaced grid that can be represented in integer values. The goal is to find the scaling factor (e.g. max and min ranges) and zero-point offset to allow such mapping [3,4,7]. Similar to [6], our approach is a data-free quantization, whereby no training data or backprogation is required. Additional methods have been proposed to deal specifically with various elements of a DNN layer such as the bias tensors [1,6]. The goal remains the same for all algorithms: reduce bit-precision and memory footprint while maintaining accuracy.

There are other methods for DNN compression including, network architecture search and quantization aware training [8]. Some training aware methods may binarize or ternarize, result in DNNs that operate with logical operations rather than multiplications and additions. These approaches are in a different algorithmic category and are not considered for discussion in this short paper.

3 Proposed Algorithm

In this paper, we use a per-tensor asymmetric PTQ scheme where the min/max in the float range is mapped to the min/max of the integer range [7]. This is performed by using a zero-point (also called quantization bias, or offset) in addition to the scale factor. For per-channel approach, the quantization scheme is done separately for each channel in the DNN [4]. Convolutional and dense layers consist of a significant number of channels. Instead of quantizing all of them as tensors, per-channel quantization can be used to quantize each channel separately to provide accuracy improvements.

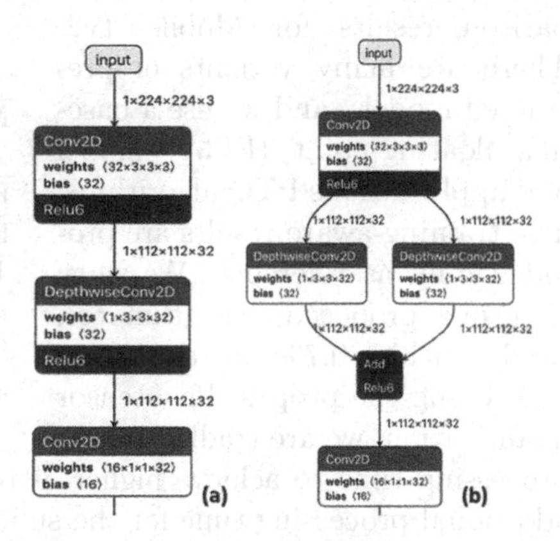

Fig. 1. Example DNN model snippet. (a) Original, (b) Proposed subtensor optimization for quantization.

We propose a new enhancement for quantizing tensors by splitting the tensors into subtensors, in order to achieve both high resolution and range in integer representation. Figure 1 shows a diagram of the original DNN graph and our proposed subtensor modification. In our approach, subtensors are better able to be quantized when done separately (e.g. splitting a problem into smaller parts makes it easier and more feasible to find the optimal mapping using asymmetric quantization). We then apply bias correction that includes weight clipping analysis and cross-layer equalization approach.

In doing so, we are increasing the bit-precision capacity of the DNN without changing the overall architecture. Previous per-tensor approaches are not able to achieve high algorithmic accuracy because there is simply not enough bits to encode both the resolution and range of a single tensor. Specifically, per-channel approaches allocates more bits along DNN channels, which requires memory access patterns that are different and less practical for most CPUs hardware.

Tensor splitting is analogous to channel splitting in [10], except that their approach duplicate tensors rather than splitting and optimizing subtensors separately. They re-scale of each duplicate tensors in order to achieve reduce floating point ranges, which loses resolution on critical portions floating point range and consequently, overall accuracy suffers.

4 Results and Analysis

For evaluation, we used the workflow [5] that builds upon a TensorFlow framework to implement all of the quantization approaches. For the DNN architecture, we use a baseline Mobilenetv2 model, trained with Imagenet 2012 dataset (1000 classes). The pre-trained model is then quantized and then compiled to generate either a native binary that can run independently or with a TFLite runtime. This workflow allows quick evaluation on embedded because the compiler targets optimal code for target embedded hardware.

Table 1 show quantization comparison results for Mobilenetv2. There are many variants of pre-trained models, and we use a baseline floating-point (FP32) model and apply various PTQ algorithms. The training-aware results are provided only as reference. We show that our proposed algorithm can reach within 0.7% of the baseline. Using the proposed subtensor quantization, we are trading tensor processing time to achieve higher algorithm accuracy. We have found that the additional processing time for the subtensors for the additional layer (the ADD layer) is minimal compared to the overall DNN processing time (only 4% of baseline). The additional overhead in memory size is also negligible (only 0.1% of baseline). Our PTQ approach using subtensors achieve a high level of accuracy, while maintaining flexibility, file size and processing speed.

Table 1. PTQ results

MobileNetv2	Accuracy (%)	Latency (ms)	Size (kB)
Baseline FP32 [2]	71.8	20.4	13651
Per-Channel [4]	69.8	11.5	4056
Training Aware [2]	70.7	8.2	3494
Bias Fine Tune [1]	70.6	8.2	3494
Proposed [5]	71.12	9.1	3508

In comparison, using per-channel quantization, each tensor is quantized separately for each channel [4]. A channel with many outlier weights will be quantized separately, resulting in improved accuracy. However, this method requires a set of quantization parameters (min-val, max-val, zero-point, scale) for each channel. Computationally, the model has to perform quantization, multiply, add, dequantization separately for each channel, for each tensor. This results in a much slower inference rate due to the more complicated memory access. Moreover, this technique of quantization is not accepted by some compilers or hardware. Specifically, our proposed subtensor approach is more hardware agnostic as the tensor memory access patterns are similar throughout the DNN.

The Mobilenet architecture, with seperable depthwise convolutions can exhibit a significant shift in the mean activation value after quantization. This shift is a result of quantization errors being unbalanced, and as such, bias correction improved overall quantized model performance. However, bias correction does not achieve near FP32 performance on its own. As shown in Table 1, subtensor level optimization improves the overall performance and brings algorithmic accuracy back to the floating point baseline.

5 Conclusion

Our goal is to change the perception that DNN quantization is a solved problem. Specifically, for post-training quantization, we show that current approaches to handle depthwise convolution layers could result in solutions that are not optimal. Through analysis of quantization effects using range and distributions of the DNN parameter values, we propose a new algorithm based on subtensor optimization that offers further DNN compression without reliance on hardware (memory and computation) support. Our work is enabled by a combined workflow that integrates a compiler to generate native binary on embedded platforms [5]. We show performance within 0.7% of accuracy for a Mobilenetv2 architecture, quantized from a floating point (FP32) baseline to integer 8-bits precision. Future work includes enhancements in the selection of subtensors to further reduce quantization errors.

References

1. Finkelstein, A., et al.: Fighting quantization bias with bias. In: ECV Workshop, CVPR2019 (2019)
2. Google: Tensorflow (2020). http://github.com/tensorflow/tensorflow/blob/v2.2.0/tensorflow/ and http://tensorflow.org/model-optimization/guide/quantization/training
3. Guo, Y.: A survey on methods and theories of quantized neural network. arXiv preprint arXiv:1808.04752 (2018)
4. Krishnamoorthi, R.: Quantizing deep convolutional networks for efficient inference: A whitepaper. arXiv preprint arXiv:1806.08342 (2018)
5. LatentAI: LEIP compress (2019). www.latentai.com/#products
6. Nagel, M., et al.: Data-free quantization through weight equalization and bias correction. In: ICCV (2019)

7. Nayak, P., et al.: Bit efficient quantization for deep neural networks. In: EMC2 Workshop, NeurIPS (2019)
8. Parajuli, S., et al.: Generalized Ternary Connect: End-to-End Learning and Compression of Multiplication-Free Deep Neural Networks. arXiv preprint arXiv:1811.04985 (2019)
9. Sandler, M., et al.: Mobilenetv 2: inverted residuals and linear bottlenecks. In: CVPR (2018)
10. Zhao, R., et al.: Improving neural network quantization without retraining using outlier channel splitting. In: ICLR (2019)

Feed-Forward On-Edge Fine-Tuning Using Static Synthetic Gradient Modules

Robby Neven[1(✉)], Marian Verhelst[2], Tinne Tuytelaars[1], and Toon Goedemé[1]

[1] KU Leuven/ESAT-PSI, Leuven, Belgium
{robby.neven,tinne.tuytelaars,toon.goedeme}@kuleuven.be
[2] KU Leuven/ESAT-MICAS, Leuven, Belgium
marian.verhelst@kuleuven.be

Abstract. Training deep learning models on embedded devices is typically avoided since this requires more memory, computation and power over inference. In this work, we focus on lowering the amount of memory needed for storing all activations, which are required during the backward pass to compute the gradients. Instead, during the forward pass, static Synthetic Gradient Modules (SGMs) predict gradients for each layer. This allows training the model in a feed-forward manner without having to store all activations. We tested our method on a robot grasping scenario where a robot needs to learn to grasp new objects given only a single demonstration. By first training the SGMs in a meta-learning manner on a set of common objects, during fine-tuning, the SGMs provided the model with accurate gradients to successfully learn to grasp new objects. We have shown that our method has comparable results to using standard backpropagation.

Keywords: Synthetic gradients · Feed-forward training · One-edge fine-tuning

1 Introduction

Most of the embedded devices currently running deep learning algorithms are used for inference only. The model gets trained in the cloud and optimized for deployment on the embedded device. Once deployed, the model is static and the device is only used for inference. If we want to incorporate new knowledge into the device, the standard way is to retrain the model in the cloud.

However, having an agile model that can be retrained on the embedded device has several advantages. Since the device is the origin of the data, no traffic between the device or the cloud has to be established, reducing latency and bandwidth. One example of this are self-driving cars, which can fine-tune their model to changing environments on the fly without the need to interact with a server or the cloud. Also, some applications like surveillance cameras deal with privacy issues which make it impossible to centralise the data. Therefore, the data has to be kept local, forcing the model to be trained on the device. One

A. Bartoli and A. Fusiello (Eds.): ECCV 2020 Workshops, LNCS 12539, pp. 131–146, 2020.
https://doi.org/10.1007/978-3-030-68238-5_11

other advantage is that each device can fine-tune its model to its own specific input data, resulting in personalized models. Instead of training one large generic model, each device can train a smaller specialist model on their specific input data, which can be beneficial (e.g., cameras only seeing one specific viewpoint).

Nevertheless, training on embedded devices is avoided for several reasons. The most determining one is that most of the embedded devices are severely resource constrained. A standard training loop requires substantially more memory, computations, precision and power than inference, which are not available on the embedded device. Another constraint is the lack of supervision, which enforces the use of unsupervised or semi-supervised methods. Having to deal with all these constraints makes training on the embedded device nearly impossible, or at the least very cumbersome.

The goal of this paper is to investigate the training cycle on embedded systems, while focusing on the memory constraint. Indeed, in small embedded platforms like [21], the amount of memory appears to be the bottleneck. More specifically, our aim is to lower the total amount of memory by training the network without backpropagation, but in a feed-forward manner. In this way, we address the update-locking problem, referring to the fact that each layer is locked for update until the network has done both a forward and a backward pass. During the forward pass all activations are computed, which are used during the backward pass to calculate the gradients. All layers are locked until their gradient is computed. Having to store all activations for the backward pass is a huge memory burden. If we can train the network in a feed-forward manner, the embedded device only has to store activations of the active layer instead of the whole network.

Figure 1 illustrates our approach. In contrast to standard backpropagation (BP), where the gradient flows backwards through the network, feed-forward training utilizes a gradient directly after computing each layer during the forward pass. This was already introduced by target propagation algorithms like [4], where they estimated this gradient by a fixed random transformation of the target. However, they showed this only worked well in a classification setting, where the target contains the sign of the global error of the network.

Instead of a fixed random transformation of the label, we propose to estimate a layer's gradient by transforming the activations of that layer together with the label. This method, called synthetic gradients, was first introduced in [9]. However, their method requires the gradient estimators (which from now on we will refer to as Synthetic Gradient Estimators or SGMs) to be trained in an online manner with a ground truth gradient. Instead, we will use static SGMs, similar to the fixed transformations of [4], which are pretrained on a set of tasks and can be used to fine-tune a model to a similar—but unseen—task in a feed-forward manner.

We tested our method on a robot grasping scenario from [16], where a robot, equipped with a camera, needs to learn to grasp new objects given only a single image. Since the CV algorithm of the robot runs on an embedded device, this application can greatly benefit from the feed-forward training method. The

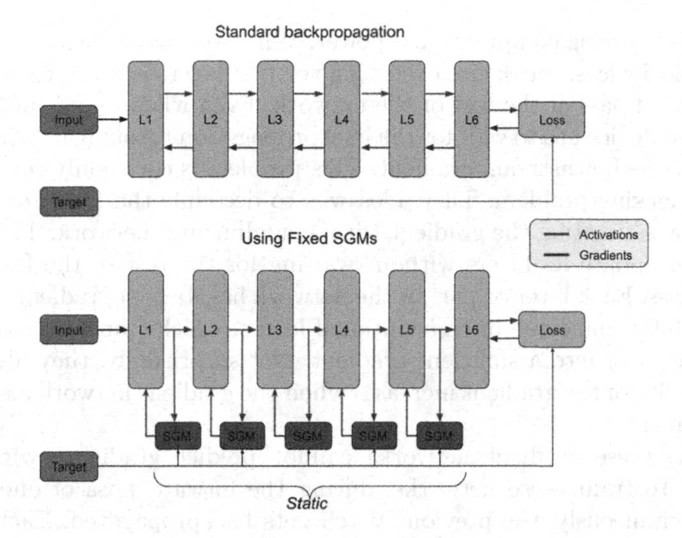

Fig. 1. Comparison between standard backpropagation (BP) and using static Synthetic Gradient Modules (SGMs). BP calculates layer activations during the forward pass, which are being used to calculate the gradients during the backward pass. In contrast, when using static SGMs, the gradients can directly be estimated based on each layer's activations and the input label. Therefore, the network can be trained in a feed-forward manner, bypassing the need to store the activations of the whole network.

objective of the model is to estimate the center and rotation of the object, meaning the setup can be seen as a multi-task setting: both a classification and regression problem. Therefore, this setup is ideal to show how our method can cope with different tasks. In the experiments and results section, we will show that we were able to successfully train the model to grasp new objects in a feed-forward manner using the static SGMs. By pretraining the SGMs in a meta-learning manner on a set of common household objects, the static SGMs were able to provide accurate updates to the model to learn new similar objects. We compared our method to standard backpropagation and showed only a slight accuracy drop.

To summarize, we used static SGMs to train a network in a feed-forward manner. Providing each layer with a gradient during the forward pass, this bypasses the update-locking problem and releases us from the memory burden of storing all activations. The SGMs were pretrained using a meta-learning setup and tested on a robot grasping scenario where we showed similar results to standard backprop.

2 Related Work

2.1 Training with Synthetic Gradients

The idea of training a network with synthetic gradients generated by small models was introduced by [9]. The goal of their work was to train a distributed network without wasting computational power. When you have a network deployed over multiple devices, the device executing the first layers is going to be idle during the forward pass of the rest of the network. Even when doing the backward pass, the first device has to wait for the backpropagation through all other devices before it gets its downstream gradient. This problem is commonly referred to as the update-locking problem. Their idea was to decouple the forward and backward pass by estimating the gradient with a small neural network. This way the device could update its layers without waiting for the rest of the forward and backward pass. Each layer or part of the network has its own gradient generator, one or two fully connected or convolutional layers, which transforms the activations of that layer into a sufficient gradient. Not surprisingly, they also showed that the quality of the gradient increases when the gradient network also receives the input label.

However, these gradient networks cannot predict gradients without any knowledge. To train these networks, during the forward pass of one batch of inputs, simultaneously, the previous batch gets backpropagated. Each transaction between devices exchange both the forward pass activations as well as the backpropagated gradient. This backpropagated gradient serves as ground truth for the gradient networks which they simply trained with an l2-loss. As shown in [9], as the target gradient distribution shifts, the gradient network follows this distribution. Therefore, there will always be some lag between the ground truth and predicted gradients, but they showed that this still is able to train the network sufficiently.

2.2 Training Without Backpropagation

Besides the use of synthetic gradients, other research explore different substitutes for standard backpropagation, more specifically the biologically plausible methods. One of these methods is the feedback alignment algorithm [13]. They argue that standard backpropagation in the brain is not possible because the forward and backward paths in the brain are physically separated (weight transport problem). To mimic this behaviour, they used fixed random feedback weights during the backward pass and showed that it could enable the network to learn.

One extension of this work is the direct feedback alignment method (DFA) [18]. Instead of backpropagating the error using fixed random weights, they use the fixed feedback weights to transform the final error of the network into a gradient for each layer. Therefore, once the error of the network is computed, each layer gets it gradient at the same time as a random transformation of the loss. One big advantage of this method is that after the forward pass, all layers can be updated in parallel. This somewhat solves the update-locking problem. However, this would require a large increase in hardware. Nevertheless, recent work showed that these algorithms scale to modern deep learning tasks [5,6,11,12,17]. Other work have also focused on reducing the fixed feedback weights with sparse connections [1] and the effect of direct feedback alignment on shallow networks [8].

All these previous methods use some transformation of the error of the network to provide a gradient to the network. [4] showed that we don't even need the final error of the network to train the upstream layers. They showed that, only the sign of the error is enough information to train the label. Having the fact that the target for classification tasks already contains the sign (one-hot encoded), they used the target as a proxy for the error of the network. Meaning, instead of computing the final error, each layer receives a random projection of the target as gradient. They showed that this indeed can support learning. This method has some huge advantages for hardware, since the network can now be trained in a feed-forward manner and reduces memory footprint since this solves the update-locking problem. Similar work have been conducted in [15] and [14].

2.3 Fine-Tuning Models with Meta-learning

If we compare the use of synthetic gradients with the bio plausible algorithms, there is not much difference between them. The first uses a network to generate gradients, which is constantly being trained with ground truth gradients. The second uses a fixed random projection of the final error or even the target. The main advantage of the synthetic gradients is that the network behaves in a feed-forward manner, solving the update-locking problem. However, these networks need to be trained. The only bio plausible algorithm from the ones discussed that really solves the update-locking problem is [4], however this works only well in classification problems and has worse performance than the synthetic gradients. For this reason, we want to combine the two by using the synthetic gradient setup, but with fixed parameters. Since this relieves us from training the synthetic gradient generators, the network can completely update during the forward pass. However, it is obvious the initialization of the gradient generators is of vital importance.

From [9], we saw that the distribution of the gradients and activations shift and that the gradient generators follow this distribution while they are trained with ground truth gradients. It is very difficult to pretrain a network which would incorporate this behaviour. Therefore, we only focus on fine-tuning a model to a new task. During fine-tuning, we expect that the distribution of the gradient does not shift that much, making it easier for the fixed gradient generators to be initialized.

To initialize the SGMs, we draw inspiration from the meta learning setting. More specifically, in [3] they introduced MAML, a model-agnostic meta learning algorithm which enables a model to be fine-tuned with fewer iterations. During training, the algorithm searches for a set of parameters which are equally distant to the optimal parameters of different tasks. If we want to fine-tune the model to a task which is similar to the ones trained on, the initialization parameters are closer to the optimal setting, resulting in fewer gradient descent steps compared to a completely random initialization. We believe that, we can use this algorithm to not only find an initialization for the model, but also for the SGMs.

3 Method

To test our method of fine-tuning a model in a feed-forward manner with static SGMs, we adapt the robot grasping setup from [16] and [2]. The goal of their setup is to fine-tune the model to learn to grasp new objects. Currently, they fine-tune the model on a GPU server. Therefore, this setup is ideal for our method, since our method can lower the total activation memory needed on the embedded device. In this section, we will first explore the robot grasping setup in more detail. Next, we will discuss the use of static SGMs and how these can be initialised by pretraining them using meta-learning.

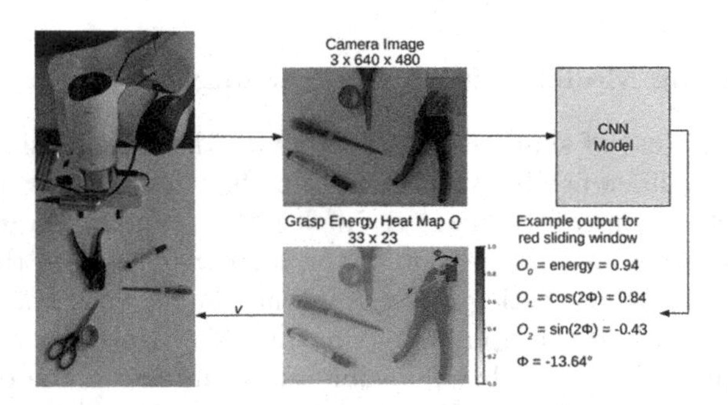

Fig. 2. Grasping setup from [16]. A cobot equipped with a camera runs a CNN model to predict a lower resolution heatmap. The heatmap contains three layers: a grasp quality layer and the rotation angle encoded in a sine and cosine layer. (Figure adapted from [16])

3.1 Robot Grasping Setup

To learn to grasp new objects, a collaborative robot (cobot) equipped with a camera is positioned above the new object. Next, the camera will take one shot of the object, having a frame where the object is centered and in the right angle. Then, to actually grasp the object, a demonstrator guides the end effector of the cobot to the correct grasping position. This uses the "program by demonstration" feature of the cobot, meaning the cobot can execute the same sequence of steps in future grasps. The goal of the computer vision algorithm is to position the cobot right above the object and rotated in the same angle as the demonstration frame. To achieve this, a fully convolutional network [20] takes in the camera image and outputs a displacement vector along with a rotation angle, which is provided to a controller as shown in Fig. 2. Instead of using a standard Cartesian controller, this results in a closed-loop "smart" controller. In this paper we will only focus on the computer vision task.

The Model. During deployment, the model takes in a $3 \times 640 \times 480$ image and outputs a heatmap of a lower resolution ($3 \times 33 \times 23$). Each pixel of the heatmap consists of a grasp quality score and a rotation angle (encoded in sine and cosine). As the network works in a fully convolutional manner (last layers are convolutional instead of dense layers), the model (Fig. 3) can be trained on crops of size $3 \times 128 \times 128$ and output a $3 \times 1 \times 1$ pixel. The grasp quality score resembles the centered position of the object in the crop (binary classification), the sine and cosine resembles the rotation angle with respect to the demonstration image (regression).

To learn both the binary classification and angle regression tasks, the grasp quality layer is trained by a log loss while both the rotation layers (sine and cosine) are trained with an l2 loss as in [7]. Since the end effector of the robot is bipodal, the rotational range of the cobot is $\pm\frac{\pi}{2}$. We incorporate this during the data generation by mapping all angles to this range. Also, to facilitate learning, we limit the output of the angle layers by adding a tanh and sigmoid layer to the sine and cosine respectively, limiting their range to (± 1) and (0, 1).

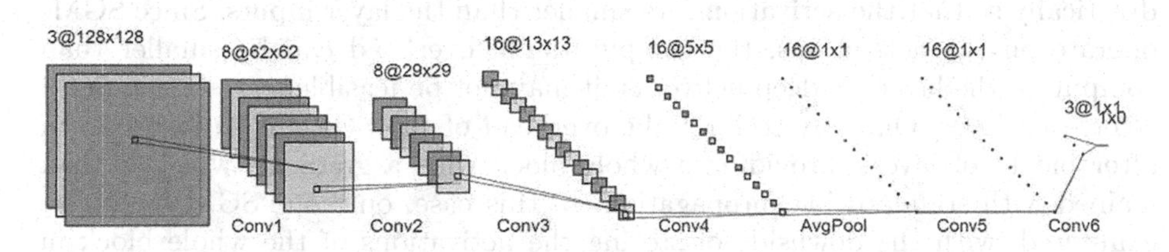

Fig. 3. The model used in [16]. The model works in a fully convolutional manner meaning the last layers are convolutional instead of dense. This way the model can be trained on crops of objects. During deployment, the model operates on full camera images and outputs a lower resolution heatmap as in Fig. 2

Dataset. In the setup from [16] they generate positive and negative samples from only one camera frame of a new object. Some frames of the dataset are depicted in Fig. 4. Positive examples are 128×128 crops, where the center of the object is in the center of the crop. Also, the object is randomly rotated so that the model can learn the rotation of the object during deployment. Negative examples are crops where the object is randomly rotated, but not situated in the center of the crop. Also, all crops use random color jitter and brightness to augment the data for better performance. In contrast to [16], instead of generating random samples on the fly, we use a fixed dataset of 1000 images per object: 800 train and 200 validation images.

3.2 Synthetic Gradient Modules

To fine-tune the model in a feed-forward manner, we will use SGMs to generate synthetic gradients for each layer as in [9]. However, during fine-tuning, these will

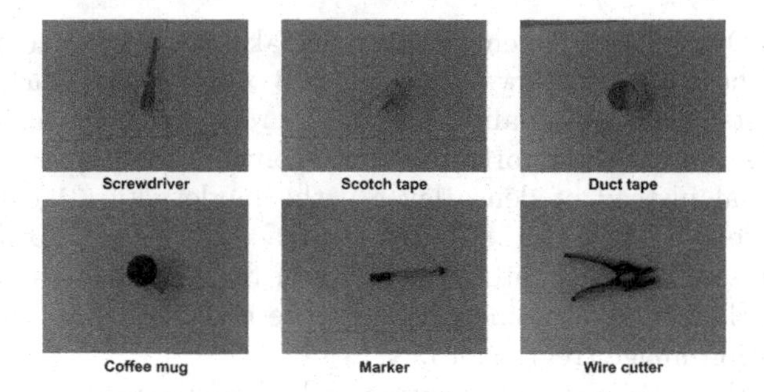

Fig. 4. Common household objects from [16]. Out of the single image of each object, positive and negative samples are generated.

remain static. The SGMs will estimate the gradient for each layer based on the activations and the target. It is obvious the SGMs will provide computational overhead. However, for the first layers of the network, the inputs are downscaled drastically so that the activations are smaller than the layer inputs. Since SGMs operate on the activations, the computational overhead can be smaller than computing the layer. In deep networks, it may not be feasible to insert an SGM after each layer. One way to limit the overhead of the SGMs is to insert them after blocks of layers, providing a whole block with a gradient, which is then trained with standard backpropagation. In this case, only one SGM has to be computed, with the downside of storing the activations of the whole block in memory.

Since the network (Fig. 3) consists of only six layers, each layer will have their own SGM. Therefore, we can use five SGMs (the last layer gets its gradient directly from the loss). Since all the activations are outputs of convolutional layers, our SGMs also consist of convolutional layers, so that they don't take up that much parameters in contrast to fully connected layers. The SGMs are implemented in Pytorch [19] as a regular layer, except that during the forward pass they compute and store the synthetic gradient. The output of the layer is the input, meaning during the forward pass it acts as a no-op. During the backward pass however, a hook function is executed where the gradient on the input is replaced with the synthetic gradient. This way, the optimizer will use the synthetic gradient to update the layer. During the pretraining step, the hook function will also update the parameters of the SGM, by using the incoming gradient of the output (which is the ground truth gradient for the activations, since the SGM acts as a no-op during the forward pass) as supervision. In all our experiments, the SGMs consist of two convolutional layers with a batchnorm and relu activation layer in between. Our experiments showed that adding the batchnorm layer resulted in a more stable training. Since during fine-tuning the SGMs are static, the batchnorm is only a shift and scale and can be folded into the convolution layer.

3.3 Training the SGMs

To train the SGMs, we will use the meta-learning algorithm MAML: model-agnostic meta learning [3]. The goal of the MAML algorithm is to find an optimal set of model parameters so that the model can be fine-tuned to a new task with as few iterations as possible. This results in finding a set of parameters which are equally distant to the optimal set of parameters for each individual task (see Fig. 5). However, instead of using standard backpropagation to train the model, we will use SGMs.

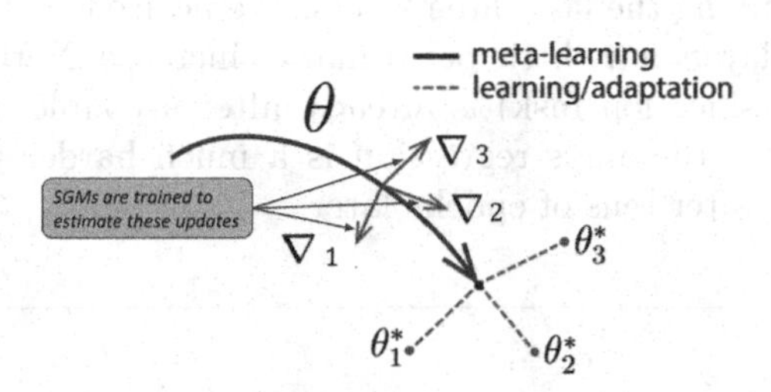

Fig. 5. The SGMs are pretrained using MAML. The difference with standard MAML is that we now use the SGMs to update the model instead of standard backpropagation. During training, the SGMs are able to train their parameters with the ground truth gradients like in [9]. This way, the SGMs learn to adapt the model to different tasks. After pretraining, we have shown that these SGMs can be left fixed during deployment to adapt the model to an unseen task, as long as the task has similarities to the training tasks. (Figure adapted from [3])

During the MAML step, the model is trained on different tasks individually, while minimizing the summed error on all tasks. Each task consists of training the model on one object like in Fig. 4. Since the model is trained by the SGMs, the SGMs learn to provide gradients to fine-tune the model towards all the different objects. Since the goal of MAML is to find parameters which can quickly adapt to new tasks, at the end of the meta-learning step, the model parameters are close to the optimal set of parameters for each task, while the SGMs have learned to provide the gradients to fine-tune the model on all these different tasks.

We will use the end state of the SGMs after the MAML step as initialization. During fine-tuning, these will remain static. Since the goal of MAML is to initialize the model with parameters close to the optimal parameters, during fine-tuning, the gradient distribution will not shift significantly, meaning the SGMs can indeed remain static. However to prove this claim, in the experiment section we will compare the performance of fine-tuning the model using static SGMs and letting them update with a ground truth gradient.

4 Experiments and Results

4.1 Fine-Tuning Using Standard Backpropagation

As a baseline, we compared the performance of our method to standard backpropagation. For each object in the dataset, we first pretrained the network on the remaining objects. Next, we fine-tuned the model on the object itself. For these experiments we used the Adam optimizer [10] with a learning rate of 1×10^{-3} with a batch size of 32. We compared the amount of fine-tuned layers ranging from only fine-tuning the last layer to all layers. Figure 6 shows the result for fine-tuning the last three layers. We noticed that fine-tuning more than the last three layers have little performance increase. Notice also that grasp quality (binary classification task) is already after a couple of epochs close to 100% accuracy, while the angle regression is a much harder task to learn and reaches a minimum after tens of epochs later.

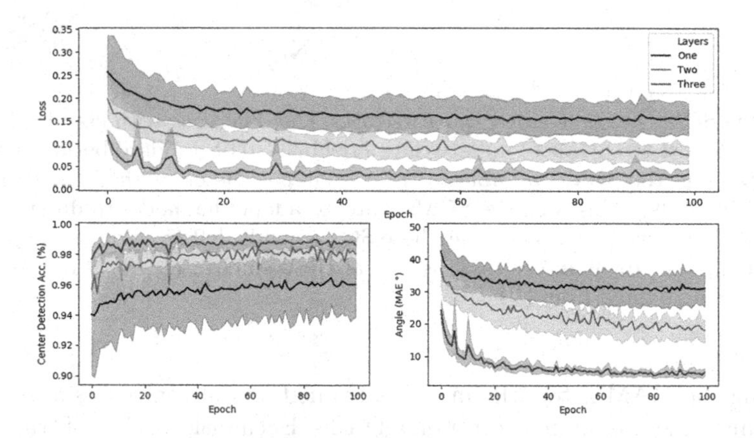

Fig. 6. Baseline using standard backpropagation. Comparison between fine-tuning the last one, two or three layers of the model. Showing the loss, detection accuracy and mean absolute error averaged for fine-tuning on six different objects' test set. For all objects, fine-tuning three layers has the highest performance. Fine-tuning more than the last three layers resulted in minimal performance increase

4.2 Fine-Tuning Using Static SGMs

We will insert SGMs after each layer (conv-bn-relu) of the network. The input of these SGMs consists of the activation of the network layer along with the label. Since in our setup the label consists of a binary grasp quality score and an angle encoded in a sine and cosine, we found out that adding three extra channels to the input activations works best: each channel filled with the grasp quality, sine and cosine. Initially, we tried to train the model without using the label as input to the SGMs. While this was able to learn the grasp quality, it was not able to learn to regress the rotation angle.

Training the SGMs. To train the SGMs during the meta-learning step, we experimented with both the l1 and l2 loss functions. While both losses were able to train the SGMs to provide accurate gradients for fine-tuning the model, we did notice an increase in accuracy when using the SGMs trained with the l2-loss. We use standard stochastic gradient descent with a learning rate of 1×10^{-1} and momentum of 0.9 to train both SGMs and the model.

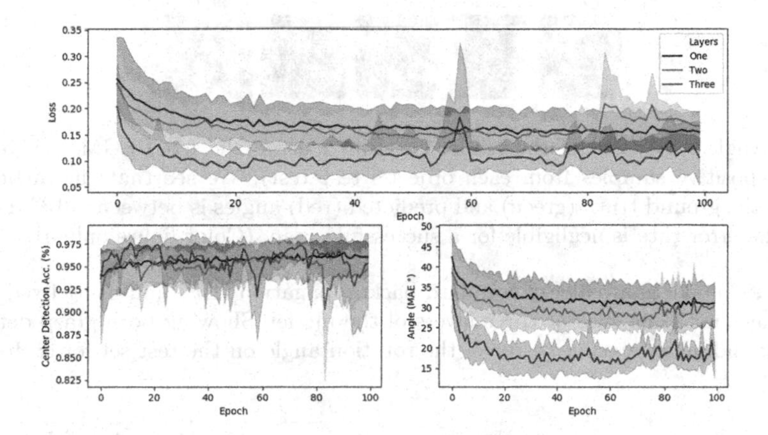

Fig. 7. Fine-tuning the model using fixed SGMs. Comparison between fine-tuning the last one, two or three layers of the model. Showing mean results for each object's test set. As with standard backprop, fine-tuning the last three layers has the highest performance. However, for all three cases, the detection accuracy is lower and angle error higher than using standard backprop (Fig. 6) by a small amount. While with standard backprop the accuracy did not increase when fine-tuning more than three layers, the performance dropped when using static SGMs (not shown in the graphs).

Fine-Tuning Using Static SGMs. In these experiments we initialized the model and SGMs with their pretrained state while keeping the SGMs static. During the forward pass, each layer gets its gradient directly from its designated SGM, fine-tuning the model in a feed-forward manner. Again, both SGD and Adam optimizer were compared with SGD performing slightly better with the same learning rate as during the pretraining step of 1×10^{-1}.

Figure 7 shows the model is able to be fine-tuned to new objects using the static SGMs. As with standard backpropagaton, the classification task (detection) is able to learn very quickly, while the angle regression needs to train considerably more epochs. The figure also shows that the combined loss converges slower than standard backprop. There also is a significant difference in the MAE of the angle. However, for grasping simple household objects with a bipodal end effector, MAE differences of around 10° will most of the time result in a successful grasp as seen in Fig. 8. Fine-tuning the last three layers resulted in the highest accuracy for both the classification and regression task. When fine-tuning more than three layers, the accuracy dropped below the three-layer experiment. Detailed results can be found in Table 1.

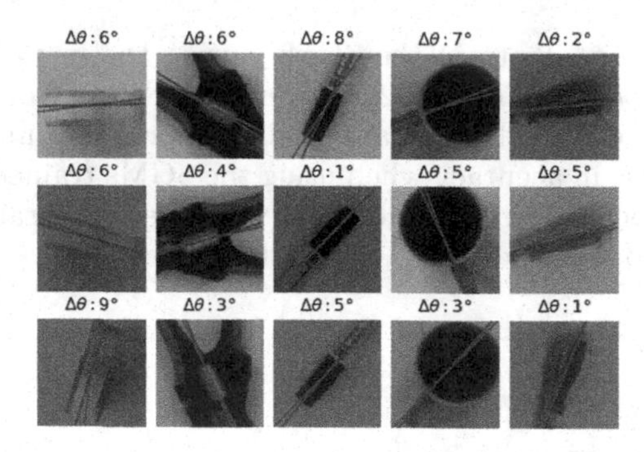

Fig. 8. Angle regression of the model fine-tuned using the fixed SGMs on different objects (positive samples from each object's test test). We see that the angle error between the ground truth (green) and predicted (red) angles is between 0–13°. For this setup, the error rate is negligible for a successful grasp. (Color figure online)

Table 1. Comparison between standard backpropagation (BP) and using fixed SGMs when fine-tuning one, two or three layers of the model. Showing both grasp detection accuracy and mean absolute error of the rotation angle on the test set for each object (a–f).

	Layers	Method	a	b	c	d	e	f	Avg	Std
Center detection (%)	1	BP	0.92	0.98	0.98	0.98	0.99	0.98	0.97	0.02
	2	BP	0.97	0.98	0.99	0.99	0.99	1.00	0.99	0.01
		SGM	0.92	0.97	0.98	0.99	0.99	0.99	0.97	0.02
	3	BP	0.99	0.99	1.00	0.99	1.00	1.00	1.0	0.0
		SGM	0.97	0.99	1.00	0.99	0.99	0.99	0.99	0.01
Angle regression (MAE °)	1	BP	37.19	23.04	33.72	25.62	35.49	16.20	28.71	6.99
	2	BP	18.19	17.65	13.55	20.56	20.76	6.63	16.44	4.57
		SGM	29.21	21.34	21.09	24.37	33.25	17.48	24.07	5.01
	3	BP	5.71	2.44	3.82	2.73	3.02	3.04	3.45	1.01
		SGM	16.82	8.59	15.32	14.35	9.99	13.71	12.65	2.94

Trained vs. Random Initialisation. In the target propagation methods like [4] and [15], the gradient is generated by a fixed transformation of the target. Since the SGMs also have the target as input, one could argue a random initialization might be able to train the network. However, to show the contribution of the trained state, we ran experiments using random static SGMs. We concluded that with a random state, the SGMs were not able to provide the model with accurate gradients, failing to converge on both the classification and regression task.

4.3 Static vs. Non-static SGMs

The main goal of this paper is to fine-tune a network with static SGMs on an edge device, eliminating the need of storing all activations during the forward pass (update-locking problem). This has the advantage that during deployment the SGMs can provide the gradients on the fly. However, since these are trained to fine-tune the model to different tasks, the gradients must generalize and differ from the ground truth gradients. Therefore, in a last experiment we wanted to show the effect of further training the SGMs during fine-tuning with the ground truth gradient. More specifically, we wanted to show if the trained initialization for the SGMs has any benefit when they are allowed to further update. Table 2 summarizes three training methods: (i) with static pretrained SGMs, (ii) with updated pretrained SGMs, and (iii) with updated randomly initialized SGMs.

Table 2. Comparison between fine-tuning two layers using static or non-static SGMs. Enabling the trained SGMs to further train during fine-tuning results in a slight accuracy improvement. However, when the SGMs are able to further train, there is no clear difference between a pretrained or random initialization.

	Init.	Static	a	b	c	d	e	f	Avg	Std
Center detection (%)	Trained	✓	0.92	0.97	0.99	0.99	0.99	0.96	0.97	0.03
		✗	0.96	0.98	0.98	0.99	0.99	0.98	0.98	0.01
	Random	✗	0.97	0.98	0.98	0.99	0.99	0.97	0.98	0.01
Angle regression (MAE °)	Trained	✓	29.21	21.34	24.37	33.25	17.48	21.77	24.57	5.25
		✗	26.20	20.05	25.54	28.95	8.30	19.58	21.44	6.76
	Random	✗	23.32	18.87	26.53	30.56	13.27	22.22	22.46	5.48

Table 2 shows that enabling the SGMs to further train during fine-tuning, the model achieves higher accuracy. This is obvious, since the SGMs can now follow the ground truth distribution of the gradients. However, the difference between the pretrained and randomly initialized SGMs is remarkable. Having the latter to perform equally well, meaning the pretrained state does not contribute that much. This possibly means that since the SGMs are only a couple of parameters, the SGMs can quickly adapt to the new target gradient distribution, both for the pretrained as well as the random initialization.

Impact on Memory. The goal of using static SGMs is to minimize the activation memory to the activation size of the largest layer. When enabling the SGMs to further train during the fine-tuning stage instead of remaining static, we also need to store its activations. Since the activations of the SGMs are the same size as the activations of the layer it provides gradients for, one can optimize which layer can update or freeze its SGM. This is something we did not investigate further in this paper.

4.4 Memory Advantage

As we have shown in this paper, having the static SGMs to provide gradients during the forward pass allows us to discard layer activations once updated. This means there is no storage and transportation to an external memory, which is responsible for a large proportion of the power consumption on embedded devices. It is clear the memory advantage increases when fine-tuning more layers. In our setup, the best performance was achieved when fine-tuning the last three layers. This resulted in saving over 81KB of transport to an external memory device. However, the use of the static SGMs introduce computational overhead which can be seen as a trade-off for the memory advantage. Table 3 shows that the computational overhead for fine-tuning the last three layers using static SGMs is around 80K MACs, which, compared to the total MAC operations for these layers during the forward pass of around 10.3M MACs, is only an increase of less than 1%. This is mainly due to the fact the SGMs operate on the layer activations, which can be significantly down scaled compared to the layer inputs. For the earlier layers of the model, where the activations are roughly the same size of the inputs, the computational overhead of the SGMs can have a greater impact. Nevertheless, this can be avoided by providing SGMs to groups of layers.

Table 3. Resource usage of model layers and SGMs.

Layer	Model			SGM	
	MAC	Param	Activations	MAC	Param
1	153,600,000	624	1,968,128	340,480,000	1400
2	87,680,000	1620	430,592	74,880,000	1400
3	35,130,240	3250	173,056	55,206,400	5100
4	10,318,080	6450	77,824	39,936	624
5	19,456	304	3072	39,936	624
6	3,264	51	192	x	x

When fine-tuning larger networks with more challenging tasks, the memory advantage will be higher, as these networks have both more layers and larger activations. Also, since the SGMs are static, it is possible to quantize these networks, lowering both the computational and memory overhead. However, this can be hard without an accurate calibration set, which can differ greatly when fine-tuning on tasks which show little resemblance to the pretrained tasks. We will investigate this in future work.

5 Conclusion

In this work towards memory-efficient on-edge training, we have been able to successfully fine-tune a model using static SGMs. By first training the SGMs on a set of tasks using MAML, while remaining static, these were able to provide accurate updates to fine-tune the model to new similar tasks. By testing our method on a multi-task robot grasping scenario, we showed comparable results to standard backpropagation both for a classification and a regression task. In further work, we will investigate the performance of our method on more challenging tasks like object detection and segmentation. Also, to lower the computational and memory overhead further, we will investigate quantizing the SGMs and the effect on the accuracy of the generated gradients.

Acknowledgement. This research received funding from the Flemish Government (AI Research Program).

References

1. Crafton, B., Parihar, A., Gebhardt, E., Raychowdhury, A.: Direct feedback alignment with sparse connections for local learning. Front. Neurosci. **13**, 525 (2019)
2. De Coninck, E., Verbelen, T., Van Molle, P., Simoens, P., Dhoedt, B.: Learning robots to grasp by demonstration. Robot. Auton. Syst. **127**, 103474 (2020)
3. Finn, C., Abbeel, P., Levine, S.: Model-agnostic meta-learning for fast adaptation of deep networks. In: Precup, D., Teh, Y.W. (eds.) Proceedings of Machine Learning Research, vol. 70, pp. 1126–1135. PMLR, International Convention Centre, Sydney (2017)
4. Frenkel, C., Lefebvre, M., Bol, D.: Learning without feedback: direct random target projection as a feedback-alignment algorithm with layerwise feedforward training (2019)
5. Han, D., Park, G., Ryu, J., jun Yoo, H.: Extension of direct feedback alignment to convolutional and recurrent neural network for bio-plausible deep learning (2020)
6. Han, D., Yoo, H.J.: Direct feedback alignment based convolutional neural network training for low-power online learning processor. In: 2019 IEEE/CVF International Conference on Computer Vision Workshop (ICCVW), pp. 2445–2452 (2019)
7. Hara, K., Vemulapalli, R., Chellappa, R.: Designing deep convolutional neural networks for continuous object orientation estimation. CoRR abs/1702.01499 (2017)
8. Illing, B., Gerstner, W., Brea, J.: Biologically plausible deep learning - but how far can we go with shallow networks? Neural Netw.: Off. J. Int. Neural Netw. Soc. **118**, 90–101 (2019)
9. Jaderberg, M., et al.: Decoupled neural interfaces using synthetic gradients. In: ICML 2017, pp. 1627–1635. JMLR.org (2017)
10. Kingma, D.P., Ba, J.: Adam: a method for stochastic optimization. In: Bengio, Y., LeCun, Y. (eds.) 3rd International Conference on Learning Representations, ICLR 2015, San Diego, CA, USA, 7–9 May 2015, Conference Track Proceedings (2015)
11. Launay, J., Poli, I., Boniface, F., Krzakala, F.: Direct feedback alignment scales to modern deep learning tasks and architectures (2020)
12. Lechner, M.: Learning representations for binary-classification without backpropagation. In: International Conference on Learning Representations (2020)
13. Lillicrap, T.P., Cownden, D., Tweed, D.B., Akerman, C.J.: Random synaptic feedback weights support error backpropagation for deep learning. Nat. Commun. **7**, 1–10 (2016)
14. Manchev, N.P., Spratling, M.W.: Target propagation in recurrent neural networks. J. Mach. Learn. Res. **21**, 7:1–7:33 (2020)
15. Meulemans, A., Carzaniga, F.S., Suykens, J.A.K., Sacramento, J., Grewe, B.F.: A theoretical framework for target propagation (2020)
16. Molle, P.V., Verbelen, T., Coninck, E.D., Boom, C.D., Simoens, P., Dhoedt, B.: Learning to grasp from a single demonstration. CoRR abs/1806.03486 (2018)

17. Moskovitz, T.H., Litwin-Kumar, A., Abbott, L.F.: Feedback alignment in deep convolutional networks. ArXiv abs/1812.06488 (2018)
18. Nøkland, A.: Direct feedback alignment provides learning in deep neural networks. In: Proceedings of the 30th International Conference on Neural Information Processing Systems, NIPS 2016, pp. 1045–1053. Curran Associates Inc., Red Hook (2016)
19. Paszke, A., et al.: PyTorch: an imperative style, high-performance deep learning library. In: Wallach, H., Larochelle, H., Beygelzimer, A., d' Alché-Buc, F., Fox, E., Garnett, R. (eds.) Advances in Neural Information Processing Systems, vol. 32, pp. 8024–8035. Curran Associates, Inc. (2019)
20. Shelhamer, E., Long, J., Darrell, T.: Fully convolutional networks for semantic segmentation. IEEE Trans. Pattern Anal. Mach. Intell. **39**(4), 640–651 (2017)
21. Verhelst, M., Murmann, B.: Machine learning at the edge. In: Murmann, B., Hoefflinger, B. (eds.) NANO-CHIPS 2030. The Frontiers Collection, pp. 293–322. Springer, Cham (2020). https://doi.org/10.1007/978-3-030-18338-7_18

W29 - Real-World Computer Vision from Inputs with Limited Quality (RLQ)

W29 - Real-World Computer Vision from Inputs with Limited Quality (RLQ)

The 2nd International Workshop and Challenge on Real-World Computer Vision from Inputs with Limited Quality (RLQ) was held virtually on August 28, 2020 in conjunction with the European Conference on Computer Vision (ECCV). The main workshop program consisted of invited talks, an invited panel discussion, and workshop papers. Alongside the workshop, we also co-hosted two challenges: Tiny Object Detection (TOD) and the Under-Display Camera (UDC). The invited talks are one of the highlights of our workshop program. We had six extraordinary invited keynote talks that covered multiple aspects of real-world computer vision:

- Imaging the Invisible, Katie Bouman, Caltech
- Recognition from Low Quality Data, Rama Chellappa, Johns Hopkins University
- Designing Cameras to Detect the "Invisible": Computational Imaging for Adverse Conditions, Felix Heide, Princeton University
- Computation + Photography: How the Mobile Phone Became a Camera, Peyman Milanfar, Google Research
- Increasing Quality of Active 3D Imaging, Srinivasa Narasimhan, CMU
- Compositional Models and Occlusion, Alan L. Yuille, Johns Hopkins University

Recorded talks can be found at: https://www.youtube.com/c/HonghuiShi.

For the workshop proceedings, we have included 11 workshop papers (peer-reviewed, acceptance rate <50%) and 8 reports on the two challenges and their top solutions (reviewed by challenge organizers, 3 for TOD and 5 for UDC).

We would like to sincerely thank all our invited keynote speakers: Katie Bouman, Rama Chellappa, Felix Heide, Peyman Milanfar, Srinivasa Narasimhan, and Alan L. Yuille, and our invited panelists: Gang Hua from Wormpex AI Research and Lei Zhang from Microsoft. We would also like to thank all authors and challenge participants for their hard work and valuable contributions. We hope that all the attendees enjoyed the workshop and found it beneficial. The organizers will continue their efforts in pushing forward real-world computer vision research through future activities including the next RLQ workshop. Lastly, our deepest appreciation to our workshop advisor Prof. Thomas S. Huang, who passed away in April 2020 following his late wife Margaret. May they rest in peace!

August 2020

Humphrey Shi
Atlas Wang
Yunchao Wei
Yifan Jiang
Ding Liu
Jiaying Liu
Zhenjun Han
Xuhui Yu
Jian Zhao
Qixiang Ye
Jie Chen
Matti Pietikäinen
Yuqian Zhou
Thomas S. Huang
Jeffrey Cohn
Nicu Sebe

Reinforcement Learning for Improving Object Detection

Siddharth Nayak(✉) and Balaraman Ravindran

Indian Institute of Technology Madras, Chennai, TN, India
siddharthnayak98@gmail.com, ravi@cse.iitm.ac.in

Abstract. The performance of a trained object detection neural network depends a lot on the image quality. Generally, images are pre-processed before feeding them into the neural network and domain knowledge about the image dataset is used to choose the pre-processing techniques. In this paper, we introduce an algorithm called ObjectRL to choose the amount of a particular pre-processing to be applied to improve the object detection performances of pre-trained networks. The main motivation for ObjectRL is that an image which looks good to a human eye may not necessarily be the optimal one for a pre-trained object detector to detect objects.

Keywords: Reinforcement learning · Object detection · Camera parameters

1 Introduction

With the advent of convolutional neural networks, object detection in images has improved significantly giving rise to several object detection algorithms like YOLO [24], SSD [16], etc. Most object detection networks work with raw image pixels as inputs. The networks are highly nonlinear in nature and thus the output predictions depend a lot on the image parameters like brightness, contrast, etc. [15,17,20,21]. In real-world scenarios, camera parameters like the shutter-speeds, gains, etc. with which the images are taken, matter a lot in the performance of an object detection network. A photographer changes a lot of parameters like the shutter speed, voltage gains, etc. [2] while capturing images according to the lighting conditions and the movements of the subject. In autonomous navigation, robotics, etc. there are several instances where the lighting conditions and the subject speed changes. In these cases, using fixed shutter speed and voltage-gain values would result in an image which would not be conducive for object detection. Most cameras rely on the built-in auto-exposure algorithms to set the exposure parameters of the camera. Although the images obtained from these auto-exposure algorithms may be *pleasing* to a human eye, they may not be the best image to perform object detection on. Also, most of the object detection networks are trained using images from a dataset which are captured either by using a single operation mode [7] or no control over the parameters of the camera

A. Bartoli and A. Fusiello (Eds.): ECCV 2020 Workshops, LNCS 12539, pp. 149–161, 2020.
https://doi.org/10.1007/978-3-030-68238-5_12

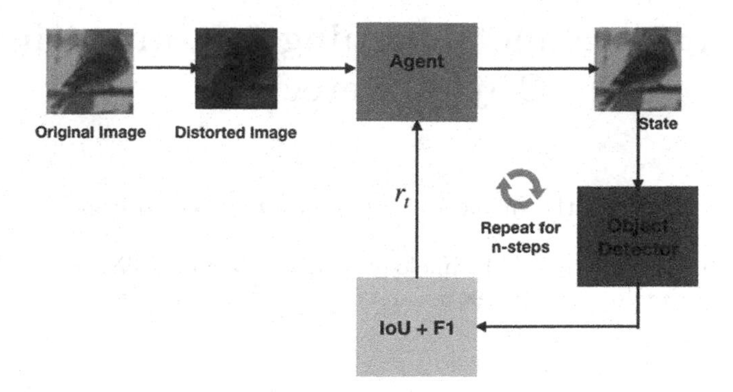

Fig. 1. The overall training procedure for *ObjectRL*. The image is randomly distorted to simulate the bad images. An episode can be carried out for n steps which we set to 1 for training stability. Thus, the agent has to take a single action on each image.

[1,5,11]. Thus, a pre-trained network may have a larger affinity towards images captured with similar parameters as the ones in the dataset it was trained on.

To tackle the problem of sudden variations in the photography conditions, we propose to train a Reinforcement Learning (RL) agent to digitally transform images in real-time such that the object detection performance is maximised. Although we perform experiments with digital transformations, this method can ideally be extended to choose the camera parameters to capture the images by using the image formation model proposed by Hassinoff et al. [9]. We train the model with images which are digitally distorted, for example: changing brightness, contrast, color, etc. It should be noted that we do not necessarily want the agent to recover the original image.

The claimed contribution of the paper is a Deep RL methodology called *ObjectRL* (Object Reinforcement Learning) to change the image digitally with rewards based on the performance of a pre-trained object detector on the agent-transformed image. An overview of the related work is provided in the next section. The proposed method for *ObjectRL* is described in detail in Sect. 4 and the experiments to validate the hypotheses along with results are provided in Sect. 5 and Sect. 6 respectively.

2 Related Works

We briefly review the literature and the existing methods related to image modifications for object detection improvement.

Bychkovsky et al. [3] present a dataset of input and retouched image pairs called MIT-Adobe FiveK, which was created by professional experts. They use this dataset to train a supervised model for color and tone adjustment in images. The main motive of this work is not inclined towards improving object detection but is more focused towards training a model to edit an image according to the user preferences.

In [28] the authors create a dataset of images taken with different combinations of shutter speeds and voltage gains of a camera. They create a performance table which is a matrix of mean average precision (mAP) for detection of objects in images taken with different combinations of shutter speed and gains. To choose the optimal parameters to capture images, they propose to choose the combination which gives the maximum precision. One of the problems with this method is that a dataset with images taken with different combinations of shutter speeds, voltage gains and illuminations has to be manually annotated with bounding boxes around the objects which is quite time-consuming. Also, the dataset consists of images with static objects. Thus, the effect of changing shutter speed is just on the overall brightness of the image. But one of the main reasons for changing shutter speed while capturing images is to increase (for artistic purposes) or (preferably) decrease motion blur in the moving objects.

In [22] the authors propose a reinforcement learning based method to recover digitally distorted images. The authors model the agent to take actions sequentially by choosing the type of modification (brightness, contrast, color saturation, etc.). The main motive of this model is to recover back the distorted images. The reward for the agent is the difference of mean square difference of the images at the current time step and the previous time step. This work is quite different from our *ObjectRL* model as our main motive is to maximise the object detection performance of a pre-trained detector.

Reinforcement Learning has been used in conjunction with computational photography in recent works by Yang et al. [29] and Hu et al. [10] where the authors train RL agents to either capture images or post-process images in such a way that the resultant image is *visually pleasing*. The agent gets a reward from the users according to their preferences of exposures on cameras in the former one whereas in the later one the agent receives a reward based on the discriminator loss of a Generative Adversarial Network [8].

Another area of research orthogonal to ours is using reinforcement learning to obtain region proposals for object-detection and object-localization [4,13,18,19]. In these works, the main motivation is to make the agent focus its attention toward candidate regions to detect objects by sequentially shifting the proposed region and rewarding the agent according to the Intersection over Union (IoU–explained in Sect. 3.2).

3 Background

3.1 Reinforcement Learning

Reinforcement learning (RL) tries to solve the sequential decision problems by learning from trial and error. Considering the standard RL setting where an agent interacts with an environment $\mathcal{E}$ over discrete time steps. In the time step t, the agent receives a state $s_t \in \mathcal{S}$ and selects an action $a_t \in \mathcal{A}$ according to its policy π, where $\mathcal{S}$ and $\mathcal{A}$ denote the sets of all possible states and actions respectively. After the action, the agent observes a scalar reward r_t and receives the next state s_{t+1}. The goal of the agent is to choose actions to maximize the cumulative sum of rewards over time. In other words, the action selection implicitly considers the future rewards. The discounted return is defined as $R_t = \sum_{\tau=t}^{\infty} \gamma^{\tau-t} r_\tau$, where $\gamma \in [0, 1]$ is a discount factor that trades-off the importance of recent and future rewards.

RL algorithms can be divided into two main sub-classes: Value-based and Policy-based methods. In value-based methods, values are assigned to states by calculating an expected cumulative score of the current state. Thus, the states which get more rewards, get higher values. In policy-based methods, the goal is to learn a map from the states to actions, which can be stochastic as well as deterministic. A class of algorithms called actor-critic methods [14] lie in the intersection of value-based methods and policy-based methods, where the critic learns a value function and the actor updates the policy in a direction suggested by the critic.

Proximal Policy Optimization (PPO): We use PPO [26] which is a type of actor-critic method for optimising the RL agent. One of the key points in PPO is that it ensures that a new update of the current policy does not change it too much from the previous policy. This leads to less variance in training at the cost of some bias, but ensures smoother training and also makes sure the agent does not go down an unrecoverable path of taking unreasonable actions. PPO uses a clipped surrogate objective function which is a first order trust region approximation. The purpose of the clipped surrogate objective is to stabilize training via constraining the policy changes at each step.

3.2 Object Detection

Object recognition is an essential research direction in computer vision. Most of the successful object recognition algorithms use deep convolutional neural networks which are trained to give the co-ordinates of the bounding boxes around the objects. To decide whether an object is detected or not, we use the Intersection over Union (IoU) criteria. Intersection over Union is the ratio of area of overlap and area of union of the predicted and the ground truth bounding boxes. Let p be the predicted box, and g be the ground truth box for the target object. Then, IoU between p and g is defined as $IoU(p, g) = Area(p \cap g)/Area(p \cup g)$. Generally, if $IoU > 0.5$ an object is said to be a True-Positive.

3.3 Image Distortions

Different parameters of an image like brightness, contrast and color can be changed digitally. We describe the formulae used to transform the pixel intensity (I) values at the co-ordinates(x, y). We assume distortion factor $\alpha \geq 0$

- Brightness: The brightness of an image can be changed by a factor α as follows:
  $I(x, y) \leftarrow \min(\alpha I(x, y), 255))$

- Color: The color of an image is changed by a factor α as follows: We evaluate the gray-scale image as:
  $gray = (I(r) + I(g) + I(b))/3$, where I(r), I(g) and I(b) are the R, G & B pixel values respectively.
  $I(x, y) \leftarrow \min(\alpha I(x, y) + (1 - \alpha)gray(x, y), 255)$
- Contrast: The contrast in an image is changed by a factor α as follows:
  $\mu_{gray} = mean(gray)$
  $I(x, y) \leftarrow \min(\alpha I(x, y) + (1 - \alpha)\mu_{gray}, 255)$.

4 Model

Given an image, the goal of *ObjectRL* is to provide a digital transformations which would be applied to the input image. This transformed image should extract maximum performance (F1 score) on object detection when given as an input to a pre-trained object detection network.

4.1 Formulation

We cast the problem of image parameter modifications as a Markov Decision Process (MDP) [23] since this setting provides a formal framework to model an agent that makes a sequence of decisions. Our formulation considers a single image as the state. To simulate the effect of *bad* images as well as increase the variance in the images in a dataset, we digitally distort the images. These digital distortions are carried out by randomly choosing α for a particular type of distortion (brightness, contrast, color). We have a pre-trained object detection network which could be trained either on the same dataset or any other dataset.

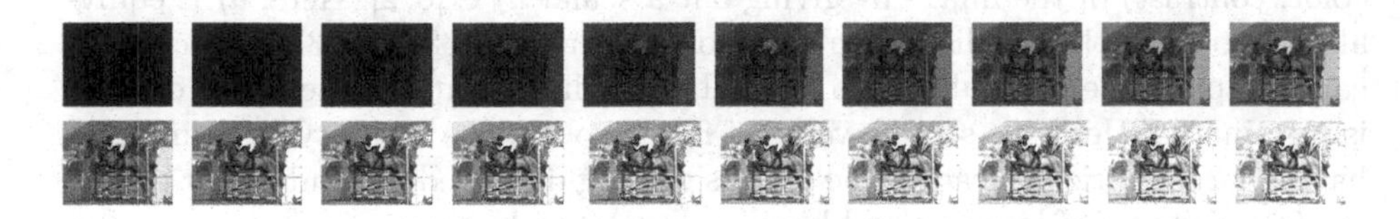

Fig. 2. Variation in images with varying brightness distortion factor α from 0 to 2 in steps of 0.1.

4.2 ObjectRL

Formally, the MDP has a set of actions $\mathcal{A}$, a set of states $\mathcal{S}$ and a reward function $\mathcal{R}$ which we define in this section.

States: The states for the agent are $128 \times 128 \times 3$ RGB images from the PascalVOC dataset [6] which are distorted by random factors α chosen according to the scale of distortion. We consider only one type of distortion (brightness, color, contrast) at a time, ie. we train different models for different types of

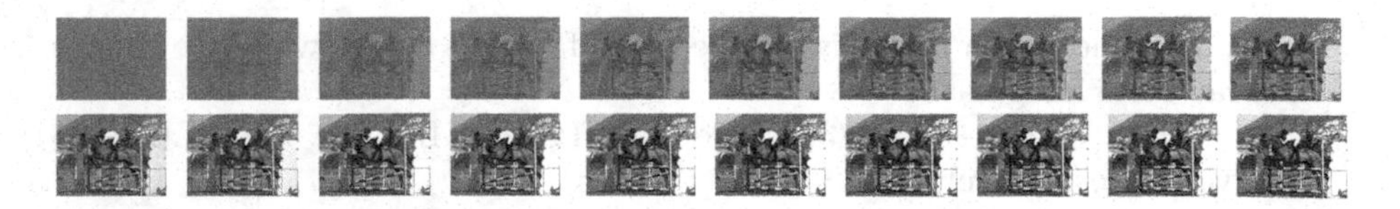

Fig. 3. Variation in images with varying contrast distortion factor α from 0 to 2 in steps of 0.1.

Fig. 4. Variation in images with varying color distortion factor α from 0 to 2 in steps of 0.1.

the distortion. Combining all the different types of distortions in a single model remains to be a key direction to explore in future work.

Scales of Distortion: We perform experiments with the following two degrees of distortion in the image:

- Full-scale distortion: The random distortion in the images $\alpha \in [0, 2]$.
- Minor-scale distortion: The random distortion in the images $\alpha \in [0.5, 1.8]$. This constraint limits the images to not have distortions which cannot be reverted back with the action space, the agent has access to.

The variation of the the distorted images can be seen in Fig. 2, 3, 4.

Actions: The agent can choose to change the global parameter (brightness, color, contrast) of the image by giving out a scalar $a_t \in [0, 2]$. Here, a_t is equivalent to α in the image distortion equations described in Sect. 3.3. The action a_t can be applied sequentially upto n number of times. After n steps the episode is terminated. Here, we set the value of $n = 1$ to achieve stability in training as having larger horizons lead to the images getting distorted beyond repair during the initial stages of learning and hence does not explore with the *better* actions.

Reward: First, we evaluate scores d_t for the images as follows:

$$d_t(x) = \gamma(IoU(x)) + (1 - \gamma)(F1(x)) \tag{1}$$

x is the input image to the pre-trained object detector. IoU is the average of all the intersection over union for the bounding boxes predicted in the image and F1 is the F1-score for the image. We set $\gamma = 0.1$ because we want to give more importance to the number of correct objects being detected.

We evaluate:

- $d_{o,t} = d_t(\text{original image})$
- $d_{d,t} = d_t(\text{distorted image})$
- $d_{s,t} = d_t(\text{state})$

where the *original image* is the one before the random distortion, *distorted image* is the image after the random distortion and *state* is the image obtained after taking the action proposed by the agent.

We define,

$$\beta_t = 2d_{s,t} - d_{o,t} - d_{d,t} \tag{2}$$

Here, β_t is positive if and only if the agent's action leads to an image which gives better detection performance than both the original image as well as the distorted image. Thus we give the reward (r_t) as follows:

$$r_t = \begin{cases} +1, & \text{if } \beta_t \geq -\epsilon \\ -1, & \text{otherwise} \end{cases}$$

Note that $d_{o,t}$ and $d_{d,t}$ do not change in an episode and only $d_{s,t}$ changes over the episode. We set the hyperparameter $\epsilon = 0.01$ as we do not want to penalise the minor shifts in bounding boxes which result in small changes in IoU in Eq. (1). Figure 1 shows the training procedure for *ObjectRL*.

4.3 Motivation for ObjectRL

In scenarios where object-detection algorithms are deployed in real-time, for example in autonomous vehicles or drones, lighting conditions and subject speeds can change quickly. If cameras use a single operation mode, the image might be quite blurred or dark and hence the image obtained may not be ideal for performing object detection. In these cases it would not be possible to create new datasets with images obtained from all the possible combinations of camera parameters along with manually annotating them with bounding-boxes. Also, due to the lack of these annotated images we cannot fine-tune the existing object-detection networks on the distorted images. Our model leverages digital distortions on existing datasets with annotations to learn a policy such that it can tackle changes in image parameters in real-time to improve the object detection performance.

One of the main motivations of *ObjectRL* is to extend it to control camera parameters to capture images which are good for object detection in real time. Thus, we propose an extension to *ObjectRL* (for future work) where we have an RL agent which initially captures images by choosing random combinations of camera parameters (exploration phase). A human would then give rewards according to the objects detected in the images in the current buffer. These rewards would then be used to update the policy to improve the choice of camera parameters. This method of assigning a $\{\pm 1\}$ reward is comparatively much faster than annotating the objects in the image to extend the dataset and training a supervised model with this extended model. This methodology is quite similar to the DAgger method (Dataset Aggregation) by Ross et al. [25] where a human labels the actions in the newly acquired data before adding it into the experience for imitation learning.

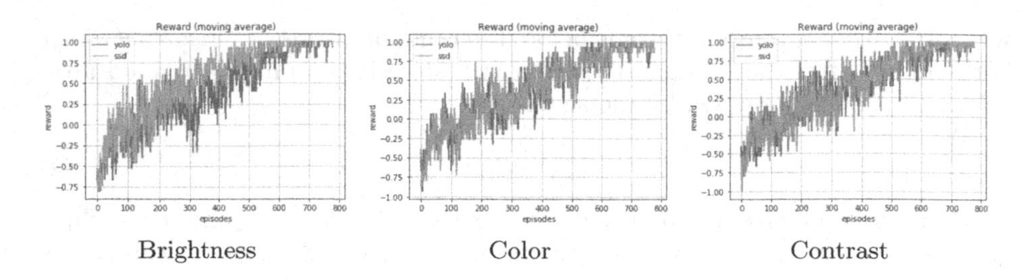

Fig. 5. Episodic return of the *ObjectRL* while training with a moving average of size 30. Each iteration represents 1K episodes.

5 Experiments

In this section, we describe the experimental setup for *ObjectRL*. We have built our network with PyTorch. For the object detector, we use a Single Shot Detector (SSD) [16] and YOLO-v3 [24] trained on the PascalVOC dataset with a VGG-base network [27] for SSD. We use Proximal Policy Optimization (PPO) [26] for optimising the *ObjectRL* agent. We train the agent network on a single NVIDIA GTX 1080Ti with the PascalVOC dataset.

Both the actor and the critic networks consist of 6 convolutional layers with (kernel size, stride, number of filters)= {(4, 2, 8), (3, 2, 16), (3, 2, 32), (3, 2, 64), (3, 1, 128), (3, 1, 256)} followed by linear layers with output size 100, 25, 1. The agent is updated after 2000 steps for 20 epochs with batch-size = 64. We use Adam Optimizer [12] with a learning rate of 10^{-3}. We use an $\epsilon-$*Greedy* method for exploration where we anneal ϵ linearly with the number of episodes until it reaches 0.05.

6 Results

6.1 Measure for Evaluation for ObjectRL: TP-Score

To the best of our knowledge, we believe no suitable measure is defined for this problem and hence we define a measure called *TP-Score(k)* (True Positive Score). This score is the number of images in which $k-$or more true positives were detected which were not detected in the image before transformation. The *TP-Score(k)* is initialised to zero for a set of images $\mathcal{I}$. For example: Let the number of true-positives detected before the transformation be 3 and let the number of true-positives detected after the transformation be 5. Then we have one image where 2 extra true-positives were detected which were not detected in the input image. Thus, we increase *TP-Score(1)* and *TP-Score(2)* by one.

6.2 Baseline for ObjectRL

To obtain the baselines, we first distort the images in the original dataset. The images are distorted with α being randomly chosen from the set $\mathcal{S}$ =

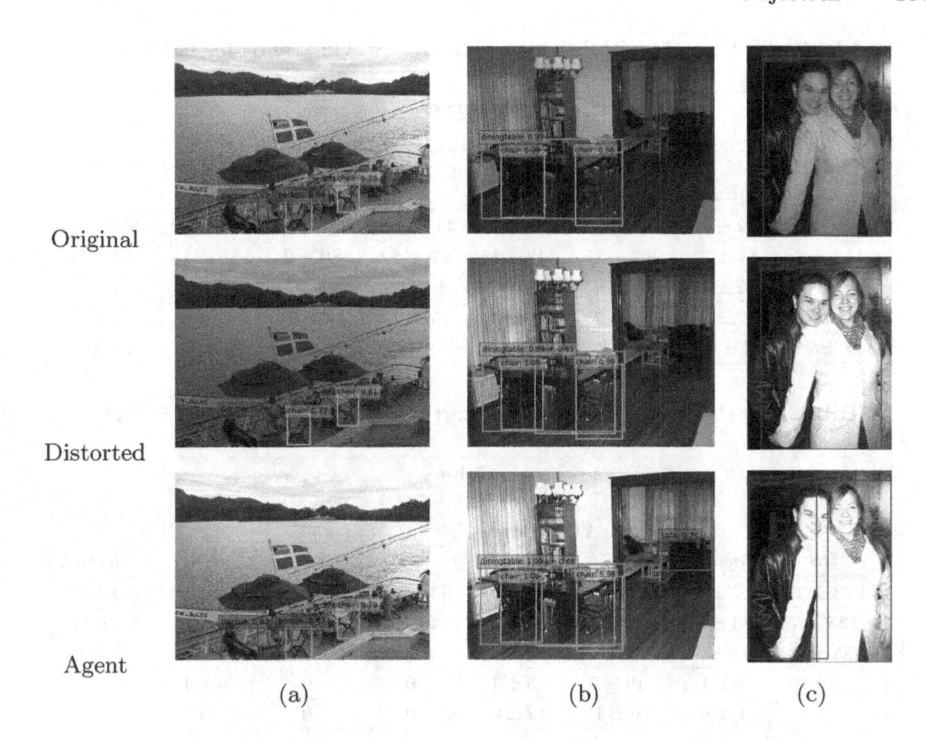

Fig. 6. A few of the outputs from *ObjectRL* with SSD and minor-scale distortion. The top row contains the original images. The second row contains the distorted images. The bottom row contains images obtained from the agent. Bounding boxes are drawn over the objects detected by the detector.

$\{0.1, \ldots, 1.9, 2.0\}$ or $\mathcal{S} = \{0.5, \ldots, 1.7, 1.8\}$ depending on the scale. The set of available actions to be applied on on these images are: $\hat{\mathcal{S}} = \{\frac{1}{s} \forall s \in \mathcal{S}\}$. We evaluate the *TP-Score(k)* on the distorted images by applying the transformations by performing a grid-search over all $\alpha \in \hat{\mathcal{S}}$ and report the scores obtained with the best-performing actions for different types and scales of distortions in Table 1, 2 and 3. We also report the *TP-Scores* obtained after applying the transformations proposed by *ObjectRL* on the images distorted using full-scale and minor-scales. The scores reported are averaged over 10 image sets $\mathcal{I}$, each containing 10,000 images. Note that the means and standard deviations are rounded to the nearest integers.

As seen in Table 1, 2 and 3, *ObjectRL* is not able to perform as well as the grid-search for full-scale distortions. The reason for this is that many of the images obtained after the full-scale distortions are not repairable with the action set provided to the agent.

But with minor-scale distortions, *ObjectRL* is able to perform as well as the grid-search. The total time taken for the grid-search over all brightness values for one image is 12.5094 ± 0.4103 s for YOLO and 15.1090 ± 0.3623 for SSD on a CPU. The advantage of using *ObjectRL* is that the time taken by the agent is

Table 1. *TP-Score(k)* with brightness distortion. GS stands for Grid-Search.

k	Brightness							
	Full-scale				Minor-scale			
	SSD		YOLO		SSD		YOLO	
	GS	*ObjectRL*	*GS*	*ObjectRL*	*GS*	*ObjectRL*	*GS*	*ObjectRL*
1	955 ± 14	532 ± 20	1360 ± 22	976 ± 18	435 ± 25	428 ± 23	1025 ± 23	883 ± 24
2	154 ± 6	87 ± 3	202 ± 15	118 ± 15	87 ± 12	80 ± 9	85 ± 15	63 ± 15
3	49 ± 3	32 ± 4	52 ± 8	18 ± 6	14 ± 5	12 ± 3	8 ± 2	5 ± 1
4	18 ± 3	7 ± 1	17 ± 2	4 ± 1	5 ± 1	3 ± 0	2 ± 0	0
5	7 ± 2	2 ± 0	4 ± 1	2 ± 0	0	0	0	0

Table 2. *TP-Score(k)* with color distortion. GS stands for Grid-Search.

k	Color							
	Full-scale				Minor-scale			
	SSD		YOLO		SSD		YOLO	
	GS	*ObjectRL*	*GS*	*ObjectRL*	*GS*	*ObjectRL*	*GS*	*ObjectRL*
1	973 ± 17	672 ± 19	1250 ± 23	1103 ± 21	561 ± 18	532 ± 22	974 ± 21	930 ± 22
2	123 ± 7	84 ± 4	210 ± 16	135 ± 13	43 ± 9	37 ± 9	83 ± 12	82 ± 12
3	53 ± 4	31 ± 3	63 ± 7	23 ± 6	1 ± 0	0	15 ± 2	10 ± 1
4	11 ± 2	3 ± 1	19 ± 2	5 ± 1	0	0	6 ± 1	3 ± 0
5	5 ± 1	1 ± 0	6 ± 1	2 ± 0	0	0	0	0

10 times less than grid-search. This latency is quite crucial in applications like surveillance drones and robots where the lighting conditions can vary quickly and the tolerance for errors in object-detection is low.

6.3 Discussion on the Outputs of *ObjectRL*

In this section, we discuss the outputs obtained from *ObjectRL* with SSD and minor-scale distortion which are shown in Fig. 6. In column (a) 4 true positives are detected in the original image, 3 true positives are detected in the distorted

Table 3. *TP-Score(k)* with contrast distortion. GS stands for Grid-Search.

k	Contrast							
	Full-scale				Minor-scale			
	SSD		YOLO		SSD		YOLO	
	GS	*ObjectRL*	*GS*	*ObjectRL*	*GS*	*ObjectRL*	*GS*	*ObjectRL*
1	955 ± 15	532 ± 20	1360 ± 21	976 ± 19	680 ± 22	663 ± 24	1038 ± 23	975 ± 24
2	163 ± 8	101 ± 4	213 ± 16	134 ± 15	62 ± 10	49 ± 9	104 ± 13	85 ± 15
3	55 ± 4	36 ± 4	67 ± 7	39 ± 6	14 ± 3	6 ± 2	19 ± 3	16 ± 2
4	21 ± 2	11 ± 1	28 ± 2	13 ± 1	1 ± 0	1 ± 0	5 ± 0	3 ± 0
5	4 ± 1	2 ± 0	5 ± 1	2 ± 0	0	0	0	0

Table 4. *TP-Score(k)* by crossing the policies.

k	Brightness		Color		Contrast	
	π_{yolo}^{ssd}	π_{ssd}^{yolo}	π_{yolo}^{ssd}	π_{ssd}^{yolo}	π_{yolo}^{ssd}	π_{ssd}^{yolo}
1	582 ± 13	1045 ± 24	800 ± 15	1249 ± 26	813 ± 15	1243 ± 26
2	36 ± 6	73 ± 11	72 ± 8	138 ± 11	65 ± 8	145 ± 12
3	2 ± 0	9 ± 4	10 ± 1	13 ± 3	2 ± 0	19 ± 4

image and 4 true positives are detected in the original image. The distorted image is slightly darker the the original one. *ObjectRL* is able to recover the object lost after distortion. In column (b) 3 true positives are detected in the original image, 4 true positives are detected in the distorted image and 5 true positives are detected in the original image. In this case, even the distorted image performs better than original image. But the agent-obtained image performs the best with 5 true-positives. In column (c) 1 true positive is detected in the original image, 1 true positive is detected in the distorted image and 2 true positives are detected in the original image. In this case the agent obtained image outperforms both the distorted and the original image. For a human eye, the agent-obtained image may not look *pleasing* as it is much brighter than the original image. Ideally for a human, the distorted image in column (c) is the most *pleasing*. Column (c) is one of the perfect examples to demonstrate the fact that whatever looks pleasing to a human eye may not necessarily be the optimal one for object-detection. Thus on an average, the agent is able to recover either as many objects as detected in the original image or more. According to our experiments, there were 8 ± 1 images with SSD and 34 ± 5 images with YOLO-v3, where the agent-obtained image had lesser number of true-positives than the original image. Although, this number of true-positives was more than the number of true-positives detected in the distorted image.

6.4 Crossing Policies

In this section we perform experiments by swapping the detectors for the learned policies. Thus, we use π_{yolo} with SSD, (denoted as π_{yolo}^{ssd}) and π_{ssd} with YOLO, (denoted as π_{ssd}^{yolo}). In Table 4, we report the number of images where $k-$or lesser true positives were detected with the swapped policy than what were detected using the original policy on their corresponding detectors. As shown in Table 4, π_{SSD} on YOLO is worse than π_{YOLO} on SSD. This is because the range of values for which SSD gives optimal performance is bigger than the range of values for which YOLO gives optimal performance. In essence, YOLO is more sensitive to the image parameters than SSD.

7 Conclusion

This paper proposes the usage of reinforcement learning to improve the object detection of a pre-trained object detector network by changing the image parameters (*ObjectRL*). We validate our approach by experimenting with distorted images and making the agent output actions necessary to improve detection. Our experiments showed that pre-processing of images is necessary to extract the maximum performance from a pre-trained detector. Future work includes combining all the different distortions in a single model and using it for controlling camera parameters to obtain images. Along with this, local image manipulations such as changing the image parameters only in certain regions of the image could be tried out.

Acknowledgements. The first author would like to thank Hannes Gorniaczyk, Manan Tomar and Rahul Ramesh for their insights on the project.

References

1. Agustsson, E., Timofte, R.: NTIRE 2017 challenge on single image super-resolution: Dataset and study. In: The IEEE Conference on Computer Vision and Pattern Recognition (CVPR) Workshops, July 2017
2. Andreopoulos, A., Tsotsos, J.K.: On sensor bias in experimental methods for comparing interest-point, saliency, and recognition algorithms. IEEE Trans. Pattern Anal. Mach. Intell. **34**(1), 110–126 (2012). https://doi.org/10.1109/TPAMI.2011.91
3. Bychkovsky, V., Paris, S., Chan, E., Durand, F.: Learning photographic global tonal adjustment with a database of input/output image pairs. In: CVPR 2011, pp. 97–104 (2011). https://doi.org/10.1109/CVPR.2011.5995332
4. Caicedo, J.C., Lazebnik, S.: Active object localization with deep reinforcement learning. CoRR abs/1511.06015 (2015). http://arxiv.org/abs/1511.06015
5. Deng, J., Dong, W., Socher, R., Li, L.J., Li, K., Fei-Fei, L.: ImageNet: a large-Scale Hierarchical Image Database. In: CVPR 2009 (2009)
6. Everingham, M., Eslami, S.M.A., Van Gool, L., Williams, C.K.I., Winn, J., Zisserman, A.: The pascal visual object classes challenge: a retrospective. Int. J. Comput. Vis. **111**(1), 98–136 (2015)
7. Geiger, A., Lenz, P., Stiller, C., Urtasun, R.: Vision meets robotics: the kitti dataset. Int. J. Robot. Res. (IJRR) **32**, 1231–1237 (2013)
8. Goodfellow, I.J., et al.: Generative adversarial nets. In: Proceedings of the 27th International Conference on Neural Information Processing Systems, NIPS 2014, vol. 2, pp. 2672–2680. MIT Press, Cambridge (2014)
9. Hasinoff, S.W., Durand, F., Freeman, W.T.: Noise-optimal capture for high dynamic range photography. In: 2010 IEEE Computer Society Conference on Computer Vision and Pattern Recognition, pp. 553–560, June 2010. https://doi.org/10.1109/CVPR.2010.5540167
10. Hu, Y., He, H., Xu, C., Wang, B., Lin, S.: Exposure: a white-box photo post-processing framework. ACM Trans. Graph. **37**(2), 1–17 (2018). https://doi.org/10.1145/3181974
11. Huiskes, M.J., Lew, M.S.: The MIR flickr retrieval evaluation. In: MIR 2008: Proceedings of the 2008 ACM International Conference on Multimedia Information Retrieval. ACM, New York (2008)
12. Kingma, D.P., Ba, J.: Adam: a method for stochastic optimization (2014). http://arxiv.org/abs/1412.6980, cite arxiv:1412.6980Comment. Published as a conference paper at the 3rd International Conference for Learning Representations, San Diego, 2015

13. Koenig, J., Malberg, S., Martens, M., Niehaus, S., Krohn-Grimberghe, A., Ramaswamy, A.: Multi-stage reinforcement learning for object detection. CoRR abs/1810.10325 (2018). http://arxiv.org/abs/1810.10325
14. Konda, V.R., Tsitsiklis, J.N.: Actor-critic algorithms. In: Solla, S.A., Leen, T.K., Müller, K. (eds.) Advances in Neural Information Processing Systems, vol. 12, pp. 1008–1014. MIT Press (2000). http://papers.nips.cc/paper/1786-actor-critic-algorithms.pdf
15. Linderoth, M., Robertsson, A., Johansson, R.: Color-based detection robust to varying illumination spectrum. In: 2013 IEEE Workshop on Robot Vision (WORV), pp. 120–125, January 2013. https://doi.org/10.1109/WORV.2013.6521924
16. Liu, W., et al.: SSD: single shot multibox detector. CoRR abs/1512.02325 (2015). http://arxiv.org/abs/1512.02325
17. Maier, W., Eschey, M., Steinbach, E.: Image-based object detection under varying illumination in environments with specular surfaces. In: 2011 18th IEEE International Conference on Image Processing, pp. 1389–1392, September 2011. https://doi.org/10.1109/ICIP.2011.6115698
18. Mathe, S., Pirinen, A., Sminchisescu, C.: Reinforcement learning for visual object detection. In: 2016 IEEE Conference on Computer Vision and Pattern Recognition (CVPR), pp. 2894–2902, June 2016. https://doi.org/10.1109/CVPR.2016.316
19. Mathe, S., Sminchisescu, C.: Multiple instance reinforcement learning for efficient weakly-supervised detection in images. CoRR abs/1412.0100 (2014). http://arxiv.org/abs/1412.0100
20. Osadchy, M., Keren, D.: Image detection under varying illumination and pose. In: Proceedings Eighth IEEE International Conference on Computer Vision, ICCV 2001, vol. 2, pp. 668–673, July 2001. https://doi.org/10.1109/ICCV.2001.937690
21. Osadchy, M., Keren, D.: Efficient detection under varying illumination conditions and image plane rotations. Comput. Vis. Image Underst. **93**, 245–259 (2004)
22. Park, J., Lee, J., Yoo, D., Kweon, I.S.: Distort-and-recover: color enhancement using deep reinforcement learning. CoRR abs/1804.04450 (2018). http://arxiv.org/abs/1804.04450
23. Puterman, M.L.: Markov Decision Processes: Discrete Stochastic Dynamic Programming, 1st edn. Wiley, New York (1994)
24. Redmon, J., Divvala, S.K., Girshick, R.B., Farhadi, A.: You only look once: unified, real-time object detection. CoRR abs/1506.02640 (2015). http://arxiv.org/abs/1506.02640
25. Ross, S., Gordon, G.J., Bagnell, J.A.: No-regret reductions for imitation learning and structured prediction. CoRR abs/1011.0686 (2010). http://arxiv.org/abs/1011.0686
26. Schulman, J., Wolski, F., Dhariwal, P., Radford, A., Klimov, O.: Proximal policy optimization algorithms. CoRR abs/1707.06347 (2017). http://arxiv.org/abs/1707.06347
27. Simonyan, K., Zisserman, A.: Very deep convolutional networks for large-scale image recognition. CoRR abs/1409.1556 (2014). http://arxiv.org/abs/1409.1556
28. Wu, Y., Tsotsos, J.K.: Active control of camera parameters for object detection algorithms. CoRR abs/1705.05685 (2017). http://arxiv.org/abs/1705.05685
29. Yang, H., Wang, B., Vesdapunt, N., Guo, M., Kang, S.B.: Personalized attention-aware exposure control using reinforcement learning. CoRR abs/1803.02269 (2018). http://arxiv.org/abs/1803.02269

Collaborative Learning with Pseudo Labels for Robust Classification in the Presence of Noisy Labels

Chanjong Park(✉), Jae-Il Jung, Dongha Bahn, Jonghee Hong, Junik Jang, Jiman Kim, and Intaik Park

Samsung Research, Seoul, Republic of Korea
{cj710.park,ji0130.jung,dongha.bahn,jonghee.hong,ji.jang,jiman14.kim, intaik.park}@samsung.com

Abstract. Supervised learning depends on labels of dataset to train models with desired properties. Therefore, data containing mislabeled samples (a.k.a. noisy labels) can deteriorate supervised learning performance significantly as it makes models to be trained with wrong targets. There are technics to train models in the presence of noise in data labels, but they usually suffer from the data inefficiency or overhead of additional steps. In this work, we propose a new way to train supervised learning models in the presence of noisy labels. The proposed approach effectively handles noisy labels while maintaining data efficiency by replacing labels of large-loss instances that are likely to be noise with newly generated pseudo labels in the training process. We conducted experiments to demonstrate the effectiveness of the proposed method with public benchmark datasets: CIFAR-10, CIFAR-100 and Tiny-ImageNet. They showed that our method successfully identified correct labels and performed better than other state-of-the-art algorithms for noisy labels.

Keywords: Deep neural network · Noisy labels · Pseudo labels

1 Introduction

Supervised learning with deep neural networks (DNNs) is a powerful method to recognize or categorize images with lots of data [7,11,18,20]. However, as labeling large-scale datasets is an expensive and error prone process, some portion of data can be mislabeled to different classes or sometimes people may label the same instance differently on marginal cases. The resulting noise in labels affect the networks to optimize toward wrong targets during training process, deteriorating their performance to correctly identify classes. We call these labels noisy labels.

Researches were conducted to deal with the noisy labels by correcting them [6,21]. These usually require additional steps, changes of network structure, or sometimes multiple trainings to correct the noisy labels.

Another way to deal with noisy labels is to detect and discard them. DNNs tend to learn and generalize simple patterns in dataset before applying brute-force memorizing on hard instances [1,25]. In the context of datasets with noisy

A. Bartoli and A. Fusiello (Eds.): ECCV 2020 Workshops, LNCS 12539, pp. 162–177, 2020.
https://doi.org/10.1007/978-3-030-68238-5_13

labels, the clean labels can be associated with small losses after initial epochs (simple patterns) and the noisy labels with large losses (hard instances). This trait can be leveraged to robustly train supervised learning models in the presence of noise as clean label instances can be identified by their small losses [5,8]. For example, Co-teaching [5] exhibited promising results by discarding large-loss instances that are likely to be instances with noise in their labels. By doing so, Co-teaching learns from only small-loss instances that are likely to be generalizable. However, as certain percentage of training data is discarded, this approach can suffer from data inefficiency.

To overcome this problem, we propose a new method that leverages large-loss instances by replacing noisy labels with labels produced by the networks (pseudo labels). This allows the networks to train with more data than the approach of utilizing only small-loss instances, but it also requires certain assurance that the pseudo labels are indeed correct. For this, a notion of certainty threshold is introduced to leave out pseudo labels that are potentially inaccurate.

We summarize contributions of the paper below.

1. We propose a method to utilize large-loss instances in training process by replacing their labels with pseudo labels.
2. We propose a method to generate and refine pseudo labels during training.
3. We present that our method not only improves performance in the initial epochs of training, but also prevents performance degradation in the late stage of training.

To demonstrate an effectiveness of our approach, we conducted experiments on CIFAR-10 [10], CIFAR-100 [10] and Tiny-ImageNet[1] datasets with added noise and compared our approach to other known technics to handle noisy labels. The data showed that our approach not only prevented the performance degradation due to noise in labels but also achieved the highest accuracy among all approaches.

2 Related Work

Many researches have been conducted to train networks with noisy labels. Among them, our research has been influenced by several recent studies, and we roughly separate them based on their approaches.

Label Correction Approaches. Label correction actively refines the quality of dataset labels that include noise. [16] proposes a method to replace the targets with a convex combination of the ground truth and the prediction of the model. [6] suggests a method to extract deep features and reassign the label by the similarity with prototypes of each class. [21] presents a two-stage method that a network alternates update their weights and labels on the first stage, then the network retrain from initial weights with the labels obtained from the first stage.

[1] https://tiny-imagenet.herokuapp.com/.

Approaches Using Small-Loss Instances. As discussed in [1] and [25], DNNs learn easy patterns first in the initial epochs. So in the presence of noisy labels which can be confusing from model perspectives (hard instances), DNNs learn from clean labels first which are easier before memorizing instances with noisy labels. Several studies have been conducted based on this concept: [8] proposed a method to supervise training of a network using another network that classifies clean labels. [12] also leverages this trait by monitoring changes of subspace dimensionality and adapting the loss function accordingly. Co-teaching [5], on which we build our method, successfully identifies and discards noisy labels using loss values at the beginning of training. Additionally, it utilizes two independent models to effectively prevent overfitting to wrong targets that can be driven by small-loss instances. In addition, co-teaching plus [24] is proposed as a complement to co-teaching. It has similar training procedures and also use small-loss instances. However, it prevents gradual convergence of two models and boosts the performance by keeping divergence of two models.

Other Related Algorithms. There are several other approaches to train DNNs with noisy labels. Two stage approach [3] is a method to train models in the second stage after removing some of labels which are suspected as noise in the first stage. A loss correction approach [15] estimates the label transition matrix that modifies loss function by multiplying itself to the prediction. This leads to a soft optimization to avoid fitting directly to the noisy labels. Decoupling [13] trains two DNNs simultaneously and use only data points that the two DNNs disagree with each other. [4] and [19] suggest adding extra layer that learns label transitions, which can identify or correct noise.

3 Proposed Approach

3.1 Overall Process

We propose a framework to train supervised learning models with noisy labels by utilizing multiple networks and pseudo labels. As done in co-teaching, it identifies instances with good labels by their small cross entropy losses and feed these instances to the other network for learning. However, instead of discarding instances with noisy labels, which can be identified with large losses, it recycles them by replacing noisy labels with pseudo labels that are generated by models. This approach allows networks to train robustly even with noisy labels while maintaining data efficiency.

A detailed training process is described in Algorithm 1. Following the co-teaching framework, we train two networks in parallel: a model with parameter w_1 and a model with parameter w_2. Each model separates mini-batch D into small-loss instances $D(x^s, y^s)$ (abbreviated as D^s) which are lower Rth percentile instances and large-loss instances $D(x^l, y^l)$ (abbreviated as D^l) which are upper $(100-R)$th percentile. R is the ratio of small-loss instances in a mini-batch and it starts at 100% and gradually decreases as the training period progresses. This is

Algorithm 1. Collaborative Learning with Pseudo Labels

Input: w_1, w_2, learning rate η , epoch E, iteration I, keep rate R, certainty threshold β, loss ℓ

for $e = 1, 2, 3, ..., E$ **do**
  <shuffle training data>
  for $i = 1, 2, 3, ..., I$ **do**
    mini batch $\grave{D}(x, y) \leftarrow D(x, y)$
    Get $\grave{D}(x_1^s, y_1^s) = \text{argmin}_{\grave{D}} \ell(f(w_1, x), y, R_t\%)$ ▷ $R_t\%$ small-loss instances
    Get $\grave{D}(x_1^l, y_1^l) = \text{argmax}_{\grave{D}} \ell(f(w_1, x), y, 100 - R_t\%)$ ▷ 100-$R_t\%$ large-loss instances
    Get $\grave{D}(x_2^s, y_2^s) = \text{argmin}_{\grave{D}} \ell(f(w_2, x), y, R_t\%)$ ▷ $R_t\%$ small-loss instances
    Get $\grave{D}(x_2^l, y_2^l) = \text{argmax}_{\grave{D}} \ell(f(w_2, x), y, 100 - R_t\%)$ ▷ 100-$R_t\%$ large-loss instances
    $z_1 = \text{argmax} f(w_1, x_1^l)$ ▷ get results produced by the network with w_1
    $z_2 = \text{argmax} f(w_2, x_2^l)$ ▷ get results produced by the network with w_2
    $\grave{D}(\hat{x}_1, \hat{y}_1) \leftarrow \grave{D}(x_1^l, z_1)$, **if** $\ell(f(w_1, x_1^l), z_1) \leq \beta$ ▷ filtering with certainty threshold
    $\grave{D}(\hat{x}_2, \hat{y}_2) \leftarrow \grave{D}(x_2^l, z_2)$, **if** $\ell(f(w_2, x_2^l), z_2) \leq \beta$ ▷ filtering with certainty threshold
    $w_1 \leftarrow w_1 - \eta \bigtriangledown \ell(f(w_1, x_2^s + \hat{x}_2), y_2^s + \hat{y}_2)$ ▷ weight(w_1) update
    $w_2 \leftarrow w_2 - \eta \bigtriangledown \ell(f(w_2, x_1^s + \hat{x}_1), y_1^s + \hat{y}_1)$ ▷ weight(w_2) update
  end for
  $R_t \leftarrow R_{t+1}$
end for

to let the models learn easy patterns at the beginning of training and to prevent them from memorizing noisy data in later iterations. The two models exchange both small and large-loss instances to each other in every iteration. Since the small-loss instances are considered as data with the potentially accurate labels, they are delivered to the peer network without any post-processing. But the large-loss instances are likely to have noisy labels and hence their labels are corrected based on the prediction results of the networks. We call these new labels as pseudo labels (denoted as z in Algorithm 1). Pseudo labels are predicted labels and they can be wrong. To reduce the chances of training models based on incorrect pseudo labels, we calculate cross entropy losses based on generated pseudo labels and deliver only the instances with small losses (denoted as $\hat{y}$) to the peer network for training, that are pseudo labeled instances with losses smaller than the certainty threshold (denoted as β).

3.2 Pseudo Labels

In this section, we introduce and analyze pseudo labels using CIFAR-10 data with 50% of symmetric noise. For clarity, we establish terms of each label category as the following. An original label is the correct label of the instance before noise is added. A pseudo label is generated label during training process and it can be correct or incorrect. A correct pseudo label matches the original label of the instance while an incorrect pseudo label is a label that is different from the original label (incorrectly predicted).

Figure 1 depicts CIFAR-10 dataset projected on 2D-space using t-SNE[22] along with their labels shown with different colors. We show four different scenarios on Fig. 1: (a) the original labels of CIFAR-10 dataset, (b) the labels after applying 50% of symmetric noise on the original labels, (c) the labels after removing large-loss instances from (b), (d) the labels after label correction process with pseudo labels (denoted by '×') on (b). Keeping only the small-loss instances removes noise effectively but it also reduces data size as in (c). On the other hand, re-categorizing large-loss instances with pseudo labels recovers the label distribution that are similar to the original labels while maintaining data volume of the original set as shown in (d). However, some of the instances end up with incorrect pseudo labels, which can be seen by the difference between (a) and (d). We discuss about these instances in the following subsections.

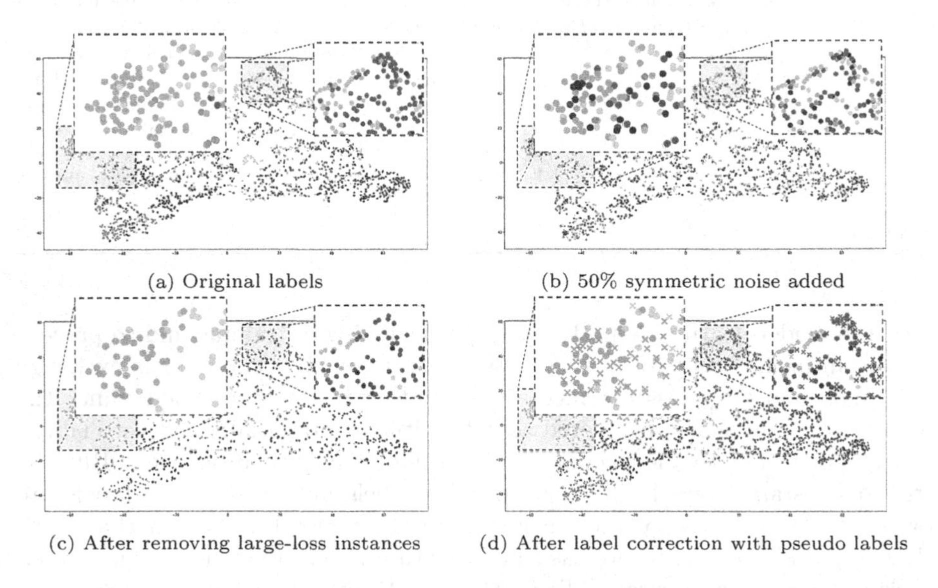

(a) Original labels (b) 50% symmetric noise added (c) After removing large-loss instances (d) After label correction with pseudo labels

Fig. 1. Feature vector distributions projected on 2D-space using t-SNE. All vectors are extracted from the inputs of the last fully-connected layer. The network trained one epoch on CIFAR-10 with symmetric 50% noise. (pseudo labels are denoted as '×'.)

The Correctness of Pseudo Labels. Pseudo labels are predicted labels by networks and they can be incorrect. Hence, there is a need for a guarantee that the generated pseudo labels are indeed correct before using them for learning. This is done by identifying potentially correct pseudo labels with the certainty threshold. Figure 2 shows two histograms of correct and incorrect labels by their losses. (a) is the histogram of entire dataset with noisy labels (incorrect labels are the noise) and it shows that the majority of correct labels are clustered in small-loss region while the most of incorrect labels have large losses. This is the reason why small-loss instances can be used for training directly without any further processing. (b) is the histogram of only the instances with pseudo labels. It also shows similar distribution in which the majority of correct pseudo labels have small losses. Likewise, using only pseudo label instances with small losses can improve the reliability of training with pseudo labels.

In this process, the certainty threshold β is used to identify correct pseudo labels for training, and it determines the trade-offs between reliability and data efficiency. Lower β will improve the reliability of pseudo labels but the more pseudo labels will not be used for training while higher β will improve data efficiency at the expense of reliability of pseudo labels. If the models converge fast, β can be set high so as to leverage more pseudo labels for training immediately. But, if the models converge slow or the performances are not good, lower β has to be used to retain quality of pseudo labels. We set the β to 0.1 empirically for general uses on various datasets.

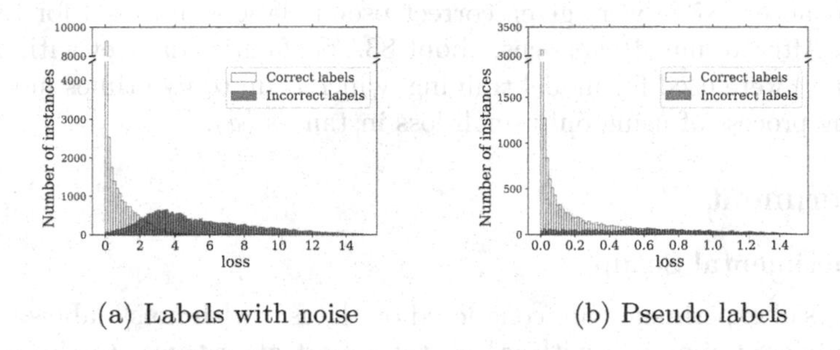

Fig. 2. Histogram of correct or incorrect labels by loss. (a) shows labels with injected noise, incorrect labels are distributed in high-loss region (symmetric 50% noise is applied). (b) shows pseudo labels. Pseudo labels also show similar distributions that correct label tend to cluster at low loss region

How Pseudo Labels Change over Iterations? Pseudo labels for large-loss instances are newly generated at each iteration in the training process, and therefore pseudo labels can change for the same instance over different iterations as the models keep changing with learning. On average, pseudo labels on CIFAR-10 changed 5.014 times during 200 epochs training.

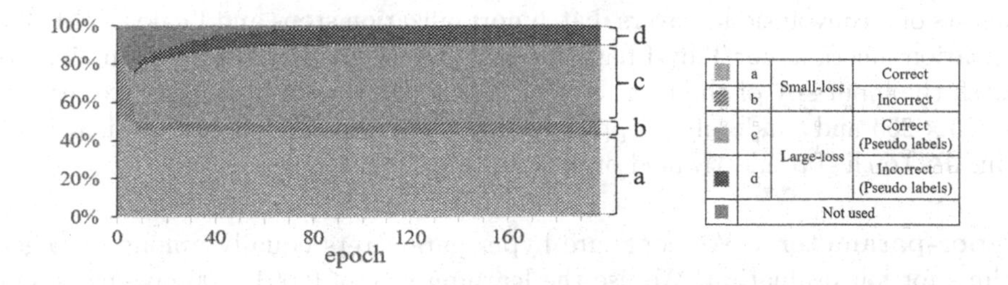

Fig. 3. Changes in the proportion of label types in the mini-batch at $\beta = 0.1$

We analyze how pseudo labels evolve over iterations and how they end up on CIFAR-10 with 50% symmetric noise. Figure 3 depicts the label changes during training process. In the very beginning when the models have not learnt much, a large portion of data instances with incorrect labels (b) were used for learning but their proportion decreases rapidly as networks learn with more epochs and incorrectness of their labels (noise) is recognized with large losses. At the same time, these large-loss data instances are given pseudo labels (c and d) and correctness of pseudo labels (the portion of correct pseudo labels among all pseudo labels) increases with more epochs as the models learn more from the dataset. The models wrongfully identified some of original label instances as noise based on their large losses(decreased portion of a). However, the majority of these instances (87%) were given correct pseudo labels and used for training in the end. After around 100 epochs, about 83.7% of entire dataset with correct labels ($a + c$) were used for model training, which is up to two times more than the training process of using only small-loss instances (a).

4 Experiment

4.1 Experimental Setup

Multiple sets of experiments are conducted on dataset with noisy labels to compare the proposed approach with other state-of-art algorithms. In this section, we describe data, setup, algorithms to compare and experimental results.

Dataset and Noisy Label. We use three representative benchmark datasets for evaluations: CIFAR-10 [10], CIFAR-100 [10], and Tiny-ImageNet which consists of a subset of ImageNet [2]. We contaminated these datasets with two types of noise: symmetric noise and pair noise [5]. Symmetric method shuffle labels in a way that the noisy labels are evenly distributed across all classes. We used three noise levels to vary the noise levels (20%, 50%, 70%). In this, 20% noise means 20% of instances in the dataset are given noisy labels and so on. Pair noise mixes labels only between similar classes.

Network Architecture. We use two types of network architecture. For the CIFAR-10 and CIFAR-100 datasets, we use the same CNN model used in [5]. It consists of 9-convolutional layers, batch normalization steps and Leaky-ReLU for activation function. For Tiny-ImageNet data, we use the ResNet-18 [7] model. To match the input size of the basic ResNet-18 model, the input images are resized to 256×256 and randomly cropped to 224×224. All models are implemented using PyTorch [14] and trained on a Nvidia M40 GPU.

Hyper-parameters. We configure hyper-parameters equally among all algorithms for fair evaluation. We use the learning rate of 0.001, 200 epochs, adam optimizer [9] with 0.9 momentum, batch size of 128, and decayed learning rate linearly from 80 epoch to 200 epoch. We repeat all experiments for five times and report performances of all algorithms using the average and standard deviation of accuracy. Certainty threshold of the proposed approach, referred as β, is set to 0.1 empirically.

4.2 Other Algorithms for Comparison

We compare our proposed approach with six different state-of-art algorithms that deal with noisy labels: standard softmax cross entropy, Boot Soft [16], Boot Hard [16], Forward [15], Co-teaching [5], Co-teaching plus [24]. We set the hyper-parameters of these algorithms following suggestions in their papers. The scaling factors for Boot Soft and Boot Hard are set to 0.95 and 0.8 respectively. For Forward, which is the loss-correction method, noise matrix estimation is carried out first before training with the estimated noise transition matrix as suggested in the paper. For all algorithms using small-loss instances, the R value decrease linearly from 100% to the percentage of the original labels in the data set (1 – noise level such as 20%, 50%, etc.) during the initial 10 epochs [5].

4.3 Experiment Results on CIFAR-10

Figure 4 shows how the test accuracies of various approaches change over training steps in different noise conditions. Most of algorithms show fast increase of accuracy in the initial epochs before reaching maximal accuracy then decrease until reaching plateaus. This seems to be associated with the trait of deep neural networks to learn and generalize with good quality labels at the beginning of training before memorizing noisy labels [1]. This phenomenon is worse for datasets with more symmetric noise or pair noise. Co-teaching is less affected by this but still show degradation after initial peak accuracy.

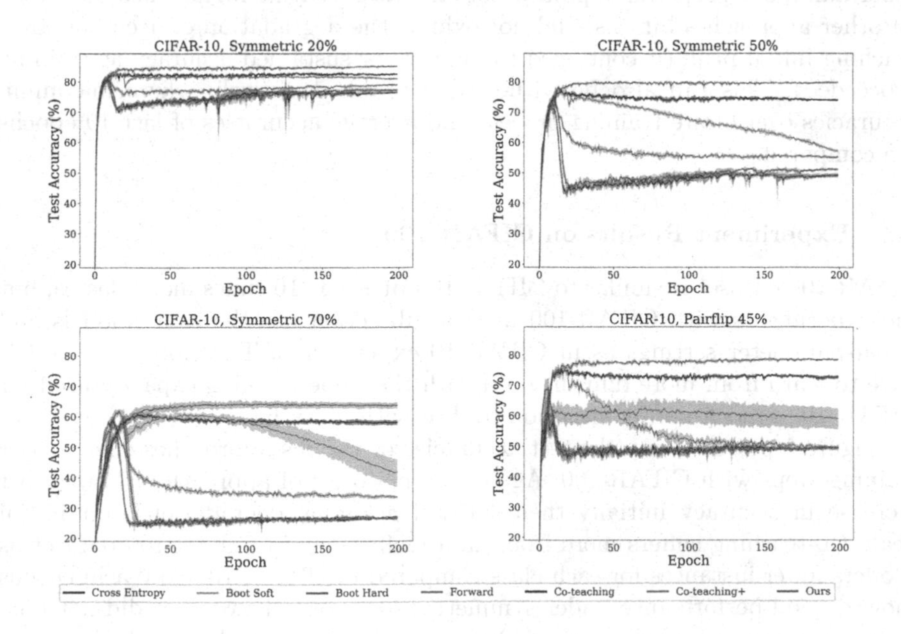

Fig. 4. Test Accuracy on CIFAR-10 dataset

Table 1. Maximal accuracy over entire training process and average accuracy of last 10 epoch on CIFAR-10 dataset

Methods	Symmetric 20%		Symmetric 50%		Symmetric 70%		Pair 45%	
	Maximum	last 10 epoch	Maximum	last 10 epoch	Maximum	last 10 epoch	Maximum	last 10 epoch
Cross Entropy	81.3 ±0.38	76.42 ±0.21	73.58 ±0.52	49.06 ±0.6	59.71 ±0.64	26.58 ±0.4	64.74 ±3.19	49.61 ±0.63
Boot Soft [16]	81.08 ±0.6	76.56 ±0.48	72.9 ±0.75	49.28 ±0.57	59.76 ±0.56	26.59 ±0.28	64.76 ±2.55	49.32 ±0.6
Boot Hard [16]	81.49 ±0.25	77.22 ±0.41	75.23 ±0.55	51.08 ±0.38	62.65 ±0.36	26.92 ±0.4	71.43 ±1.72	49.42 ±0.29
Forward [15]	81.35 ±0.33	78.97 ±0.34	73.77 ±0.28	55.13 ±0.4	57.66 ±0.64	33.78 ±0.55	65.28 ±3.12	59.23 ±3.46
Co-teaching [5]	83.68 ±0.21	82.46 ±0.3	77.84 ±0.22	73.96 ±0.23	61.27 ±0.82	57.87 ±0.65	76.49 ±0.43	72.69 ±0.44
Co-teaching+ [24]	83.48 ±0.27	81.01 ±0.31	77.24 ±0.5	57.8 ±0.51	61.58 ±0.65	42.17 ±4.35	76.32 ±0.65	49.43 ±0.4
Ours	**85.64 ±0.16**	**85.09 ±0.15**	**80.57 ±0.17**	**79.72 ±0.15**	**64.67 ±1.2**	**63.58 ±0.98**	**79.32 ±0.24**	**78.54 ±0.36**

Co-teaching and co-teaching plus also show more degradation with more noises which can be due to data inefficiency as more data is discarded. On the other hand, our proposed approach not only achieved the highest accuracy over all other approaches but also did not exhibit the degradation of accuracy after reaching initial peak (it continued to improve or sustained accuracy as training proceeded). This can also be witnessed in Table 1 as well, where maximum accuracies over entire training process and average accuracies of last 10 epochs are compared.

4.4 Experiment Results on CIFAR-100

CIFAR-100 dataset is similar to CIFAR-10 but it has 10 times more classes, but the experiments with CIFAR-100 in this subsection use the same models and hyper-parameter settings as in CIFAR-10 experiments. Therefore, the models have to learn from more difficult data with the same learning capacity used for CIFAR-10; a much harder situation to deal with noisy labels.

Figure 5 presents how the test accuracies of various approaches change over training steps with CIFAR-100. Again, the majority of approaches showed fast increase in accuracy initially then suffered accuracy degradation after initial peak. Co-teaching suffers more from data inefficiency in CIFAR-100 dataset as it offers fewer instances for each class compared to CIFAR-10. Co-teaching plus showed good performance under symmetric 20% noise, however it did not hold the performance with more noises. Our proposed approach showed the best performance over all approaches in CIFAR-100 experiments as well.

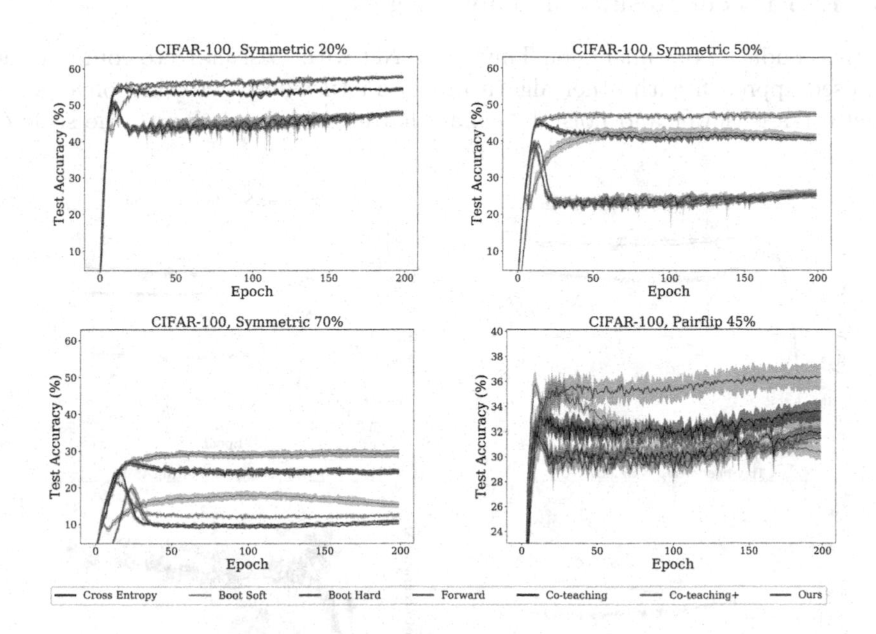

Fig. 5. Test Accuracy on CIFAR-100 dataset

Table 2. Maximal accuracy over entire training process and average accuracy of last 10 epoch on CIFAR-100 dataset

Methods	Symmetric 20%		Symmetric 50%		Symmetric 70%		Pair 45%	
	Maximum	last 10 epoch	Maximum	last 10 epoch	Maximum	last 10 epoch	Maximum	last 10 epoch
Cross Entropy	50.63 ±0.98	47.54 ±0.45	39.65 ±0.94	25.42 ±0.56	24.86 ±0.4	10.7 ±0.46	32.93 ±0.55	31.97 ±0.38
Boot Soft [16]	50.85 ±0.42	47.58 ±0.39	39.8 ±0.76	25.73 ±0.57	24.93 ±0.62	10.86 ±0.225	33.41 ±0.61	32.15 ±0.35
Boot Hard [16]	51.11 ±0.75	47.57 ±0.83	39.66 ±0.83	24.92 ±0.225	20.67 ±0.55	9.96 ±0.12	32.92 ±0.55	31.48 ±0.39
Forward [15]	51.28 ±0.59	47.52 ±0.42	38.82 ±1.25	25.42 ±1.29	21.92 ±0.41	12.46 ±0.51	36.2 ±0.71	32.95 ±0.69
Co-teaching [5]	55.62 ±0.31	54.04 ±0.225	45.89 ±1.12	41.01 ±0.87	27.33 ±0.78	24.35 ±0.82	34.38 ±0.81	33.63 ±0.82
Co-teaching+ [24]	57.88 ±0.43	**57.36 ±0.14**	43.88 ±1.52	41.18 ±1.25	18.58 ±1.14	15.65 ±0.97	36.78 ±0.33	30.35 ±0.51
Ours	**57.96 ±0.23**	57.23 ±0.14	**48.37 ±0.56**	**47.45 ±0.75**	**29.96 ±1.07**	**29.32 ±1.03**	**36.9 ±0.93**	**36.17 ±1.1**

4.5 Experiment Results on Tiny-ImageNet

We also conduct experiments on Tiny-ImageNet with ResNet-18 to compare the proposed approach with other algorithms in a situation where the convergence of networks is slower and noisy labels are harder to correct due to the scale of

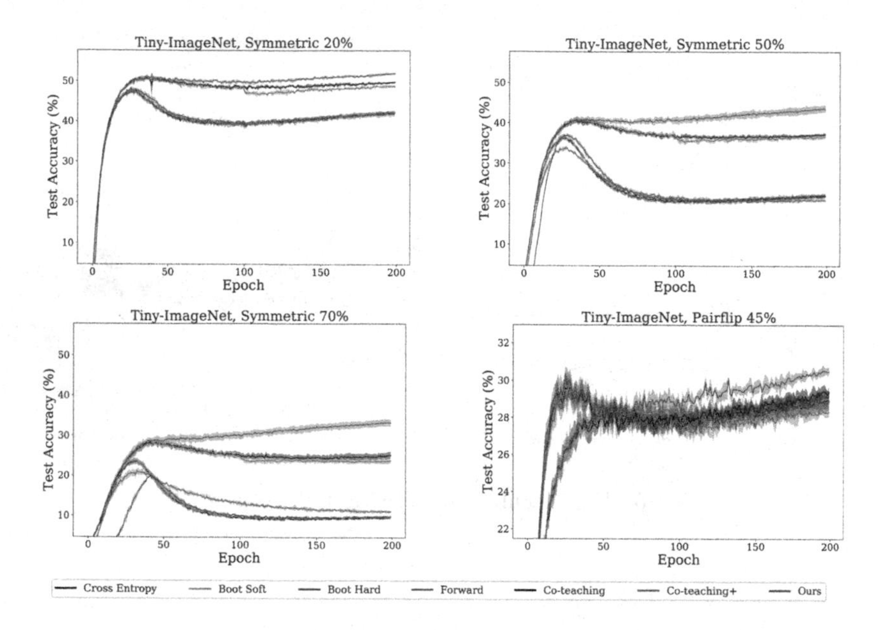

Fig. 6. Test Accuracy on Tiny-Imagenet dataset

Table 3. Maximal accuracy over entire training process and average accuracy of last 10 epoch on Tiny-ImageNet

Methods	Symmetric 20%		Symmetric 50%		Symmetric 70%		Pair 45%	
	Maximum	last 10 epoch	Maximum	last 10 epoch	Maximum	last 10 epoch	Maximum	last 10 epoch
Cross Entropy	47.93 ±0.27	42.17 ±0.41	36.54 ±0.53	21.96 ±0.36	23.93 ±0.63	9.62 ±0.34	30.41 ±0.26	28.9 ±0.53
Boot Soft [16]	47.94 ±0.49	42.31 ±0.41	36.64 ±0.3	21.77 ±0.5	23.9 ±0.56	9.71 ±0.29	30.44 ±0.48	28.79 ±0.22
Boot Hard [16]	48.53 ±0.35	42.22 ±0.33	37.38 ±0.36	22.21 ±0.43	19.94 ±0.24	9.85 ±0.29	30.66 ±0.45	28.83 ±0.4
Forward [15]	47.85 ±0.27	41.88 ±0.54	34.12 ±0.33	20.98 ±0.34	21.35 ±0.8	11.15 ±0.34	30.36 ±0.25	28.88 ±0.42
Co-teaching [5]	51.4 ±0.39	49.69 ±0.35	41.19 ±0.59	37.09 ±0.36	28.71 ±0.65	25.1 ±0.71	29.76 ±0.34	29.3 ±0.42
Co-teaching+ [24]	51.21 ±0.2	48.91 ±0.3	41.28 ±0.53	36.55 ±0.225	28.55 ±0.4	23.74 ±0.38	29.47 ±0.2	28.15 ±0.2
Ours	**52.33 ±0.19**	**51.94 ±0.21**	**43.89 ±0.78**	**43.48 ±0.79**	**33.44 ±0.88**	**33.12 ±0.89**	**30.82 ±0.13**	**30.35 ±0.16**

dataset and depth of the neural networks. Even in this setting, the proposed approach achieved the best performance over all approaches and its accuracy curve did not flatten at the end of 200 epochs suggesting that the higher accuracy is achievable with more epochs.

4.6 Experiment on Open-Set Noisy Labels

So far, we considered symmetric and pair noise that are created within the boundaries of existing dataset labels (close-set). In this subsection, we introduce a different type of noise; an open-set noise dataset, which contains external images that are not related to the labels of original dataset in scope [23]. This is to reflect more realistic environment where noise can be introduced from many different sources simultaneously. We create open-set noise dataset off CIFAR-10 with 40% of noise ratio and take mislabeled images from CIFAR-100 as noise. Additionally, we mix close-set and open-set noise in datasets with different ratios: 0:40, 10:30, 20:20, 30:10 (close-set:open-set ratio). We compare cross entropy, Co-teaching, and the proposed approach to demonstrate their effectiveness under various noise situations. All algorithms are repeated five times and Table 4 reports the results include maximum accuracies over all time periods and average accuracies of last 10 epochs as done in previous subsections.

Table 4. Maximal accuracy and average accuracy of last 10 epoch on mixture dataset of open-set noisy labels and close-set noisy labels

Methods	0:40		10:30		20:20		30:10	
	Maximum	last 10 epoch	Maximum	last 10 epoch	Maximum	last 10 epoch	Maximum	last 10 epoch
Cross Entropy	81.15 ±0.56	78.71 ±0.45	79.16 ±0.26	72.38 ±0.65	77.88 ±0.1	67.08 ±0.48	76.81 ±0.26	62.98 ±0.45
Co-teaching	**84.4 ±0.28**	**83.99 ±0.2**	83.71 ±0.2	83.38 ±0.18	82.57 ±0.36	81.95 ±0.3	81.58 ±0.41	80.23 ±0.21
Ours	83.68 ±0.18	83.37 ±0.19	**83.96 ±0.34**	**83.58 ±0.34**	**83.97 ±0.32**	**83.4 ±0.25**	**83.56 ±0.16**	**82.86 ±0.27**

Overall, the proposed approach achieved the best performance except the case with only open-set noise, in which co-teaching achieved the highest accuracy. This is because the pseudo labels cannot be correctly identified under open-set noise (as there is no correct labels for them); Discarding noisy labels and facing data inefficiency (co-teaching) seems to be a better strategy rather than trying to correct them (our approach) in this situation. However, our approach performed almost in par with the best performing approach of co-teaching even in this case as the generated pseudo labels are pruned with certainty threshold that can filter out incorrect ones.

5 Ablation Study

An ablation study was conducted to show the impact of pseudo labels on the performance of models. Certainty threshold β controls the quantity and quality of pseudo labels, and we changed β from 0 to infinity; $\beta = 0$ prevents generation of pseudo labels and discards all large-loss instances just like co-teaching. On the other hand, $\beta = \infty$ uses all the generated pseudo labels for training regardless of their quality represented by cross entropy loss. In other words, lower β guarantees fewer pseudo labels with high quality while higher β generates more pseudo labels with potentially lower quality.

Table 5. Maximal accuracy and average accuracy of last 10 epoch on ablation test on dataset with symmetric noise 50%

β	CIFAR-10		CIFAR-100	
	Maximum	last 10 epoch	Maximum	last 10 epoch
0	77.77 ± 0.34	74.12 ± 0.26	45.81 ± 0.86	41.28 ± 0.94
0.01	79.87 ± 0.34	79.21 ± 0.31	47.3 ± 0.85	46.91 ± 0.91
0.1	80.76 ± 0.23	79.78 ± 0.39	48.3 ± 0.92	47.4 ± 0.86
0.5	81.19 ± 0.15	79.93 ± 0.25	$\mathbf{49.41 \pm 0.99}$	$\mathbf{48.42 \pm 0.96}$
1.0	$\mathbf{81.28 \pm 0.23}$	$\mathbf{80.21 \pm 0.37}$	49.33 ± 1.23	48.39 ± 1.11
5.0	80.98 ± 0.19	80.0 ± 0.21	45.55 ± 0.62	44.73 ± 0.7
10.0	81.07 ± 0.21	80.03 ± 0.28	44.95 ± 0.69	44.45 ± 0.7
∞	81.19 ± 0.14	80.07 ± 0.07	45.38 ± 1.03	44.56 ± 1.27

Figure 7 shows test accuracy, label precision (the correctness of pseudo labels), and the number of correct pseudo labels over training steps while changing β with CIFAR-10. As expected, lower β results in better label precision in the early stage of training. However, the label precision decreases after peaks as more epochs progress, which makes the networks to fit more to dataset and lowers overall loss of pseudo labels producing more pseudo label instances with loss below beta. Eventually, the label precision and the number of correct pseudo labels converge to certain values for all betas in later stage of training. This tells that β controls training process with the trade-off between fewer labels with high quality and more labels with lower quality. Table 5 shows model performances with different β values on CIFAR-10 and CIFAR-100. As it can be seen, the optimal β values for different settings are different and hence should be empirically identified. In this paper, we suggested absolute value of cross entropy loss as β thresholds. However, as loss distributions of pseudo labels change (gradually decrease) over training steps, it may be worth-while to try the percentiles of pseudo label losses as certainty thresholds.

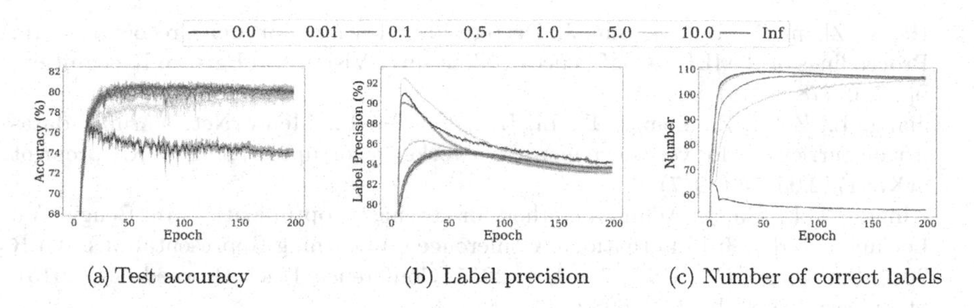

(a) Test accuracy (b) Label precision (c) Number of correct labels

Fig. 7. Ablation study on CIFAR-10 with symmetric noise 50%. Labels indicate β values. Total batch size is 128 in graph (c).

6 Conclusion

In this paper, we propose an end-to-end training framework which allows networks to train robustly in the presence of noise in data labels while keeping data efficiency. It identifies good instances with their small losses and train multiple models to prevent over-fitting to the small set of data. Additionally, it corrects noisy labels with pseudo labels and leverages them in training rather than discarding them. It is shown this approach is extremely robust against noisy labels with experiments on CIFAR-10, CIFAR-100, and Tiny-ImageNet datasets with added noise.

References

1. Arpit, D., et al.: A closer look at memorization in deep networks. In: Precup, D., Teh, Y.W. (eds.) Proceedings of the 34th International Conference on Machine Learning, ICML 2017, Sydney, NSW, Australia, 6–11 August 2017. Proceedings of Machine Learning Research, vol. 70, pp. 233–242. PMLR (2017). http://proceedings.mlr.press/v70/arpit17a.html
2. Deng, J., Dong, W., Socher, R., Li, L.J., Li, K., Fei-Fei, L.: ImageNet: a large-scale hierarchical image database. In: 2009 IEEE Conference on Computer Vision and Pattern Recognition, pp. 248–255. IEEE (2009)
3. Ding, Y., Wang, L., Fan, D., Gong, B.: A semi-supervised two-stage approach to learning from noisy labels. In: 2018 IEEE Winter Conference on Applications of Computer Vision (WACV), pp. 1215–1224. IEEE (2018)
4. Goldberger, J., Ben-Reuven, E.: Training deep neural-networks using a noise adaptation layer. In: 5th International Conference on Learning Representations, ICLR 2017, Toulon, France, 24–26 April 2017, Conference Track Proceedings. OpenReview.net (2017)
5. Han, B., et al.: Co-teaching: robust training of deep neural networks with extremely noisy labels. In: Advances in Neural Information Processing Systems, pp. 8527–8537 (2018)
6. Han, J., Luo, P., Wang, X.: Deep self-learning from noisy labels. In: Proceedings of the IEEE International Conference on Computer Vision, pp. 5138–5147 (2019)

7. He, K., Zhang, X., Ren, S., Sun, J.: Deep residual learning for image recognition. In: Proceedings of the IEEE Conference on Computer Vision and Pattern Recognition, pp. 770–778 (2016)
8. Jiang, L., Zhou, Z., Leung, T., Li, L.J., Fei-Fei, L.: MentorNet: learning data-driven curriculum for very deep neural networks on corrupted labels. arXiv preprint arXiv:1712.05055 (2017)
9. Kingma, D.P., Ba, J.: Adam: a method for stochastic optimization. In: Bengio, Y., LeCun, Y. (eds.) 3rd International Conference on Learning Representations, ICLR 2015, San Diego, CA, USA, 7–9 May 2015, Conference Track Proceedings (2015). http://arxiv.org/abs/1412.6980
10. Krizhevsky, A., Hinton, G., et al.: Learning multiple layers of features from tiny images (2009)
11. Krizhevsky, A., Sutskever, I., Hinton, G.E.: ImageNet classification with deep convolutional neural networks. In: Bartlett, P.L., Pereira, F.C.N., Burges, C.J.C., Bottou, L., Weinberger, K.Q. (eds.) Advances in Neural Information Processing Systems 25: 26th Annual Conference on Neural Information Processing Systems 2012, 3–6 December 2012, Lake Tahoe, pp. 1106–1114 (2012). http://papers.nips.cc/paper/4824-imagenet-classification-with-deep-convolutional-neural-networks
12. Ma, X., et al.: Dimensionality-driven learning with noisy labels. In: International Conference on Machine Learning, pp. 3355–3364 (2018)
13. Malach, E., Shalev-Shwartz, S.: Decoupling "when to update" from "how to update". In: Advances in Neural Information Processing Systems, pp. 960–970 (2017)
14. Paszke, A., et al.: Pytorch: an imperative style, high-performance deep learning library. In: Wallach, H.M., Larochelle, H., Beygelzimer, A., d'Alché-Buc, F., Fox, E.B., Garnett, R. (eds.) Advances in Neural Information Processing Systems 32: Annual Conference on Neural Information Processing Systems 2019, NeurIPS 2019, 8–14 December 2019, Vancouver, pp. 8024–8035 (2019). http://papers.nips.cc/paper/9015-pytorch-an-imperative-style-high-performance-deep-learning-library
15. Patrini, G., Rozza, A., Krishna Menon, A., Nock, R., Qu, L.: Making deep neural networks robust to label noise: a loss correction approach. In: Proceedings of the IEEE Conference on Computer Vision and Pattern Recognition, pp. 1944–1952 (2017)
16. Reed, S.E., Lee, H., Anguelov, D., Szegedy, C., Erhan, D., Rabinovich, A.: Training deep neural networks on noisy labels with bootstrapping. In: Bengio, Y., LeCun, Y. (eds.) 3rd International Conference on Learning Representations, ICLR 2015, San Diego, CA, USA, 7–9 May 2015, Workshop Track Proceedings (2015)
17. Ren, M., Zeng, W., Yang, B., Urtasun, R.: Learning to reweight examples for robust deep learning. In: Dy, J., Krause, A. (eds.) Proceedings of the 35th International Conference on Machine Learning. Proceedings of Machine Learning Research, vol. 80, pp. 4334–4343. PMLR (2018)
18. Simonyan, K., Zisserman, A.: Very deep convolutional networks for large-scale image recognition. In: Bengio, Y., LeCun, Y. (eds.) 3rd International Conference on Learning Representations, ICLR 2015, San Diego, CA, USA, 7–9 May 2015, Conference Track Proceedings (2015). http://arxiv.org/abs/1409.1556
19. Sukhbaatar, S., Estrach, J.B., Paluri, M., Bourdev, L., Fergus, R.: Training convolutional networks with noisy labels. In: 3rd International Conference on Learning Representations, ICLR 2015 (2015)

20. Szegedy, C., et al.: Going deeper with convolutions. In: IEEE Conference on Computer Vision and Pattern Recognition, CVPR 2015, Boston, MA, USA, 7–12 June 2015, pp. 1–9. IEEE Computer Society (2015). https://doi.org/10.1109/CVPR.2015.7298594
21. Tanaka, D., Ikami, D., Yamasaki, T., Aizawa, K.: Joint optimization framework for learning with noisy labels. In: Proceedings of the IEEE Conference on Computer Vision and Pattern Recognition, pp. 5552–5560 (2018)
22. Van Der Maaten, L.: Accelerating T-SNE using tree-based algorithms. J. Mach. Learn. Res. **15**(1), 3221–3245 (2014)
23. Wang, Y., et al.: Iterative learning with open-set noisy labels. In: Proceedings of the IEEE Conference on Computer Vision and Pattern Recognition, pp. 8688–8696 (2018)
24. Yu, X., Han, B., Yao, J., Niu, G., Tsang, I.W., Sugiyama, M.: How does disagreement help generalization against label corruption? In: Proceedings of the 36th International Conference on Machine Learning, ICML. Proceedings of Machine Learning Research, vol. 97, pp. 7164–7173 (2019)
25. Zhang, C., Bengio, S., Hardt, M., Recht, B., Vinyals, O.: Understanding deep learning requires rethinking generalization. arXiv preprint arXiv:1611.03530 (2016)

Addressing Neural Network Robustness with Mixup and Targeted Labeling Adversarial Training

Alfred Laugros[1,2(✉)], Alice Caplier[2(✉)], and Matthieu Ospici[1(✉)]

[1] Atos, Bezons, France
{alfred.laugros,matthieu.ospici}@atos.net
[2] Univ. Grenoble Alpes, Grenoble, France
alice.caplier@grenoble-inp.fr

Abstract. Despite their performance, Artificial Neural Networks are not reliable enough for most of industrial applications. They are sensitive to noises, rotations, blurs and adversarial examples. There is a need to build defenses that protect against a wide range of perturbations, covering the most traditional common corruptions and adversarial examples. We propose a new data augmentation strategy called M-TLAT and designed to address robustness in a broad sense. Our approach combines the Mixup augmentation and a new adversarial training algorithm called Targeted Labeling Adversarial Training (TLAT). The idea of TLAT is to interpolate the target labels of adversarial examples with the ground-truth labels. We show that M-TLAT can increase the robustness of image classifiers towards nineteen common corruptions and five adversarial attacks, without reducing the accuracy on clean samples.

Keywords: Neural network · Robustness · Common corruptions · Adversarial training · Mixup

1 Introduction

Artificial neural networks have been proven to be very efficient in various image processing tasks [26,38,43]. Unfortunately, in real-world computer vision applications, a lot of common corruptions may be encountered such as blurs, colorimetry variations, or noises, etc. Such corruptions can dramatically decrease neural network efficiency [7,11,23,48]. Besides, deep neural networks can be easily attacked with adversarial examples [42]. These attacks can reduce the performance of most of the state-of-the-art neural networks to zero [1,37].

Some techniques have been proposed to make neural networks more robust. When a specific perturbation is considered, we can build a defense to protect

Electronic supplementary material The online version of this chapter (https://doi.org/10.1007/978-3-030-68238-5_14) contains supplementary material, which is available to authorized users.

A. Bartoli and A. Fusiello (Eds.): ECCV 2020 Workshops, LNCS 12539, pp. 178–195, 2020.
https://doi.org/10.1007/978-3-030-68238-5_14

a neural network against it, whether it is a common corruption [25,48,58] or an adversarial attack [33,39,54]. However, increasing robustness towards a specific perturbation generally does not help with another kind of perturbation. For instance, geometric transformation robustness is orthogonal with worst-case additive noise [10]. Fine tuning on blur does not increase the robustness to Gaussian noise [8]. Even worse, making a model robust to one specific perturbation can make it more sensitive to another one. For instance, a data augmentation with corruptions located in the high frequency domain tends to decrease the robustness to corruptions located in the low frequency domain [55]. Besides, increasing adversarial robustness often implies a diminution of the *clean accuracy* (the accuracy of a model on not-corrupted samples) [27,47]. Therefore, a company or a public organism may feel reluctant to use neural networks in their projects since it is hard to make them robust to a large diversity of corruptions, and it is difficult to predict how a neural network will react to unexpected corruptions. There is a need to build new defenses that address robustness in a broad sense, covering the most encountered common corruptions and adversarial examples.

We propose to address this issue with a new data augmentation approach called M-TLAT. M-TLAT is a combination of Mixup [57] and a new kind of adversarial training called Targeted Labeling Adversarial Training (TLAT). The idea of this adversarial training is to label target adversarial examples with soft labels that contain information about the used target. We show that M-TLAT can increase the robustness of image classifiers to nineteen common corruptions and five adversarial attacks, without reducing the accuracy on clean samples. This algorithm is easy to implement and to integrate to an existing training process. It intends to make the neural networks used in real-world applications more reliable.

2 Related Works

2.1 Protecting Neural Networks Against Common Corruptions

Neural Networks are known to be sensitive to a lot of perturbations such as noises [25], rotations [11], blurs [48] or colorimetry variations [23], etc. We call these perturbations common corruptions. They are often encountered in industrial applications, but generally absent from academic datasets. For instance the faces in the dataset celeba are always well illuminated with a consistent eyes positioning [32], yet those conditions are not always guaranteed in the industrial applications. Because of common corruptions, the performance of a neural network can be surprisingly low [14].

There are a few methods that succeded in increasing the robustness to several common corruptions simultaneously. Among them, the robust pre-training algorithm proposed by Liu et al. can make classifiers more robust to noises, occlusions and blurs [31]. In [13], it is proposed to change the style of the training images using style transfer [12]. Neural networks trained this way are obliged to focus more on shapes than textures. The Augmix algorithm proposes to interpolate a clean image with perturbed versions of this image [18]. The obtained images are used to augment the training set of a neural network. These methods are useful to make neural networks more robust to common corruptions but they do not address the case of adversarial examples.

2.2 Protecting Neural Networks Against Adversarial Examples

Adversarial Examples are another threat that can make neural networks give unexpected answers [42]. Unlike common corruptions they are artificial distortions. They are crafted by humans so as to especially fool neural networks. Adversarial examples can completely fool even the state-of-the-art models [27]. Those perturbations are even more dangerous because humans can hardly see if an image has been adversarially corrupted or not [1,37]. In other words, a system can be attacked without anyone noticing it. They are two kinds of adversarial examples called white-box and black-box attacks.

White-Box Attacks. The attacker has access to the whole target network: its parameters and its architecture. White-box attacks are tailored so as to especially fool a specific network. White-box adversarial examples are very harmful, defending a model against it is a tough task [3,40,47].

Black-Box Attacks. An adversarial example crafted with a limited access to the targeted network is called a black-box attack. When only the training set of a neural network is known, it is still possible to make a transfer attack. Considering a dataset and two neural networks trained with it, it has been shown that an adversarial example crafted using one of the models, can harm the other one [15,42]. This phenomenon occurs even when the two models have distinct architectures and parameters.

A lot of methods have been proposed to protect against adversarial examples. Adversarial training uses adversarial examples to augment the training set of a neural network [33,46]. Defense-Gan [39] and feature squeezing [54] are used to remove adversarial patterns from images. Stochastic Activation Pruning [6] and defensive dropout [51] make the internal operations of neural networks more difficult to access in order to make these networks more difficult to attack. These methods can significantly increase the robustness of neural networks to adversarial examples but they do not provide any protection towards common corruptions.

2.3 Addressing Robustness in a Broad Sense

The methods mentioned above increase either the adversarial robustness or the robustness to common corruptions. Unfortunately, increasing the robustness to common corruptions generally does not imply increasing the robustness to adversarial examples and conversely [29]. The experiments carried out in [55] show that data augmentation with traditional adversarial examples makes models less robust to low frequency corruptions. Robustness to translations and rotations is independent from robustness to the L_p bounded adversarial examples [10].

A natural approach to address robustness in a broad sense is to combine defenses that address common corruptions with defenses that address adversarial examples. Unfortunately, it is possible that two defenses are not compatible and do not combine well [21,45,55].

A few standalone methods have been recently proposed to address adversarial robustness and robustness to common corruptions at the same time. In [34,52], a large set of unlabeled data are leveraged to get a significant increase in common corruption robustness and a limited increase in adversarial robustness. However, using these methods has a prohibitive computational cost. Adversarial Noise Propagation adds adversarial noise into the hidden layers of neural networks during the training phase to address both robustnesses [30]. Adversarial Logit Pairing (ALP) encourages trained models to output similar logits for adversarial examples and their clean counterparts [22]. In addition to the adversarial robustness provided, it has been reported that ALP increases the robustness to some common corruptions [17]. The drawback of these methods is that they reduce the clean accuracy of trained models.

To be useful in real-world applications, we want our method to preserve the clean accuracy of the trained models, to increase the robustness to both adversarial examples and common corruptions, and to be easy to integrate into an existing training framework.

3 Combining Mixup with Targeted Labeling Adversarial Training: M-TLAT

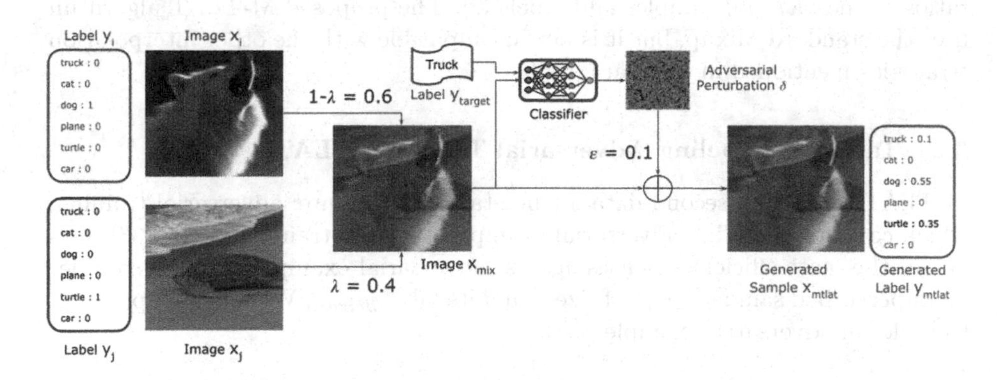

Fig. 1. Visualising of the generation of a new training couple using M-TLAT

Our approach called M-TLAT is a combination of two data augmentation algorithms, which are Mixup [57] and a new adversarial training strategy called Targeted Labeling Adversarial Training (TLAT). Basically, Mixup aims to increase the common corruption robustness while TLAT aims to increase the adversarial robustness. We go more into details in Sect. 5.3 to understand the contribution of each component. In practice, we observe that those augmentations combine well to address robustness in a broad sense.

3.1 Mixup

Let us consider the couples (x_i, y_i) and (x_j, y_j), where x_i and x_j are images of the training set and y_i and y_j are their associated one-hot encoding labels. Mixup [57] is a data augmentation algorithm that interpolates linearly samples and labels (see Fig. 1):

$$\begin{aligned} x_{mix} &= \lambda * x_i + (1 - \lambda) * x_j \\ y_{mix} &= \lambda * y_i + (1 - \lambda) * y_j \end{aligned} \tag{1}$$

where λ is drawn from a Beta distribution defined by an hyperparameter α: $\lambda \sim Beta(\alpha, \alpha)$. This augmentation strategy encourages the trained models to have a linear behavior in-between training samples [49,57]. In practice, Mixup reduces the generalization error of classifiers, makes neural networks less sensitive to corrupted labels and slightly improves the robustness to adversarial examples. In the ablation study carried out in Sect. 5.3, we detail the influence of Mixup on neural network robustness.

Augmenting datasets by interpolating samples of training sets have been largely studied [20,44]. Among the most successful, the Manifold Mixup interpolates hidden representations instead of interpolating the inputs directly [49]. The Mixup Inference proposes to use Mixup in the inference phase to degrade the perturbations that may corrupt the input images [35]. Directional Adversarial Training and Untied Mixup are alternative policies to pick the interpolation ratios of the *mixuped* samples and labels [2]. The proposed M-TLAT algorithm uses the standard Mixup, but it is not incompatible with the other interpolation strategies mentioned in this paragraph.

3.2 Targeted Labeling Adversarial Training: TLAT

M-TLAT relies on a second data augmentation procedure: adversarial training, which consists in adding adversarial examples into the training set [15,33]. It is one of the most efficient defenses against adversarial examples [3]. We consider an unperturbed sample x_{clean} of size S and its label y_{clean}. We can corrupt x_{clean} to build an adversarial example x_{adv}:

$$\begin{aligned} x_{adv} &= x_{clean} + \underset{\delta \in [-\epsilon, \epsilon]^S}{arg\,max} \{L(x_{clean} + \delta, y_{clean}, \theta)\} \\ y_{adv} &= y_{clean} \end{aligned} \tag{2}$$

Where θ are the parameters of the attacked model. L is a cost function like the cross-entropy function. The value ϵ defines the amount of the introduced adversarial perturbation. As suggested in [1], adversarial training is even more efficient when it uses adversarial examples that target a specific class y_{target}. The augmentation strategy becomes:

$$
\begin{aligned}
x_{adv} &= x_{clean} - \underset{\delta \in [-\epsilon,\epsilon]^S}{arg\,max}\{L(x_{clean}+\delta, y_{target}, \theta)\} \\
y_{adv} &= y_{clean}
\end{aligned} \tag{3}
$$

One advantage to use target adversarial examples during training is to prevent label leaking [27]. We propose to improve this augmentation strategy by using y_{target} in the labeling of the adversarial examples. In particular, we propose to mix the one-hot encoding ground-truth labels of the original samples with the one-hot encoding target labels used to craft the adversarial examples:

$$
\begin{aligned}
x_{adv} &= x_{clean} - \underset{\delta \in [-\epsilon,\epsilon]^S}{arg\,max}\{L(x_{clean}+\delta, y_{target}, \theta)\} \\
y_{adv} &= (1-\epsilon) * y_{clean} + \epsilon * y_{target}
\end{aligned} \tag{4}
$$

We call this augmentation strategy Targeted Labeling Adversarial Training (TLAT). The *arg max* part is approximated with an adversarial example algorithm such as target FGSM [1]:

$$
x_{adv} = x_{clean} - \epsilon * sign(\nabla_{x_{clean}} L(x_{clean}, y_{target}, \theta)) \tag{5}
$$

FGSM is used because it is a computationally efficient way to craft adversarial examples [15,46]. As for traditional adversarial trainings, models trained with TLAT should be trained on both clean samples and adversarial samples [1]. The advantage of TLAT is to make models have a high clean accuracy compared to the models trained with a standard adversarial training algorithm. More details can be found in Sect. 5.4.

TLAT uses soft labels instead of one-hot encoding labels. Using soft labels is a recurrent idea in the methods that address robustness. As mentioned above, Mixup interpolates training labels to generate soft labels [57]. Label smoothing replaces the zeros of one-hot encoding labels by a smoothing parameter $s > 0$ and normalizes the high value so that the distribution still sums to one [41]. Distillation learning uses the logits of a trained neural network to train a second neural network [36]. The second network is enforced to make smooth predictions by learning on soft labels. Bilateral training generates soft labels by adversarially perturb training labels [50]. It uses the gradient of the cost function of an attacked model to generate adversarial labels. Models trained on both adversarial examples and adversarial labels are encouraged to have a small gradient magnitude which makes them more robust to adversarial examples.

The originality of TLAT is to use the target labels of adversarial attacks as a component of the soft labels. Intuitions about why this labeling strategy works are provided in Sect. 5.4.

3.3 M-TLAT

The idea of M-TLAT is to generate new training couples by applying sequentially the TLAT perturbations (4) after the Mixup interpolations (1):

$$
\begin{aligned}
x_{mtlat} &= x_{mix} - \underset{\delta \in [-\epsilon,\epsilon]^S}{arg\,max}\{L(x_{mix} + \delta, y_{target}, \theta)\} \\
y_{mtlat} &= (1-\epsilon) * y_{mix} + \epsilon * y_{target}
\end{aligned}
\tag{6}
$$

As displayed in Fig. 1, x_{mtlat} contains features that come from three distinct sources: two clean images and an adversarial perturbation that targets a specific class. The label y_{mtlat} contains the class and the weight associated with each source of features. These weights are determined by the values λ and ϵ. A model trained with M-TLAT is not only constraint to predict the classes that correspond to the three sources. It also has to predict the weight of each source within the features of x_{mtlat}. We believe that being able to predict the class and the weighting of the three sources requires a subtle understanding of the features present in images. In practice, being trained with augmented couples (x_{mtlat}, y_{mtlat}) makes neural networks more robust.

The expressions (6) are the essence of the algorithm. The whole process of one training step with M-TLAT is provided in the algorithm description 1. As recommended in [1], the training minibatches contain both adversarial and non-adversarial samples. In our algorithm, the non-adversarial samples are obtained using a standard Mixup interpolation, the adversarial samples are crafted by combining Mixup and TLAT.

Another recently proposed approach combines Mixup and adversarial training [28]. Their method called Interpolated Adversarial Training (IAT) is different from M-TLAT for two main reasons. Most importantly, they do not use the labeling strategy of TLAT. Basically, their adversarially corrupted samples are labeled using the standard Mixup interpolation while our labels contain information about the amount and the target of the used adversarial examples. Secondly, we interpolate images before adding the adversarial corruptions. On the contrary they adversarially corrupt images before mixing them up. In practice, we get better adversarial robustness when we proceed in our order. Besides, proceeding in their order doubles the number of adversarial perturbations to compute: it increases the training time. In Sect. 5.2, we compare the robustness of two models trained with these approaches.

4 Experiment Set-Up

4.1 Perturbation Benchmark

We want to evaluate the robustness of neural networks in a broad sense, covering the most encountered common corruptions and adversarial examples. To achieve this, we gather a large set of perturbations that contains nineteen common corruptions and five adversarial attacks.

Common Corruptions. The benchmark includes the common corruptions of ImageNet-C [17]. The ImageNet-C set of perturbations contains diverse kinds of common corruptions such as 1) noises: Gaussian noise, shot noise and impulse

Algorithm 1. One training step of the M-TLAT algorithm

Require: θ the parameters of the trained neural network
Require: L a cross entropy function
Require: $(x_1, y_1), (x_2, y_2), (x_3, y_3), (x_4, y_4) \sim$ Dataset
Require: $\lambda_1, \lambda_2 \sim Beta(\alpha, \alpha)$
Require: $\epsilon \sim U[0, \epsilon_{max}]$ where U is a uniform distribution and ϵ_{max} is the maximum perturbation allowed
Require: $y_{target} \sim U[0, N]$ where N is the number of classes
Require: *optim* an optimizer like Adam or SGD

$x_{mix1} = \lambda_1 * x_1 + (1 - \lambda_1) * x_2$
$y_{mix1} = \lambda_1 * y_1 + (1 - \lambda_1) * y_2$

$x_{mix2} = \lambda_2 * x_3 + (1 - \lambda_2) * x_4$
$y_{mix2} = \lambda_2 * y_3 + (1 - \lambda_2) * y_4$

$x_{mtlat} = x_{mix2} - \epsilon * sign(\nabla_{x_{mix2}} L(x_{mix2}, y_{target}, \theta))$
$y_{mtlat} = (1 - \epsilon) * y_{mix2} + \epsilon * y_{target}$

$loss_1 = L(x_{mix1}, y_{mix1}, \theta)$
$loss_2 = L(x_{mtlat}, y_{mtlat}, \theta)$
$loss = loss_1 + loss_2$
$gradients = \nabla_\theta loss$
$optim.update(gradient, \theta)$ The optimizer updates θ according to the gradients

noise 2) blurs: defocus blur, glass blur, motion blur and zoom blur 3) weather related corruptions: snow, frost, fog and brightness 4) digital distortions: contrast, elastic, pixelate and jpeg compression. Each corruption is associated with five severity levels. We use the corruption functions[1] provided by the authors to generate those perturbations.

The rotation, translation, color distortion and occlusion common corruptions are absent from ImageNet-C, yet they are often encountered in industrial applications. We decided to add those four perturbations to the pool of common corruptions. The occlusion perturbation is modeled by masking a randomly selected region of images with a grey square. For the images corrupted with color distortion, one of the RGB channel is randomly chosen and a constant is added to all the pixels of this channel. For the rotation and translation perturbations, the pixel values of the area outside the transformed images are set to zero.

These four corruptions are associated with a severity range. The lower bound of the severity range has been set to reduce the accuracy of the standard ResNet-50 by five percent on the test set. The upper bound has been set to reduce its accuracy by thirty percent. As a result, the rotations can turn images from 8 to 40 degrees. In other words, when the images of the test set are rotated by eight degrees, the accuracy of the standard ResNet-50 is five percent lower than for a not-corrupted test set. The side of the square mask used in the occlusions varies

[1] https://github.com/hendrycks/robustness.

from 60 to 127 pixels. The translations can move images from 15 to 62 pixels. For the color distortion corruption, the values of the pixels of the perturbed channel are increased from 8% to 30% of the maximum possible pixel value.

Adversarial Examples. Following the recommendations in [5], we carefully choose the adversarial examples to be added to the benchmark. Firstly, we use adversarial examples in both white box and black box settings. Secondly, we use two different metrics, the L_∞ and the L_2 norms, to compute the bound of adversarial attacks. Thirdly, we employ targeted and untargeted attacks. Fourthly, several amounts of adversarial perturbations are used. Finally, the selected adversarial examples are not used during trainings. We build a set of adversarial examples to cover all the conditions mentioned above.

We note ϵ the amount of the introduced perturbation in adversarial examples. We use PGD with $\epsilon = 0.04$ as a white-box L_∞ bounded attack [33]. We generate targeted adversarial attacks by using PGD_LL, which targets the least likely class according to attacked models [1]. We use MI_FGSM with $\epsilon = 0.04$ and $\epsilon = 0.08$ as black-box attacks [9]. A VGG network trained on the same training set is used to craft these black-box attacks. PGD, PGD_LL and MI_FGSM are computed over ten iterations. We use the Carlini-Wagner attack (CW_2) as a L_2 white-box bounded attack [4]. We perform the optimization process of this attack with 40 iterations and a confidence score of 50.

The gathered common perturbations and adversarial examples constitute the perturbation benchmark. It is used in the experimental section in order to evaluate and compare the robustness of models.

4.2 Training Details

The trained models are either a ResNet-50 [16] or a ResNeXt-50 with the 32×4d template [53]. We use a batch size of 256 and 90 training epochs. The Adam optimizer [24] is used with a learning rate of 0.002 and a weight decay of 10^{-4}. At the end of the epochs 30, 60 and 80, the learning rate is divided by 10. The cost function is the cross entropy. We use the same hyperparameters for all trainings.

All models are trained and tested using ImageNet. Because of a limited computational budget, we used a subset of ImageNet built on 100 randomly chosen classes. For each class, ten percent of the images are preserved for the test set. Then, the training set and the test set contain respectively 10^5 and 10^4 images. The only pre-processing used is the resizing of images to the 224 * 224 format.

We call the models trained without any data augmentation the standard models. We observed that the highest clean accuracy for the models trained with mixup is reached when $\alpha = 0.4$, so we used this value in all experiments. The adversarial examples used in trainings are crafted using FGSM with $\epsilon \sim U[0, 0.025]$. The range of pixel values of images in our experiments is $[0, 1]$.

5 Performance Evaluation

5.1 Robustness Score

To measure the robustness of a model to a perturbation, we compare the performance of this model on clean samples with its performance on perturbed samples:

$$R_N^{\phi} = \frac{A_{\phi}}{A_{clean}} \tag{7}$$

We call R_N^{ϕ} the robustness score of a neural network N towards a perturbation ϕ. A_{clean} is the accuracy of the model on the clean test set and A_{ϕ} is its accuracy on the test set corrupted with the ϕ perturbation. ϕ can be either a common corruption or an adversarial attack.

This robustness score metric should be used carefully because it masks the clean accuracy of neural networks. Indeed, an untrained model that always makes random outputs, would have the same accuracy for clean samples than for corrupted samples. Its robustness scores would be equal to 1, so it could be considered as completely robust to any common corruption. Therefore, in this study, before comparing the robustness scores of two neural networks we always make sure that their clean accuracies are also comparable.

5.2 Performances of M-TLAT on the Perturbation Benchmark

For the first experiment we train one Resnet-50 and one ResNeXt-50, using the M-TLAT algorithm. The training of these models took a dozen of hours using a single GPU Nvidia Tesla V100. We also train one Resnet-50 and one ResNeXt-50 with the IAT algorithm [28]. We compute the robustness scores of the trained models towards all the perturbations of the benchmark. The results are reported in Table 1.

In Tables 1 and 2, the *Clean* column contains the accuracy of the models on the not-corrupted test set. Each of the other columns contains the robustness scores to a perturbation of the benchmark. For the corruptions of the ImageNet-C benchmark, the displayed scores correspond to the mean robustness score of the corruptions computed with their five severity levels. To better visualize the effect of the augmentation algorithms, we use either a "-" index or a "+" index, to signify if a model is less or more robust than the standard model to a perturbation.

We observe in Table 1 that the models trained with M-TLAT are slightly more accurate than the standard models on clean images. They are also more robust than the standard models to every single tested common corruption. We see that using M-TLAT makes neural networks much more robust to the CW_2 and PGD_LL attacks. It also makes models less sensitive to black-box adversarial examples. We observe that the robustness gain for the PGD attack is less important.

For comparison, the IAT algorithm tends to reduce the clean accuracy. It does not increase the robustness to all the common corruption of the benchmark. In particular, it significantly decreases the robustness towards the Jpeg perturbation. Besides, the IAT models are significantly less robust to adversarial examples than the M-TLAT models.

Using FGSM during training is known to poorly defend models against iterative adversarial attack such as PGD [27]. The robustness of the M-TLAT models towards PGD can likely be increased by replacing FGSM by an iterative adversarial attack [33]. But this would increase significantly the training time. That is the reason why this option has not been tested yet.

To our knowledge, M-TLAT is the first data augmentation approach that is able to increase the robustness to every single common corruption and adversarial example of a large set of diverse perturbations, without reducing the clean accuracy.

Table 1. Effect on robustness of M-TLAT and comparison with IAT

(a) Robustness scores towards the common corruptions of ImageNet-C

		Clean	Gauss	Shot	Impul	Defocus	Glass	Motion	Zoom	Snow	Fog	Frost	Bright	Contr	Elastic	Pixelate	Jpeg
ResNet	standard	73.3	0.17	0.17	0.12	0.25	0.35	0.41	0.47	0.31	0.48	0.34	0.78	0.34	0.73	0.65	0.80
	IAT	73.2^-	0.47^+	0.46^+	0.42^+	0.51^+	0.65^+	0.58^+	0.68^+	0.49^+	0.78^+	0.66^+	0.79^+	0.78^+	0.84^+	0.82^+	0.63^-
	M-TLAT	73.9^+	0.56^+	0.54^+	0.52^+	0.41^+	0.61^+	0.58^+	0.63^+	0.49^+	0.59^+	0.61^+	0.82^+	0.58^+	0.85^+	0.94^+	0.95^+
ResNeXt	standard	76.4	0.25	0.25	0.20	0.28	0.37	0.44	0.48	0.36	0.53	0.37	0.79	0.36	0.75	0.72	0.74
	IAT	74.7^-	0.46^+	0.44^+	0.43^+	0.53^+	0.67^+	0.62^+	0.70^+	0.51^+	0.73^+	0.69^+	0.81^+	0.80^+	0.85^+	0.83^+	0.59^-
	M-TLAT	76.5^+	0.57^+	0.55^+	0.54^+	0.44^+	0.64^+	0.60^+	0.66^+	0.52^+	0.68^+	0.67^+	0.86^+	0.70^+	0.85^+	0.95^+	0.95^+

(b) Robustness scores towards our additional common corruptions

		Obstru	Color	Trans	Rot
ResNet	standard	0.74	0.76	0.78	0.68
	IAT	0.71^-	0.89^+	0.75^-	0.72^+
	M-TLAT	0.75^+	0.86^+	0.79^+	0.74^+
ResNeXt	standard	0.75	0.82	0.82	0.72
	IAT	0.72^-	0.90^+	0.77^-	0.71^-
	M-TLAT	0.76^+	0.89^+	0.82	0.74^+

(c) Robustness scores towards adversarial examples

		pgd ϵ=0.04	pgd_ll ϵ=0.04	cw_l2	mi_fgsm ϵ=0.04	mi_fgsm ϵ=0.08
ResNet	standard	0.00	0.00	0.00	0.58	0.34
	IAT	0.01^+	0.08^+	0.84^+	0.87^+	0.78^+
	M-TLAT	0.08^+	0.45^+	1.00^+	0.96^+	0.87^+
ResNeXt	standard	0.00	0.00	0.00	0.58	0.33
	IAT	0.01^+	0.11^+	0.95^+	0.87^+	0.81^+
	M-TLAT	0.09^+	0.38^+	0.99^+	0.95^+	0.87^+

5.3 Complementarity Between Mixup and TLAT

To better understand the effect of each constituent of the M-TLAT algorithm, we proceed to an ablation study. Two ResNet-50 are respectively trained with the Mixup and TLAT data augmentations. We report their robustness to the perturbation benchmark in Table 2.

First, we notice that Mixup causes an increase of the clean accuracy, which is coherent with observations made in [57]. On the contrary, TLAT makes the trained model less accurate on the clean data. But those two effects seem to cancel each other because the M-TLAT model and the standard model have comparable clean accuracies as observed in Table 1a.

In Tables 2a and 2b, we observe that Mixup makes the trained model more robust than the standard model to all the common corruptions but the *Motion Blur*, *Pixelate* and *Jpeg* corruptions. We observe in Table 2c that Mixup has a little influence on adversarial robustness, with either a slight increase or decrease of the robustness depending on the considered attack.

Fortunately, TLAT makes models much more robust to any adversarial attacks. Indeed, the TLAT model is much more difficult to attack with the black box adversarial examples or with the CW_2 attack. It is also significantly more robust to PGD_LL and slightly more robust to PGD. For common corruptions, the effect of TLAT is very contrasted. Concerning the noise and blur corruptions (the seven first corruptions of the Table 2a), the TLAT model is much more robust than the standard model. For some other common corruptions like *Fog* or *Contrast*, the TLAT augmentation decreases significantly the robustness scores.

It is clear that the M-TLAT models are much more robust to adversarial examples thanks to the contribution of TLAT. However, for common corruptions, Mixup and TLAT are remarkably complementary. Concerning the few corruptions for which Mixup has a very negative effect on robustness (*Jpeg* and *Pixelate*), TLAT has a strong positive effect. Similarly, for the *Fog* and *Contrast* corruptions, TLAT makes models less robust while Mixup makes them much more robust.

The ablation study indicates that both components are important to increase the robustness to a large diversity of perturbations.

Table 2. Influence on robustness of the Mixup and TLAT data augmentations

(a) Robustness scores towards the corruptions of ImageNet-C

		Clean	Gauss	Shot	Impul	Defocus	Glass	Motion	Zoom	Snow	Fog	Frost	Bright	Contr	Elastic	Pixelate	Jpeg
ResNet	standard	73.3	0.17	0.17	0.12	0.25	0.35	0.41	0.47	0.31	0.48	0.34	0.78	0.34	0.73	0.65	0.80
	Mixup	74.9^+	0.28^+	0.28^+	0.24^+	0.25	0.38^+	0.39^-	0.53^+	0.38^+	0.79^+	0.53^+	0.78	0.75^+	0.75^+	0.58^-	0.61^-
	TLAT	69.4^-	0.57^+	0.54^+	0.51^+	0.43^+	0.60^+	0.56^+	0.60^+	0.41^+	0.15^-	0.41^+	0.78	0.13^-	0.84^+	0.94^+	0.97^+

(b) Robustness scores towards our additional common corruptions

		Obstru	Color	Trans	Rot
ResNet	standard	0.74	0.76	0.78	0.68
	Mixup	0.75^+	0.88^+	0.79^+	0.71^+
	TLAT	0.69^-	0.76	0.67^-	0.66^-

(c) Robustness scores towards adversarial examples

		pgd $\epsilon=0.04$	pgd_ll $\epsilon=0.04$	cw_l2	mi_fgsm $\epsilon=0.04$	mi_fgsm $\epsilon=0.08$
ResNet	standard	0.00	0.00	0.00	0.58	0.34
	Mixup	0.00	0.00	0.202^+	0.61^+	0.22^-
	TLAT	0.10^+	0.74^+	0.97^+	0.98^+	0.93^+

5.4 Labeling in Adversarial Trainings

Comparison of the Labeling Strategies

Adversarial trainings increase the adversarial robustness of the trained models, but they also reduce their accuracy on clean samples [27,47]. In this section, we want to show that TLAT decreases less the clean accuracy of the trained models than traditional adversarial trainings.

To achieve it, we trained four ResNet-50 with different kinds of adversarial training algorithms. The first model is trained using untarget FGSM and the second is trained using target FGSM with randomly chosen target. Both use adversarial examples labeled with the ground-truth labels. We train another model with target FGSM but regularized via label smoothing (LS) [41], we call it the LS model. For this model, we use a smoothing parameter equal to ϵ, where ϵ is the amount of the FGSM perturbation. In other words, the one values of the one-hot encoding vectors are replaced by $1 - \epsilon$ and the zeros are replaced by ϵ/N where N is the number of classes. The fourth model is trained using the TLAT algorithm. All models are trained with minibatches that contain both clean samples and adversarial examples. We measure the clean accuracy of those models: results are displayed in Table 3.

We see that TLAT is the adversarial training method that reduces the less the clean accuracy. This result shows that TLAT is important to preserve the clean accuracy of the models trained with M-TLAT, all the while making them more robust to adversarial examples.

Using soft labels in trainings is known to help models to generalize [36,41]. Here we want to make sure that the usage of soft labels is not the main reason of high clean accuracy of TLAT. To achieve it, we compare the performances of the TLAT and LS models. Even if the LS model also uses soft labels during training, it performs worse than the TLAT model. Consequently, the good performances of TLAT are not due to the usage of soft labels. We believe TLAT performs well because it uses labels that contain information about the target of the adversarial examples.

Table 3. Comparison of the performances of the TLAT augmentation with the performances of other kinds of adversarial trainings

	Standard	FGSM	target-FGSM	LS	TLAT
Clean accuracy	73.3	65.8	68.3	67.1	69.4

Interpretation

The TLAT augmentation is motivated by the works of Ilyas et al. [19]. In their study, they reveal the existence of brittle yet highly predictive features in the data. Neural networks largely depend on those features even if they are nearly invisible to human eye. They are called non-robust features. Ilyas et al. show that a model that only uses non-robust features to complete properly a task still generalize well on unseen data.

They use this phenomenon to interpret adversarial vulnerability. Adversarial attacks are small perturbations, so they mainly affect non-robust features. They can especially make those brittle features anti-correlated with the true label. As neural networks largely rely on non-robust features, their behaviour is completely disturbed by adversarial attacks.

In Adversarial trainings, neural networks encounter adversarial examples that can have non-robust features uncorrelated with the true label. The trained models are then constrained to less rely on non-robust features. This can explain the success of adversarial training: adversarially trained models give less importance to non-robust features so they are much more difficult to attack with small perturbations.

Despite its efficiency, adversarial training generally causes a decrease in clean accuracy [27,47,56]. One possible reason could be that adversarial patterns in the training adversarial examples are not coherent with the ground-truth label. Consequently, an adversarially trained model encounters samples for which the features are not completely coherent with labelling.

The proposed method tries to make the labeling of target adversarial examples more correlated with its non-robust features. Targeted adversarial attacks make non-robust features of a sample correlated with a target class. So instead of labelling only with the ground-truth class of the attacked sample, the method introduces a part related to the target class. Therefore, the trained model still learns the true original class of the attacked sample, but the label used for learning is more correlated with the non-robust features of this sample. We believe this could be the reason why our method has better performance than traditional target adversarial training in practice.

6 Conclusion

We propose a new data augmentation strategy that increases the robustness of neural networks to a large set of common corruptions and adversarial examples. The experiments carried out suggest that the effect of M-TLAT is always positive: basically, it increases the robustness to any corruption without reducing the clean accuracy. We believe using M-TLAT can be particularly useful to help industrials to increase the robustness of their neural networks without being afraid of any counterpart.

As part of the M-TLAT algorithm, we use the new adversarial augmentation strategy TLAT. We show that models trained with TLAT have a better accuracy on clean samples than the models trained with a standard adversarial training algorithm. The idea of TLAT is to interpolate the target labels of adversarial examples with the ground-truth labels. This operation is computationally negligible and can be used in any trainings with target adversarial examples in order to improve the clean accuracy.

In future works, we would like to replace FGSM by an iterative adversarial attack in our algorithm and observe how this would influence the clean accuracy and the adversarial robustness of the models trained with M-TLAT. It would be also interesting to replace Mixup by a different interpolation strategy such as Manifold Mixup [49], and test if it further improves the performances of the algorithm.

References

1. Kurakin, A., Ian, J., Goodfellow, S.B.: Adversarial examples in the physical world. In: International Conference on Learning Representations (2017)
2. Archambault, G.P., Mao, Y., Guo, H., Zhang, R.: Mixup as directional adversarial training. arXiv preprint arXiv:1906.06875 (2019)
3. Athalye, A., Carlini, N., Wagner, D.: Obfuscated gradients give a false sense of security: circumventing defenses to adversarial examples. In: Proceedings of the 35th International Conference on Machine Learning, pp. 274–283. PMLR (2018)
4. Carlini, N., Wagner, D.: Towards evaluating the robustness of neural networks. In: 2017 IEEE Symposium on Security and Privacy (SP), pp. 39–57, May 2017
5. Carlini, N., et al.: On evaluating adversarial robustness. arXiv preprint arXiv:1902.06705 (2019)
6. Dhillon, G.S., et al.: Stochastic activation pruning for robust adversarial defense. In: International Conference on Learning Representations (2018)
7. Dodge, S., Karam, L.: Understanding how image quality affects deep neural networks (2016)
8. Dodge, S.F., Karam, L.J.: Quality resilient deep neural networks. CoRR abs/1703.08119 (2017)
9. Dong, Y., et al.: Boosting adversarial attacks with momentum. In: The IEEE Conference on Computer Vision and Pattern Recognition (CVPR), June 2018
10. Engstrom, L., Tran, B., Tsipras, D., Schmidt, L., Madry, A.: Exploring the landscape of spatial robustness. In: Chaudhuri, K., Salakhutdinov, R. (eds.) Proceedings of the 36th International Conference on Machine Learning. Proceedings of Machine Learning Research, vol. 97, pp. 1802–1811. PMLR (2019)
11. Engstrom, L., Tsipras, D., Schmidt, L., Madry, A.: A rotation and a translation suffice: fooling CNNs with simple transformations. ArXiv abs/1712.02779 (2017)
12. Gatys, L.A., Ecker, A.S., Bethge, M.: Image style transfer using convolutional neural networks. In: 2016 IEEE Conference on Computer Vision and Pattern Recognition (CVPR), pp. 2414–2423, June 2016
13. Geirhos, R., Rubisch, P., Michaelis, C., Bethge, M., Wichmann, F.A., Brendel, W.: ImageNet-trained CNNs are biased towards texture; increasing shape bias improves accuracy and robustness. In: International Conference on Learning Representations (2019)
14. Geirhos, R., Temme, C.R.M., Rauber, J., Schütt, H.H., Bethge, M., Wichmann, F.A.: Generalisation in humans and deep neural networks. In: Advances in Neural Information Processing Systems, vol. 31, pp. 7538–7550. Curran Associates Inc. (2018)
15. Goodfellow, I.J., Shlens, J., Szegedy, C.: Explaining and harnessing adversarial examples. CoRR abs/1412.6572 (2015)
16. He, K., Zhang, X., Ren, S., Sun, J.: Deep residual learning for image recognition. In: 2016 IEEE Conference on Computer Vision and Pattern Recognition (CVPR), pp. 770–778, June 2016

17. Hendrycks, D., Dietterich, T.: Benchmarking neural network robustness to common corruptions and perturbations. In: Proceedings of the International Conference on Learning Representations (2019)
18. Hendrycks, D., Mu, N., Cubuk, E.D., Zoph, B., Gilmer, J., Lakshminarayanan, B.: AugMix: a simple method to improve robustness and uncertainty under data shift. In: International Conference on Learning Representations (2020)
19. Ilyas, A., Santurkar, S., Tsipras, D., Engstrom, L., Tran, B., Madry, A.: Adversarial examples are not bugs, they are features. In: Advances in Neural Information Processing Systems, vol. 32, Curran Associates Inc. (2019)
20. Inoue, H.: Data augmentation by pairing samples for images classification. arXiv preprint arXiv:1801.02929 (2018)
21. Kang, D., Sun, Y., Hendrycks, D., Brown, T., Steinhardt, J.: Testing robustness against unforeseen adversaries. arXiv preprint arXiv:1908.08016 (2019)
22. Kannan, H., Kurakin, A., Goodfellow, I.: Adversarial logit pairing. arXiv preprint arXiv:1803.06373 (2018)
23. Karahan, S., Kilinc Yildirum, M., Kirtac, K., Rende, F.S., Butun, G., Ekenel, H.K.: How image degradations affect deep CNN-based face recognition? In: 2016 International Conference of the Biometrics Special Interest Group (BIOSIG), pp. 1–5, September 2016
24. Kingma, D.P., Ba, J.: Adam: a method for stochastic optimization. In: Bengio, Y., LeCun, Y. (eds.) 3rd International Conference on Learning Representations, ICLR (2015)
25. Koziarski, M., Cyganek, B.: Image recognition with deep neural networks in presence of noise - dealing with and taking advantage of distortions. Integ. Comput.-Aid. Eng. **24**, 1–13 (2017)
26. Krizhevsky, A., Sutskever, I., Hinton, G.E.: ImageNet classification with deep convolutional neural networks. In: Proceedings of the 25th International Conference on Neural Information Processing Systems, NIPS 2012, vol. 1. pp. 1097–1105. Curran Associates Inc. USA (2012)
27. Kurakin, A., Goodfellow, I.J., Bengio, S.: Adversarial machine learning at scale. In: International Conference on Learning Representations (2017)
28. Lamb, A., Verma, V., Kannala, J., Bengio, Y.: Interpolated adversarial training: achieving robust neural networks without sacrificing too much accuracy. In: Proceedings of the 12th ACM Workshop on Artificial Intelligence and Security, pp. 95–103 (2019)
29. Laugros, A., Caplier, A., Ospici, M.: Are adversarial robustness and common perturbation robustness independant attributes ? In: The IEEE International Conference on Computer Vision (ICCV) Workshops, October 2019
30. Liu, A., Liu, X., Zhang, C., Yu, H., Liu, Q., He, J.: Training robust deep neural networks via adversarial noise propagation. arXiv preprintarXiv:1909.09034 (2019)
31. Liu, D., Cheng, B., Wang, Z., Zhang, H., Huang, T.S.: Enhance visual recognition under adverse conditions via deep networks. IEEE Trans. Image Process. **28**(9), 4401–4412 (2019)
32. Liu, Z., Luo, P., Wang, X., Tang, X.: Deep learning face attributes in the wild. In: 2015 IEEE International Conference on Computer Vision (ICCV), pp. 3730–3738, December 2015
33. Madry, A., Makelov, A., Schmidt, L., Tsipras, D., Vladu, A.: Towards deep learning models resistant to adversarial attacks. In: International Conference on Learning Representations (2018)
34. Orhan, A.E.: Robustness properties of Facebook's resnext WSL models. CoRR abs/1907.07640 (2019)

35. Pang, T., Xu, K., Zhu, J.: Mixup inference: better exploiting mixup to defend adversarial attacks. In: International Conference on Learning Representations (2020)
36. Papernot, N., McDaniel, P., Wu, X., Jha, S., Swami, A.: Distillation as a defense to adversarial perturbations against deep neural networks. In: 2016 IEEE Symposium on Security and Privacy (SP) (2016)
37. Papernot, N., McDaniel, P.D., Goodfellow, I.J., Jha, S., Celik, Z.B., Swami, A.: Practical black-box attacks against machine learning. In: ACM on Asia Conference on Computer and Communications Security (2017)
38. Ren, S., He, K., Girshick, R., Sun, J.: Faster R-CNN: towards real-time object detection with region proposal networks. IEEE Trans. Pattern Anal. Mach. Intell. **39**, 1137–1149 (2015)
39. Samangouei, P., Kabkab, M., Chellappa, R.: Defense-GAN: protecting classifiers against adversarial attacks using generative models. In: International Conference on Learning Representations (2018)
40. Schmidt, L., Santurkar, S., Tsipras, D., Talwar, K., Madry, A.: Adversarially robust generalization requires more data. In: Advances in Neural Information Processing Systems, vol. 31, pp. 5014–5026. Curran Associates Inc. (2018)
41. Szegedy, C., Vanhoucke, V., Ioffe, S., Shlens, J., Wojna, Z.: Rethinking the inception architecture for computer vision. In: The IEEE Conference on Computer Vision and Pattern Recognition (CVPR), June 2016
42. Szegedy, C., et al.: Intriguing properties of neural networks. In: International Conference on Learning Representations (2014)
43. Taigman, Y., Yang, M., Ranzato, M., Wolf, L.: DeepFace: closing the gap to human-level performance in face verification. In: 2014 IEEE Conference on Computer Vision and Pattern Recognition, pp. 1701–1708, June 2014
44. Tokozume, Y., Ushiku, Y., Harada, T.: Between-class learning for image classification. In: 2018 IEEE/CVF Conference on Computer Vision and Pattern Recognition, pp. 5486–5494, June 2018
45. Tramer, F., Boneh, D.: Adversarial training and robustness for multiple perturbations. In: Advances in Neural Information Processing Systems, vol. 32. Curran Associates Inc. (2019)
46. Tramèr, F., Kurakin, A., Papernot, N., Goodfellow, I., Boneh, D., McDaniel, P.: Ensemble adversarial training: attacks and defenses. In: International Conference on Learning Representations (2018)
47. Tsipras, D., Santurkar, S., Engstrom, L., Turner, A., Madry, A.: Robustness may be at odds with accuracy. In: International Conference on Learning Representations (2019)
48. Vasiljevic, I., Chakrabarti, A., Shakhnarovich, G.: Examining the impact of blur on recognition by convolutional networks. arXiv preprintarXiv:1611.05760 (2016)
49. Verma, V., et al.: Manifold mixup: Better representations by interpolating hidden states. In: Proceedings of the 36th International Conference on Machine Learning. pp. 6438–6447. PMLR (2019)
50. Wang, J., Zhang, H.: Bilateral adversarial training: towards fast training of more robust models against adversarial attacks. In: The IEEE International Conference on Computer Vision (ICCV), October 2019
51. Wang, S., et al.: Defensive dropout for hardening deep neural networks under adversarial attacks. In: 2018 IEEE/ACM International Conference on Computer-Aided Design (ICCAD), pp. 1–8 (2018)
52. Xie, Q., Hovy, E.H., Luong, M.T., Le, Q.V.: Self-training with noisy student improves imagenet classification. ArXiv abs/1911.04252 (2019)

53. Xie, S., Girshick, R., Dollár, P., Tu, Z., He, K.: Aggregated residual transformations for deep neural networks. In: 2017 IEEE Conference on Computer Vision and Pattern Recognition (CVPR) (2017)
54. Xu, W., Evans, D., Qi, Y.: Feature squeezing: detecting adversarial examples in deep neural networks. In: 25th Annual Network and Distributed System Security Symposium (2018)
55. Yin, D., Lopes, R.G., Shlens, J., Cubuk, E.D., Gilmer, J.: A Fourier perspective on model robustness in computer vision. In: ICML Workshop on Uncertainty and Robustness in Deep Learning (2019)
56. Zhang, H., Yu, Y., Jiao, J., Xing, E., Ghaoui, L.E., Jordan, M.: Theoretically principled trade-off between robustness and accuracy. In: Proceedings of the 36th International Conference on Machine Learning, pp. 7472–7482 (2019)
57. Zhang, H., Cisse, M., Dauphin, Y.N., Lopez-Paz, D.: Mixup: beyond empirical risk minimization. In: International Conference on Learning Representations (2018)
58. Zhou, Y., Song, S., Cheung, N.: On classification of distorted images with deep convolutional neural networks. In: 2017 IEEE International Conference on Acoustics, Speech and Signal Processing (ICASSP), pp. 1213–1217, March 2017

What Does CNN Shift Invariance Look Like? A Visualization Study

Jake Lee[1(✉)], Junfeng Yang[1], and Zhangyang Wang[2]

[1] Columbia University, New York, NY 10027, USA
jake.h.lee@jpl.nasa.gov, junfeng@cs.columbia.edu
[2] The University of Texas at Austin, Austin, TX 78712, USA
atlaswang@utexas.edu

Abstract. Feature extraction with convolutional neural networks (CNNs) is a popular method to represent images for machine learning tasks. These representations seek to capture global image content, and ideally should be independent of geometric transformations. We focus on measuring and visualizing the shift invariance of extracted features from popular off-the-shelf CNN models. We present the results of three experiments comparing representations of millions of images with exhaustively shifted objects, examining both local invariance (within a few pixels) and global invariance (across the image frame). We conclude that features extracted from popular networks are not globally invariant, and that biases and artifacts exist within this variance. Additionally, we determine that anti-aliased models significantly improve local invariance but do not impact global invariance. Finally, we provide a code repository for experiment reproduction, as well as a website to interact with our results at https://jakehlee.github.io/visualize-invariance.

Keywords: Feature extraction · Shift invariance · Robust recognition

1 Introduction

Convolutional neural networks (CNNs) are able to achieve state-of-the-art performance on computer vision tasks such as object classification [10,12] and image segmentation [9]. Transfer learning methods [24] allow tasks to leverage models pre-trained on large, generic datasets such as ImageNet [5] and MIT Places [26] instead of training a CNN from scratch, which can be costly. One popular method is extracting neuron activations of a layer in the pre-trained CNN and treating them as feature representations of the input image [22]. Using these feature vectors with machine learning methods result in competitive performance for classification in different domains [21,22], content based image retrieval [2], and novel image detection [13].

However, it is known that CNNs, and therefore CNN features, lack geometric invariance. Simple, small transformations such as translations and rotations can significantly impact classification accuracy [1,7,17]. For feature extraction,

A. Bartoli and A. Fusiello (Eds.): ECCV 2020 Workshops, LNCS 12539, pp. 196–210, 2020.
https://doi.org/10.1007/978-3-030-68238-5_15

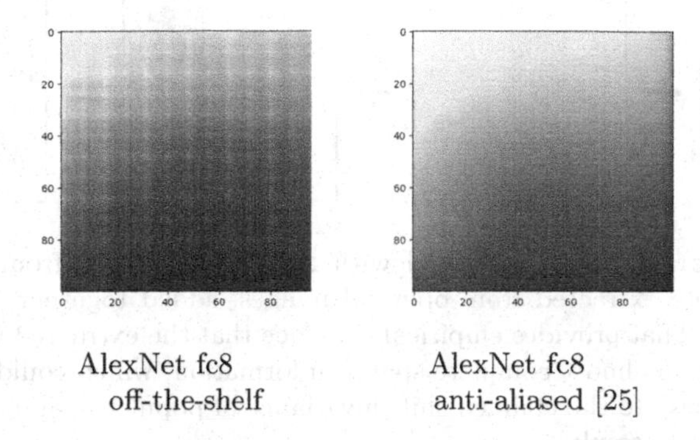

Fig. 1. An example heatmap of cosine similarities as an indicator for shift invariance, using features extracted from the last fully-connected layer of AlexNet. Features from each shift location are compared to the features of an image with the object at the top left. Brighter colors at each shift location indicate more similar features. Refer to Sect. 3.3 for more results.

geometric invariance may be desired to retrieval a global descriptor of the image robust to minor changes in image capture. This goal is especially relevant for content based image retrieval tasks, as images only a small shift or rotation apart should result in similar features [2]. However, any task that relies on CNN feature extraction would benefit from models more robust to geometric transformations, as the extracted features would better represent the content of the images.

Several methods to improve geometric invariance have been proposed, including geometric training dataset augmentation [23], spatial transformer networks [11], and anti-aliasing [25]. For improving the invariance of extracted features, methods have been proposed in the context of image retrieval [8] by post-processing the extracted features instead of making the model itself more robust. Despite these proposals, it remains questionable to what extent we can trust state-of-the-art classifier networks as translation-invariant, even with those add-ons, and there is a lack of principled study on examining that either qualitatively or quantitatively.

In this work, we examine the shift invariance of features extracted from off-the-shelf pre-trained CNNs with visualizations and quantitative experiments. These experiments take advantage of exhaustive testing of the input space, extracting features from millions of images with objects at different locations in the image frame. Such fine-grained experiments allow for observing invariance from *two different lenses*: both **global shift invariance**, defined the maximal coverage of object location translations that the feature can stay relatively invariant to (based some feature similarity threshold); and **local shift invariance**, defined as the relative invariance changes when perturbing the object location for just a few pixels. Note that most literature discussed the translation invari-

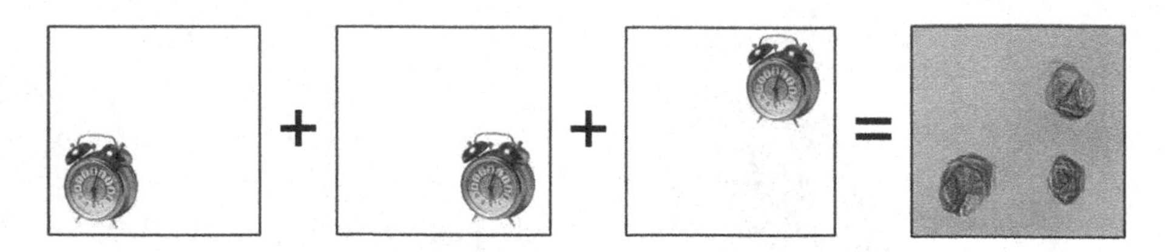

Fig. 2. "Feature arithmetic" performed with features extracted from AlexNet's fc7 layer. Features were extracted from operand images, added together, and visualized with DeepSiM [6]. That provides empirical evidence that the extracted fully-connected features still preserve almost complete spatial information, which could be considered to "counterexamples" to the claimed shift invariance of popular deep models. Refer to Sect. 3.5 for complete results.

ance through the latter lens [23,25], while we advocate both together for more holistic and fine-grained understanding.

We focus on features extracted from fully-connected layers of the models, as they are often assumed to be more geometrically invariant [14] than convolutional feature maps. We also compare the robustness of standard pre-trained models [4], and those models trained by a state-of-the-art anti-aliasing technique to boost translation invariance [25]. We draw the following main observations:

- Visualizing cosine similarities of features from shifted images show that almost all existing pre-trained classifiers' extracted features, even from high-level fully-connected layers, are brittle to input translations. Those most accurate models, such as ResNet-50, suffer even more from translation fragility than simpler ones such as AlexNet.
- Interestingly, we observe an empirical bias towards greater similarity for horizontally shifted images, compared to vertical translations. Also, a grid pattern was universally observed in vanilla pre-trained models, showing significant local invariance fluctuations and concurring the observation in [25]. An example is shown in Fig. 1a.
- Antialiased models [25] are able to suppress such local fluctuations and improve local invariance, but do not visibly improve the global invariance. An example is shown in Fig. 1b.
- A side product that we create is a "feature arithmetic" demonstration: adding or subtracting features extracted from images with shifted object locations, then visualizing the result hidden feature to the pixel domain. That results in images with objects added or removed spatially correctly, despite the lack of any latent space optimization towards that goal. An example is shown in Fig. 2, which we suggest may serve as another "counterexamples" to the claimed shift invariance of popular deep models.

2 Methods

We perform three sets of experiments to describe and quantify the robustness of extracted features to object translation. First, we measure the sensitivity of extracted features to translation by calculating cosine similarities between features of translated patches (Sect. 2.1). Next, we train a linear SVM on extracted features of translated patches to measure basic separability (Sect. 2.2). Finally, we demonstrate that extracted features can be added and subtracted, and coherent spatial information can still be recovered (Sect. 2.3).

2.1 Feature Similarity

We adopt Zhang's definition of shift invariance [25]: an extracted feature is shift invariant if shifting the input image results in an identical extracted feature. To quantify the invariance, we exhaustively test the input space. We define a segmented object image patch and generate all translations of the patch across a white background with a stride of one pixel. Compared to an alternative method—exhaustively cropping a smaller image from a larger image (similarly to data augmentation to prevent overfitting [12]), our method ensures that the pixel content is exactly the same for every image and the only variation is the geometric transformation.

After feature extraction, we calculate the cosine similarity between vectors. Since similarity is comparative, we define five anchor patch locations to compare vectors against: top left, top right, center, bottom left, and bottom right. By our definition, a completely shift-invariant model would have identical extracted features for every image, and a cosine similarity of 1 for every comparison. However, a less shift-invariant model would have lower cosine similarities. By observing the results for each anchor, we can determine if the features are more sensitive to shifts in a certain axis, or if certain corners of the image are more sensitive than others.

2.2 Feature Separability

To evaluate the shift invariance of features beyond consine similarity, we also measure the separability of the extracted features. While the previous experiments quantify feature similarity, it does not show whether spatial information is well-encoded by the features. It is possible that the features are not shift invariant, but the differences due to shifting are random. It is also possible that the changes are well-correlated with patch location, and therefore separable.

To measure this separability, we train a linear SVM classifier on the extracted features to determine whether the patch is on the top or bottom half of the image, and another to determine whether the patch is on the left or right half of the image. We perform 5-fold stratified cross-validation for each patch object.

A completely shift invariant model would generate identical features for every image, resulting in random classifier accuracy (50%). A less shift invariant model may still generate features not correlated with the patch location. In this case, we can still expect near-random classifier accuracy. Finally, a less shift invariant model may generate features well-correlated with the patch location, resulting in higher classifier accuracy.

2.3 Feature Arithmetic

Finally, we consider a case in which extracted features have encoded shift information and spatial information very well. It may be possible to manipulate extracted features and still recover information, similarly to latent space interpolation or arithmetic in autoencoders and generative adversarial networks (GANs) [3,18]. Dosovitskiy and Brox have previously shown that images can be recovered from features extracted from layers of AlexNet, as well as from interpolated features [6]. Performing feature arithmetic with these features may give us further insight into the shift invariance of extracted features.

We perform feature arithmetic by extracting features of images with objects in different locations. We then add or subtract the extracted features, hoping to add or remove images of said objects in different locations. Finally, we visualize these features with the GAN model proposed by Dosovitskiy and Brox (referred to as DeePSiM) [6]. We only perform these experiments with features extracted from the fc6, fc7, and fc8 layers of AlexNet [12], as DeePSiM architectures were only provided for those layers. However, we make a slight adjustment and use features extracted prior to ReLU activation to take advantage of the additional information. We used re-trained DeePSim model weights provided by Lee and Wagstaff [13] for this modification.

If the extracted features encode location information well, then the visualizations of the results of arithmetic should be similar to what we would have expected had we performed the arithmetic in pixel space. It should be noted that, in contrast to unsupervised GANs trained to learn an efficient latent space that can be recovered [18], AlexNet had no motivation to learn such a latent space when training for ImageNet object classification. Therefore, it is already impressive that the input image can be successfully recovered from features extracted from the fully connected layers. If meaningful images can be recovered from added or subtracted features, it would further show that the features encode spatial information in addition to class and object information.

3 Results

3.1 Datasets

For our feature similarity and separability experiments, we use 20 segmented objects each from 10 classes in the COCO 2017 training set as patches [15]: `car`, `airplane`, `boat`, `dog`, `bird`, `zebra`, `orange`, `banana`, `clock`, and `laptop`. These classes were selected because of their range of object types, and because the same classes also exist in ImageNet [5], on which all of our models being evaluated are trained. This does not mean that the results of the experiment are exclusive to ImageNet, however; feature extraction is known to be applicable domains outside of the training set [22].

The white background images are 224×224 in size. We resize each segmented and cropped object to 75×75 (small) and 125×125 (large) to explore translating patches of different sizes. Each patch is then translated with stride 1 to every possible location in the 224×224 frame. This results in 22.5k images for the smaller patch and 10k images for the larger patch. In total, for 200 unique patches, there are 4.5 million small-patch and 2 million large-patch images.

For our feature arithmetic experiments, we use a handcrafted dataset of object image patches against a white background. We place some of the same image patches at specific locations so that, when added or subtracted, multiple patches can appear or patches can be removed. These images are described in further detail in Sect. 3.5.

3.2 Feature Extraction Models

For our feature similarity and separability experiments, we evaluate the following popular pre-trained models provided by Pytorch [16]: AlexNet [12], ResNet-50 [10], and MobileNetV2 [20].

We extract features from all three fully-connected layers of AlexNet to determine if and how shift invariance changes deeper into the network. ResNet-50 has only one fully-connected layer (the final classification layer prior to softmax activation), but is far deeper overall than AlexNet, enabled by its skip connections. It also has a global average pooling layer prior to the final fully connected layer. Finally, MobileNetV2 also only has one fully-connected layer as the final classification layer prior to softmax activation.

We also evaluate the anti-aliased versions of these models by Zhang [25], as they claim to improve shift invariance and classification accuracy. We use the Bin-5 filter, as they reported the best "consistency" compared to other filters.

All models were pre-trained on ILSVRC2012 [5] with data augmentation.[1] First, a crop of random size (from 0.08 to 1.0 of the original size) and random aspect ratio (3/4 to 4/3) is made, and is resized to 224×224. Second, the image is flipped horizontally with 0.5 probability. For each model, we extract features from fully connected layers prior to ReLU or Softmax activation.

3.3 Feature Similarity

First, observe the similarities of features extracted from the large-patch dataset. Fig. 3 shows heatmaps visualizing the average cosine similarities of 200 object patches at each shift location. Only comparisons with the center (C) and top-left (TL) anchor points are shown. Heatmaps for all comparison anchor points can be seen at https://jakehlee.github.io/visualize-invariance.

Overall, features extracted from the fc7 layer of AlexNet seem to be the most shift invariant, with the anti-aliased ResNet-50 close behind. This is also supported by Table 1, which reports the mean and standard deviation of the mean cosine similarity for each object patch.

Interestingly, while similarities with the center anchor remain fairly consistent (with lower similarities at the corners), results for the top-left anchor show that

[1] https://github.com/pytorch/vision/blob/master/references/classification/train.py#L96-L103.

extracted features remain more similar when the patch is shifted horizontally than vertically. This is most clearly seen in Fig. 3f. Additionally, the heatmap in Fig. 3a takes the form of an oval stretched horizontally, indicating the same bias. We hypothesized that this may be due to random horizontal flipping included in the training data augmentation. The model may have encountered more objects varying horizontally than vertically during training, resulting in greater invariance along that axis.

To test this hypothesis, we trained AlexNet models on the ILSVRC2012 ImageNet training set [19] with and without random horizontal flip augmentation for 5 epochs using PyTorch-provided training hyperparameters.[2] While the original models were trained for 90 epochs, we believe 5 epochs are sufficient for comparison. Heatmaps generated with these two models are shown in Fig. 5. The heatmaps reveal no significant differences between the two models, rejecting our hypothesis. The bias must be a result of the dataset or architecture, not the training set augmentation.

Additionally, we can observe the general improvement that anti-aliasing provides to shift invariance and overall consistency. In the heatmaps for all layers of off-the-shelf models, there is a clear grid pattern visible, most distinct in AlexNet's fc8, ResNet-50, and MobileNetV2. The anti-aliased models either completely eliminate or significantly suppress this grid pattern, improving the consistency and local invariance. Additionally, for some layers, the antialiased model also slightly improves overall similarity and global invariance.

Similar patterns are visible in the heatmaps for the small-patch dataset in Fig. 4 and the aggregate cosine similarities in Table 2. In fact, these phenomena are even more pronounced, as smaller patches can be shifted further in the image frame.

3.4 Feature Separability

Next, we investigate the separability of extracted features. After we perform 5-fold cross validation for each classifier on each object patch, we report the mean of the mean accuracies and the mean of the standard deviations for the cross validations. These metrics are reported for both the left-right classifier and the top-bottom classifier. Table 3 reports the metrics for the large-patch dataset while Table 4 reports the metrics for the small-patch dataset.

The results clearly show that extracted features are well-correlated with patch location. AlexNet's fc6 layer, for example, had a 99.9% accuracy for left-right classification. However, the top-bottom classifiers show a significantly lower accuracy by about 10%, as well as a larger standard deviation in cross validation. This seems to conflict with our observations in feature similarity, in which horizontally shifted images had more similar features. Intuitively, more similar features should be less separable.

We propose that features extracted from horizontally shifted patches are more similar and more correlated. While less values differ between horizontally shifted

[2] https://github.com/pytorch/vision/tree/master/references/classification.

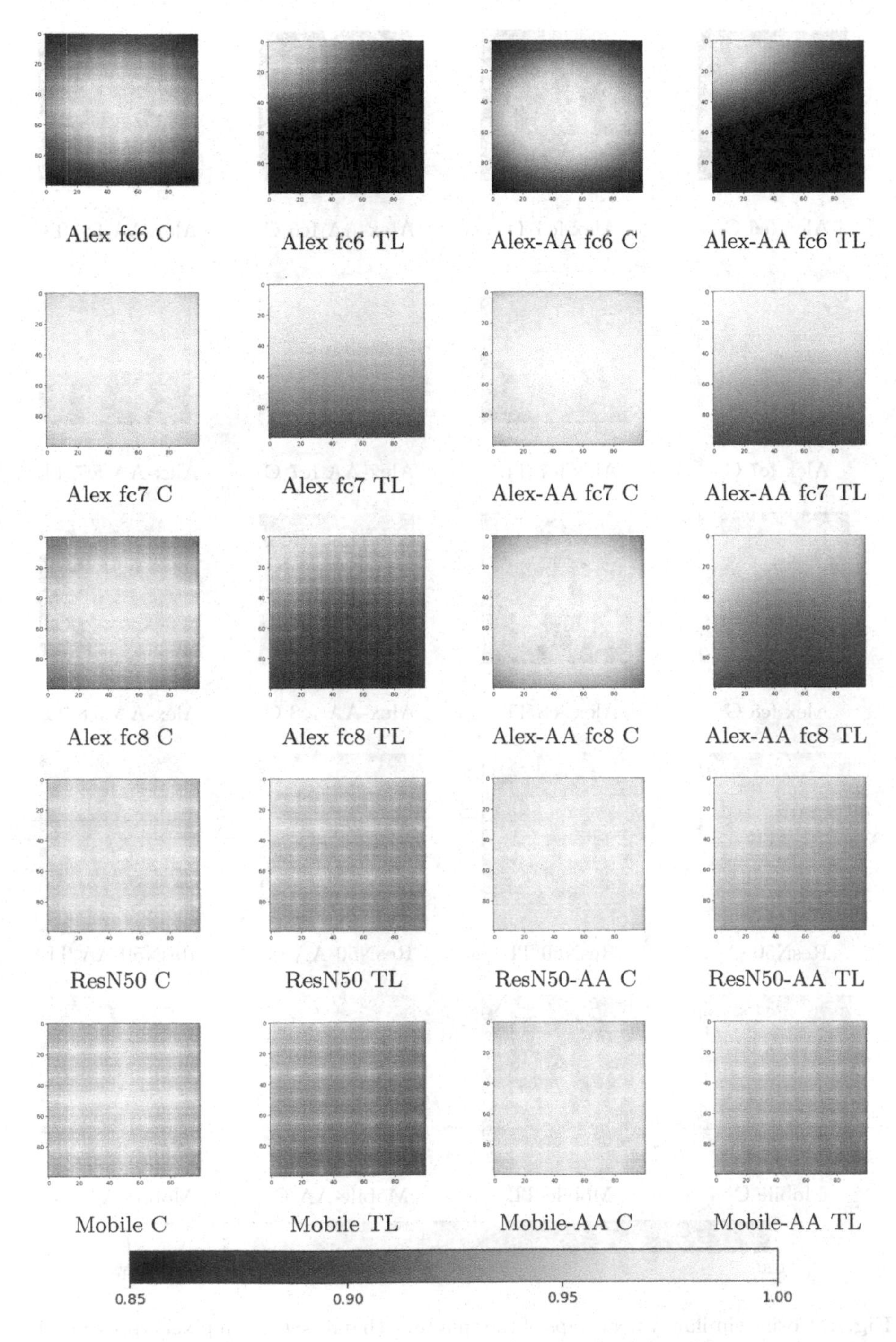

Fig. 3. Cosine similarity heatmaps of the large-patch dataset. Each pixel represents the average cosine similarity of different object patches at that shift location. C indicates the center anchor and TL indicates the top-left anchor.

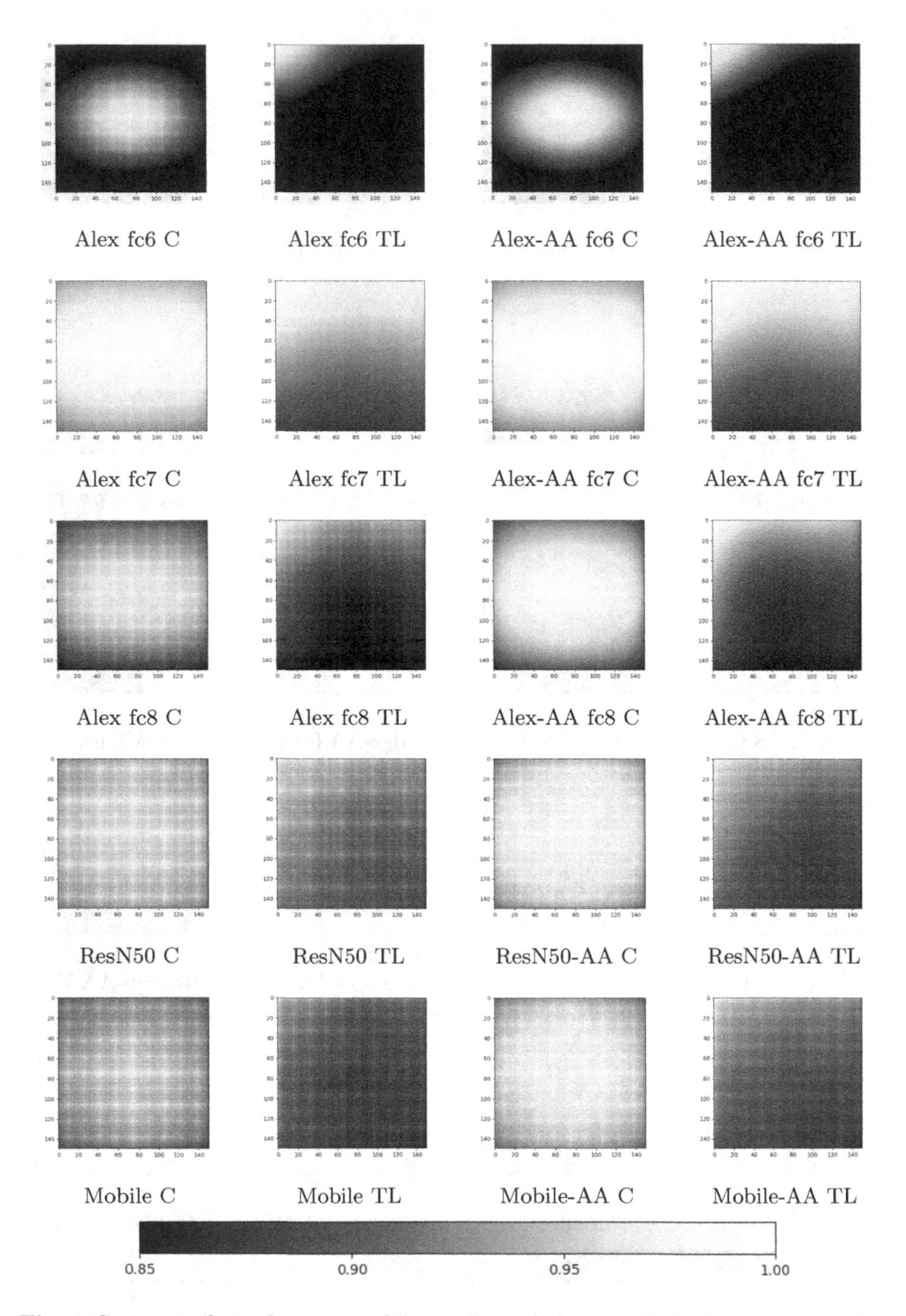

Fig. 4. Cosine similarity heatmaps of the small-patch dataset. Each pixel represents the average cosine similarity of different object patches at that shift location. C indicates the center anchor and TL indicates the top-left anchor.

Table 1. Feature similarity experiment metrics for the large-patch dataset. $\bar{x}$ is the mean of the mean accuracies and σ is the standard deviation of the mean accuracies.

model	layer	top-left		top-right		center		bottom-left		bottom-right	
		$\bar{x}$	σ	$\bar{x}$	σ	$\bar{x}$	σ	$\bar{x}$	σ	$\bar{x}$	σ
AlexNet	fc6	0.830	0.030	0.823	0.031	0.928	0.012	0.822	0.031	0.816	0.032
	fc7	**0.949**	0.016	**0.946**	0.018	0.977	0.007	0.945	0.015	**0.944**	0.017
	fc8	0.918	0.032	0.908	0.043	0.955	0.016	0.912	0.032	0.908	0.035
ResNet-50	fc	0.945	0.022	0.935	0.044	0.961	0.015	0.942	0.025	0.930	0.042
MobileV2	fc	0.931	0.029	0.920	0.040	0.954	0.017	0.930	0.027	0.917	0.041
AlexNet-AA	fc6	0.826	0.029	0.819	0.032	0.936	0.009	0.819	0.031	0.811	0.032
	fc7	0.947	0.016	0.945	0.020	**0.981**	0.006	0.944	0.016	0.941	0.019
	fc8	0.930	0.028	0.916	0.051	0.970	0.010	0.925	0.029	0.911	0.045
ResNet-50-AA	fc	0.948	0.023	0.943	0.031	0.975	0.011	**0.948**	0.028	0.938	0.035
MobileV2-AA	fc	0.943	0.024	0.934	0.034	0.970	0.012	0.941	0.025	0.934	0.033

Table 2. Feature similarity experiment metrics for the small-patch dataset. $\bar{x}$ is the mean of the mean accuracies and σ is the standard deviation of the mean accuracies.

model	layer	top-left		top-right		center		bottom-left		bottom-right	
		$\bar{x}$	σ	$\bar{x}$	σ	$\bar{x}$	σ	$\bar{x}$	σ	$\bar{x}$	σ
AlexNet	fc6	0.754	0.046	0.744	0.046	0.874	0.021	0.737	0.048	0.731	0.049
	fc7	**0.931**	0.015	**0.928**	0.018	0.967	0.007	**0.923**	0.017	**0.923**	0.019
	fc8	0.889	0.039	0.877	0.053	0.937	0.022	0.865	0.051	0.863	0.058
ResNet-50	fc	0.921	0.031	0.908	0.050	0.949	0.019	0.913	0.039	0.892	0.063
MobileV2	fc	0.897	0.043	0.883	0.058	0.935	0.026	0.882	0.047	0.868	0.066
AlexNet-AA	fc6	0.740	0.043	0.733	0.042	0.882	0.017	0.724	0.046	0.720	0.044
	fc7	0.927	0.017	0.926	0.019	**0.971**	0.006	0.921	0.018	0.920	0.019
	fc8	0.898	0.039	0.885	0.067	0.956	0.013	0.879	0.051	0.877	0.058
ResNet-50-AA	fc	0.909	0.040	0.906	0.052	0.962	0.015	0.906	0.046	0.900	0.052
MobileV2-AA	fc	0.908	0.033	0.900	0.044	0.956	0.016	0.904	0.039	0.898	0.042

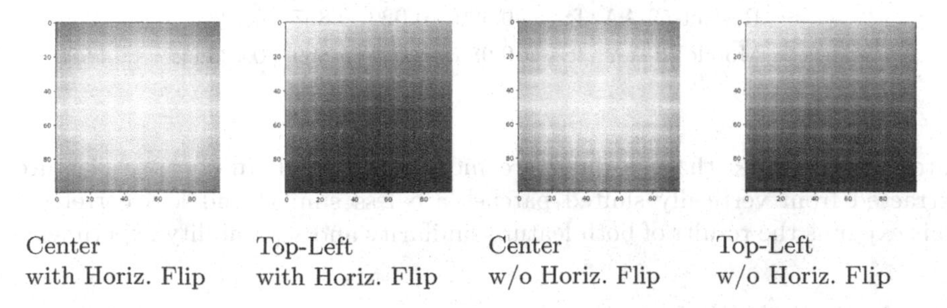

Fig. 5. Cosine similarity heatmaps of AlexNet fc8 features on the large-patch dataset. Models were trained with and without random horizontal flip augmentation for 5 epochs.

Table 3. Feature separability experiment metrics for the large-patch dataset. $\overline{Acc}$ is the mean of the mean accuracies and $\bar{\sigma}$ is the mean of the standard deviations of cross validations.

model	layer	left-right		top-bottom	
		$\overline{Acc}$	$\bar{\sigma}$	$\overline{Acc}$	$\bar{\sigma}$
AlexNet	fc6	0.999	0.002	0.898	0.126
	fc7	0.995	0.009	0.896	0.128
	fc8	0.984	0.023	0.896	0.129
ResNet-50	fc	0.992	0.012	0.889	0.118
MobileV2	fc	0.989	0.015	**0.884**	0.119
AlexNet-AA	fc6	0.998	0.004	0.898	0.125
	fc7	0.988	0.020	0.899	0.124
	fc8	**0.976**	0.036	0.899	0.124
ResNet-50-AA	fc	0.990	0.016	0.893	0.124
MobileV2-AA	fc	0.989	0.017	0.891	0.119

Table 4. Feature separability experiment metrics for the small-patch dataset. $\overline{Acc}$ is the mean of the mean accuracies and $\bar{\sigma}$ is the mean of the standard deviations of cross validations.

model	layer	left-right		top-bottom	
		$\overline{Acc}$	$\bar{\sigma}$	$\overline{Acc}$	$\bar{\sigma}$
AlexNet	fc6	0.997	0.004	0.897	0.127
	fc7	0.985	0.022	0.896	0.129
	fc8	0.966	0.043	0.895	0.130
ResNet-50	fc	0.980	0.029	**0.893**	0.124
MobileV2	fc	0.984	0.022	0.899	0.121
AlexNet-AA	fc6	0.995	0.008	0.900	0.123
	fc7	0.977	0.034	0.899	0.124
	fc8	**0.959**	0.056	0.898	0.127
ResNet-50-AA	fc	0.980	0.029	0.897	0.128
MobileV2-AA	fc	0.986	0.021	0.899	0.123

patches, the values that do differ are more meaningful. In contrast, features extracted from vertically shifted patches are less similar and less correlated. This explains the results of both feature similarity and separability experiments.

3.5 Feature Arithmetic

Finally, we report the results of the feature arithmetic experiments. We performed feature arithmetic with several different image patches and operands,

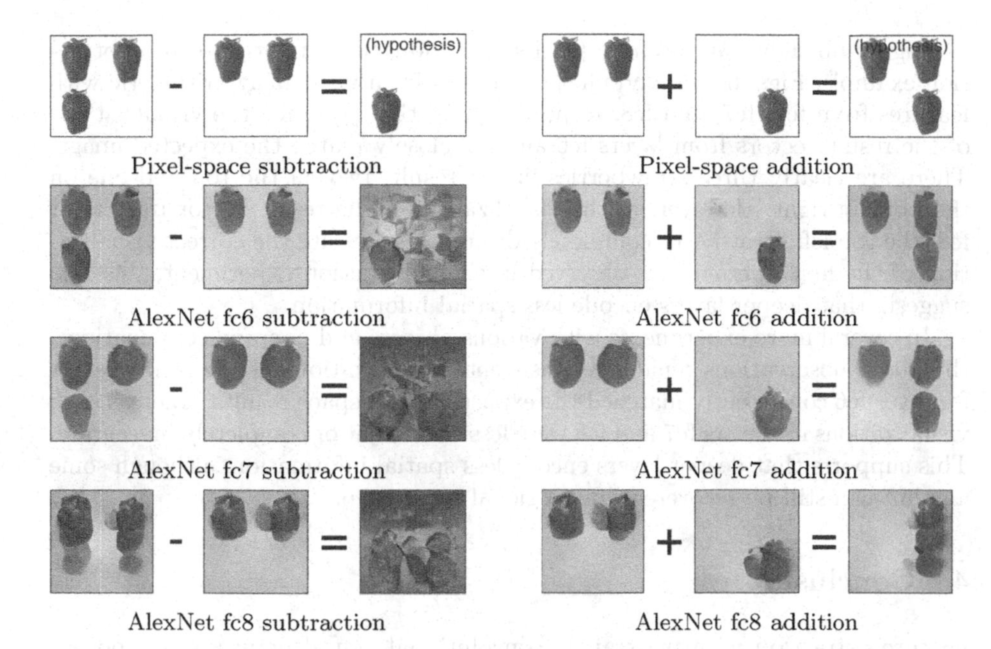

Fig. 6. Feature arithmetic visualizations

but we have chosen the examples in Fig. 6 for discussion. Figure 6a shows an example of a pixel-space subtraction, where we expect some strawberries in an image to be removed via the subtraction. To perform this extraction in feature space, we extract the features of the two operand images from AlexNet's fc6, fc7, and fc8 layers. Then, for each layer, we calculate a result vector by subtracting the operands' feature vectors. Finally, using DeePSiM [6], we visualize the operand feature vectors and the result vector for each layer. The results are shown in Figs. 6c, 6e, and 6g.

Since DeePSiM was trained to reconstruct and visualize extracted feature vectors, the visualizations of the operand feature vectors were expected to be close to the original image. However, DeePSiM was not trained to visualize modified feature vectors, so the visualizations of the result vectors are of interest. For the fc6 layer, despite the noisy background, the visualization of the result vector is similar to the expected image: the top two strawberries have been removed, and only a single strawberry remains in the lower left corner.

For the fc7 and fc8 layers, however, the results are less clear. The visualizations for both result vectors contain an extra strawberry in the bottom right corner, which did not exist in any of the original images. In fact, for fc8, the visualization of the extracted operand features is not accurate, displaying four (or more) strawberries when there were only three originally. Features at these deeper layers seem to encode less spatial information, which is consistent with the results of previous experiments.

Figure 6b shows an intuitive pixel-space addition. Similarly to the subtraction example, Figs. 6d, 6f, and 6h show results from visualizing arithmetic with features from fc6, fc7, and fc8, respectively. In this example, the visualizations of the result vectors from layers fc6 and fc7 closely match the expected image. There are clearly three strawberries in the result, two on the top and one on the bottom right. However, in the visualization of the result vector from layer fc8, the top left strawberry completely disappears, despite the correct visualization of the first operand. As observed in the subtraction experiment, this also suggests that deeper layers encode less spatial information.

In several more experiments with various objects and operand combinations, the above observations remained consistent. Visualizations of the result vector for layer fc6 consistently matched the expected pixel-space results, whereas such visualizations for layers fc7 and fc8 were less consistent or completely inaccurate. This supports that deeper layers encode less spatial information, although some amount can still be recovered due to global invariance.

4 Conclusion

Feature extraction with pre-trained convolutional neural networks is a powerful computer vision tool. However, some tasks may require that these extracted features be shift invariant, to be more robust to camera movement or other geometric perturbations. In this work, we measure the shift invariance of extracted features from popular CNNs with three simple experiments and visualizations.

Heatmaps of cosine similarities between extracted features of shifted images reveal that, while various models and layers have different degrees of invariance, none are globally shift invariant. Patterns in the heatmaps show a bias of extracted features towards horizontally shifted images, which remain more similar than vertically shifted images. A grid pattern of local fluctuations is visible in all off-the-shelf models, but anti-aliasing is able to significantly suppress this effect and improve local shift invariance. However, anti-aliasing does not significantly improve global shift invariance.

Results of the features separability experiments suggest that, while features of horizontally shifted images are more similar, they are also more correlated with the geometric transformation. Linear SVM classifiers were significantly better at classifying shifts between left and right than top and bottom. Features of vertically shifted images are more different, but the differences are less correlated and less separable.

Finally, features extracted from some layers of AlexNet can be added and subtracted, and spatial information can still be visualized and recovered. Visualizations of added and subtracted extracted features match the expected pixel-space result for earlier fully connected layers of AlexNet, such as fc6 and fc7. Features extracted from the deepest layer, fc8, are not added and subtracted as easily, as visualizations often do not match the expected pixel-space result.

Further work would be even more valuable to examine the robustness and sensitivity of extracted features. Larger scale experiments with more patch sizes and different background colors and textures may provide better understanding. A more thorough investigation of the separability of extracted features is needed, perhaps from a metric learning approach. Finally, applying the lessons learned to improve the training or architecture of the models is crucial for improving shift invariance, robustness to geometric perturbations, and transferability of models to other tasks.

References

1. Azulay, A., Weiss, Y.: Why do deep convolutional networks generalize so poorly to small image transformations? J. Mach. Learn. Res. (2018)
2. Babenko, A., Lempitsky, V.: Aggregating local deep features for image retrieval. In: Proceedings of the IEEE International Conference on Computer Vision, pp. 1269–1277 (2015)
3. Bojanowski, P., Joulin, A., Lopez-Paz, D., Szlam, A.: Optimizing the latent space of generative networks. In: Dy, J.G., Krause, A. (eds.) Proceedings of the 35th International Conference on Machine Learning, ICML 2018, Stockholmsmässan, Stockholm, Sweden, 10–15 July 2018. Proceedings of Machine Learning Research, vol. 80, pp. 599–608. PMLR (2018). http://proceedings.mlr.press/v80/bojanowski18a.html
4. Chen, T., Liu, S., Chang, S., Cheng, Y., Amini, L., Wang, Z.: Adversarial robustness: from self-supervised pre-training to fine-tuning. In: Proceedings of the IEEE/CVF Conference on Computer Vision and Pattern Recognition, pp. 699–708 (2020)
5. Deng, J., Dong, W., Socher, R., Li, L.J., Li, K., Fei-Fei, L.: ImageNet: a large-scale hierarchical image database. In: 2009 IEEE Conference on Computer Vision and Pattern Recognition, pp. 248–255. IEEE (2009)
6. Dosovitskiy, A., Brox, T.: Generating images with perceptual similarity metrics based on deep networks. In: Advances in Neural Information Processing Systems, pp. 658–666 (2016)
7. Engstrom, L., Tsipras, D., Schmidt, L., Madry, A.: A rotation and a translation suffice: fooling CNNS with simple transformations. In: Proceedings of the International Conference on Machine Learning (2019)
8. Gong, Y., Wang, L., Guo, R., Lazebnik, S.: Multi-scale orderless pooling of deep convolutional activation features. In: Fleet, D., Pajdla, T., Schiele, B., Tuytelaars, T. (eds.) ECCV 2014. LNCS, vol. 8695, pp. 392–407. Springer, Cham (2014). https://doi.org/10.1007/978-3-319-10584-0_26
9. He, K., Gkioxari, G., Dollár, P., Girshick, R.: Mask R-CNN. In: Proceedings of the IEEE International Conference On Computer Vision, pp. 2961–2969 (2017)
10. He, K., Zhang, X., Ren, S., Sun, J.: Deep residual learning for image recognition. In: Proceedings of the IEEE Conference on Computer Vision and Pattern Recognition, pp. 770–778 (2016)
11. Jaderberg, M., Simonyan, K., Zisserman, A., et al.: Spatial transformer networks. In: Advances in Neural Information Processing Systems, pp. 2017–2025 (2015)
12. Krizhevsky, A., Sutskever, I., Hinton, G.E.: ImageNet classification with deep convolutional neural networks. In: Advances in Neural Information Processing Systems, pp. 1097–1105 (2012)
13. Lee, J.H., Wagstaff, K.L.: Visualizing image content to explain novel image discovery. In: Data Mining and Knowledge Discovery, pp. 1–28 (2020)

14. Lenc, K., Vedaldi, A.: R-CNN minus R. ArXiv abs/1506.06981 (2015)
15. Lin, T.-Y., et al.: Microsoft COCO: common objects in context. In: Fleet, D., Pajdla, T., Schiele, B., Tuytelaars, T. (eds.) ECCV 2014. LNCS, vol. 8693, pp. 740–755. Springer, Cham (2014). https://doi.org/10.1007/978-3-319-10602-1_48
16. Paszke, A., et al.: Pytorch: an imperative style, high-performance deep learning library. In: Wallach, H., Larochelle, H., Beygelzimer, A., d'Alché-Buc, F., Fox, E., Garnett, R. (eds.) Advances in Neural Information Processing Systems, vol. 32, pp. 8024–8035. Curran Associates, Inc. (2019). http://papers.neurips.cc/paper/9015-pytorch-an-imperative-style-high-performance-deep-learning-library.pdf
17. Pei, K., Cao, Y., Yang, J., Jana, S.: Towards practical verification of machine learning: the case of computer vision systems. arXiv preprint arXiv:1712.01785 (2017)
18. Radford, A., Metz, L., Chintala, S.: Unsupervised representation learning with deep convolutional generative adversarial networks. In: Bengio, Y., LeCun, Y. (eds.) 4th International Conference on Learning Representations, ICLR 2016, San Juan, Puerto Rico, 2–4 May 2016, Conference Track Proceedings (2016). http://arxiv.org/abs/1511.06434
19. Russakovsky, O., et al.: ImageNet large scale visual recognition challenge. Int. J. Comput. Vis. **115**(3), 211–252 (2015). https://doi.org/10.1007/s11263-015-0816-y
20. Sandler, M., Howard, A., Zhu, M., Zhmoginov, A., Chen, L.C.: MobileNetv 2: inverted residuals and linear bottlenecks. In: Proceedings of the IEEE Conference on Computer Vision and Pattern Recognition, pp. 4510–4520 (2018)
21. Schwarz, M., Schulz, H., Behnke, S.: RGB-D object recognition and pose estimation based on pre-trained convolutional neural network features. In: 2015 IEEE International Conference on Robotics and Automation (ICRA), pp. 1329–1335. IEEE (2015)
22. Sharif Razavian, A., Azizpour, H., Sullivan, J., Carlsson, S.: CNN features off-the-shelf: an astounding baseline for recognition. In: Proceedings of the IEEE Conference on Computer Vision and Pattern Recognition Workshops, pp. 806–813 (2014)
23. Shorten, C., Khoshgoftaar, T.M.: A survey on image data augmentation for deep learning. J. Big Data **6**(1), 60 (2019)
24. Yosinski, J., Clune, J., Bengio, Y., Lipson, H.: How transferable are features in deep neural networks? In: Advances in Neural Information Processing Systems, pp. 3320–3328 (2014)
25. Zhang, R.: Making convolutional networks shift-invariant again. In: Proceedings of the International Conference on Machine Learning (2019)
26. Zhou, B., Lapedriza, A., Xiao, J., Torralba, A., Oliva, A.: Learning deep features for scene recognition using places database. In: Advances in Neural Information Processing Systems, pp. 487–495 (2014)

Challenges from Fast Camera Motion and Image Blur: Dataset and Evaluation

Zunjie Zhu[1], Feng Xu[2(✉)], Mingzhu Li[1], Zheng Wang[1], and Chenggang Yan[1(✉)]

[1] Department of Automation, Hangzhou Dianzi University, Hangzhou, China
{zunjiezhu,mingzhuli,zhengwang,cgyan}@hdu.edu.cn
[2] BNRist and School of Software, Tsinghua University, Beijing, China
feng-xu@tsinghua.edu.cn

Abstract. To study the impact of the camera motion speed for image/video based tasks, we propose an extendable synthetic dataset based on real image sequences. In our dataset, image sequences with different camera speeds featuring the same scene and the same camera trajectory. To synthesize a photo-realistic image sequence with fast camera motions, we propose an image blur synthesis method that generates blurry images by their sharp images, camera motions and the reconstructed 3D scene model. Experiments show that our synthetic blurry images are more realistic than the ones synthesized by existing methods. Based on our synthetic dataset, one can study the performance of an algorithm in different camera motions. In this paper, we evaluate several mainstream methods of two relevant tasks: visual SLAM and image deblurring. Through our evaluations, we draw some conclusions about the robustness of these methods in the face of different camera speeds and image motion blur.

Keywords: Dataset · Fast camera motion · Image motion blur · SLAM · Deblurring

1 Introduction

Along with the rapid development of information technology, images and videos will play an increasingly important role to enable a lot of applications. Tasks applied to those applications are all affected by the quality of images, while the image motion blur induced by fast camera motions is a common factor that severely degrades the image quality.

It is crucial to study the performance of those tasks when images/videos have different motion blur with fast camera motions. In theory, a dataset that provides image sequences with different camera speeds is required for the above

Electronic supplementary material The online version of this chapter (https://doi.org/10.1007/978-3-030-68238-5_16) contains supplementary material, which is available to authorized users.

A. Bartoli and A. Fusiello (Eds.): ECCV 2020 Workshops, LNCS 12539, pp. 211–227, 2020.
https://doi.org/10.1007/978-3-030-68238-5_16

study. To be more specific, these image sequences should be recorded in the same scene with the same camera trajectory. However, it is rather difficult to make this kind of dataset in real world, since the image sequences are required to be recorded in just the same camera trajectory and capture positions.

In this work, we overcome the above-mentioned difficulties by proposing a synthetic method. The synthetic method utilizes real image sequences to generate sequences with faster camera motions. We call these synthetic sequences together as fast camera motion dataset (FMDataset). Our FMDataset meets the requirement that provides different sequences with the same scene information as well as the same camera trajectory. A core to generate FMDataset is our blurry image synthesis method which generates photo-realistic image motion blur in an image sequence by their sharp images, scene structures and camera motions.

By quantitatively comparing against existing blurry image synthesis methods, we see that our blurry images are more similar to the real blurry images. Besides, we further evaluate two relevant tasks (image deblurring and visual SLAM) based on our synthetic dataset, i.e., evaluating the performance of public deblurring methods in our blurry images with different blur degrees; evaluating three kinds of visual SLAM methods with different camera speeds and with/without image blur. Through our evaluation, we get some conclusions on the robustness of each method for camera speed and image motion blur. Our main contributions are summarized as follows:

- A synthetic dataset that provides image sequences recorded in the same scene and same camera trajectory with different camera speeds.
- A synthesis method that generates images sequences with different camera speeds based on real captured sequences.
- Comprehensive evaluations and analyses for SLAM methods and deblurring methods, concerning different camera speeds and different image blur degrees.

2 Related Work

2.1 Datasets

SLAM Datasets. There are several well known video-based SLAM datasets, including ICL-NUIM [11], EuRoC [2], TUM RGB-D [25], and ScanNet [3]. ICL-NUIM is a virtual dataset that renders color and depth image sequences by 3D virtual scenes as well as camera trajectories. EuRoC is recorded by an UAV with color cameras and IMUs. TUM RGB-D uses a Kinect sensor to capture color and depth images as well as accelerations in several real scenes. ScanNet provides RGB-D and full IMU information of various indoor scenes without ground truth. All these datasets are recorded under slow camera motions, which limit the evaluation of SLAM methods for different camera speeds.

Synthetic Blur Datasets. We classify the synthetic blur datasets by their image blur synthesis methods in Table 1. The Convolution method synthesizes blurry images by convolving sharp images with pre-defined blur kernels.

Levin et al. [16] generate dataset by 255 × 255 uniform blur kernels. Sun et al. [28] built a higher resolution dataset that has natural images of diverse scenes. [27] and [15] generate non-uniform blurry images based on blur kernels which are calculated by 6 DOF camera motions. However, the Convolution method essentially ignores the impact of the 3D scene structure, and synthesizes unreal blur for images that have rich structure information.

To mimic natural blur, the Integration method synthesizes blurry images by integrating latent sharp images which are captured in image exposure time. [19,20,24,26] utilize high-speed cameras to capture high frame-rate videos, and synthesize blurry images by averaging successive frames in a video. The successive frames implicitly involve the 3D scene structures and camera/object motions. For instance, Nah et al. [19] generate a blurry image by averaging 7–11 successive frames from a 240 frame-rate video which is recorded by the GoPro camera. However, these techniques require high frame-rate input sequences, which limits the size of the dataset. Techniques that can utilize a large amount of ordinary videos are required to generate a large dataset.

Table 1. Summary of existing and our proposed datasets for blurry image synthesis.

Method	Public datasets	Blur model	Motion description	3D scene	Extra device
Convolution	[15,16,27,28]	(Non-)Uniform	3/6 DOF Cam. Pose		
Integration	[19,20,24,26]	Non-Uniform	Dynamic Scene	✓	✓
Ours	FMDataset	Non-Uniform	Cam. Trajectory	✓	

2.2 Algorithms

SLAM Methods. Feature point based SLAM methods such as ORB-SLAM2 [18] and VINS [21] extract point features from each frame and estimate camera poses by matching point features of different frames. In addition, direct methods such as SVO [8], DSO [5] and LSD [6] directly match the image intensities of two consecutive frames. While in dense SLAM, InfiniTAM [12], BundleFusion [4] and ElasticFusion [29] involve depth images to calculate camera poses as well as scene models by the ICP algorithm [1].

However, none of the above methods have evaluated their performance in different camera speeds. In theory, the feature point method cannot detect sufficient features on blurry images which are generated by fast camera motions. The SE effect introduced by Zhu et al. [33] always happens with fast camera motions, results in inaccurate feature tracking of the direct method. The ICP algorithm used in dense SLAM is easily trapped in local solutions when camera perspectives change a lot in inter-frames.

Deblurring Methods. There has been enormous research efforts in image quality restoration from image motion blur, and can be roughly divided into single-image deblurring and video deblurring. The sparsity prior has proved effective in single-image deblurring for blur kernel estimation. For example, Krishnan et al. [14] apply the normalized sparsity in their MAP framework to estimate the blur kernel. Xu et al. [32] propose an approximation of the *L0*-norm as a sparsity before jointly estimate sharp image and blur kernel. Xu et al. [31] propose a two-phase method that first estimates the blur kernel by the selected image edges and then refines it by iterative support detection (ISD) optimization, the latent sharp image is finally restored by total-variation(TV)-*L1* deconvolution.

As for the video image deblurring, Whyte et al. [30] propose a stereo-based method with the geometry model, and handle the non-uniform image blur in a monocular video. Kim et al. [13] utilize optical flow for the blur kernel estimation, which works well for certain object motion blur. Sellent et al. [22] further used piece-wise rigid scene representation to estimate 3D scene flow from blurry images.

3 Proposed FMDataset

In this section, we first introduce our definition of the sequence speed and difficulty. Then, we introduce our synthesis method which generates image sequences with fast camera motions and blurry images based on real slow sequences.

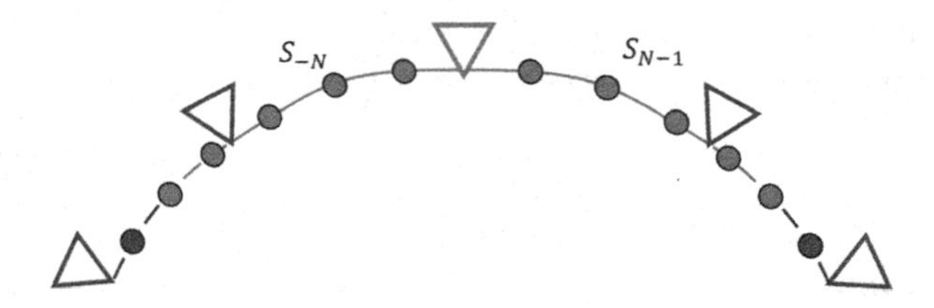

Fig. 1. This figure shows the camera frames (triangles) and grund truth camera motion values (dots) used for the camera speed calculation. The camera speed of a frame (red) is calculated by the motions of $2N$ nearby GT values (red). Note that N is small enough ($N = 2$ in our experiment) to adapt the assumption of uniformly acceleration motion in nearby frames. (Color figure online)

3.1 Speed and Difficulty Definition

For an image sequence, we calculate the camera motion speed of each frame by the ground truth camera motion (GTCM) values of the sequence captured by a Vicon system. Figure 1 illustrates the relationship between the camera frames and the GTCM values in a camera motion trajectory. Since we find there still exist high-frequency noises in GTCM values captured by a Vicon system, so we use $2N$ nearby GTCM values to calculate a smoother camera speed for an image frame. We assume the kinematical model in nearby frames as uniformly accelerated motion model. Then, the speed $\mathbf{V}$ and the acceleration $\mathbf{a}$ of each frame are easily calculated as:

$$\mathbf{V} = \frac{\omega}{2N} \cdot \sum\nolimits_{i=-N}^{N-1} \mathbf{S}_i, \tag{1}$$

$$\mathbf{a} = \left(\frac{\omega}{N}\right)^2 \cdot \sum\nolimits_{i=0}^{N-1} \mathbf{S}_i - \mathbf{S}_{i-N}, \tag{2}$$

where the ω is the capturing frequency of the Vicon system. $\mathbf{V}$, $\mathbf{a}$ and $\mathbf{S}$ are six-dimensional vectors, while $\mathbf{S}$ represents the relative camera motion between two consecutive GT values.

Figure 2 shows the influences of camera motions and scene depths for image blur. According to the influences illustrated in this figure, we define the difficulty G of an image frame as the moving distance of the center pixel in inter-frame. The difficulty is calculated as:

$$G(\mathbf{V}) = \frac{f}{\mathrm{FPS}} \cdot \left\| \mathbf{V}_r + \frac{\mathbf{V}_t}{D} \right\|_2 \tag{3}$$

where f is the camera focal length, FPS is the camera frame-rate, and D is the depth value of the center pixel. $\mathbf{V}_t$ and $\mathbf{V}_r$ are acquired from the calculated camera speed $\mathbf{V}$, where $\mathbf{V}_t$ is the 3×1 translation speed and $\mathbf{V}_r = \left[\mathbf{V}_{\mathrm{pitch}} \; -\mathbf{V}_{\mathrm{yaw}} \; 0\right]^{\mathsf{T}}$ is the rotation speeds of pitch and yaw. The derivations of the Eq. 3 are detailed in our supplementary material.

Figure 3 shows the speeds and difficulties of three image sequences from TUM RGB-D dataset [25], ICL-NUIM [11] and FastFusion [33], respectively. In this figure, we find that the difficulties vary a lot in sequences with fast camera motions. Therefore, we empirically set the value at the top 10% to represent the difficulty of a sequence, as the value (dotted line in Fig. 3) better represents the real difficulty of a sequence (for video-based tasks such as SLAM) than mean or medium value. Note that the speed of a sequence is also represented by the value at the top 10%. We calculate the speed and difficulty of several public SLAM datasets, and the results are listed in Table 2.

3.2 Synthesis Method

Firstly, we down-sampling a real image sequence by a fixed interval λ to get an accelerated sequence. Since the ground truth camera motions of the accelerated sequence are same with the ones of original sequence, the camera speed $\hat{\mathbf{V}}$ and acceleration $\hat{\mathbf{a}}$ of a frame in the accelerated sequence are easily deduced by Eqs. 1 and 2, as $\hat{\mathbf{V}} = \lambda \cdot \mathbf{V}$ and $\hat{\mathbf{a}} = \lambda^2 \cdot \mathbf{a}$, where $\mathbf{V}$ and $\mathbf{a}$ are the values of the same image in the original sequence.

The fast camera motion commonly blurs the images which are captured by consumer cameras, therefore we also need to synthesize blurry images for the accelerated image sequence. As is shown in Table 1, traditional blurry image synthesis methods either ignore the influence of scene structure or require an extra high-speed camera. In this paper, we do not rely on high-speed cameras,

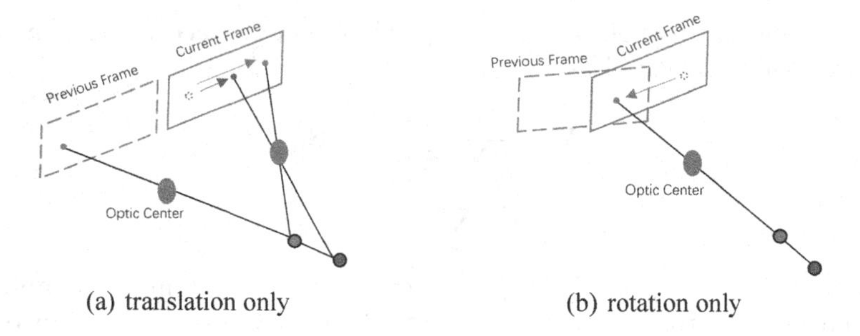

(a) translation only (b) rotation only

Fig. 2. This figure shows how scene depths and camera motions affect the image blur. In an image, the blur degree of an object equals to the distance it moves in the image domain during the exposure time. Assuming the exposure time is constant, then in the translation only camera motion (a), the moving distances of the red and green objects in the image domain are both affected by camera translation speed and scene depth. Therefore, the green object which is closer to the camera has a more severe blur than the red one. While in rotation only camera motion (b), the moving distances only depend on the camera rotation speed. (Color figure online)

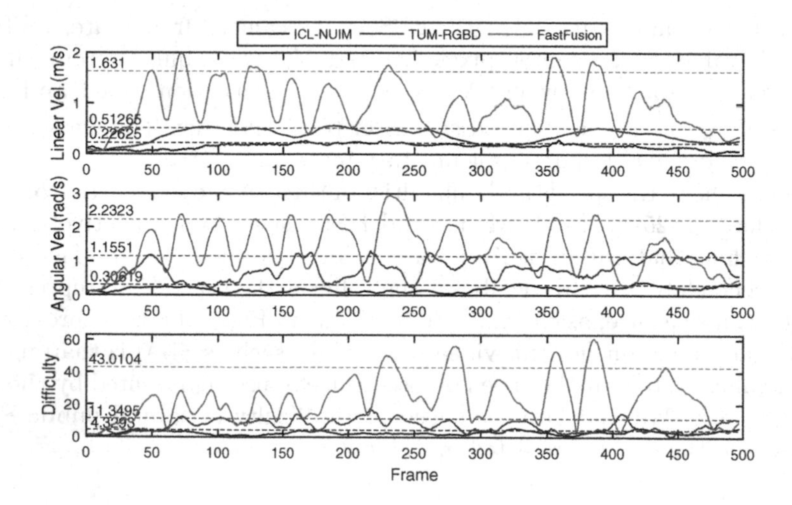

Fig. 3. This figure shows velocities and difficulties of top 500 frames of three sequences, including or_kt3 of ICL-NUIM [11], fr1_desk of TUM RGB-D [25], and Dorm2_fast1 of FastFusion [33]. The solid lines denote the velocities and difficulties of each frame, the numbers over the dotted lines represent the sequence rotation/translation speed and sequence difficulty.

instead, we synthesize blurry images by utilizing normal frame-rate sequences which provide color and depth images as well as camera motions. The scene structures of these sequences are easily acquired by a 3D reconstruction algorithm such as BundleFusion [4]. Note that slow image sequences used for the synthesis can be found from public SLAM datasets, or captured by oneself.

Table 2. The velocities and difficulties of sequences in ICL-NUIM [11], TUM RGB-D [25], VaFRIC [10] and FastFusion [33].

Dataset	Linear Vel. (m/s)	Angular Vel. (rad/s)	Difficulty				
ICL_lr_kt1	0.087	0.312	6.1	TUM_fr2_xyz	0.1	0.156	4.0
ICL_lr_kt2	0.362	0.348	7.3	TUM_fr3_office	0.361	0.433	7.4
ICL_or_kt3	0.242	0.311	5.7	Dorm1_slow	0.598	1.082	19.3
VaFRIC (20FPS)	1.81	0.22	34.3	Dorm1_fast1	0.954	2.617	53.5
TUM_fr1_desk	0.585	0.849	15.9	Dorm1_fast2	0.93	2.43	49.2
TUM_fr1_desk2	0.69	1.165	20.3	Dorm2_fast1	1.61	2.201	42.1
TUM_fr1_room	0.555	1.133	16.4	Dorm2_fast2	1.413	2.528	42.4
TUM_fr1_rpy	0.128	1.655	27.3	Hotel_slow	0.668	1.19	21.6
TUM_fr1_360	0.331	1.141	21.3	Hotel_fast1	1.294	2.356	43.7
TUM_fr2_desk	0.318	0.348	8.7	Hotel_fast2	1.566	2.183	43.0

Similar to Nah et al. [19], the generation of a blurry image is modeled as the averaging of images which are latently observed during the blurry image exposure time. In the following, we call these latently observed images as latent images. Nah et al. [19] acquire latent images by a high-speed camera, which limits the latent image number and causes the ghost effect in blurry images. For example, 11 latent images are used by Nah et al. to generate a blurry image with 50 ms exposure time, which are far from enough to mimic the camera exposure. Different from them, we use sharp images to mimic any number of latent images. To be specific, given a sharp image $\mathbf{C}$ of our accelerated sequence, a nearby latent image $\mathbf{L}$ is generated as follows.

Firstly, a pixel (u, v) of the latent image is back-projected to 3D by its depth d. Note the depth d is acquired from the 3D geometry model by the position of the latent image. Then, we re-project the 3D point of the pixel to the image coordinate of $\mathbf{C}$ by the relative camera motion $\mathbf{T}$ between the sharp and latent images. Finally, the value of pixel (u, v) is acquired from the sharp image by the inverse distance weighted interpolation. Note that the relative camera motion $\mathbf{T}$ is calculated by the camera speed and the time node t of the latent image with the uniformly accelerated motion model. The latent image is formulized as:

$$\mathbf{L}(u, v) = \mathbf{C}\Big(\mathbf{\Psi}\big(\mathbf{T} \cdot \mathbf{\Psi}^{-1}(u, v, d)\big)\Big) \tag{4}$$

where $\mathbf{\Psi}(\cdot)$ is to project a 3D point into image domain, and $\mathbf{\Psi}^{-1}(\cdot)$ is the inverse operation. Note the pixel $\mathbf{L}(u, v)$ will be invalid once the projection is out of the image region of $\mathbf{C}$ or is covered by another pixel which is more close to the camera. In practice, to better solve the occlution problem, we generate latent images by 3 nearby sharp images, i.e. the value of a pixel in a latent image is acquired by averaging the corresponding valid values of the 3 sharp images (Eq. 4). Finally, the blurry image $\mathbf{B}$ is synthesized by averaging its latent images. The latent image number $\eta = 6\lambda + 1$ is changed from the acceleration degree λ, i.e. when the acceleration (blur) degree is $\lambda = 5$, then the latent image number is $\eta = 31$.

3.3 Synthetic Dataset

As is described above, our synthesis method can generate image sequences with different camera speeds based on existing image sequences. So we create a synthetic dataset by implementing our synthesis method on public SLAM datasets and our recorded slow sequences. Firstly, we choose 3 sequences (lr_kt1, lr_kt2, or_kt3) of ICL-NUIM [11], and 3 sequences (fr2_desk, fr2_xyz, fr3_office) of TUM-RGBD [25] for our dataset synthesis. All these sequences have slow camera motions and small difficulties, which can be viewed in Table 2. Then, for each sequence, we synthesize 9 faster sequences with the down-sampling intervals λ from 2 to 10 and the blurry image exposure time $\tau = 20$ ms. Each synthetic sequence provides blurry color images, ground truth sharp color images, depth images corresponding to sharp color images, and ground truth camera motions. The size of each image is 640×480, and the frame-rate of each sequence is 30 FPS. Besides, we also record 4 slow sequences by a Microsoft Azure Kinect, and generate their synthetic sequences in the same way. However, different from the above public datasets, the camera motions of our recorded sequences are estimated by the state-of-the-art SLAM method ORB-SLAM2 [18] which in practice shows accurate estimation results in slow sequences.

In summary, our synthetic dataset contains 90 synthetic sequences which can be used to evaluate the performance of an algorithm in different camera motion speeds. It is worth mentioning that our synthetic dataset is extendable. One may use our synthesis method to generate more sequences by his recorded sequence or other public datasets.

4 Evaluation and Analysis

In this section, we first compare our blurry image synthesis method against existing synthesis methods. Then, we evaluate the performance of deblurring methods on our images with different blur degrees. Finally, we evaluate and analyze different SLAM methods on our synthetic sequences.

4.1 Evaluation of Synthetic Image Blur

To evaluate the performance of our blurry image synthesis method, we quantitatively compare against the existing synthesis methods which are classified in Table 1 and are introduced in Sect. 2.1.

We compare on a public dataset VaFRIC [10] which provides corresponding sharp and blurry image sequences with multiple frame-rate. VaFRIC renders color and depth images by the acknowledged tool POV-Ray, and the sequences with different frame-rates are rendered by the same virtual scene with the same camera trajectory. The sharp and blurry images rendered at the same frame-rate

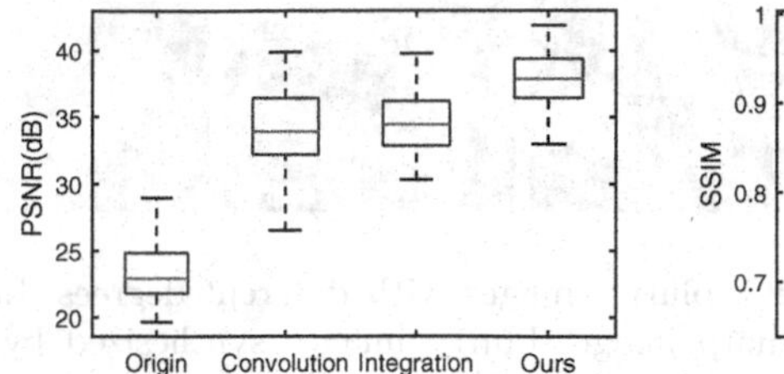

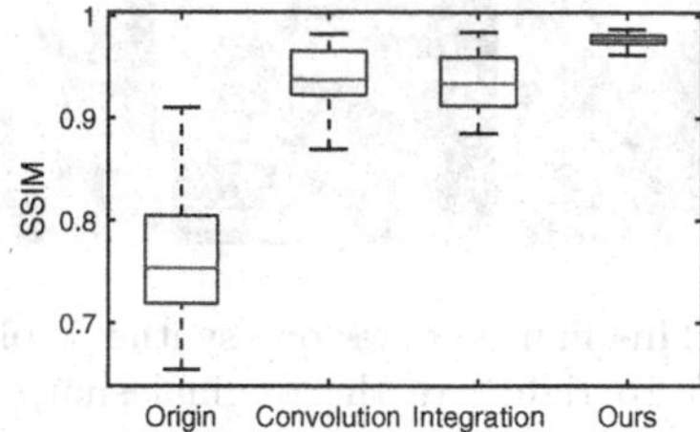

Fig. 4. This figure shows the results of 100 synthetic blurry images of each synthesis method, measured by PSNR and SSIM. The "Origin" represents the results that we directly compare the original sharp images with the ground truth blurry images.

perfectly correspond to each other, as a sharp image is directly rendered at the center time node of the blurry image exposure time. The speed and difficulty of the 20 FPS sequence in VaFRIC are shown in Table 2.

We choose the 20 FPS sharp sequence as the input of the Convolution method and our method. While the input of the Integration method is the corresponding 200 FPS sharp sequence, because this method has to use a high frame-rate sequence for the synthesis. Each method finally outputs a blurry sequence with 100 images. To measure the similarity between the synthetic blurry images and the ground truth blurry images, we calculate image quality evaluation metrics PSNR and SSIM for each synthetic blurry image.

The performance of each method is shown in Fig. 4. The results of the Convolution method have greater variance than the Integration method and ours, because the ignorance of scene structure in the Convolution method causes the unreal blur for images which have rich 3D structure information. As is shown in Fig. 4, our method outperforms the others on both evaluation metrics. The reason Integration method inferiors to ours is that the Integration method cannot generate a good approximation for the real blur. Because the image exposure time of the 20 FPS sequence is 20 ms, the Integration method can only average at most 5 sharp images for the synthesis of a blurry image. Note that the Integration methods in Table 1 always set the exposure time of a blurry image to be more than 200 ms which rarely happens in videos.

Also, our method is independent of the high-speed camera, which not only decreases the experiment cost but also contributes to the efficient collection of various training datasets for deep learning methods. Besides, our method is also able to control the blur degree of an image, which can be used to evaluate the robustness of deblurring methods concerning different degrees of image blur.

4.2 Evalution of Deblurring Methods

We choose 7 deblurring methods described in Sect. 2.2, and evaluate their performance on our synthetic blurry images. In this experiment, these deblurring methods are implemented by their public source code with appropriate parameters. We randomly choose 5 sharp images from the public sequence TUM_fr3_office,

Fig. 5. This figure shows our synthetic blurry images with different degrees. Images from left to right are the original sharp image, blurry images synthesized by $\lambda = 3, 5, 7, 9$.

and synthesize 8 blurry images with different degrees for each sharp image. The blur degree of each image is controlled by the λ (mentioned in Sect. 3.2) which denotes the magnification times of the camera speed. Different blur degrees are named directly by their λ. Figure 5 displays our synthetic blurry images with different degrees.

Figure 6 shows the deblurring results of each method. In this figure, most methods have the same phenomenon that the deblurring ability degrades severely as the blur degree increases. The reason is that the severe image blur caused by the larger blur kernel is much harder to be estimated by a deblurring method. Besides, We also find that the rankings of Xu [32] at the smallest and the largest degrees are significantly different. Meanwhile, the results of Xu [32] and Levin [16] are smoother than others in different blur degrees.

The robustness of deblurring methods for different blur degrees is an interesting ability, which intuitively reflects whether a deblurring method able to work stably in real applications. However, none of the 7 deblurring methods have measured their robustness in terms of different blur degrees. Therefore, we further calculate the coefficient of variation (CV) for each method:

$$\mathrm{CV}(A_i) = \frac{\sqrt{\frac{1}{N}\sum_{i=1}^{N}(A_i - r)^2}}{r} \cdot 100\% \tag{5}$$

where A_i denotes the deblurring accuracy corresponding to a blur degree i, N is the number of blur degrees, and r is the average value of all A_i.

Table 3 lists the CVs of each method. The ranking of each method is similar in both PSNR CV and SSIM CV. In this table, Levin [17] and Xu [32] have smaller CVs than others, which is consistent with the performance in Fig. 6. As a consequence, we hold the opinion that the robustness for image blur degree is as important as the deblurring accuracy for monitoring deblurring methods.

4.3 Evaluation and Analysis of SLAM Methods

The study of visual SLAM always assumes the camera motions are slow and the images are sharp. Therefore, based on the advantage of our synthetic dataset, we evaluate three kinds of SLAM methods, including feature point method, direct method, and ICP method, to see their performance on our sequences with different camera motions and with/without image blur. The reason we use

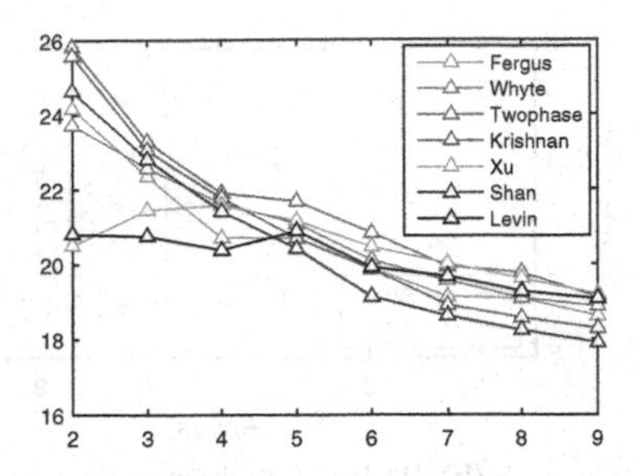

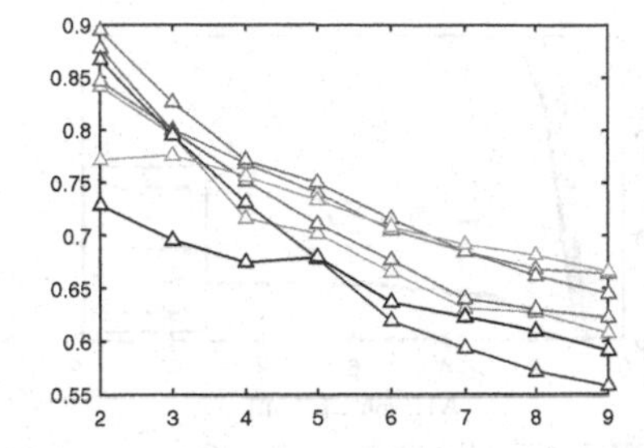

Fig. 6. Accuracy of deblurred images in different blur degrees (x-coordinate), measured by PSNR (left) and SSIM (right). The accuracy of each degree is calculated by averaging the results of 5 deblurred images.

Table 3. The coefficient of variation [%] calculated by the results of Fig. 6. The numbers in parentheses are the rankings of deblurring methods.

Method	Fergus [7]	Whyte [30]	Twophase [31]	Krishnan [14]	Xu [32]	Shan [23]	Levin [17]
PSNR CV	8.5 (4)	7.8 (3)	9.5 (5)	11.3 (7)	3.9 (2)	11 (6)	**3.3** (1)
SSIM CV	11.3 (5)	8.5 (3)	10.8 (4)	11.9 (6)	**5.5** (1)	15.6 (7)	6.8 (2)

sequences without image blur is that, the evaluations on these sequences can directly examine the impact of camera motion speed for a method. Further, by comparing the results between sequences with and without blur, the impacts of image blur are clearly observed.

In the following experiments, synthetic sequences with blurry images are named by their down-sampling interval λ with a label "-Blur", while their corresponding sequences with sharp images are named with another label "-Sharp". The original sequence used for the dataset synthesis is directly named as "1-Sharp".

Feature Point Method. We choose the well known open-source method ORB-SLAM2 [18] to represent the feature point based SLAM method in our experiment, and we close the loop closure function for the better evaluation of the feature tracking as well as the camera pose estimation.

The accuracy of camera pose estimation is normally measured by the ATE RMSE between the estimated camera trajectory and the ground truth trajectory. So we design an index "repeatability" which is inspired by Gauglitz et al. [9], to further quantitatively measure the performance of the feature tracking. The repeatability REP is the ratio of the successfully tracked features between the previous frame and the current frame:

$$\text{REP}_i = \frac{m_i}{M_{i-1}} \tag{6}$$

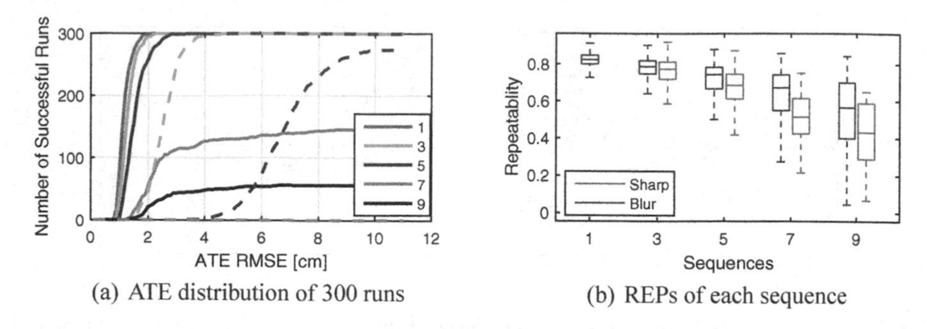

(a) ATE distribution of 300 runs (b) REPs of each sequence

Fig. 7. The results of ORB-SLAM2 on original sequence TUM_fr3_office (1) and its synthetic sharp/blurry sequences (3, 5, 7, 9). In plot (a), the solid lines represent the results of sharp sequences; The dotted lines: blurry sequences. In plot (b), the REPs of each sequence belong to the one which is coverged in plot (a). Note that the runs in 7-Blur and 9-Blur are all failed, so we draw their box-plots only by the successfully tracked frames.

where M is the number of 3D feature points (also named as mappoints in ORB-SLAM2) in the previous frame $i-1$, and m represents the number of tracked 3D feature points in the current frame i. The overall reapeatability of a sequence is calculated by the root mean square error (RMSE):

$$\text{REP RMSE} = \Big(\frac{1}{n}\sum_{i=1}^{n}(\text{REP}_i - 1)^2\Big)^{\frac{1}{2}} \tag{7}$$

where n is the number of frames after the system initialization. REP RMSE can be used to measure the robustness of a feature point method, beacuse the system with lower repeatability will be more sensitive to the unfriendly situations such as texture-less or low-light scenes as well as the image motion blur.

Figure 7 shows the accuracy of camera pose estimation and the repeatabilities of each sequence. The fast camera motions cause the great changes of camera perspectives, leading to smaller overlap in two consecutive frames. As a consequence, by comparing the results of sharp sequences and the results of blurry sequences respectively, we find that the repeatability decreases as the sequence speed increases, and the convergence of ATE is slower in faster sequences. Besides, the comparison between the results of sharp and blurry sequences indicates that, the image motion blur will lead to further degeneration of ORB-SLAM2.

Then, we calculate the REP RMSE for each sequence in Table 4. Although 7-Blur and 9-Sharp have approximate REP RMSE, runs on 7-Blur are all failed while 9-Sharp still has more than 50 successful runs. This is because the lower repeatability makes the feature point method more vulnerable to the image blur.

As a result, the robustness of the feature point method is affected by the camera motion speed, and the affection will be more severe when images are blurred.

Table 4. The REP RMSE of sharp (S) and blurry (B) image sequences.

	1-S	3-S	5-S	7-S	9-S	3-B	5-B	7-B	9-B
REP RMSE	0.187	0.233	0.296	0.402	0.500	0.245	0.342	0.506	0.602

Table 5. ATE RMSE [cm] of three direct methods on sequences with different speeds and with/without image blur, each value is averaged over five runs. Note that the scale error is neglected for the ATE calculation. The symbol – indicates that the method is failed in a sequence.

	Origin	Sharp				Blur			
		3	5	7	9	3	5	7	9
LSD [ECCV'14]	18.7	22.6	59.0	105.3	141.2	27.0	80.0	–	–
SVO [TOR'16]	3.6	8.8	10.9	14.5	–	12.8	–	–	–
DSO [TPAMI'18]	1.9	2.1	14.8	36.9	96.5	3.4	37.5	55.2	105.0

Direct Method. We evaluate three direct methods including LSD [6], SVO [8] and DSO [5] by their open source code. We still use TUM_fr3_office and its synthetic sequences in this evaluation. Table 5 lists the ATE RMSE of the three direct methods. On each successful sequence, SVO has smaller ATE than LSD, because SVO strictly selects interest pixel points and accurately estimates depths for those pixels. However, similar to the feature point method, the strict selection strategy is also more sensitive to the image motion blur and the fast camera motions, which makes SVO easily fail in fast and blurry sequences.

As is shown in Fig. 8(c), DSO effectively optimizes the photometric errors for tracked pixels, and successfully runs in all sequences. Nonetheless, DSO still has larger ATE in sequences with fast camera motions and severe image blur, which is mainly caused by the lack of successfully tracked pixels. Note that the lack of tracked pixel/patch also appears on SVO and LSD (shown in Fig. 8(a) and (b)).

Comparing to the results of the feature point method, although the image blur is also hard for the direct method, it will not exercise a decisive influence if the direct method shows a good ability for optimization, such as DSO. One possible explanation is that, because of the similar camera speed between two consecutive frames, the blur degrees of the two frames are close to each other, then the intensities of the tracked pixels are also approximate in two consecutive frames, leading to the reasonable values of photometric errors.

ICP Method. To see the impact of camera motion speed to the iterative closest point (ICP) method, we implement the real-time point-to-plane ICP algorithm similar to [4,12,29]. Besides, we also perform the pyramid strategy to optimize the camera poses from coarse to fine by the Levenberg-Marquardt method. The ICP algorithm only requires the inputs of depth images. Depth images captured by a structured-light sensor are less sensitive to the image blur, such as Intel Realsense ZR300. Therefore, we choose synthetic sequences of TUM_fr3_office,

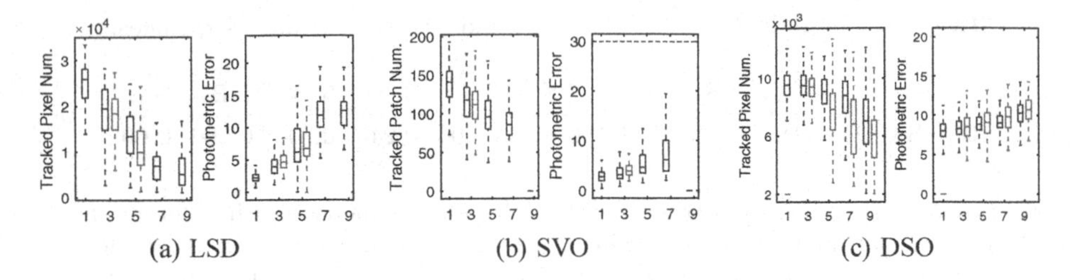

Fig. 8. Box-plots of the tracked pixel/patch numbers and the average pixel photometric errors of each frame. The X-axis represents the name of each sequence. The blue and red boxes represent the results of sharp and blurry sequences, respectively. (Color figure online)

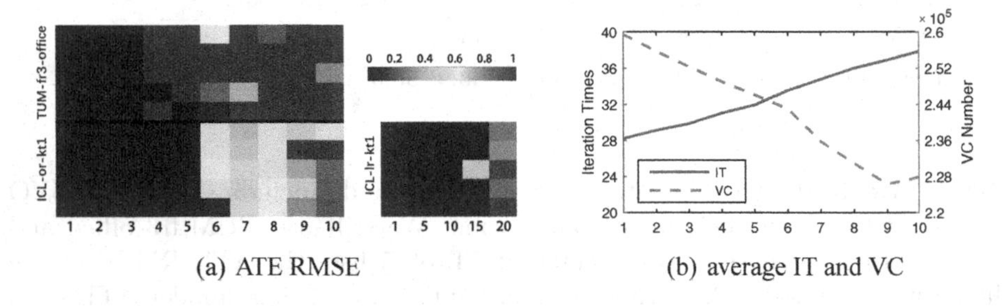

Fig. 9. The results of the ICP method. In plot (a), the numbers at the bottom represent each sequence, and the squares represent the colored ATE RMSE of estimated trajectories. We run five times for each sequence. In plot (b), the X-axis represents synthetic sequences of ICL_or_kt1, and the Y-axis on the left denotes the average iteration times (IT) of ICP; right Y-axis: the valid correspondence (VC) pixel number in two consecutive frames. The values in the plot (b) are the results of one run. Note that each run has similar IT and VC.

ICL_or_kt1 and ICL_lr_kt1, and only use their depth images for the evaluation. Note that ICL_or_kt1 and ICL_lr_kt1 are rendered in different virtual scenes with the same camera trajectory.

Figure 9 shows the results of each sequence. In plot (a), we find that ICP cannot accurately track the camera once the sequence speed exceeds a threshold. And the threshold of each original sequence is different from the others. By comparing the results of ICL_or_kt1 and ICL_lr_kt1 which share the same camera trajectory, we hold that the threshold is affected by the different scene structures of each sequence. Plot (b) shows that the average valid correspondence number of a sequence decreases as the sequence speed increases, while the iteration times of the ICP algorithm is the opposite. The reason is that the camera speed determines the size of the inter-frame camera parallax, and the larger parallax size will cause a smaller overlap between two frames and more iterations for the convergence of the ICP algorithm.

Therefore, when the parallax size in two consecutive frames is greater than a threshold, the initialization of the real-time ICP algorithm will greatly deviate from the solution. Then, the algorithm will prematurely converge to a local optimum, leading to the bad ATE, while the iteration times remains normal.

Besides, by comparing the above three SLAM methods on synthetic sequences of TUM_fr3_office (where the results are respectively shown in Fig. 7(a), Table 5 and Fig. 9(a)), we find that the ICP method is more sensitive to the camera motion speed. The reason is that, during the tracking process of the ICP method, the information of a previous frame is directly acquired from the map, while the bad mapping caused by an inaccurate convergence will severely affects the following camera pose estimations, leading to the bad ATE of the ICP method.

5 Conclusion and Discussion

In this paper, we presented a synthesis method to generate a synthetic dataset for the evaluation of image/video based tasks concerning fast camera speed and severe image blur. To this end, we first defined the speed and difficulty of an image sequence, and synthesized faster sequences based on slow image sequences. Our synthetic sequences with different camera speeds featuring the same scene information and the same camera trajectory. To mimic the real image motion blur, we proposed to synthesize blurry images by real sharp images, scene structures and camera speeds.

The dataset has been designed to be extendable. New videos or images can be synthesized for relevant tasks by public SLAM datasets or self-recorded slow sequences. We aim to extend our dataset for tasks where current methods lose sight of the challenges from fast camera motions and image blur.

In this work, we have evaluated several mainstream methods of two relevant tasks: SLAM and image deblurring, on our sequences synthesized by public SLAM datasets. In our experiments, we evaluated the performance of 7 deblurring methods on our synthetic blurry images with different blur degrees, while the results indicate that most of them are not robust to the different degrees of image blur. As for the visual SLAM methods, we designed separation experiments to evaluate the impact of the camera motion speed and the image blur. As a result, the three kinds of visual SLAM methods, including feature point method, direct method and ICP method, are all severely affected by the fast camera motions. Meanwhile, the direct method is more robust than the feature point method when images are severely blurred, and the ICP method is most vulnerable to the impacts of fast camera motions.

Acknowledgement. Thiswork was funded by National Science Foundation of China (61931008, 61671196, 61701149, 61801157, 61971268, 61901145, 61901150, 61972123), National Science Major Foundation of Research Instrumentation of PR China under Grants 61427808, Zhejiang Province NatureScienceFoundationofChina (R17F030006, Q19F010030), Higher Education Discipline Innovation Project 111 Project D17019.

References

1. Besl, P.J., McKay, N.D.: Method for registration of 3-D shapes. In: Sensor fusion IV: control paradigms and data structures, vol. 1611, pp. 586–606. International Society for Optics and Photonics (1992)
2. Burri, M., et al.: The euroc micro aerial vehicle datasets. Int. J. Robot. Res. **35**(10), 1157–1163 (2016)
3. Dai, A., Chang, A.X., Savva, M., Halber, M., Funkhouser, T., Nießner, M.: ScanNet: richly-annotated 3D reconstructions of indoor scenes. In: Proceedings of Computer Vision and Pattern Recognition (CVPR). IEEE (2017)
4. Dai, A., Nießner, M., Zollhöfer, M., Izadi, S., Theobalt, C.: BundleFusion: Real-time globally consistent 3D reconstruction using on-the-fly surface reintegration. ACM Trans. Graph. (ToG) **36**(3), 24 (2017)
5. Engel, J., Koltun, V., Cremers, D.: Direct sparse odometry. IEEE Trans. Pattern Anal. Mach. Intell. **40**(3), 611–625 (2017)
6. Engel, J., Schöps, T., Cremers, D.: LSD-SLAM: large-scale direct monocular SLAM. In: Fleet, D., Pajdla, T., Schiele, B., Tuytelaars, T. (eds.) ECCV 2014. LNCS, vol. 8690, pp. 834–849. Springer, Cham (2014). https://doi.org/10.1007/978-3-319-10605-2_54
7. Fergus, R., Singh, B., Hertzmann, A., Roweis, S.T., Freeman, W.T.: Removing camera shake from a single photograph. In: ACM SIGGRAPH 2006 Papers, pp. 787–794 (2006)
8. Forster, C., Pizzoli, M., Scaramuzza, D.: SVO: fast semi-direct monocular visual odometry. In: 2014 IEEE international Conference on Robotics and Automation (ICRA), pp. 15–22. IEEE (2014)
9. Gauglitz, S., Höllerer, T., Turk, M.: Evaluation of interest point detectors and feature descriptors for visual tracking. Int. J. Comput. Vis. **94**(3), 335–360 (2011)
10. Handa, A., Newcombe, R.A., Angeli, A., Davison, A.J.: Real-time camera tracking: when is high frame-rate best? In: Fitzgibbon, A., Lazebnik, S., Perona, P., Sato, Y., Schmid, C. (eds.) ECCV 2012. LNCS, vol. 7578, pp. 222–235. Springer, Heidelberg (2012). https://doi.org/10.1007/978-3-642-33786-4_17
11. Handa, A., Whelan, T., McDonald, J., Davison, A.: A benchmark for RGB-D visual odometry, 3D reconstruction and SLAM. In: IEEE International Conference on Robotics and Automation, ICRA, Hong Kong, China, May 2014
12. Kahler, O., Prisacariu, V.A., Ren, C.Y., Sun, X., Torr, P.H.S., Murray, D.W.: Very high frame rate volumetric integration of depth images on mobile device. IEEE Trans. Vis. Comput. Graph. **22**(11), 1241–1250 (2015)
13. Kim, T.H., Lee, K.M.: Segmentation-free dynamic scene deblurring. In: Proceedings of the IEEE Conference on Computer Vision and Pattern Recognition, pp. 2766–2773 (2014)
14. Krishnan, D., Tay, T., Fergus, R.: Blind deconvolution using a normalized sparsity measure. In: CVPR 2011, pp. 233–240. IEEE (2011)
15. Lai, W.S., Huang, J.B., Hu, Z., Ahuja, N., Yang, M.H.: A comparative study for single image blind deblurring. In: IEEE Conferene on Computer Vision and Pattern Recognition (2016)
16. Levin, A., Weiss, Y., Durand, F., Freeman, W.T.: Understanding and evaluating blind deconvolution algorithms. In: 2009 IEEE Conference on Computer Vision and Pattern Recognition, pp. 1964–1971. IEEE (2009)
17. Levin, A., Weiss, Y., Durand, F., Freeman, W.T.: Efficient marginal likelihood optimization in blind deconvolution. In: CVPR 2011, pp. 2657–2664. IEEE (2011)

18. Mur-Artal, R., Tardós, J.D.: ORB-SLAM2: an open-source SLAM system for monocular, stereo, and RGB-D cameras. IEEE Trans. Rob. **33**(5), 1255–1262 (2017)
19. Nah, S., Kim, T.H., Lee, K.M.: Deep multi-scale convolutional neural network for dynamic scene deblurring. In: 2017 IEEE Conference on Computer Vision and Pattern Recognition (CVPR) (2017)
20. Noroozi, M., Chandramouli, P., Favaro, P.: Motion deblurring in the wild. In: Roth, V., Vetter, T. (eds.) GCPR 2017. LNCS, vol. 10496, pp. 65–77. Springer, Cham (2017). https://doi.org/10.1007/978-3-319-66709-6_6
21. Qin, T., Li, P., Shen, S.: VINS-mono: a robust and versatile monocular visual-inertial state estimator. IEEE Trans. Rob. **34**(4), 1004–1020 (2018)
22. Sellent, A., Rother, C., Roth, S.: Stereo video deblurring. In: Leibe, B., Matas, J., Sebe, N., Welling, M. (eds.) ECCV 2016. LNCS, vol. 9906, pp. 558–575. Springer, Cham (2016). https://doi.org/10.1007/978-3-319-46475-6_35
23. Shan, Q., Jia, J., Agarwala, A.: High-quality motion deblurring from a single image. ACM Trans. Graph. (TOG) **27(3)**, 1–10 (2008)
24. Shen, Z., et al.: Human-aware motion deblurring. In: Proceedings of the IEEE International Conference on Computer Vision, pp. 5572–5581 (2019)
25. Sturm, J., Engelhard, N., Endres, F., Burgard, W., Cremers, D.: A benchmark for the evaluation of RGB-D slam systems. In: Proceedings of of the International Conference on Intelligent Robot Systems (IROS), October 2012
26. Su, S., Delbracio, M., Wang, J., Sapiro, G., Heidrich, W., Wang, O.: Deep video deblurring for hand-held cameras. In: Proceedings of the IEEE Conference on Computer Vision and Pattern Recognition, pp. 1279–1288 (2017)
27. Sun, J., Cao, W., Xu, Z., Ponce, J.: Learning a convolutional neural network for non-uniform motion blur removal. In: Proceedings of the IEEE Conference on Computer Vision and Pattern Recognition, pp. 769–777 (2015)
28. Sun, L., Cho, S., Wang, J., Hays, J.: Edge-based blur kernel estimation using patch priors. In: IEEE International Conference on Computational Photography (ICCP), pp. 1–8. IEEE (2013)
29. Whelan, T., Salas-Moreno, R.F., Glocker, B., Davison, A.J., Leutenegger, S.: ElasticFusion: real-time dense slam and light source estimation. Int. J. Robot. Res. **35**(14), 1697–1716 (2016)
30. Whyte, O., Sivic, J., Zisserman, A., Ponce, J.: Non-uniform deblurring for shaken images. Int. J. Comput. Vis. **98**(2), 168–186 (2012)
31. Xu, L., Jia, J.: Two-phase kernel estimation for robust motion deblurring. In: European Conference on Computer Vision, pp. 236–252. Springer (2014)
32. Xu, L., Zheng, S., Jia, J.: Unnatural l0 sparse representation for natural image deblurring. In: Proceedings of the IEEE Conference on Computer Vision and Pattern Recognition, pp. 1107–1114 (2013)
33. Zhu, Z., Xu, F., Yan, C., Hao, X., Ji, X., Zhang, Y., Dai, Q.: Real-time indoor scene reconstruction with RGBD and inertial input. In: 2019 IEEE International Conference on Multimedia and Expo (ICME), pp. 7–12. IEEE (2019)

Self-supervised Attribute-Aware Refinement Network for Low-Quality Text Recognition

Younkwan Lee[1(✉)], Heongjun Yoo[1], Yechan Kim[1], Jihun Jeong[1], and Moongu Jeon[1,2]

[1] Gwangju Institute of Science and Technology, Gwangju, South Korea
{brightyoun,jhdf1234,yechankim,jihunj1111,mgjeon}@gist.ac.kr
[2] Korea Culture Technology Institute, Gwangju, South Korea

Abstract. Scene texts collected from unconstrained environments encompass various types of degradation, including low-resolution, cluttered backgrounds, and irregular shapes. Training a model for text recognition with such types of degradations is notoriously hard. In this work, we analyze this problem in terms of two attributes: semantic and a geometric attribute, which are crucial cues for describing low-quality text. To handle this issue, we propose a new Self-supervised Attribute-Aware Refinement Network (SAAR-Net) that addresses these attributes simultaneously. Specifically, a novel text refining mechanism is combined with self-supervised learning for multiple auxiliary tasks to solve this problem. In addition, it can extract semantic and geometric attributes important to text recognition by introducing mutual information constraint that explicitly preserves invariant and discriminative information across different tasks. Such learned representation encourages our method to evidently generate a clear image, thus leading to better recognition performance. Extensive results demonstrate the effectiveness in refinement and recognition simultaneously.

Keywords: Scene text recognition · Self-supervised learning · Mutual information

1 Introduction

Scene text recognition is one of the fundamental problems in a range of computer vision applications including intelligent transportation systems, self-driving. While various visual recognition research has made great success in recent years, most text recognition methods are proposed, applied, and evaluated on high-quality visual data captured from constrained scenarios. However, in many real-world applications, the performance of text recognition is largely impaired by

Electronic supplementary material The online version of this chapter (https://doi.org/10.1007/978-3-030-68238-5_17) contains supplementary material, which is available to authorized users.

A. Bartoli and A. Fusiello (Eds.): ECCV 2020 Workshops, LNCS 12539, pp. 228–244, 2020.
https://doi.org/10.1007/978-3-030-68238-5_17

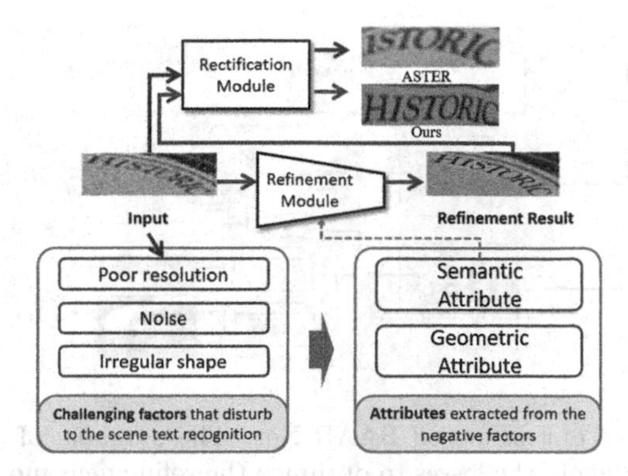

Fig. 1. Illustration of various factors disturbing the scene text recognition. There are various negative factors in low-quality data. Under these factors, comparison of detection results between ASTER [51] and proposed method.

challenging scene texts obtained from complex uncontrolled environments, suffering from various kinds of degradations such as poor resolution, out-of-focus, arbitrary shape, etc.

To address these difficulties, previous works [4,8,20,30,33,38,47,50,51,57–59,62–64] have been made to tackle a vast amount of challenging factors that disturb the scene text recognition, leveraging deep neural networks that can exploit the potential feature representation. However, one of its fatal limitations is that they focus on the existing datasets which rarely contain extremely low-quality visual data so that the performances of text recognition among low-quality data remain a challenging problem. As a result, such an approach on these datasets may lead to poorly optimized representation deviating from their pure original-purpose.

Even if it is not possible to collect scene texts of all environments, we assume that the limited annotated dataset has inherent attributes of the scene text. More clearly, all scene texts maintain specific intra-information even if the raw images are embedded in arbitrary feature space by refinement. Therefore, the basic idea behind our assumption is to learn refinement feature representation in which all scene texts consist of two attributes: semantic and geometric attributes. It also enables robust representation learning on low-quality samples that do not have any existing ground-truth annotations. Figure 1 illustrates our motivation by observing three challenging factors of the scene text. However, it is difficult to define valid embedding tasks that actually help and lead to unbiased representations.

To deal with the above problems, we propose a new Self-supervised Attribute-Aware Refinement Network (SAAR-Net) that to integrate self-supervised learning into the attribute-aware refinement without additional human effort. We then concatenate the latent space of each encoder that contains attribute information and propagate it as input to the main task network for final refinement output. Besides, we introduce mutual information constraint losses that enforce the

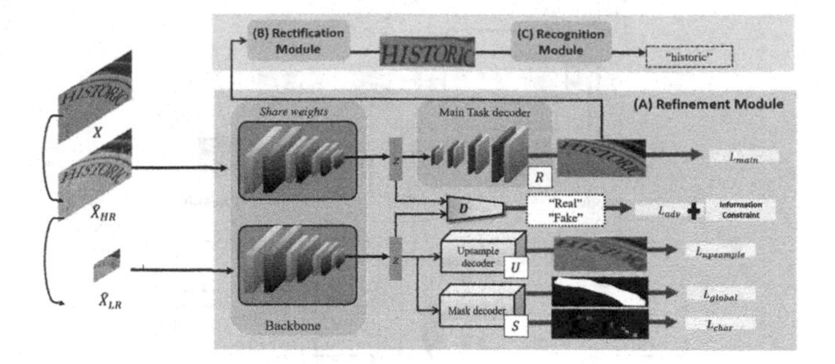

Fig. 2. Illustration of our proposed SAAR-Net. The symbols of L_{main}, $L_{upsample}$, L_{global}, L_{char}, L_{adv} denote the losses to optimize the refinement module of refinement, upsample, global region, individual character region, and discriminator, respectively. The symbol of information constraint is the loss for preserving domain-specific features.

refinement network to preserve inherent information. Motivated by information-theoretical approach for deep learning [1,5], a variational upper bound and a contrastive lower bound of mutual information are computed to approximately optimize the learning model. Consequently, we enforce the auxiliary task networks to capture those task-dependent factors while excluding task-independent factors.

To sum up, this work makes the following three main contributions. **(1)** First, we propose a novel Self-supervised Attribute-Aware Refinement Network (SAAR-Net) for scene text recognition, which combines a robust text refinement module. **(2)** Second, we propose a new information constraint loss that promotes the main task network to generate clearer texts for easier recognition. **(3)** Finally, we conduct experiments on the scene-text datasets ICDAR 2013 [26], 2015 [25], SVTP [47], and CUTE [48], as well as on license plate datasets SSIG [12] and VTLP [31] and obtain encouraging text recognition performance.

2 Related Works

2.1 Scene Text Recognition

Early works employed hand-crafted features to extract pattern features for character-level detection and recognition one by one with the sliding window methods [45,54,55,60]. However, they were also not efficient to the unconstrained environment due to the limited representation capacity of hand-crafted feature. In recent years, the deep neural architectures were introduced into this field. [22,61] adopted a convolutional neural networks (CNN) to address unconstrained recognition. With the advancement of recurrent neural networks (RNNs) [10,19], many researchers combined CNN and RNN for better representation learning of context. [49] integrated the CNN and RNN into an end-to-end trainable network with CTC loss [14] that learns conditional probability between sequential

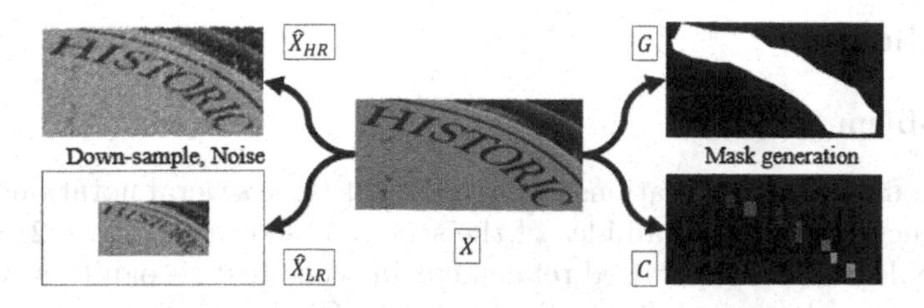

Fig. 3. Label generation for the training of the proposed method. From a high-quality image as ground truth, the noisy images can be obtained using a Gaussian Distribution, and the low-quality image is processed through downsampling $\times$ 1/2 of noisy image. The segmented image is inferred through region generation [43] of the original image.

labels and predictions. Though useful for estimating straight or slightly distorted texts, most existing methods are still facing problems in coping the irregular text geometry. [37] developed an affine transformation network for rotated and scaled text rectification, and extracted features for handling more complex background noise using residual network [17]. In addition, [50] conducted a regression task for obtaining fiducial transformation points on distorted text. [62] proposed an iterative rectification approach with perspective and curvature distortions. [9] designs an arbitrary orientation text recognition network for identifying characters in special locations.

Our proposed model adopts the top-down approach. Although numerous methods are using the refinement module for real-world text recognition, our approach differs in the disentanglement of the essential attributes emerging from the scene text, especially for the handling of limited annotations, which has not fully addressed in existing methods (Fig. 3).

2.2 Mutual Information Criterion

Mutual information has been widely researched in the literature since it captures the relationships between different distributions. [52] formulated the objectives of optimal feature space using mutual information for the domain similarity and classification error. Although they proposed an information-theoretic framework to solve their formulas, their approach could not be scaled into deep neural network. Inspired by the success of deep learning in various applications, mutual information has been used to disentangle feature representation [1,5,29,32]. In particular, [7] captured the semantic attribute without supervision while maximizing mutual information. Contrary to their approach, [27] reduced bias information while minimizing mutual information. Different from the existing methods, in this paper, we use information constraints to establish disentangling representation between auxiliary tasks and the main task.

3 Method

3.1 Problem Settings

In order to make the formulations clearer, we illustrate several notations prior to the introduction of the overall idea of the study, as shown in Figure 2. We focus on the problem of self-supervised refinement in scene text recognition, where we have a training dataset and a testing dataset. Given a training dataset of N input pairs $\{X_i, Y_i\}_i^N$, X_i is the ground truth sample and $Y_i = \{y_{i,1}, ..., y_{i,T}\}$ is its character information, where T is the number of characters. For the self-supervised labeling, we have corresponding targets for main task decoder R, upsampling decoder U, mask decoder S, and text recognition network respectively. For the refinement task, we define the input $\hat{X}_{HR}$ of main task decoder in the refinement module:

$$\hat{X}_{HR} = (X) + (\eta \sim \mathcal{N}(0, \sigma^2 \mathbf{I})), \tag{1}$$

where the noise η is obtained from the Gaussian distribution with standard deviation σ. This noisy image $\hat{X}_{HR}$ makes R robust against noise and distortions of the input images X at the training phase. For the upsampling task, we perform bicubic downsampling the $\hat{X}_{HR}$ to a $1/2 \times$ low-resolution image $\hat{X}_{LR}$. For the mask task, we automatically generate a global text G instance and a set of character instance C ($C = \{C_t\}_{t=1}^T$) for each original image X (see Figrue 3 and paper [43] in label generation for further details).

After performing the refinement task with the auxiliary tasks, we also introduce the rectification and recognition module. Using the refined image, the rectification module extracts the text geometric attributes based on pixel-wise prediction, with which the input image is expected to be rectified through Thin-Plate-Spline (TPS). Finally, the rectified image is propagated into the recognition network for text recognition. Even though different kinds of supervised models are involved, our approach does not require any additional annotation with human effort.

3.2 Refinement Module

Functionally, our refinement module consists of five components: the front-end backbone network F, the main task decoder R, the upsampling decoder U, the discriminator D and the mask decoder S. For the high-level representations at all scales, we adopt FPN [35] a backbone with ResNet-50 [17]. We describe the details for tasks of each component as follows.

Upsampling Decoder. Due to the complex environments by heavily corrupted conditions by noise, blur and clutter background in the wild, the scene text is usually lacking fine details and blurring. Therefore, appearance information of an image is usually easily damaged, which causes semantic properties to be lost. Toward this problem, we employ the upsampling network as an auxiliary task, which can lead backbone network to eliminate the artifact and thus acquire restored semantic attributes of original scene text image. With the training pairs of $\{X, \hat{X}_{LR}\}$, the upsampling network can be trained using a pixel-wise L2 loss $l_2(\cdot, \cdot)$:

$$\mathcal{L}_{upsample}(F, U) = \mathbb{E}_{(x,y)\sim p(\hat{X}_{LR}, X)}[l_2(U(F(x)), y)]. \quad (2)$$

However, while calculating less loss between the generated and the original image in pixel-wise prediction, the geometric attributes such as the arrangement or position of the scene text in the image are still not taken into account. Even if the semantic properties of the image are captured, it is difficult to recognize characters without the geometric properties.

Mask Decoder. To gain more geometric properties, we introduce the mask decoder to guide the backbone to capture geometric attribute. There are two branches in our mask decoder, the global text instance and the character instances. The global text instance defines the binary localization of a text spot, despite the arbitrary shape of the text. With the training pairs of $\{G, \hat{X}_{LR}\}$, this is acquired by L_{global}:

$$\mathcal{L}_{global}(F, S) = \mathbb{E}_{(x,y)\sim p(\hat{X}_{LR}, G)}[ylog(S(x)) + (1 - y)log(1 - S(x))], \quad (3)$$

where $S(\dot{})$ denotes a sigmoid function.

Besides the character instances consist of 36 maps, including 26 letters and 10 Arabic numbers. To calculate the instances, we introduce a spatial softmax loss to the objective loss, defined as:

$$\mathcal{L}_{char}(F, S) = \mathbb{E}_{(x,y)\sim p(\hat{X}_{LR}, C)}[\sum_{t=1}^{T} y_t log(\frac{e^{x_t}}{\sum_{k=1}^{T} e^{x_k}}))], \quad (4)$$

where, C is a set of instance ($C = \{C_t\}_{t=1}^{T}$). Note that, with these auxiliary tasks, our backbone encoder will produce rich feature representations adapted to the semantic and geometric attributes of the input image.

Main Task Decoder. The main task decoder is built on the features emitted from the backbone encoder. In addition, the main task treats the $\hat{X}$ with no noise added as input. Therefore, the decoder for the main task is optimized using the L1 loss l_1 for the original image X:

$$\mathcal{L}_{main}(F, R) = \mathbb{E}_{(x,y)\sim p(\hat{X}_{LR}, X)}[l_1(R(F(x)), y)]. \quad (5)$$

Although a sophisticated task for upsampling is proposed to refine some missing details in the low-quality image, it still does not recover sharp features due to scene-texts of too low resolutions. Therefore, we perform our main task with the L1 loss as well as high-resolution input, which yields sharp features based on the rich backbone features updated from auxiliary tasks.

Discriminator. Because we separate the inputs of the auxiliary tasks and the main task from each other, latent space can suffer from domain-specific features that learn the domain class of the input image. In essence, during the training process, the classifier that distinguishes the extracted features between $\hat{X}_{HR}$ and $\hat{X}_{LR}$ should confuse to obtain domain-invariant features. To achieve more robust feature maps of backbone encoder, we introduce the discriminator network, and it is achieved by adversarial minimax loss:

$$\mathcal{L}_{adv}(F, D) = -\mathbb{E}_{x \sim p(\hat{X}_{LR})}[log(D(F(x)))] - \mathbb{E}_{x \sim p(\hat{X}_{HR})}[log(1 - D(F(x)))]. \quad (6)$$

3.3 Mutual Information Constraint

Although domain-specific patterns are ignored by Eq. (6), forcing the similarity in the marginal distribution does not directly affect the capture of valid information in each domain. In particular, low-quality scene text requires domain-specific information to be maintained, as the geometric and semantic properties change drastically with small changes in image quality. Therefore, maintaining domain information that is distinguished by image quality can contribute to the abundant preservation of the essential characteristics of the image. To solve this issue, we limit the minimum value of MI $I(z_i, X_i)$ for all domains and this maintains the representation specified by all domains.

Let z_i and X_i denote the random variables of latent representations and input images from i-th domain (assume the number of domains is at hand M For $i = 0, ..., M$. $i = 0$ is $\hat{X}_{LR}$ and $i = 1$ is $\hat{X}_{HR}$), respectively. Rewriting the adversarial loss, we obtain the following constrained problem:

$$\begin{aligned} \mathcal{L}_{adv}(F, D) = -\mathbb{E}_{x \sim p(\hat{X}_{LR})}[log(D(F(x)))] - \mathbb{E}_{x \sim p(\hat{X}_{HR})}[log(1 - D(F(x)))] \\ s.t. \quad I(z_i, X_i) > \epsilon_1, \forall \in \{0, 1\}. \end{aligned} \quad (7)$$

Mutual information is lower bounded by Noise Contrastive Estimation [46]:

$$I(z_i, X_i) \geq \hat{I}^{(NCE)}(z_i, X_i) := \mathbb{F}_{p_{i(X)}}[D(F(X_i), X_i) - \mathbb{F}_{\tilde{p}_{i(X)}}[log \sum_{x_i^{'}} e^{D(F(X_i), X_i)}]], \quad (8)$$

where $X_i^{'}$ is sampled from the distribution $\tilde{p}_i(X) = p_i(X)$. By providing stable approximation results [18], mutual information can be maximized by replacing maximization in Jensen-Shannon divergence (JSD):

$$\hat{I}^{JSD}(z_i, X_i) := \mathbb{F}_{p_{i(X)}}[-sp(-D(F(X_i), X_i))] - \mathbb{F}_{p_{i(X)} \times p_{i(X)}}[sp(D(F(X_i), X_i^{'}))], \quad (9)$$

where $sp(\cdot)$ is a softplus function. As discussed in [18], the discriminator D can share backbone encoder F so that maximizing Eq. (9) will maximize the MI $I(z_i, X_i)$. With tractable approximation by using lower bound of MI, we can equivalently optimize Eq. (7) as follows:

$$\begin{aligned} \mathcal{L}_{adv}(F, D) = -\mathbb{E}_{x \sim p(\hat{X}_{LR})}[log(D(F(x)))] \\ - \mathbb{E}_{x \sim p(\hat{X}_{HR})}[log(1 - D(F(x)))] + \lambda_l \sum_{i=0}^{M} \hat{I}^{JSD}(z_i, X_i). \end{aligned} \quad (10)$$

3.4 Rectification Module

We followed the rectification network on geometric attribute prediction [40]. Then, in Supplementary, we will discuss both how to rectify the text image based on the given geometric properties and the necessity of character orientation for accurate rectification.

3.5 Recognition Module

For character recognition, we utilize the backbone encoder F for extracting features. It consists of a CNN-LSTM encoder and a decoder. In the encoder, the input is an output from the rectification module of the scene text image. Empirically feature maps from the lower layers are known to improve semantic segmentation performance [39]. In the same vein, the area of the scene text is mostly very small relative to the input image. Therefore, only seven lower convolutional layers of the encoder are used to extract features with two 2×2 max-pooling operations. The encoder network is followed by Bi-directional LSTM [15] each of which uses 256 hidden units that explicitly control data flow. For the decoder, we employ the attentional mechanism with GRU [42] and LSTMs. In the inference phase, the decoder predicts an individual text class y_k at step k until the last step of scene text.

3.6 Training

For SAAR-Net, The full objective function is a weighted sum of all the losses from main task to the auxiliary tasks:

$$\begin{aligned} \mathcal{L}(F,U,S,D) = \lambda_1\mathcal{L}_{upsample} + \lambda_2\mathcal{L}_{global} + \lambda_3\mathcal{L}_{char} + \\ \lambda_4\mathcal{L}_{main} + \lambda_5\mathcal{L}_{adv} + \lambda_l\sum_{i=1}^{M}\hat{I}^{JSD}(z_i, X_i). \end{aligned} \tag{11}$$

We employ a stage-wise training strategy to optimize main tasks with auxiliary tasks and empirically set the weights of each loss as detailed in Sect. 4.5.

3.7 Inference

At the testing phase, the auxiliary tasks are removed. Given a scene-text test image I_{test}, R outputs the recovered image via backbone encoder and main task decoder. Rectification module *Rect* based on the TPS [58] takes the recovered image and generates a rectification result $Rect_{result}$ of I_{test}. Finally, recognition module *Recog* based on Conv-LSTM takes the rectified image and outputs a character recognition result $Recog_{result}$ and it is denoted as Eq. (12):

$$Recog_{result} = Recog(Rect(R(F(I_{test})))). \tag{12}$$

4 Experimental Results

4.1 Implementation Details

All the reported implementations are based on the PyTorch framework, and our method has done on one NVIDIA TITAN X GPU and one Intel Core i7-6700K CPU. In all the experiments, we resize all images to 256 $\times$ 64. The proposed refinement module is trained in 1 million iterations with a batch size of 16. All models adopt SynthText [16] and Synth90k [23] as the training set and are trained for the first 10 epochs with a learning rate of 10^{-1} despite higher values, and then for the remaining epochs at the learning rate of 10^{-2}. Batch normalization [21] and LeakyReLU [44] are used in all layers of our networks. Also, backbone network as baseline, we use the FPN with ResNet-50 model pre-trained on ImageNet [11]. In our experiment, we set the weights for λ_1, λ_2, λ_3, λ_4, λ_5 and λ_l to 0.1, 0.1, 0.1, 0.2, 0.3 and 0.2, respectively.

4.2 Ablation Studies

To evaluate SAAR-Net, we first compare our SAAR-Net with the baseline scene text recognition researches to prove the effectiveness of image refinement performance. Both results on two types of the dataset (1) scene text and 2) license plate) are reported for the following five types of our methods where each module is optionally added: a) without proposed refinement method (Baseline); b) adding the main task to (a); c) adding one auxiliary task to (b); d) adding the main task and all auxiliary task to (c); e) adding discriminator network to (d); f) adding information constraint loss to (e); g) adding all of the modules (namely, proposed method).

We present the scene text recognition accuracy for each type on six datasets in Table 1, and the visual comparisons are shown in Fig. 4. From Table 1, we can find that adding the main task (R and the auxiliary tasks (U and S), respectively, significantly improves the scene text recognition performance (type b, c). In addition, we observe that recognition performance improves more when both tasks are applied at the same time (type d). As shown in Fig. 4. (f), noise and blurring effect are removed from the low-quality image (a), and characters are enhanced well compared to (c). This confirms that performing two auxiliary tasks at the same time is more helpful to recover high-quality images. Despite showing better scene text recognition performance, we still find that the output image contains elements that interfere with recognition performance. For example, there are still challenges to detect the suitable text region, including a region that is unnecessary for recognition, such as blurry and noise (see in Fig. 4. (e, f)), and ambiguity that not well detected between consecutive characters. Therefore, when discriminator is added to the proposed method, recovered image quality can be better (Fig. 4. (g)) and we observe some improvements on scene text recognition performance (Table 1. e)). Finally, we incorporate all the tasks, perform experiments on it and observe the best performance improvement in scene text recognition (Table 1. (f)). Furthermore, the recovered image in Fig. 4. (h) is the most realistic of all results (Fig. 6).

Table 1. Ablation study on the effectiveness of different components. Main task (refinement), upsampling decoder, mask decoder, discriminator, and MI constraint loss represent the R, U, S, D, and *Eq.* (9), respectively.

Type	Method	ICDAR2013	ICDAR2015	SVTP	CUTE	SSIG	VTLP
(A)	Baseline	91.6	76.0	74.6	77.4	85.6	85.2
(B)	add R	93.1	76.5	76.9	82.6	88.3	88.4
(C)	add R, U	93.3	76.5	79.1	82.6	88.6	90.1
	add R, S	93.5	76.7	79.2	85.1	88.9	89.0
(D)	add R, U, S	93.8	76.9	80.2	85.1	89.1	90.4
(E)	add R, U, S, D	93.8	77.0	80.6	86.8	89.1	90.5
(F)	Ours (R, U, S, D, *Eq.*(9))	94.0	77.3	81.2	88.2	89.2	90.9

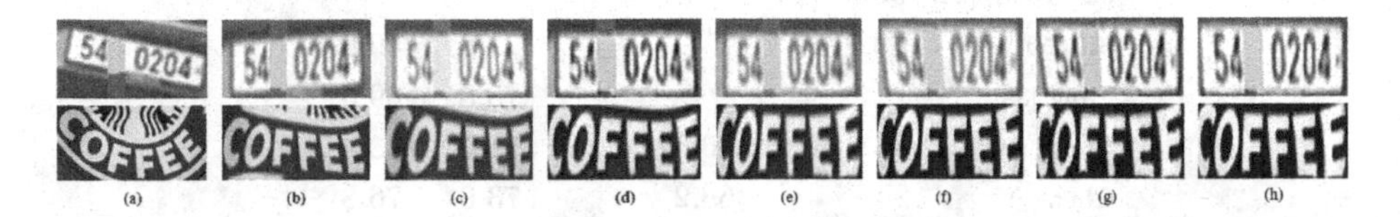

Fig. 4. Ablation Study. (a) shows noisy input; (b) baseline rectification; (c) adds refinement decoder; (d) add upsample decoder to (c); (e) adds mask decoder to (c); (f) add all auxiliary tasks from (c); (g) adds discriminator to (f); (h) all of tasks, namely our proposed model.

4.3 Comparisons with State-of-the-Art Methods

We compare the SAAR-Net with some state-of-the-art scene text recognition and license plate recognition methods, respectively. For the baseline, the proposed method has been evaluated over the six datasets as described in Sect. 4.2 that contain scene text images and low-quality license plates with a variety of geometric variations.

Table 2 shows the recognition performance consistently outperforms the existing methods across all datasets due to the use of the proposed refinement network. However, ICDAR2015 is also evaluated to be more effective than most methods, and if not, it shows a relatively small performance difference. Therefore, it can be explained that our proposed method is more useful for LPR than other methods.

Scene Text Dataset Results. For the ICDAR13, ICDAR15, SVTP, and CUTE, SAAR-Net demonstrates that our recovery image can significantly improve the performance of recognition on real-world images. This is mainly due to the fact that scene text samples which usually have geometrically tilted cases and semantically damaged are processed into a well-rectified image. The results are listed in Table 2, and our method obtains the highest performance **(94.0%)**, **(81.2%)**, **(88.2%)** in ICDAR13, SVTP, and CUTE. Note that what we want to illustrate in all the benchmarks (especially see the difference between Baseline and ours in Table 2) is that our method can benefit from the refinement

Table 2. Scene text recognition performance comparison of method with existing methods on ICDAR13, ICDAR15, SVTP and CUTE.

Methods	ICDAR13	ICDAR2015	SVTP	CUTE
Wang [56]	–	–	–	–
Bissacco [6]	87.6	–	–	–
Yao [60]	–	–	–	–
Almazan [2]	–	–	–	–
Gordo [13]	–	–	–	–
Jaderberg [22]	81.8	–	–	–
Jaderberg [24]	90.8	–	–	–
Shi [50]	88.6	–	71.8	59.2
Yang [59]	–	–	75.8	69.3
Cheng [8]	93.3	70.6	71.5	63.9
Cheng [9]	–	68.2	73.0	76.8
Shi [51]	91.8	76.1	78.5	79.5
Zhan [62]	91.3	76.9	79.6	83.3
Liao [34]	91.4	–	–	79.9
Bai [3]	94.4	73.9	–	–
Luo [41]	92.4	68.8	76.1	77.4
Liu [36]	91.1	74.2	78.9	–
Yang [58]	93.9	**78.7**	80.8	87.5
Ours-Baseline	91.6	76.0	74.6	77.4
Ours	**94.0**	77.3	**81.2**	**88.2**

module, which captures the semantic and geometric attributes of images despite low-quality image.

License Plate Dataset Results. The quantitative results for SSIG and VTLP datasets are shown in Tables 3 and 4, and the visual comparisons are illustrated in Fig. 5. Our approach shows superior performance to other license plate recognition (LPR) algorithms on LPR accuracy and image refinement. Furthermore, we achieve comparable results with state-of-the-art LPR method [28,31,53]. From Table 3, our method obtains the highest performance **(89.18%)**, and outperforms the state-of-the-art methods by more than 0.62% (88.56% *vs* 89.18%). Moreover, Table 3, our method obtains the highest performance **(90.87%)**, and outperforms the state-of-the-art methods by more than 4.21% (86.66% *vs* 90.87%). Note that our proposed method achieves robust performance in VTLP that are collected in low-resolution environments rather than other datasets.

Original
Refine+Rectify
Original Recog
Refine+Rectify Recog
mnudser
historic
newschesups
manchester
asgatify
vacation
srs
boohstorf
36 0746
36 8746
32 7910
32 2910

Fig. 5. Illustration of scene text refinement, rectification and recognition by our proposed method. Red characters are wrongly recognized characters. Green characters are correctly recognized characters. Best viewed on the computer, in color and zoomed in.

Table 3. Full LPR performance (percentage) comparison of our method with the existing methods on **SSIG**.

Method	SSIG Full LPR accuracy (%)
Laroca *et al.* [28]	85.45
Silva *et al.* [53]	88.56
Ours - Baseline	85.63
Ours	**89.18**

Fig. 6. Visual comparison of different scene text refinement and rectification results. For the four original images in the first column, columns 2–4 show the refined and rectified images by RARE, ESIR, ours, respectively. The proposed refinement method contributes to getting the better rectification results. Best viewed on the computer, in color and zoomed in.

Table 4. Full LPR performance (percentage) comparison of our method with the existing methods on **VTLP**.

Method	VTLP Full LPR accuracy (%)
YOLO v3	80.45
Laroca *et al.* [28]	87.34
Silva *et al.* [53]	84.73
Lee *et al.* [31]	86.66
Ours - Baseline	85.20
Ours	**90.87**

5 Conclusion

In this paper, we have proposed a novel Self-supervised Attribute-Aware Refinement Network (SAAR-Net), which refines and recognizes scene texts and can be trained end-to-end thoroughly. Comparing with existing approaches, we design a novel refinement module that incorporates self-supervised learning for upsampling and segmentation tasks. Moreover, the mutual information constraint losses are brought to the refinement module to exploit invariant and discriminative information with unbiased description. Extensive experiments on five benchmarks demonstrated the effectiveness of our proposed method in rectifying arbitrary shape scene text accurately and showed that the recognition performance outperforms other state-of-the-art algorithms. As future research, we would like to extend the proposed method to domain adaptive recognition which can deal with text instances of unseen domains.

Acknowledgements. This work was partly supported by Institute of Information & Communications Technology Planning & Evaluation (IITP) grant funded by the Korea government (MSIT) (No.2014-3-00077, AI National Strategy Project) and Ministry of Culture, Sports and Tourism and Korea Creative Content Agency(Project Number: R2020070004).

References

1. Alemi, A.A., Fischer, I., Dillon, J.V., Murphy, K.: Deep variational information bottleneck. arXiv preprint arXiv:1612.00410 (2016)
2. Almazán, J., Gordo, A., Fornés, A., Valveny, E.: Word spotting and recognition with embedded attributes. IEEE Trans. Pattern Anal. Mach. Intell. **36**(12), 2552–2566 (2014)
3. Bai, F., Cheng, Z., Niu, Y., Pu, S., Zhou, S.: Edit probability for scene text recognition. In: Proceedings of the IEEE Conference on CVPR, pp. 1508–1516 (2018)
4. Bartz, C., Yang, H., Meinel, C.: See: towards semi-supervised end-to-end scene text recognition. In: Thirty-Second AAAI Conference on Artificial Intelligence (2018)
5. Belghazi, M.I., Baratin, A., Rajeswar, S., Ozair, S., Bengio, Y., Courville, A., Hjelm, R.D.: Mine: mutual information neural estimation. arXiv preprint arXiv:1801.04062 (2018)

6. Bissacco, A., Cummins, M., Netzer, Y., Neven, H.: Photoocr: reading text in uncontrolled conditions. In: Proceedings of the IEEE International Conference on Computer Vision, pp. 785–792 (2013)
7. Chen, X., Duan, Y., Houthooft, R., Schulman, J., Sutskever, I., Abbeel, P.: Infogan: interpretable representation learning by information maximizing generative adversarial nets. In: NIPS, pp. 2172–2180 (2016)
8. Cheng, Z., Bai, F., Xu, Y., Zheng, G., Pu, S., Zhou, S.: Focusing attention: towards accurate text recognition in natural images. In: ICCV, pp. 5076–5084 (2017)
9. Cheng, Z., Xu, Y., Bai, F., Niu, Y., Pu, S., Zhou, S.: Aon: towards arbitrarily-oriented text recognition. In: Proceedings of the IEEE Conference on CVPR, pp. 5571–5579 (2018)
10. Cho, K., et al.: Learning phrase representations using RNN encoder-decoder for statistical machine translation. arXiv preprint arXiv:1406.1078 (2014)
11. Deng, J., Dong, W., Socher, R., Li, L.J., Li, K., Fei-Fei, L.: ImageNet: a large-scale hierarchical image database. In: 2009 IEEE Conference on CVPR, pp. 248–255. Ieee (2009)
12. Gonçalves, G.R., da Silva, S.P.G., Menotti, D., Schwartz, W.R.: Benchmark for license plate character segmentation. J. Electron. Imaging **25**(5), 053034 (2016)
13. Gordo, A.: Supervised mid-level features for word image representation. In: Proceedings of the IEEE conference on CVPR, pp. 2956–2964 (2015)
14. Graves, A., Fernández, S., Gomez, F., Schmidhuber, J.: Connectionist temporal classification: labelling unsegmented sequence data with recurrent neural networks. In: Proceedings of the 23rd International Conference on Machine Learning, pp. 369–376. ACM (2006)
15. Graves, A., Liwicki, M., Fernández, S., Bertolami, R., Bunke, H., Schmidhuber, J.: A novel connectionist system for unconstrained handwriting recognition. IEEE Trans. Pattern Anal. Mach. Intell. **31**, 855–868 (2009)
16. Gupta, A., Vedaldi, A., Zisserman, A.: Synthetic data for text localisation in natural images. In: Proceedings of the IEEE Conference on CVPR, pp. 2315–2324 (2016)
17. He, K., Zhang, X., Ren, S., Sun, J.: Deep residual learning for image recognition. In: Proceedings of the IEEE Conference on CVPR, pp. 770–778 (2016)
18. Hjelm, R.D., et al.: Learning deep representations by mutual information estimation and maximization. arXiv preprint arXiv:1808.06670 (2018)
19. Hochreiter, S., Schmidhuber, J.: Long short-term memory. Neural Comput. **9**(8), 1735–1780 (1997)
20. Hradiš, M., Kotera, J., Zemcık, P., Šroubek, F.: Convolutional neural networks for direct text deblurring. In: Proceedings of BMVC, vol. 10, p. 2 (2015)
21. Ioffe, S., Szegedy, C.: Batch normalization: accelerating deep network training by reducing internal covariate shift. arXiv preprint arXiv:1502.03167 (2015)
22. Jaderberg, M., Simonyan, K., Vedaldi, A., Zisserman, A.: Deep structured output learning for unconstrained text recognition. arXiv preprint arXiv:1412.5903 (2014)
23. Jaderberg, M., Simonyan, K., Vedaldi, A., Zisserman, A.: Synthetic data and artificial neural networks for natural scene text recognition. arXiv preprint arXiv:1406.2227 (2014)
24. Jaderberg, M., Simonyan, K., Vedaldi, A., Zisserman, A.: Reading text in the wild with convolutional neural networks. Int. J. Comput. Vision **116**(1), 1–20 (2016)
25. Karatzas, D., et al.: Icdar 2015 competition on robust reading. In: 2015 13th International Conference on Document Analysis and Recognition (ICDAR), pp. 1156–1160. IEEE (2015)

26. Karatzas, D., et al.: Icdar 2013 robust reading competition. In: 2013 12th International Conference on Document Analysis and Recognition, pp. 1484–1493. IEEE (2013)
27. Kim, B., Kim, H., Kim, K., Kim, S., Kim, J.: Learning not to learn: training deep neural networks with biased data. In: Proceedings of the IEEE Conference on CVPR, pp. 9012–9020 (2019)
28. Laroca, R., et al.: A robust real-time automatic license plate recognition based on the yolo detector. In: 2018 IJCNN. IEEE (2018)
29. Lee, Y., Jeon, J., Yu, J., Jeon, M.: Context-aware multi-task learning for traffic scene recognition in autonomous vehicles. arXiv preprint arXiv:2004.01351 (2020)
30. Lee, Y., Jun, J., Hong, Y., Jeon, M.: Practical license plate recognition in unconstrained surveillance systems with adversarial super-resolution. arXiv preprint arXiv:1910.04324 (2019)
31. Lee, Y., Lee, J., Ahn, H., Jeon, M.: Snider: Single noisy image denoising and rectification for improving license plate recognition. In: Proceedings of the IEEE International Conference on Computer Vision Workshops (2019)
32. Lee, Y., Lee, J., Hong, Y., Ko, Y., Jeon, M.: Unconstrained road marking recognition with generative adversarial networks. In: 2019 IEEE Intelligent Vehicles Symposium (IV) pp. 1414–1419. IEEE (2019)
33. Lee, Y., Yun, J., Hong, Y., Lee, J., Jeon, M.: Accurate license plate recognition and super-resolution using a generative adversarial networks on traffic surveillance video. In: 2018 IEEE International Conference on Consumer Electronics-Asia (ICCE-Asia), pp. 1–4. IEEE (2018)
34. Liao, M., Zhang, J., Wan, Z., Xie, F., Liang, J., Lyu, P., Yao, C., Bai, X.: Scene text recognition from two-dimensional perspective. In: Proceedings of the AAAI Conference on Artificial Intelligence, vol.33, pp. 8714–8721 (2019)
35. Lin, T.Y., Dollar, P., Girshick, R., He, K., Hariharan, B., Belongie, S.: Feature pyramid networks for object detection. In: The IEEE Conference on CVPR, July 2017
36. Liu, W., Chen, C., Wong, K.Y.K.: Char-net: a character-aware neural network for distorted scene text recognition. In: AAAI (2018)
37. Liu, W., Chen, C., Wong, K.Y.K., Su, Z., Han, J.: Star-net: a spatial attention residue network for scene text recognition. In: BMVC, vol. 2, p. 7 (2016)
38. Liu, Y., Wang, Z., Jin, H., Wassell, I.: Synthetically supervised feature learning for scene text recognition. In: Ferrari, V., Hebert, M., Sminchisescu, C., Weiss, Y. (eds.) ECCV 2018. LNCS, vol. 11209, pp. 449–465. Springer, Cham (2018). https://doi.org/10.1007/978-3-030-01228-1_27
39. Long, J., Shelhamer, E., Darrell, T.: Fully convolutional networks for semantic segmentation. In: CVPR
40. Long, S., Ruan, J., Zhang, W., He, X., Wu, W., Yao, C.: TextSnake: a flexible representation for detecting text of arbitrary shapes. In: Ferrari, V., Hebert, M., Sminchisescu, C., Weiss, Y. (eds.) ECCV 2018. LNCS, vol. 11206, pp. 19–35. Springer, Cham (2018). https://doi.org/10.1007/978-3-030-01216-8_2
41. Luo, C., Jin, L., Sun, Z.: Moran: a multi-object rectified attention network for scene text recognition. Pattern Recogn. **90**, 109–118 (2019)
42. Luong, T., Pham, H., Manning, C.D.: Effective approaches to attention-based neural machine translation. In: EMNLP (2015)

43. Lyu, P., Liao, M., Yao, C., Wu, W., Bai, X.: Mask TextSpotter: an end-to-end trainable neural network for spotting text with arbitrary shapes. In: Ferrari, V., Hebert, M., Sminchisescu, C., Weiss, Y. (eds.) Computer Vision – ECCV 2018. LNCS, vol. 11218, pp. 71–88. Springer, Cham (2018). https://doi.org/10.1007/978-3-030-01264-9_5
44. Maas, A.L., Hannun, A.Y., Ng, A.Y.: Rectifier nonlinearities improve neural network acoustic models. In: Proceedings of icml, vol. 30, p. 3 (2013)
45. Neumann, L., Matas, J.: Real-time scene text localization and recognition. In: CVPR, pp. 3538–3545 (2012)
46. Oord, A.v.d., Li, Y., Vinyals, O.: Representation learning with contrastive predictive coding. arXiv preprint arXiv:1807.03748 (2018)
47. Quy Phan, T., Shivakumara, P., Tian, S., Lim Tan, C.: Recognizing text with perspective distortion in natural scenes. In: ICCV, pp. 569–576 (2013)
48. Risnumawan, A., Shivakumara, P., Chan, C.S., Tan, C.L.: A robust arbitrary text detection system for natural scene images. Expert Syst. Appl. **41**(18), 8027–8048 (2014)
49. Shi, B., Bai, X., Yao, C.: An end-to-end trainable neural network for image-based sequence recognition and its application to scene text recognition. IEEE Trans. Pattern Anal. Mach. Intell. **39**(11), 2298–2304 (2016)
50. Shi, B., Wang, X., Lyu, P., Yao, C., Bai, X.: Robust scene text recognition with automatic rectification. In: Proceedings of the IEEE Conference on CVPR, pp. 4168–4176 (2016)
51. Shi, B., Yang, M., Wang, X., Lyu, P., Yao, C., Bai, X.: Aster: an attentional scene text recognizer with flexible rectification. IEEE Trans. Pattern Anal. Mach. Intell. **PP**(99), 1 (2018)
52. Shi, Y., Sha, F.: Information-theoretical learning of discriminative clusters for unsupervised domain adaptation. arXiv preprint arXiv:1206.6438 (2012)
53. Silva, S.M., Jung, C.R.: License plate detection and recognition in unconstrained scenarios. In: Ferrari, V., Hebert, M., Sminchisescu, C., Weiss, Y. (eds.) ECCV 2018. LNCS, vol. 11216, pp. 593–609. Springer, Cham (2018). https://doi.org/10.1007/978-3-030-01258-8_36
54. Wang, K., Babenko, B., Belongie, S.: End-to-end scene text recognition. In: ICCV, pp. 1457–1464. IEEE (2011)
55. Wang, K., Belongie, S.: Word spotting in the wild. In: Daniilidis, K., Maragos, P., Paragios, N. (eds.) ECCV 2010. LNCS, vol. 6311, pp. 591–604. Springer, Heidelberg (2010). https://doi.org/10.1007/978-3-642-15549-9_43
56. Wang, T., Wu, D.J., Coates, A., Ng, A.Y.: End-to-end text recognition with convolutional neural networks. In: Proceedings of the 21st International Conference on Pattern Recognition (ICPR2012), pp. 3304–3308. IEEE (2012)
57. Xue, C., Lu, S., Zhan, F.: Accurate scene text detection through border semantics awareness and bootstrapping. In: Ferrari, V., Hebert, M., Sminchisescu, C., Weiss, Y. (eds.) ECCV 2018. LNCS, vol. 11220, pp. 370–387. Springer, Cham (2018). https://doi.org/10.1007/978-3-030-01270-0_22
58. Yang, M., et al.: Symmetry-constrained rectification network for scene text recognition. In: ICCV, pp. 9147–9156 (2019)
59. Yang, X., He, D., Zhou, Z., Kifer, D., Giles, C.L.: Learning to read irregular text with attention mechanisms. In: IJCAI, vol. 1, p. 3 (2017)
60. Yao, C., Bai, X., Shi, B., Liu, W.: Strokelets: a learned multi-scale representation for scene text recognition. In: Proceedings of the IEEE Conference on CVPR, pp. 4042–4049 (2014)

61. Yin, F., Wu, Y.C., Zhang, X.Y., Liu, C.L.: Scene text recognition with sliding convolutional character models. arXiv preprint arXiv:1709.01727 (2017)
62. Zhan, F., Lu, S.: Esir: End-to-end scene text recognition via iterative image rectification. In: Proceedings of the IEEE Conference on CVPR, pp. 2059–2068 (2019)
63. Zhan, F., Lu, S., Xue, C.: Verisimilar image synthesis for accurate detection and recognition of texts in scenes. In: Ferrari, V., Hebert, M., Sminchisescu, C., Weiss, Y. (eds.) ECCV 2018. LNCS, vol. 11212, pp. 257–273. Springer, Cham (2018). https://doi.org/10.1007/978-3-030-01237-3_16
64. Zhan, F., Xue, C., Lu, S.: Ga-dan: geometry-aware domain adaptation network for scene text detection and recognition. In: ICCV, pp. 9105–9115 (2019)

Two-Stage Training for Improved Classification of Poorly Localized Object Images

Sravanthi Bondugula(✉), Gang Qian, and Allison Beach

Object Video Labs, LLC, McLean, USA
{sbondugula,gqian,abeach}@objectvideo.com

Abstract. State-of-the-art object classifiers finetuned from a pretrained (e.g. from ImageNet) model on a domain-specific dataset can accurately classify well-localized object images. However, such classifiers often fail on poorly localized images (images with lots of context, heavily occluded/partially visible, and off-centered objects). In this paper, we propose a two-stage training scheme to improve the classification of such noisy detections, often produced by low-compute algorithms such as motion based background removal techniques that run on the edge. The proposed two-stage training pipeline first trains a classifier from scratch with extreme image augmentation, followed by finetuning in the second stage. The first stage incorporates a lot of contextual information around the objects, given access to the corresponding full images. This stage works very well for classification of poorly localized input images, but generates a lot of false positives by classifying non-object images as objects. To reduce the false positives, a second training is done on the tight ground-truth bounding boxes (as done traditionally) by using the trained model in the first stage as the initial model and very slowly adjusting its weights during the training. To demonstrate the efficacy of our approach, we curated a new classification dataset for poorly localized images - noisy PASCAL VOC 2007 test dataset. Using this dataset, we show that the proposed two-stage training scheme can significantly improve the accuracy of the trained classifier on both well-localized and poorly-localized object images.

Keywords: Poorly-localized objects · Low-quality bounding boxes · Limited quality · Noisy · Occluded bounding boxes · Context · Image classification

1 Introduction

Object detection using deep learning models is a popular research topic with many practical applications. In many real-life scenarios, for latency and privacy reasons, deep object detection must be carried out on the edge directly on embedded devices and platforms such as smart phones and surveillance cameras,

A. Bartoli and A. Fusiello (Eds.): ECCV 2020 Workshops, LNCS 12539, pp. 245–260, 2020.
https://doi.org/10.1007/978-3-030-68238-5_18

Fig. 1. Accurately (top) and Poorly (bottom) localized bounding boxes of objects in images. Bottom row may be examples of low quality detections of objects produced by background removal techniques

instead of sending the images to the backend cloud for deep model inference. However, deep object detectors are both memory and computationally intensive (e.g., the YOLOv2 [23] object detector takes 250 MB to just host the model parameters, and it takes few seconds on CPU to run the model on a single image). Such compute requirements go way beyond embedded devices capacities. There has been a surge of research in the field of optimization of deep neural networks with memory and energy saving constraints for low power devices. Optimizing neural networks entails distilling the neural network with techniques such as pruning [4,10], quantization [16], knowledge distillation [11,21], early exit, and finding networks automatically that meet the budget constraints using Neural Architecture Search (NAS) [22,28,38]. Researchers have made strides of improvement in both reducing the memory and obtaining faster inference times, but the optimal models performance is no match to that of the full models. State of the art classification model [29] has 88.5% top-1 accuracy on ImageNet, while state-of-the-art optimized model for mobile devices [2] has just hit 80% top-1 accuracy (for Pixel 1 phone with 590 M FLOPS). Further, additional training and fine-tuning times are incurred for the development of optimized neural networks for each optimization procedure.

Security cameras is one of the most commonly used edge devices for event surveillance and activity monitoring in residential/commercial settings. These cameras continuously monitor an area and send alert notifications to users when pre-defined events/activities occur. Examples of alerts could be a person/car entering a driveway, or human activities detected in a protected area. As the security cameras are usually mounted at a fixed location, one can efficiently employ background modeling and subtraction techniques to find regions of motion in the input video feed. Although these motion-based foreground extraction techniques can be efficiently deployed on the edge, they often produce many false positives and poorly localized bounding boxes of objects. To reduce the false positives, one can employ classification models optimized for edge devices. Figure 1 shows some examples of accurately and poorly localized bounding boxes of objects. In this paper, we refer to the images that are off-centered, partially visible, heavily occluded, and contains a large portion of background (context) as poorly-localized object images. State-of-the-art classification models can accu-

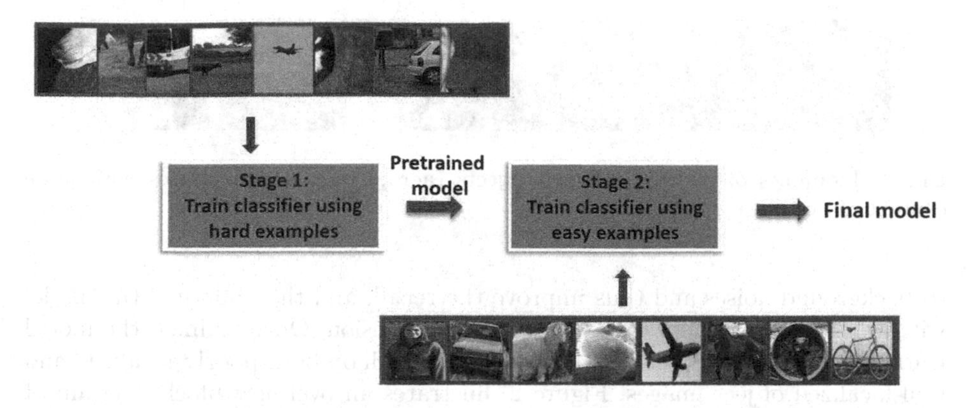

Fig. 2. Our two-stage training framework for improving classification performance of poorly localized images

rately detect well-localized objects in images, but often fail at classifying poorly-localized object images. Our proposed approach is designed to improve classification of these poorly-localized object images, without loosing performance on the well-localized object images and can still run efficiently on the camera.

Poorly-localized object images often contain large portions of contextual background information that outweighs the object presence. Traditional classifiers are robust enough to handle small linear transformations of the objects, as the training process employs such data augmentation techniques to boost its classification performance. However, they cannot handle large displacements of the objects in the images. When objects are off-centered and occupy only a small portion of the image (see for example the car in the bottom right image in Fig. 1), off-the-shelf classifiers usually perform poorly on these images. In addition, other hard cases include images with heavily occluded objects. To improve the classification of such poorly localized object images, in this paper we propose a two-stage training approach. Our first stage trains the classifier with images with extreme augmentation including a mixture of poorly localized bounding boxes and accurately localized bounding boxes. This allows the classifier to learn objects in the presence of heavy context. The goal of this stage is to allow the classifier not only to learn a strong representation of the target objects but also to ignore irrelevant features from the context in the mix of strong background noise. As the training inclines towards doing better at classifying poorly localized objects, the first stage classifier may still generate a lot of false positives of classifying non-objects as objects due to low SNR in the first stage training. To address this issue, the second stage training will follow in which the model from the first stage training is carefully finetuned (with low learning rates) using well-localized training object image. The goal of the second stage training is to rectify the model and reinforce the model's response to true target object classes. The intuition behind the proposed two-stage training approach is to first train the model using hard examples to make the model more acceptable and resilient

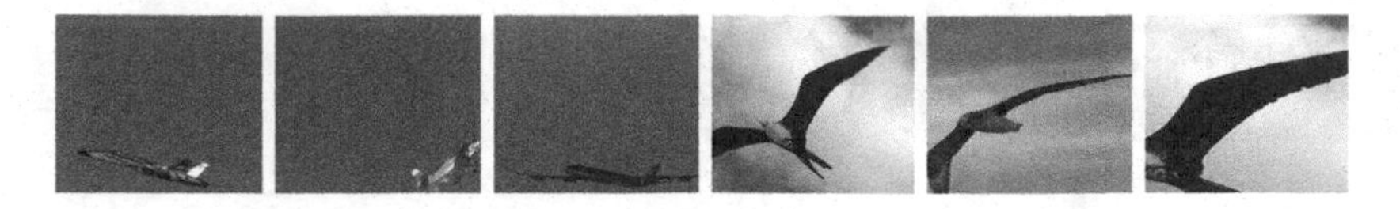

Fig. 3. Examples of poorly localized objects (aeroplane/bird) in images with large context of sky

to background noises and thus improve the recall, and then finetune the model with well-localized examples to improve the precision. Once trained, the model learned using the proposed approach performs well on both poorly-localized and well-localized object images. Figure 2 illustrates an overview block diagram of the proposed two-stage training approach.

Our main contributions of this paper are:

1. Propose a two-stage classifier training approach to improve classification performance of poorly-localized object images.
2. Utilize contextual information in training classifiers.
3. Ability to train a model from scratch without the need for a pretrained model (typically trained on ImageNet [6]) to obtain good performance.

In the following sections of the paper, we will talk about related work in Sect. 2. In Sect. 3, we explain the proposed framework to improve the classification of low quality detections. Then, we discuss datasets and experiments in Sect. 4. In Sect. 5, we provide ablation study showing the importance of two stage framework with extreme augmentation compared to the traditional one-stage fine tuning with different augmentations. We will also analyze how hard the examples should be, what is the right mixture of contextual information in the first stage and what is the right learning rate in the second stage that leads to optimal performance. Finally, we conclude in Sect. 6.

2 Related Work

Optimized neural networks for image classification on the Edge, specifically for mobile devices has been recently a focused research in the computer vision community. MobileNet series [12,13,24] pioneered this research by introducing mobile optimized models and showed improved accuracies that can run efficiently on mobile devices. MobileNetV1 [13] introduced depthwise separable convolutions, while MobileNetV2 [24] introduced inverted residuals and linear bottlenecks and MobileNetV3 [12] utilized an architecture similar to MNasNet [27] that employed Neural Architecture Search methods [38] using Reinforcement Learning to learn an optimized architecture for mobile devices. In MobileNetV3, Howard et al. reduced the number of filters in the input layer and rearranged the layers to keep the overall FLOPS low and introduced a new activation function called swish activation function to deal with the no gradient update for negative values. Away from the MobileNet series, several optimized networks like ShuffleNet [36], CondenseNet [14], ShiftNet [33] have been proposed that are also optimized for mobile devices. These methods try to keep the number of operations (MAdds) and the actual measured latency low instead of reducing the number of parameters like SqueezeNet [15]. Neural Architecture Search has shown remarkable results in finding optimal architectures for the task at hand and outperform state of the art hand-crafted architectures. Among those promising are NasNet [39], AmoebaNet [22], MNasNet [27], PNASNet [18], DARTS [19], FBNet series [5,31,32], EfficientNets B0-B7 [28] and the most recent state-of-the art method called Once-For-All Networks by Cai et al. [2]. We utilize the architectures of MobileNetV2 [24] and the original FBNet [32] as our base network architectures in this paper. More optimizations like pruning, quantization and mixed precision techniques can be applied to further compress and fit on edge devices. Discussing such optimizations is out of scope for our paper, as these methods only strive to maintain performance but not improve it further. Furthermore, they are only bound to drop performance on the poorly localized object images.

Data augmentation is a popular technique used to increase generalizability and reduce overfitting of deep neural networks. This comes at no cost during inference and is referred to as one of the bag of freebies method that only increase training time but does not increase testing time as described in YOLO v4 by Bockovskiy et al. [1]. Two types of data augmentations exist: pixel-wise data augmentation and region-wise data augmentation. Pixel wise data augmentation modifies each input pixel independently while region wise data augmentation modifies regions of images simultaneously. Examples of pixel-wise data augmentation include photometric distortions like adjusting the brightness, contrast, hue, saturation and noise of an image while geometric distrortions include random scaling, cropping, flipping and rotating. These data augmentations increase the variability of the input images to generalize to scenes and environments that the original training data does not include. Examples of region-wise data augmentation include methods like RandomErase [37], CutOut [7], Hide-and-Seek [26], Grid Mask [3], MixUp [35], CutMix [34], Mosaic [1] and many more. In methods such as Random Erase [37] and CutOut [7], a single image is manipulated by choosing a rectangular region randomly and is filled with a random value or zero. On the other hand, in methods such as Mixup [35], CutMix [34] and Mosaic [1], multiple images are combined together to generate a single training image. Region wise data augmentation techniques learn to improve classification of objects by attending on less discriminative parts of the object (e.g. leg as opposed to head of a person) by learning more from local features. Such augmentation increases robustness of the models by serving as a regularizer. Although data augmentation (both pixel wise or region wise) increases generalizability and improves object localization capabilities than being dependent on context, our method proposed learns from both local features to support occlusion and also from context available from full images to improve classification of poorly localized objects in images.

Efficient performance can be obtained by training both easy and hard examples in one-stage using novel loss functions likes Focal loss [17] introduced by Lin

Fig. 4. Few examples of poorly localized objects in images of VOC07noisy test datasets

et al. Focal loss enables training to focus on hard, misclassified samples while reducing the relative loss for well-classified examples. Such loss function boosts the overall performance but does not indicate how to utilize contextual information during training and there is no guarantee on the improvement on poorly local detections. The first stage in our two stage approach can employ any loss function including Focal loss. Another popular technique to improve training is using Bootstrapping [9,30], where the model is exposed to more hard negatives over time. Our method differs from this as it aims to learn from hard examples first and then train on easy examples later.

3 A Two-Stage Training Framework Utilizing Context

We propose a two-stage classification training system to improve the performance of accurately classifying images with poorly localized detections. In the first stage, we learn an object-focused model that can correctly classify images with

Table 1. Table comparing the top-1 accuracies of MobileNetV2 models finetuned with the traditional finetuning and our two-stage approach on the on the VOC07noisy-0.1, 0.2, 0.3, 0.4 datasets

Training scheme	VOCnoisy-0.1	VOCnoisy-0.2	VOCnoisy-0.3	VOCnoisy-0.4
Finetuning (traditional)	64	68.52	73.31	84.78
Ours	**66.5**	**72.05**	**77.41**	**87.48**

any amount of object present in it. The learned object-focused model is very good in detecting objects but has poor accuracies for background class. In the second stage, we refine the model trained in the first stage very slowly by providing only clean bounding boxes (accurately localized bounding boxes). By finetuning the model trained in the first stage with clean bounding boxes, we adjust the weights to learn to clearly distinguish between background and objects. This pipeline improves the overall classification performance on both well and poorly localized objects in images.

The intuition behind this idea is to first learn a model that generalizes well by associating objects with context. For example, we want to train a model on a dataset containing an aeroplane or a bird image in the sky, by including images of aeroplane or bird with large portions of sky in the image as shown in Fig. 3. The learned model associates sky with these two object categories rather than any random object. During inference, when the classifier sees a full or partial object in the sky, it most likely predicts it to contain aeroplane or bird rather than any other object. We expect that the context with which the classifier has been trained helps in aiding correct classification of object, even if only a portion of the object is present in the image. However, the downside is that now a sky image with no object present in it can also be mistaken to contain an object with high probability i.e. sometimes too much association of context with objects also hurts performance [25]. Instead of answering the riddled question of "Is context helpful or not?", we ask the question of "How to make context helpful?". To this end, we propose to approach this problem with a novel solution by first training with too much context and then finetune with little or no context as available in tight ground truth boxes. We think that the second stage finetuning adjusts the model to perform well on all easy, hard and background samples. This is different from the traditional ensemble-based approaches that improves performance by providing hard-samples incrementally [9]. We feel that the local minima achieved by learning only on hard examples is very different from the local minima achieved by introducing hard examples incrementally and therefore can generalize well when trained on easy examples later. We move the local minima slowly i.e. use very low learning rate in the second stage to refine the pretrained model.

Finally, we want to emphasize that we only use the model generated in the second stage for inference and this adds no additional cost in inference times.

4 Dataset and Experiments

4.1 Dataset

Traditional classification datasets like ImageNet, CIFAR-10, MNIST comprise majority of well-localized bounding boxes. They do not comprise many poorly localized bounding boxes. Further, to utilize additional context to improve classification accuracy of poorly localized bounding boxes, we want access to full images, i.e we want detection datasets. We curated a new VOC 2007 noisy test dataset[1], based on Pascal VOC datasets [8] to evaluate models on poorly localized object images.

Dataset Generation: We combine VOC 2007 train+valid, 2012 train+test datasets to form our train set. We use the 2007 VOC test images as our test dataset, as used for testing commonly by the researchers. We obtain a total of 16551 full train images and 4952 full test images. To generate clean and well localized classification test set, we use all the 12029 ground truth bounding box images from the full test set. We refer to this dataset as the clean VOC 2007 classification test set consisting of 21 classes: 20 objects and 1 background class. To generate the noisy test sets, for each full test image, we generate 2 poorly localized bounding boxes and 2 background images. A poorly localized bounding box is generated by finding a random bounding box that overlaps with any of the object bounding boxes in the full test image and has atleast an Intersection over Union (IoU) overlap of IoU with the object. The generated bounding box may contain multiple objects, but we are interested in detecting the largest object present in it. Therefore, we assign a single ground truth label (i.e. a hard assignment) for each image, which is the label of the object that has the maximum IoU with the selected poorly localized bounding box. We coin 4 noisy datasets, referred to as VOC07noisy-IoU consisting of 21 classes, where IoU in $\{0.1, 0.2, 0.3, 0.4\}$. To generate background images, we also randomly selected bounding boxes with no IoU overlap with any of the object bounding box. In total, we muster 9721 images. For some test images, we would not be able to extract background images as the objects occupy the whole image. For VOC07noisy-0.1, we obtained 9904 poorly localized images and 9890, 9860, 9860 images for VOC07noisy-0.2, VOC07noisy-0.3 and VOC07noisy-0.4 respectively. Figure 4 shows some poorly localized images of the VOC07noisy-0.1, highlighting the challenges of classification.

4.2 Experiments: Two-Stage Framework on Poorly Localized Object Images

Training Stage 1 Details: We use a similar augmentation strategy proposed in SSD [20] to generate images for first stage training, but here we train a classifier instead of a detector. We use all the full train images (i.e. 16551 images) for training in Stage 1. In each batch of training, a set of 11 annotation samplers

[1] https://github.com/sravbond/voc2007noisy.

Fig. 5. Figure showing correct classifications of poorly localized images by our approach on some of the VOC classes, that are missed by traditional finetuning with one stage.

are applied on each full input image. Each annotation sampler is defined by specifying the range for jaccard, sample coverage and object coverage overlap. A set of random boxes are generated m number of times, each time checking if any of the criteria specified by the sampler is met. If the criteria is met, a random box is chosen and a ground truth label is assigned to the object with maximum IoU in the bounding box selected by the sampler. If no critera is met,

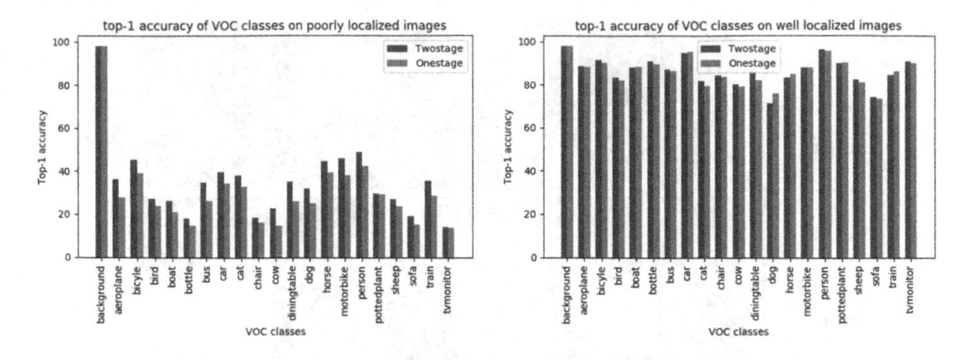

Fig. 6. Top-1 accuracy of our model with one stage model on both well-localized and poorly-localized VOC 2007 classification test sets.

Table 2. Table comparing the performance of the architectures trained from scratch using our approach to the one-stage training on the VOC 2007 classification test set.

Model architecture	Training Pipeline	Top-1 acc	Top-5 acc
MobileNetV2 [24]	One Stage	79.35	94.75
MobileNetV2 [24]	Two Stage (Ours)	**81.56**	**96.65**
FBNet [32]	One Stage	78.81	94.85
FBNet [32]	Two Stage (Ours)	**82.33**	**96.97**

a background image is chosen. So, for each annotation sampler, we gather 11 randomly selected bounding boxes, out of which only one of them is chosen in random to train for that batch. We use $m = 50$. We train for 90 epochs and use MobileNetV2 model trained on ImageNet as a pretrained model. We train for a small input resolution of 96×96, instead of traditional input resolution of 224×224 used in training classifier models so as to fit the models on the edge. The initial learning rate we use for finetuning is 1e-3 and we use SGD with momentum and a step wise decay of 0.1 every $\frac{1}{3}$ of the training time.

Training Stage 2 Details: We use the model trained in the first stage to initialize the weights for training the second stage. In this stage, only the perfectly cropped ground truth bounding boxes of the VOC 2007 + 2012 train datasets are used for training. We use a very low learning rate of 1e-5 and ADAM optimizer with step decay of 0.1 at every $\frac{1}{3}$ of the training time. We again train for 90 epochs and the input resolution is 96×96.

We compare the one-stage finetuned model with our two-stage approach for all the VOC07noisy-IoU datasets with a minimum overlap IoU threshold of $\{0.1, 0.2, 0.3, 0.4\}$ in the Table 1. We observe that the top-1 accuracies on poorly localized bounding boxes over all the IoU thresholds, our approach performs significantly better than the traditional finetuned model. In particular, for VOC07noisy-0.2, we improve performance by an absolute 3.53% in top-1 accuracy, and for VOCnoisy-0.3, we improve by +4.1%. Figure 5 shows qualitative results of our two-stage approach from VOCnoisy-0.2 test set. Each row in the figure shows the poorly localized images of a class that are correctly classified by our approach. We want to emphasize that none of these images were classified correctly using one-stage approach that uses traditional finetuning. As we can see, heavily occluded and off-centered images with large context are accurately classified by our two-stage training framework showing the effectiveness of our method.

Figure 6 compares the top-1 accuracies of one-stage model with two stage models for each of the 21 classes for both VOC07noisy-0.1 and VOC07 clean classification test sets. We observe that our two-stage trained model significantly outperforms the model finetuned on the classification boxes for many of the classes in poorly localized boxes, while maintaining the performance(or slightly more) on the perfectly localized VOC 2007 classification test set. We also observe that the generalization ability of the model trained in first stage is better and can be used as a pretrained model to improve performance, specifically when the pretrained model on ImageNet is not available. The next section elaborates this idea.

4.3 Improving the Model Performance Without a Pretrained Model

Training a state-of-the-art deep learning model on a domain-specific dataset is usually a two-step process. First, we train the model on a larger dataset like ImageNet [6]. We call this model a pretrained model. Second, we finetune the pretrained model on a domain-specific dataset by changing the number of the outputs of the last layer. This pipeline is the de-facto standard for obtaining the best performance of a model for a domain-specific application. The first step is to obtain a pretrained model that is well generalized and the second step is to tailor the model's responses to the objects in the customized domain.

Often, for a given budget, an architecture may required to be altered (i.e. remove layers, reduce number of filters in each layer etc.) to meet the runtime needs during deployment. And there are only limited set of pretrained models available, and if the new architecture deviates from the original one, it is hard to train a pretrained model on ImageNet. In this work, we propose that with our two stage training strategy, when no pretrained model is available, we obtain a boost in performance compared to having trained the model from scratch. This is because our first stage training yields a better generalized model that gives a better training initialization than having learned from scratch.

Training Details: We employ a similar training strategy as done in the previous section, except that we do not initialize the first stage model training with any weights. We use learning rate of 0.016 with SGD and step-wise decay to train the model. Since, we are training from scratch, we train for 200 epochs. We use the first stage model as a pretrained model in the second stage and train with learning rate of 1e-5. For training the second stage, we use the clean ground truth bounding boxes for training as done in the previous experiment.

Table 3. Table comparing the performance of the MobileNetV2 models trained and finetuned with various training mechanism in one stage and two stage on the test set of VOC 2007 classification dataset and noisy VOC 2007 classification set with IoU threshold 0.2.

Model type (finetuned)	Training details	Top-1 acc on VOC07 classification test	Top-1 acc on VOC07 noisy test set
Traditional	One-stage	93.35	68.52
with DA	One-Stage	89.29	70.5
with CutOut [7]	One-Stage	93.36	65.76
with CA (Ours,Stage-1 only)	One-Stage	86.87	**78.04**
with Two-Stage (Ours)	Two-Stage	**93.48**	72.05

Table 4. Table showing the effect of using different amounts of contextual information in the first stage training and compares the models performance on the clean and noisy test set with IoU Threshold 0.3 . * denotes here in this case, only one stage training is done with easy examples as ground truth boxes of objects (i.e. IoU is 1.0).

% of samples min IoU 0.1	% of samples min IoU 0.5	Training scheme	Top-1 acc on VOC07 clean test set	Top-1 acc on VOC07 noisy test set
0	100*	One-stage*	93.35	73.31
0	100	Two-stage	93.67	76.85
5	95	Two-stage	93.49	76.84
10	90	Two-stage	93.42	77
25	75	Two-stage	93.39	77.36
50	50	Two-stage	**93.71**	**77.59**
75	25	Two-stage	93.62	77.53
100	0	Two-stage	93.6	77.53
random	random	Two-stage	93.48	77.41

Results: Here, our test set is the clean classification set. Table 2 shows improved performance results of our method compared to model trained from scratch for both architectures: MobileNetV2 [24] and FBNet [32] demonstrating better generalization ability.

5 Ablation Study

Why Not Just One Stage with Extreme Data Augmentation? A natural question to ask is, how is the two-stage training framework better than applying traditional augmentation strategy in one stage? Table 3 compares the performance of finetuning the pretrained model of MobileNetV2 [24] trained with various augmentation techniques in one stage with our proposed two-stage

Table 5. Table showing the performance on the well localized and noisy test sets with IoU Threshold 0.4 with decreasing learning rates in the second stage

learning rate of second stage	Top-1 acc on VOC07 classification	Top-1 acc on VOC07 noisy set
1e–3	94.3	79.53
1e–4	94.0	81.95
1e–5	**93.48**	**87.48**
1e–6	92.22	87.3

approach on the VOC07noisy dataset[2] and VOC 2007 classification test set. In particular, we compared with traditional data augmentation using random crops and flips denoted by DA, CutOut [7] and with our model trained in the first stage denoted as CA. For all the trainings, we used Digits[3] framework with nvcaffe backend[4] to maintain fair comparison. For CutOut augmentation, we generated 10 augmented images with randomly masked regions of length that is 40% of the maximum of each side. CutOut augmentation maintained the top-1 accuracies on the VOC07 clean classification test set, but remarkably reduced the performance on the noisy test sets. While, finetuning with DA only improved performance by a little amount on the noisy set, it significantly dropped the performance on the clean test set. Incorporating context while training a classifier using augmentation strategy in SSD [20], i.e. our stage 1 model significantly improves the performance on VOC07noisy test set, but looses performance drastically on the clean test set. So, one stage training with both data and context augmentations alone cannot be employed for classifying well on both poorly localized and accurately localized object images. Our approach employing two-stage training scheme significantly improves the performance on poorly localized object images and also maintains performance on classifying accurately localized object images as shown in the last row of Table 3. If we were to only classify poorly localized object images, our stage-1 model gives the best performance on the VOC07noisy test set.

How Much Context is Actually Helpful? So, the next natural question is: How much context should we include to improve performance? In Table 4, we compare performance of using different amounts of contextual information. For this experiment, we generate two different training samples: one sample with a lot of context i.e. with a minimum IoU of 0.1 with the ground truth object bounding box and other sample with moderate amount of context i.e. object occupies atleast 50% of the image having a minimum IoU of 0.5. We report

[2] https://github.com/sravbond/voc2007noisy.

[3] https://developer.nvidia.com/digits.

[4] https://docs.nvidia.com/deeplearning/frameworks/caffe-user-guide/index.html.

performance of 8 different combination of mixing these sample sets i.e. in one training, we use 50% of samples containing more context and 50% of samples with moderate context. In another, we use 25% and 75% of images with more and some amount of context. We observe that all the training combinations utilizing contextual information in first stage boosts the performance of the final model on the noisy test set, compared to the one-stage training that uses no contextual information. Further, we observe that we get the best performance using equal amount of images with more context and moderate context in the first stage. For the stage 1 model trained with equal split, we get 81% (+4% over the best performance) top-1 accuracy on the VOC07noisy-0.3 test set, and only 86% (−7% of the best performance) top-1 accuracy on the clean test set, thereby emphasizing the importance of finetuning in the second stage to do better on both the clean and noisy test sets.

Slow or Fast Training? In our two-stage training pipeline, the second stage training uses the model trained in the first stage to initialize its weights. It is important that we train the second stage with very low learning rate, to achieve the right balance of performance between well-localized and poorly-localized objects in images. Table 5 emphasizes the importance that the weights need to be adjusted using a small learning rate of 1e-5 to boost the performance on clean classification test set, while maintaining performance on the noisy test set.

6 Conclusion

We proposed a two-stage training framework to improve the accuracy of the traditional classifiers on poorly localized objects in images. We leverage contextual information from full images in the first stage and train with hard noisy examples in the first stage. In the second stage, we use the classifier trained in the first stage as the pretrained model and train on perfectly localized images. We use a very low learning rate to adjust the weights of the model trained in the first stage to perform well on both accurately and poorly localized objects in images. We introduce a new VOC07noisy-IoU with different level of difficulties with IoU in $\{0.1, 0.2, 0.3, 0.4\}$ dataset that releases poorly localized images containing a minimum of IoU overlap with any of the ground truth boxes, generated from Pascal VOC 2007 test images [8]. VOC07noisy-0.1 is a much challenging dataset than VOC07noisy-0.4, as the test images of the former may contain only small amount of object compared to the background. We demonstrate that our approach with MobileNetV2 [24] architecture outperformed over the state of the art finetuned models trained in one stage using state of the art data augmentation procedures on all the VOC07noisy-IoU datasets. We also empirically observed that, using our approach, we get a better generalized model when trained from scratch. We show improved results with both MobileNetV2 and FBNet architectures when trained from scratch using the proposed two stage training framework on the well-localized VOC 2007 classification test set.

References

1. Bochkovskiy, A., Wang, C.Y., Liao, H.Y.M.: Yolov4: optimal speed and accuracy of object detection (2020)
2. Cai, H., Gan, C., Wang, T., Zhang, Z., Han, S.: Once-for-all: train one network and specialize it for efficient deployment (2019)
3. Chen, P., Liu, S., Zhao, H., Jia, J.: Gridmask data augmentation (2020)
4. Chin, T., Ding, R., Zhang, C., Marculescu, D.: Legr: filter pruning via learned global ranking. CoRR abs/1904.12368 (2019). http://arxiv.org/abs/1904.12368
5. Dai, X., et al.: Fbnetv3: joint architecture-recipe search using neural acquisition function (2020)
6. Deng, J., Dong, W., Socher, R., Li, L.J., Li, K., Fei-Fei, L.: ImageNet: a large-scale hierarchical image database. In: CVPR09 (2009)
7. Devries, T., Taylor, G.W.: Improved regularization of convolutional neural networks with cutout. CoRR abs/1708.04552 (2017). http://arxiv.org/abs/1708.04552
8. Everingham, M., Eslami, S.M.A., Van Gool, L., Williams, C.K.I., Winn, J., Zisserman, A.: The pascal visual object classes challenge: a retrospective. Int. J. Comput. Vision **111**(1), 98–136 (2015)
9. Freund, Y., Schapire, R.E.: A decision theoretic generalization of on-line learning and an application to boosting. In: Vitányi, P.M.B. (ed.) Second European Conference on Computational Learning Theory (EuroCOLT-95), pp. 23–37 (1995). citeseer.nj.nec.com/freund95decisiontheoretic.html
10. He, Y., Han, S.: AMC: automated deep compression and acceleration with reinforcement learning. CoRR abs/1802.03494 (2018). http://arxiv.org/abs/1802.03494
11. Hinton, G., Vinyals, O., Dean, J.: Distilling the knowledge in a neural network (2015)
12. Howard, A., et al.: Searching for mobilenetv3. CoRR abs/1905.02244 (2019). http://arxiv.org/abs/1905.02244
13. Howard, A.G., et al.: Mobilenets: efficient convolutional neural networks for mobile vision applications. CoRR abs/1704.04861 (2017). http://arxiv.org/abs/1704.04861
14. Huang, G., Liu, S., van der Maaten, L., Weinberger, K.Q.: Condensenet: an efficient densenet using learned group convolutions. CoRR abs/1711.09224 (2017). http://arxiv.org/abs/1711.09224
15. Iandola, F.N., Moskewicz, M.W., Ashraf, K., Han, S., Dally, W.J., Keutzer, K.: Squeezenet: Alexnet-level accuracy with 50x fewer parameters and ¡1mb model size. CoRR abs/1602.07360 (2016), http://arxiv.org/abs/1602.07360
16. Krishnamoorthi, R.: Quantizing deep convolutional networks for efficient inference: a whitepaper. CoRR abs/1806.08342 (2018), http://arxiv.org/abs/1806.08342
17. Lin, T.Y., Goyal, P., Girshick, R., He, K., Dollár, P.: Focal loss for dense object detection (2017)
18. Liu, C., et al.: Progressive neural architecture search. CoRR abs/1712.00559 (2017), http://arxiv.org/abs/1712.00559
19. Liu, H., Simonyan, K., Yang, Y.: DARTS: differentiable architecture search. CoRR abs/1806.09055 (2018), http://arxiv.org/abs/1806.09055
20. Liu, W., Anguelov, D., Erhan, D., Szegedy, C., Reed, S., Fu, C.Y., Berg, A.C.: SSD: single shot multibox detector (2015), http://arxiv.org/abs/1512.02325, cite arxiv:1512.02325Comment: ECCV 2016

21. Polino, A., Pascanu, R., Alistarh, D.: Model compression via distillation and quantization. In: International Conference on Learning Representations (2018). https://openreview.net/forum?id=S1XolQbRW
22. Real, E., Aggarwal, A., Huang, Y., Le, Q.V.: Regularized evolution for image classifier architecture search. CoRR abs/1802.01548 (2018), http://arxiv.org/abs/1802.01548
23. Redmon, J., Farhadi, A.: YOLO9000: better, faster, stronger. CoRR abs/1612.08242 (2016), http://arxiv.org/abs/1612.08242
24. Sandler, M., Howard, A.G., Zhu, M., Zhmoginov, A., Chen, L.: Inverted residuals and linear bottlenecks: Mobile networks for classification, detection and segmentation. CoRR abs/1801.04381 (2018), http://arxiv.org/abs/1801.04381
25. Shetty, R., Schiele, B., Fritz, M.: Not using the car to see the sidewalk: quantifying and controlling the effects of context in classification and segmentation. CoRR abs/1812.06707 (2018), http://arxiv.org/abs/1812.06707
26. Singh, K.K., Yu, H., Sarmasi, A., Pradeep, G., Lee, Y.J.: Hide-and-seek: a data augmentation technique for weakly-supervised localization and beyond. CoRR abs/1811.02545 (2018), http://arxiv.org/abs/1811.02545
27. Tan, M., et al.: Mnasnet: platform-aware neural architecture search for mobile. In: Proceedings of the IEEE/CVF Conference on Computer Vision and Pattern Recognition (CVPR) (June 2019)
28. Tan, M., Le, Q.V.: Efficientnet: rethinking model scaling for convolutional neural networks. CoRR abs/1905.11946 (2019), http://arxiv.org/abs/1905.11946
29. Touvron, H., Vedaldi, A., Douze, M., Jégou, H.: Fixing the train-test resolution discrepancy: Fixefficientnet (2020)
30. Viola, P.A., Jones, M.J.: Rapid object detection using a boosted cascade of simple features. In: CVPR (1), pp. 511–518. IEEE Computer Society (2001)
31. Wan, A., et al.: Fbnetv2: differentiable neural architecture search for spatial and channel dimensions (2020)
32. Wu, B., et al.: Fbnet: hardware-aware efficient convnet design via differentiable neural architecture search. CoRR abs/1812.03443 (2018), http://arxiv.org/abs/1812.03443
33. Wu, B., et al.: Shift: a zero flop, zero parameter alternative to spatial convolutions. CoRR abs/1711.08141 (2017), http://arxiv.org/abs/1711.08141
34. Yun, S., Han, D., Oh, S.J., Chun, S., Choe, J., Yoo, Y.: Cutmix: regularization strategy to train strong classifiers with localizable features. CoRR abs/1905.04899 (2019), http://arxiv.org/abs/1905.04899
35. Zhang, H., Cissé, M., Dauphin, Y.N., Lopez-Paz, D.: mixup: beyond empirical risk minimization. CoRR abs/1710.09412 (2017), http://arxiv.org/abs/1710.09412
36. Zhang, X., Zhou, X., Lin, M., Sun, J.: Shufflenet: an extremely efficient convolutional neural network for mobile devices. CoRR abs/1707.01083 (2017), http://arxiv.org/abs/1707.01083
37. Zhong, Z., Zheng, L., Kang, G., Li, S., Yang, Y.: Random erasing data augmentation. CoRR abs/1708.04896 (2017), http://arxiv.org/abs/1708.04896
38. Zoph, B., Le, Q.V.: Neural architecture search with reinforcement learning. CoRR abs/1611.01578 (2016), http://arxiv.org/abs/1611.01578
39. Zoph, B., Vasudevan, V., Shlens, J., Le, Q.V.: Learning transferable architectures for scalable image recognition. CoRR abs/1707.07012 (2017), http://arxiv.org/abs/1707.07012

Face Mask Invariant End-to-End Face Recognition

I. Putu Agi Karasugi and Williem(✉)

Verihubs Inteligensia Nusantara, Jakarta, Indonesia
{agikarasugi,williem}@verihubs.com

Abstract. This paper introduces an end-to-end face recognition network that is invariant to face images with face masks. Conventional face recognition networks have degraded performance on images with face masks due to inaccurate landmark prediction and alignment results. Thus, an end-to-end network is proposed to solve the problem. We generate face mask synthesized datasets by properly aligning the face mask to images on available public datasets, such as CASIA-Webface, LFW, CALFW, CPLFW, and CFP. Then, we utilize those datasets as training and testing datasets. Second, we introduce a network that contains two modules: alignment and feature extraction modules. These modules are trained end-to-end, which makes the network invariant to face images with a face mask. Experimental results show that the proposed method achieves significant improvement from state-of-the-art face recognition network in face mask synthesized datasets.

Keywords: Face mask · Face recognition · End-to-end network · Face alignment

1 Introduction

For the last decade, there has been a significant increase in the performance of face recognition algorithms in unconstrained and open-set conditions, where considerable variation such as pose and illumination exists in the data. These increasing performance trends are driven by the utilization of Deep Convolutional Neural Networks (DCNNs) architectures in modern face recognition methods [3,14]. Typically, modern face recognition techniques involve three steps: face detection, face feature extraction, and classification. In face detection, face images often go through pose normalization by transforming detected face landmarks to reference landmark location, as in the case of [22]. These pose normalized face images are then fed to CNNs to obtain face features, where these features are finally compared to one another with metrics such as cosine similarity or euclidean distance to perform classification. The performance of the classification correlates with the discriminating nature of the extracted face features. These steps are applied in two categories of face recognition: face identification, which is comparing a face with a collection of faces to get the most

A. Bartoli and A. Fusiello (Eds.): ECCV 2020 Workshops, LNCS 12539, pp. 261–276, 2020.
https://doi.org/10.1007/978-3-030-68238-5_19

similar results; and face verification, which checks if a face is identical or has the same identity with another face. Works such as Sphereface [8], CosFace [17], and ArcFace [1] achieve state-of-the-art performances in face verification and identification by utilizing angular margin-based losses. These losses maximize inter-class variance and minimize the intra-class variance of face feature, which yields discriminative feature space that improves the face recognition model's discriminative power.

Beside algorithms and network architectures, the rise of publicly available large-scale face recognition datasets such as CASIA-Webface [21] and MS-Celeb-1M [2] also supports the advancement of face recognition. According to [9], the underlying ability of CNNs to learn from massive training sets allows techniques using CNNs to be effective alongside the development of network architectures. These massive datasets are possible because of the vast availability of public information on internet sites that can be gathered automatically.

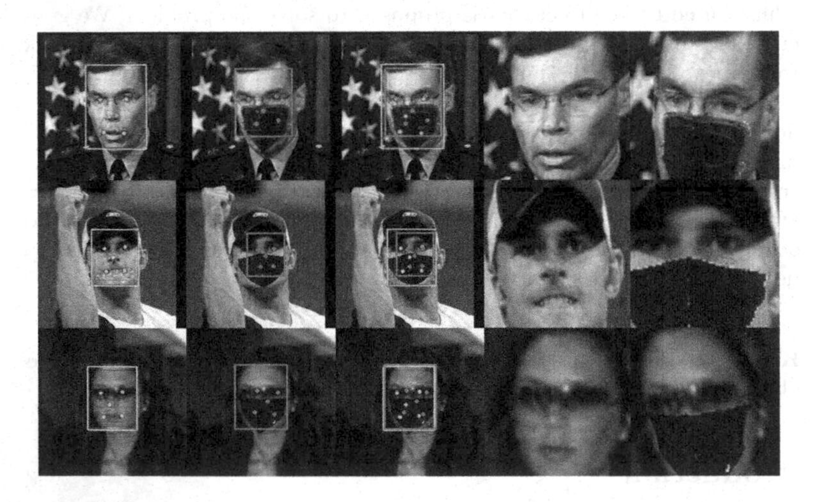

Fig. 1. The comparison between MTCNN [22] detected face landmarks and pose normalization results between image samples in LFW [4] (*green, first column*) and our face mask augmented LFW (*yellow, second column*). The presence of a face mask covering the face results in unreliable and inconsistent pose normalization shown in the *last column.* (Color figure online)

Despite these achievements on face recognition, there is still limited progress in face recognition performance on obstructed face images. The recent COVID-19 pandemic around the world prompted many countries to recommend or even enforce their citizens to wear a face mask to limit the spread of the outbreak. The obstruction of many face areas such as mouth, nose, and cheeks can cause conventional face detection and face recognition methods to have severely degraded performance. Pose normalization methods, such as [22], have difficulties in predicting face landmark locations covered by face mask, which results in inaccurate face landmark predictions. Figure 1 shows that [22] produces unreliable face alignment output, which gives a performance penalty to face recognition networks that depend on pose normalized inputs.

Moreover, the availability of large-scale face recognition datasets with face masks is still quite rare or insufficient in number. Gathering and annotating the images and identities of face images with a face mask requires a tremendous effort and precision to ensure minimal labeling error. A small number of visible face areas can lead to a high number of labeling errors from human annotators or an automated system if the labeling process is not done correctly.

In this paper, we explore the impact of face mask on face verification performance. Specifically, we introduce a face mask augmentation technique to enable the creation of face mask synthesized datasets from currently available datasets. We analyze the effects of the face mask synthesized datasets in the alignment result and verification performance on state-of-the-art method ArcFace [1]. We propose using an end-to-end trainable joint face alignment and recognition architecture to align and recognize faces with mask images. The code for this work can be accessed at https://github.com/agikarasugi/Face-Mask-Invariant-End-to-End-Face-Recognition.

2 Related Works

2.1 Deep Face Recognition and Alignment

DCNNs have come a long way toward improving face recognition algorithms' performance. DeepFace [16] pioneered the use of CNN in their work, where they adopted CNN with locally connected layers and used a 3D alignment model to transform all detected faces to frontal view before the face images are fed into the network. After that, the DeepID series [15] introduced a method that learns an ensemble of neural networks with different local face patches aligned in different ways, which provides a rich set of features for face verification. FaceNet [12] then introduced Triplet Loss, which learns directly from the embedding of face images to a compact Euclidean space, where distances directly corresponded to a measure of face similarity. The rise of loss functions that aimed to enhance the discriminative power of the softmax loss such as Center Loss [19], Sphereface [8], CosFace [17], and ArcFace [1] enabled Deep Face Recognition methods to achieve very high face recognition performance surpassing human-level performance.

Most of the face recognition methods required a face alignment technique to provide pose-normalized face images that minimized variance in the location of the face and its attributes, which helped to improve the robustness of the recognition. In recent CNN-based face recognition methods, the model's inputs were 2D aligned face images during both training and testing. One of the most common face alignment methods utilized by [1,17] was by transforming detected facial landmarks from a face alignment network such as MTCNN [22] to fix reference positions. Recent research showed that proper face alignment positively affected face recognition performance in the testing stage, as reported by [10]. However, in real-life scenarios, face images often had variations such as occlusions and pose, which could hinder facial landmarks detection. In the case of a person using a face mask, important landmarks such as nose, mouth, and possibly the jawline were obstructed, causing inaccurate landmark detection in these areas. Therefore, it could produce sub-optimal aligned images that caused a drop in recognition performance.

2.2 Spatial Transformer Networks

Spatial Transformer Networks [5] are differentiable modules that applied spatial transformation such as affine or projective transformation to a feature map. Generally, spatial transformers consisted of three parts: *localization network*, which took a feature map and generated transformation parameters; *grid generator* that took the predicted transformation parameters and generated sampling grid; *sampler* that sampled the input feature map at sampling grid that produced the output map. Spatial Transformers Networks could be inserted into existing convolutional architectures, allowing the learning of optimal transformation for feature maps. It also gave underlying architecture robustness to translation, scale, and rotation without any extra supervision, which had allowed state-of-the-art performance in digits recognition and several classification problems. Spatial Transformer Networks had been applied in the face recognition problem. [25] explored the possibility of alignment learning in end-to-end face recognition by inserting spatial transformer layers in front of the feature extraction layers to align the feature map using identical, similarity, affine, and projective transformation. Another work that utilized Spatial Transformer Network in face recognition is [20], where the work proposed the use of Recursive Spatial Transformer that allowed face alignment and recognition to be jointly optimized in one network in an end-to-end fashion.

Those works showed that Spatial Transformer Networks could be integrated into face recognition networks to enable face alignment inside the architecture without prior landmark detection, making face alignment more feasible in conditions where a face detector could not accurately detect face landmarks. Furthermore, Spatial Transformer Networks also provided the convenience of end-to-end training, easing, and simplifying the training process.

3 Dataset Preparation

To train and evaluate face mask invariant face recognition networks, we synthesize publicly available face recognition datasets with face masks. We use the same procedures for the training and evaluation datasets. To synthesize the face mask occluded face image, we use 15 face mask images gathered from the internet, which includes cloth masks, N95 masks, and surgical masks. These face mask images are used together with the binary mask and stored in a face mask dictionary.

3.1 Training Data

For the training dataset used in our experiments, we choose CASIA-Webface [21] and augment it using the face mask augmentation. CASIA-Webface comprises 500K face images of 10K subjects semi-automatically collected from the internet. Each image from the dataset is augmented with a random face mask image. First, face landmarks are extracted using Dlib's [6] 68-face-landmarks detector to guide the boundary of the face mask. Note that we only utilize 6 of the detected landmarks, denoted as l_n $(0 \leq n \geq 5)$, each contains point $(x,\ y)$ coordinates, where n corresponds to the position number that can be seen in Fig. 2. We randomly select the face mask from the dictionary and split the face mask into two parts, left and right regions shown in Fig. 2.

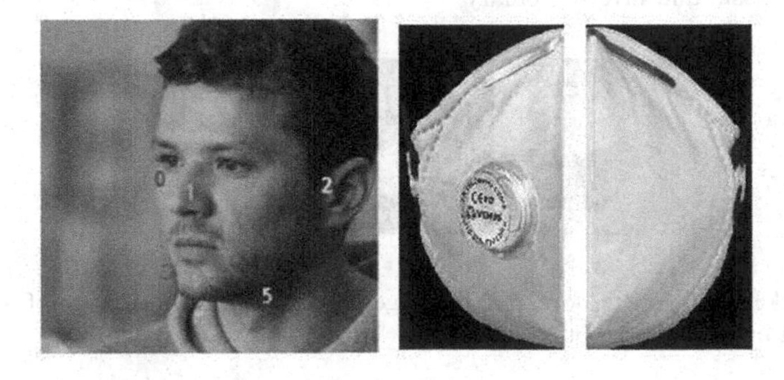

Fig. 2. Face image with $(l_0, l_1, \ldots, l_5)$ landmark points, left half, and right half of the face mask image. Top left corner of each *number* is the location of the landmark point.

To naturally synthesize the face mask onto the image, we measure the homography matrix between the four corners of the half-face mask and the desired face landmark points. Each half 4 image corners (clockwise, from top left) is used as source point, where (l_0, l_1, l_4, l_3) and (l_1, l_2, l_4, l_5) are used as destination points for the left and right half respectively. Then, we project the face mask onto the face image using the estimated homography matrix. Figure 3 shows the projection result of each half mask with the binary mask and inverse binary mask.

The transformed half images, binary and inverted binary masks are then combined into one. With the inverted binary masks, the background image is obtained by only keeping pixels outside the desired face mask area. The combined face mask image acts as the foreground image. By adding the foreground and the background image together, the transformed face mask image is applied to the face image, as shown in Fig. 4. This process is used to each face image on the dataset to generate a fully face-mask synthesized version. The result of our augmentation method closely resembles real-life person wearing a face mask, and the use of perspective transformation allows our face mask augmentation method to be applied in most face poses. Figure 5 shows samples obtained from the face masked synthesized CASIA-Webface.

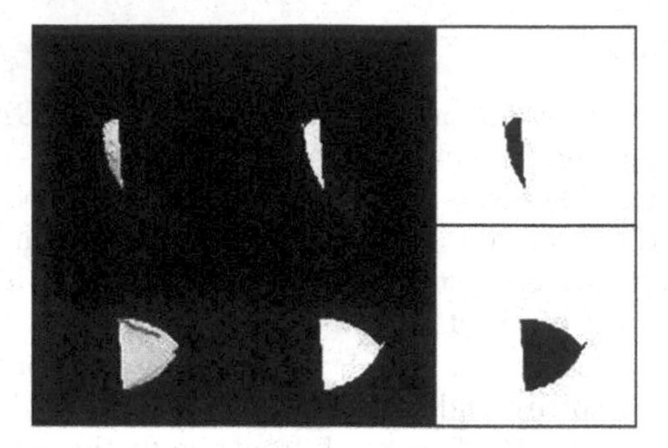

Fig. 3. The resulting perspective transformation applied to each half of face mask image. *left*, *middle*, and *right column* shows the transformed half images, transformed half binary mask, and inverted binary

Fig. 4. Background image, foreground image, and the resulting image of adding the background image to foreground image.

3.2 Synthesized Testing Data

We generate the face mask synthesized evaluation dataset using augmentation method explained in the previous subsection on several popular verification datasets: LFW [4], CALFW [24], CPLFW [23], and CFP [13]. Labeled Faces in the Wild (LFW) is a popular public benchmark for face verification that consists of 13K images with 5.7K identities. Cross-Age LFW (CALFW) is a renovation of LFW that further emphasizes the age gap of positive pairs to enlarge intra-class variance. Cross-Pose LFW (CPLFW) is a renovation of LFW that emphasizes pose differences to enlarge intra-class variance. Celebrities in

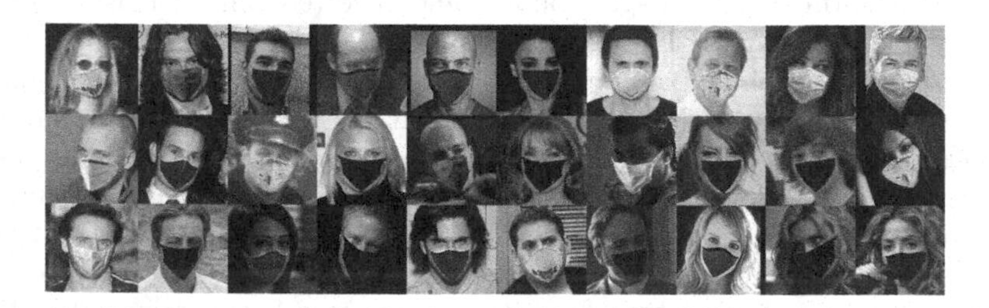

Fig. 5. Random samples obtained from our face mask augmented CASIA-Webface dataset.

Frontal-Profile (CFP) is a face recognition dataset that is designed to facilitate research in the problem of frontal to profile face verification in the wild. We used both Frontal-Frontal and Frontal-Profile verification pairs from CFP. Because of large numbers of failed landmark detection in profile images, only frontal face images are synthesized with face masks. Figure 6 shows the example of face mask augmented images.

(a) Masked CALFW samples

(b) Masked CPLFW samples

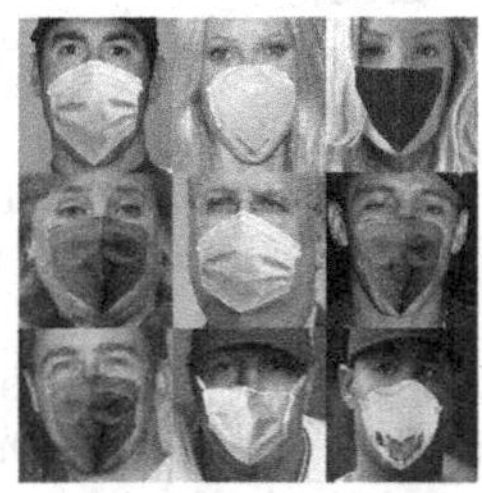

(c) Masked CFP frontal samples

Fig. 6. Face mask augmented/synthesized samples of several testing datasets

3.3 Real-World Testing Data

Besides only using the synthesized versions of standard face recognition testing datasets, we also utilize Real-World Masked Face Dataset (RMFD) [18]. At the time of the writing, RMFD is proclaimed as the world's largest masked face dataset. The dataset contains 5,000 masked faces of 525 people and 90,000 normal faces from cleaned and labeled images crawled from the internet.

4 Proposed Method

Most face recognition methods utilize face alignment to normalize face pose to minimize the spatial variance of the face images fed into the network, which improves recognition accuracy. Standard alignment method by warping detected landmarks into reference position is unreliable and inconsistent when key face landmarks such as nose, mouth, and possibly jawline are occluded from the usage of face mask, as shown in Fig. 1. To mitigate this problem, we propose using a localization network that utilizes the Spatial Transformer Network to learn the optimal alignment of face masked images. Furthermore, the localization network also acts to localize face in the image. The Spatial Transformer Network is used in conjunction with a DCNN feature extractor to facilitate an end-to-end trainable network.

Our proposed method is divided into two modules: face localization and alignment module, which localizes and aligns the face to an optimal position,

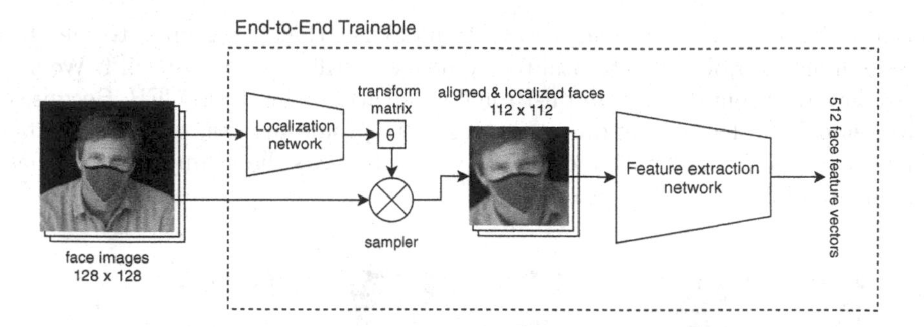

Fig. 7. The overall architecture of the proposed network.

and face feature extraction module to produce face feature vector. The overview of the system can be seen in Fig. 7. For a clear and concise explanation of our proposed method, we use notations to refer to different parts of our method. N_{loc} and N_{recog} denote localization and recognition networks, respectively; I_o and I_a denote original and aligned face images. During the training phase, N_{loc} takes I_o input images with 128×128 size, and it generates affine transformation parameter for the grid generator, creating a sampling grid for the sampler to output the aligned I_a. The I_a image is then fed into N_{recog}, which outputs 512 face feature vectors. In order for the network to learn and produce discriminative face feature vectors, Li-ArcFace [7] loss is utilized, which is a variant of ArcFace [1] loss that takes the value of the angle through a linear function as the target logits rather than through cosine function that offers better convergence and performance on low dimensional embedding feature learning for face recognition.

To generate the affine transformation parameter for face image alignment, we designed a compact localization network N_{loc}, which utilizes the bottleneck blocks from IR-SE50 [1]. The localization network takes RGB input image I_o with 128×128 size and produces θ. The θ consists of 6 elements representing

Table 1. The configuration of localization network N_{loc}. S and c denote the number of stride and output channels respectively. All spatial convolutions in Bottleneck use 3×3 kernel. AvgPool2d kernel is of the same size as the input (8×8).

Input	Operator	S	c
$128^2 \times 3$	Bottleneck	2	16
$64^2 \times 16$	Bottleneck	2	32
$32^2 \times 32$	Bottleneck	2	32
$16^2 \times 32$	Bottleneck	2	64
$8^2 \times 64$	Bottleneck	1	64
$8^2 \times 64$	AvgPool2d	1	1
$1^2 \times 64$	FC	1	6

affine transformation parameters that allow rotation, translation, scaling, and shearing. Table 1 shows the configuration of the localization network.

$$\theta = N_{align}(I_o) = \begin{bmatrix} \theta_{11} & \theta_{12} & \theta_{13} \\ \theta_{21} & \theta_{22} & \theta_{23} \end{bmatrix} \quad (1)$$

The θ is then used by grid sampler to produce 2D sampling grids. Finally, the sampling grids are used to interpolate the output value, where it specifies the sampling pixel locations normalized by the input I_o spatial dimension. This process yields aligned I_a images. We choose 112×112 as the dimension of I_a. Figure 8 show the progression of the face alignment on each iteration.

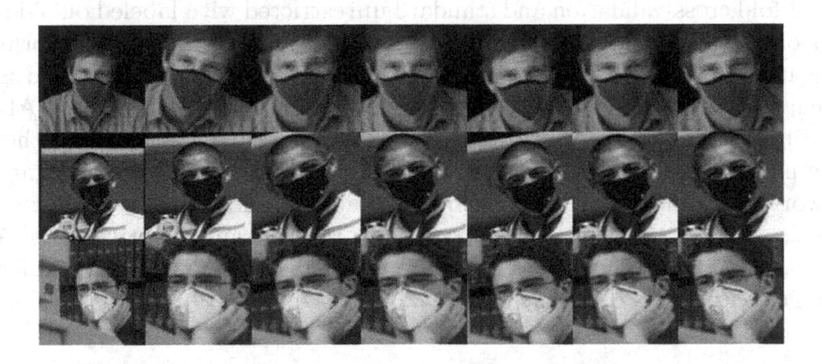

Fig. 8. Progression of the alignment result from sample LFW image generated by localization network and spatial transformer inside our proposed method. The *first column* on the image is initial image before being forwarded to the network, the rest of the *columns* is the output from epoch 3, 6, 9, 12, 15, and 18.

Face recognition network can be any of the state-of-the-art deep learning networks designed for face feature extraction. In our experiments, we utilize the same underlying architecture as the state-of-the-art IR-SE50 [1] for the face recognition network N_{recog}, where it takes I_a to output 512 face feature vectors. In the training stage, face feature vectors are forwarded to Li-ArcFace loss. Because of the continuous and uninterrupted flow of information from the alignment network to the face recognition network and the loss, the whole system can be trained end-to-end, simplifying the training process and eliminating the need for separate face detectors and landmark predictors degraded by the face masked images.

5 Experimental Results

For the experiments, we set the batch size to 512 images for each training iteration. The learning rate of the localization N_{loc} and face recognition network N_{recog} is set to $5e-4$ and $1e-1$ respectively and reduced by a factor of 0.1 when accuracy has stopped improving in 5 accumulated epochs. Weight decay is set to $5e-4$ set to all layers except Batch Normalization. The training process is done on 4 Nvidia Tesla V100. We use PyTorch [11] as the deep learning framework for all of our training and evaluations.

5.1 Comparison with ArcFace

Evaluation of ArcFace is done by aligning the images with affine transform using face landmarks detected by MTCNN [22]. For the ArcFace model trained on face-mask synthesized CASIA-Webface, we used the same hyperparameters as described in [1] for training on the dataset. For evaluation in the LFW dataset, we used 6000 face mask synthesized image pairs derived from the original pairs, with 10-fold cross-validation and standard unrestricted with labeled outside data protocol. We also created half-synthesized pairs to test verification performance between normal images and face-masked images, where only the second image in each pair is face mask synthesized. The evaluation in CPLFW, CALFW, CFP-FF, and CFP-FP also utilized synthesized image pairs and half-synthesized image pairs based on original pairs provided by the dataset. The experiment on real-world dataset RMFD is done by randomly generating 800 mask-to-mask and mask-to-non mask pairs, with equivalent positive and negative pairs. Additionally, non-mask-to-non-mask pairs are also generated and used for reference in images without face masks.

Table 2. Averaged accuracy on face mask synthesized, half-synthesized, and original image pairs on LFW dataset. LFW^{s} and LFW^{hs} refers to synthesized and half-synthesized LFW pairs. C denotes combined face mask synthesized and original CASIA-Webface.

Method	LFW^{s}	LFW^{hs}	LFW
ArcFace (R50-SE, MS1M)	89.06%	91.86%	99.75%
ArcFace (R50-SE, *m*CASIA)	93.83%	93.66%	92.08%
STN-IR-SE50 (Ours, *m*CASIA)	**98.25%**	97.98%	97.88%
STN-IR-SE50 (Ours, *m*CASIA, C)	98.20%	**98.41%**	98.69%

Table 2 shows that the ArcFace method trained in large-scale MS-Celeb-1M dataset struggles to achieve good performance in both synthesized and half-synthesized LFW pairs. ArcFace method trained on synthesized datasets has better verification performance in both types of pairs, but only by a little. Meanwhile, our method can achieve significantly higher accuracy in both types of pairs. By looking at Fig. 9, our method is positioned above the ArcFace methods with a sizable gap, signifying better performance. Combining the face-mask synthesized dataset with the original helped to increase verification performance between face images with a face mask and normal face images, with a small drop on verification performance in fully synthesized pairs. Sample predictions of the evaluated methods with alignment applied are shown in Fig. 10

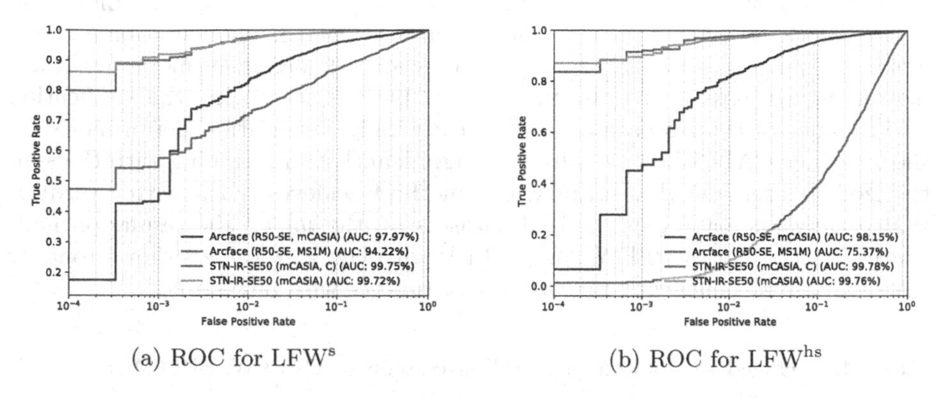

(a) ROC for LFW^{s} (b) ROC for LFW^{hs}

Fig. 9. ROC curves of face verification on fully face-mask synthesized LFW^{s} pairs and half face-mask synthesized LFW^{hs} pairs

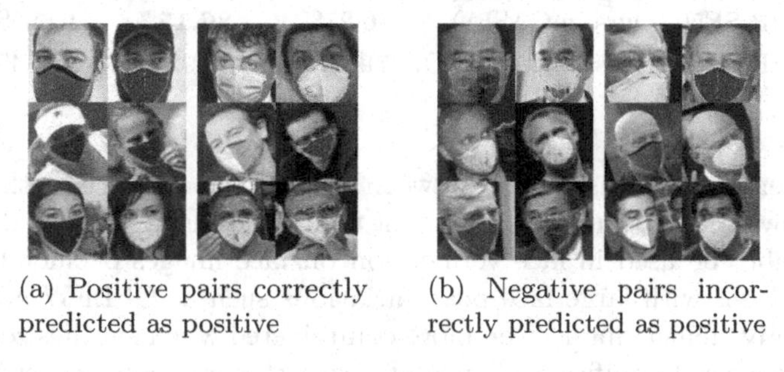

(a) Positive pairs correctly predicted as positive (b) Negative pairs incorrectly predicted as positive

Fig. 10. True positive and false positive predictions on synthesized LFW dataset. Top, middle, and bottom row corresponds to image aligned and predicted by ArcFace (R50-SE, mCASIA), STN-IR-SE50 (Ours, mCASIA), and STN-IR-SE50 (Ours, mCASIA, C). Top row images are pre-aligned using MTCNN, the rest is aligned by respective methods' STN. Note the variability of alignment result in the top row.

Table 3. Averaged accuracy on face mask synthesized pairs on CALFW, CPLFW, CFP-FF, and CFP-FP. C denotes combined face mask synthesized and original CASIA-Webface.

Method	sCPLFW	sCALFW	sCFP-FF	sCFP-FP
ArcFace (R50-SE, MS1M)	49.83%	72.40%	83.72%	69.32%
ArcFace (R50-SE, mCASIA)	57.61%	66.51%	92.47%	81.60%
STN-IR-SE50 (Ours, mCASIA)	75.63%	**87.08%**	**95.84%**	**84.24%**
STN-IR-SE50 (Ours, mCASIA, C)	**77.05%**	86.13%	93.68%	82.82%

Table 3 shows even more gaps in verification performance in CPLFW and CALFW. Method trained in a dataset without face mask augmentation is hopeless in recognizing face-masked images with standard alignment methods. Training on the synthesized dataset only helps a little in improving the verification performance of the ArcFace method. Surprisingly, verification performance on the cross-age CALFW dataset dropped significantly from training with the synthesized dataset. On the contrary, our method achieves much better verification performance in all synthesized datasets on the table. The results on half-synthesized pairs of CPLFW and CALFW displayed in Table 4 shows roughly the same improvement in verification performance with our method.

Table 4. Averaged accuracy on half-synthesized pairs on CALFW, CPLFW, and CFP-FF.

Method	hsCPLFW	hsCALFW	hsCFP-FF
ArcFace (R50-SE, MS1M)	69.93%	79.57%	94.83%
ArcFace (R50-SE, *m*CASIA)	52.55%	65.20%	92.79%
STN-IR-SE50 (Ours, *m*CASIA)	76.27%	**86.15%**	94.06%
STN-IR-SE50 (Ours, *m*CASIA, C)	**79.42%**	85.92%	94.44%

These performance disparities between the regular ArcFace method and our method shows that current face alignment techniques using predicted landmarks cannot reliably be used in face verification on face images occluded with face-mask. A dataset with large face pose variations such as CPLFW and CFP-FP that is already challenging is even more complicated when synthesized with face mask, which caused significant alignment errors that subsequently cause low verification accuracy. Our method shows that learned face alignment using Spatial

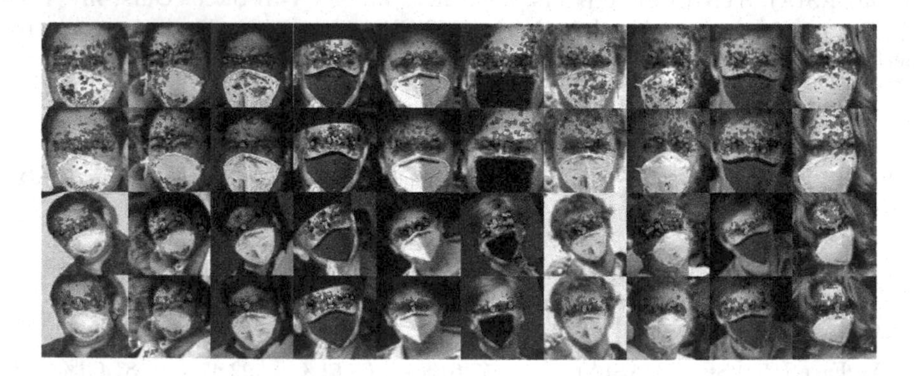

Fig. 11. Overlay of saliency map obtained from gradients backpropagated from the face embeddings. *First* and *second rows* are from ArcFace method trained with MS1M and synthesized CASIA-Webface. *Third* and *last rows* are from our methods trained from synthesized and combined training data.

Transformer Networks can have more robust alignment on heavily occluded face wearing a face mask, which helps face recognition performance significantly.

The performance disparities can be explained by looking at the saliency map in Fig. 11. ArcFace method trained on regular MS1M dataset factors the face mask to generate face feature vectors, which is terrible considering the face mask is not the descriptive part of one's face and will have different vector if the face mask is different. ArcFace method trained on the synthesized train datasets concentrates on the eye and forehead regions but not as localized as our methods, which have a tighter area that contributes more for discriminative feature vectors.

Table 5. Accuracy on Real-World Masked Face Dataset (RMFD) pairs. Notation *m2m*, *m2nm*, *nm2nm* refers to mask-to-mask pair, mask-to-nonmask pair, and nonmask-to-nonmask pair respectively.

Method	RMFD (*m2m*)	RMFD (*m2nm*)	RMFD (*nm2nm*)
ArcFace (R50-SE, MS1M)	51.00%	35.62%	82.75%
ArcFace (R50-SE, *m*CASIA)	54.88%	63.38%	65.75%
STN-IR-SE50 (Ours, *m*CASIA)	68.38%	**64.38%**	65.38%
STN-IR-SE50 (Ours, *m*CASIA, C)	**75.50%**	58.00%	65.00%

In the mask-to-mask evaluation of RMFD pairs, our method can get a sizable performance increase from ArcFace. In mask-to-non-mask pairs, our method trained on a fully synthesized dataset only achieve minor performance improvement. Interestingly, our method trained on the combined dataset has lower performance on this pair, which may be caused by the challenging nature of the dataset's images. The verification performance on RMFD pairs is shown in Table 5.

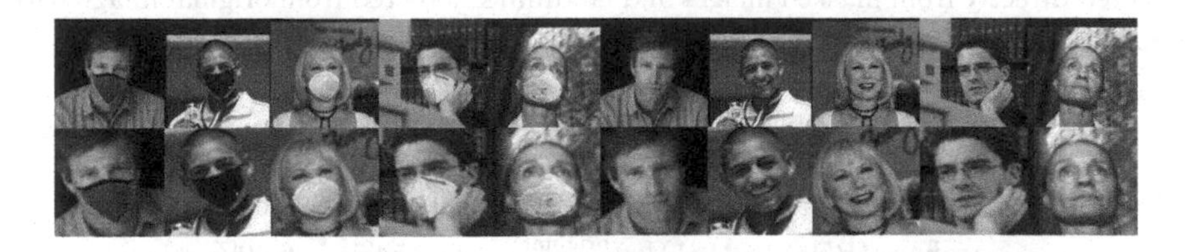

Fig. 12. Overview of alignment result from Spatial Transformer Network in our method. Top row shows unaligned original and augmented images. Bottom row shows the alignment results from the Spatial Transformer Network.

The Spatial Transformer Network used in our method can also align original face images without augmentation, as shown in Fig. 12 where there is little difference between image aligned from the original and synthesized sample, which shows that the Spatial Transformer Network has learned general features that enables it to perform alignment in both face-masked and normal face scenario.

5.2 Ablation Study

Our experiment shows that face alignment using landmarks detected by face detector degrades verification performance because of often inconsistent results. However, what if face landmark detected from unmodified images is used to align augmented images? After all, more consistent face alignment should, in theory, yield better verification performance. To test the theory, we perform an ablation study to determine if alignment using the original landmarks rather than directly predicting the landmarks can yield better results on the ArcFace method. The comparison of direct alignment and alignment using the original face landmarks is shown in Fig. 13.

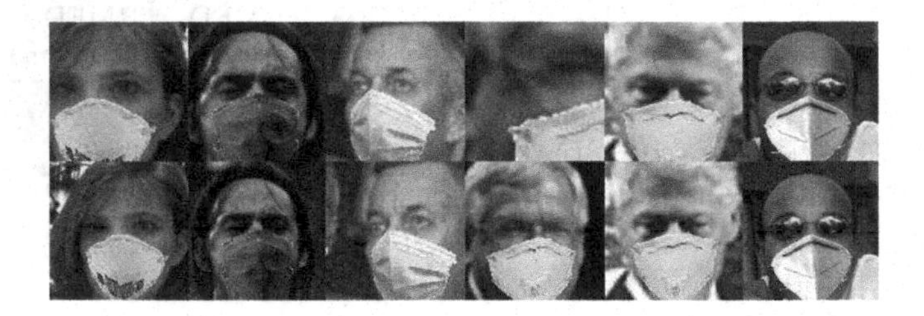

Fig. 13. Overview of alignment result on the sample of synthesized LFW using face landmark predicted by MTCNN directly (*top row*) and from original unmodified image without augmentation (*bottom row*).

Table 6. Averaged accuracy on face mask synthesized, and half-synthesized image pairs on LFW dataset. a^{direct} and $a^{original}$ refers to alignment method using landmark detected directly from masked images and landmark detected from original image without mask.

Method	LFW^{s}	LFW^{hs}
ArcFace (R50-SE, MS1M, a^{direct})	89.06%	91.86%
ArcFace (R50-SE, *m*CASIA, a^{direct})	93.83%	93.66%
ArcFace (R50-SE, MS1M, $a^{original}$)	96.54%	96.40%
ArcFace (R50-SE, *m*CASIA, $a^{original}$)	95.45%	94.13%
STN-IR-SE50 (Ours, *m*CASIA)	**98.25%**	97.98%
STN-IR-SE50 (Ours, *m*CASIA, C)	98.20%	**98.41%**

Results in Table 6 show a significant improvement in verification accuracy when the images are aligned with landmarks detected from the original image without face mask augmentation in the ArcFace method. Nevertheless, this method still yields results that are lower than the performance of using the Spatial Transformer Network in our method. This result proves that in the case of a heavily occluded face using a face mask, a different alignment approach is needed for the facial recognition network to learn the still-exposed part of the face more effectively and have more discriminative power. Moreover, it is impractical or impossible to separately get unobstructed face landmarks of the same image in a real-life scenario. Therefore, learned joint face alignment is still the best way to tackle the problem of alignment on heavily occluded face images wearing a face mask.

6 Conclusion

In this paper, we explored the impact of face mask occlusion in face verification performance. We proposed an augmentation method to generate face mask synthesized datasets that could be used as training or testing dataset. We also explored the effect of the synthesized dataset to face alignment used in ArcFace and proposed an end-to-end face recognition network that is invariant to face mask. The proposed system was able to achieve significantly better verification performance on the synthesized and real-world testing dataset.

References

1. Deng, J., Guo, J., Xue, N., Zafeiriou, S.: ArcFace: additive angular margin loss for deep face recognition. In: Proceedings of the IEEE Conference on Computer Vision and Pattern Recognition, pp. 4690–4699 (2019)
2. Guo, Y., Zhang, L., Hu, Y., He, X., Gao, J.: MS-Celeb-1M: a dataset and benchmark for large-scale face recognition. In: Leibe, B., Matas, J., Sebe, N., Welling, M. (eds.) ECCV 2016. LNCS, vol. 9907, pp. 87–102. Springer, Cham (2016). https://doi.org/10.1007/978-3-319-46487-9_6
3. He, K., Zhang, X., Ren, S., Sun, J.: Deep residual learning for image recognition. In: 2016 IEEE Conference on Computer Vision and Pattern Recognition (CVPR), pp. 770–778 (2016)
4. Huang, G.B., Ramesh, M., Berg, T., Learned-Miller, E.: Labeled faces in the wild: a database for studying face recognition in unconstrained environments. Tech. rep. 07–49, University of Massachusetts, Amherst (October 2007)
5. Jaderberg, M., Simonyan, K., Zisserman, A., et al.: Spatial transformer networks. In: Advances in Neural Information Processing Systems, pp. 2017–2025 (2015)
6. King, D.E.: Dlib-ml: a machine learning toolkit. J. Mach. Learn. Res. **10**, 1755–1758 (2009)
7. Li, X., Wang, F., Hu, Q., Leng, C.: AirFace: lightweight and efficient model for face recognition. In: Proceedings of the IEEE International Conference on Computer Vision Workshops (2019)
8. Liu, W., Wen, Y., Yu, Z., Li, M., Raj, B., Song, L.: Sphereface: deep hypersphere embedding for face recognition. In: Proceedings of the IEEE Conference on Computer Vision and Pattern Recognition, pp. 212–220 (2017)

9. Masi, I., Tran, A.T., Hassner, T., Leksut, J.T., Medioni, G.: Do we really need to collect millions of faces for effective face recognition? In: Leibe, B., Matas, J., Sebe, N., Welling, M. (eds.) ECCV 2016. LNCS, vol. 9909, pp. 579–596. Springer, Cham (2016). https://doi.org/10.1007/978-3-319-46454-1_35
10. Parkhi, O.M., Vedaldi, A., Zisserman, A.: Deep face recognition. In: Xie, X., Jones, M.W., Tam, G.K.L. (eds.) Proceedings of the British Machine Vision Conference (BMVC), pp. 41.1–41.12. BMVA Press (September 2015). https://doi.org/10.5244/C.29.41
11. Paszke, A., et al.: Automatic differentiation in PyTorch. In: NIPS-W (2017)
12. Schroff, F., Kalenichenko, D., Philbin, J.: FaceNet: a unified embedding for face recognition and clustering. In: Proceedings of the IEEE Conference on Computer Vision and Pattern Recognition, pp. 815–823 (2015)
13. Sengupta, S., Cheng, J., Castillo, C., Patel, V., Chellappa, R., Jacobs, D.: Frontal to profile face verification in the wild. In: IEEE Conference on Applications of Computer Vision (February 2016)
14. Simonyan, K., Zisserman, A.: Very deep convolutional networks for large-scale image recognition. In: Bengio, Y., LeCun, Y. (eds.) 3rd International Conference on Learning Representations, ICLR 2015, San Diego, CA, USA, May 7–9, 2015, Conference Track Proceedings (2015). http://arxiv.org/abs/1409.1556
15. Sun, Y., Wang, X., Tang, X.: Deeply learned face representations are sparse, selective, and robust. In: The IEEE Conference on Computer Vision and Pattern Recognition (CVPR) (June 2015)
16. Taigman, Y., Yang, M., Ranzato, M., Wolf, L.: DeepFace: closing the gap to human-level performance in face verification. In: The IEEE Conference on Computer Vision and Pattern Recognition (CVPR) (June 2014)
17. Wang, H., et al.: CosFace: large margin cosine loss for deep face recognition. In: Proceedings of the IEEE Conference on Computer Vision and Pattern Recognition, pp. 5265–5274 (2018)
18. Wang, Z., et al.: Masked face recognition dataset and application (2020)
19. Wen, Y., Zhang, K., Li, Z., Qiao, Y.: A discriminative feature learning approach for deep face recognition. In: Leibe, B., Matas, J., Sebe, N., Welling, M. (eds.) ECCV 2016. LNCS, vol. 9911, pp. 499–515. Springer, Cham (2016). https://doi.org/10.1007/978-3-319-46478-7_31
20. Wu, W., Kan, M., Liu, X., Yang, Y., Shan, S., Chen, X.: Recursive spatial transformer (rest) for alignment-free face recognition. In: 2017 IEEE International Conference on Computer Vision (ICCV), pp. 3792–3800 (2017)
21. Yi, D., Lei, Z., Liao, S., Li, S.Z.: Learning face representation from scratch (2014)
22. Zhang, K., Zhang, Z., Li, Z., Qiao, Y.: Joint face detection and alignment using multitask cascaded convolutional networks. IEEE Signal Process. Lett. **23**(10), 1499–1503 (2016). https://doi.org/10.1109/lsp.2016.2603342
23. Zheng, T., Deng, W.: Cross-pose LFW: a database for studying cross-pose face recognition in unconstrained environments. Tech. rep. 18–01, Beijing University of Posts and Telecommunications (February 2018)
24. Zheng, T., Deng, W., Hu, J.: Cross-age LFW: a database for studying cross-age face recognition in unconstrained environments. CoRR abs/1708.08197 (2017). http://arxiv.org/abs/1708.08197
25. Zhong, Y., Chen, J., Huang, B.: Toward end-to-end face recognition through alignment learning. IEEE Signal Process. Lett. **24**(8), 1213–1217 (2017). https://doi.org/10.1109/lsp.2017.2715076

Visible Feature Guidance for Crowd Pedestrian Detection

Zhida Huang(✉), Kaiyu Yue, Jiangfan Deng, and Feng Zhou

Algorithm Research, Aibee Inc., Beijing, China
hzhida@pku.edu.cn

Abstract. Heavy occlusion and dense gathering in crowd scene make pedestrian detection become a challenging problem, because it's difficult to guess a precise full bounding box according to the invisible human part. To crack this nut, we propose a mechanism called Visible Feature Guidance (VFG) for both training and inference. During training, we adopt visible feature to regress the simultaneous outputs of visible bounding box and full bounding box. Then we perform NMS only on visible bounding boxes to achieve the best fitting full box in inference. This manner can alleviate the incapable influence brought by NMS in crowd scene and make full bounding box more precisely. Furthermore, in order to ease feature association in the post application process, such as pedestrian tracking, we apply Hungarian algorithm to associate parts for a human instance. Our proposed method can stably bring about 2–3% improvements in mAP and AP_{50} for both two-stage and one-stage detector. It's also more effective for MR^{-2} especially with the stricter IoU. Experiments on Crowdhuman, Cityperson, Caltech and KITTI datasets show that visible feature guidance can help detector achieve promisingly better performances. Moreover, parts association produces a strong benchmark on Crowdhuman for the vision community.

1 Introduction

Pedestrian detection is widely applied in a number of tasks, such as autonomous driving, robot-navigation and video surveillance. Many previous efforts [1,21, 22,28,30,33,34,36] have been made to improve its performance. Although reasonably good performance has been achieved on some benchmark datasets for detecting non-occluded, slightly occluded and crowd pedestrians, it's still far from being satisfactory for detecting heavily occluded and crowded pedestrians. To push the cutting-edge boundary, this work aims to propose a novel and simple method to precisely detect pedestrians in the heavily crowd scene.

In detection, bounding box (or box) is commonly used to represent the object location in an image. According to its application scenario, it generally has two distinct types: visible one and full one. As shown in Fig. 1(a), a visible box only covers the visible part of an object. In contrast, the full box covers the whole expected region of an object even though it's occluded by the other objects. However, the full bounding box needs detector to estimate the invisible border.

A. Bartoli and A. Fusiello (Eds.): ECCV 2020 Workshops, LNCS 12539, pp. 277–290, 2020.
https://doi.org/10.1007/978-3-030-68238-5_20

Fig. 1. Visible bounding box *vs.* Full bounding box. (a) We illustrate the visible and full bounding box of a same human instance in a crowd scene. (b) Conspicuously, the full bounding box has a dramatically larger overlap (IoU) than visible bounding box.

This property hinders the detector to produce highly precise outputs. We analyze this problem in two folds: (1) First, intuitively, it's unreasonable to use invisible feature to guide a regressor for producing the full bounding box. This will easily plunge the detector into a random process to guess the border for an occluded object. (2) Second, for two heavily crowed and occluded instances, it's hard to perform Non-Maximum-Suppression (NMS) well on their full boxes to separate them precisely, because there is a large overlap between the full bounding boxes of them, which would be easily merged into a same box by NMS, as shown in Fig. 1(b).

To overcome these two obstacles, we propose an effective mechanism named Visible Feature Guidance (VFG). For training, we restrict RPN only produces the visible bounding-box proposals. Then we adopt the visible feature to simultaneously regress the visible bounding box and full bounding box. This is motivated by the visible part of object that represents the clear appearance to pinpoint the area for guiding the regressor where should be focused on. In inference, we use the visible feature as well to generate the coupled pair of the visible box and full box. But we only perform NMS on visible bounding-boxes to shave the confused ones, then achieving the corresponding full bounding box in the end. The insight of this behavior is that we found the overlap of visible bounding boxes between heavily crowed instances is smaller than that of full bounding boxes, as illustrated in Fig. 1. Using visible boxes to perform NMS can effectively prevent the boxes from being merged into a same box for the common case of two heavily crowed instances (e.g., the IoU of full boxes is larger than threshold, but the IoU of visible boxes is smaller than threshold).

In addition, some application scenario not only requires the simple output of the whole object bounding box, also requires the precise localization for its part (e.g., head, face). They need detector to stably generate the coupled localization of the main object and its semantic parts. In this paper, we transform the parts association task into an assignment problem solved by Hungarian algorithm [13]. This simple post process can effectively produce the combined output of full bounding box, visible bounding box and part localizations for a same instance. Experiments on Crowdhuman [25] introduce a strong baseline and benchmark for the parts association task.

Overall, our main contributions in this work are summarized into three folds:

- We present visible feature guidance (VFG) mechanism. It regresses visible bounding box and full bounding box simultaneously during training. Then we perform NMS on visible bounding-boxes to achieve the best fitting full box in inference in order to mitigate the influence of NMS on full boxes.
- We transform parts association into a linear assignment problem, including body-head, body-face or head-face association. We build a strong benchmark on Crowdhuman [25] dataset for this parts association task.
- Experiments are carried out on four challenging pedestrian and car detection datasets, including KITTI [8], Caltech [5], CityPerson [32], and Crowdhuman [25], to demonstrate that our proposed method works stably better than previous methods.

2 Related Work

With the development of deep convolution neural network (CNN), object detection has made a dramatic improvement in both performance and efficiency. Recently, detection methods can be roughly divided into two types: two-stage and one-stage. Two-stage detector, such as Faster R-CNN [24], R-FCN [3] and Mask R-CNN [10] first generates the region proposals and then refine these coarse boxes after an scale-invariant feature aggregating operation named ROI-Pool [9] or RoI-Align [10]. This coarse-to-fine process leads to achievements of top performance. One-stage detector, such as YOLO [23], SSD [18] and RetinaNet [16], predicts locations directly in an unified one-shot structure which is fast and efficient. Recently, many novel anchor-free detectors have emerged [6,12,14,27], these methods cancel the hand-craft tuning of pre-defined anchors, and simplify the training process. In the area of pedestrian detection, previous works such as such as Adapted Faster R-CNN [20,32] design an anchor-free method to predict the body center and scale of instance box.

However, crowd occlusion is still one of the most important and difficult problems in pedestrian detection. Repulsion loss [28] is designed to penalize the predicted box to avoid shifting to the wrong objects and push it far from the other ground-truth targets. Bi-box regression [36] aims to use two branches for regressing the full and visible boxes simultaneously. Occlusion-Aware RCNN [35] proposes an aggregation loss to enforce proposals closely for objects. It also uses the occlusion-aware region of interest (RoI) pooling with structure information. Compared with [36] and [35], our proposed method is more concise with using visible box to produce full box directly.

Non-Maximum-Suppression (NMS) is mostly used by object detection in post-processing stage. Firstly, it sorts all detection boxes based on their scores.

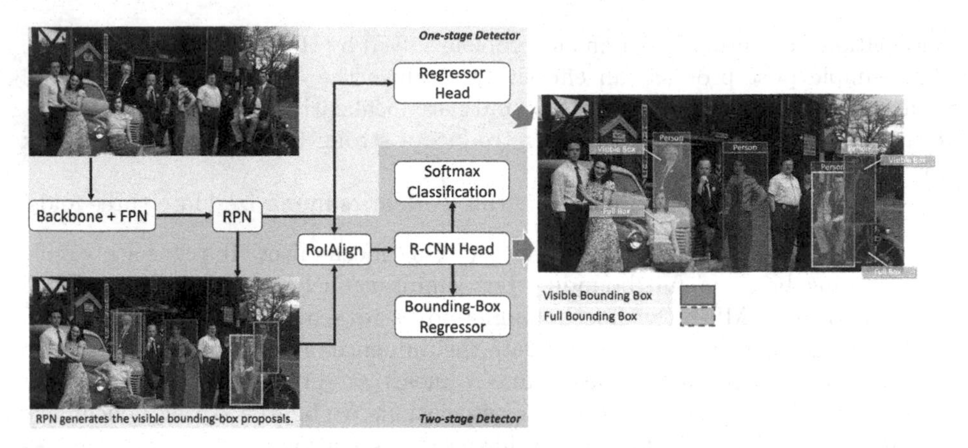

Fig. 2. Visible feature guidance (VFG) for one-stage and two-stage detector. We let RPN to regress visible box and RCNN to regress pairs of visible and full boxes simultaneously by using visible feature for two-stage detector. For one-stage detector, regression head directly regresses the pairs of visible and full boxes.

Then the detection box with maximum score is selected and all other detection boxes which have large overlap with the selected one are suppressed. This process is running recursively until there are no remaining boxes. Soft-NMS [1] decreases detection scores using a continuous function of their overlaps such as Guassian Kernel. It drops the candidate boxes progressively and carefully. Adaptive-NMS [17] designs a detector sub-network that learns the target density to decide what dynamic suppression threshold should be applied to an instance.

However, NMS will remove valid boxes in crowded scene. Soft-NMS would remove valid boxes and introduce false positive boxes. Adaptive-NMS need a additional network to generate the threshold value. In this work, we propose VFG-NMS which simply performs NMS with visible boxes and achieves the corresponding full boxes according to the indexes of remaining visible boxes. This manner make detection more robust and less sensitive to NMS threshold.

3 Methodology

Our VFG method is general and independent of detector types. Considering the existence of two typical frameworks: two-stage and one-stage, we go into the details of building the VFG module with these two models respectively. Then we discuss the method of parts association, which is the downstream process of pedestrian detection.

Architecture. The proposed method can be easily implemented into current detectors. Figure 2 shows the main structure of VFG method. An input image is firstly filled into backbone and FPN (if exists) to extract features. Then, for two-stage detector like Faster R-CNN [24], the Region Proposal Network (RPN) [24] generates visible bounding-box proposals and sends them into RoIPool [9] or

RoIAlign [10], from which the RoI feature is aggregated. After that, the R-CNN head uses these visible RoI features to regress pairs of visible and full boxes in the *parallel* manner. For one-stage detector, following similar scheme, visible bounding boxes are generated as base proposals, which are used by the final submodule to regress visible and full bounding boxes at the same time. Compared to the typical pipeline of detector, this structure includes two key differences: (1) The detector regresses eight coordinate values for both visible and full bounding box, instead of the conventional four values. (2) The positive and negative sampling during training is only based on the IoU of visible proposals and ground truth.

Visible Feature Guidance. As shown in Fig. 3, red and green points represent the feature region of visible and full box of an object (a pedestrian or a car in the image) respectively. For the two-stage detector, RoIAlign could be applied with one of two potential feature regions: visible region (in red points) and full region (in green points). Previous methods in default choose full feature regions to perform RoIAlign for guiding regression. However, the features of full box is half occupied by the background, which is harmful that plunges detector into a random process to regress the invisible border. Fortunately, the feature region of visible box focus on the valid part of an object, which has a better potential to estimate entire structure precisely. Thus we utilize visible feature region to apply RoIAlign to produce a clear and effective feature represents for guiding the bounding-box regressor and classifier. For one-stage detector, in the same sense, taking visible box to regress full box can be considered as adding attention factors directly to the visible region, facilitating and guiding the network to learn the object shape and estimate the full box location.

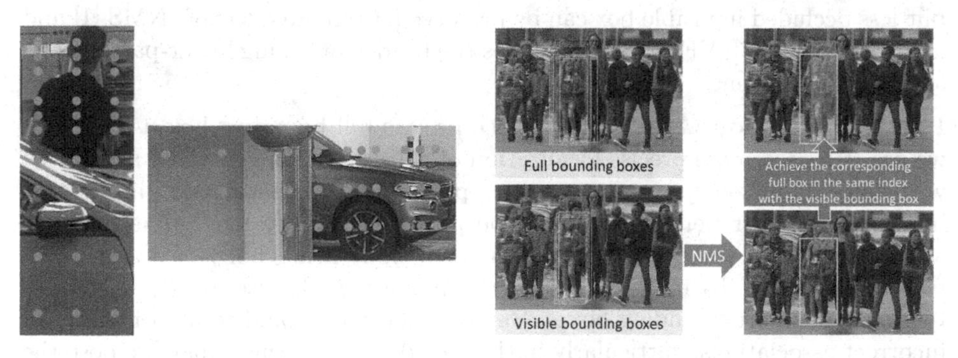

Fig. 3. **Feature regions** produced by visible (red) and full boxes (green). (Color figure online)

Fig. 4. **VFG-NMS** is done with visible boxes.

Multi-task Loss. In the procedure of training, we propose multi-task training to supervise and guide bounding box regressor and classifier using visible feature regions. The bounding-box regressor produces two outputs for a instance in parallel: visible box and full box. Let the center coordination of visible bounding-box proposal for an object[1] be (x_v^p, y_v^p) within width P_w and height P_h. The ground-truth of visible box is (x_v^g, y_v^g, V_w, V_h) and which of full box is (x_f^g, y_f^g, F_w, F_h). So the visible and full box regression formula is:

$$\triangle_{x_v}^* = \frac{x_v^g - x_v^p}{P_w}, \triangle_{y_v}^* = \frac{y_v^g - y_v^p}{P_h}, \triangle_{w_v}^* = \log\frac{V_w}{P_w}, \triangle_{h_v}^* = \log\frac{V_h}{P_h} \tag{1}$$

$$\triangle_{x_f}^* = \frac{x_f^g - x_f^p}{P_w}, \triangle_{y_f}^* = \frac{y_f^g - y_f^p}{P_h}, \triangle_{w_f}^* = \log\frac{F_w}{P_w}, \triangle_{h_f}^* = \log\frac{F_h}{P_h} \tag{2}$$

The multi-task loss L on each labeled RoI to jointly train for classification and bounding-box regression, including visible box and full box:

$$L = L_{cls}(c, c^*) + \lambda_{loc} L_{loc}(t, t^*) \tag{3}$$

where c is predicted output and c^* is ground-truth. L_{cls} denotes the loss for classification: commonly softmax cross-entropy loss for two-stage detector and focal loss [16] for one-stage detector. $t = (\triangle_{x_v}, \triangle_{y_v}, \triangle_{w_v}, \triangle_{h_v}, \triangle_{x_f}, \triangle_{y_f}, \triangle_{w_f}, \triangle_{h_f})$ and $t^* = (\triangle_{x_v}^*, \triangle_{y_v}^*, \triangle_{w_v}^*, \triangle_{h_v}^*, \triangle_{x_f}^*, \triangle_{y_f}^*, \triangle_{w_f}^*, \triangle_{h_f}^*)$ represent eight predicted and ground-truth parameterized targets. L_{loc} is the loss function for regression, such as smooth-L_1 loss or IoU loss [29]. λ_{loc} is a loss-balancing parameter and we set $\lambda_{loc} = 3$.

VFG-NMS. In the procedure of inference, NMS is a common post-processing method for object detection. However, it may remove valid bounding boxes improperly in the heavily crowded scene. To address this issue, we propose VFG-NMS: using the less crowded visible boxes to guide full boxes filtering. In Fig. 4, the detector outputs visible and full boxes in pairs, we send all visible boxes into a standard NMS and get the corresponding full boxes according to remaining indices. Through this operation, instances which are highly occluded in full box but less occluded in visible box can be preserved. Compared to Soft-NMS [1] and adaptive-NMS [17], VFG-NMS alleviates the burden of tuning hyper-parameters, and is more efficient.

Parts Association. After achieving the precise full bounding box using VFG, we are still need more precise localization of the semantic parts to be associated with full box in some scenarios, such as pedestrian re-identification and tracking. Because their performances can be promoted by the fine-grained feature aggregation from the parts association (e.g., body-head, body-face and head-face association). To reach this goal, the pioneer works use greedy algorithm to couple the part bounding boxes. But this behavior would recall a number of incorrect associations, particularly in the crowd scene. In this paper, we pose the body-parts association problem as a Linear Assignment problem. We resolve it by the efficient Hungarian algorithm [13] to minimize the assignment cost. The cost matrix is constructed by one of two measurements[2]: geometry distance and

[1] Here all the notations ignore the object classes. In default, the notation indicates the representation for a instance of one object class.

[2] Generally, for the association of which body-part is in a small overlap, e.g. body-head association, we select one of distances to build the cost matrix. Vice in versa, for the association which the overlap is too large, e.g. head-face association, we prefer to use the IoU metric to constrain the parts' boxes.

IoU metric. With R bodies and C heads overall the instances in an image, we can use appropriate distance metric and IoU as the constrain condition to construct an $N_R \times N_C$ ($N_R \leq N_C$) cost matrix D according to the spatial relations of body and part, which are belonging to the same human instance. The standard assignment problem can be optimized by minimizing the cost function:

$$X^* = \arg\min \sum_{i=1}^{N_R} \sum_{j=1}^{N_C} \rho_{i,j} x_{i,j}; \; st. \sum_{j=1}^{N_C} x_{i,j} = 1, \forall i; \; \sum_{i=1}^{N_R} \leq 1, \forall j; x_{i,j} \in \{0,1\} \quad (4)$$

where $\rho_{i,j}$ is the cost function to measure the relation of i−th body and j−th part, which can be distance function (e.g. Euclidean distance) or IoU metric. After solving the optimization problem of unknown x_{ij}, we can get body-head association.

4 Experiment

We have done with massive experiments on four datasets: CityPerson [32], Crowdhuman [25], Caltech [5] for pedestrian detection and KITTI [8] for car detection. All of three pedestrian datasets provide visible and full bounding-box annotations. Specially, Crowdhuman [25] has extra head bounding-box annotations for human instances. Since KITTI only provides full boxes, we use Mask R-CNN pre-trained on COCO to generate the visible boxes of car and associate it with labeled full boxes by using Hungarian Algorithm.

4.1 Dataset and Evaluation Metrics

The CityPerson [32] dataset is originated from the semantic segmentation dataset Cityscape [2] for crowed pedestrian detection. Visible bounding boxes are auto-generated from instance segmentation of each pedestrian. The Crowdhuman [25] dataset is targeted for addressing the task of crowd pedestrian detection. It provides complete annotations: full bounding box, visible box, and head box. Furthermore, compared to other human detection datasets, it's a larger-scale dataset with much higher crowdness. Caltech [5] is one of several predominant datasets for pedestrian detection and Zhang *et al.* [31] provided refined full box annotations. KITTI [8] is the dataset for autonomous driving scene and it only provides full box annotations for the classes of pedestrian, car, tram, truck and van.

In our experiments, we follow the standard Caltech evaluation [5] metrics. The log miss-rate averaged over FPPI (false positive per-image) range of $[10^{-2}, 10^0]$ (denoted as MR^{-2}) is used to evaluate the pedestrian detection performance (lower is better). We also use the standard evaluation of object detection, including mAP, recall and AP_{50} to investigate the performance of pedestrian detection. For car detection on KITTI, unlike previous works follow PASCAL [7] criteria, we use COCO evaluation criteria with the stricter IoU to fully report performances.

4.2 Implementation Details

We implement our VFG method in Faster R-CNN [24] (FRCNN) and RetinaNet [16] with FPN [15] structure. ResNet-50 [11] pre-trained on ImageNet [4] is adopted as backbone. We set the anchor ratios for full box into {1.0, 1.5, 2.0, 2.5, 3.0} and our VFG method of visible box into {0.5, 1.0, 2.0}. During training, we use a total batch size of 8 and 16 for Faster R-CNN and RetinaNet respectively. Stochastic Gradient Descent (SGD) solver is used to optimize the networks on 8 1080Ti GPUs. We set the weight decay to 0.0001 and momentum to 0.9. The threshold of greedy-NMS and soft-NMS with the linear method is 0.5.

When training on Crowdhuman, images are resized so that the short side is 800 pixels while the long side does not exceed 1400 pixels. For Faster R-CNN, models are trained for 72k iters in total, the initial learning rate is set to 0.02 and decreased by a factor of 10 for 48k and 64k iters. For RetinaNet, models are trained for 36k iters in total, the initial learning rate is set to 0.01 and decreased by a factor of 10 for 24k and 32k iters. Multi-scale training/testing are not applied to ensure fair comparisons. On Cityperson, we train Faster R-CNN for 9k iters in total using the base learning rate of 0.02 and decrease it by a factor of 10 after 6k and 8k iters. On KITTI, we train Faster R-CNN for 18k iters in total using the base learning rate of 0.02 and decrease it by a factor of 10 after 12k and 16k iters. Original input image size was used to evaluate on Cityperson and KITTI.

4.3 Experimental Results

Crowdhuman Dataset. We evaluate Faster R-CNN and RetinaNet with VFG on Crowdhuman. Table 1 shows that, in Faster R-CNN, our proposed VFG method outperforms the baseline, bringing about 3.5% and 4.5% improvement in mAP and recall while reducing MR^{-2} by 0.84% and 0.74% in `Reasonable` and `All` sets respectively. As comparison, soft-NMS can improve mAP obviously but showing little effect in MR^{-2}. In the case of RetinaNet, soft-NMS decreases performance a lot for introducing false positives in crowded scene. On the contrary, our VFG method can still bring improvement in mAP (Table 3).

Cityperson Dataset. Experiments on Cityperson follow the evaluation standard in RepLoss [28] and OR-CNN [35], in which the `Reasonable` part (occlusion $< 35\%$) of the validation set is divided into `Partial` ($10\% <$ occlusion $< 35\%$) and `Bare` (occlusion $\leq 10\%$) subsets.

In Table 2, our VFG method has obvious better performance compared to baseline, improving the mAP by 2.1% and reduce the MR^{-2} by 1.8%. Compared to Soft-NMS, VFG has competitive performances in mAP and better results in MR^{-2}. Especially, when we set the confidence threshold to 0.3, which is often applied in real usage, our VFG method suggests strong superiority.

Caltech Dataset. Whereas the work of Zhang *et al.* [31] has provided refined annotations, the visible box annotations on Caltech are still noisy. So we don't

Table 1. Performances of full bounding box on Crowdhuman

Method	mAP	AP_{50}^{bbox}	Recall	MR^{-2}			
				Reasonable	Small	Heavy	All
FRCNN+greedy-NMS(offical)	–	84.95	90.24	–	–	–	50.42
FRCNN+adaptive-nms(offical)	–	84.71	91.27	–	–	–	49.73
FRCNN+greedy-NMS	47.9	83.1	87.8	25.49	25.29	46.76	49.02
FRCNN+soft-NMS	50.8	85.8	**92.8**	25.49	24.98	**46.65**	49.02
FRCNN+VFG w/o VFG-NMS	49.5	84.5	89.3	24.95	24.94	47.97	48.33
FRCNN+VFG	**51.4**	**86.4**	92.3	**24.65**	**24.95**	47.25	**48.28**
RetinaNet+greedy-NMS(official)	–	80.83	93.28	–	–	–	63.33
RFBNet+adaptive-nms(official)	–	79.67	94.77	–	–	–	63.03
RetinaNet+greedy-NMS	42.1	77.7	85.1	33.70	35.45	**52.85**	**57.90**
RetinaNet+soft-NMS	40.2	70.1	87.8	58.0	49.37	67.76	77.37
RetinaNet+VFG	**47.0**	**82.3**	**91.0**	**33.41**	**32.08**	53.71	58.13

Table 2. Performances of full bounding box on Cityperson.

Method	Score	mAP	AP_{50}^{bbox}	Recall	MR^{-2}			
					Reasonable	Heavy	Partial	Bare
FRCNN+greedy-NMS	0.05	52.8	80.6	84.5	12.87	50.99	13.15	7.73
FRCNN+soft-NMS	0.05	**54.9**	**82.9**	**89.1**	12.76	**50.91**	13.14	7.42
FRCNN+VFG w/o VFG-NMS	0.05	53.8	81.0	84.6	11.63	51.41	12.36	6.95
FRCNN+VFG	0.05	54.8	82.3	86.9	**11.04**	50.94	**11.56**	**6.69**
FRCNN+greedy-NMS	0.3	51.4	77.3	79.7	12.88	51.69	13.44	7.82
FRCNN+soft-NMS	0.3	51.9	78.1	80.3	12.76	51.61	13.40	7.59
FRCNN+VFG w/o VFG-NMS	0.3	52.5	77.6	79.8	11.67	52.01	12.47	6.98
FRCNN+VFG	0.3	**53.4**	**79.1**	**81.2**	**11.04**	**51.46**	**11.68**	**6.73**

use this dataset for training. Instead, we evaluate models trained on Crowdhuman on the Caltech, verifying the generality of the proposed method. In Table 4, compared to Faster R-CNN with greedy-NMS, the VFG method boosts mAP, AP_{50} and recall with 3.2%, 4.5% and 8.4%, reduces 0.6%, 1.41%, 4.76% and 2.13% MR^{-2} in `Reasonable`, `Small`, `Heavy` and `All` subsets respectively. These results indicate generality of our VFG method.

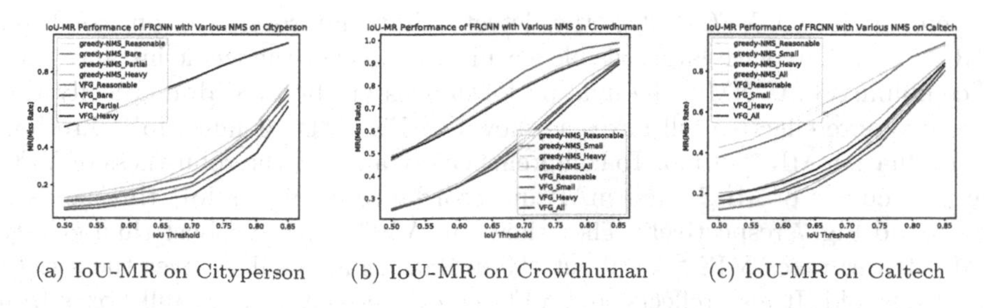

(a) IoU-MR on Cityperson (b) IoU-MR on Crowdhuman (c) IoU-MR on Caltech

Fig. 5. IoU influence of MR evaluation. With stricter IoU to calculate MR^{-2}, it shows VFG method is more robust.

Table 3. Comparison with state-of-the-art methods on Cityperson. All the experiments are done with the original input size.

Method	Backbone	Reasonable	Heavy	Partial	Bare
Adapted-FRCNN [32]	VGG-16	15.40	–	–	–
FRCNN Adaptive-NMS [17]	VGG-16	12.90	56.40	14.40	7.00
OR-CNN [35]	VGG-16	12.80	55.70	15.30	6.70
RepLoss [28]	ResNet-50	13.20	56.90	16.80	7.60
TLL [26]	ResNet-50	15.5	53.60	17.20	10.0
ALFNet [19]	ResNet-50	12.00	51.90	**11.4**	8.40
FRCNN+greedy	ResNet-50	12.87	50.99	13.15	7.73
FRCNN+VFG	ResNet-50	**11.04**	**50.94**	11.56	**6.69**

Table 4. Performances of full bounding box on Caltech.

Method	mAP	AP_{50}	Recall	MR^{-2}			
				Reasonable	Small	Heavy	All
FRCNN+greedy	26.4	53.5	57.0	12.49	16.35	42.64	46.47
FRCNN+soft	27.2	53.6	58.7	12.54	16.49	42.19	46.68
FRCNN+VFG w/o VFG-NMS	29.5	57.7	65.8	11.72	14.96	37.95	44.19
FRCNN+VFG	**29.6**	**58.0**	**66.4**	**11.89**	**14.94**	**37.88**	**44.34**

KITTI Dataset. For KITTI dataset, we choose the category of car to evaluate our proposed VFG method. Since the KITTI dataset provides full boxes only. In this experiment, we use Mask R-CNN pre-trained on COCO to produce the visible boxes of car and associate them with full boxes by Hungarian Algorithm. As Table 5 shown, VFG method outperforms the FRCNN with greedy-NMS 1.9%, 2.8% and 4.9% in mAP, AP80 and AP90 respectively with 0.05 detection score. To evaluate the performance further, we set the confidence threshold to 0.3 and get consistent result.

Ablation Study and Robustness Analysis. To better investigate our VFG, we decompose its pipeline into two isolated flows: regression guided by visible feature and VFG-NMS. We experiment ablations on these two aspects in the whole of above four datasets. When using visible feature guidance for regressor only (without VFG-NMS), the detector have improvements on mAP and MR^{-2}, especially with larger visible feature region over the whole instance area. If combining VFG-NMS, the final performances are boosted dramatically that remains more effective full boxes as shown in Fig. 6. By default, IoU threshold for evaluation MR^{-2} is 0.5. To comprehensively analyze the robustness of VFG, we plot curves of MR^{-2} and mAP in accordance of higher IoU thresholds in Fig. 5 and Fig. 7 respectively. They show our VFG is more robust than greedy-NMS. It decreases MR^{-2} by about 2% in `Reasonable` subset over the stricter IoU threshold. It also reflects that VFG can effectively prevent full boxes from being filtered out.

(a) FRCNN+greedy-NMS (b) FRCNN+VFG

Fig. 6. Visualization of detection result. Red and green boxes mean ground-truth and predicted box respectively. The left (a) is the result of FRCNN with greedy-NMS method while right (b) is the result of our FRCNN with VFG method. The top, middle and bottom images are from Crowdhuman, Cityperson and KITTI datasets respectively. (Color figure online)

4.4 Parts Association and Benchmark

To completely investigate the algorithm for post parts association, we do experiments with two tasks on Crowdhuman: (1) we associate body-head parts; (2) we decompose the pair output of detector and to associate the visible and full bounding boxes. The experimental results build a simple and strong benchmark on Crowdhuman.

First, we describe the evaluation method for the task of parts associations using an example, e.g., body-head. The pair of ground truth for body-head is indicated by (B_{gt}, H_{gt}). The predicted pair of it by our proposed algorithm is (B, H). The validation condition is $\frac{area(B_{gt} \cup B)}{area(B_{gt} \cap B)} \geq thresh_B$ and $\frac{area(H_{gt} \cup H)}{area(H_{gt} \cap H)} \geq thresh_H$ where $thresh_B$ is the threshold for body and $thresh_H$ for head to measure their relationship. When (B, H) meets the above condition, it will match the ground truth. Otherwise it won't.

Table 6 and 7 show the Hungarian algorithm can achieve the optimal solution with *un-associated* ground-truth boxes on Crowdhuman. The results of recall and precision are fully correct. This upper bound verify the correctness and effect of our matching algorithm. In addition, we have trained the detectors for regressing out head, visible and full box respectively for performing association as shown in

Table 5. Performances of full bounding box on KITTI Car.

Method	Score	mAP	AP70	AP80	AP90
FRCNN+greedy	0.05	69.8	86.4	75.9	29.5
FRCNN+VFG w/o VFG-NMS	0.05	70.9	87.5	77.9	34.4
FRCNN+VFG	0.05	**71.7**	**87.6**	**78.7**	**34.4**
FRCNN+greedy	0.3	69.2	85.6	75.9	29.5
FRCNN+VFG w/o VFG-NMS	0.3	69.8	85.9	77.1	34.0
FRCNN+VFG	0.3	**70.5**	**86.8**	**78.0**	**34.4**

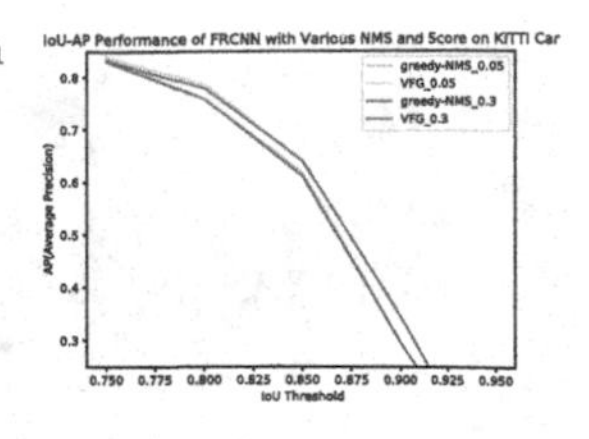

Fig. 7. IoU-AP car performance

Table 6. Body-head association

Box-type	Score	Recall	Precision
GT box	–	1.0	1.0
Pred box	0.05	73.83	64.58
Pred box	0.3	65.01	83.20
Pred box	0.7	56.99	92.14

Table 7. Visible-full body association

Box-type	Score	Recall	Precision
GT box	–	1.0	1.0
Pred box	0.05	87.23	61.58
Pred box	0.3	81.91	80.35
Pred box	0.7	73.45	90.33

(a) Full Body-Head Association

(b) Visible-Full Body Association

Fig. 8. Visualization of parts association result. Same color of body-parts box means belonging to the same person

Fig. 8. In contrast, the predicted box (Pred box) output by deep learning model produces bad results due to the unsatisfied detected bounding boxes in crowd scene. As shown in Table 6 and 7, increasing the detection score threshold can make precision of association increase but recall decrease. This suggests that parts association largely depends on parts detection performance and we can conclude better detection could bring better association.

5 Discussion

We have evaluated our proposed VFG on four popular datasets and show it works stably well in two-stage and one-stage detector. All the experimental results verify that VFG can stably promote 2–3% in mAP and AP_{50} and also be more effective for MR^{-2} especially with stricter IoU. Moreover, we have benchmarked strongly for the task of parts association using Hungarian algorithm on Crowdhuman. Although all the ablations show the effect of VFG, they suggest that there is an interesting topic on reducing false positives further without loss of recall in the future research.

References

1. Bodla, N., Singh, B., Chellappa, R., Davis, L.S.: Soft-NMS-improving object detection with one line of code. In: ICCV (2017)
2. Cordts, M., et al.: The cityscapes dataset for semantic urban scene understanding. In: CVPR (2016)
3. Dai, J., Li, Y., He, K., Sun, J.: R-FCN: object detection via region-based fully convolutional networks (2016)
4. Deng, J., Dong, W., Socher, R., Li, L.J., Li, K., Fei-Fei, L.: ImageNet: a large-scale hierarchical image database. In: CVPR (2009)
5. Dollár, P., Wojek, C., Schiele, B., Perona, P.: Pedestrian detection: a benchmark. In: CVPR (2009)
6. Duan, K., Bai, S., Xie, L., Qi, H., Huang, Q., Tian, Q.: CenterNet: keypoint triplets for object detection (2019)
7. Everingham, M., Van Gool, L., Williams, C.K., Winn, J., Zisserman, A.: The pascal visual object classes (VOC) challenge. ICCV **88**, 303–338 (2010)
8. Geiger, A., Lenz, P., Urtasun, R.: Are we ready for autonomous driving? The KITTI vision benchmark suite. In: CVPR (2012)
9. Girshick, R.: Fast R-CNN. In: ICCV (2015)
10. He, K., Gkioxari, G., Dollár, P., Girshick, R.: Mask R-CNN. In: ICCV (2017)
11. He, K., Zhang, X., Ren, S., Sun, J.: Deep residual learning for image recognition. In: CVPR (2016)
12. Kong, T., Sun, F., Liu, H., Jiang, Y., Shi, J.: FoveaBox: beyond anchor-based object detector. arXiv preprint arXiv:1904.03797 (2019)
13. Kuhn, H.W.: The Hungarian method for the assignment problem. Naval Res. Logist. Q. **2**, 83–97 (1955)
14. Law, H., Deng, J.: CornerNet: detecting objects as paired keypoints (2018)
15. Lin, T.Y., Dollár, P., Girshick, R., He, K., Hariharan, B., Belongie, S.: Feature pyramid networks for object detection. In: CVPR (2017)
16. Lin, T.Y., Goyal, P., Girshick, R., He, K., Dollár, P.: Focal loss for dense object detection. In: ICCV (2017)
17. Liu, S., Huang, D., Wang, Y.: Adaptive NMS: refining pedestrian detection in a crowd. In: CVPR (2019)
18. Liu, W., et al.: SSD: single shot multibox detector. In: Leibe, B., Matas, J., Sebe, N., Welling, M. (eds.) ECCV 2016. LNCS, vol. 9905, pp. 21–37. Springer, Cham (2016). https://doi.org/10.1007/978-3-319-46448-0_2
19. Liu, W., Liao, S., Hu, W., Liang, X., Chen, X.: Learning efficient single-stage pedestrian detectors by asymptotic localization fitting. In: ECCV (2018)
20. Liu, W., Liao, S., Ren, W., Hu, W., Yu, Y.: High-level semantic feature detection: a new perspective for pedestrian detection. In: CVPR (2019)
21. Ouyang, W., Wang, X.: A discriminative deep model for pedestrian detection with occlusion handling. In: CVPR (2012)
22. Pang, Y., Xie, J., Khan, M.H., Anwer, R.M., Khan, F.S., Shao, L.: Mask-guided attention network for occluded pedestrian detection. In: ICCV (2019)

23. Redmon, J., Divvala, S., Girshick, R., Farhadi, A.: You only look once: unified, real-time object detection. In: CVPR (2016)
24. Ren, S., He, K., Girshick, R., Sun, J.: Faster R-CNN: towards real-time object detection with region proposal networks. In: NIPS (2015)
25. Shao, S., et al.: Crowdhuman: a benchmark for detecting human in a crowd. arXiv preprint arXiv:1805.00123 (2018)
26. Song, T., Sun, L., Xie, D., Sun, H., Pu, S.: Small-scale pedestrian detection based on topological line localization and temporal feature aggregation. In: ECCV (2018)
27. Tian, Z., Shen, C., Chen, H., He, T.: FCOS: fully convolutional one-stage object detection (2019)
28. Wang, X., Xiao, T., Jiang, Y., Shao, S., Sun, J., Shen, C.: Repulsion loss: detecting pedestrians in a crowd. In: CVPR (2018)
29. Yu, J., Jiang, Y., Wang, Z., Cao, Z., Huang, T.: UnitBox: an advanced object detection network. In: Multimedia (2016)
30. Zhang, L., Lin, L., Liang, X., He, K.: Is faster R-CNN doing well for pedestrian detection? In: Leibe, B., Matas, J., Sebe, N., Welling, M. (eds.) ECCV 2016. LNCS, vol. 9906, pp. 443–457. Springer, Cham (2016). https://doi.org/10.1007/978-3-319-46475-6_28
31. Zhang, S., Benenson, R., Omran, M., Hosang, J., Schiele, B.: How far are we from solving pedestrian detection? In: CVPR (2016)
32. Zhang, S., Benenson, R., Schiele, B.: CityPersons: a diverse dataset for pedestrian detection. In: CVPR (2017)
33. Zhang, S., Benenson, R., Schiele, B., et al.: Filtered channel features for pedestrian detection. In: CVPR (2015)
34. Zhang, S., Yang, J., Schiele, B.: Occluded pedestrian detection through guided attention in CNNs. In: CVPR (2018)
35. Zhang, S., Wen, L., Bian, X., Lei, Z., Li, S.Z.: Occlusion-aware R-CNN: detecting pedestrians in a crowd (2018)
36. Zhou, C., Yuan, J.: Bi-box regression for pedestrian detection and occlusion estimation. In: ECCV (2018)

The Impact of Real Rain in a Vision Task

Iago Breno Araujo(✉), Eric K. Tokuda, and Roberto M. Cesar Jr.

Institute of Mathematics and Statistics, University of São Paulo, São Paulo, Brazil
{ibaraujo,keiji,cesar}@ime.usp.br

Abstract. Single image deraining has made impressive progress in recent years. However, the proposed methods are heavily based on high-quality synthetic data for supervised learning which are not representative of practical applications with low-quality real-world images. In a real setting, the rainy images portray a scene with a complex degradation caused by the rain weather and the low-quality factors. The goal of this paper is to investigate the impact of two visual factors that affect vision tasks: *image quality and rain effect*. To evaluate this, an image dataset with images varying these factors has been created. Aiming to evaluate them, different object detection algorithms are applied and evaluated on the dataset. Our findings indicate that the fine-tuned models can efficiently cope with this problem regardless of the rain intensity of the scene, however it is greatly affected by the image quality gap.

Keywords: Rain · Object detection · Transfer learning · Deraining

1 Introduction

Single image deraining has received increasing attention in the last few years. The goal is to enhance the visual quality of an original rainy image so that the rain artifacts are removed. Recently proposed methods have achieved impressive results on this endeavor [14,22]. However, the common deraining framework has presented unsatisfactory aspects.

Despite the success, the proposed methods are largely based on the image restoration perspective and trained using synthetic images. Therefore, the rain removal problem is being approached disconnected from its intrinsic nature, where it usually co-occurs with other complex perceptual degradation problems. Besides, they do not explore the potential of deraining in benefiting high-level computer vision tasks [14].

In this paper, we study the impact of the rain on a practical application, with varying rain intensities and image qualities. Our aim is to understand how state-of-the-art object detection methods deal with a real-world rainy scene. We created a dataset according to the image quality and to the intensity of the rain, which resulted in four subsets: high-quality sunny, low-quality sunny, low-quality with light rain and low-quality with heavy rain. We gain insights on how and to what extent the rain and the low-quality factors impact on the object detection task. Our results indicate a need towards task-driven approaches beyond the image restoration view.

A. Bartoli and A. Fusiello (Eds.): ECCV 2020 Workshops, LNCS 12539, pp. 291–298, 2020.
https://doi.org/10.1007/978-3-030-68238-5_21

2 Related Work

2.1 Deraining Algorithms

Deraining methods can be categorized into video and single image classes. Traditional approaches for video deraining are based on a multi-frame view, exploring the temporal relationship in frames to achieve a derained result [8,9]. This temporal information is not available in the single image case making it a more challenging problem. This work investigate challenges related to the current state-of-the-art single image deraining methods.

In single image deraining, the traditional approaches are based on priors regarding the rain degradation in an image. Kang et al. [12] approaches the problem in the frequency domain and model the rain streaks as a high frequency layer. Similarly, Chen and Hsu [3] decomposes the image into a background and a rain streak layer but based on low-rank priors. Kang et al. [12] approaches the problem in the frequency domain and model the rain streaks as a high frequency layer. Similarly, Chen and Hsu [3] decomposes the image into a background and a rain streak layer but based on low-rank priors. Li et al. [15] propose to learn from image patches using Gaussian mixture models for both the background and the rain layers. The patch corresponding to the rain allows the representation of multiple orientation and scale of the rain streaks. Zhu et al. [26] method identify rain-dominated regions to estimate the rain direction and then apply a joint layers optimization to iteratively modify the image.

In the last few years, deep learning approaches achieved remarkable results in the field [21,24]. The majority of them deals only with a single rain type, the rain streaks, and they are heavily based on synthetic images [7,21,24]. More recently, some progress has been made in considering deraining in a real context and connecting it with high-level computer vision tasks [14]. Deng et al. [6] proposed a method for rain removal using detail recovery and evaluated the impact of the deraining in a vision application, but it only evaluated on 30 high-quality real-world rainy images. Jiang et al. [11] also evaluated its proposal on a vision application, but on synthetic images. Yasarla et al. [23] demonstrate an interesting progress using real-world rainy images for training, but the method was only evaluated on high-quality real rainy images for restoration purposes.

2.2 Deraining Datasets

The single-image deraining field contains a broad set of synthetic rainy datasets and a few real-world rainy datasets. It is common to find new datasets accompanying a proposed deraining method in the literature. These characteristics are a challenge in terms of evaluating how the field has evolved over time. As with the progress in the methods, the deraining datasets has experimented a huge improvement on the quality of datasets ranging from datasets composed of a few images with simulated rain to large benchmarks comprising diverse rain types and annotated real-world rainy images for practical applications.

One of the pioneer public datasets in the field [15] consists in 12 images synthesized with photo-realistic rendering techniques. A larger set was subsequently proposed in [21], which included two synthetic sets for rain streaks, each one containing 100 images. Zhang et al. [25] synthesized rainy images from clean images in UCID [19] and BSD [1] datasets, totaling 800 clean-rainy image pairs. Additionally, they collected 50 real-world rainy images from the Internet. The datasets kept growing in size [10, 24]. Zhang and Patel [24] proposed a synthetic dataset consisting of 12,000 images categorized by their rain intensity and Hu et al. [10] proposed the *RainCityscapes* dataset, synthetizing 9,432 training images and 1,188 testing images from *Cityscapes* [5] simulating the effect of rain streaks and fog.

3 Proposed Approach

Image Quality and Rain Effect. The rain degradation may pose different impact in the high-level vision tasks according to the image quality. To better understand the impact of each degradation factor in a practical application, we analyze in isolation the image quality and the rain effect. For that, we categorize each image into low-quality (LQ) and high-quality (HQ) and into sunny, light rain and heavy rain to finally obtain the partitions of Fig. 1. State-of-the-art detection algorithms, Faster R-CNN [18] and RetinaNet [16], are applied over each partition. Faster R-CNN [18] is a two-stage detector that uses deep convolutional networks to propose regions. RetinaNet [16] is an one-stage detector that includes the focal loss to deal with the imbalance of classes. The detection results are evaluated considering the mean average precision (mAP). These results allow a comparison of the degradation impact of the quality and of the rain.

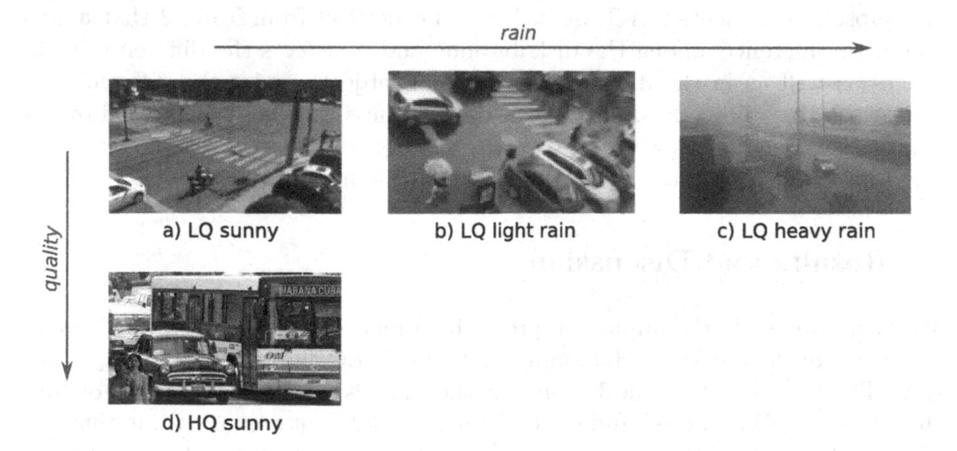

Fig. 1. Sample of the partitions of the dataset. Image may have low and high resolution (quality) and may present or not rain.

Table 1. Number of images in each subset. HQ sunny was sampled from [17], LQ sunny was obtained from personal collection, and LQ light and heavy rain were sampled from [13].

	Number of images
HQ sunny	1388
LQ sunny	2326
LQ light-rain	1230
LQ heavy-rain	818

Table 2. Number of *objects* in each subset.

	Car	Person	Truck	Bus	Motorcycle
HQ sunny	992	1416	330	231	110
LQ sunny	12804	1988	588	363	409
LQ light-rain	6682	1877	419	298	141
LQ heavy-rain	4733	810	254	190	134

Dataset. We created a dataset composed of 5,762 urban scene images, partitioned into 4 subsets according to the image quality and the rain intensity: HQ sunny, LQ sunny, LQ light rain, and LQ heavy rain. Table 1 shows statistics for each subset. These images were originated from COCO [17] datasets, MPID [14] datasets and also from public surveillance cameras [20]. The MPID [14] have a wide variety of images, and for the purposes of this work we categorized them according to the rain intensity. Besides, we manually annotated the images from the public surveillance cameras. The objects considered were car, person, bus, truck, and motorcycle and the distribution of the number of these objects across the subsets is presented in Table 2. It can be noticed from Table 2 that a considerable difference among the distributions and it reflects the difference in the scenes as well as in the distributions of these objects under the different rain scenarios. To avoid confusion between the darkness caused by night and by the rain, night-time images were filtered out. Cloudy images were also not included due to an analogous reason between sunny and light rain images.

4 Results and Discussion

We first investigate the impact on pre-trained models on scenarios with increasing levels of degradation: HQ sunny, LQ sunny, LQ light rain, and LQ heavy rain. The rows in the Table 3 represent the models (trained on COCO or fine-tuned on the LQ datasets) and the columns, the different scenarios of evaluation. The results are divided into two blocks, each representing an object detection method [16,18].

We can observe that the model's performance decreases as the scenarios increase in difficulty. For instance, the Faster-RCNN model pre-trained on COCO achieves 70.67, 30.27, 23.42 and 10.95 on mAP, when evaluated on scenarios of increasing levels of degradation. Therefore, the low-quality and the distinct rain intensities contributed as expected to the impairment of the detection quality. We can observe examples of detected results in these scenarios in Fig. 2.

High-quality sunny.

Low-quality sunny.

Low-quality light rain.

Low-quality heavy rain

Fig. 2. The average and standard deviation of critical parameters: Region R4

Our results also show that the object detection accuracy is impacted by image quality and rain intensity in different ways. Fined-tuned models evaluated on respective scenarios achieved the best results as expected with similar values (67.96, 65.87, 67.44) regardless of the rain intensity. There is a minor variation between the scenarios and we believe this is due to the variation in the scenes between them. Even with the LQ sunny containing more images, the models were able to improve to similar performances for all scenarios with the RetinaNet achieving mAP values on the LQ heavy rain just 0.52 less than the LQ sunny.

A large gap was observed when comparing the results of pre-trained model on the high-quality scenario (79.89). These results indicate that the image quality is a more dominant factor in the task when compared to the rain intensity. The results of Faster-RCNN corroborate these findings. One can also observe a large gap when the models are finetuned on the LQ datasets with a huge decrease in the performance on the HQ sunny scenario. We believe that the image quality poses a hard challenge on the adaptation suffering from catastrophic forgetting [2,4] with the disruption of the previous weights learned for the high-quality sunny scenario.

Table 3. Detection results (mAP) for Faster R-CNN and RetinaNet pre-trained on COCO 2017 and fine-tuned (FT) on the low-quality datasets. The scenarios are ordered in increasing degradation.

		HQ sunny	LQ sunny	LQ light rain	LQ heavy rain
Faster R-CNN	Pre-train	**70.67**	30.27	23.42	10.95
	FT on LQ sunny	5.41	**54.79**	21.21	16.30
	FT on LQ light rain	4.14	25.34	**49.61**	31.59
	FT on LQ heavy rain	2.30	19.47	31.36	**50.94**
RetinaNet	Pre-train	**79.89**	33.52	26.84	15.43
	FT on LQ sunny	14.67	**67.96**	31.53	28.36
	FT on LQ light rain	11.77	44.84	**65.87**	46.99
	FT on LQ heavy rain	11.11	38.82	46.59	**67.44**

5 Conclusions

The single-image deraining field currently relies heavily on synthetic data for supervised learning and this simplifies the task, but does not accommodate the realness to solve the actual problem. In this paper we analyzed the impact of rain in a real scenario considering also the image quality factor. The results show that the pre-trained models suffer a decreasing performance degradation as more rain factors are found in the scene. The fine-tuning overcomes this performance degradation and the fine-tuned models achieve good and similar results for images with same quality, regardless of the weather scenario. However, the image quality poses a more challenging factor to the models, which is shown in this work by the large gap between the high-quality results and the low-quality ones. We believe the challenge in the rainy scenes is posed on finding representations robust to the weather change. The complexity involved in a real rainy scene, regarding the variety on rain types and intensities, makes a supervised approach costly due to the manual labelling involved. Therefore, unsupervised learning would be a potential research direction. An alternative and complementary procedure to this work would be apply domain adaptation techniques to compare the performances of the fine-tuned and the *adapted* models. In the future, we expect to propose an unsupervised domain adaptation approach that can cope with the degradation imposed by the rain and the quality.

Acknowledgements. This work was supported by FAPESP - Fundação de Amparo à Pesquisa do Estado de São Paulo (grants #15/22308-2, #2019/01077-3). This study was financed in part by the Coordenação de Aperfeiçoamento de Pessoal de Nível Superior - Brasil (CAPES) - Finance Code 001.

References

1. Arbelaez, P., Maire, M., Fowlkes, C., Malik, J.: Contour detection and hierarchical image segmentation. IEEE Trans. Pattern Anal. Mach. Intell. **33**(5), 898–916 (2010)
2. Borthakur, A., Einhorn, M., Dhawan, N.: Preventing catastrophic forgetting in an online learning setting
3. Chen, Y.L., Hsu, C.T.: A generalized low-rank appearance model for spatio-temporally correlated rain streaks. In: IEEE International Conference on Computer Vision, pp. 1968–1975 (2013)
4. Chen, Z., Liu, B.: Lifelong machine learning. Synth. Lect. Artif. Intell. Mach. Learn. **12**(3), 1–207 (2018)
5. Cordts, M., et al.: The cityscapes dataset for semantic urban scene understanding. In: Proceedings of the IEEE Conference on Computer Vision and Pattern Recognition, pp. 3213–3223 (2016)
6. Deng, S., et al.: Detail-recovery image deraining via context aggregation networks. In: Proceedings of the IEEE Conference on Computer Vision and Pattern Recognition, pp. 14560–14569 (2020)
7. Fu, X., Huang, J., Zeng, D., Huang, Y., Ding, X., Paisley, J.: Removing rain from single images via a deep detail network. In: IEEE Conference on Computer Vision and Pattern Recognition (2017)
8. Garg, K., Nayar, S.K.: Detection and removal of rain from videos. In: IEEE Conference on Computer Vision and Pattern Recognition (2004)
9. Garg, K., Nayar, S.K.: When does a camera see rain? In: IEEE International Conference on Computer Vision (2005)
10. Hu, X., Fu, C.W., Zhu, L., Heng, P.A.: Depth-attentional features for single-image rain removal. In: Proceedings of the IEEE Conference on Computer Vision and Pattern Recognition, pp. 8022–8031 (2019)
11. Jiang, K., et al.: Multi-scale progressive fusion network for single image deraining. In: Proceedings of the IEEE Conference on Computer Vision and Pattern Recognition, pp. 8346–8355 (2020)
12. Kang, L.W., Lin, C.W., Fu, Y.H.: Automatic single-image-based rain streaks removal via image decomposition. IEEE Trans. Image Process. **21**(4), 1742 (2012)
13. Li, B., et al.: Benchmarking single-image dehazing and beyond. IEEE Trans. Image Process. **28**(1), 492–505 (2019)
14. Li, S., et al.: Single image deraining: a comprehensive benchmark analysis. In: Proceedings of the IEEE Conference on Computer Vision and Pattern Recognition, pp. 3838–3847 (2019)
15. Li, Y., Tan, R.T., Guo, X., Lu, J., Brown, M.S.: Rain streak removal using layer priors. In: IEEE Conference on Computer Vision and Pattern Recognition, pp. 2736–2744 (2016)
16. Lin, T.Y., Goyal, P., Girshick, R., He, K., Dollár, P.: Focal loss for dense object detection. IEEE Trans. Pattern Anal. Mach. Intell. **39**, 2999–3007 (2018)
17. Lin, T.-Y., et al.: Microsoft COCO: common objects in context. In: Fleet, D., Pajdla, T., Schiele, B., Tuytelaars, T. (eds.) ECCV 2014. LNCS, vol. 8693, pp. 740–755. Springer, Cham (2014). https://doi.org/10.1007/978-3-319-10602-1_48
18. Ren, S., He, K., Girshick, R., Sun, J.: Faster R-CNN: towards real-time object detection with region proposal networks. In: Advances in Neural Information Processing Systems, pp. 91–99 (2015)

19. Schaefer, G., Stich, M.: UCID: an uncompressed color image database. In: Storage and Retrieval Methods and Applications for Multimedia 2004, vol. 5307, pp. 472–480. International Society for Optics and Photonics (2003)
20. Tokuda, E.K., Ferreira, G.B., Silva, C., Cesar, R.M.: A novel semi-supervised detection approach with weak annotation. In: 2018 IEEE Southwest Symposium on Image Analysis and Interpretation (SSIAI), pp. 129–132. IEEE (2018)
21. Yang, W., Tan, R.T., Feng, J., Liu, J., Guo, Z., Yan, S.: Deep joint rain detection and removal from a single image. In: IEEE Conference on Computer Vision and Pattern Recognition (2017)
22. Yang, W., Tan, R.T., Wang, S., Fang, Y., Liu, J.: Single image deraining: from model-based to data-driven and beyond. IEEE Trans. Pattern Anal. Mach. Intell. (2020)
23. Yasarla, R., Sindagi, V.A., Patel, V.M.: Syn2Real transfer learning for image deraining using gaussian processes. In: Proceedings of the IEEE Conference on Computer Vision and Pattern Recognition, pp. 2726–2736 (2020)
24. Zhang, H., Patel, V.M.: Density-aware single image de-raining using a multi-stream dense network. In: IEEE Conference on Computer Vision and Pattern Recognition (2018)
25. Zhang, H., Sindagi, V., Patel, V.M.: Image de-raining using a conditional generative adversarial network. arXiv preprint arXiv:1701.05957 (2017)
26. Zhu, L., Fu, C.W., Lischinski, D., Heng, P.A.: Joint bilayer optimization for single-image rain streak removal. In: IEEE International Conference on Computer Vision (2017)

Hard Occlusions in Visual Object Tracking

Thijs P. Kuipers(✉), Devanshu Arya(✉), and Deepak K. Gupta(✉)

Informatics Institute, University of Amsterdam, Amsterdam, The Netherlands
kuiperthijs@gmail.com, d.arya@ua.ul, gupta.deepak@gmail.com

Abstract. Visual object tracking is among the hardest problems in computer vision, as trackers have to deal with many challenging circumstances such as illumination changes, fast motion, occlusion, among others. A tracker is assessed to be good or not based on its performance on the recent tracking datasets, *e.g.*, VOT2019, and LaSOT. We argue that while the recent datasets contain large sets of annotated videos that to some extent provide a large bandwidth for training data, the hard scenarios such as occlusion and in-plane rotation are still underrepresented. For trackers to be brought closer to the real-world scenarios and deployed in safety-critical devices, even the rarest hard scenarios must be properly addressed. In this paper, we particularly focus on hard occlusion cases and benchmark the performance of recent state-of-the-art trackers (SOTA) on them. We created a small-scale dataset (Dataset can be accessed at https://github.com/ThijsKuipers1995/HTB2020) containing different categories within hard occlusions, on which the selected trackers are evaluated. Results show that hard occlusions remain a very challenging problem for SOTA trackers. Furthermore, it is observed that tracker performance varies wildly between different categories of hard occlusions, where a top-performing tracker on one category performs significantly worse on a different category. The varying nature of tracker performance based on specific categories suggests that the common tracker rankings using averaged single performance scores are not adequate to gauge tracker performance in real-world scenarios.

Keywords: Visual object tracking · Occlusion · Benchmarks · Metrics · Dataset

1 Introduction

Visual object tracking remains a challenging problem, even though it has been studied for several decades. A visual object tracker has to account for many different and varying circumstances. For instance, changes in the illumination may alter the appearance of the target object. The object could also blend in with the background environment, or it might get occluded, resulting in the object, or part of the object, being obstructed from view. Because of all the possible

A. Bartoli and A. Fusiello (Eds.): ECCV 2020 Workshops, LNCS 12539, pp. 299–314, 2020.
https://doi.org/10.1007/978-3-030-68238-5_22

different circumstances, visual object trackers have to account for, visual tracking is considered a hard problem [1].

Commonly, a tracker is assessed to be good or not based on its performance on established benchmarking datasets such as OTB100 [2], UAV123 [3], VOT2019 [4], GOT-10k [5], LaSOT [6], and TrackingNet [7]. These datasets comprise large sets of video sequences spanning across different challenges of tracking. For every dataset, the performance of a tracker is averaged over all sequences.

For a tracker to be used in real-world scenarios and embedded in safety-critical equipment, it should be able to handle even the rarest and hardest instances of tracking. Therefore, the evaluation datasets should also contain such instances. We argue that this is not yet the case, and scenarios such as occlusion, in-plane rotation, and out-of-plane rotation are still underrepresented in these datasets. Moreover, most performance metrics compute an average score across all sequences, thereby overshadowing the poor performance of the subjected tracker on a certain specific challenge. For a deeper study on these aspects, we tailor the focus of this paper to only hard occlusions.

Occlusion refers to the phenomenon where parts of the target object are blocked from the field of view. For instance, an object can be occluded when it either moves partially or fully out of frame. Occlusion also occurs when another object fully or partially blocks the target object. When the target object is partially blocked, either certain specific features of the target object can disappear, or part of the entire target object appearance will disappear. Unlike other challenges of tracking, learning a distribution for occlusion is hard - no distribution exists in parts of the object that are occluded. This makes occlusion a hard problem. Some methods for handling occlusion do exist, but they are often focused on very specific aspects of tracking such as solely tracking pedestrians [8]. A major problem in evaluating visual object trackers on occlusion is the lack of data containing hard occlusions in current datasets. While the above-mentioned datasets do contain samples of occlusion, they often do not represent the challenging cases of occlusion that can occur when tracking in the wild [2,6,7]. Therefore, the available benchmarks might not accurately evaluate tracker performance on hard occlusions.

This work aims to evaluate a set of current state-of-the-art (SOTA) visual object trackers on the occurrences of hard occlusions. To perform this evaluation, a small dataset containing various samples of hard occlusions is compiled. Our preliminary results show that the performance of the SOTA trackers is, in general, lower on the hard occlusion scenarios. Further, we analyze whether the leading tracker among the ones chosen in this study performs superior on different scenarios of occlusion. Our results reveal interesting insights on whether even the best tracker, decided based on the existing evaluation strategies, could be considered a safe choice for deployment in real-world scenarios, especially for safety-critical devices. Further, it raises the question of whether the current performance metrics, averaging the score over all the sequences, are the right choice to assess the performance of trackers.

2 Related Work

In the following, we present an overview of some previous works that are relevant to this study. First, we present an overview of the recent visual object tracking algorithms, followed by works related to tackling occlusion in tracking. Finally, an overview of different tracking benchmarks is presented.

2.1 Object Tracking

The task of object tracking can refer to tracking multiple objects simultaneously (multi-object tracking) or tracking a single instance of an object (single-object tracking) [7]. This work will only consider *generic* single-object trackers. To perform successful tracking, trackers must be able to create a strong appearance model for the target object and be capable of reliable and fast localization of the target object.

To address the above-mentioned challenge, various methods have been proposed. One such category is Correlation Filters (CF), which forms the basis of several SOTA single-object trackers [7]. CF uses circular correlation, which results in all shifted variants of input samples being implicitly included. This enables the construction of a powerful appearance model with very limited data [9]. Furthermore, CF allows for fast run-times as computations are performed in the Fourier domain [10]. The MOSSE [11] tracker paved the way for more advanced CF-based approaches such as the use of multi-dimensional features [12–14], improved robustness to variations in scale and deformations [15,16], mitigating boundary effects [9,17], and the use of deep convolutional filters [18,19]. The advancements made in CF trackers resulted in large and complex models, which significantly increases the computational cost. ECO [20] improves upon the CF-framework by reducing the model complexity by proposing a factorized convolution operation to increase running speeds.

Another category is deep learning-based trackers. Recurrent Neural Networks have been proposed for tracking [21,22], but do not yield competitive performance compared to the SOTA trackers. MDNet [23] implements a deep convolutional network that is trained offline, and performs Stochastic Gradient Descent during tracking but is not able to operate in real-time. GOTURN [24] utilizes a convolutional network to learn a function between image pairs. SiamFC [25] introduces a fully-connected Siamese architecture to perform general similarity learning. The goal is to learn a similarity function offline, which can then be evaluated during tracking to locate an exemplar image within a larger candidate image. SiamRPN [26] implements a Siamese network that is extended with a Region Proposal Network (RPN) which allows for improved bounding-box predictions. SiamRPN++ [27] improves upon the Siamese architecture by enabling the use of deep networks in Siamese trackers. ATOM [28] improves upon the Siamese trackers by introducing an approach that consists of a target estimation module that is learned offline, and a target classification module that is learned online. DiMP [29] consists of a discriminative model prediction architecture derived from a discriminative learning loss for visual tracking which can fully exploit background information.

2.2 Occlusion in Tracking

Occlusion remains largely unexplored in generic single-target object tracking. Some tracking architectures that do explicitly handle occlusion have been proposed. ROT [30] utilizes occlusion-aware real-time object tracking by overcoming target model decay which can occur when the target object is being occluded. In [31] SiamRPN++ and SiamFC are equipped with structured dropouts to handle occlusion. Other methods include more experimental strategies such as analyzing the occurrence of occlusion by utilizing spatiotemporal context information [32]. By further analyzing motion constraints and the target reference the strategy allows for better discrimination between the target object and background distractors causing occlusion. Another experimental strategy uses layer-based strategies, extending it by specific background occluding layers [33]. Strategies that focus on handling occlusion have also been proposed in more specific object tracking tasks, such as tracking pedestrians [8] and cars [34]. In [13] a multi-object tracker approach that handles occlusion is proposed, which is built on the idea of object permanence, using a region-level association and object-level localization process to handle long periods of occlusion. In [35] alternative SOTA methods for handling occlusion are presented such as depth-analysis [36,37] and fusion methods such as a Kalman filter for predicting target object motion and location [38,39].

2.3 Tracking Datasets

To evaluate the robustness of single-target visual object trackers, many datasets have been proposed. ALOV300 [1] contains 314 short sequences. ALOV300 does include 14 different attributes, including occlusion. However, it does not differentiate between different kinds of occlusion. OTB [2] is another well-known dataset. The full dataset (OTB100) contains 98 sequences, while OTB50 is a subset of OTB100 containing 51 of the most challenging sequences. OTB offers 11 attributes, including both partial and full occlusion. Since the rise of deep trackers, the demand for large-scale datasets has increased. TrackingNet [7] is introduced to accommodate these demands. TrackingNet consists of over 30 thousand sequences with varying frame rates, resolutions, and lengths. TrackingNet includes 15 attributes, including both partial and full occlusion, as well as out-of-frame occlusion. GOT-10k [5] consists of over 10 thousand sequences. GOT-10k offers an impressive 563 different object classes, and offers a train and test-set with zero overlaps between classes, resulting in more accurate evaluations. GOT-10k offers several attributes, including occlusion. Many of the sequences contained in the above-mentioned datasets are of relatively short duration. However, when tracking in the wild, tracking often occurs for sustained periods. LaSOT [6] introduces a dataset consisting of 1400 longer duration sequences with an average sequence length of over 2500 frames. Similarly to the previously mentioned datasets, LaSOT includes a variety of attributes, including both partial and full occlusion, and out-of-frame occlusion. UAV123 [3] contains 123 videos captured from a UAV, as well as artificial sequences generated by a flight-simulator.

3 Benchmarking on Hard Occlusions

While most datasets described in Sect. 2.3 contain cases of occlusion, the chosen instances are still very simple and do not account for the hard scenarios. These datasets do not take specific sub-categories within occlusion into account, as often solely general cases of occlusion, such as partial or full, are considered. Furthermore, many of the sequences contained in the datasets are of relatively short duration. As a result, occlusion often occurs for only short amounts of time. These short durations are not enough to accurately assess tracker performance on occlusion. Another issue is that often the occlusions that do occur involve simple cases. The occluded target object often possesses barely any movement relative to the camera, or the target object remains stationary throughout the sequence (see Fig. 1a). The challenging LaSOT [6] dataset does contain more challenging cases of occlusion, including longer durations and more extreme movement of the target object (see Fig. 1b). However, the set of sequences containing these hard occlusions remains very limited.

Occlusion examples in OTB.

Occlusion examples in LaSOT.

Fig. 1. Examples of occlusion from the OTB and LaSOT datasets.

Here, we present our Hard Occlusion Benchmark[1] (HOB), a small-scale dataset containing 20 long-duration sequences that encompass a variety of different hard occlusion scenarios. For the sake of demonstration, Fig. 2 shows the first frame of some of the sequences with the corresponding ground-truth. Each sequence is of similar length, with an average of 2760 frames per sequence. Each sequence in HOB is annotated every 15th frame. Despite the lack of fully annotated sequences, with an average of 185 annotated frames per sequence, there exist ample ground-truths to perform an accurate evaluation. Naturally, HOB contains the general cases of hard occlusion, such as *partial occlusion*, *full occlusion*, and *out-of-frame occlusion*. The cases of occlusion occur for long periods and are combined with strong movement and scale-variations of the target object

[1] The dataset can be accessed at https://github.com/ThijsKuipers1995/HTB2020.

relative to the camera. Also, these general cases are complemented with more specific attributes, to obtain a more precise evaluation of the SOTA tracker implementations on hard occlusions.

Fig. 2. First frame of the sequences in HOB with ground-truth.

In its current form, HOB dataset comprises the following occlusion types.

- *Full out of frame occlusion (FOC).* The target object moves fully out of the frame for extended periods. The target object may enter the frame at a different location compared to where it exited the frame.
- *Feature occlusion (FO).* Some specific features of the target are omitted from view. It is still possible for the entire target object to be in view in this case.
- *Occlusion by transparent object (OCT).* The target object is being occluded by a transparent object. This means the target object can still be visible through the occluding object, although the occluding object does alter the appearance of the target object.
- *Occlusion by similar object (OCS).* The target object is being occluded by a similar-looking object.

4 Experiments and Evaluations

The current section introduces the set of trackers that are evaluated. Next, an overview of the metrics used for evaluation is given. Finally, the performance of the selected set of trackers is evaluated. The selected set of trackers aims to cover a variety of common state-of-the-art (SOTA) tracking principles. The following trackers are chosen for the evaluation: ECO [20], SiamFC [25], SiamRPN++ [27], ATOM [28], DiMP [29]. ECO proposes a general framework to improve upon the discriminant correlation filter tracking. SiamFC proposed similarity matching by using a fully connected Siamese network. SiamRPN++ uses a deep network for more sophisticated features, as well as a region proposal network. Three variants of SiamRPN++ are evaluated: the original version using the ResNet50 (r50) [40] (r50) backbone, a version using the shallow AlexNet [41] backbone, and the long-term version which uses the long-term update strategy as described in [42]. ATOM proposes a tracking architecture consisting of dedicated target estimation and target classification components. DiMP [29] proposes a tracking architecture that can utilize both target and background information. Both ATOM and DiMP utilize a memory model to update the appearance model to take into account the change of appearance over time. Solely trackers with publicly available code are used in this work.

4.1 Evaluation Methodology

We perform a One Pass Evaluation (OPE) measuring precision, success rate, area-under-curve, and the least-subsequence-metric [43] on the 20 hard occlusion sequences in HOB. A brief overview of these metrics is presented below.

Precision. When tracking precision, the center localization error is calculated. The center localization error is defined as the Euclidean distance between the center of the ground-truth bounding box and the prediction bounding box. A frame is deemed successfully tracked if the calculated distance is below a certain threshold, say t. While precision does not take the predicted bounding box size into account, it does correctly measure how close the position of the prediction is to the ground truth, which is not always the case when using success rates, as only the overlap between prediction and ground truth is considered. In the case of occlusion, where the target object is not entirely visible, precision can depict to what extend the tracker manages to correctly predict the location of the occluded target object. The issue with using precision as a metric is its sensitiveness to resolution, which in the case of HOB is avoided since every sequence is of the same resolution. The final precision score for each tracker is evaluated using a threshold of 20 pixels such as in [44].

Success Rate. The success rate makes use of the average bounding box overlap, which measures performance by calculating the overlap between ground-truth bounding boxes and prediction bounding boxes. It takes both position accuracy and accuracy of the predicted bounding box size and shape into account. Therefore, the success rate can offer a solid measurement of tracker performance. Bounding box overlap is calculated using the intersection-over-union (IoU) score. Similar to precision, a frame is deemed successfully tracked when the calculated IoU meets a certain threshold, say t. By calculating the success rate at a range of different thresholds, a success plot can be formed. A threshold of $t > 0.5$ is often used to measure success. However, this does not necessarily represent a successfully tracked frame [2]. Because of this, the area-under-curve (AuC) of the success plot is calculated instead, which takes the entire range of thresholds into account. Furthermore, frames in which the absence of the target object is correctly predicted are given an IoU score of 1.

Least Subsequence Metric. The least-subsequence-metric (LSM) quantifies long term tracking behavior by computing the ratio between the length of the longest continuously "successfully" tracked sequence of frames and the full length of the sequence. A sequence of frames is deemed as successfully tracked if at least a certain percentage p of frames within this is successfully tracked [43]. The representative LSM score is calculated at a threshold of $p = 95\%$ as in [43]. A frame is considered correctly tracked when the IoU of that frame is greater than 0.5. Because LSM calculates the ratio between the longest continuously tracked subsequence and the length of the entire sequence, it can introduce a bias towards extremely long and short sequences. However, all sequences used in this work are of similar length, therefore this is not an issue for accurate evaluation.

4.2 Baseline Dataset

For the sake of comparison with HOB, we use LaSOT as the baseline dataset. LaSOT is a large benchmark that focuses on long-term tracking, and it includes many sequences containing occlusions and out-of-frame occlusions. Due to this reason, it is one of the more difficult tracking benchmarks. Evaluating the selected visual object trackers on LaSOT will, therefore, offer a great baseline for comparing to HOB. HOB is a relatively small dataset containing only 20 sequences. To keep the comparison between HOB and LaSOT fair, only the top 20 occlusion heavy sequences from LaSOT are selected. Furthermore, while LaSOT offers per-frame annotations of ground-truths, HOB contains a ground-truth annotation every 15th frame. Therefore only every 15th frame of LaSOT will be used during the evaluation procedure.

4.3 Overall Performance

Figure 4 and Fig. 5 depicts the predictions for each of the evaluated trackers on four sequences corresponding to the mentioned attributes.

Figure 3 shows the precision, success rate, and LSM of each evaluated tracker on both HOB and LaSOT. In Table 1, the representative scores of each of the metrics are shown. The results show that on average, the performance of the evaluated trackers is worse on HOB compared to LaSOT. SiamRPN++(r50) is the top-performing tracker on HOB on all metrics. SiamRPN++(r50) outperforms SiamRPN++(lt) by a small margin on the AuC metric, and similar observations can be made for the other two metrics as well. This is an interesting

Table 1. Representative scores for precision (t = 20), area under curve (AuC) and LSM (x = 0.95) scores for each of the evaluated trackers on HOB and LaSOT. Best scores are shown in bold.

Dataset	Precision		AuC		LSM	
	HOB	LaSOT	HOB	LaSOT	HOB	LaSOT
ATOM	0.142	0.342	0.243	0.317	0.122	0.228
DiMP	0.173	0.421	0.324	**0.391**	0.126	**0.292**
ECO	0.070	0.191	0.149	0.199	0.098	0.166
SiamFC	0.093	0.225	0.205	0.191	0.090	0.135
SiamRPN++ (alex)	0.154	0.320	0.300	0.300	0.125	0.185
SiamRPN++ (lt)	0.192	**0.437**	0.343	0.383	0.127	0.248
SiamRPN++ (r50)	**0.195**	0.318	**0.359**	0.278	**0.133**	0.245

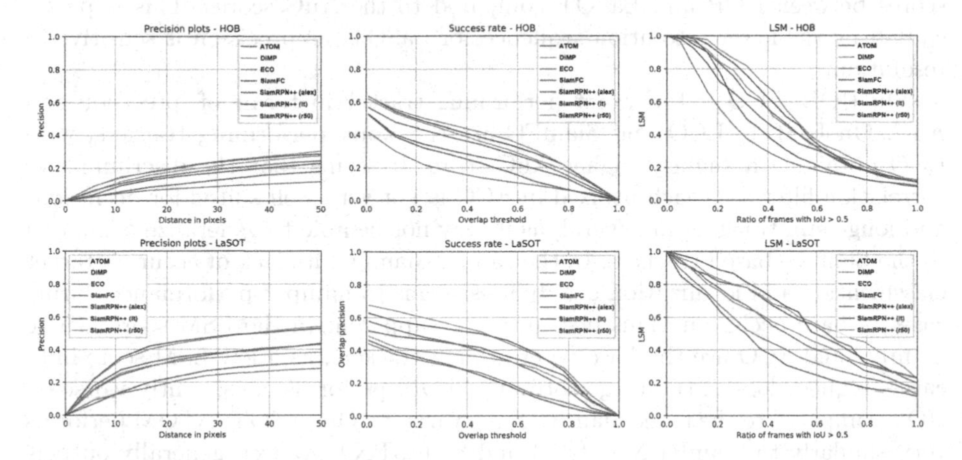

Fig. 3. Overall results on HOB (top) and LaSOT (bottom) on the precision rate, success rate, and LSM (left, middle, and right respectively).

result, as SiamRPN++(lt) is specifically tailored to handling long-term tracking which includes occlusion of the target object. These results imply that even the re-detection module of SiamRPN++(lt) can occasionally drift the tracker model to false targets. This could be attributed to SiamRPN++(lt) re-detecting the wrong target object and sticking to it during long and heavy stretches of occlusion, which would result in lower overall performance. Contrary to the results obtained on HOB, performance on LaSOT seems significantly different. On LaSOT, DiMP is the top-performing tracker on the AuC and LSM metrics, and second-best on precision. Only SiamRPN++(lt) shows comparative performance, and as shown in Table 1. The performance of the remaining trackers is significantly lower.

It is interesting to note that DiMP consistently underperforms compared to both SiamRPN++(r50) and SiamRPN++(lt) on HOB. While HOB dataset contains only 20 sequences, the sequences of HOB have been chosen with no intended bias towards Siamese and SiamRPN-type trackers. Thus, we would ideally expect DiMP to perform the best, as has been seen on LaSOT. One possible reason for the reduced performance on HOB could be attributed to the fact that the training set of DiMP comprised the training set of LaSOT sequences as well. This could mean that DiMP tends to overfit on objects similar to those observed in the LaSOT training set. Since most other datasets contain similar tracking instances, DiMP performs well on them. On the contrary, the scenarios observed in HOB are quite different, and that leads to reduced performance on this dataset. Another possible reason for the performance decay of DiMP on HOB could be attributed to the bias added to the tracking model due to frequent model updates happening even under the scenarios of occlusion [45]. This is not the case for the SiamRPN++ variants, as they do not perform any model update. Note that Table 1 shows a relatively large difference in precision scores between HOB and LaSOT compared to the AuC scores. This is partly caused by the lower resolution sequences of LaSOT, as precision is sensitive to resolution.

On HOB, ECO is the worst performing tracker in terms of, precision and AuC. On LaSOT, ECO and SiamFC are the worst-performing trackers, with ECO obtaining a slightly higher AuC score. It seems that the discriminative correlation filter approach utilized in ECO is not very well suited for occlusions and long-term tracking in general, as it may not be able to generalize compared to the Siamese based trackers. In the case of SiamFC, its lack of accurate target classification and localization capabilities seems to hamper performance during cases of hard occlusion. This becomes more apparent in the LSM score, where SiamFC and ECO are the lowest-performing trackers, as a low LSM score indicates frequent loss of the target object. ATOM performs consistently worse on HOB compared to the three SiamRPN++ variants. On LaSOT, ATOM performs very similarly to SiamRPN++(r50) and SiamRPN++(alex), generally outperforming them by a slight margin. ATOM also utilizes a model update strategy, which could result in the decay of the appearance model during cases of hard occlusion.

4.4 Attribute Evaluation

While the results from the previous section have shown that most trackers struggle in the presence of hard occlusions, it is of interest to analyze further how different occlusion types affect the overall performance. In this section, we study the trackers for the different categories of occlusion that we have defined earlier. The success plots for each of the categories are shown in Fig. 4. Figure 5 depicts the predictions for each of the evaluated trackers on sequences corresponding to each of the categories.

Full Out-of-Frame Occlusion (FOC). FOC seems to be a very challenging problem for the visual object trackers, with SiamRPN++(lt) being the top-performing tracker in this category. This is most likely attributed to its re-

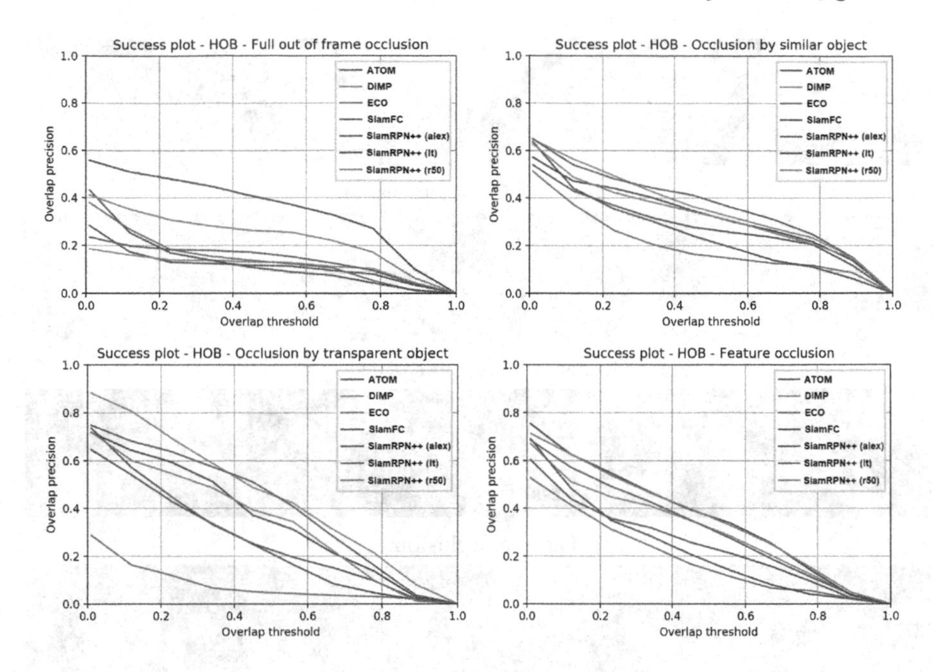

Fig. 4. Success plots for full out of frame occlusion (top left), occlusion by similar object (top right), occlusion by transparent object (bottom left), and feature occlusion (bottom right).

initialization strategy when detecting target object loss. The second-best performing tracker on FOC is SiamRPN++(r50), performing considerably better compared to the rest. Having access to rich features at different scales seems to aid its re-detection capabilities when the target object moves within the localization area. When observing the predictions, only SiamRPN++(lt) seems to consistently be able to re-detect the target object (see Fig. 5a). Overall, the SiamRPN++ variants outperform the other evaluated trackers. DiMP performs considerably worse during FOC. Interestingly, even trackers with weak discriminative power, such as SiamFC and ECO, perform on par with DiMP and ATOM for cases of FOC. ATOM, DiMP, and ECO update their appearance model during tracking. In the case of FOC, this could result in the appearance updating on samples that do not contain the target object causing strong appearance model decay. This is not the case for the Siamese trackers, as their appearance model remains fixed during tracking.

Occlusion by Similar Object (OCS). In the case of OCS, SiamRPN++(alex) has the highest overall performance, while SiamRPN++(r50) performs the worst of the SiamRPN++ trackers. Interestingly, the use of the shallow AlexNet as a backbone results in better performance compared to using the deep ResNet, even outperforming the long-term SiamRPN++ variant. Re-initialising on target loss does not offer an advantage during OCS, as the performance of SiamRPN++(lt) is similar to the performance of SiamRPN++(alex). DiMP is the second-best

Full out of frame occlusion.

Occlusion by similar object.

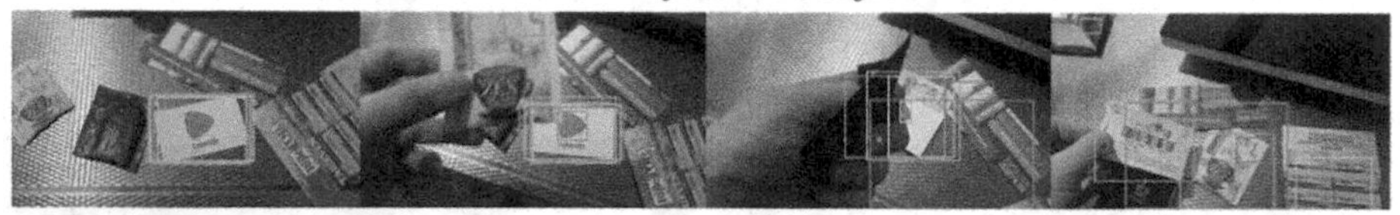

Feature occlusion.

Occlusion by transparent object.

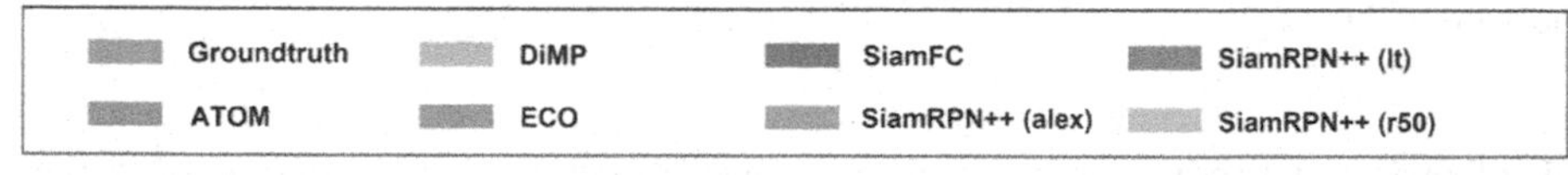

Fig. 5. Depiction of four frames including prediction and groundtruth bounding-boxes for each the categories *full out-of-frame occlusion*, *occlusion by similar object*, *feature occlusion*, and *occlusion by transparent object.*

performing tracker, with ATOM, SiamFC and, ECO being the lowest-performing trackers. ECO performs considerably lower compared to the other trackers. When observing the predictions during OCS, trackers struggle to accurately keep track of the target object (see Fig. 5b).

Feature Occlusion (FO). During FO, SiamRPN++(r50) and SiamRPN++(lt) are the top performing trackers, with near-identical performance. As objects tend to stay at least partially visible in this category, the re-initialization strategy of SiamRPN++(lt) does not offer much benefit in tracking performance. SiamRPN++(lt) and SiamRPN++(r50) are closely followed by SiamRPN++-(alex) and DiMP. Once again, ECO is the worst performing tracker. The results of the FO category are very similar to the overall performance on occlusion as shown in Fig. 3, although on average the trackers seem to perform slightly worse on feature occlusion specifically at higher thresholds.

Occlusion by Transparent Object (OCT). DiMP is the best performing tracker on OCT, SiamRPN++(alex) being a very close second. Both DiMP and SiamRPN++(alex) perform considerably better than to the other SiamRPN++ variants, similar to the OCS category. SiamRPN++(r50) and SiamRPN++(lt) have very similar performance. In the cases of OCS and OCT, it seems that the ability to generalize for objects that are not seen during training could play an important role. The appearance model predictor implemented in DiMP contains few parameters, leading to better generalization as less overfitting to observed classes occurs during the offline training phase [29]. Likewise, SiamRPN++(alex) using AlexNet contains less parameters compared to SiamRPN++(r50) [27]. The performance of ATOM is considerably lower compared to DiMP, while both use the same IoU maximization based architecture for updating the appearance model, suggesting the appearance model update is of less importance during OCT and OCS. It is interesting to note that when observing the predictions on OCT and FE, DiMP and ATOM tend to strongly focus on striking target object features, as can be observed in Fig. 5c and Fig. 5d.

5 Conclusion

In this work, we presented an evaluation of the current state-of-the-art (SOTA) visual object trackers on hard occlusions. We compiled a small dataset containing sequences that encompass several samples of hard occlusions to assess the performance of these trackers in such occluded scenarios. Furthermore, we evaluated the trackers on a subset of the most occlusion-heavy LaSOT sequences. From the results, we show that on average the trackers perform worse on hard occlusion scenarios, suggesting that occlusion is still a relatively unsolved problem in tracking. While DiMP is the best performing tracker on the LaSOT benchmark, it is consistently outperformed by SiamRPN++ using the ResNet backbone architecture (r50) and its long-term tracking variant (lt) on instances of hard occlusions. Furthermore, we show that the top-performing tracker can vary drastically between specific scenarios of hard occlusions. For example, while DiMP seems the best for handling occlusions caused by semi-transparent objects, it performs the worst for full out-of-frame occlusion scenarios.

The set of results presented in this paper hint towards the fact that even the best performing tracker based on the current benchmark datasets might not be suited for real-world deployment, especially in safety-critical applications, such as self-driving cars. Real-world problems do not promise the presence of a uniform set of challenges, and at any random instance, a different tracking challenge could be the most important. Correspondingly, we focused on the challenge of hard occlusions in this paper, and trackers behaved differently than they did on LaSOT. This implies two important things for future research. First, tracking datasets need to incorporate more instances of difficult tracking challenges. Second, evaluation methodologies need to be designed that give more importance to instances where a certain tracker performs the worst. To summarize on a high-level, a model that handles even the most difficult challenges of tracking sufficiently well should be considered a better visual object tracker.

References

1. Smeulders, A.W., Chu, D.M., Cucchiara, R., Calderara, S., Dehghan, A., Shah, M.: Visual tracking: an experimental survey. IEEE Trans. Pattern Anal. Mach. Intell. **36**(7), 1442–1468 (2013)
2. Wu, Y., Lim, J., Yang, M.H.: Object tracking benchmark. IEEE Trans. Pattern Anal. Mach. Intell. **37**, 1 (2015)
3. Mueller, M., Smith, N., Ghanem, B.: A benchmark and simulator for UAV tracking. In: Leibe, B., Matas, J., Sebe, N., Welling, M. (eds.) ECCV 2016. LNCS, vol. 9905, pp. 445–461. Springer, Cham (2016). https://doi.org/10.1007/978-3-319-46448-0_27
4. Kristan, M., et al.: The seventh visual object tracking vot2019 challenge results (2019)
5. Huang, L., Zhao, X., Huang, K.: GOT-10k: a large high-diversity benchmark for generic object tracking in the wild. IEEE Trans. Pattern Anal. Mach. Intell. (2019)
6. Fan, H., et al.: LaSOT: a high-quality benchmark for large-scale single object tracking. In: Proceedings of the IEEE Conference on Computer Vision and Pattern Recognition, pp. 5374–5383 (2019)
7. Muller, M., Bibi, A., Giancola, S., Alsubaihi, S., Ghanem, B.: TrackingNet: a large-scale dataset and benchmark for object tracking in the wild. In: Proceedings of the European Conference on Computer Vision (ECCV), pp. 300–317 (2018)
8. Noh, J., Lee, S., Kim, B., Kim, G.: Improving occlusion and hard negative handling for single-stage pedestrian detectors. In: Proceedings of the IEEE Conference on Computer Vision and Pattern Recognition, pp. 966–974 (2018)
9. Danelljan, M., Hager, G., Shahbaz Khan, F., Felsberg, M.: Learning spatially regularized correlation filters for visual tracking. In: Proceedings of the IEEE International Conference on Computer Vision, pp. 4310–4318 (2015)
10. Bolme, D.S., Beveridge, J.R., Draper, B.A., Lui, Y.M.: Visual object tracking using adaptive correlation filters. In: 2010 IEEE Computer Society Conference on Computer Vision and Pattern Recognition, pp. 2544–2550. IEEE (2010)
11. Bolme, D.S., Beveridge, J.R., Draper, B.A., Lui, Y.M.: Visual object tracking using adaptive correlation filters (2010)
12. Danelljan, M., Shahbaz Khan, F., Felsberg, M., Van de Weijer, J.: Adaptive color attributes for real-time visual tracking. In: Proceedings of the IEEE Conference on Computer Vision and Pattern Recognition, pp. 1090–1097 (2014)
13. Huang, Y., Essa, I.: Tracking multiple objects through occlusions. In: 2005 IEEE Computer Society Conference on Computer Vision and Pattern Recognition (CVPR 2005), vol. 2, pp. 1051–1058. IEEE (2005)
14. Henriques, J.F., Caseiro, R., Martins, P., Batista, J.: High-speed tracking with kernelized correlation filters. IEEE Trans. Pattern Anal. Mach. Intell. **37**(3), 583–596 (2014)
15. Li, Y., Zhu, J.: A scale adaptive kernel correlation filter tracker with feature integration. In: Agapito, L., Bronstein, M.M., Rother, C. (eds.) ECCV 2014. LNCS, vol. 8926, pp. 254–265. Springer, Cham (2015). https://doi.org/10.1007/978-3-319-16181-5_18
16. Danelljan, M., Häger, G., Khan, F., Felsberg, M.: Accurate scale estimation for robust visual tracking. In: British Machine Vision Conference, Nottingham, September 1–5, 2014. BMVA Press (2014)
17. Kiani Galoogahi, H., Sim, T., Lucey, S.: Correlation filters with limited boundaries. In: Proceedings of the IEEE Conference on Computer Vision and Pattern Recognition, pp. 4630–4638 (2015)

18. Danelljan, M., Robinson, A., Shahbaz Khan, F., Felsberg, M.: Beyond correlation filters: learning continuous convolution operators for visual tracking. In: Leibe, B., Matas, J., Sebe, N., Welling, M. (eds.) ECCV 2016. LNCS, vol. 9909, pp. 472–488. Springer, Cham (2016). https://doi.org/10.1007/978-3-319-46454-1_29
19. Ma, C., Huang, J.B., Yang, X., Yang, M.H.: Hierarchical convolutional features for visual tracking. In: Proceedings of the IEEE International Conference on Computer Vision, pp. 3074–3082 (2015)
20. Danelljan, M., Bhat, G., Shahbaz Khan, F., Felsberg, M.: ECO: efficient convolution operators for tracking. In: Proceedings of the IEEE Conference on Computer Vision and Pattern Recognition, pp. 6638–6646 (2017)
21. Gan, Q., Guo, Q., Zhang, Z., Cho, K.: First step toward model-free, anonymous object tracking with recurrent neural networks. arXiv preprint arXiv:1511.06425 (2015)
22. Kahou, S.E., Michalski, V., Memisevic, R., Pal, C., Vincent, P.: RATM: recurrent attentive tracking model. In: 2017 IEEE Conference on Computer Vision and Pattern Recognition Workshops (CVPRW), pp. 1613–1622. IEEE (2017)
23. Nam, H., Han, B.: Learning multi-domain convolutional neural networks for visual tracking. In: Proceedings of the IEEE Conference on Computer Vision and Pattern Recognition, pp. 4293–4302 (2016)
24. Held, D., Thrun, S., Savarese, S.: Learning to track at 100 FPS with deep regression networks. In: Leibe, B., Matas, J., Sebe, N., Welling, M. (eds.) ECCV 2016. LNCS, vol. 9905, pp. 749–765. Springer, Cham (2016). https://doi.org/10.1007/978-3-319-46448-0_45
25. Bertinetto, L., Valmadre, J., Henriques, J.F., Vedaldi, A., Torr, P.H.S.: Fully-convolutional Siamese networks for object tracking. In: Hua, G., Jégou, H. (eds.) ECCV 2016. LNCS, vol. 9914, pp. 850–865. Springer, Cham (2016). https://doi.org/10.1007/978-3-319-48881-3_56
26. Li, B., Yan, J., Wu, W., Zhu, Z., Hu, X.: High performance visual tracking with Siamese region proposal network. In: Proceedings of the IEEE Conference on Computer Vision and Pattern Recognition, pp. 8971–8980 (2018)
27. Li, B., Wu, W., Wang, Q., Zhang, F., Xing, J., Yan, J.: SiamRPN++: evolution of Siamese visual tracking with very deep networks. In: Proceedings of the IEEE Conference on Computer Vision and Pattern Recognition, pp. 4282–4291 (2019)
28. Danelljan, M., Bhat, G., Khan, F.S., Felsberg, M.: ATOM: accurate tracking by overlap maximization. In: Proceedings of the IEEE Conference on Computer Vision and Pattern Recognition, pp. 4660–4669 (2019)
29. Bhat, G., Danelljan, M., Gool, L.V., Timofte, R.: Learning discriminative model prediction for tracking (2019)
30. Yilmaz, A., Li, X., Shah, M.: Contour-based object tracking with occlusion handling in video acquired using mobile cameras. IEEE Trans. Pattern Anal. Mach. Intell. **26**(11), 1531–1536 (2004)
31. Gupta, D.K., Gavves, E., Smeulders, A.W.: Tackling occlusion in Siamese tracking with structured dropouts. arXiv:2006.16571 (2020)
32. Pan, J., Hu, B.: Robust occlusion handling in object tracking. In: 2007 IEEE Conference on Computer Vision and Pattern Recognition, pp. 1–8. IEEE (2007)
33. Zhou, Y., Tao, H.: A background layer model for object tracking through occlusion. In: Proceedings Ninth IEEE International Conference on Computer Vision, pp. 1079–1085. IEEE (2003)
34. Koller, D., Weber, J., Malik, J.: Robust multiple car tracking with occlusion reasoning. In: Eklundh, J.-O. (ed.) ECCV 1994. LNCS, vol. 800, pp. 189–196. Springer, Heidelberg (1994). https://doi.org/10.1007/3-540-57956-7_22

35. Lee, B.Y., Liew, L.H., Cheah, W.S., Wang, Y.C.: Occlusion handling in videos object tracking: a survey. In: IOP Conference Series: Earth and Environmental Science, vol. 18, p. 012020. IOP Publishing (2014)
36. Greenhill, D., Renno, J., Orwell, J., Jones, G.A.: Occlusion analysis: learning and utilising depth maps in object tracking. Image Vis. Comput. **26**(3), 430–441 (2008)
37. Ma, Y., Chen, Q.: Depth assisted occlusion handling in video object tracking. In: Bebis, G., et al. (eds.) ISVC 2010. LNCS, vol. 6453, pp. 449–460. Springer, Heidelberg (2010). https://doi.org/10.1007/978-3-642-17289-2_43
38. Ali, A., Terada, K.: A framework for human tracking using Kalman filter and fast mean shift algorithms. In: 2009 IEEE 12th International Conference on Computer Vision Workshops, ICCV Workshops, pp. 1028–1033. IEEE (2009)
39. Zhao, J., Qiao, W., Men, G.Z.: An approach based on mean shift and Kalman filter for target tracking under occlusion. In: 2009 International Conference on Machine Learning and Cybernetics, vol. 4, pp. 2058–2062. IEEE (2009)
40. He, K., Zhang, X., Ren, S., Sun, J.: Deep residual learning for image recognition. In: Proceedings of the IEEE Conference on Computer Vision and Pattern Recognition, pp. 770–778 (2016)
41. Krizhevsky, A., Sutskever, I., Hinton, G.E.: ImageNet classification with deep convolutional neural networks. In: Advances in Neural Information Processing Systems, pp. 1097–1105 (2012)
42. Kristan, M., et al.: The sixth visual object tracking vot2018 challenge results. In: Proceedings of the European Conference on Computer Vision (ECCV) (2018)
43. Moudgil, A., Gandhi, V.: Long-term visual object tracking benchmark. arXiv preprint arXiv:1712.01358 (2017)
44. Wu, Y., Lim, J., Yang, M.: Online object tracking: a benchmark. In: 2013 IEEE Conference on Computer Vision and Pattern Recognition, pp. 2411–2418 (2013)
45. Gavves, E., Tao, R., Gupta, D.K., Smeulders, A.W.M.: Model decay in long-term tracking. arXiv:1908.01603 (2019)

The 1st Tiny Object Detection Challenge: Methods and Results

Xuehui Yu[1], Zhenjun Han[1](✉), Yuqi Gong[1], Nan Jan[1], Jian Zhao[2], Qixiang Ye[1], Jie Chen[3], Yuan Feng[4], Bin Zhang[4], Xiaodi Wang[4], Ying Xin[4], Jingwei Liu[5], Mingyuan Mao[6], Sheng Xu[6], Baochang Zhang[6], Shumin Han[4], Cheng Gao[7], Wei Tang[7], Lizuo Jin[7], Mingbo Hong[8], Yuchao Yang[8], Shuiwang Li[8], Huan Luo[8], Qijun Zhao[8], and Humphrey Shi[9]

[1] UCAS, Beijing, China
hanzhj@ucas.ac.cn
[2] NUS, Singapore, Singapore
[3] Pengcheng Lab, Shenzhen, China
[4] Baidu Inc., Beijing, China
[5] ICT, CAS, Beijing, China
[6] Beihang University, Beijing, China
[7] Southeast University, Nanjing, China
[8] Sichuan University, Chengdu, China
[9] University of Oregon, Eugene, USA

Abstract. The 1st Tiny Object Detection (TOD) Challenge aims to encourage research in developing novel and accurate methods for tiny object detection in images which have wide views, with a current focus on tiny person detection. The TinyPerson dataset was used for the TOD Challenge and is publicly released. It has 1610 images and 72651 box-level annotations. Around 36 participating teams from the globe competed in the 1st TOD Challenge. In this paper, we provide a brief summary of the 1st TOD Challenge including brief introductions to the top three methods.The submission leaderboard will be reopened for researchers that are interested in the TOD challenge. The benchmark dataset and other information can be found at: https://github.com/ucas-vg/TinyBenchmark.

Keywords: Tiny Object Detection · Visual recognition

1 Introduction

Object detection is an important topic in the computer vision community. With the rise of deep convolutional neural networks, research in object detection has seen unprecedented progress [1–3,9,11,13,20]. Nevertheless, detecting tiny objects remains challenging and far from well-explored. One possible reason for this is because there is a lack of datasets and benchmarks for tiny object detection, and researchers thus pay much less attention to challenges in tiny object detection compared to general object detection.

A. Bartoli and A. Fusiello (Eds.): ECCV 2020 Workshops, LNCS 12539, pp. 315–323, 2020.
https://doi.org/10.1007/978-3-030-68238-5_23

Tiny object detection is very important for real-world vision application and differs from general object detection in several aspects. For example, since the objects are extremely small while the whole input image has relatively large field-of-view, there is much less information from the targeting objects and much more from background distractions. In addition, the large field-of-view characteristic of input images usually means that the tiny objects are imaged from a long distance, this makes detection of tiny objects with various poses and viewpoints even more difficult. All these distinctions make tiny object detection a uniquely challenging task.

To encourage researchers to develop better methods to solve the tiny object detection problem, we held the first Tiny Object Detection Challenge. The TinyPerson dataset [18] was adopted in this challenge. The dataset contains 1610 images with 72651 box-level annotations and is collected from real-world scenarios. The persons in the challenge dataset are very small, and their aspect ratios have a large variance so that they are representative of different type of objects. In the following section, we will summarize the detail information about the challenge, the methods and results.

2 The TOD Challenge

2.1 Challenge Dataset

The images used in the challenge are collected from some real-world videos to build the TinyPerson dataset [18]. We sample images from video every 50 frames and delete images with a certain repetition for homogeneity. And finally, 72651 objects with bounding boxes are annotated. TinyPerson have four important properties: 1) The persons in TinyPerson are quite tiny compared with other representative datasets, and the size of image are mainly 1920 * 1080, which is the main characteristics; 2) The aspect ratio of persons in TinyPerson has a large variance. TinyPerson has the various poses and viewpoints of persons, it brings more complex diversity of the persons. 3) In TinyPerson, we mainly focus on person around seaside. 4) There are many images with dense objects (more than 200 persons per image) in TinyPerson. In TinyPerson, we classify persons as "sea person" (persons in the sea) or "earth person" (persons on the land). Some annotation rules in TinyPerson are defined to determine which label a person belongs to(as an example show in Fig. 1(a)): 1) Persons on boat are treated as "sea person"; 2) Persons lying in the water are treated as "sea person"; 3) Persons with more than half body in water are treated as "sea person"; 4) others are treated as "earth person". In addition, there are three conditions where persons are labeled as "ignore": 1) Crowds, which we can recognize as persons. But the crowds are hard to be separated one by one when labeled with standard rectangles; 2) Ambiguous regions, which are hard to clearly distinguish whether there is one or more persons, and 3) Reflections in water.Some objects are hard to be recognized as human beings, we directly labeled them as "uncertain".

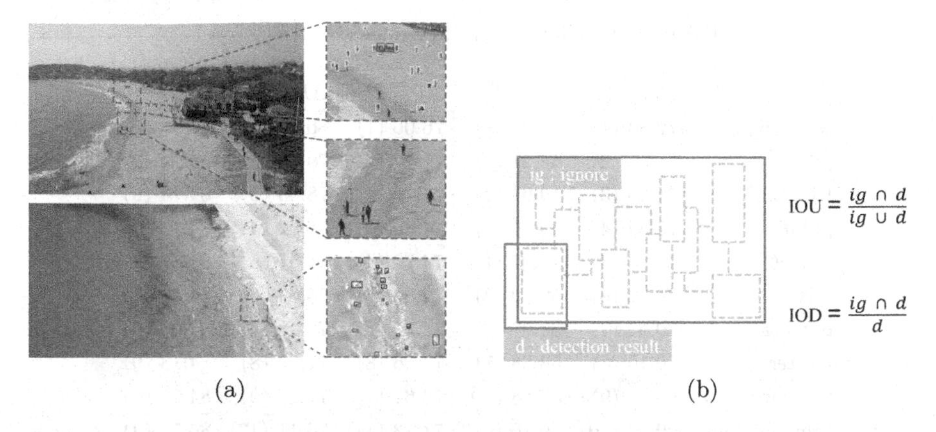

Fig. 1. (a): the annotation examples. "sea person", "earth person","uncertain sea person", "uncertain earth person", "ignore region" are represented with red, green, blue, yellow, purple rectangle, respectively. The regions are zoomed in and shown on right. (b): IOU (insertion of union) and IOD (insertion of detection). IOD is for ignored regions for evaluation. The outline (inviolet) box represents a labeled ignored region and the dash boxes are unlabeled and ignored persons. The red box is a detection's result box that has high IOU with one ignored person. (Color figure online)

2.2 Evaluation Metric

We use both AP (average precision) and MR(miss rate) for performance evaluation. The size range is divided into 3 intervals: tiny [2, 20], small [20, 32] and all [2, inf]. And for tiny [2, 20], it is partitioned into 3 sub-intervals: tiny1 [2, 8], tiny2 [8, 12], tiny3 [12, 20]. And the IOU threshold is set to 0.25 and 0.5 for performance evaluation. Same as pedestrian detection, 'ignore' regions do not participate in the evaluation, which means detection bounding boxes match them do not make a false positive. However in TinyPerson, most of ignore regions are much larger than that of a person. Therefore, we change IOU criteria to IOD(show in Fig. 1(b)) for ignore regions (IOD criteria only applies to ignore region, for other classes still use IOU criteria). In this challenge, we also treat uncertain same as ignore while evaluation.

3 Result and Methods

The 1st TOD challenge was held between April 20, 2020 and July 25, 2020. Around 36 teams submitted their final results in this challenge. Submission are evaluated on 786 images with 13787 person boxes and 1989 ignore regions, the images' resolution is mainly 1920 * 1080, some is even 3840 * 2160. The results of the first TOD challenge show in Table 1. In this section, we will briefly introduce the methodologies of the top 3 submissions.

Table 1. Challenge results ranked by AP_{50}^{tiny}

Team	AP_{50}^{tiny}	AP_{50}^{tiny1}	AP_{50}^{tiny2}	AP_{50}^{tiny3}	AP_{25}^{tiny}
baidu_ppdet	72.33 (1)	58.87 (8)	76.06 (1)	80.23 (1)	87.28 (1)
pilafsama	71.53 (2)	59.21 (5)	75.22 (2)	79.68 (2)	85.27 (2)
BingBing	71.36 (3)	59.91 (2)	74.70 (4)	78.63 (3)	84.74 (6)
pplm	71.35 (4)	59.89 (4)	74.69 (5)	78.62 (4)	84.75 (5)
tod666	71.34 (5)	59.97 (1)	74.70 (4)	78.57 (6)	84.61 (7)
mix	71.32 (6)	59.90 (3)	74.68 (6)	78.60 (5)	84.61 (7)
potting	70.91 (7)	58.87 (8)	74.73 (3)	78.18 (7)	84.93 (3)
matter	70.10 (8)	58.88 (7)	73.06 (8)	77.49 (8)	84.13 (9)
tiantian12414	69.71 (9)	57.88 (10)	73.67 (7)	77.26 (9)	84.25 (8)
Mingbo_Hong	69.34 (10)	59.10 (6)	71.73 (11)	76.11 (13)	84.76 (4)
dilidili	69.32 (11)	58.28 (9)	72.71 (10)	76.55 (11)	83.38 (11)
LHX	69.20 (12)	57.14 (11)	72.80 (9)	77.25 (10)	84.09 (10)
Washpan	68.73 (13)	57.12 (12)	71.52 (12)	76.21 (12)	82.93 (12)
liw	67.87 (14)	56.63 (13)	70.82 (13)	75.33 (14)	82.76 (13)
ZhangYuqi	65.31 (15)	49.34 (20)	69.57 (15)	75.13 (15)	80.86 (17)
xieyy	65.27 (16)	49.59 (18)	69.65 (14)	74.56 (16)	81.23 (16)
times	64.92 (17)	49.49 (19)	69.37 (16)	73.98 (17)	81.58 (15)
Michealz	63.34 (18)	53.55 (14)	65.58 (21)	69.66 (22)	81.77 (14)
yujia	62.94 (19)	50.66 (16)	67.33 (19)	69.76 (21)	78.61 (21)
LLP	62.88 (20)	50.97 (15)	66.64 (20)	69.78 (19)	79.58 (19)
ctt	62.83 (21)	46.53 (21)	68.54 (17)	71.76 (18)	79.62 (18)
Lee_Pisces	62.58 (22)	50.55 (17)	67.59 (18)	69.77 (20)	78.93 (20)
xuesong	58.79 (23)	44.81 (22)	61.46 (22)	68.81 (23)	76.73 (22)
alexto	57.52 (24)	43.04 (23)	60.05 (23)	66.64 (24)	75.61 (23)
stevehsu	54.34 (25)	35.74 (27)	59.04 (24)	65.94 (25)	74.91 (24)
fisty	52.88 (26)	42.54 (24)	55.83 (25)	61.49 (29)	71.51 (26)
Evali	51.38 (27)	37.06 (26)	55.09 (27)	62.51 (27)	72.87 (25)
mmeendez	51.07 (28)	31.98 (28)	55.20 (26)	63.11 (26)	70.24 (28)
bobson	50.72 (29)	38.82 (25)	53.73 (28)	58.84 (30)	70.52 (27)
daavoo	49.45 (30)	30.36 (29)	52.78 (29)	61.52 (28)	68.08 (30)
divyanshahuja	46.84 (31)	29.77 (31)	51.96 (30)	57.74 (31)	68.46 (29)
xie233	44.67 (32)	30.01 (30)	45.07 (31)	54.78 (32)	64.38 (31)
yingling	39.57 (33)	24.30 (32)	43.98 (32)	50.80 (34)	61.91 (32)
zhaoxingjie	33.83 (34)	5.09 (34)	34.17 (33)	52.91 (33)	57.49 (34)
suntinger	32.98 (35)	16.88 (33)	32.11 (34)	47.43 (35)	60.30 (33)
Sugar_bupt	13.61 (36)	1.79 (35)	13.04 (35)	22.92 (36)	36.40 (35)

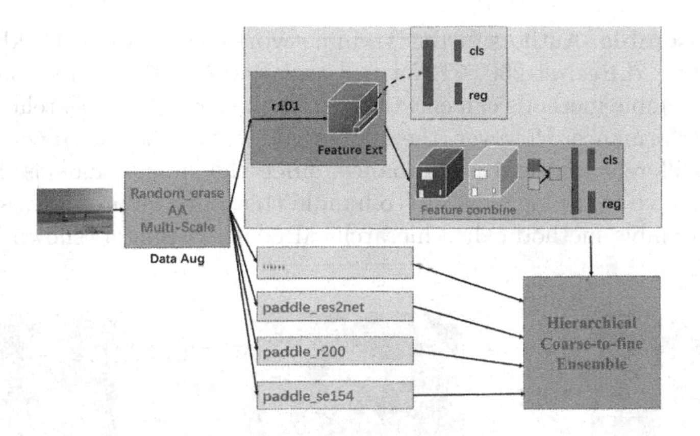

Fig. 2. Framwork of team baidu_ppdet.

3.1 Team baidu_ppdet

Yuan Feng, Bin Zhang, Xiaodi Wang, Ying Xin, Jingwei Liu, Mingyuan Mao, Sheng Xu, Baochang Zhang, Shumin Han. (Baidu & Beihang University)

Authors build detector based on the two-stage detection framework. They examine the performance of different components of detectors, leading to a large model pond for ensemble. The two-stage detectors include Faster R-CNN [11], FPN [8], Deformable R-CNN [4], and Cascade R-CNN [1]. The training datasets are separated into two parts: 90% images for training and the remaining for validation. The framework is illustrated in Fig. 2.

Data Augmentation. Authors pretrained their models on MSCOCO [9] and Object365 [12] dataset for better performance. To address the scale variance issue, authors implement scale match [18] on MSCOCO by rescaling training images to match the size distribution of images in TinyPerson, which helps achieve 2% to 3% improvement in terms of AP 50%.

Training Strategy. Authors train Faster R-CNN with ResNet-101 [6] as baseline, and the AP_{50}^{tiny} is 57.9%. After applying multi-scale training tricks, the AP_{50}^{tiny} reaches 60.1%. Additionally, erase ignore regions while validate on valid set to keep same as evaluation also comes with earnings with nearly 2%. Optimize the threshold of NMS, sample ratio and quantity. Finally, the AP_{50}^{tiny} of FRCNN-Res101 reaches 65.38%.

Model Refinement. Feature representation is always the key to tiny object detection. A new feature fusion method is designed to improve the feature representation ability of networks. For Faster R-CNN and FPN, the P3 layer can better represent tiny object. In contrast, the lack of P3 layer brings the loss of the semantic information. Thus, authors use PAFPN [15] to replace FPN in detector, which improves the mAP about 1.5%.

Model Ensemble. Authors further train networks with various backbones such as SENet-154 [7],ResNet-200, CBNet [10] and Res2Net-200 [5] for combination. Existing ensemble methods can effectively fuse the networks with relatively close size and performance. However, the results get worse when it comes to models with very different size and performance, since the smaller models deteriorate the performance of the bigger ones. To handle this, authors propose a simple and effective ensemble method called hierarchical coarse-to-fine as shown in Fig. 3.

Fig. 3. Illustration of hierarchical coarse-to-fine.

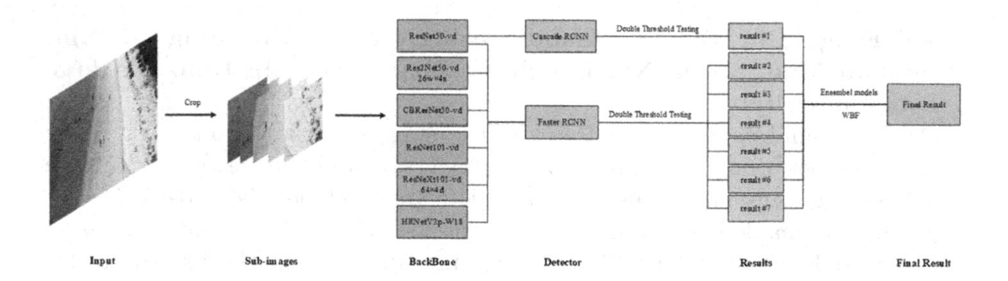

Fig. 4. Illustration of pipline of team STY-402.

3.2 Team STY-402

Cheng Gao, Wei Tang, Lizuo Jin (Southeast University)

Authors adopt Faster R-CNN with ResNet-50, FPN, DCNv2 as the baseline model. All models are pre-trained on MS COCO. The baseline employs feature pyramid levels from P_2 to P_6 and the area of anchors are defined as (12^2,24^2,48^2,96^2,192^2) pixels. Deformable convolutions are applied in all convolution layers in stages 3–5 in ResNet-50.

Backbone. In baseline, the weights of the first stage are frozen. Since the difference between MS COCO and TinyPerson is obvious, all the convolution layers in the backbone are unfreezed. In addition, Batch Normalization layer is added after each convolution layer of FPN. Authors replace ResNet with ResNet-vd. Moreover, Res2Net is a new multi-scale backbone architecture, which can further improve the performance of several representative computer vision tasks with no effort. Authors also train Res2Net-50 with 26w×4s, and the performance improves by 3% compared to ResNet-50-vd.

Multi-scale Training. The scale of short side is randomly sampled from 832, 896, 960, 1024, 1088, 1152, 1216, 1280, 1344, 1408, 1472, 1500 and the longer edge is fixed to 2000 in PaddleDetection. In particular, due to the limited GPU memory, the maximum value of the short side is set to 1280 when training ResNeXt101 (64 × 4d) [17]. In MMDetection, the scale of short side is randomly sampled from 480, 528, 576, 624, 672, 720, 768, 816, 912, 960 and the longer edge is fixed to 1280.

Training Tricks. In the training stage, the number of proposals before NMS is changed from 2000 to 12000. And the data is changed to 6000 in testing stage.

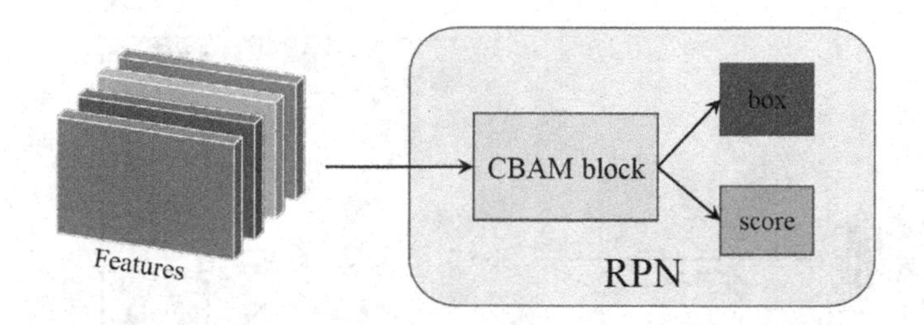

Fig. 5. Illustration of improved CBAM in Region Proposal Network

Data Augmentation. Random horizontal flip, random cropping, random expanding and CutMix are adopt to augment training data. VisDrone is also used as additional data, which only use categories 0, 1, 2, and delete categories 3–11.

Large Scale Testing. A large scale (1280 × 960) is adopted for testing. In order to get better performance, another large scale (1024 × 768) is also used for testing at the same time.

Double Threshold Testing. If the highest confidence of a sub-image detection results is less than a certain threshold (0.5), the sub-image will be regarded as a pure background image and ignored.

Model Ensemble. They train 7 models using different backbones as shown in Fig. 4. Except HRNetV2p-W18 is trained on MMDetection, the rest of the models are trained on PaddleDetection. Finally, the final ensemble results is obtained by weighted boxes fusion (the IoU threshold is 0.6).

3.3 Team BRiLliant

Mingbo Hong, Yuchao Yang, Huan Luo, Shuiwang Li, Qijun Zhao
(College of Computer Science, Sichuan University)

To explore more detail features in tiny objects, authors utilize High Resolution Net (HRNet) [14] as backbone network, which allows the network to extract high-resolution representation. To simultaneously detect objects of varying scales, authors introduce an improved Convolutional Block Attention Module

(CBAM) [16] in Region Proposal Network to guide network to "Look Where" as shown in Fig. 5. Unlike the traditional CBAM, improved CBAM adds a suppression block to balance the attention value between objects of different scales. Furthermore, in order to raise different numbers of proposals for different scale objects, authors use "Top k sampler" instead of fixed threshold to select positive samples as shown in Fig. 6, and the selection is based on the IOU metric rather than center distance that was utilized in ATSS [19]. The proposed sampler is adaptive to the scale of objects, which can be more accurate in detecting tiny objects, whereas ATSS may not generate any positive samples for tiny objects at all.

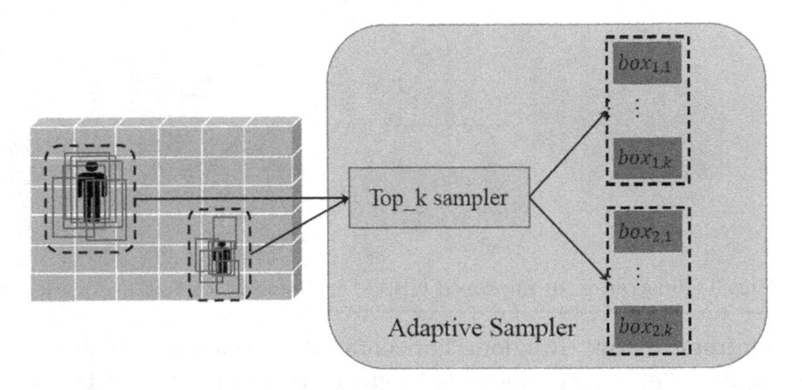

Fig. 6. Illustration of Adaptive Sampler. The blue solid line indicates low-quality proposals, the red solid line indicates high-quality proposals, and Adaptive Sampler will adaptively select positive sample according to the quality of proposals. (Color figure online)

4 Conclusions

We held the 1st TOD Challenge to encourage novel visual recognition research into tiny object detection. We used TinyPerson as the challenge dataset, and both AP and MR as evaluation metric. Approximately 36 teams around the globe participated in this competition, in which top-3 leading teams, together with their methods, are briefly introduced in this paper. It is our vision that tiny object detection should extend far beyond person detection. Tiny object detection related tasks are important for many real-world computer vision applications, and the advancement of addressing its technical challenges can help general object detection research as well. We hope our 1st TOD Challenge is a useful initial step in this promising research direction.

References

1. Cai, Z., Vasconcelos, N.: Cascade R-CNN: delving into high quality object detection. In: CVPR, pp. 6154–6162 (2018)
2. Cheng, B., Wei, Y., Shi, H., Feris, R., Xiong, J., Huang, T.: Decoupled classification refinement: hard false positive suppression for object detection. arXiv preprint arXiv:1810.04002 (2018)
3. Cheng, B., Wei, Y., Shi, H., Feris, R., Xiong, J., Huang, T.: Revisiting RCNN: on awakening the classification power of faster RCNN. In: ECCV, pp. 453–468 (2018)
4. Dai, J., Qi, H., Xiong, Y., Li, Y., Zhang, G., Hu, H., Wei, Y.: Deformable convolutional networks. In: ICCV, pp. 764–773 (2017)
5. Gao, S., Cheng, M.M., Zhao, K., Zhang, X.Y., Yang, M.H., Torr, P.H.: Res2net: a new multi-scale backbone architecture. IEEE Trans. Pattern Anal. Mach. Intell. (2019)
6. He, K., Zhang, X., Ren, S., Sun, J.: Deep residual learning for image recognition. In: CVPR, pp. 770–778 (2016)
7. Hu, J., Shen, L., Sun, G.: Squeeze-and-excitation networks. In: CVPR, pp. 7132–7141 (2018)
8. Lin, T.Y., Dollár, P., Girshick, R., He, K., Hariharan, B., Belongie, S.: Feature pyramid networks for object detection. In: CVPR, pp. 2117–2125 (2017)
9. Lin, T.-Y., et al.: Microsoft COCO: common objects in context. In: Fleet, David, Pajdla, Tomas, Schiele, Bernt, Tuytelaars, Tinne (eds.) ECCV 2014. LNCS, vol. 8693, pp. 740–755. Springer, Cham (2014). https://doi.org/10.1007/978-3-319-10602-1_48
10. Liu, Y., et al.: CBNet: a novel composite backbone network architecture for object detection. In: AAAI, pp. 11653–11660 (2020)
11. Ren, S., He, K., Girshick, R., Sun, J.: Faster R-CNN: towards real-time object detection with region proposal networks. In: Advances in Neural Information Processing Systems, pp. 91–99 (2015)
12. Shao, S., et al.: Objects365: a large-scale, high-quality dataset for object detection. In: ICCV, pp. 8430–8439 (2019)
13. Shen, Z., et al.: Improving object detection from scratch via gated feature reuse. In: BMVC (2019)
14. Sun, K., Xiao, B., Liu, D., Wang, J.: Deep high-resolution representation learning for human pose estimation. In: CVPR, pp. 5693–5703 (2019)
15. Tan, M., Pang, R., Le, Q.V.: EfficientDet: scalable and efficient object detection. In: CVPR, pp. 10781–10790 (2020)
16. Woo, S., Park, J., Lee, J.Y., So Kweon, I.: CBAM: convolutional block attention module. In: ECCV, pp. 3–19 (2018)
17. Xie, S., Girshick, R., Dollár, P., Tu, Z., He, K.: Aggregated residual transformations for deep neural networks. In: CVPR, pp. 1492–1500 (2017)
18. Yu, X., Gong, Y., Jiang, N., Ye, Q., Han, Z.: Scale match for tiny person detection. In: WACV, pp. 1257–1265 (2020)
19. Zhang, S., Chi, C., Yao, Y., Lei, Z., Li, S.Z.: Bridging the gap between anchor-based and anchor-free detection via adaptive training sample selection. In: CVPR, pp. 9759–9768 (2020)
20. Zhang, X., et al.: SkyNet: a hardware-efficient method for object detection and tracking on embedded systems. In: MLSys (2020)

Effective Feature Enhancement and Model Ensemble Strategies in Tiny Object Detection

Yuan Feng[1], Xiaodi Wang[1], Ying Xin[1], Bin Zhang[1], Jingwei Liu[2], Mingyuan Mao[3], Sheng Xu[3], Baochang Zhang[3,4](✉), and Shumin Han[1](✉)

[1] Baidu Inc., Beijing, China
hanshumin@baidu.com
[2] ICT, CAS, Beijing, China
[3] Beihang University, Beijing, China
bczhang@buaa.edu.cn
[4] Shenzhen Academy of Aerospace Technology, Shenzhen, China

Abstract. We introduce a novel tiny-object detection network that achieves better accuracy than existing detectors on TinyPerson dataset. It is an end-to-end detection framework developed on PaddlePaddle. A suit of strategies are developed to improve the detectors performance including: 1) data augmentation based on scale-match that aligns the object scales between the existing large-scale dataset and TinyPerson; 2) comprehensive training methods to further improve detection performance by a large margin; 3) model refinement based on the enhanced PAFPN module to fully utilize semantic information; 4) a hierarchical coarse-to-fine ensemble strategy to improve detection performance based on a well-designed model pond.

Keywords: Data augmentation · Feature pyramid · Model ensemble

1 Introduction

In this competition, we build our detector based on the two-stage detection framework. We examine the performance of different components of detectors, leading to a large model pond for ensemble. The two-stage detectors we investigated include Faster R-CNN [2], FPN [4], and Cascade R-CNN [1]. The training datasets are separated into two parts: 90% images for training and the remaining for validation. The mAP between our validation and test set is positive correlation, that means we can apply experiment results on validation set to the test set. Our framework is illustrated in Fig. 1.

The performance of each detector on public leaderboard evolved as follows: Faster R-CNN [2] with ResNet-101 backbone (Faster RCNN-Res101) was trained to get a baseline AP_{50}^{tiny} as 57.9%. After applying multi-scale training tricks, the mAP reached 60.1%. And random erasing also comes with earnings with nearly two points. We also adjust the threshold of NMS and in the model and

A. Bartoli and A. Fusiello (Eds.): ECCV 2020 Workshops, LNCS 12539, pp. 324–330, 2020.
https://doi.org/10.1007/978-3-030-68238-5_24

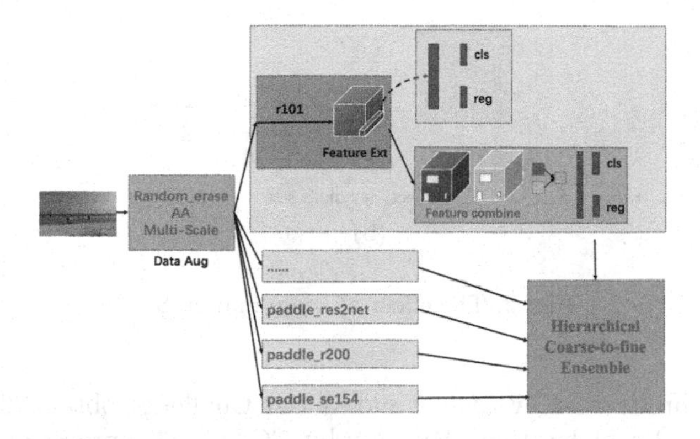

Fig. 1. The framework of our tiny-object detector

sample ratio and number. Finally, the AP_{50}^{tiny} of Faster RCNN-Res101 reached 65.38%. Otherwise, we design a new feature combine method the improve the feature of tiny object. We further trained different backbone networks such as SENet-154 [11], ResNet-200 [9], CBNet [12], Res2Net-200 [10]. Then the AP_{50}^{tiny} of ensemble of different backbone models increased to 72.23%. We found that feature enhancement and model ensemble play a significant role in tiny object detection.

2 Our Method

Our work is mainly separate into four parts: 1) data augmentation, 2) training strategy, 3)feature enhancement, 4) model ensemble. The performance of each detector evolved as follows:

2.1 Data Augmentation

Considering TinyPerson dataset only contains 794 images for training, we pre-train our models on MSCOCO dataset or OpenImage365 dataset for better performance. We implement scale match [8] to align the object scales between the two datasets for favorable tiny-object representation, as shown in Fig. 2, which helps achieve 2% to 3% improvement in terms of AP_{50}^{tiny}. As we observe that scale match method [1] can effectively provide a subtly pre-trained model with prior knowledge of similarly sized objects, it occurs us that prior knowledge of similarly sized objects of the same class may work better. As the person objects have very similar features, we have the instinct that the pre-train model person detection dataset on will work better in theory.

We calculate the absolute size, relative size and aspect ratio following the same pipeline in [8]. We take an ablation study on the dataset for pre-train, and the results are listed in Table 1. We employ Faster RCNN [2] -FPN [4] with ResNet-50 backbone (FRCNN-Res50) for experiments. As listed in Table 1,

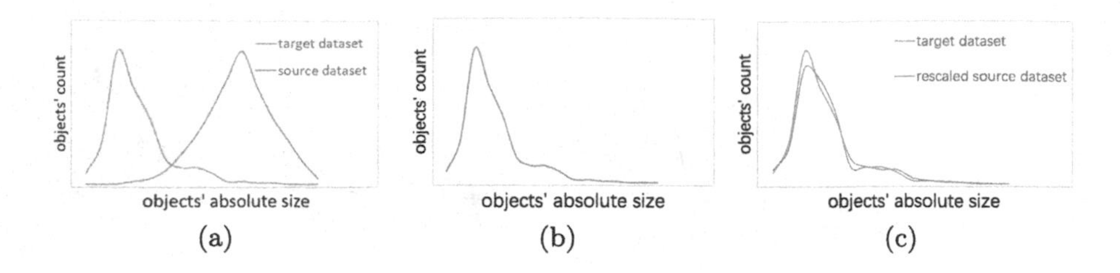

Fig. 2. The effect of scale match [8]

FRCNN-R50 pre-trained by COCO and Wider Challenge obtains 62% mAP in test, which is 4.1% higher than that only by COCO. This proves our instinct of pre-train dataset. Hence, we finally decide to obtain the pre-trained model via MS COCO and Wider Challenge with scale match for the following experiments on larger models such as Cascade R-CNN [1] and Res2Net [5].

Table 1. The results on different dataset for pre-train

Dataset for pre-train	AP_{50}^{tiny}
COCO	61.59%
Wider challenge	61.31%
COCO+Wider challenge	62%

2.2 Training Strategy

We train Faster R-CNN [2] with ResNet-101 backbone (FRCNN-Res101) as our baseline, and the AP_{50}^{tiny} is 57.9%.

(a) **Multi-scale Training:** We randomly rescale input images to 0.5, 1, 1.5 times of the original size while training to help address the scale variance issue, the AP_{50}^{tiny} reaches 60.1%.

(b) **Larger-Scale Training:** Considering the objects in TinyPerson are extremely small, based on the multi-scale training strategy, we upsample the training images for higher resolution and gain better results. However, this method has its bottleneck. Upsampling training images to larger scales helps detector extract more detailed features, but at the same time more distortion. The results are shown in Table 2. Note that the results in Table 2 are based on Faster RCNN-Res50.

(c) **Fine-tuned Scale Match:** Rescaling input images while training detectors on TinyPerson brings scale mismatch problem between pre-training and training process. Thus, we need to adjust scale parameters when pre-training models on COCO with scale match according to the scale we set while training them on TinyPerson. By implementing this we improve AP_{50}^{tiny} from 57.31% to 59.64%.

(d) **2x Training Schedule:** The open source baseline implements 1x training schedule(12epochs). Nevertheless, we find that the model does not converge well. Instead, we choose 2x training schedule (24 epochs) and gain 0.8% more mAP.

Additionally, considering some blurry areas in the images are erased and treated as ignored area during training, we test our model on the corresponding test set, whose blurry areas are already erased, rather than the original test set with no erased areas. This trick also comes with earnings with nearly 2%. We also optimize the threshold of NMS, sample ratio and quantity. Finally, the AP_{50}^{tiny} of Faster-RCNN-Res101 reaches 65.38%.

Table 2. Scale selection while training

Scale	1	2	3
AP_{50}^{tiny}	52.48%	57.31%	55.29%

2.3 Feature Enhancement

Feature representation is always the key to tiny object detection. We design a new feature fusion method to achieve feature enhancement in our networks. For Faster-RCNN [2] and FPN [4], the P3 layer can better represent tiny object. In contrast, the lack of P3 layer brings the loss of the semantic information. Thus, we use PAFPN [6] to replace FPN in our detector, which improves the mAP about 1.5%. The architecture of PAFPN is shown in Fig. 3.

Table 3. The area distribution of bounding boxes TinyPerson.

Area	[6,720]	(720, 1438]	(1438, 2158]	(2158, 2877]	(2877, 3596]
Box num	31965	289	89	40	29
Area	(3596, 4315]	(4315.5034]	(5034, 5754]	(5754, 6473]	(6473, 7192]
Box num	7	4	1	4	2

In addition, as for a better performance, we statistics the distribution of the tiny dataset, as the Table 3 and 4 shows as the decrease of area, instead, the number of objects are increasing. Small object and aspect ratio variable are the key challenges in object detection. In object detection, task is divided into two pillars: classification and localization. However, the CNN-based architecture is insensitive to localization and the modules are optimized relying on the presuppose anchor. We redesign our anchor size and ratio, which improving the mAP about 1.2%. In the inference, we evaluate the model with the IoU of 0.5. Hence, we adjust the rpn-positive-overlap (paddle detection parametric) from 0.7 to 0.5 and RPN-negative-overlap from 0.3 to 0.25, which gain 1% mAP improvement in all models. Besides, we adjust the sample ratio and sample number between positive and negative anchors/proposal to ensure the positive sample exceed the evaluation number.

Table 4. The area (from 0 to 720) distribution of bounding boxes of TinyPerson.

Area	[0, 72]	(72, 144]	(144, 216]	(216, 288]	(288, 360]
Box num	8569	6812	3808	2432	1683
Area	(360, 432]	(432, 504]	(504, 576]	(576, 648]	(648, 720]
Box num	1258	879	767	502	408

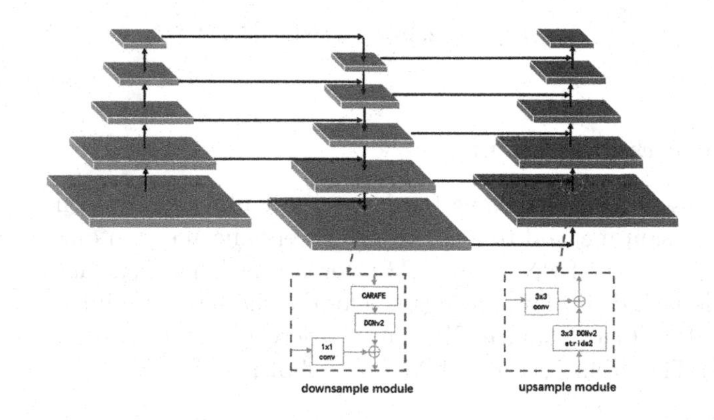

Fig. 3. Architecture of PAFPN.

2.4 Hierarchical Coarse-to-fine Ensemble

PaddleDetection provides varied object detection architectures in modular design, and wealthy data augmentation methods, network components, loss functions, etc. We further train our networks with various backbones such as SENet-154 [3], ResNet-200, CBNet [7] and Res2Net-200 for combination. Existing ensemble methods can effectively fuse the networks with relatively close size and performance. However, the results get worse when it comes to models with very different size and performance, since the smaller models deteriorate the performance of the bigger ones. To handle this problem, we propose a simple and effective ensemble method, called hierarchical coarse-to-fine, as shown in Fig. 4.

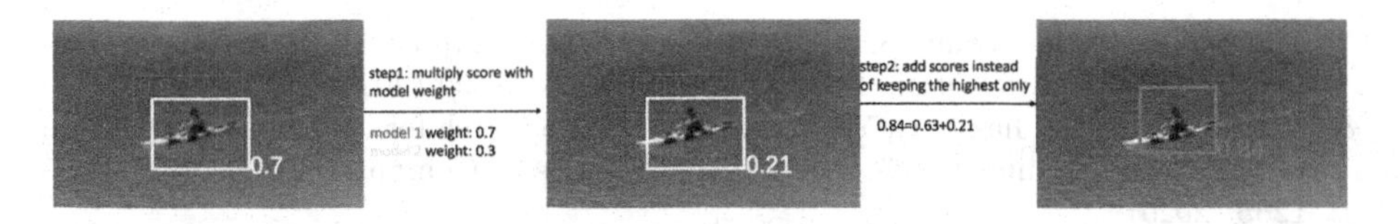

Fig. 4. Ensemble strategy.

The steps are as follows:

(a) Give each model a scalar weight from 0 to 1. All weights sum up to 1.
(b) The confidence score of boxes from each model is multiplied by its weight.
(c) Merge boxes from all models and run the original NMS, except that we add scores from different models instead of keeping the highest one only. The IOU threshold is 0.5 in this step.

We use an adaptive weighting scheme to combine the different detectors based on the boosting method. By implementing hierarchical coarse-to-fine ensemble strategy, the AP_{50}^{tiny} increased to 72.23%.

3 Conclusions

In this paper, several novel strategies are proposed for tiny object detection. By implementing scale match in pre-training, multiscale training strategy, feature enhancement (PAFPN) and hierarchical coarse-to-fine ensemble to the target dataset, we improve the detector performance on tiny object dataset with a large margin. We hope these novel strategies provide fresh insights to future works on tiny object detection.

Acknowledgements. The work was supported in part by National Natural Science Foundation of China under Grants 62076016. This work is supported by Shenzhen Science and Technology Program KQTD2016112515134654. Baochang Zhang and Shumin Han are the correspondence authors.

References

1. Cai, Z., Vasconcelos, N.: Cascade R-CNN: delving into high quality object detection. In: IEEE CVPR, pp. 6154–6162 (2018)
2. Girshick, R.B.: Fast R-CNN. In: IEEE ICCV, pp. 1440–1448 (2015)
3. J, H., L, S., J, S.: Squeeze-and-excitation networks. In: IEEE CVPR, pp. 7132–7141 (2014)
4. Lin, T., Dollár, P., Girshick, R.B., He, K., Hariharan, B., Belongie, S.J.: Feature pyramid networks for object detection. In: IEEE CVPR, pp. 936–944 (2017)
5. Gao, S., Cheng, M. M., Zhao, K.: Res2net: a new multi-scale backbone architecture. IEEE Trans. Pattern Anal. Mach. Intell. **43**(2) (2019)
6. Liu, S., Qi, L., Qin, H., Shi, J., Jia, J.: Path aggregation network for instance segmentation. In: IEEE CVPR, pp. 8759–8768 (2018)

7. Liu, Y., Wang, Y., Wang, S.: CBNet: a novel composite backbone network architecture for object detection. In: AAAI, pp. 11653–11660 (2020)
8. Yu, X., Gong, Y., Jiang, N., Ye, Q., Han, Z.: Scale match for tiny person detection. In: The IEEE Winter Conference on Applications of Computer Vision, pp. 1257–1265 (2020)
9. He, K., Zhang, X., Ren, S., Sun, J.: Identity mappings in deep residual networks. In: ECCV, pp. 630–645 (2016)
10. PddlePaddle. https://github.com/PaddlePaddle/PaddleClas/
11. Hu, J., Shen, L., Albanie, S,, Sun, G., Wu, E.: Squeeze-and-excitation networks. In: IEEE CVPR, pp. 7132–7141 (2018)
12. Liu, Y., Wang, Y., Wang, S.: CBNet: a novel composite backbone network architecture for object detection. In: AAAI, pp. 11653–11660 (2020)

Exploring Effective Methods to Improve the Performance of Tiny Object Detection

Cheng Gao[1], Wei Tang[1], Lizuo Jin[1(✉)], and Yan Jun[2]

[1] School of Automation, Southeast University, Nanjing, China
{gaocheng,w.tang}@seu.edu.cn,
jinlizuo@gmail.com
[2] Department of Geriatrics,
Nanjing Brain Hospital Affiliated to Nanjing Medical University, Nanjing, China
2519002414@qq.com

Abstract. In this paper, we present our solution of the 1st Tiny Object Detection (TOD) Challenge. The purpose of the challenge is to detect tiny person objects (2–20 pixels) in large-scale images. Due to the extreme small object size and low signal-to-noise ratio, the detection of tiny objects is much more challenging than objects in other datasets such as COCO and CityPersons. Based on Faster R-CNN, we explore some effective and general methods to improve the detection performance of tiny objects. Since the model architectures will not be changed, these methods are easy to implement. Accordingly, we obtain the 2nd place with the AP_{50}^{tiny} score of 71.53 in the challenge.

Keywords: Pedestrian detection · Tiny Object Detection

1 Introduction

Tiny Object Detection (TOD) is a task of detecting extreme small objects in large-scale images, which is a challenging task. Firstly, it is difficult to distinguish small objects from the background. Secondly, low signal-to-noise ratio can seriously damage the features used for classification and localization. In TinyPerson [16], the size range is divided into 3 intervals: tiny ([2, 20]), small ([20, 32]), and all ([2, inf]). And for tiny scale, it is partitioned into 3 sub-intervals: tiny 1 ([2, 8]), tiny 2 ([8, 12]) and tiny 3 ([12, 20]). Both AP (average precision) and MR (miss rate) are employed as evaluation criteria for TOD Challenge with the three IoU thresholds (0.25, 0.50, 0.75). But only AP_{50}^{tiny} are utilized to score and rank.

In this paper, we propose some methods (*e.g.* Feature Pyramid Network [9], Deformable ConvNets v2 [20], multi-scale training and testing) that can simply and effectively improve the performance of tiny object detection based Faster RCNN [12]. In TOD Challenge, our ensemble model achieves 2nd place with the AP_{50}^{tiny} score of 71.53.

A. Bartoli and A. Fusiello (Eds.): ECCV 2020 Workshops, LNCS 12539, pp. 331–336, 2020.
https://doi.org/10.1007/978-3-030-68238-5_25

2 Methods

2.1 Baseline

We adopt Faster R-CNN with ResNet-50 [6], FPN, DCNv2 as the baseline model. To extract more precise features, we adopt RoI Align [5] instead of RoI Pooling. Most of the experiments are conducted on PaddleDetection[1], and a few are conducted on MMDetection [2]. All models are pretrained on MS COCO [10]. Our baseline employs feature pyramid levels from P_2 to P_6 and the area of anchors are set as $\{12^2, 24^2, 48^2, 96^2, 192^2\}$ pixels on $\{P_2, P_3, P_4, P_5, P_6\}$. DCNv2 has the ability to focus on pertinent image regions, through increased modeling power and stronger training. Deformable convolutions are applied in all 3×3 convolution layers in stages conv3-5 in ResNet-50.

For TinyPerson, the two categories (sea person and earth person) are merged into one category (person). The ignore region is replaced with the mean value of the images. The origin images are cut into some sub-images with the size of 640×480 and the 20% overlap rate during training and testing. Then the NMS strategy is used to merge all results of the sub-images in one same image, with the threshold of 0.5.

We train the models for 12 epochs with the linear scaling rule [4] ($lr = 0.0025$, $batch\ size = 2$) on one Tesla V100 (32 GB), and adopt SGD with momentum 0.9 and weight decay 0.0001 for training.

2.2 Bag of Tricks for Detector

Overview. In this section, as shown in Table 1, the step-by-step improvements of different tricks adopted in our baseline will be presented.

Backbone Tricks. In baseline, the weights in the first stage are fronzen. Since the difference between MS COCO and TinyPerson is obvious, we unfreeze all the convolution layers in the backbone. In addition, we add batch normalization (BN) [8] after each convolution layer of FPN. According to [7], ResNet-vd is generally better than original ResNet in classification and object detection tasks. Therefore, we replace ResNet with ResNet-vd. Res2Net [3] is a new multi-scale backbone architecture, which can further improve the performance of several representative computer vision tasks with no effort. We also train Res2Net50-vd with $26\,\text{w} \times 4\,\text{s}$, and the performance improves by 3% compared to ResNet50-vd.

Multi-scale Training. As a simple data augmentation method, multi-scale training is commonly used. The scale of shorter side is randomly sampled from {832, 896, 960, 1024, 1088, 1152, 1216, 1280, 1344, 1408, 1472, 1500} and the longer edge is fixed to 2000 in PaddleDetection. In particular, due to the limited GPU memory, the maximum value of the shorter side is set to 1280 when training

[1] https://github.com/PaddlePaddle/PaddleDetection/.

ResNeXt-101($64 \times 4d$) [15]. In MMDetection, the scale of short side is randomly sampled from {480, 528, 576, 624, 672, 720, 768, 816, 912, 960} and the longer edge is fixed to 1280.

Training Tricks. In training stage, the number of proposals before NMS is changed from 2000 to 12000. And the data is changed to 6000 in testing stage.

Table 1. The performances of step-by-step improvements on the TinyPerson test set.

Method	AP_{50}^{tiny}
Baseline	46.83
+ResNet50-vd	47.85
+Multi-scale training	49.28
+Training tricks	51.63
+Anchor scale = 4	52.12
+Test flip	53.10
+Unfreeze backbone	53.62
+Double threshold testing	53.80
+Res2Net50-vd	55.54
+RandomCrop	56.75
+RandomExpand	56.99
+VisDrone	62.22
+CutMix	63.05
+Anchor scale = 3	64.43
+FPN BN	65.27
+Large scale testing	66.06

Data Augmentation. By default, we augment training data by random horizontal flip with the probability of 0.5. Furthermore, we also use random cropping and random expanding. CutMix [17] is another augmentation strategy to improve the model robustness against input corruptions and its out-of-distribution detection performances. VisDrone [19] detection training set is used as additional data. We only use categories 0, 1, 2, and delete categories 3–11.

Label Smoothing. During training, we use cross entropy loss as shown in formula (1). Label smoothing [18] can reduce classification confidence of a model, measured by the difference between the largest and smallest logits. We utilize formula (2) to smooth the ground truth distribution q in Faster R-CNN RoI classification head, where $\varepsilon = 0.001$ and K is the total number of classes. We only exploit label smoothing on the model trained by MMDetetection, which increases AP_{50}^{tiny} by 0.1.

$$CE = -\sum_{i=1}^{K} q_i \log p_i \tag{1}$$

$$q_i = \begin{cases} 1-\varepsilon & \text{if } i = y \\ \varepsilon/(K-1) & \text{otherwise} \end{cases} \tag{2}$$

Large Scale Testing. We adopt a large scale 1280 × 960 for testing instead of 640 × 480. In order to get better performance, we also use another large scale 1024 × 768 for testing at the same time.

Double Threshold Testing. As shown in Algorithm 1, if the highest confidence of a sub-image detection results is less than a certain threshold (0.5), the sub-image is regarded as a pure background image and ignored.

Algorithm 1: Double Threshold Testing

Input: $T_{filter}, T_{nms}, \mathcal{B} = \{b_1, \ldots, b_N\}$
T_{filter} is the threshold for filtering pure background sub-images.
T_{nms} is the NMS threshold for merging all results of the sub-images in one same image.
B is the detection results of all sub-images in one same image, where b_i is the results of the ith sub-image of the image.
Output: D: The final detection result of one image.

```
1 D ← {};
2 for b_i in B do
3     m ← argmax b_i[4];
4     if m >= T_filter then
5         D ← D ∪ b_i;
6     end
7 end
8 D ← NMS(D, T_nms);
9 return D
```

Model Ensemble. We train 7 models using different backbones. Except HRNet-V2p-W18 [14] is trained on MMDetection, the rest of the models are trained on PaddleDetection. Cascade R-CNN [1] is trained with ResNet50-vd. Then, we get the final ensemble results by weighted boxes fusion [13] with the IoU threshold of 0.6. WBF uses information from all boxes, which can fix some cases where all boxes are predicted inaccurate by all models. As shown in Table 2, the ensemble method significantly improves the performance with more than 6% improvement on AP_{50}^{tiny}.

Table 2. The performances of different models and final ensemble model on the TinyPerson test set.

Model	AP_{50}^{tiny}	AP_{50}^{tiny1}	AP_{50}^{tiny2}	AP_{50}^{tiny3}	AP_{25}^{tiny}	AP_{75}^{tiny}	MR_{50}^{tiny}
HRNetV2p-W18	61.73	47.06	65.88	72.22	76.10	12.42	82.27
Cascade R-CNN	62.35	50.94	65.78	70.05	79.04	12.00	80.84
CBResNet50-vd [11]	64.03	51.66	67.82	71.98	79.69	12.08	81.61
ResNet50-vd	65.94	54.69	69.50	73.58	81.58	12.37	80.29
ResNet101-vd	66.04	53.17	69.14	74.36	81.94	12.61	80.03
Res2Net50-vd	66.06	53.21	70.10	74.40	81.64	12.33	80.54
ResNeXt101-vd	66.95	55.47	70.06	75.20	83.01	12.85	80.51
Ensemble Result	**71.53**	**59.21**	**75.22**	**79.68**	**85.27**	**14.84**	**74.69**

3 Conclusions

In this paper, we present some effective and general methods to improve the detection performance of tiny objects, which helps us win the 2nd place in the 1st TOD Challenge.

Acknowledgement. This paper was financial supported partially by Special Funds of the Jiangsu Provincial Key Research and Development Projects (grant No. BE2019612) and Jiangsu Provincial Cadre Health Research Projects (grant No. BJ17006).

References

1. Cai, Z., Vasconcelos, N.: Cascade R-CNN: delving into high quality object detection. In: Proceedings of the IEEE Conference on Computer Vision and Pattern Recognition (CVPR), June 2018
2. Chen, K., et al.: MMDetection: open MMLab detection toolbox and benchmark. arXiv preprint arXiv:1906.07155 (2019)
3. Gao, S., Cheng, M.M., Zhao, K., Zhang, X.Y., Yang, M.H., Torr, P.H.: Res2net: a new multi-scale backbone architecture. IEEE Trans. Pattern Anal. Mach. Intell. (2019)
4. Goyal, P., et al.: Accurate, large minibatch SGD: training ImageNet in 1 hour. arXiv preprint arXiv:1706.02677 (2017)
5. He, K., Gkioxari, G., Dollar, P., Girshick, R.: Mask R-CNN. In: Proceedings of the IEEE International Conference on Computer Vision (ICCV), October 2017
6. He, K., Zhang, X., Ren, S., Sun, J.: Deep residual learning for image recognition. In: Proceedings of the IEEE Conference on Computer Vision and Pattern Recognition (CVPR), June 2016
7. He, T., Zhang, Z., Zhang, H., Zhang, Z., Xie, J., Li, M.: Bag of tricks for image classification with convolutional neural networks. In: Proceedings of the IEEE/CVF Conference on Computer Vision and Pattern Recognition (CVPR), June 2019
8. Ioffe, S., Szegedy, C.: Batch normalization: accelerating deep network training by reducing internal covariate shift. arXiv preprint arXiv:1502.03167 (2015)

9. Lin, T.Y., Dollar, P., Girshick, R., He, K., Hariharan, B., Belongie, S.: Feature pyramid networks for object detection. In: Proceedings of the IEEE Conference on Computer Vision and Pattern Recognition (CVPR), July 2017
10. Lin, T.Y., et al.: Microsoft COCO: common objects in context. In: European Conference on Computer Vision, pp. 740–755. Springer (2014). https://doi.org/10.1007/978-3-319-10602-1_48
11. Liu, Y., et al.: CBNet: a novel composite backbone network architecture for object detection. In: AAAI, pp. 11653–11660 (2020)
12. Ren, S., He, K., Girshick, R., Sun, J.: Faster R-CNN: towards real-time object detection with region proposal networks. IEEE Trans. Pattern Anal. Mach. Intell. **39**(6), 1137–1149 (2017)
13. Solovyev, R., Wang, W.: Weighted boxes fusion: ensembling boxes for object detection models. arXiv preprint arXiv:1910.13302 (2019)
14. Sun, K., et al.: High-resolution representations for labeling pixels and regions. arXiv preprint arXiv:1904.04514 (2019)
15. Xie, S., Girshick, R., Dollar, P., Tu, Z., He, K.: Aggregated residual transformations for deep neural networks. In: Proceedings of the IEEE Conference on Computer Vision and Pattern Recognition (CVPR), July 2017
16. Yu, X., Gong, Y., Jiang, N., Ye, Q., Han, Z.: Scale match for tiny person detection. In: Proceedings of the IEEE/CVF Winter Conference on Applications of Computer Vision (WACV), March 2020
17. Yun, S., Han, D., Oh, S.J., Chun, S., Choe, J., Yoo, Y.: CutMix: regularization strategy to train strong classifiers with localizable features. In: Proceedings of the IEEE/CVF International Conference on Computer Vision (ICCV), October 2019
18. Zhang, Z., He, T., Zhang, H., Zhang, Z., Xie, J., Li, M.: Bag of freebies for training object detection neural networks. arXiv preprint arXiv:1902.04103 (2019)
19. Zhu, P., Wen, L., Du, D., Bian, X., Hu, Q., Ling, H.: Vision meets drones: past, present and future. arXiv preprint arXiv:2001.06303 (2020)
20. Zhu, X., Hu, H., Lin, S., Dai, J.: Deformable ConvNets v2: more deformable, better results. In: Proceedings of the IEEE/CVF Conference on Computer Vision and Pattern Recognition (CVPR), June 2019

UDC 2020 Challenge on Image Restoration of Under-Display Camera: Methods and Results

Yuqian Zhou[1(✉)], Michael Kwan[1], Kyle Tolentino[1], Neil Emerton[2], Sehoon Lim[2], Tim Large[2], Lijiang Fu[1], Zhihong Pan[3], Baopu Li[3], Qirui Yang[4], Yihao Liu[4], Jigang Tang[4], Tao Ku[4], Shibin Ma[5], Bingnan Hu[5], Jiarong Wang[5], Densen Puthussery[6], P. S. Hrishikesh[6], Melvin Kuriakose[6], C. V. Jiji[6], Varun Sundar[7], Sumanth Hegde[7], Divya Kothandaraman[7], Kaushik Mitra[7], Akashdeep Jassal[8], Nisarg A. Shah[9], Sabari Nathan[10], Nagat Abdalla Esiad Rahel[11], Dafan Chen[5], Shichao Nie[5], Shuting Yin[5], Chengconghui Ma[5], Haoran Wang[5], Tongtong Zhao[12], Shanshan Zhao[12], Joshua Rego[13], Huaijin Chen[13], Shuai Li[13], Zhenhua Hu[13], Kin Wai Lau[14,15], Lai-Man Po[14], Dahai Yu[15], Yasar Abbas Ur Rehman[14,15], Yiqun Li[14,15], and Lianping Xing[15]

[1] University of Illinois at Urbana-Champaign, Champaign, USA
yuqian2@illinois.edu
[2] Applied Science Group, Microsoft Corporation, Albuquerque, USA
tlarge@microsoft.com
[3] Baidu Research, Sunnyvale, USA
zhihongpan@baidu.com
[4] Chinese Academy of Sciences, Shenyang, China
yangqirui18@mails.ucas.ac.cn
[5] Xidian University, Xi'an, China
dfchen@stu.xidian.edu.cn
[6] College of Engineering, Trivandrum, India
puthusserydensen@gmail.com
[7] Indian Institute of Technology Madras, Chennai, India
ee16b068@smail.iitm.ac.in
[8] Punjab Engineering College, Chandigarh, India
akash.deep.jassal@hotmail.com
[9] IIT Jodhpur, Karwar, India
[10] Couger Inc., Tokyo, Japan
sabarinathantce@gmail.com
[11] Al-Zintan University, Zintan, Libya
[12] Dalian Maritime University, Dalian, China
daitoutiere@gmail.com
[13] SenseBrain Technology, San Jose, USA
joshua.rego@sensebrain.ai
[14] City University of Hong Kong, Hong Kong, Hong Kong, China
kinwailau6-c@my.cityu.edu.hk
[15] TCL Research Hong Kong, Hong Kong, Hong Kong SAR, China

A. Bartoli and A. Fusiello (Eds.): ECCV 2020 Workshops, LNCS 12539, pp. 337–351, 2020.
https://doi.org/10.1007/978-3-030-68238-5_26

Abstract. This paper is the report of the first Under-Display Camera (UDC) image restoration challenge in conjunction with the RLQ workshop at ECCV 2020. The challenge is based on a newly-collected database of Under-Display Camera. The challenge tracks correspond to two types of display: a 4k Transparent OLED (T-OLED) and a phone Pentile OLED (P-OLED). Along with about 150 teams registered the challenge, eight and nine teams submitted the results during the testing phase for each track. The results in the paper are state-of-the-art restoration performance of Under-Display Camera Restoration. Datasets and paper are available at https://yzhouas.github.io/projects/UDC/udc.html.

Keywords: Under-display camera · Image restoration · Denoising · Debluring

1 Introduction

Under-Display Camera (UDC) [34] is specifically designed for full-screen devices as a new product trend, eliminating the need for bezels. Improving the screen-to-body ratio will enhance the interaction between users and devices. Mounting the display in front of a camera imaging lens will cause severe image degradation like low-light and blur. It is then desirable to propose proper image restoration algorithms for a better imaging quality identical to the original lens. It will also potentially benefit the downstream applications like object detection [27] and face analysis [33].

Obtaining such algorithms can be challenging. First, most existing methods leverage the advantages of deep learning to resolve multiple image degradation problems, such as image denoising [1,2,13,31,32], deblurring [8], super-resolution [18,19] etc. Restoration of UDC images, as a problem of recovering combined degradation, requires the joint modeling of methods resolving different optical effects caused by the displays and camera lens. It also requires the researchers to understand inter-disciplinary knowledge of optics and vision. Second, data acquisition process can be challenging due to variant types of displays and cameras. Collecting data sets consisting of pairs of degraded and undegraded images that are in other respects identical is challenging even using special display-camera combination hardware. Furthermore, the trained model may not be easily generalized to other devices.

In this paper, we report the methods and results from the participants of the first Under-Display Camera Challenge in conjunction with the Real-world Recognition from Low-quality Inputs (RLQ) workshop of ECCV 2020. We held this image restoration challenge to seek an efficient and high-performance image restoration algorithm to be used for recovering under-display camera images. Participants greatly improved the restoration performance compared to the baseline paper. More details will be discussed in the following sections.

2 Under-Display Camera (UDC) Challenge

2.1 UDC Dataset

The UDC dataset is collected using a monitor-based imaging system as illustrated in the baseline paper [34]. Totally 300 images from DIV2K [3] dataset are displayed on the monitor screen, and paired data is recorded using a FLIR Flea camera. In this challenge, we only use the 3-channel RGB data for training and validation. The training data consists of 240 pairs of 1024×2048 images, totally 480 images. Validation and Testing inputs each consist of 30 images of the same resolution. The challenge is organized in two phases: validation and testing. We only release the ground truth of the validation set after the end of the validation phase, while the ground truth of the testing partition is kept hidden.

2.2 Challenge Tracks and Evaluation

The challenge had two tracks: **T-OLED** and **P-OLED** image restoration. Participants were encouraged to submit results on both of them, but only attending one track was also acceptable. For both tracks, we evaluated and ranked the algorithms using the standard Peak Signal To Noise Ratio (PSNR). Additional measurements like Structural Similarity (SSIM) and inference time are also reported for reference. Although an algorithm with high efficiency is extremely important for portable devices, we did not rank the participants based on the inference time. In total of 83 teams took part in the T-OLED track, and 73 teams registered the P-OLED track. Finally, 8 teams submitted the testing results to T-OLED track, and 9 teams to the P-OLED track.

3 Teams and Methods

In this section, we summarize all the methods from the participants who submitted the final results and reports for each track.

3.1 Baidu Research Vision

Members: Zhihong Pan, Baopu Li
Affiliations: Baidu Research (USA)
Track: T-OLED
Title: Dense Residual Network with Shade-Correction for UDC Image Restoration. The architecture proposed by Team Baidu Research Vision is shown in Fig. 1. The team branches off of prior work of a dense residual network for raw image denoising and demosaicking [1] with a newly added novel shade-correction module. The novel shade-correction module consists of a set of learned correction-coefficients with the same dimension as the full-size image. Matching coefficients of the input patch are multiplied with the input for shade correction. The proposed shade-correction module could learn patterns related to the specific T-OLED screen, so additional learning might be needed for different set-ups. However, the team believes this fine-tuning process tends to be efficient. The model is trained on patches of size 128×128.

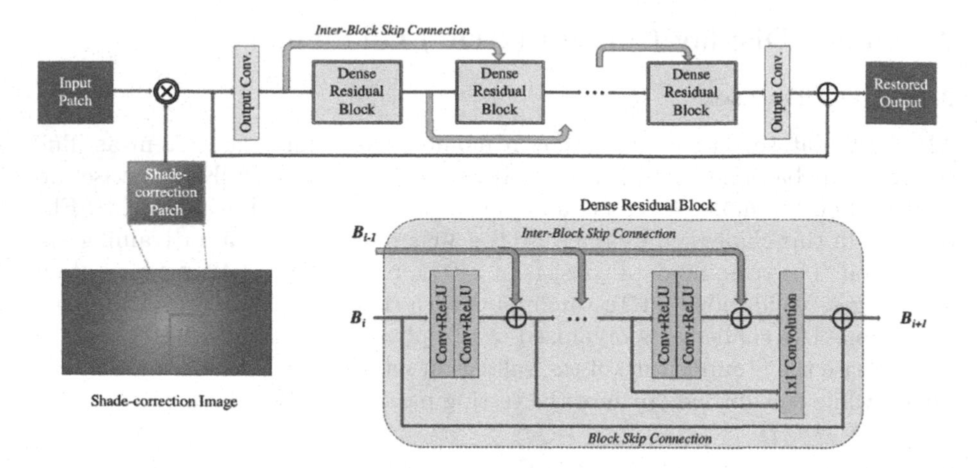

Fig. 1. The dense residual network architecture proposed by team Baidu Research Vision.

3.2 BigGuy

Members: Qirui Yang, Yihao Liu, Jigang Tang, Tao Ku
Affiliations: Chinese Academy of Sciences, Shenyang Institutes of Advanced Technology
Track: T-OLED and P-OLED
Title: Residual and Dense U-Net. [21] **for Under-display Camera Restoration.** The team's method is based on the U-net network. The U-Net encoder network consists of several residual dense blocks. A decoder network is constructed by skip connection and pixel-shuffle upsampling [22]. Their experiments show that T-OLED and P-OLED have different adaptability to the model and the patch size during training. Therefore, they proposed two different U-Net structures. For the T-OLED track, they used Residual Dense Blocks as the basic structure and proposed a Residual-Dense U-Net model (RDU-Net) as shown in Fig. 2.

For the P-OLED track, they found P-OLED panels allow small amounts of light to enter the camera so that P-OLED images present a dull feature. The difference between input and output images of P-OLED dataset is mainly reflected in color and high-frequency texture information. They thus explored residual dense blocks [29], ResNet [9], and RCAB [28] and found residual block achieved the best validation PSNR. The model structure for the P-OLED track is shown in Fig. 2.

3.3 BlackSmithM

Members: Shibin Ma, Bingnan Hu, Jiarong Wang
Affiliations: Xidian University, China
Track: P-OLED
Title: P-OLED Image Reconstruction Based on GAN Method. In view of the poor quality blurred image of the P-OLED track, the team focused on

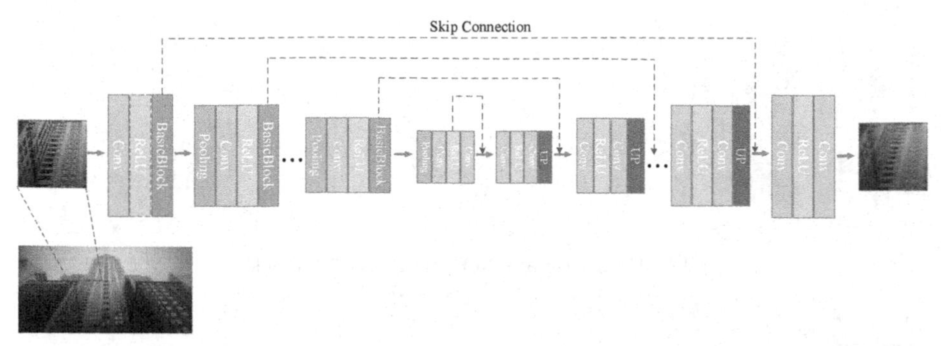

(a) The overall architecture is similar with UNet. We can choose or design different "basic blocks" (e.g. residual block,dense block, residual dense block) to obtain better performance.

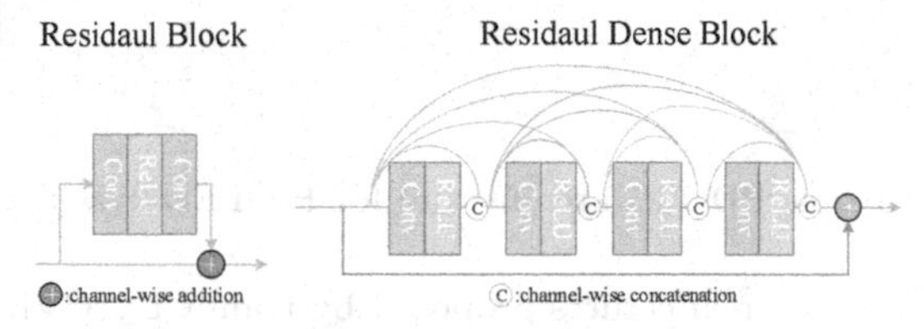

(b) Left: residual block. Right: the proposed residual dense block. It consists of a residual connection and a dense module composed of four convolutional layers.

Fig. 2. The architectures proposed by team BigGuy.

adjusting the light first. After the image is dimmed, they can better remove the blur. The team used the pix2pix [12,35] model to adjust the poor image light, and at the same time, it was possible to recover the image information. Before the image passed the pix2pix model, the team preprocessed the single scale Retinex (SSR) [25], and croped the 1024 data set to the left and right images of 1024 × 1024. The image after the pix2pix network contained noise, so a Gaussian filter was used to process the image to make the resulting image more smooth and real, thus improving the PSNR value of the image.

3.4 CET_CVLab

Members: Densen Puthussery, Hrishikesh P S, Melvin Kuriakose, Jiji C V
Affiliations: College of Engineering, Trivandrum, India
Track: T-OLED and P-OLED
Title: Dual Domain Net (T-OLED) and Wavelet Decomposed Dilated Pyramidal CNN (P-OLED). The team proposed encoder-decoder structures to learn the restoration. For the T-OLED track, they proposed a dual-domain net (DDN) inspired by [30]. In the dual domain method, the image features are processed in both pixel domain and frequency domain using implicit discrete cosine transform. This enables the network to correct the image degradation in both frequency and pixel domain, thereby enhancing the restoration. The DDN architecture is shown in Fig. 3(a).

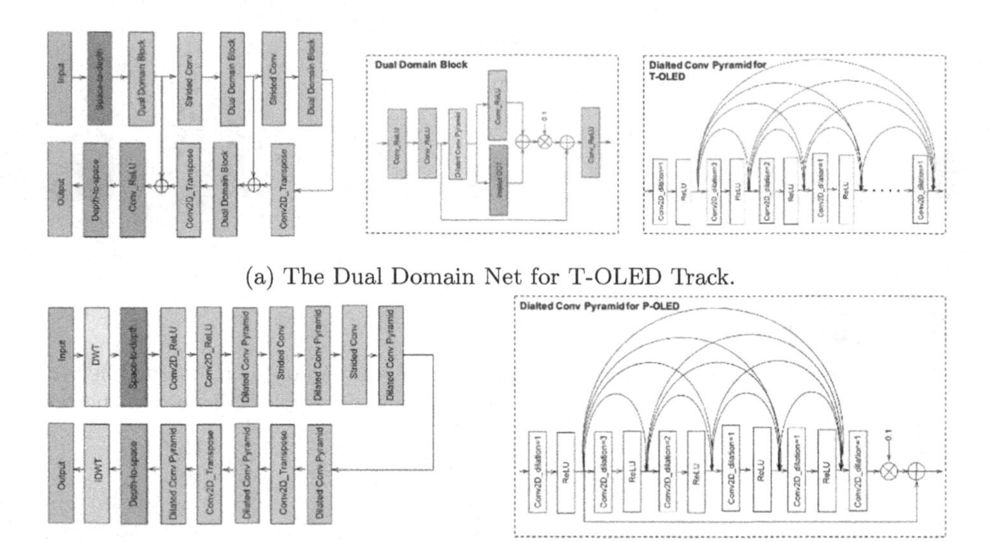

(a) The Dual Domain Net for T-OLED Track.

(b) The PDCRN architecture for P-OLED Track

Fig. 3. The architectures proposed by team CET_CVLab.

For the P-OLED track, inspired by the multi-level wavelet-CNN (MWCNN) proposed by Liu et al. [14], they proposed Pyramidal Dilated Convolutional RestoreNet (PDCRN) which follows an encoder-decoder structure as shown in Fig. 3(b). In the proposed network, the downsampling operation in the encoder is discrete wavelet transform (DWT) based decomposition instead of downsampling convolution or pooling. Similarly, in the decoder network, inverse discrete wavelet transform (IDWT) is used instead of upsampling convolution. In the wavelet based decomposition used here, the information from all the channels are combined in the downsampling process to minimize information loss when compared to that of convolutional downsampling. The feature representation for both tracks is made efficient using a pyramidal dense dilated convolutional block. The dilation rate is gradually decreased as the dilated convolution pyramid is moved up. This is to compensate for information loss that may occur with highly dilated convolution due to a non-overlapping moving window in the convolution process.

3.5 CILab IITM

Members: Varun Sundar, Sumanth Hegde, Divya Kothandaraman, Kaushik Mitra
Affiliations: Indian Institute of Technology Madras, India
Track: T-OLED and P-OLED
Title: Deep Atrous Guided Filter for Image Restoration in Under Display Cameras. The team uses two-stage pipeline for the task as shown in Fig. 4. The first stage is a low-resolution network (LRNet) which restores image quality at low-resolution. The low resolution network retains spatial resolution and

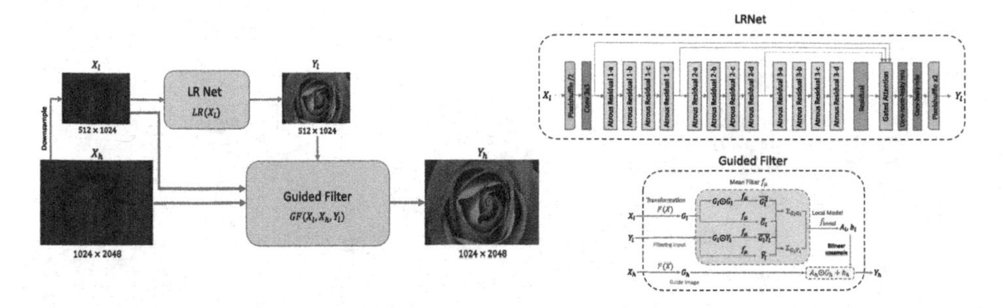

Fig. 4. The deep atrous guided filter architectures of the LRNet and the guided filter proposed by team CILab IITM.

emulates multi-scale information fusion with multiple atrous convolution blocks [5,6] stacked in parallel. In the second stage, they leverage a guided filter to produce a high resolution image from the low resolution refined image obtained from stage one. They further propose a simulation scheme to augment data and boost performance. More details are in the team's challenge report [23].

3.6 Hertz

Members: Akashdeep Jassal[1], Nisarg A. Shah[2]
Affiliations: Punjab Engineering College, India[1], IIT Jodhpur, India[2]
Track: P-OLED
Title: P-OLED Reconstruction Using GridDehazeNet. Based on GridDehazeNet which aims to clear haze from a low resolution image [16], the team uses the network which contains a pre-processing residual dense block, a grid-like backbone of the same residual dense blocks interconnected with convolutional downsampling and upsampling modules, and a post-processing stage of another residual dense block.

3.7 Image Lab

Members: Sabari Nathan[1], Nagat Abdalla Esiad Rahel[2]
Affiliations: Couger Inc., Japan[1], Al-Zintan University, Libya[2]
Track: T-OLED and P-OLED
Title: Image Restoration Using Light weight Multi Level Supervision Model. The team proposes a Lightweight Multi Level Supervision Model inspired by [20]. The architecture is shown in Fig. 5. The input image is first passed to the coordinate convolutional layer to map the pixels to a Cartesian coordinate space [15], and then fed into the encoder. The encoder composed of 3×3 convolution layers, two Res2Net [7] blocks, and a downsampling layer, while the decoder block replaces the last component with a subpixel scaling layer [22]. A convolution block attention module (CBAM) [26] in the skip connection is concatenated with the encoding block as well.

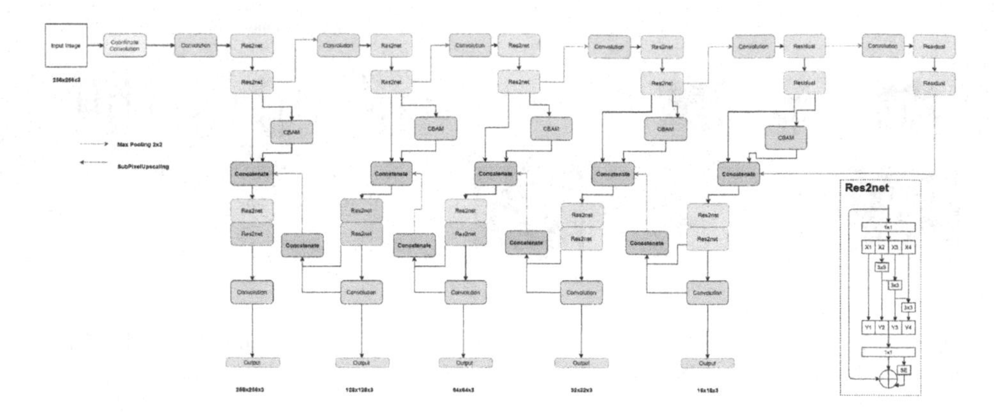

Fig. 5. The lightweight multi level supervision model architecture proposed by team Image Lab.

3.8 IPIUer

Members: Dafan Chen, Shichao Nie, Shuting Yin, Chengconghui Ma, Haoran Wang
Affiliations: Xidian University, China
Track: T-OLED
Title: Channel Attention Image Restoration Networks with Dual Residual Connection. The team proposes a novel UNet model inspired by Dual Residual Networks [17] and Scale-recurrent Networks (SRN-DeblurNet) [24]. As shown in Fig. 6, in the network, there are 3 EnBlocks, 3 DeBlocks and 6 DRBlocks. The entire network has three stages and one bridge between encoder and decoder. Every stage consists of one EnBlock/DeBlock and a residual group (ResGroup). The ResGroup has seven residual blocks (ResBlock). Between the encoder and decoder, a dual residual block (DRBlock) is used to extract high level semantic features effectively. The skip connection uses squeeze-and-exitation blocks [10] which aims to highlight the features of some dimensions.

3.9 Lyl

Members: Tongtong Zhao, Shanshan Zhao
Affiliations: Dalian Maritime University, China
Track: T-OLED and P-OLED
Title: Coarse to Fine Pyramid Networks for Progressive Image Restoration. The team proposes a coarse to fine network (CFN) for progressive reconstruction. Specifically, in each network level, the team proposes a lightweight upsampling module (LUM), also named FineNet as in Fig. 7, to process the input, and merge it with the input features. Such progressive cause-and-effect process helps to achieve the principle for image restoration: high-level information can guide an image to recover a better restored image. The authors claim that they can achieve competitive results with a modest number of parameters.

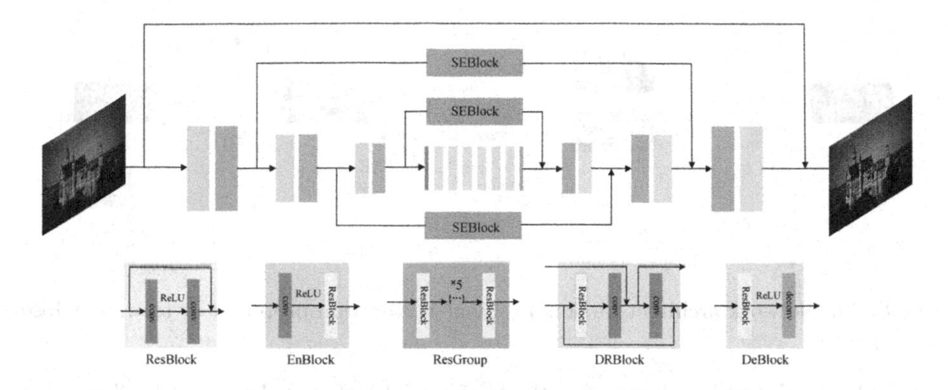

Fig. 6. The novel Unet structure proposed by team IPIUer.

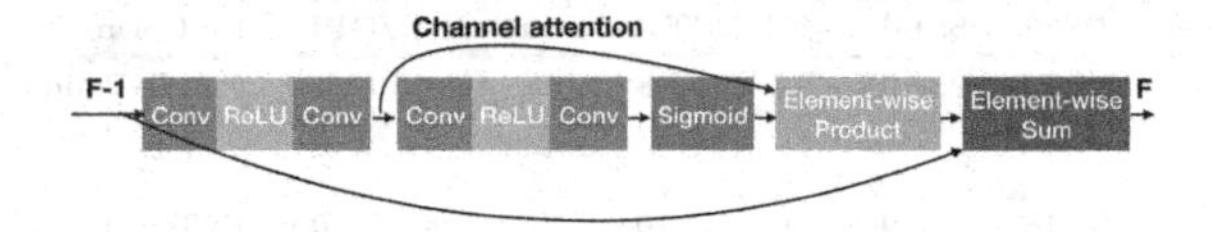

Fig. 7. The Lightweight Upsampling Module (LUM) named FineNet proposed by team lyl.

3.10 San Jose Earthquakes

Members: Joshua Rego, Huaijin Chen, Shuai Li, Zhenhua Hu
Affiliations: SenseBrain Technology, USA
Track: T-OLED and P-OLED
Title: Multi-stage Network for Under-Display Camera Image Restoration. The team proposes multiple stage networks as shown in Fig. 8 to solve different issues caused by under-display cameras. For T-OLED, the pipeline is two-staged. The first stage uses a multi-scale network PyNET [11] to recover intensity and largely apply deblur to the input image. The second stage is a U-Net fusion network that uses the output of the PyNET as well as a sharper, but noisier, alternating direction method of multipliers (ADMM) [4] reconstruction as the inputs of the network and outputs weights used to combine the two inputs for a sharper, de-noised result. P-OLED pipeline uses an additional third stage color correction network to improve color consistency with the target image by pixel-wise residual learning.

The authors mentioned that training solely through the first-stage network, while greatly restoring towards the target image, was unable to restore sharpness completely, especially in higher frequency textures. The ADMM output, on the other hand, was able to retain high-frequency sharpness, but suffered largely from noise and some additional artifacts. The fusion network blends the desirable characteristics from the two to slightly improve the results. However, one drawback of the method is that ADMM takes about 2.5 mins to solve each

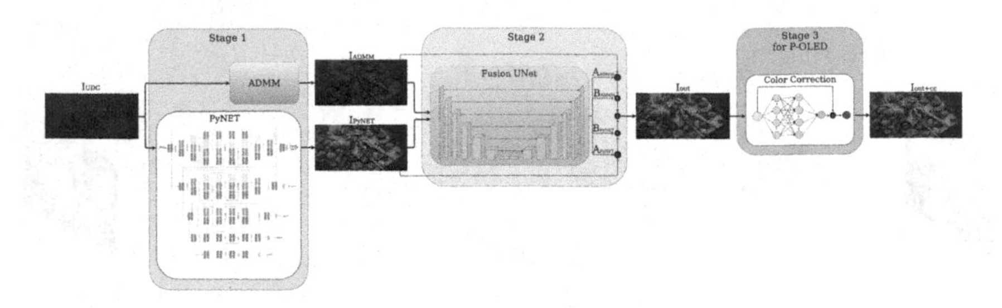

Fig. 8. The network architecture of multi-stage restoration proposed by team San Jose earthquakes.

Table 1. Results and rankings of methods of T-OLED Track. TT: Training Time. IT: Inference Time

Team	Username	PSNR	SSIM	TT(h)	IT (s/frame)	CPU/GPU	Platform	Ensemble	Loss
Baidu Research Vision	zhihongp	$38.23_{(1)}$	$0.9803_{(1)}$	12	11.8	Tesla M40	PaddlePaddle, PyTorch	flip(×4)	L_1, SSIM
IPIUer	TacoT	$38.18_{(2)}$	$0.9796_{(3)}$	30	1.16	Tesla V100	PyTorch	None	L_1
BigGuy	JaydenYang	$38.13_{(3)}$	$0.9797_{(2)}$	23	0.3548	Tesla V100	PyTorch	-	L_1
CET_CVLAB	hrishikeshps	$37.83_{(4)}$	$0.9783_{(4)}$	72	0.42	Tesla P100	Tensorflow	-	L_2
CILab IITM	ee15b085	$36.91_{(5)}$	$0.9734_{(6)}$	96	1.72	GTX 1080 Ti	PyTorch	flip(×4) model(×8)	L_1
lyl	tongtong	$36.72_{(6)}$	$0.9776_{(5)}$	72	3	-	PyTorch	model(-)	L_1
Image Lab	sabarinathan	$34.35_{(7)}$	$0.9645_{(7)}$	-	1.6	GTX 1080 Ti	Keras	model(-)	L_2, SSIM
San Jose earthquakes	jdrego	$33.78_{(8)}$	$0.9324_{(8)}$	18	180	-	PyTorch	model(-)	-

image, which is longer for inference. Nevertheless, the method is a novel approach fusing the implementation of a traditional and deep-learning method.

3.11 TCL Research HK

Members: Kin Wai Lau[1,2], Lai-Man Po[1], Dahai Yu[2], Yasar Abbas Ur Rehman[1,2], Yiqun Li[1,2], Lianping Xing[2]
Affiliations: City University of Hong Kong [1], TCL Research Hong Kong [2]
Track: P-OLED
Title: Self-Guided Dual Attention Network for Under Display Camera Image Restoration. The team proposes a multi-scale self-guidance neural architecture containing (1) multi-resolution convolutional branch for extracting multi-scale information, (2) low-resolution to high-resolution feature extraction for guiding the intermediate high-resolution feature extraction process, (3) spatial and channel mechanisms for extracting contextual information, (4) Dilated Residual Block (DRB) to increase the receptive field for preserving the details, (5) local and global feature branch for adjusting local information (e.g., contrast,

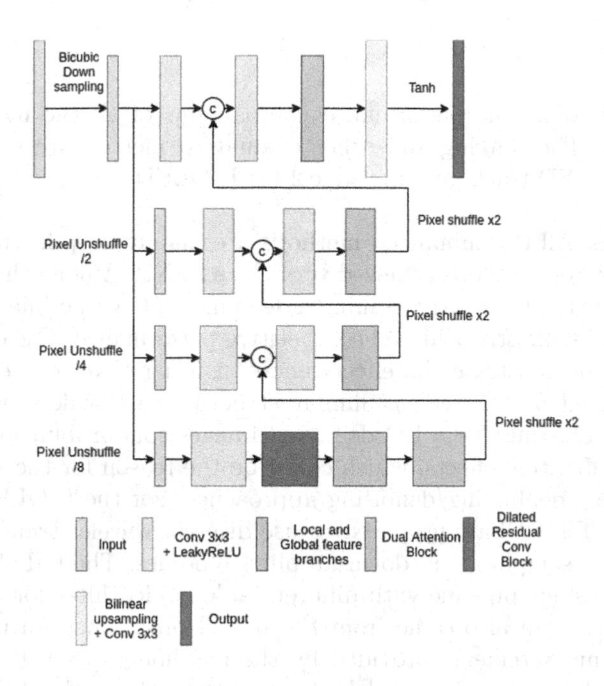

Fig. 9. The network architecture proposed by team TCL Research HK.

Table 2. Results and rankings of methods of P-OLED Track. TT: Training Time. IT: Inference Time

Team	Username	PSNR	SSIM	TT(h)	IT (s/frame)	CPU/GPU	Platform	Ensemble	Loss
CET_ -CVLAB	Densen	$32.99_{(1)}$	$0.9578_{(1)}$	72	0.044	Tesla T4	Tensorflow	-	L_2
CILab IITM	varun19299	$32.29_{(2)}$	$0.9509_{(2)}$	96	1.72	GTX 1080 Ti	PyTorch	flip(×4) model(×8)	L_1
BigGuy	JaydenYang	$31.39_{(3)}$	$0.9506_{(3)}$	24	0.2679	Tesla V100	PyTorch	model	L_1
TCL Research HK	stevenlau	$30.89_{(4)}$	$0.9469_{(5)}$	48	1.5	Titan Xp	PyTorch	-	L_1, SSIM
BlackSmithM	BlackSmithM	$29.38_{(5)}$	$0.9249_{(6)}$	-	2.43	Tesla V100	PyTorch	-	L_1
San Jose Earthquakes	jdrego	$28.61_{(6)}$	$0.9489_{(4)}$	18	180	-	PyTorch	model	-
Image Lab	sabarinathan	$26.60_{(7)}$	$0.9161_{(7)}$	-	1.59	-	-	None	L_2, SSIM
Hertz	akashdeepjassal	$25.72_{(8)}$	$0.9027_{(8)}$	-	2.29	Tesla K80	PyTorch	None	VGG
lyl	tongtong	$25.46_{(9)}$	$0.9015_{(9)}$	72	3	-	PyTorch	model(-)	L_1

detail, etc.) and global information (e.g., global intensity, scene category, color distribution, etc.) The network architecture proposed by Team TCL Research HK for Self-Guided Dual Attention Network is shown in Fig. 9.

4 Results

This section presents the performance comparisons of all the methods in the above sections. The ranking and other evaluation metrics are summarized in Table 1 for T-OLED track, and in Table 2 for P-OLED.

Top Methods. All the submitted methods are mainly deep-learning oriented. On both tracks, top methods achieved very close PSNR. Among the top-3 methods for T-OLED track, directly training a deep model (e.g. modified UNet) in an end-to-end fashion mostly achieved competitive performance. The outperformed results further demonstrate the effectiveness of using UNet embedding Residual Blocks shared by the teams. Similar structures are widely used in image denoising or deblurring. T-OLED degraded images contain blur and noisy patterns due to diffraction effects, which could be the reason for the superiority of directly applying deblurring/denoising approaches. For the P-OLED track, the winner team, CET_CVLab, proposed to use discrete wavelet transform (DWT) to replace the upsampling and downsampling modules. The CILab IITM team proposed a two-stage pipeline with differentiable guided filter for training. The performance gain can also come from the model pre-trained on the simulated data by using measurements provided by the baseline paper [34]. The BigGuy team conducted an extensive model search to find the optimal structures for P-OLED images. Some methods proposed by other teams, though not ranked on top-3, are also novel and worthy to mention. The team San Jose Earthquakes proposed to combine the results of the deep-learning and traditional methods by leveraging the benefits from both ends. The multi-level supervision model proposed by the Image Lab team restores the image in a progressive way. Similarly, the lyl team also share the progressive idea.

In addition to module design, the model depth, parameter amounts, data augmentation and normalization, or training strategies can also cause performance differences. Most teams also report the performance gains from model or inputs ensemble strategies.

T-OLED v.s. P-OLED. According to the experiment results, T-OLED demonstrates an easier task than P-OLED. The T-OLED restoration problem itself resembles a combination of deblurring and denonising tasks. However, imaging through P-OLED suffers heavily from lower light transmission and color shift. Some teams like CILab IITM, Image Lab and lyl, which participated in both tracks, chose to use the same models for two tracks, while other teams tend to use different model structures. Team BigGuy explored different module options to better resolve the low-light and color issues of P-OLED inputs. Team CET_CVLab addresses the information loss issues of downsampling from P-OLED inputs by using a wavelet-based decomposition. Team San Jose Earthquakes added an additional color correction stage for P-OLED.

Inference Efficiency. We did not rank the methods by the inference time due to device and platform difference of different methods. Most methods run about

1 to 3 s per image of size 1024 × 2048 on GPUs. Without further optimization, these models may not be easily applied in mobile devices or laptops for real-time inference of live streams or videos. However, it is still feasible to restore degraded images or videos in an offline way. Team San Jose Earthquakes run a longer inference time since their method involves an additional ADMM optimization process. Team CET_CVLab claimed to achieve 0.044s inference time on a better GPU, which makes the method both outperformed and high-efficient.

5 Conclusions

We summarized and presented the methods and results of the first image restoration challenge on Under-Display Camera (UDC). The testing results represent state-of-the-art performance on Under-Display Camera Imaging. Participants extensively explored state-of-the-art deep-learning based architectures conventionally used for image restoration. Some additional designs like shade and color correction are also proven beneficial to be adaptive to the through-display imaging tasks. However, the results are specifically limited to two display types, P-OLED and T-OLED, and a single camera. This further suggests the need for exploring more display types and hardware set-ups so the model can be generalized.

Acknowledgment. We thank the UDC2020 challenge and RLQ workshop Sponsor: Microsoft Applied Science Group.

References

1. Abdelhamed, A., Afifi, M., Timofte, R., Brown, M.S.: Ntire 2020 challenge on real image denoising: Dataset, methods and results. In: Proceedings of the IEEE/CVF Conference on Computer Vision and Pattern Recognition Workshops, pp. 496–497 (2020)
2. Abdelhamed, A., Timofte, R., Brown, M.S.: Ntire 2019 challenge on real image denoising: methods and results. In: Proceedings of the IEEE Conference on Computer Vision and Pattern Recognition Workshops (2019)
3. Agustsson, E., Timofte, R.: Ntire 2017 challenge on single image super-resolution: dataset and study. In: Proceedings of the IEEE Conference on Computer Vision and Pattern Recognition Workshops, pp. 126–135 (2017)
4. Boyd, S., Parikh, N., Chu, E.: Distributed Optimization and Statistical Learning via the Alternating Direction Method of Multipliers. Now Publishers Inc., Breda (2011)
5. Brehm, S., Scherer, S., Lienhart, R.: High-resolution dual-stage multi-level feature aggregation for single image and video deblurring. In: Proceedings of the IEEE/CVF Conference on Computer Vision and Pattern Recognition Workshops, pp. 458–459 (2020)
6. Chen, D., et al.: Gated context aggregation network for image dehazing and deraining. In: 2019 IEEE Winter Conference on Applications of Computer Vision (WACV), pp. 1375–1383. IEEE (2019)

7. Gao, S., Cheng, M.M., Zhao, K., Zhang, X.Y., Yang, M.H., Torr, P.H.: Res2net: a new multi-scale backbone architecture. IEEE Trans. Pattern Anal. Mach. Intell. (2019)
8. Guo, Q., et al.: Effects of blur and deblurring to visual object tracking. arXiv preprint arXiv:1908.07904 (2019)
9. He, K., Zhang, X., Ren, S., Sun, J.: Deep residual learning for image recognition. In: Proceedings of the IEEE Conference on Computer Vision and Pattern Recognition, pp. 770–778 (2016)
10. Hu, J., Shen, L., Sun, G.: Squeeze-and-excitation networks. In: Proceedings of the IEEE Conference on Computer Vision and Pattern Recognition, pp. 7132–7141 (2018)
11. Ignatov, A., Van Gool, L., Timofte, R.: Replacing mobile camera isp with a single deep learning model. In: Proceedings of the IEEE/CVF Conference on Computer Vision and Pattern Recognition Workshops, pp. 536–537 (2020)
12. Isola, P., Zhu, J.Y., Zhou, T., Efros, A.A.: Image-to-image translation with conditional adversarial networks. In: Proceedings of the IEEE Conference on Computer Vision and Pattern Recognition, pp. 1125–1134 (2017)
13. Liu, J., Wu, C.H., et al.: Learning raw image denoising with Bayer pattern unification and bayer preserving augmentation. In: Proceedings of the IEEE Conference on Computer Vision and Pattern Recognition Workshops (2019)
14. Liu, P., Zhang, H., Zhang, K., Lin, L., Zuo, W.: Multi-level wavelet-cnn for image restoration. In: Proceedings of the IEEE Conference on Computer Vision and Pattern Recognition Workshops, pp. 773–782 (2018)
15. Liu, R., et al.: An intriguing failing of convolutional neural networks and the CoordConv solution. In: Advances in Neural Information Processing Systems, pp. 9605–9616 (2018)
16. Liu, X., Ma, Y., Shi, Z., Chen, J.: Griddehazenet: attention-based multi-scale network for image dehazing. In: Proceedings of the IEEE International Conference on Computer Vision, pp. 7314–7323 (2019)
17. Liu, X., Suganuma, M., Sun, Z., Okatani, T.: Dual residual networks leveraging the potential of paired operations for image restoration. In: Proceedings of the IEEE Conference on Computer Vision and Pattern Recognition, pp. 7007–7016 (2019)
18. Mei, Y., et al.: Pyramid attention networks for image restoration. arXiv preprint arXiv:2004.13824 (2020)
19. Mei, Y., Fan, Y., Zhou, Y., Huang, L., Huang, T.S., Shi, H.: Image super-resolution with cross-scale non-local attention and exhaustive self-exemplars mining. In: Proceedings of the IEEE/CVF Conference on Computer Vision and Pattern Recognition, pp. 5690–5699 (2020)
20. Nathan, D.S., Beham, M.P., Roomi, S.: Moire image restoration using multi level hyper vision net. arXiv preprint arXiv:2004.08541 (2020)
21. Ronneberger, O., Fischer, P., Brox, T.: U-Net: convolutional networks for biomedical image segmentation. In: Navab, N., Hornegger, J., Wells, W.M., Frangi, A.F. (eds.) MICCAI 2015, Part III. LNCS, vol. 9351, pp. 234–241. Springer, Cham (2015). https://doi.org/10.1007/978-3-319-24574-4_28
22. Shi, W., et al.: Real-time single image and video super-resolution using an efficient sub-pixel convolutional neural network. In: Proceedings of the IEEE Conference on Computer Vision and Pattern Recognition, pp. 1874–1883 (2016)
23. Sundar, V., Hegde, S., Kothandaraman, D., Mitra, K.: Deep atrous guided filter for image restoration in under display cameras. arXiv preprint arXiv:2008.06229 (2020)

24. Tao, X., Gao, H., Shen, X., Wang, J., Jia, J.: Scale-recurrent network for deep image deblurring. In: Proceedings of the IEEE Conference on Computer Vision and Pattern Recognition, pp. 8174–8182 (2018)
25. Wei, C., Wang, W., Yang, W., Liu, J.: Deep retinex decomposition for low-light enhancement. arXiv preprint arXiv:1808.04560 (2018)
26. Woo, S., Park, J., Lee, J.-Y., Kweon, I.S.: CBAM: convolutional block attention module. In: Ferrari, V., Hebert, M., Sminchisescu, C., Weiss, Y. (eds.) ECCV 2018, Part VII. LNCS, vol. 11211, pp. 3–19. Springer, Cham (2018). https://doi.org/10.1007/978-3-030-01234-2_1
27. Yu, X., Gong, Y., Jiang, N., Ye, Q., Han, Z.: Scale match for tiny person detection. In: The IEEE Winter Conference on Applications of Computer Vision, pp. 1257–1265 (2020)
28. Zhang, Y., Li, K., Li, K., Wang, L., Zhong, B., Fu, Y.: Image super-resolution using very deep residual channel attention networks. In: Ferrari, V., Hebert, M., Sminchisescu, C., Weiss, Y. (eds.) ECCV 2018, Part VII. LNCS, vol. 11211, pp. 294–310. Springer, Cham (2018). https://doi.org/10.1007/978-3-030-01234-2_18
29. Zhang, Y., Tian, Y., Kong, Y., Zhong, B., Fu, Y.: Residual dense network forimage restoration. IEEE Trans. Pattern Anal. Mach. Intell. (2020)
30. Zheng, B., Chen, Y., Tian, X., Zhou, F., Liu, X.: Implicit dual-domain convolutional network for robust color image compression artifact reduction. IEEE Trans. Circuits Syst. Video Technol. (2019)
31. Zhou, Y., Jiao, J., Huang, H., Wang, J., Huang, T.: Adaptation strategies for applying AWGN-based denoiser to realistic noise. In: Proceedings of the AAAI Conference on Artificial Intelligence, vol. 33, pp. 10085–10086 (2019)
32. Zhou, Y., et al.: When awgn-based denoiser meets real noises. arXiv preprint arXiv:1904.03485 (2019)
33. Zhou, Y., Liu, D., Huang, T.: Survey of face detection on low-quality images. In: 2018 13th IEEE International Conference on Automatic Face and Gesture Recognition (FG 2018), pp. 769–773. IEEE (2018)
34. Zhou, Y., Ren, D., Emerton, N., Lim, S., Large, T.: Image restoration for under-display camera. arXiv preprint arXiv:2003.04857 (2020)
35. Zhu, J.Y., Park, T., Isola, P., Efros, A.A.: Unpaired image-to-image translation using cycle-consistent adversarial networks. In: Proceedings of the IEEE International Conference on Computer Vision, pp. 2223–2232 (2017)

A Dual Residual Network with Channel Attention for Image Restoration

Shichao Nie(✉), Chengconghui Ma, Dafan Chen, Shuting Yin, Haoran Wang, LiCheng Jiao, and Fang Liu

Key Laboratory of Intelligent Perception and Image Understanding of Ministry of Education, International Research Center for Intelligent Perception and Computation, Joint International Research Laboratory of Intelligent Perception and Computation School of Artificial Intelligence, Xidian University, Xi'an 710071, Shaanxi, China
{scnie,cchma,dfchen,styin,hrwang}@stu.xidian.edu.cn,
lchjiao@mail.xidian.edu.cn, f63liu@163.com

Abstract. Deep learning models have achieved significant performance on image restoration task. However, restoring the images with complicated and combined degradation types still remains a challenge. For this purpose, we proposed a dual residual network with channel attention (DRANet) to address complicated degradation in the real world. We further exploit the potential of encoder-decoder structure. To fuse feature more efficiently, we adopt the channel attention module with skip-connections. To better process low- and high-level information, we introduce the dual residual connection into the network architecture. And we explore the effect of multi-level connection to image restoration. Experimental results demonstrate the superiority of our proposed approach over state-of-the-art methods on the UDC T-OLED dataset.

Keywords: Image restoration · Channel attention · Dual residual connection · Multi-level connection

1 Introduction

Image restoration refers to the task that aims to restore a high-quality image from its degraded version. As a classic computer vision problem, it has a long history of research. According to the types of degradation, it is generally categorized into different subtasks such as image de-noising, de-blurring, de-hazing, super-resolution and so on. Previously, researchers tackled the problems by modeling clean natural images. The main methods are filtering [3,6–8], sparse coding [14,21,25], effective prior [22,29,30] or modeling based on physical models [2,13]. However, these methods often have a large number of parameters and lack universality.

Recently, with the development of deep learning, deep convolutional models have achieved excellent performance in dealing with image degradation as in

S. Nie, C. Ma, D. Chen, S. Yin—Equal contribution.

A. Bartoli and A. Fusiello (Eds.): ECCV 2020 Workshops, LNCS 12539, pp. 352–363, 2020.
https://doi.org/10.1007/978-3-030-68238-5_27

many other computer vision tasks. Kai Zhang *et al.* [26] demonstrate the ability of the residual block to the image restoration tasks. Since then, many researchers adopt the residual block as the feature learning module. Inspired by other studies, Xing Liu *et al.* [11] proposed the dual residual networks as shown in Fig. 1. By exploiting the potential of paired operations, this method achieves a better feature extraction.

Furthermore, the encoder-decoder structure has been proven to be effective in many vision tasks [1,17], and it is widely used as the model framework [12,18,23]. Based on the study of Xin Tao *et al.* [19], it can be seen that using both encoder-decoder structure and the residual block is more effective than using any of them alone in image restoration tasks.

In this study, we pay attention to the Under-Display Camera (UDC) image restoration. It is a new problem in the fields of image restoration. UDC is a new imaging system that mounts a display screen on top of a traditional digital camera lens. It is of great significance for the development of full screen equipment, which can provide better user perceptive and intelligent experience. Moreover, it also provides a better human-computer interaction. However, because the imaging sensor is mounted behind a display, there are complex degradation types in the captured map image. The main low-quality problems are noisy and blurry caused by low light transmission rates and diffraction effects.

However, aiming to restore the data with single degradation types, most of the existing methods are hard to achieve good results on the images with complicated and combined degradation types. Different from them, our study is aimed to address the complicated real degradation problem.

In this paper, we propose a more effective encoder-decoder network for complicated real degraded image restoration, which we call dual residual network with channel attention (DRANet). We make some modifications to adapt encoder-decoder networks into our architecture. In our network, we adopt the channel-wise attention mechanism into skip-connections. This operation can combine different levels of information and enhance essential features, which can help to get more accurate results. Besides, we add the dual connection as the bridge between encoder and decoder. This method has a more dense connection than general connection so that it can help to process high-level semantic features more effectively. Moreover, multi-level residual blocks are introduced into our networks, which are composed of stacked residual blocks. By using multi-level residual blocks, our networks can get the better capability of feature extraction.

We evaluate the effectiveness of our proposed method through several experiments. Furthermore, the contributions of this study are summarized as follows:

1) We propose a novel encoder-decoder network to restore real-world low-quality images with complicated and combined degradation types. Our method outperforms the state-of-the-art image restoration approaches for the task.
2) We show that simple channel-wise attention is effective for image restoration and adopt it in the skip-connections.

3) We further exploit the potential of dual residual connection in the field of image restoration.
4) We show how the multi-level connection structure might affect performance.

2 Proposed Method

In this paper, we propose a novel UNet-based [17] end-to-end model, which we call dual residual network with channel attention (DRANet). We will describe the architecture of our proposed DRANet in this section. The overall architecture of the network is shown in Fig. 1. It takes a sequence of degraded images as input, and produces corresponding sharp images.

2.1 Network Architecture

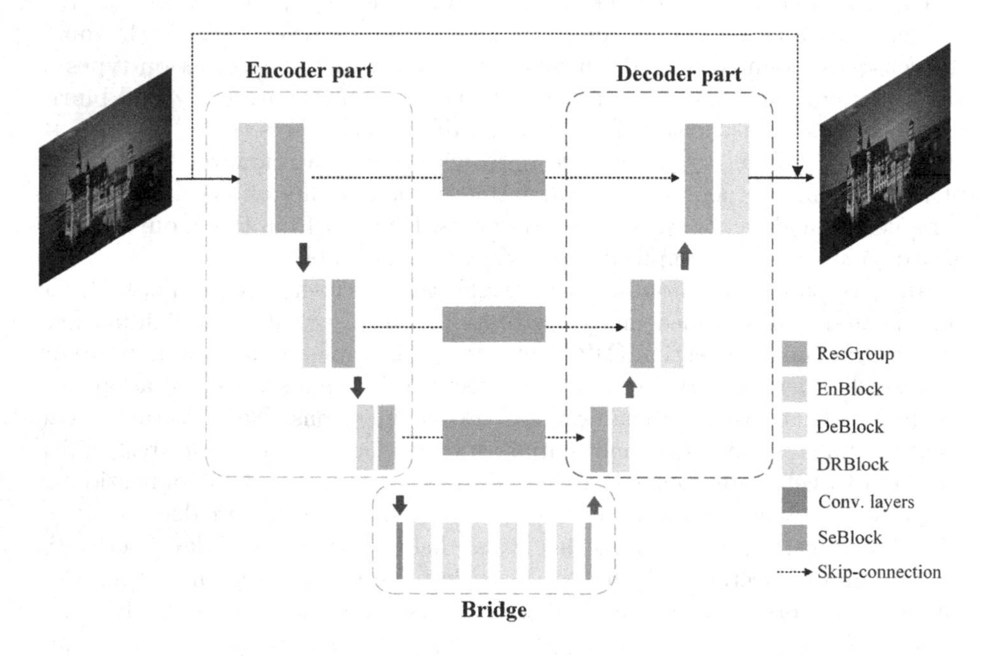

Fig. 1. Our proposed dual residual network with channel attention (DRANet). It consists of an encoder part, a decoder part and a bridge between encoder and decoder.

As explained above, we adopt the encoder-decoder structure with several skip-connections in our network [10,15]. The encoder-decoder structure is a symmetrical network. Through the encoder-decoder network, the input data is first converted to smaller feature maps with more channels, and then converted to original size in the end. It can get a strong feature that combines pixel-level information and high-level semantic information, which has been proven to be effective in many image processing tasks. This kind of structure still has much potential.

As shown in Fig. 1, our entire network includes three components, an encoder part, a decoder part and a bridge between encoder and decoder. The encoder part is a set of downsamplers that extract the features of different scales from degraded, and the decoder part is used to reconstruct the sharp images by using a set of upsamplers.

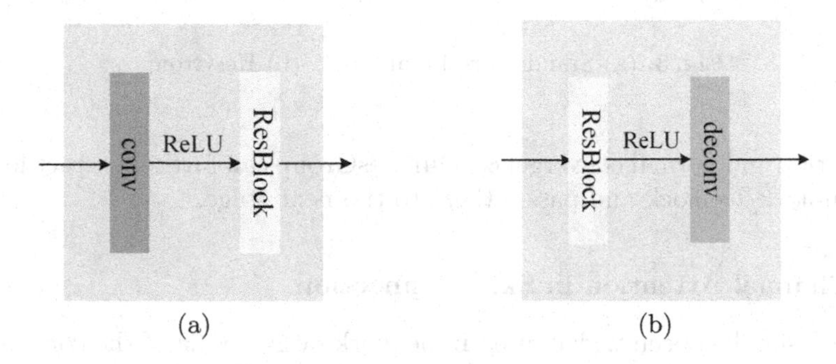

Fig. 2. (a) EnBlock. (b) DeBlock.

For an encoder-decoder structure, our method has three stages (refer to the numbers of the upsampler/downsampler). In every stage, there are an EnBlock(DeBlock) and a ResGroup. As illustrated in Fig. 2(a), our EnBlock includes one convolution layer with stride = 2 followed by one standard ResBlock. The DeBlock is symmetric to EnBlock, and the convolutional layer is replaced by the deconvolution layer in it, as shown in Fig. 2(b). In the same stage, we choose the same kernel size and kernel numbers. Besides, the number of kernels in the back stages is the twice it in the preceding stages.

Furthermore, we adopt the dual residual connection as the bridge part. And there are three skip-connections with SeBlock for two features with the same size of feature map between encoder and decoder. Moreover, we also add a global shortcut connection from input to output.

2.2 Multi-level Connection Structure

Most general encoder-decoder structures tend to have a few layers in every stage. Unlike the general method, the multi-level residual blocks are introduced into our networks. As mentioned earlier, in order to enhance the networks, there is a ResGroup in every stage (as illustrated in Fig. 3(b)). The ResGroup consists of 7 standard residual blocks (as shown in Fig. 3(a)). In every residual block, there are two convolution layers with 5×5 kernels followed by a ReLU.

In some way, image restoration is a pixel-level task. To get better results, we want to process different information accurately. By using multi-level residual blocks, our networks can get better fitting capability while ensuring gradient

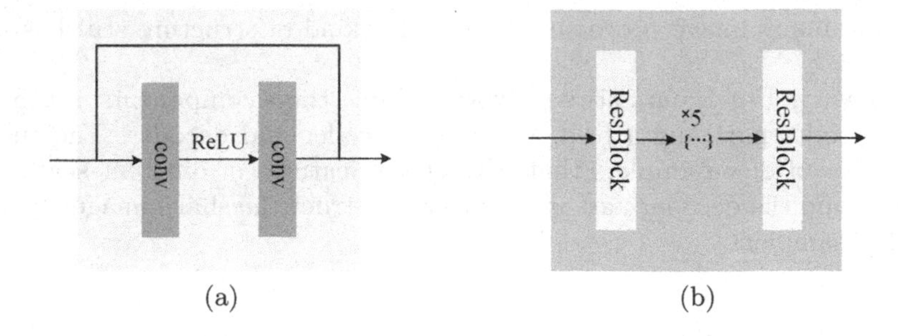

(a) (b)

Fig. 3. (a) Standard residual block. (b) ResGroup.

propagation and normal convergence. Our ResGroup can further extract features from Enblock/Deblock and passes them to the next stage.

2.3 Channel Attention in Skip-Connection

Feature fusion has been widely used in network design. One of the conventional methods is reusing the features from previous stages. Between the encoder part and the decoder part, skip-connection is usually used to combine the feature maps with the same spatial size. We adopt an element-wise summation in skip-connections rather than concatenation in our method.

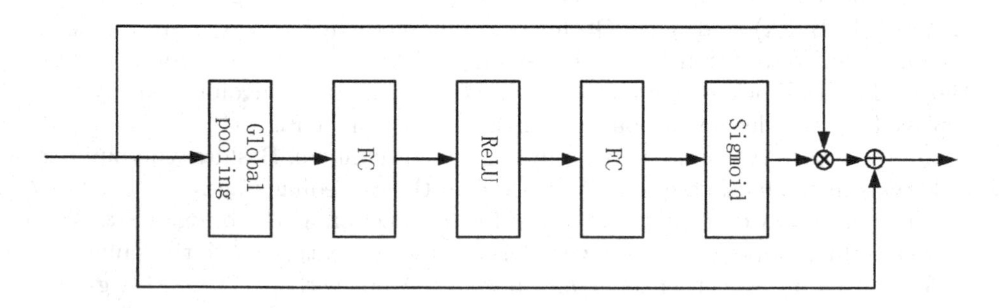

Fig. 4. SeBlock module.

Moreover, to better assist upsampler to reconstruct a sharp date, the channel-wise attention mechanism is introduced into our skip-connection. As shown in Fig. 1, we adopt the SeBlock module [5] in skip-connections, which can enhance the essential features and weaken others. Besides, we add a connection between the two ends of the SeBlock to keep original features from encoder while refining features. As illustrated in Fig. 4, the SeBlock contains a global average pooling layer, two full connection layers followed by rectified linear unit (ReLU) and Sigmoid, respectively.

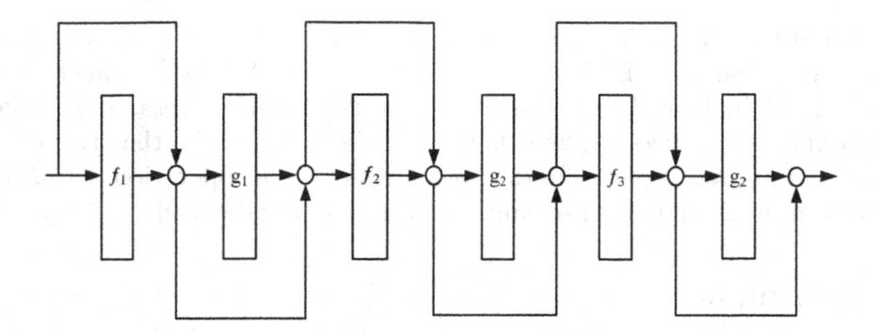

Fig. 5. Dual residual connection

By using these operations, our networks reuse the features of the same scale from the previous layer and offer the upsamplers more high-frequency information, which can remedy the missing feature during the downsampling process. Moreover, our SeBlock module can highlight the features of some dimensions and provides more refined high-resolution information. After fusing different features, it can help to get more accurate results.

2.4 Dual Residual Connection

Between encoder and decoder, we add some operations to further process the feature map gained in the encoder part and to enhance overall performance. For the effectiveness of the residual networks, a favoured explanation is from [20]: a sequential connection of residual blocks is regarded as an ensemble of many sub-networks. There are more paths of sub-networks and it can still be improved. Besides, we find that most of the encoder-decoder structures usually lack enough connection between the features from non-adjacent layers.

To fully exploiting the features from non-adjacent, a more dense connection is introduced into our method. We use dual residual connection like [11] between encoder and decoder (as shown in Fig. 5), rather than ResBlock. The dual residual exploits the potential of paired operations and has more paths of sub-networks as mentioned before.

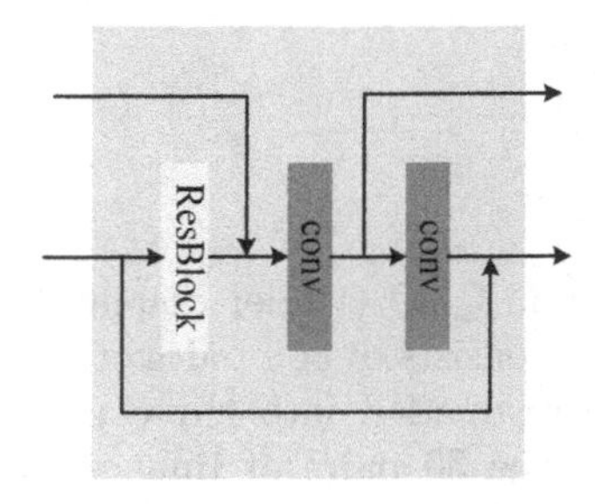

Fig. 6. DRBlock

We adopt the dual residual modules into the bridge between encoder part and decoder part, named as DRBlock (as shown in Fig. 6). The DRBlock includes a standard residual block followed by two convolutional layers. This structure adds a connection between different blocks, which guarantee that two convolutional layers are always paired in the possible path. This operations can help our network to process high-level semantic features more effectively.

3 Experiment

Our experiment is conducted on two devices, a workstation with Intel Xeon E5 CPU and two NVIDIA GTX 1080 GPU and a workstation with Intel Xeon E5 CPU and two Tesla-V100 GPU. We implement our method on the PyTorch platform. In this section, we present our experimental results to validate the effectiveness of the proposed model.

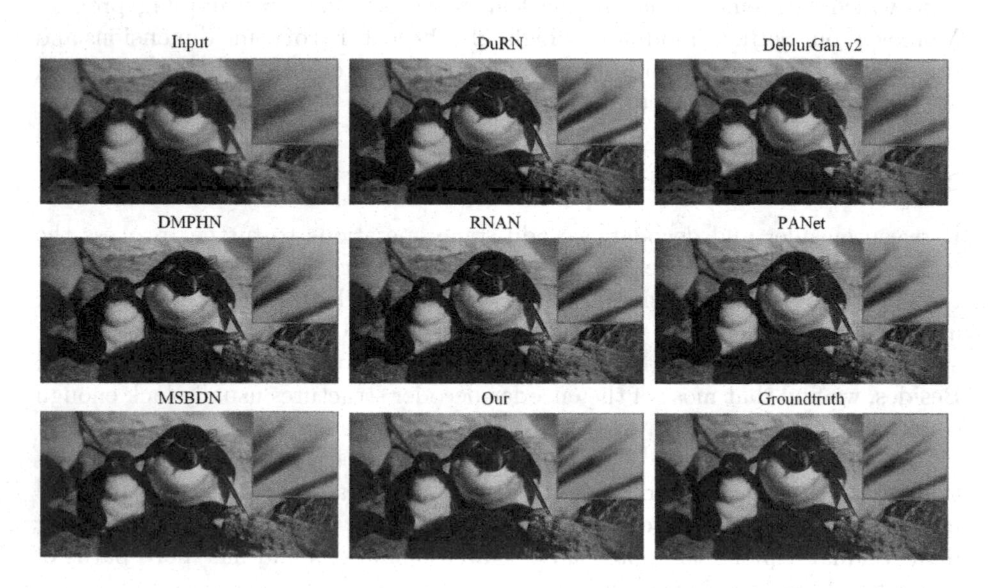

Fig. 7. Visualization of the selected parts restored by different methods. Our method has better ability to restore the real-world low-quality images than other methods.

3.1 Data Preparation

We train/test our methods on UDC (Under-display Camera) T-OLED dataset [28]. UDC T-OLED dataset consists of 300 pairs of 1024×2048 training images. The 300 pairs of images are divided into three parts: (1) the first 240 pairs of images for training, (2) the 30 pairs of images for validation, and (3) the remaining 30 pairs of images for testing. Each pair of images includes a sharp image (ground truth) and a degraded image.

3.2 Implementation Details

As is known to all, large amounts of data need to be prepared to a deep neural network model, and a dataset which contains 240 pairs of images is too small to train a network fully. Because of the large size of the images in the dataset, we randomly crop 256×256-pixel patches as training input. In other words, we get a much larger dataset, and the amount of data is enough to train a deep learning model. Moreover, to prevent the overfitting problem, we adopt the data augmentation, including horizontal, vertical flipping and rotations.

For model training, we use Adam optimization with $\beta_1 = 0.9$, $\beta_2 = 0.999$, and $\varepsilon_1 = 10^{-8}$. The initial learning rate is set to 0.0001 and decayed to $1e^{-5}$ and $1e^{-6}$ in epoch 1600 and 3200, respectively. We train our model for 4000 epochs with batch size 8. For training loss, we use *l1* loss between the restored images and their ground truths.

3.3 Comparisons

We compare our method with the state-of-the-arts image restoration approaches including DuRN [11], SRN [19], DeblurGAN v2 [9], DMPHN [24], RNAN [27], PANet [16], MSBDN [4]. The peak signal-to-noise ratio (PSNR) are employed as the evaluation metrics.

Table 1. Quantitative results for baseline models. All the methods are evaluated on the UDC T-OLED dataset using the same training setting as the proposed method.

Methods	PSNR
SRN	36.86
DuRN	35.9
DeblurGan v2	31.73
DMPHN	30.78
RNAN	36.11
PANet	36.64
MSBDN	37.72
Ours	**38.93**

Table 1 shows the PSNR values of the several methods on validation data of UDC T-OLED dataset. It is seen that our method outperforms others in PSNR metric and achieves 1.21 dB performance improvement over them at least. As mentioned earlier, these state-of-the-arts image restoration approaches can achieve significant performance on restoring images with single degradation type. But it's difficult for them to enhance real-world low-quality images with complicated and combined degradation types.

We select some examples for visualization, as shown in Fig. 7. It is observed that there is a color shift issue in some of other methods. And compared with other approaches, our proposed method can produce superior results with sharper structures and clear details. That indicates that our method can address the complicated real degradation more effectively.

3.4 Ablation Study

Our proposed dual residual network with channel attention (DRANet) can be broken down into a base model and three parts: multi-level residual block, channel-wise attention and dual residual connection. We conduct some ablation studies to evaluate how they contribute to the performance. All methods below are trained using the same setting as the proposed algorithm for fair comparisons.

Table 2. Effects of the num of stages and global short cut.

Num of stages	Global short cut	PSNR
2		37
3		37.5
4		37.3
3	√	37.9

Base Model. As explained earlier, our base model can be regarded as an encoder-decoder-UNet structure with the short cut, and we replace normal convolutional layers with standard residual block. One of the vital ingredients of this structure is the number of stages. To demonstrate how the number of stages performs for our task, we conduct a comparison. In our experiments, we replace the ResGroup with standard residual block and remove the DRBlock and SeBlock. Table 2 shows that when the number of stages is set to 3, it achieves 0.5 dB performance improvement over the network with stage = 2. If the number of stages is set to 4, it will get a lower PSNR. So our method can get the best results when the number of stages is set to 3. With the number of stage reaches 4 and more, the size of feature maps will be too small to keep effective information.

Table 3. Effects of the multi-level connection.

Num of level	PSNR
3	38.02
4	38.34
5	38.65
6	38.74
7	38.80
8	38.78

Multi-level. In every stages, the number of ResBlock in the ResGroup is a hyperparameter. To investigate the effect of it, we evaluate the ResGroup with different layers under the same configurations. At first, as shown in Table 3, there are 3 ResBlocks in the ResGroup and PSNR reaches 38.02 dB. As we add more ResBlocks, the PSNR value climbs step by step with the number of ResBlocks. When the number is set to 7, our method achieves the best PSNR, 38.80 dB, while the PSNR decreases to 38.78 dB when we add one more ResBlock. It is shown that there are limits on the feature processing in the same stage. Too many layers in the same stage will has negative effect on feature extraction, instead of positive.

Table 4. Effects of the dual residual connection and channel attention.

Dual residual connection	Channel attention	PSNR
√	√	38.93
√		38.84
	√	38.85

Dual Residual Connection and Channel Attention. Here we analyze how the dual residual connection and channel attention perform for our task, DRBlock and SeBlock in our networks. We evaluate the following alternatives to demonstrate the effectiveness of them. Begining with the baseline DRANet, we first remove the SeBlock and add the DRBlock as the bridge. Next, we re-introduce the SeBlock and remove the DRBlock. The evaluation results are shown in Table 4. It is shown that the network with both dual residual connection and channel attention can get the best PSNR. The network with both of them can outperform others by 0.08 dB at least.

4 Conclusions

In this paper, we propose an end-to-end dual residual network with channel attention (DRANet) to enhance real-world low-quality images. Our DRANet is a deep UNet-based model. By introducing the channel attention with skip-connections, the network efficiently refines information while fusing the feature. To fully exploit the features from non-adjacent layers, we adopt the dual residual connection as the bridge between encoder and decoder. And experimental results demonstrate that multi-level connection can further bring improvement in this task. Moreover, our method achieves the state-of-the-art performance on UDC T-OLED dataset. In the future, we expect a broader range of applications for it, especially in the image restoration task.

References

1. Chen, L.C., Papandreou, G., Kokkinos, I., Murphy, K., Yuille, A.L.: Semantic image segmentation with deep convolutional nets and fully connected CRFs. arXiv preprint arXiv:1412.7062 (2014)
2. Cho, T.S., Paris, S., Horn, B.K., Freeman, W.T.: Blur kernel estimation using the radon transform. In: CVPR 2011, pp. 241–248. IEEE (2011)
3. Dabov, K., Foi, A., Katkovnik, V., Egiazarian, K.: Image restoration by sparse 3D transform-domain collaborative filtering. In: Image Processing: Algorithms and Systems VI. vol. 6812, p. 681207. International Society for Optics and Photonics (2008)
4. Dong, H., et al.: Multi-scale boosted dehazing network with dense feature fusion. In: Proceedings of the IEEE/CVF Conference on Computer Vision and Pattern Recognition, pp. 2157–2167 (2020)
5. Hu, J., Shen, L., Sun, G.: Squeeze-and-excitation networks. In: Proceedings of the IEEE Conference on Computer Vision and Pattern Recognition, pp. 7132–7141 (2018)
6. Isola, P., Zhu, J.Y., Zhou, T., Efros, A.A.: Image-to-image translation with conditional adversarial networks. In: Proceedings of the IEEE Conference on Computer Vision and Pattern Recognition, pp. 1125–1134 (2017)
7. Knaus, C., Zwicker, M.: Progressive image denoising. IEEE Trans. Image Process. **23**(7), 3114–3125 (2014)
8. Shreyamsha Kumar, B.K.: Image denoising based on non-local means filter and its method noise thresholding. Signal Image Video Process. **7**(6), 1211–1227 (2012). https://doi.org/10.1007/s11760-012-0389-y
9. Kupyn, O., Martyniuk, T., Wu, J., Wang, Z.: Deblurgan-v2: Deblurring (orders-of-magnitude) faster and better. In: Proceedings of the IEEE International Conference on Computer Vision, pp. 8878–8887 (2019)
10. Li, G., He, X., Zhang, W., Chang, H., Dong, L., Lin, L.: Non-locally enhanced encoder-decoder network for single image de-raining. In: Proceedings of the 26th ACM international conference on Multimedia, pp. 1056–1064 (2018)
11. Liu, X., Suganuma, M., Sun, Z., Okatani, T.: Dual residual networks leveraging the potential of paired operations for image restoration. In: Proceedings of the IEEE Conference on Computer Vision and Pattern Recognition, pp. 7007–7016 (2019)
12. Liu, Z., Yeh, R.A., Tang, X., Liu, Y., Agarwala, A.: Video frame synthesis using deep voxel flow. In: Proceedings of the IEEE International Conference on Computer Vision, pp. 4463–4471 (2017)
13. Mai, L., Liu, F.: Kernel fusion for better image deblurring. In: Proceedings of the IEEE Conference on Computer Vision and Pattern Recognition (CVPR), June 2015
14. Mairal, J., Bach, F., Ponce, J., Sapiro, G., Zisserman, A.: Non-local sparse models for image restoration. In: 2009 IEEE 12th International Conference on Computer Vision, pp. 2272–2279. IEEE
15. Mao, X., Shen, C., Yang, Y.B.: Image restoration using very deep convolutional encoder-decoder networks with symmetric skip connections. In: Advances in Neural Information Processing Systems, pp. 2802–2810 (2016)
16. Mei, Y., et al.: Pyramid attention networks for image restoration. arXiv preprint arXiv:2004.13824 (2020)

17. Ronneberger, O., Fischer, P., Brox, T.: U-Net: convolutional networks for biomedical image segmentation. In: Navab, N., Hornegger, J., Wells, W.M., Frangi, A.F. (eds.) MICCAI 2015, Part III. LNCS, vol. 9351, pp. 234–241. Springer, Cham (2015). https://doi.org/10.1007/978-3-319-24574-4_28
18. Tao, X., Gao, H., Liao, R., Wang, J., Jia, J.: Detail-revealing deep video super-resolution. In: Proceedings of the IEEE International Conference on Computer Vision, pp. 4472–4480 (2017)
19. Tao, X., Gao, H., Shen, X., Wang, J., Jia, J.: Scale-recurrent network for deep image deblurring. In: Proceedings of the IEEE Conference on Computer Vision and Pattern Recognition, pp. 8174–8182 (2018)
20. Veit, A., Wilber, M.J., Belongie, S.: Residual networks behave like ensembles of relatively shallow networks. In: Advances in Neural Information Processing Systems, pp. 550–558 (2016)
21. Wen, B., Ravishankar, S., Bresler, Y.: Structured overcomplete sparsifying transform learning with convergence guarantees and applications. Int. J. Comput. Vis. **114**(2–3), 137–167 (2015). https://doi.org/10.1007/s11263-014-0761-1
22. Xu, J., Zhang, L., Zuo, W., Zhang, D., Feng, X.: Patch group based nonlocal self-similarity prior learning for image denoising. In: Proceedings of the IEEE International Conference on Computer Vision, pp. 244–252 (2015)
23. Xu, N., Price, B., Cohen, S., Huang, T.: Deep image matting. In: Proceedings of the IEEE Conference on Computer Vision and Pattern Recognition, pp. 2970–2979 (2017)
24. Zhang, H., Dai, Y., Li, H., Koniusz, P.: Deep stacked hierarchical multi-patch network for image deblurring. In: Proceedings of the IEEE Conference on Computer Vision and Pattern Recognition, pp. 5978–5986 (2019)
25. Zhang, J., Zhao, D., Gao, W.: Group-based sparse representation for image restoration. IEEE Trans. Image Process. **23**(8), 3336–3351 (2014)
26. Zhang, K., Zuo, W., Chen, Y., Meng, D., Zhang, L.: Beyond a Gaussian denoiser: residual learning of deep CNN for image denoising. IEEE Trans. Image Process. **26**(7), 3142–3155 (2017)
27. Zhang, Y., Li, K., Li, K., Zhong, B., Fu, Y.: Residual non-local attention networks for image restoration. arXiv preprint arXiv:1903.10082 (2019)
28. Zhou, Y., Ren, D., Emerton, N., Lim, S., Large, T.: Image restoration for under-display camera. arXiv preprint arXiv:2003.04857 (2020)
29. Zoran, D., Weiss, Y.: From learning models of natural image patches to whole image restoration. In: 2011 International Conference on Computer Vision, pp. 479–486. IEEE (2011)
30. Zuo, W., Zhang, L., Song, C., Zhang, D.: Texture enhanced image denoising via gradient histogram preservation. In: Proceedings of the IEEE Conference on Computer Vision and Pattern Recognition, pp. 1203–1210 (2013)

Transform Domain Pyramidal Dilated Convolution Networks for Restoration of Under Display Camera Images

Hrishikesh Panikkasseril Sethumadhavan(✉), Densen Puthussery, Melvin Kuriakose, and Jiji Charangatt Victor

College of Engineering, Trivandrum, India
{hrishikeshps,puthusserydenson,memelvin,jijicv}@cet.ac.in

Abstract. Under-display camera (UDC) is a novel technology that can make digital imaging experience in handheld devices seamless by providing large screen-to-body ratio. UDC images are severely degraded owing to their positioning under a display screen. This work addresses the restoration of images degraded as a result of UDC imaging. Two different networks are proposed for the restoration of images taken with two types of UDC technologies. The first method uses a pyramidal dilated convolution within a wavelet decomposed convolutional neural network for pentile-organic LED (P-OLED) based display system. The second method employs pyramidal dilated convolution within a discrete cosine transform based dual domain network to restore images taken using a transparent-organic LED (T-OLED) based UDC system. The first method produced very good quality restored images and was the winning entry in European Conference on Computer Vision (ECCV) 2020 challenge on image restoration for Under-display Camera - Track 2 - P-OLED evaluated based on PSNR and SSIM. The second method scored 4th position in Track-1 (T-OLED) of the challenge evaluated based on the same metrics.

Keywords: Under-display camera · Image restoration · Wavelet · T-OLED · P-OLED

1 Introduction

In recent years there has been a growing demand for casual photography and video conferencing using hand-held devices, especially in smartphones. To address such demands various imaging systems were introduced such as pop-up cameras, punch-hole cameras etc. Under-display camera (UDC) is such a digital data acquisition technique or an imaging system that is a by-product of the growing demand and greater hunger for new technologies. UDC is an imaging system in which a display screen is mounted atop the lens of a conventional digital camera. The major attraction of a UDC device is that the highest possible display-to-body ratio can be realized with a bezel-free design. UDC devices

P. S. Hrishikesh and D. Puthussery—The authors contributed equally.

A. Bartoli and A. Fusiello (Eds.): ECCV 2020 Workshops, LNCS 12539, pp. 364–378, 2020.
https://doi.org/10.1007/978-3-030-68238-5_28

can provide a seamless digital experience through perfect gaze tracking in teleconferences. Leading smartphone makers like OPPO and XIAOMI have already developed and integrated under-display camera technology into some of their flagship products.

The quality of a UDC image is poor due to severe degradation resulting from diffraction and low light transmission. Under-display camera image restoration is relatively a new task for the computer vision community since the technology is new. However, it has a high correlation with many image restoration tasks such as image denoising, deblurring, dehazing and low light image enhancement. The degradations in UDC images can be considered as a combination of degradations addressed in such restoration problems.

Two different techniques are predominantly used for display purpose in a UDC device. They are transparent organic light emitting diode (T-OLED) and the pentile organic light emitting diode (P-OLED). The type of degradation is different in either of these technologies. The major degradations in T-OLED are blur and noise while in P-OLED they are color shift, low light and noise [20]. Figure 1 shows degraded images taken using P-OLED and T-OLED UDC cameras. The clean image taken without mounting a display screen is also depicted for reference.

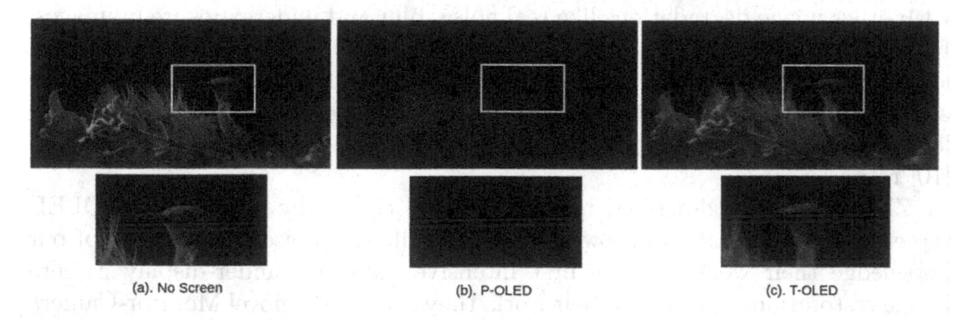

Fig. 1. (a) Original image without display screen (b) degraded image with P-OLED screen mounted on the camera (c) degraded image when using T-OLED screen.

This work addresses the restoration of under-display camera images taken with P-OLED and T-OLED configurations. Recently, deep convolutional neural network based techniques have proven to be effective for most image restoration tasks. Two methods are proposed in this work to address the restoration of under-display camera images. The first method employs a pyramidal dilated convolution block within a wavelet decomposed convolutional neural network for the restoration of P-OLED degraded images. The second method integrates a discrete cosine transform (DCT) based dual domain sub-net within a fully convolutional encoder-decoder architecture and performs well on T-OLED degraded images.

The first method is a state-of-the-art that won the first position in the European Conference on Computer Vision (ECCV) 2020 challenge on image restoration for Under-display Camera - Track 2 - P-OLED [19] organized in conjunction with RLQ-TOD workshop. The second method achieved 4th position in Track -1 - T-OLED in the same challenge and has comparable performance with the winning methodology in that challenge track. The main contributions of the proposed work are :

- A pyramidal dilated convolutional block has been integrated into a wavelet decomposed encoder-decoder network for state-of-the-art restoration results on P-OLED under-display camera images.
- An encoder-decoder network with DCT based dual domain sub-net has been developed to address T-OLED under-display camera image restoration.

Rest of the paper is organized as follows : in Sect. 2 we review related works and in Sect. 3 we describe the two methodologies that we propose for UDC image restoration. Section 4 details our experiments and in Sect. 5 we present our result analysis. Finally, Sect. 6 concludes the work.

2 Related Work

Degradation caused by the under-display camera has a significant correlation with other image degradations like real noise, blur and underexposure in images. Recently, learning based methods have brought a significant improvement in low level vision tasks like super-resolution [7,8], deblurring [14], denoising [2] etc., and have produced promising results. A collection of state of the art deep learning techniques on super-resolution and deblurring can be found in [15] and [10] respectively.

Zhou *et al.* [20] described the degradation caused by UDC with P-OLED screens as a combination of low-light, color shift and noise. To the best of our knowledge their work was the first intensive study on under-display camera image restoration problem. In their work, they proposed a novel Monitor-Camera Imaging System (MCIS) for paired data acquisition for studying UDC problem. They demonstrated a data synthesizing pipeline for generating UDC images from natural images using only the camera measurements and display pattern. Additionally, they proposed a dual encoder variant of UNet [11] to restore the under-display camera images they generated using the proposed data pipeline.

Extensive works have been done in tasks like low-light enhancement and denoising using deep learning techniques. Earlier such tasks were addressed using synthetic data generated using Additive White Gaussian Noise (AWGN) to develop training set for deep learning based networks. The DnCNN [16] network proposed by Zang *et al.* performs denoising on noisy images of unknown noise levels. This network follows a modified VGG19 [4] architecture. Recently, there is a shift in the dataset used for training deep learning models. Datasets containing images pairs of original image and images with natural degradation are becoming more relevant. Abdelhamed *et al.* [1] proposed novel smartphone image denoising dataset (SIDD) containing images with real noise in smartphone images. Several denoising methods like WNNM [3] and DnCNN [16] were benchmarked based on this dataset. Kim *et al.* [6] proposed a network that uses adaptive instance normalisation (AIN) to learn a denoiser for real noisy images from the models trained on synthetic training pair. Their network has better generalisation and has produced state-of-the-art results on Darmstadt Noise Dataset (DND) which contains real noisy images.

Liu *et al.* [9] proposed a multi-level Wavelet-CNN (MWCNN) for general image restoration tasks. MWCNN follows an encoder-decoder architecture and uses wavelet decomposition for a computationally efficient network for subsampling. In their work, they demonstrated the effectiveness of their method for denoising and super-resolution. The winning team of CVPR NTIRE 2020 Challenge on single image demoireing [13] used a wavelet decomposed network similar to MWCNN and achieved state-of-the-art results.

Zhang *et al.* [17] proposed the use of dilated convolutions for removing compression artifacts from images. In their work, diverse dilation rates were used, with which they reached a receptive field of up to 145×145. Their theory of receptive field was adopted by Zheng *et al.* [18] wherein they used a pyramidal dilated convolution structure and implicit domain transform to remove compression artifacts. In their work, the dilation rate was gradually increased along with the dilated convolution stack and proved the effectiveness of dilated convolution in removing artifacts with a wide spatial range.

In this work to restore P-OLED degraded images, we propose a fully convolutional network with wavelet decomposition namely Pyramidal Dilated Convolutional RestoreNet (PDCRN) inspired by MWCNN [9]. Here, the feature extraction in the encoder and decoder levels has been improved by integrating the pyramidal dilated convolution blocks. A different approach was followed to address restoration of T-OLED degraded images wherein, a discrete cosine transform (DCT) based dual domain CNN was employed.

3 Proposed Methods

In this section, we describe the two different networks that we propose for the restoration of under-display camera images. Section 3.1 details the solution for P-OLED while Sect. 3.2 details the solution for T-OLED images. Although both networks follow a fully convolutional encoder-decoder architecture, the encoder and decoder sub-nets are different in either case.

3.1 Restoring P-OLED Degraded Images

The proposed Pyramidal Dilated Convolutional RestoreNet (PDCRN) has a fully convolutional encoder-decoder architecture similar to a UNet [11] as shown in Fig. 2. There are three different operation levels in terms of the spatial resolution of the features being processed. The major degradations in a P-OLED under-display camera image are low light, color shift and noise as mentioned earlier. The task of the encoder is to extract the image features including contrast, semantic information and illumination from the degraded image. The decoder then uses this information to predict the clean image that mimics the one taken without a display screen mounted on the lens.

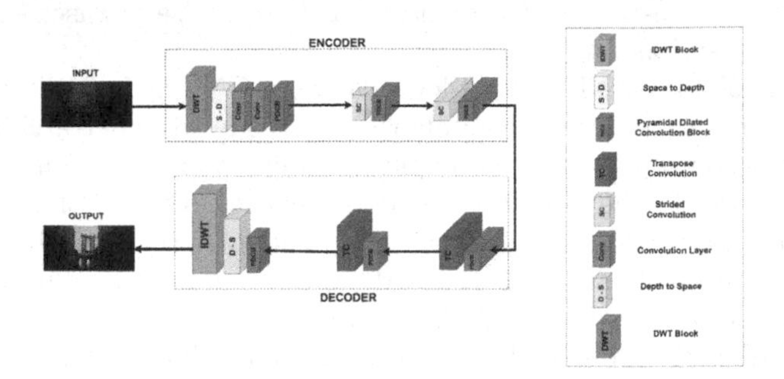

Fig. 2. Proposed PDCRN architecture. The network has an encoder and a decoder sub-net as shown in the figure. The network uses PDCB for efficient feature extraction.

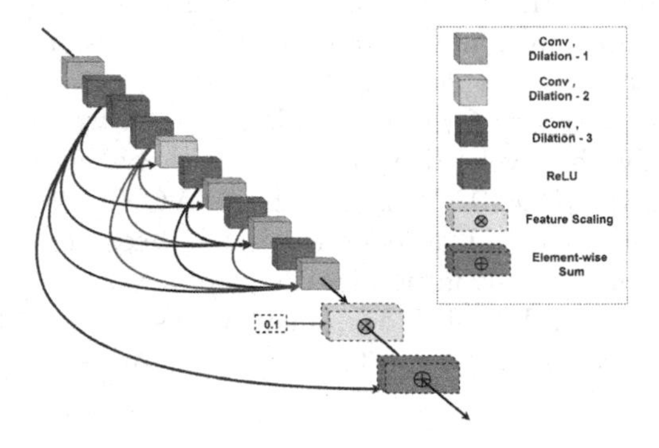

Fig. 3. The PDCB sub-net used in PDCRN. The block has a series of densely connected convolutional layers with varying dilation rates for feature extraction.

PDCRN Encoder. In the encoder section, the input image is processed at 3 different spatial scales obtained by progressively downsampling the inputs from the previous blocks. At the initial spatial scales of the encoder, the features such as colour distribution, brightness are captured. As encoder progress to smaller spatial scales, higher frequency details like texture, edges etc. are learned from the image. The initial downsampling is done using discrete wavelet transform (DWT) based on Haar wavelet as it envelops a larger receptive field which in turn helps in acquiring maximum spatial information from the input image. The downsampling in the subsequent layers is done using strided convolutions as it is empirically found to be more effective than using DWT based downsampling along the whole of encoder. At each level in the encoder, a pyramidal dilated convolution block (PDCB) is used for efficient feature extraction. A detailed description of PDCB is discussed later.

PDCRN Decoder. The decoder section consists of a series of upsampling operations with PDCB blocks in between to learn the representation of the upscaled image. The decoder progressively upsamples the feature maps from the encoder with 3 levels of upsampling until the required spatial resolution is reached. The first two upsampling processes are done using transposed convolutions and inverse discrete wavelet transform (IDWT) is used to upsample the image to the original resolution. To make the training process more efficient space-to-depth and depth-to-space [12] blocks are used at both ends of the network.

Pyramidal Dilated Convolutional Block (PDCB). To capture the degradations that have wide spatial extension, convolutional filters of large receptive field is necessary. But this would potentially increase the network complexity and will require a huge dataset to avoid the risk of overfitting. Another solution is to have a sequence of convolutional layers with low filter size which effectively provide a larger receptive field with lower learnable parameters. However, this would also increase the number of trainable parameters and may lead to vanishing gradient problem for deep networks. Dilated convolutions with large dilation factors can overcome these limitations. However, large dilation rates may lead to information loss as the aggregated features in two adjacent pixels come from completely non-overlapping input feature set [9]. Additionally, these may lead to gridding effect in the reconstructed images. To tackle this problem, a pyramidal dilated convolutional block (PDCB) is used for feature extraction in each encoder and decoder levels.

PDCB is a residual densely [5] connected convolutional network having varying dilatation rates in different convolution layers as shown in Fig. 3. The block is said to be pyramidal because the dilation rate used in the convolution layers is progressively reduced within the block in which the layers are stacked. Each convolutional layer in the block receives the activation outputs of every preceding convolutional layer as well as the input to the block. At the bottom of the pyramid, the information from far away pixels can be aggregated by using a high dilation rate. The information leakage that occurs with the highly dilated convolution is captured in the top of the pyramid. Thus the network achieves a large receptive field without information loss.

The proposed PDCRN produced excellent results for P-OLED degraded images and the same network was also employed for the restoration of T-OLED degraded images and achieved considerable performance. To further improve the performance for T-OLED images, the network was modified using a dual domain approach with implicit Discrete Cosine Transform (DCT) and obtained superior results for T-OLED compared to PDCRN. This network is explained in detail in the next section.

3.2 Restoring T-OLED Degraded Images

In a T-OLED display, the size of horizontal gratings or openings for light transmission have a size comparable to visible light wavelength [20]. Due to this, the light is heavily spread in the horizontal axis and results in blurry images. Thus the restoration of T-OLED degraded images is majorly a deblurring problem for horizontal distortion apart from removing the added noise during the acquisition process.

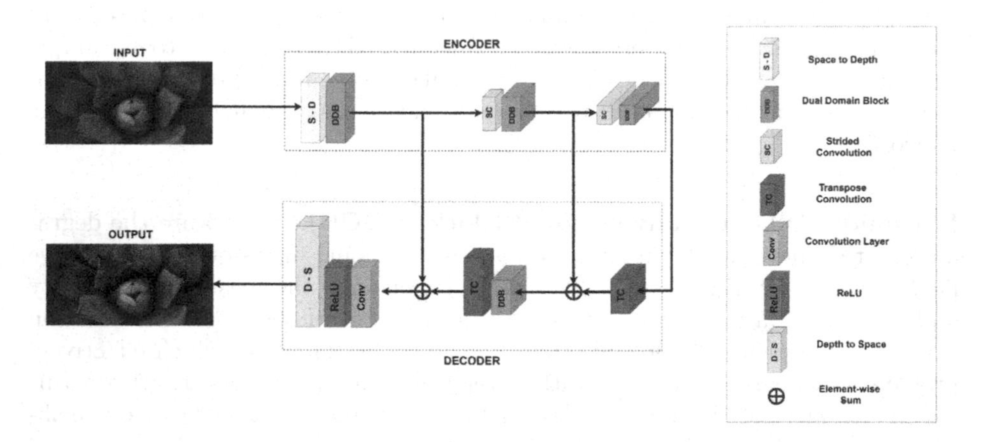

Fig. 4. Modified PDCRN architecture with dual domain block (DDB) for T-OLED degraded image restoration.

The modified PDCRN with dual domain block also follows an encoder-decoder architecture that process the image features at 3 different scales. The architecture of the modified PDCRN is depicted in Fig. 4. Each encoder and decoder level in the network uses a dual domain block with implicit DCT inspired from Implicit Dual-domain Convolutional Network (IDCN) proposed by Zheng *et al.* [18]. IDCN was introduced for restoration of compressed images. This restoration involves correction of details in the high frequency band and removal of distortion artifacts. In a T-OLED degraded image, the restoration task is to remove the horizontally spread blur distortion. since the blur leads to smoothening of edges and other sharp details, the correction is to be done in the mid and high frequency bands. Due to this reason, the IDCN methodology was adopted and improved in this work.

Dual Domain Block. Here we use a simplified version of the IDCN proposed in Zheng *et al.* [18]. In this block, the image features are processed in the pixel domain using convolution layers and in the frequency domain using implicit discrete cosine transform (DCT). The DCT domain branch in the dual domain block takes the feature inputs and convert them into the DCT domain using a quantization rectified unit (QRU) and an implicit translation unit (ITU). The

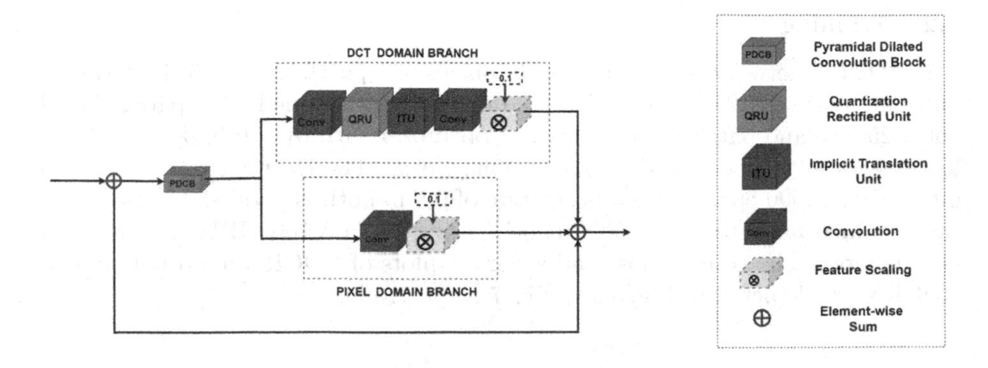

Fig. 5. Expansion of the dual domain block in modified PDCRN

QRU thresholds the feature values into a range of -0.5 to $+0.5$. This output is given as the input to the ITU where the quantized DCT coefficients of the feature maps are obtained. The DCT domain features are then fused with the pixel domain features obtained from the parallel pixel domain branch as shown in Fig. 5. Thus the dual domain approach performs efficient feature corrections fusing the multiple domain features. The major task of this network is to make corrections in the mid and high frequency details of the image. When DCT is applied to the features from PDCB, the correction in the subsequent layers are happening in frequency domain. This correction will aid the network in learning the edges and other sharp details of the ground truth.

3.3 Loss Function

The network is trained based on mean squared error (MSE) and the MSE loss is the mean squared error between the ground-truth and the predicted images. It is incorporated to generate high fidelity images and is formulated as :

$$L_{MSE} = \frac{1}{W \times H \times C} \sum_{i=0}^{W-1} \sum_{j=0}^{H-1} \sum_{k=0}^{C-1} (Y_{i,j,k} - \hat{Y}_{i,j,k})^2 \tag{1}$$

where, W, H and C are the width, height and number of channels of the output, Y is the ground truth image and $\hat{Y}$ is the predicted image.

4 Experiments

4.1 Dataset

The dataset used in the experiments is the challenge dataset provided by the organizers of RLQ-TOD workshop 2020 for the image restoration for Under-display Camera - Track 1 - T-OLED and Track 2- P-OLED. Both the datasets contains 240 pairs of degraded under-display camera images and their corresponding ground truth images for training, 30 pairs for validation and 30 pairs for testing. Each image is of resolution $1024 \times 2048 \times 3$.

4.2 Training

The PDCRN network was trained with images of a patch size of $1024 \times 2048 \times 3$ and batch size of 12. For T-OLED, the images were trained with patch size of $256 \times 256 \times 3$ and batch size of 16. Adam optimizer with $\beta_1 = 0.9$, $\beta_2 = 0.99$ and learning rate $2e^{-4}$ was used in both the networks. The learning rate is decayed after every 10000 epoch with a decay rate of 0.5 in both networks. The networks were trained for a total of 30000 epochs using Tesla V100 GPU card with 32 Gib memory. The training loss and accuracy plots of PDCRN and dual domain PDCRN are depicted in Fig. 6 and Fig. 7 respectively.

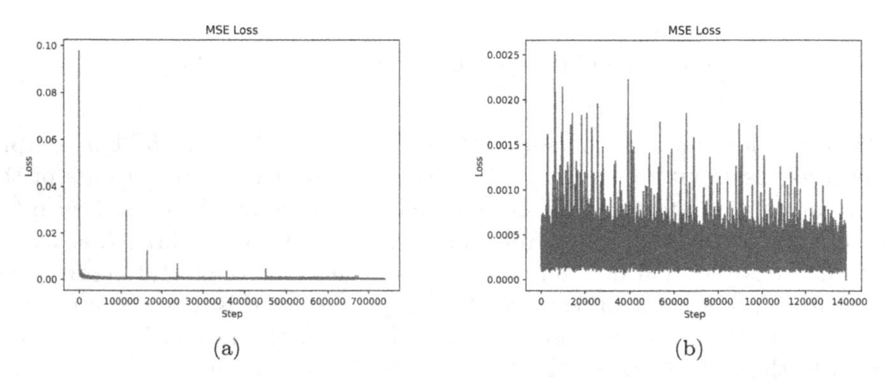

Fig. 6. Training loss plot of (a) PDCRN for P-OLED (b) PDCRN with dual domain block for T-OLED. The spikes in loss function resulted from discontinuous training.

4.3 Evaluation Metrics

To evaluate the performance of the methodology, the evaluation metric used is the peak signal to noise ratio (PSNR) and structural similarity (SSIM).

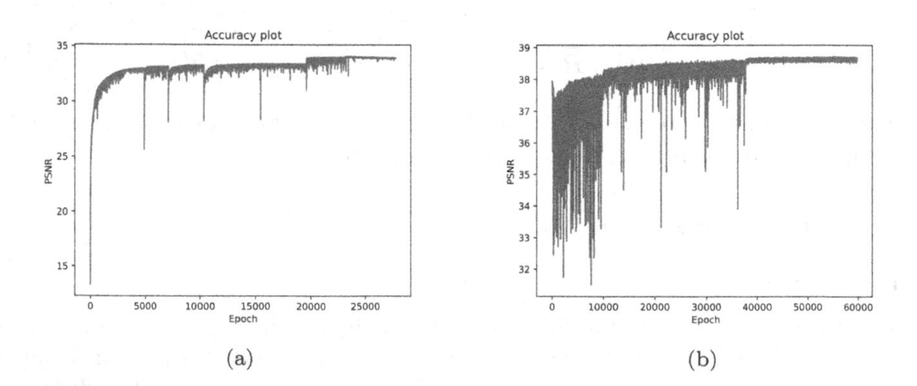

Fig. 7. Accuracy plot of (a) PDCRN for P-OLED (b) PDCRN with dual domain block for T-OLED. The accuracy plot in (a) is not smooth due to discontinuous training. The accuracy plot in (b) is high in initial epochs because of a high learning rate.

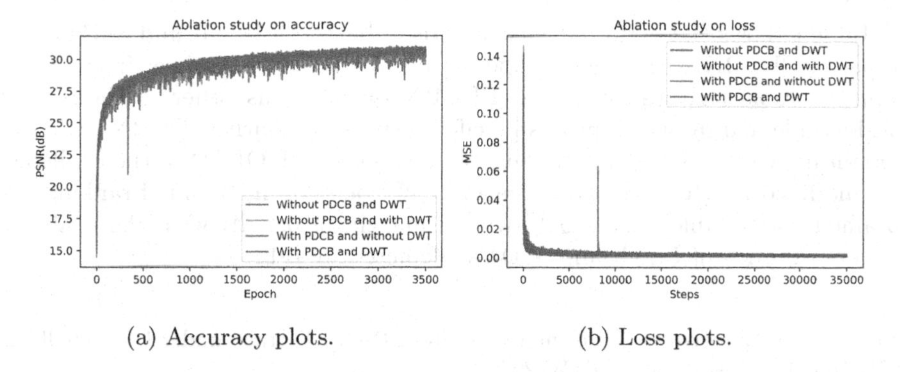

(a) Accuracy plots.

(b) Loss plots.

Fig. 8. Study of the effects of various components of the proposed PDCRN on accuracy in Fig.(a) and mean squared error(MSE) loss in Fig.(b)

4.4 Ablation Study for PDCRN

The Fig. 8 shows the accuracy and loss curves generated that shows the effectiveness of various components of the proposed network. The accuracy plot is generated based on the PSNR metric in every epoch of training and loss plot is based on MSE loss of every mini-batch of training. For standardisation, all the networks were trained up to 3500 epochs were the plots reaching a comparative saturation. From the accuracy plots, we can understand that network without pyramid dilation and DWT performed the worse, attaining a PSNR of 30.81 dB. The introduction of dilated convolutions in the pixel domain improved the results to 31 dB. The network of the proposed method obtained best PSNR of 30.25 dB out of all. So with the addition of both DWT and pyramid dilation, there was performance gain of 0.5 dB when compared to network with a no dilation and DWT. Loss plot doesn't give much insights into the performance comparison.

5 Result Analysis

The Fig. 9 and 10 shows six sets of examples for P-OLED and T-OLED degraded image restoration. The proposed PDCRN was able to attain superior results in PSNR scores and from the visual comparison shows the high fidelity of the restored images with the ground truth. Similarly, the modified version of PDCRN with dual domain network also achieves high perceptual quality and fidelity scores for restored images degraded by T-OLED display. Two different networks have been proposed in this work to address the restoration of two types of under-display camera images. The PDCRN proposed for P-OLED image restoration was the winning entry in European conference on computer vision (ECCV) 2020 challenge on image restoration for Under-display Camera - Track 2 - P-OLED organized in conjunction with RLQ-TOD workshop.

Table 1 shows the performance comparison based on PSNR and SSIM of the proposed PDCRN with other competing entries in the challenge. It is evident from the table that the proposed PDCRN outperforms other methods by a considerable margin which proves its effectiveness. Similarly, PDCRN with dual domain network was a competing entry in Track -1 - T-OLED of the challenge. The modified PDCRN was able to achieve 4th position in the final ranking and obtained comparable results only with a margin of 0.4 dB with the challenge winners in terms of PSNR and SSIM as depicted in Table 2.

Table 1. Comparative study of the results from P-OLED track under UDC challenge in RLQ-TOD'20 workshop, ECCV-2020

Teams	**Ours**	Team 1	Team 2	Team 3	Team 4	Team 5	Team 6	Team 7	Team 8
PSNR	**32.99(1)**	32.29(2)	31.39(3)	30.89(4)	29.38(5)	28.61(6)	26.6(7)	25.72(8)	25.46(9)
SSIM	**0.9578(1)**	0.9509(2)	0.9506(3)	0.9469(5)	0.9249(6)	0.9489(4)	0.9161(7)	0.9027(8)	0.9015(9)

Table 2. Comparative study of the results from T-OLED track under UDC challenge in RLQ-TOD'20 workshop, ECCV-2020

Teams	**Ours**	Team 1	Team 2	Team 3	Team 4	Team 5	Team 6	Team 7
PSNR	**37.83 (4)**	38.23 (1)	38.18 (2)	38.13 (3)	36.91 (5)	36.72 (6)	34.35 (7)	33.78 (8)
SSIM	**0.9783(4)**	0.9803 (1)	0.9796 (2)	0.9797 (3)	0.9734(6)	0.9776 (5)	0.9645 (7)	0.9324 (8)

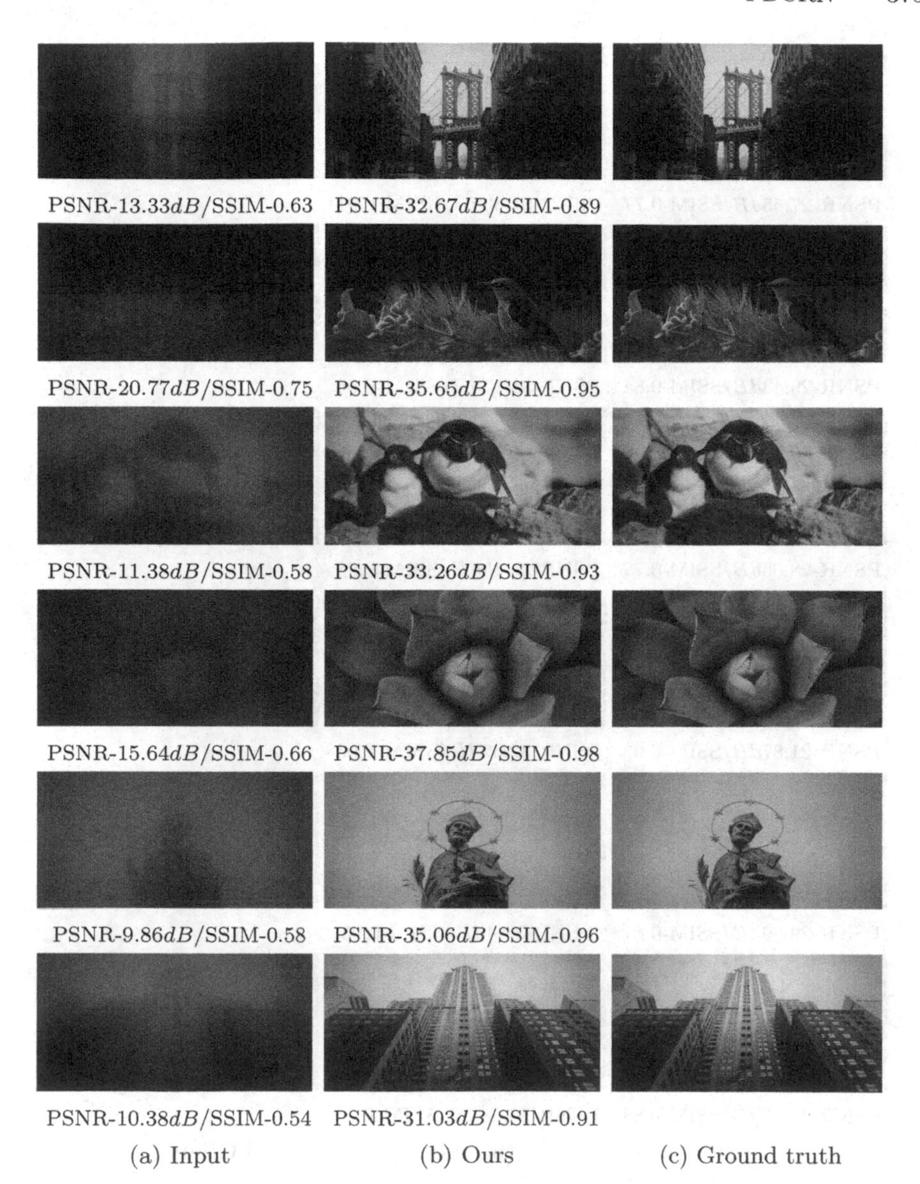

Fig. 9. Sample results of the proposed PDCRN for (a). P-OLED degraded images, (b). images restored using PDCRN shows high fidelity with (c). the ground-truth captured without mounting any display.

Fig. 10. Sample results of the modified PDCRN with DDB for (a). T-OLED degraded images, (b). images restored using the network shows high fidelity with (c). the ground-truth captured without mounting any display.

6 Conclusions

In this paper, we proposed two different fully convolutional networks for the restoration of images degraded due to under-display imaging. Pyramidal Dilated Convolutional RestoreNet proposed for pentile-organic LED images has obtained state-of-the-art restoration performance in terms of standard evaluation metrics PSNR and SSIM. The network proposed for transparent-organic LED based imaging used a dual domain approach with implicit DCT and obtained considerable performance when compared with state-of-the-art methodologies. An intensive study may be conducted to devise a methodology that can generalize both P-OLED and T-OLED degradations equally well.

Acknowledgements. We gratefully acknowledge the support of NVIDIA PSG Cluster and Trivandrum Engineering Science and Technology Research Park (TrEST) with the donation of the Tesla V100 GPUs used for this research.

References

1. Abdelhamed, A., Lin, S., Brown, M.S.: A high-quality denoising dataset for smartphone cameras. In: 2018 IEEE/CVF Conference on Computer Vision and Pattern Recognition, pp. 1692–1700 (2018)
2. Anwar, S., Barnes, N.: Real image denoising with feature attention. In: 2019 IEEE/CVF International Conference on Computer Vision (ICCV), pp. 3155–3164 (2019)
3. Gu, S., Zhang, L., Zuo, W., Feng, X.: Weighted nuclear norm minimization with application to image denoising. In: 2014 IEEE Conference on Computer Vision and Pattern Recognition, pp. 2862–2869 (2014)
4. He, K., Zhang, X., Ren, S., Sun, J.: Deep residual learning for image recognition. In: 2016 IEEE Conference on Computer Vision and Pattern Recognition (CVPR), pp. 770–778 (2016)
5. Huang, G., Liu, Z., Van Der Maaten, L., Weinberger, K.Q.: Densely connected convolutional networks. In: 2017 IEEE Conference on Computer Vision and Pattern Recognition (CVPR), pp. 2261–2269 (2017)
6. Kim, Y., Soh, J.W., Park, G.Y., Cho, N.I.: Transfer learning from synthetic to real-noise denoising with adaptive instance normalization. In: IEEE/CVF Conference on Computer Vision and Pattern Recognition (CVPR), June 2020
7. Ledig, C., et al.: Photo-realistic single image super-resolution using a generative adversarial network. In: 2017 IEEE Conference on Computer Vision and Pattern Recognition (CVPR), pp. 105–114 (2017)
8. Lim, B., Son, S., Kim, H., Nah, S., Lee, K.M.: Enhanced deep residual networks for single image super-resolution. In: 2017 IEEE Conference on Computer Vision and Pattern Recognition Workshops (CVPRW), pp. 1132–1140 (2017)
9. Liu, P., Zhang, H., Zhang, K., Lin, L., Zuo, W.: Multi-level wavelet-cnn for image restoration. In: 2018 IEEE/CVF Conference on Computer Vision and Pattern Recognition Workshops (CVPRW), pp. 886–88609 (2018)
10. Nah, S., et al.: Ntire 2020 challenge on image and video deblurring. In: 2020 IEEE/CVF Conference on Computer Vision and Pattern Recognition Workshops (CVPRW), pp. 1662–1675 (2020)
11. Ronneberger, O., Fischer, P., Brox, T.: U-Net: convolutional networks for biomedical image segmentation. In: Navab, N., Hornegger, J., Wells, W.M., Frangi, A.F. (eds.) MICCAI 2015, Part III. LNCS, vol. 9351, pp. 234–241. Springer, Cham (2015). https://doi.org/10.1007/978-3-319-24574-4_28

12. Shi, W., et al.: Real-time single image and video super-resolution using an efficient sub-pixel convolutional neural network. In: 2016 IEEE Conference on Computer Vision and Pattern Recognition (CVPR), pp. 1874–1883 (2016)
13. Yuan, S., et al.: Ntire 2020 challenge on image demoireing: ethods and results. In: 2020 IEEE/CVF Conference on Computer Vision and Pattern Recognition Workshops (CVPRW), pp. 1882–1893 (2020)
14. Zhang, H., Dai, Y., Li, H., Koniusz, P.: Deep stacked hierarchical multi-patch network for image deblurring. In: 2019 IEEE/CVF Conference on Computer Vision and Pattern Recognition (CVPR), pp. 5971–5979 (2019)
15. Zhang, K., et al.: Ntire 2020 challenge on perceptual extreme super-resolution: methods and results. In: 2020 IEEE/CVF Conference on Computer Vision and Pattern Recognition Workshops (CVPRW), pp. 2045–2057 (2020)
16. Zhang, K., Zuo, W., Chen, Y., Meng, D., Zhang, L.: Beyond a Gaussian denoiser: Residual learning of deep CNN for image denoising. IEEE Trans. Image Process. **26**(7), 3142–3155 (2017)
17. Zhang, X., Yang, W., Hu, Y., Liu, J.: DMCNN: dual-domain multi-scale convolutional neural network for compression artifacts removal. In: 2018 25th IEEE International Conference on Image Processing (ICIP), pp. 390–394 (2018)
18. Zheng, B., Chen, Y., Tian, X., Zhou, F., Liu, X.: Implicit dual-domain convolutional network for robust color image compression artifact reduction. IEEE Trans. Circuits Syst. Video Technol. 1 (2019)
19. Zhou, Y., et al.: UDC 2020 challenge on image restoration of under-display camera: Methods and results. arXiv abs/2008.07742 (2020)
20. Zhou, Y., Ren, D., Emerton, N., Lim, S., Large, T.: Image restoration for under-display camera (2020)

Deep Atrous Guided Filter for Image Restoration in Under Display Cameras

Varun Sundar(✉), Sumanth Hegde, Divya Kothandaraman, and Kaushik Mitra

Indian Institute of Technology Madras, Chennai, India
{varunsundar,sumanth,ee15b085}@smail.iitm.ac.in, kmitra@ee.iitm.ac.in

Abstract. Under Display Cameras present a promising opportunity for phone manufacturers to achieve bezel-free displays by positioning the camera behind semi-transparent OLED screens. Unfortunately, such imaging systems suffer from severe image degradation due to light attenuation and diffraction effects. In this work, we present Deep Atrous Guided Filter (DAGF), a two-stage, end-to-end approach for image restoration in UDC systems. A Low-Resolution Network first restores image quality at low-resolution, which is subsequently used by the Guided Filter Network as a filtering input to produce a high-resolution output. Besides the initial downsampling, our low-resolution network uses multiple, parallel atrous convolutions to preserve spatial resolution and emulates multi-scale processing. Our approach's ability to directly train on megapixel images results in significant performance improvement. We additionally propose a simple simulation scheme to pre-train our model and boost performance. Our overall framework ranks 2nd and 5th in the RLQ-TOD'20 UDC Challenge for POLED and TOLED displays, respectively.

Keywords: Under-display camera · Image restoration · Image enhancement

1 Introduction

Under Display Cameras (UDC) promise greater flexibility to phone manufacturers by altering the traditional location of a smartphone's front camera. Such systems place the camera lens behind the display screen, making truly bezel-free screens possible and maximising screen-to-body ratio. Mounting the camera at the centre of the display also offers other advantages such as enhanced video call experience and is more relevant for larger displays found in laptops and TVs. However, image quality is greatly degraded in such a setup, despite the superior light efficiency of recent display technology such as OLED screens [57].

V. Sundar and S. Hegde—Equal Contribution.

Electronic supplementary material The online version of this chapter (https://doi.org/10.1007/978-3-030-68238-5_29) contains supplementary material, which is available to authorized users.

A. Bartoli and A. Fusiello (Eds.): ECCV 2020 Workshops, LNCS 12539, pp. 379–397, 2020.
https://doi.org/10.1007/978-3-030-68238-5_29

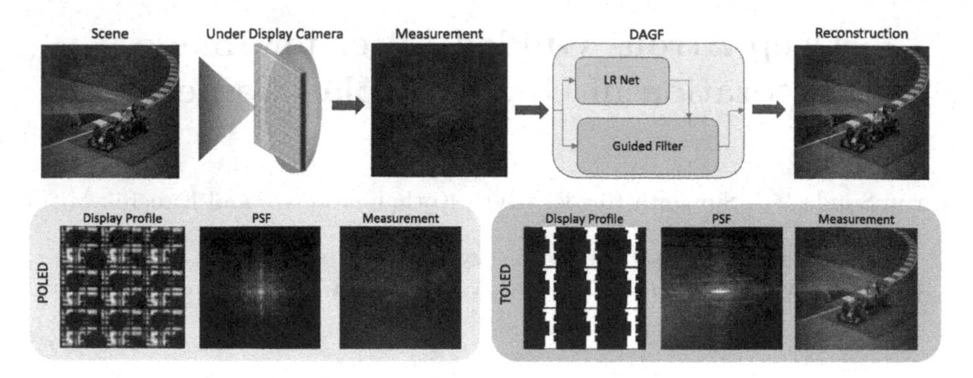

Fig. 1. Under display ameras [75] mount lenses behind semi-transparent OLED displays leading to image degradation. In this work, we introduce DAGF, which performs image restoration at megapixel resolution for both POLED and TOLED displays.

As illustrated in Fig. 1, UDC imaging systems suffer from a range of artefacts including colour degradation, noise amplification and low-light settings. This creates a need for restoration algorithms which can recover photorealistic scenes from UDC measurements.

Learning based methods, accentuated by deep learning, have achieved state-of-the-art performance on a variety of image restoration tasks including deblurring [5,41,47], dehazing [3,6,44], denoising [1,66,68], deraining [6,30] and image enhancement [8,13]. However, deep learning techniques face two main drawbacks with regard to UDC imaging systems. First, such methods are do not scale computationally with input image resolution, and are typically run on much smaller patches. This is problematic for restoring severely degraded images such as UDC measurements, since small patches lack sufficient context. Second, common Convolutional Neural Networks (CNNs) employed in image restoration use multiple down-sampling operations to stack more layers and expand their receptive field without blowing up their memory footprint. Down-sampling leads to a loss of spatial information and affects performance in pixel-level dense prediction tasks such as image restoration [5,10,38]. An alternative is to simply omit such sub-sampling and resort to atrous (or dilated) convolutions. Owing to memory constraints, this is not feasible since we deal with high-resolution images in UDC systems.

To overcome these drawbacks, we propose a two-stage, end-to-end trainable approach utilizing atrous convolutions in conjunction with guided filtering. The first stage performs image restoration at low-resolution using multiple, parallel atrous convolutions. This allows us to maximally preserve spatial information without an exorbitant memory requirement. The guided filter then uses the low-resolution output as the filtering input to produce a high-resolution output via joint upsampling. Our approach makes it possible to directly train on high resolution images, and results in significant performance gains. Our contributions are as follows:

- We propose a novel image restoration approach for UDC systems utilizing atrous convolutions in conjunction with guided filters (Sect. 3).
- We show that directly training on megapixel inputs allows our approach to significantly outperform existing methods (Sect. 4.3).
- We propose a simple simulation scheme to pre-train our model and further boost performance (Sect. 4.2).

Our code and simulated data is publicly available at varun19299.github.io/deep-atrous-guided-filter/.

2 Related Work

Image restoration encompasses tasks like image denoising, dehazing, deblurring and super resolution [1,3,36,41]. In recent years, deep learning has been the go-to tool in the field, with fully convolutional networks at the forefront of this success [37,45,50,65]. Of these, residual dense connections [71] exploiting hierarchical features has garnered interest with subsequent works in specific restoration tasks [6,25,44,64,71]. Another class of techniques use a GAN [15] based setting. Methods like [23,28] fall in this category. Finally, there exist recent work exploiting CNNs as an effective image prior [29,53,67]. However, the above-mentioned methods operate on small patches of the input image and do not scale to larger input dimensions.

Joint upsampling seeks to generate a high-resolution output, given a low-resolution input and a high-resolution guidance map. Joint Bilateral Upsampling [27] uses a bilateral filter towards this goal, obtaining a piecewise-smooth high-resolution output, but at a large computational cost. Bilateral Grid Upsampling [7] greatly alleviates this cost by fitting a grid of local affine models on low-resolution input-output pairs, which is then re-used at high resolution. Deep Bilateral Learning [13] integrates bilateral filters in an end-to-end framework, with local affine grids that can be learnt for a particular task.

Guided filters [19] serve as an alternative to Joint Bilateral Upsampling, with superior edge-preserving properties at a lower computational cost. Deep Guided Filtering [59] integrates this with fully convolutional networks and demonstrates it for common image processing tasks, with recent interest in the hyperspectral [17], remote [60] and medical imaging [14]. Guided filters have been mainly explored in the context of accelerating image processing operators. In our work, we present a different application of image restoration.

Atrous or dilated convolutions incorporate a larger receptive field without an increase in the number of parameters or losing spatial resolution. Yu *et al.* [62] proposed a residual network using dilated convolutions. Dilated convolutions have found success in semantic segmentation [8,73], dehazing [6,44] and deblurring [5] tasks as well as general image processing operations [11]. However, a major challenge in atrous networks is keeping memory consumption in check. Multi-scale fusion via pyramid pooling or encoder-decoder networks [9,10,32,40,67] can offload intensive computation to lower scales, but can lead

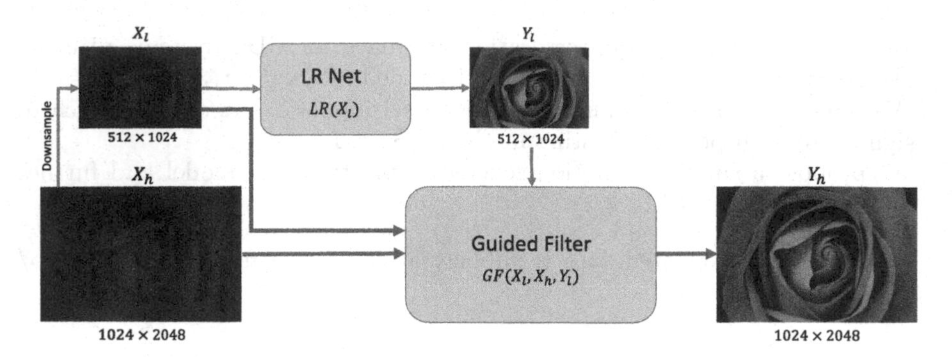

Fig. 2. Framework overview of DAGF. Our architecture seeks to operate directly on megapixel images by performing joint upsampling. A low resolution network (LRNet) restores a downsampled version X_l of input X_h to produce Y_l. The guided filter then uses this to yield the final high-resolution output Y_h.

to missing fine details. Instead, we include channel and pixel attention [44] to have a flexible receptive field at each stage while better tending to severely degraded regions (Fig. 3).

Compared to prior work, **our main novelty** lies in directly training on megapixel images by incorporating multiple, parallel smoothed atrous convolutions in a guided filter framework. This adapts the proposed framework in Wu *et al.* [59]- primarily developed for image processing tasks- to handle the challenging scenario of image restoration for Under Display Cameras.

3 Deep Atrous Guided Filter

To address the challenges posed by Under Display Cameras, we employ a learning based approach that directly trains on megapixel images. We argue that since UDC measurements are severely degraded, it is imperative to train models with large receptive fields on high-resolution images [26,47,48].

Our approach, **Deep Atrous Guided Filter Network (DAGF)**, consists of two stages: (a) Low Resolution Network (LRNet), which performs image restoration at a lower resolution, and (b) Guided Filter Network, which uses the restored low-resolution output and the high-resolution input to produce a high-resolution output. Our guided filter network, trained end-to-end with LRNet, restores content using the low-resolution output while preserving finer detail from the original input.

We design our approach to perform image restoration for two types of OLED displays: Pentile OLED (POLED) and Transparent OLED (TOLED). As seen in Fig. 1, TOLED has a stripe pixel layout, while POLED has a pentile pixel layout with a much lower light transmittance. Consequently, TOLED results in a blurry image, while POLED results in a low-light, colour-distorted image.

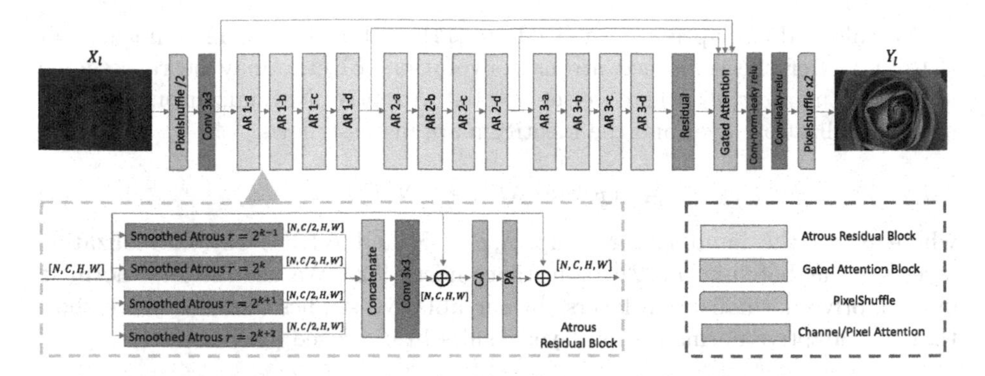

Fig. 3. Overview of LRNet. LRNet operates on a low-resolution version X_l of original input X_h. The input image X_l is downsampled via pixelshuffle, encoded via many atrous residual blocks and finally a gated attention mechanism aggregates contextual information to produce low-resolution output Y_l.

3.1 LR Network

LRNet comprises of three key components: i) PixelShuffle [46] ii) atrous residual blocks, and iii) a gated attention mechanism [6,49]. We first use PixelShuffle [46] to lower input spatial dimensions while expanding channel dimensions. This affords us a greater receptive field at a marginal memory footprint [16,31]. We further encode the input into feature maps via successive atrous residual blocks and then aggregate contextual information at multiple levels by using a gated attention mechanism. We now describe each component of LRNet.

Smoothed Atrous Convolutions. Unlike common fully convolutional networks employed in image restoration [16,31,39], which use multiple downsampling blocks, we opt to use atrous (or dilated) convolutions [61] instead. This allows us expand the network's receptive field without loss in spatial resolution, which is beneficial for preserving fine detail in dense prediction tasks.

Atrous convolutions, however, lead to gridding artefacts in their outputs [18,54,56]. To alleviate this, we insert a convolution layer before each dilated convolution, implemented via shared separable kernels for computational and parameter efficiency [56]. Concretely, for a input feature map F^{in} with C channels, the smoothed atrous convolution layer produces output feature map F^{out} with C channels as follows:

$$F_i^{\text{out}} = \sum_{j \in [C]} \left((F_j^{\text{in}} * K^{\text{sep}} + b_i) *_r K_{ij} \right) \tag{1}$$

where F_i^{out} is the ith output channel, b_i is a scalar bias, $*$ is a 2D convolution and $*_r$ is a dilated convolution with dilation r. K_{ij} is a 3×3 convolution kernel and K^{sep} is the shared separable convolution kernel, shared among all input feature channels. For dilation rate r, we use a shared separable kernel of size $2r - 1$.

We also add adaptive normalization [11] and leaky rectified linear unit (LReLU) after the smoothed atrous convolution. LReLU may be represented as: $\Phi(x) = \max(\alpha x, x)$, where we set $\alpha = 0.2$. Adaptive Normalization combines any normalization layer and the identity mapping as follows:

$$AN(F^{\text{in}}) = \lambda F^{\text{in}} + \mu N(F^{\text{in}}) \tag{2}$$

where F^{in} is the input feature map, $\lambda, \mu \in \mathbb{R}$ and $N(.)$ is any normalization layer such as batch-norm [22] or instance-norm [52]. We use instance-norm in our adaptive normalization layers. In our ablative studies (Sect. 5.2), we show that our adaptive normalization layer results in improved performance.

Atrous Residual Blocks. As depicted in Fig. 3, we propose to use multiple, parallel, smoothed atrous convolutions with various dilation rates in our residual blocks, following its recent success in image deblurring [5]. For atrous residual block AR-k, belonging to the kth group, we use four smoothed atrous convolutions with dilation rates $\{2^{k-1}, 2^k, 2^{k+1}, 2^{k+2}\}$. Each convolution outputs a feature map with $C/2$ channels, which we concatenate to obtain $2C$ channels. These are subsequently reduced to C channels via a 1×1 convolution. Our atrous residual blocks also utilize channel and pixel attention mechanisms, which are described below.

Channel Attention. We use the channel attention block proposed by Qin *et al.* [44]. Specifically for a feature map F^{in} of dimensions $C \times H \times W$, we obtain channel-wise weights by performing global average pooling (GAP) and further encode it via two 1×1 conv layers. We multiply F^{in} with these channel weights CA to yield output F^{out}:

$$GAP_i = \frac{1}{HW} \sum_{u \in [H], v \in [W]} F_i^{\text{in}}(u, v) \tag{3}$$

$$CA_i = \sigma\Big(\sum_{j \in [C]} \Phi\big(\sum_{k \in [C/8]} GAP_k * K_{jk} + b_j \big) * K'_{ij} + b_i \Big) \tag{4}$$

$$F_i^{\text{out}} = CA_i \odot F_i^{\text{in}} \tag{5}$$

where, σ is the sigmoid activation, Φ is LReLU described earlier and $\odot$ is element-wise multiplication.

Pixel Attention. To account for uneven context distribution across pixels, we use a pixel attention module [44] that multiplies the input feature map F^{in} of shape $C \times H \times W$ with an attention map of shape $1 \times H \times W$ varying across pixels, but constant across channels. We obtain the pixel attention map PA by using two 1×1 conv layers:

$$PA = \sigma\Big(\sum_{j \in [C/8]} \Phi\big(\sum_{k \in [C]} F_k^{\text{in}} * K_{jk} + b_j \big) * K'_j + b \Big) \tag{6}$$

$$F_i^{\text{out}} = PA \odot F_i^{\text{in}} \tag{7}$$

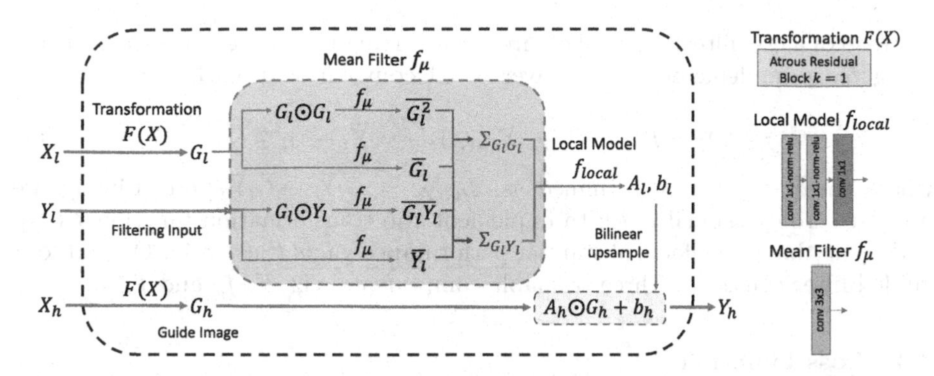

Fig. 4. Computational graph of guided filter stage. The guided filter first transforms the high-resolution input X_h to guide image G_h, and then yields the final output Y_h via joint upsampling. Our guided filter network is differentiable and end-to-end trainable [59].

Gated Attention. We utilise a gated attention mechanism [6,49] to aggregate information across several atrous residual blocks. Fusing features from different levels is beneficial for both low-level and high-level tasks [33,63,73]. We extract feature maps before the first atrous residual block (F^0), and right after each atrous residual group ($F^1, ..., F^k$). For k atrous groups, we concatenate these $k+1$ feature maps and output $k+1$ masks, using $\mathcal{G}$, a 3×3 conv layer:

$$(\mathcal{M}^0, \mathcal{M}^1, ..., \mathcal{M}^k) = \mathcal{G}(F^0, F^1, ..., F^k) \tag{8}$$

$$F^{\text{out}} = \mathcal{M}^0 \odot F^0 + \sum_{l \in [k]} \mathcal{M}^l \odot F^l \tag{9}$$

3.2 Guided Filter Network

Given a high-resolution input X_h, low-resolution input X_l and low-resolution output Y_l, we seek to produce a high-resolution output Y_h, which is perceptually similar to Y_l while preserving fine detail from X_h. We adopt the guided filter proposed by He *et al.* [19,20] and use it in an end-to-end trainable fashion [59]. As illustrated in Fig. 4, the guided filter formulates Y_h as:

$$Y_h = A_h \odot G_h + b_h \tag{10}$$

where $G_h = F(X_h)$ is a transformed version of input X_h. We bilinear upsample A_h and b_h from low-resolution counterparts A_l and b_l, such that:

$$\overline{Y}_l = A_l \odot \overline{G}_l + b_l \tag{11}$$

where $\overline{G}_l, \overline{Y}_l$ are mean filtered versions of G_l, Y_l, i.e., $\overline{G}_l = f_\mu(G_l)$ and $\overline{Y}_l = f_\mu(Y_l)$. Compared to Wu *et al.* [59], we implement f_μ by a 3×3 convolution (instead of a box-filter). Instead of directly inverting Eq. 11, we obtain its solution using f_{local}, implemented by a 3 layer, 1×1 convolutional block:

$$A_l = f_{\text{local}}(\Sigma_{G_l Y_l}, \Sigma_{G_l G_l}), b_l = \overline{Y}_l - A_l \odot \overline{G}_l \tag{12}$$

where covariances are determined as, $\Sigma_{G_l Y_l} = \overline{G_l Y_l} - \overline{G}_l \overline{Y}_l$, etc. Finally, we use our atrous residual block to implement the transformation function $F(X)$, and show that it confers substantial performance gains (Sect. 5.1). Overall, our guided filter consists of three trainable components, viz. F, f_μ and f_{local}.

3.3 Loss Function

L1 Loss. We employ Mean Absolute Error (or L1 loss) as our objective function. We empirically justify our choice L1 loss over other loss formulations (including MS-SSIM [72], perceptual [24] and adversarial [15] losses) in Sect. 5.2.

4 Experiments and Analysis

4.1 Dataset

Our network is trained on the POLED and TOLED datasets [75] provided by the UDC 2020 Image Restoration Challenge. Both datasets comprise of 300 images of size 1024×2048, where 240 images are used for training, 30 for validation and 30 for testing in each track. We do not have access to any specific information of the forward model (such as the PSF or display profile), precluding usage of non-blind image restoration methods such as Wiener Filter [42].

4.2 Implementation Details

Model Architecture. LRNet comprises of 3 atrous residual groups, with 4 blocks each. The intermediate channel size in LRNet is set to 48. The training data is augmented with random horizontal flips, vertical flips and 180° rotations. All images are normalized to a range between -1 and 1. The AdamW [35] optimizer, with initial learning rate $\eta = 0.0003$, $\beta_1 = 0.9$ and $\beta_2 = 0.999$ is used. The learning rate is varied with epochs as per the cosine annealing scheduler with warm restarts [34]. We perform the first warm restart after 64 epochs, post which we double the duration of each annealing cycle. The models are trained using PyTorch [43] on 4 NVIDIA 1080Ti GPUs with a minibatch size of 4, for 960 epochs each.

Pre-training Strategy. To aid in faster convergence and boost performance, we pre-train our model with simulated data. The UDC dataset is created using monitor acquisition [75], where images from the DIV2K dataset [2] are displayed on a LCD monitor and captured by a camera mounted behind either glass (considered ground-truth) or POLED/TOLED panels (low-quality images). To simulate data, we need to transform clean images from the DIV2K dataset to various display measurements (POLED, TOLED or glass).

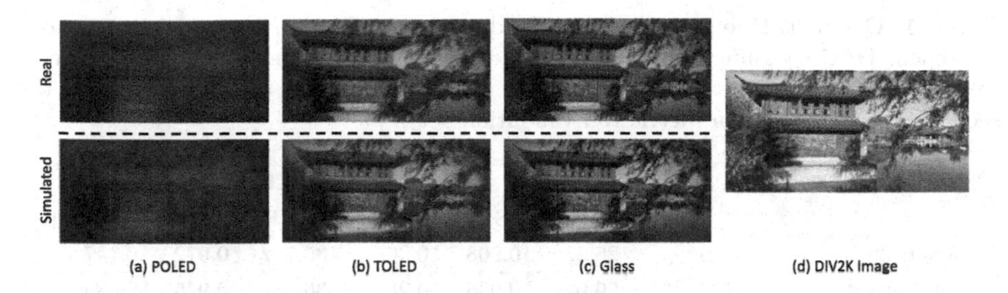

Fig. 5. Pre-training using simulated data enhances performance. We transform 800 DIV2K [2] images via a simulation network to various display measurements (Glass, TOLED and POLED). To train our simulation network, we use the misalignment tolerant CoBi [70] loss.

Using Fresnel Propagation to simulate data with either the display profile or calibrated PSF can be inaccurate [75]. Instead, a shallow variant of our model is trained to transform 800 images from the DIV2K dataset to each measurement. Since DIV2K images do not align with display measurements, we leverage the Contextual Bilateral (CoBi) Loss [70], which can handle moderately misaligned image pairs. For two images P and Q, with $\{\boldsymbol{p}_{ij}\}$ and $\{\boldsymbol{q}_{ij}\}$ ($i \in [H], j \in [W]$) representing them as a grid of RGB intensities, CoBi loss can be written as:

$$\text{CoBi}(P, Q) = \frac{1}{HW} \sum_{i,j} \min_{k,l} \left[\mathbb{D}(\boldsymbol{p}_{ij}, \boldsymbol{q}_{kl}) + \gamma \left((i-k)^2 + (j-l)^2 \right) \right] \tag{13}$$

where $\mathbb{D}$ is any distance metric (we use cosine distance). γ allows CoBi to be flexible to image-pair misalignment. As seen in Fig. 5, our simulated measurements closely match real measurements. Such an initialisation procedure gives our model (DAGF-PreTr) around 0.3 to 0.5 dB improvement in PSNR (Table 1). More simulation results can be found in the supplementary material.

4.3 Quantitative and Qualitative Analysis

Baseline Methods. Our method is compared against four image restoration methods: DGF [59], PANet [39], UNet [45] and FFA-Net [44]. DGF utilises a trainable guided filter for image transformation. For DGF, we use 9 layers in the CAN [11] backbone (instead of 5) for better performance. UNet is a popular architecture in image restoration. A variant of UNet with a double encoder [75] and 64 intermediate channels in the first block is used. PANet and FFA-Net are specifically designed architectures for image denoising and dehazing, respectively. Small patch sizes often provide little information for faithful restoration. Hence, to make a fair comparison, a much larger patch size of 96×96 for PANet (compared to 48×48 in Mei *et al.* [39]), 256×512 for UNet (256×256 in Zhou *et al.* [75]) and 256×512 for FFA-Net (240×240 in Qin *et al.* [44]) is used.

Table 1. Quantitative comparison. By directly training on megapixel images, our approach, DAGF significantly outperforms baselines. To further boost performance, we pre-train on simulated data (DAGF-PreTr). Red indicates the best and Blue the second best in the chosen metric (on validation set).

Method	#Params ↓	POLED			TOLED		
		PSNR ↑	SSIM ↑	LPIPS ↓	PSNR ↑	SSIM ↑	LPIPS ↓
PANet [39]	6.0M	26.22	0.908	0.308	35.712	0.972	0.147
FFA-Net [44]	1.6M	29.02	0.936	0.256	36.33	0.975	0.126
DGF [59]	0.4M	29.93	0.931	0.362	34.43	0.956	0.220
Unet [45]	8.9M	29.98	0.932	0.251	36.73	0.971	0.143
DAGF (Ours)	1.1M	33.29	0.952	0.236	37.27	0.973	0.141
DAGF-PreTr (Ours)	1.1M	33.79	0.958	0.225	37.57	0.973	0.140

Quantitative and Qualitative Discussion. All our methods are evaluated on PSNR, SSIM and the recently proposed LPIPS [69] metrics. Higher PSNR and SSIM score indicate better performance, while lower LPIPS indicates better perceptual quality. As seen in Table 1, our approach (DAGF) significantly outperforms the baselines, with an improvement of 3.2 dB and 0.5 dB over the closest baseline on POLED and TOLED measurements, respectively.

Our approach's ability to directly train on megapixel images and hence aggregate contextual information over large receptive fields leads to a significant improvement. This is more evident on the POLED dataset, where patch based methods such as PANet, UNet and FFA-Net lack sufficient context despite using larger patch-sizes. With the exception of DGF, our approach also uses much lesser parameters. Visual comparisons in Fig. 6 are consistent with our quantitative results. Our approach closely resembles groundtruth, having lesser artefacts and noise. Notably, in Fig. 6a, we can observe line artefacts in patch based methods (further detailed in Sect. 5.1).

Challenge Results. This work is initially proposed for participating in the UDC 2020 Image Restoration Challenge [74]. For the challenge submission, geometric self-ensembling [47,51] is incorporated in DAGF-PreTr to boost performance, denoted in Table 2 as DAGF-PreTr+. Self-ensembling involves feeding various rotated and flipped versions of the input image to the network, and performing corresponding inverse transforms before averaging their outputs.

Quantitatively, our method ranks 2nd and 5th on the POLED and TOLED tracks, respectively (Table 2), proving that DAGF is effective at image restoration, especially in the severe degradation setting of POLED. While our approach is competitive on both tracks, there is scope to better adapt our model to moderate image degradation scenarios such as TOLED measurements.

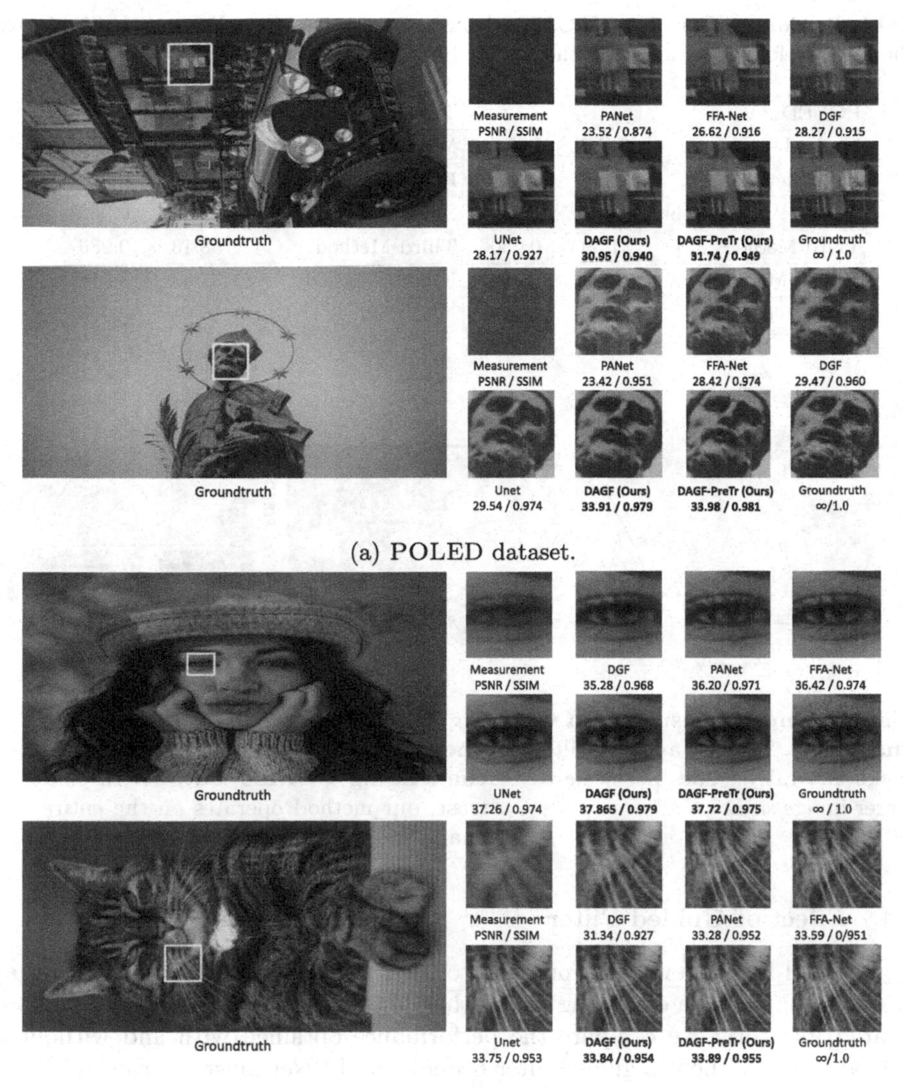

(b) TOLED dataset.

Fig. 6. Qualitative results. DAGF is considerably superior to patch based restoration methods [39,44,45], more evident on the severely degraded POLED measurements. Metrics evaluated on entire image. Zoom in to see details.

5 Further Analysis

To understand the role played by various components in DAGF, extensive ablative studies have been conducted. These experiments have been performed on downsized measurements, i.e., 512 × 1024, in order to reduce training time.

Table 2. Comparison on UDC2020 image restoration challenge. Red indicates the best performance and Blue the second best (on challenge test set).

POLED			TOLED		
Method	PSNR ↑	SSIM ↑	Method	PSNR ↑	SSIM ↑
First Method	32.99	0.957	First Method	38.23	0.980
DAGF-PreTr+ (Ours)	32.29	0.951	Second Method	38.18	0.980
Third Method	31.39	0.950	Third Method	38.13	0.980
Fourth Method	30.89	0.947	Fourth Method	37.83	0.978
Fifth Method	29.38	0.925	DAGF-PreTr+ (Ours)	36.91	0.973

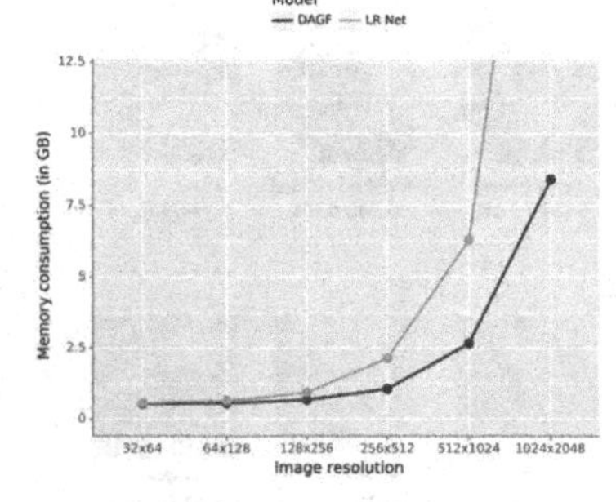

Fig. 7. Memory consumption vs image size. Without a guided filter backbone, LRNet does not scale to larger image sizes.

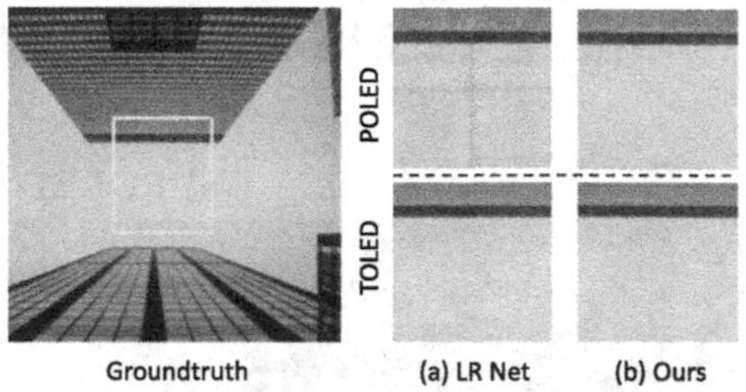

Fig. 8. Patch based methods lead to line artefacts, evident in the more challenging POLED output. In contrast, our method operates on the entire image and produces no such artefacts.

5.1 Effect of Guided Filter

The guided filter allows our approach to directly use high resolution images as input, as opposed to operating on patches or downsampled inputs. To demonstrated its utility, we compare the performance obtained with and without a guided filter. Without a guided filter framework, LRNet must be trained patch-wise, due to memory constraints (Fig. 7). At test time, we assemble the output patch-wise.

Using a guided filter provides a significant benefit of 2.5 dB and 1.8 dB on POLED images and TOLED images, respectively (Table 3). Although marginally better LPIPS metrics indicate that LRNet produces more visually pleasing outputs, line artefacts can be observed in the outputs (Fig. 8). Such artefacts are prominent in the more challenging POLED dataset. Alternative evaluation strategies for LRNet such as using overlapping patches followed by averaging, or feeding the entire high-resolution input results in blurry outputs and degrades performance.

Table 3. Using a **trainable guided filter** provides greater context by scaling to larger image dimensions.

Backbone	POLED			TOLED		
	PSNR ↑	SSIM ↑	LPIPS ↓	PSNR ↑	SSIM ↑	LPIPS ↓
No guided filter	30.14	0.938	**0.212**	33.92	0.963	**0.132**
Conv 1 × 1 guided Filter	32.39	0.940	0.223	35.605	0.963	0.141
Conv 3 × 3 guided filter	32.50	0.942	0.220	35.84	0.965	0.150
Smoothed atrous block	**32.87**	**0.946**	0.216	**35.87**	**0.966**	0.147

Table 3 also features comparisons against other transformation functions $F(X)$. Experiments indicate a clear advantage in using our atrous residual block over either 1 × 1 or 3 × 3 conv layers proposed in Wu *et al.* [59].

5.2 Other Ablative Studies

All ablative results are presented in Table 4.

Smoothed Dilated Convolutions. Using either 3 × 3 convolutions or exponentially growing dilation rates [6,11] with the same number blocks leads to inferior performance. In contrast, parallel atrous convolutions lead to a larger receptive field at similar depth [5] and improves performance. We also verify that introducing a smoothing operation before atrous convolutions is beneficial, and qualitatively leads to fewer gridding (or checkerboard) artefacts [18,54,56].

Residual and Gated Connections. Consistent with Zhang *et al.* [71], removing local residual connections leads to a considerable degradation in performance. Similarly, gated attention, which can be perceived as a global residual connection with tunable weights, provides a noticeable performance gain.

Adaptive Normalisation. Using adaptive equivalents of batch-norm [22] or instance-norm [52] improves performance. Experiments indicate a marginal increase of adaptive batch-norm over adaptive instance-norm. However, since we use smaller minibatch sizes while training on 1024 × 2048, we prefer adaptive instance-norm.

Channel Attention. Unlike recent variants of channel attention [12,21,55,58], our implementation learns channel-wise weights, but does not capture inter-channel dependency. Experimenting with Efficient Channel Attention (ECA) [55] did not confer any substantial benefit, indicating that modelling inter-channel dependencies may not be important in our problem.

Loss Functions. Compared to L1 loss, optimising MS-SSIM [72] loss can improve PSNR marginally, but tends to be unstable during the early stages of training. Perceptual [4] and adversarial [15] losses improve visual quality, reflected in better LPIPS scores, but degrade PSNR and SSIM metric performance [4]. Overall, L1 loss is a simple yet superior choice.

Table 4. Ablative studies. We experiment with various components present in our approach to justify our architecture choices.

Conditions				POLED			TOLED		
Smooth atrous convolutions									
Atrous	Parallel atrous	Smooth atrous		PSNR ↑	SSIM ↑	LPIPS ↓	PSNR ↑	SSIM ↑	LPIPS ↓
-	-	-		31.26	0.936	0.257	34.76	0.960	0.157
✓	-	-		31.11	0.928	0.311	32.78	0.945	0.240
✓	✓	-		32.39	0.943	0.233	35.46	0.963	0.157
✓	✓	✓		**32.87**	**0.946**	**0.216**	**35.87**	**0.966**	**0.147**
Residual and gated connections									
Residual		Gated		PSNR ↑	SSIM ↑	LPIPS ↓	PSNR ↑	SSIM ↑	LPIPS ↓
-		-		29.59	0.907	0.382	32.03	0.938	0.267
✓		-		32.19	0.941	0.255	35.14	0.961	0.162
✓		✓		**32.87**	**0.946**	**0.216**	**35.87**	**0.966**	**0.147**
Normalization layers									
BN [22]	IN [52]	ABN [11]	AIN	PSNR ↑	SSIM ↑	LPIPS ↓	PSNR ↑	SSIM ↑	LPIPS ↓
-	-	-	-	31.78	0.937	0.278	35.09	0.96	0.165
✓	-	-	-	30.75	0.919	0.268	33.20	0.943	0.191
-	✓	-	-	30.17	0.918	0.289	30.54	0.92	0.224
-	-	✓	-	32.73	0.945	0.225	**36.02**	**0.966**	**0.14**
-	-	-	✓	**32.87**	**0.946**	**0.216**	35.87	**0.966**	0.147
Channel attention									
ECA [55]		FFA [6]		PSNR ↑	SSIM ↑	LPIPS ↓	PSNR ↑	SSIM ↑	LPIPS ↓
-		-		32.62	0.944	0.225	35.72	0.964	0.152
✓		-		32.66	0.944	0.232	**35.98**	**0.966**	**0.143**
-		✓		**32.87**	**0.946**	**0.216**	35.87	**0.966**	0.147
Loss functions									
L1	MS-SSIM [72]	Percep. [24]	Adv. [15]	PSNR ↑	SSIM ↑	LPIPS ↓	PSNR ↑	SSIM ↑	LPIPS ↓
✓	-	-	-	**32.87**	**0.946**	0.216	35.87	0.966	0.147
✓	✓	-	-	32.55	**0.946**	0.208	**36.20**	**0.968**	0.125
✓	-	✓	-	31.75	0.936	0.189	35.45	0.963	0.112
✓	-	✓	✓	31.81	0.94	**0.178**	34.59	0.922	**0.086**

6 Conclusions

In this paper, we introduce a novel architecture for image restoration in Under Display Cameras. Deviating from existing patch-based image restoration methods, we show that there is a significant benefit in directly training on megapixel images. Incorporated in an end-to-end manner, a guided filter framework alleviates artefacts associated with patch based methods. We also show that a carefully designed low-resolution network utilising smoothed atrous convolutions and various attention blocks is essential for superior performance. Finally, we develop a simple simulation scheme to pre-train our model and boost performance. Our overall approach outperforms current models and attains 2nd place in the UDC 2020 Challenge- Track 2:POLED.

As evidenced by our superlative performance on POLED restoration, the proposed method is more suited for higher degree of image degradation. Future work could address modifications to better handle a variety of image degradation tasks. Another promising perspective is to make better use of simulated data, for instance, in a domain-adaptation framework.

Acknowledgements. The authors would like to thank Genesis Cloud for providing additional compute hours.

References

1. Abdelhamed, A., Afifi, M., Timofte, R., Brown, M.S.: Ntire 2020 challenge on real image denoising: Dataset, methods and results. In: Proceedings of the IEEE/CVF Conference on Computer Vision and Pattern Recognition (CVPR) Workshops, June 2020
2. Agustsson, E., Timofte, R.: Ntire 2017 challenge on single image super-resolution: dataset and study. In: Proceedings of the IEEE Conference on Computer Vision and Pattern Recognition (CVPR) Workshops, July 2017
3. Ancuti, C.O., Ancuti, C., Vasluianu, F.A., Timofte, R.: Ntire 2020 challenge on nonhomogeneous dehazing. In: Proceedings of the IEEE/CVF Conference on Computer Vision and Pattern Recognition (CVPR) Workshops, June 2020
4. Blau, Y., Mechrez, R., Timofte, R., Michaeli, T., Zelnik-Manor, L.: The 2018 PIRM Challenge on Perceptual Image Super-Resolution. In: Leal-Taixé, L., Roth, S. (eds.) ECCV 2018, Part V. LNCS, vol. 11133, pp. 334–355. Springer, Cham (2019). https://doi.org/10.1007/978-3-030-11021-5_21
5. Brehm, S., Scherer, S., Lienhart, R.: High-resolution dual-stage multi-level feature aggregation for single image and video deblurring. In: Proceedings of the IEEE/CVF Conference on Computer Vision and Pattern Recognition (CVPR) Workshops, June 2020
6. Chen, D., et al.: Gated context aggregation network for image dehazing and deraining. In: Proceedings of the IEEE/CVF Winter Conference on Applications of Computer Vision (WACV) Workshops, March 2018
7. Chen, J., Adams, A., Wadhwa, N., Hasinoff, S.W.: Bilateral guided upsampling. ACM Trans. Graph. (TOG) **35**(6), 1–8 (2016)
8. Chen, L.C., Papandreou, G., Kokkinos, I., Murphy, K., Yuille, A.L.: DeepLab: semantic image segmentation with deep convolutional nets, atrous convolution, and fully connected CRFs. IEEE Trans. Pattern Anal. Mach. Intell. **40**(4), 834–848 (2017)
9. Chen, L.C., Papandreou, G., Schroff, F., Adam, H.: Rethinking atrous convolution for semantic image segmentation. arXiv preprint arXiv:1706.05587 (2017)
10. Chen, L.-C., Zhu, Y., Papandreou, G., Schroff, F., Adam, H.: Encoder-decoder with atrous separable convolution for semantic image segmentation. In: Ferrari, V., Hebert, M., Sminchisescu, C., Weiss, Y. (eds.) ECCV 2018, Part VII. LNCS, vol. 11211, pp. 833–851. Springer, Cham (2018). https://doi.org/10.1007/978-3-030-01234-2_49
11. Chen, Q., Xu, J., Koltun, V.: Fast image processing with fully-convolutional networks. In: Proceedings of the IEEE International Conference on Computer Vision (ICCV), October 2017
12. Chen, Y., Kalantidis, Y., Li, J., Yan, S., Feng, J.: Aˆ 2-nets: double attention networks. In: Advances in Neural Information Processing Systems, December 2018
13. Gharbi, M., Chen, J., Barron, J.T., Hasinoff, S.W., Durand, F.: Deep bilateral learning for real-time image enhancement. ACM Trans. Graph. (TOG) **36**(4), 118 (2017)

14. Gong, E., Pauly, J., Zaharchuk, G.: Boosting SNR and/or resolution of arterial spin label (ASL) imaging using multi-contrast approaches with multi-lateral guided filter and deep networks. In: Proceedings of the Annual Meeting of the International Society for Magnetic Resonance in Medicine, Honolulu, Hawaii (2017)
15. Goodfellow, I., et al.: Generative adversarial nets. In: Advances in Neural Information Processing Systems, December 2014
16. Gu, S., Li, Y., Gool, L.V., Timofte, R.: Self-guided network for fast image denoising. In: Proceedings of the IEEE/CVF International Conference on Computer Vision (ICCV), October 2019
17. Guo, Y., Han, S., Cao, H., Zhang, Y., Wang, Q.: Guided filter based deep recurrent neural networks for hyperspectral image classification. Procedia Computer Science **129**, 219–223 (2018)
18. Hamaguchi, R., Fujita, A., Nemoto, K., Imaizumi, T., Hikosaka, S.: Effective use of dilated convolutions for segmenting small object instances in remote sensing imagery. In: Proceedings of the IEEE/CVF Winter Conference on Applications of Computer Vision (WACV) Workshops, March 2018
19. He, K., Sun, J., Tang, X.: Guided image filtering. IEEE Trans. Pattern Anal. Mach. Intell. **35**(6), 1397–1409 (2013)
20. He, K., Sun, J.: Fast guided filter. arXiv preprint arXiv:1505.00996 (2015)
21. Hu, J., Shen, L., Sun, G.: Squeeze-and-excitation networks. In: Proceedings of the IEEE Conference on Computer Vision and Pattern Recognition (CVPR), June 2018
22. Ioffe, S., Szegedy, C.: Batch normalization: Accelerating deep network training by reducing internal covariate shift. arXiv preprint arXiv:1502.03167 (2015)
23. Jiang, Y., et al.: Enlightengan: Deep light enhancement without paired supervision. arXiv preprint arXiv:1906.06972 (2019)
24. Johnson, J., Alahi, A., Fei-Fei, L.: Perceptual losses for real-time style transfer and super-resolution. In: Leibe, B., Matas, J., Sebe, N., Welling, M. (eds.) ECCV 2016, Part II. LNCS, vol. 9906, pp. 694–711. Springer, Cham (2016). https://doi.org/10.1007/978-3-319-46475-6_43
25. Kim, D.W., Ryun Chung, J., Jung, S.W.: GRDB: grouped residual dense network for real image denoising and GAN-based real-world noise modeling. In: Proceedings of the IEEE/CVF Conference on Computer Vision and Pattern Recognition (CVPR) Workshops, June 2019
26. Kim, J., Lee, J.K., Lee, K.M.: Accurate image super-resolution using very deep convolutional networks. In: Proceedings of the IEEE Conference on Computer Vision and Pattern Recognition (CVPR), June 2016
27. Kopf, J., Cohen, M.F., Lischinski, D., Uyttendaele, M.: Joint bilateral upsampling. ACM Trans. Graph. (ToG) **26**(3), 96-es (2007)
28. Kupyn, O., Martyniuk, T., Wu, J., Wang, Z.: Deblurgan-v2: deblurring (orders-of-magnitude) faster and better. In: Proceedings of the IEEE/CVF International Conference on Computer Vision (ICCV), October 2019
29. Lehtinen, J., etal.: Noise2noise: learning image restoration without clean data. arXiv preprint arXiv:1803.04189 (2018)
30. Li, X., Wu, J., Lin, Z., Liu, H., Zha, H.: Recurrent squeeze-and-excitation context aggregation net for single image deraining. In: Proceedings of the European Conference on Computer Vision (ECCV) (September 2018)
31. Lim, B., Son, S., Kim, H., Nah, S., Mu Lee, K.: Enhanced deep residual networks for single image super-resolution. In: Proceedings of the IEEE Conference on Computer Vision and Pattern Recognition (CVPR) Workshops, July 2017

32. Lin, D., Ji, Y., Lischinski, D., Cohen-Or, D., Huang, H.: Multi-scale context intertwining for semantic segmentation. In: Ferrari, V., Hebert, M., Sminchisescu, C., Weiss, Y. (eds.) ECCV 2018, Part III. LNCS, vol. 11207, pp. 622–638. Springer, Cham (2018). https://doi.org/10.1007/978-3-030-01219-9_37
33. Lin, T.Y., Dollar, P., Girshick, R., He, K., Hariharan, B., Belongie, S.: Feature pyramid networks for object detection. In: Proceedings of the IEEE Conference on Computer Vision and Pattern Recognition (CVPR), July 2017
34. Loshchilov, I., Hutter, F.: SGDR: Stochastic gradient descent with warm restarts. In: International Conference on Learning Representations, April 2017
35. Loshchilov, I., Hutter, F.: Decoupled weight decay regularization. In: International Conference on Learning Representations, April 2019
36. Lugmayr, A., Danelljan, M., Timofte, R.: Ntire 2020 challenge on real-world image super-resolution: Methods and results. In: Proceedings of the IEEE/CVF Conference on Computer Vision and Pattern Recognition (CVPR) Workshops, June 2020
37. Mao, X., Shen, C., Yang, Y.B.: Image restoration using very deep convolutional encoder-decoder networks with symmetric skip connections. In: Advances in Neural Information Processing Systems, December 2016
38. Marin, D., et al.: Efficient segmentation: learning downsampling near semantic boundaries. In: Proceedings of the IEEE/CVF International Conference on Computer Vision (ICCV), October 2019
39. Mei, Y., et al.: Pyramid attention networks for image restoration. arXiv preprint arXiv:2004.13824 (2020)
40. Sarker, M.M.K., Rashwan, H.A., Talavera, E., Banu, S.F., Radeva, P., Puig, D.: MACNet: multi-scale atrous convolution networks for food places classification in egocentric photo-streams. In: Leal-Taixé, L., Roth, S. (eds.) ECCV 2018, Part V. LNCS, vol. 11133, pp. 423–433. Springer, Cham (2019). https://doi.org/10.1007/978-3-030-11021-5_26
41. Nah, S., Son, S., Timofte, R., Lee, K.M.: Ntire 2020 challenge on image and video deblurring. In: Proceedings of the IEEE/CVF Conference on Computer Vision and Pattern Recognition (CVPR) Workshops, June 2020
42. Orieux, F., Giovannelli, J.F., Rodet, T.: Bayesian estimation of regularization and point spread function parameters for wiener-hunt deconvolution. JOSA A **27**(7), 1593–1607 (2010)
43. Paszke, A., et al.: Pytorch: an imperative style, high-performance deep learning library. In: Advances in Neural Information Processing Systems, December 2019
44. Qin, X., Wang, Z., Bai, Y., Xie, X., Jia, H.: FFA-Net: feature fusion attention network for single image dehazing. In: AAAI, February 2020
45. Ronneberger, O., Fischer, P., Brox, T.: U-Net: convolutional networks for biomedical image segmentation. In: Navab, N., Hornegger, J., Wells, W.M., Frangi, A.F. (eds.) MICCAI 2015, Part III. LNCS, vol. 9351, pp. 234–241. Springer, Cham (2015). https://doi.org/10.1007/978-3-319-24574-4_28
46. Shi, W., et al.: Real-time single image and video super-resolution using an efficient sub-pixel convolutional neural network. In: Proceedings of the IEEE Conference on Computer Vision and Pattern Recognition (CVPR), June 2016
47. Sim, H., Kim, M.: A deep motion deblurring network based on per-pixel adaptive kernels with residual down-up and up-down modules. In: Proceedings of the IEEE/CVF Conference on Computer Vision and Pattern Recognition (CVPR) Workshops, June 2019
48. Simonyan, K., Zisserman, A.: Very deep convolutional networks for large-scale image recognition. In: International Conference on Learning Representations (ICLR), May 2015

49. Tai, Y., Yang, J., Liu, X., Xu, C.: MemNet: A persistent memory network for image restoration. In: Proceedings of the IEEE International Conference on Computer Vision (ICCV), October 2017
50. Tao, X., Gao, H., Shen, X., Wang, J., Jia, J.: Scale-recurrent network for deep image deblurring. In: Proceedings of the IEEE Conference on Computer Vision and Pattern Recognition (CVPR), June 2018
51. Timofte, R., Rothe, R., Van Gool, L.: Seven ways to improve example-based single image super resolution. In: Proceedings of the IEEE Conference on Computer Vision and Pattern Recognition (CVPR), June 2016
52. Ulyanov, D., Vedaldi, A., Lempitsky, V.: Improved texture networks: Maximizing quality and diversity in feed-forward stylization and texture synthesis. In: Proceedings of the IEEE Conference on Computer Vision and Pattern Recognition (CVPR), July 2017
53. Ulyanov, D., Vedaldi, A., Lempitsky, V.: Deep image prior. In: Proceedings of the IEEE Conference on Computer Vision and Pattern Recognition (CVPR), June 2018
54. Wang, P., et al.: Understanding convolution for semantic segmentation. In: Proceedings of the IEEE/CVF Winter Conference on Applications of Computer Vision (WACV) Workshops, March 2018
55. Wang, Q., Wu, B., Zhu, P., Li, P., Zuo, W., Hu, Q.: Eca-net: Efficient channel attention for deep convolutional neural networks. In: Proceedings of the IEEE/CVF Conference on Computer Vision and Pattern Recognition (CVPR), June 2020
56. Wang, Z., Ji, S.: Smoothed dilated convolutions for improved dense prediction. In: Proceedings of the 24th ACM SIGKDD International Conference on Knowledge Discovery and Data Mining. ACM, August 2018
57. Wenke, I.G.: Organic light emitting diode (OLED). Research gate (2016)
58. Woo, S., Park, J., Lee, J.-Y., Kweon, I.S.: CBAM: convolutional block attention module. In: Ferrari, V., Hebert, M., Sminchisescu, C., Weiss, Y. (eds.) ECCV 2018, Part VII. LNCS, vol. 11211, pp. 3–19. Springer, Cham (2018). https://doi.org/10.1007/978-3-030-01234-2_1
59. Wu, H., Zheng, S., Zhang, J., Huang, K.: Fast end-to-end trainable guided filter. In: Proceedings of the IEEE Conference on Computer Vision and Pattern Recognition (CVPR), June 2018,
60. Xu, Y., Wu, L., Xie, Z., Chen, Z.: Building extraction in very high resolution remote sensing imagery using deep learning and guided filters. Remote Sens. **10**(1), 144 (2018)
61. Yu, F., Koltun, V.: Multi-scale context aggregation by dilated convolutions. In: International Conference on Learning Representations (ICLR), May 2016
62. Yu, F., Koltun, V., Funkhouser, T.: Dilated residual networks. In: Proceedings of the IEEE Conference on Computer Vision and Pattern Recognition (CVPR), July 2017
63. Zhang, H., Patel, V.M.: Densely connected pyramid dehazing network. In: Proceedings of the IEEE Conference on Computer Vision and Pattern Recognition (CVPR), June 2018
64. Zhang, H., Patel, V.M.: Density-aware single image de-raining using a multi-stream dense network. In: Proceedings of the IEEE Conference on Computer Vision and Pattern Recognition (CVPR), June 2018
65. Zhang, J., Pan, J., Lai, W.S., Lau, R.W.H., Yang, M.H.: Learning fully convolutional networks for iterative non-blind deconvolution. In: Proceedings of the IEEE Conference on Computer Vision and Pattern Recognition (CVPR), July 2017

66. Zhang, K., Zuo, W., Chen, Y., Meng, D., Zhang, L.: Beyond a Gaussian denoiser: residual learning of deep CNN for image denoising. IEEE Trans. Image Process. **26**(7), 3142–3155 (2017)
67. Zhang, K., Zuo, W., Gu, S., Zhang, L.: Learning deep CNN denoiser prior for image restoration. In: Proceedings of the IEEE Conference on Computer Vision and Pattern Recognition (CVPR), July 2017
68. Zhang, K., Zuo, W., Zhang, L.: FFDNet: toward a fast and flexible solution for CNN-based image denoising. IEEE Trans. Image Process. **27**(9), 4608–4622 (2018)
69. Zhang, R., Isola, P., Efros, A.A., Shechtman, E., Wang, O.: The unreasonable effectiveness of deep features as a perceptual metric. In: Proceedings of the IEEE Conference on Computer Vision and Pattern Recognition (CVPR), June 2018
70. Zhang, X., Chen, Q., Ng, R., Koltun, V.: Zoom to learn, learn to zoom. In: Proceedings of the IEEE Conference on Computer Vision and Pattern Recognition (CVPR), June 2019
71. Zhang, Y., Tian, Y., Kong, Y., Zhong, B., Fu, Y.: Residual dense network for image restoration. IEEE Trans. Pattern Anal. Mach. Intell. 1 (2020). https://ieeexplore.ieee.org/document/8964437
72. Zhao, H., Gallo, O., Frosio, I., Kautz, J.: Loss functions for image restoration with neural networks. IEEE Trans. Comput. Imaging **3**(1), 47–57 (2016)
73. Zhao, H., Shi, J., Qi, X., Wang, X., Jia, J.: Pyramid scene parsing network. In: Proceedings of the IEEE Conference on Computer Vision and Pattern Recognition (CVPR), July 2017
74. Zhou, Y., et al.: UDC 2020 challenge on image restoration of under-display camera: Methods and results. arXiv preprint arXiv:2008.07742 (2020)
75. Zhou, Y., Ren, D., Emerton, N., Lim, S., Large, T.: Image restoration for under-display camera. arXiv preprint arXiv:2003.04857 (2020)

Residual and Dense UNet for Under-Display Camera Restoration

Qirui Yang[1,2,4(✉)], Yihao Liu[3,4], Jigang Tang[5], and Tao Ku[1,2]

[1] Shenyang Institute of Automation, Chinese Academy of Sciences, Shenyang 110016, China
kutao@sia.cn

[2] Institutes for Robotics and Intelligent Manufacturing, Chinese Academy of Sciences, Shenyang 110169, China

[3] Shenzhen Institutes of Advanced Technology, Chinese Academy of Sciences, Shenzhen 518055, China

[4] University of Chinese Academy of Sciences, Beijing 100049, China
{yangqirui18,liuyihao14}@mails.ucas.ac.cn

[5] Institute of Acoustics, Chinese Academy of Sciences, Beijing 100190, China
tangjigang18@mails.ucas.ac.cn

Abstract. With the rapid development of electronic products, the increasing demand for full-screen devices has become a new trend, which facilitates the investigation of Under-Display Cameras (UDC). UDC can not only bring larger display-to-body ratio but also improve the interactive experience. However, when imaging sensor is mounted behind a display, existing screen materials will cause severe image degradation due to lower light transmission rate and diffraction effects. In order to promote the research in this field, RLQ-TOD 2020 held the Image Restoration Challenge for Under-Display Camera. The challenge was composed of two tracks – 4k Transparent OLED (T-OLED) and phone Pentile OLED (P-OLED) track. In this paper, we propose a UNet-like structure with two various basic building blocks to tackle this problem. We discover that T-OLED and P-OLED have different preferences with the model structure and the input patch size during training. With the proposed model, our team won the third place in the challenge on both T-OLED and P-OLED tracks.

Keywords: Under-display camera · Image restoration · Deblurring · Enhancement · Denoising

1 Introduction

Under-Display Camera (UDC) [27] is a new imaging system. It places the camera under the screen to provide users with better human-computer interaction experience, and brings a larger display-to-body ratio. However, due to lower

Q. Yang and Y. Liu—The first two authors are co-first authors.

A. Bartoli and A. Fusiello (Eds.): ECCV 2020 Workshops, LNCS 12539, pp. 398–408, 2020.
https://doi.org/10.1007/978-3-030-68238-5_30

light transmission rate and diffraction effects, the images taken by UDC severely degraded with blur, noise and other unsatisfactory defects. Therefore, recovering under-display camera images is highly desired in the industry. This problem can be regarded as an image restoration task, which commonly includes image denoising, deblurring, decompression, enhancement, etc. Recently, plenty of learning-based methods have been proposed for image restoration and achieved promising results, especially in the fields of image denoising [12,22,23,26], deraining [21], defogging [5,9,14], and super-resolution [10,18].

In order to compensate for the unavoidable degradation caused by under-display camera, we propose a Residual and Dense UNet for under-display camera restoration. Built upon UNet [16] structure, we devise two different basic building blocks to further enhance the network capacity and representation capability – Residual Block (RB) and Residual Dense Block (RDB). Experiments show that T-OLED and P-OLED have different preferences with the model structure and the input patch size during training. T-OLED images favor complex network structures with more dense connections and smaller patch size, while P-OLED images favor structures with only residual connections and larger patch size. Hence, for the two screen materials (T-OLED and P-OLED), we propose two variants of the enhanced UNet – ResUNet and RDUNet, for under-camera image restoration. With the proposed model, our team won the third place in the ECCV Challenge on Image Restoration for Under-Display Camera on both T-OLED and P-OLED tracks.

2 Related Work

The denoising [5,12,22,23,26] problem of under-display camera imaging has always been the main obstacle to the launch of full-screen mobile phones. [27] take a 4k Transparent OLED (T-OLED) and a phone Pentile OLED (P-OLED) to analyze the optical performance of the two display screens, and designed a new imaging system. On this basis, a learning-based method propose to restore the full-screen image to a high-quality image. At the same time, single image restoration [5,21] has always been a research hotspot in the field of computer vision and image processing, especially in many fields such as security monitoring and medical diagnosis, smartphone camera. However, due to camera shake, moving objects, dark light, noise and other reasons, the real image will often appear fuzzy, virtual shadow and other phenomena. Therefore, the research of image denoising technology has important theoretical and practical application value, especially the image restoration in the real world [1,6] is becoming a new challenge in low-level vision.

At the same time, with the rapid development of deep neural networks, the noise reduction effect of neural networks on synthetic data has made significant progress, but these models are not effective for images with real degradation. Therefore, we need to have different ideas for the image restoration task of real-world images. Many datasets [20,24] about real noise have been proposed recently. Literature [2] first focused on the problem of using real HR to obtain natural low-resolution (LR) images, and designed a GAN-based High-to-Low network to obtain natural LR images from high-resolution (HR) images to simulate real low-resolution data. [19] established a real-world super-resolution dataset by adjusting the focal length of a digital camera to shoot a pair of LR-HR images in the same scene. [3] studied SR from the perspective of the camera, reducing the inherent trade-off between resolution and field of view (focal length), using a more realistic LR-HR dataset, and CameraSR [4] can be extended to different content and devices. [11] proposed a method combining two GAN models, one model learns to simulate the process of image blur, and the other model learns how to restore the image. At the same times, in order to reduce the difference between real blur and synthetic blur, a new set of real world blurred images.

In this article, we study the image restoration of the under-display camera based on the existing truly degraded natural image data set. We propose a new "under-display camera" image recovery method. According to experiments, T-OLED and P-OLED have different adaptability to model and patch size during training. Therefore, for these two screen materials, we propose two different UNet structures to meet different task requirements.

3 Method

In this section, we will introduce in detail the proposed network architecture – Residual Dense UNet (RDUNet) for the T-OLED display screen, and Residual UNet (ResUNet) for the P-OLED display screen.

3.1 Network Architecture

The overall structure of RDUNet and ResUNet is shown in Fig. 1. Based on the architecture of UNet [16], we devise two different basic building blocks, residual dense block and residual block, to enhance the network learning capability (see Fig. 2).

Similar to the original UNet, the proposed architecture contains an encoder and a decoder, with skip connections propagated from the encoder path. The main modification is that we expand the encoder with grouping a series of plain convolutional and ReLU layers into basic blocks. Then we could devise and select different basic blocks with various structures for exploring better performance.

The encoder is composed of several encoding blocks. Each block is composed of a convolutional layer, a basic block and a pooling layer. The residual block is the same as [10]. The proposed residual dense block is shown in Fig. 2, which consist of five layers of 3×3 convolution to form a dense network structure with a skip connection. Such residual and dense connections can extract richer context information and facilitate the information flow from shallow layers to deep layers. Based on these two building blocks, we develop the Residual UNet (ResUNet) and the Residual Dense UNet (RDUNet).

In our implementation, ResUNet and RDUNet contains five and four encoding blocks, respectively. As ResUNet has less complex connections, we deepen its depth for improving the learning capacity. For upsampling in the decoding path, ResUNet adopts PixelShuffle [17] operation while RDUNet utilizes bilinear upsampling. Further, in order to meet the needs of different tasks and reduce the overall parameters, we replaced part of the 3×3 convolution in the decoder stage with 1×1 convolution.

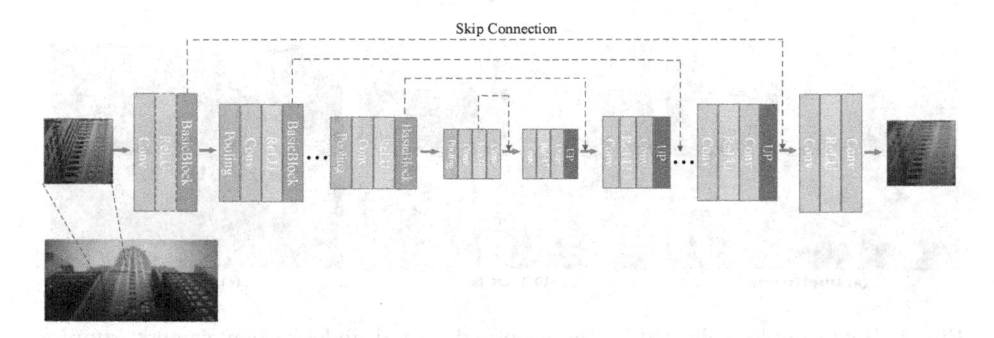

Fig. 1. We employ the basic architecture of UNet [16], and most of the calculations are done in the low-resolution feature space. We can choose or design "basic blocks" (for e.g., residual blocks [8], dense blocks [25], residual dense block) to obtain better performance.

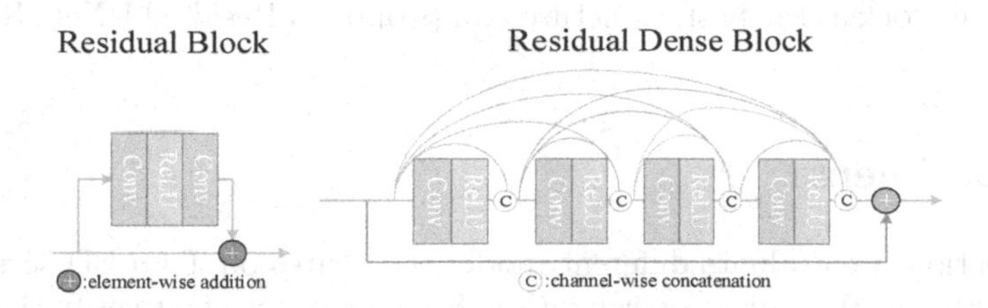

Fig. 2. Left: residual block. Right: the proposed residual dense block. It consists of a residual connection and a dense module composed of four convolutional layers.

3.2 Adaptability

Through the analysis of the images captured by the two screens, there is a big difference in the degree of light transmission and image degradation of the two screens (as shown in Table 1).

Based on the original UNet model, we propose two UNet models with different infrastructures, RDUNet and ResUNet models. Through cross-validation, our experiments show that T-OLED display and P-OLED display have different adaptability to the model structure and patch size during training. Hence, we propose two different UNet structures to meet different mission needs. For

Table 1. Comparison of two displays, PSNR obtained by calculating degraded images and labels in the training set.

Metrics	T-OLED	P-OLED
PSNR	29.06	15.51
SSIM	0.8405	0.6011
Open area [27]	21%	23%
Transmission rate [27]	20%	2.9%

Fig. 3. Real samples collected by the proposed paired under-display camera samples collected by [27].

T-OLED, we propose to apply residual dense block as the basic structure and propose a Residual-Dense UNet model (RDUNet). For P-OLED, we propose to use residual block as the basic structure and propose a Residual UNet (ResUNet) (Fig. 2).

4 Experiments

In this section, we evaluate different model structures on T-OLED screens and P-OLED screens through cross-validation. Specifically, we first verify that the T-OLED screen and P-OLED have different preferences with the model structure, and then verify the huge impact of patch size on the improvement of model performance during the training process. At the same time, we compare with several common image restoration models to verify the superiority of our method.

4.1 Implementation Details

In order to evaluate the performance of our RDUNet and ResUNet on image restoration, we use the datasets provided by 'ECCV Challenge on Image Restoration for Under-display Camera', which are T-OLED Dataset and P-OLED Dataset. The T-OLED Dataset and P-OLED Dataset [27] respectively contain a total of 270 pairs of images, of which the training set contains 240 pairs of 1024×2048 images, a total of 480 images, and the verification set is the remaining 30 pairs of 1024×2048 images. We train the model using MultiStepLR optimizer, and the momentum is 0.9. The learning rate is initialized as 1×10^{-4}. The whole algorithm developed based on PyTorch framework.

4.2 Adaptability

We designed a crossover experiment to verify the adaptability of different screens to different algorithms. Therefore, in this part, we designed four experiments, using UNet as the basic network, RB (Residual Block) and RDB (Residual Dense Block) as the basic modules to construct a ResUNet-5 network and an RDUNet-5 network with 5 encoder blocks. We apply these two networks to perform image restoration on two datasets to verify our ideas. The experimental results are depicted in Table 2.

Experiments show that RDUNet has better recovery performance for low-quality T-OLED UDC images; ResUNet more is friendly to the low-quality P-OLED UDC images.

Table 2. Performance comparison of two models on two datasets. ResUNet-n and RDUNet-n, n represents the number of basic blocks, which can indicate the depth of the network.

Data	Method	PSNR
Toled	ResUNet-5	37.95
	RDUNet-5	**38.22**
Poled	ResUNet-5	**30.62**
	RDUNet-5	30.12

4.3 Ablation Study

In this part, we will analyze the T-OLED track and P-OLED track separately to verify the influence of network depth, patch size, and upsampling mechanism on model performance.

Exploration on ResUNet. In this section, we explored the ResUNet from the following three aspects on the P-OLED dataset: 1. the choice of encoding module; 2. the choice of different upsampling and downsampling mechanisms; 3. patch size.

- Encoder Block: There are two main network architectures in image-to-image conversion: encoder-decoder and plain CNN. The comparsion of MWCNN [13] and MSRResNet [7] show that the architecture is more suitable for this challenge. Due to better texture transfer characteristics, residual and dense mechanism has been widely used in the field of image restoration and de-noising. On our experiments, we make an overall comparsion of Residual-Dense [25], ResNet [8] and common convolution. The results show that residual block achieved the highest PSNR.

- Up and Down Sampling: For upsampling, we adopt Bilinear and PixelShuffle [17]. For down sampling, haar wavelet transform [13], and global average pooling are tried. As shown in Table 3, the results show that PixelShuffle and global average pooling get the best performance.
- Patch Size: The task shows a strong sensitivity to patch size. Based on the same network structure and parameters, larger patch size generally shows better performance.

Finally, ResUNet-5-PS achieves the best performance of 32.54 dB on the validation set, when the patch size is 1024 × 1024. The visual results can be seen in Fig. 4.

Table 3. A comparison of different methods and mechanisms on P-OLED dataset.The i in $ResUNet - i$ represent the number of encoder blocks. The BL and PS means Bilinear and PixelShuffle Upsampling, respectively. The W_2 stands for haar wavelet transform and inverse wavelet transform are used after encoder block 2 and before the corresponding second last decoder, respectively.

Methods	Patch size	Parameter	PSNR
Camera output			15.51
Wiener filter [15]			16.97
MWCNN [13]	256	16.17M	28.09
ResNet-RAW [8]	256	**1.37M**	27.42
MSRResNet [7]	256	1.52M	25.25
UNet-SynRAW [27]	256	8.93M	25.88
UNet-RAW [27]	256	8.93M	30.45
RDUNet-5-PS	128	64.96M	28.18
RDUNet-5-PS	256	64.96M	29.96
ResUNet-5-PS	512	82.21M	30.62
ResUNet-5-BL	512	86.17M	30.44
ResUNet-5-W2	640	82.21M	31.53
ResUNet-5-PS	640	82.21M	31.63
ResUNet-5-PS	832	82.21M	32.23
ResUNet-5-PS	896	82.21M	32.45
ResUNet-5-PS	**1024**	82.21M	**32.54**

Exploration on RDUNet. In this part, we carry out experiments on T-OLED dataset from the following two aspects: 1. The impact of network depth on performance. 2. The influence of patch size.

- Network Depth: Residual and residual dense block play a different effects on different datasets. From the comparsion of Table 4, four residual dense blocks get the best performance on T-OLED dataset.

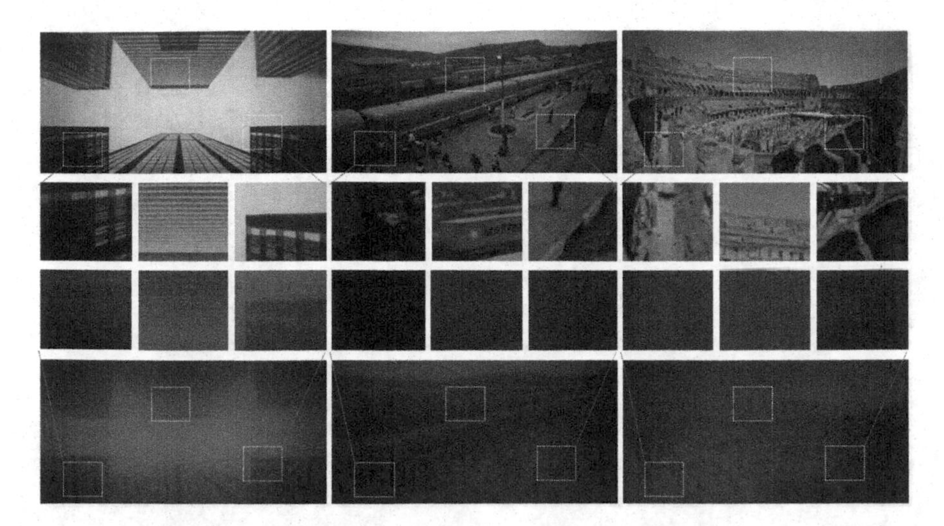

Fig. 4. A comparsion of low-quality P-OLED images and the predicted results of our method. The first line: predicted results of our method. Second line: three local samples of generated image. Third line: three local samples of input image. Fourth line: input low-quality images of P-OLED dataset.

Table 4. A comparison of different methods and mechanisms on T-OLED dataset.

Methods	Patch size	Parameter	PSNR
Camera output			29.06
Wiener filter [15]			28.50
ResNet-RAW [8]	256	**1.37M**	36.26
UNet-SynRAW [27]	256	8.93M	32.42
UNet-RAW [27]	256	8.93M	36.71
MSRResNet [7]	256	1.52M	36.45
ResUNet-5	512	82.21M	37.81
RDUNet-3	256	8.68M	38.10
RDUNet-3	512	8.68M	37.82
RDUNet-5	128	64.96M	38.12
RDUNet-5	192	64.96M	37.97
RDUNet-4-PS	**192**	31.7M	**38.98**

- Patch Size: Different from P-OLED dataset, the T-OLED datset is not sensitive to patch size. When the patch size has a large change, the PSNR performance does not change significantly.

Finally, RDUNet-4-PS achieves the best performance of 38.89dB on the validation set, when the patch size is 192 × 192. The visual results can be seen in Fig. 5.

Fig. 5. A comparsion of low-quality T-OLED images and the predicted results of our method. The first line: predicted results of our methods. Second line: three local samples of generated image. Third line: three local samples of input image. Fourth line: input low-quality images of T-OLED dataset.

5 Conclusions

In this paper, we study the image restoration of the under-display camera based on the existing truly degraded natural image datasets. We proposes a new under-display camera image recovery method. And according to experiments, T-OLED and P-OLED have different adaptability to model and patch size during training. Therefore, for these two screen materials, we respectively proposed the RDUNet network suitable for T-OLED screens and the ResUNet network suitable for P-OLED screens based on the UNet network. The proposed method won the third place in the ECCV Challenge on Image Restoration for Under-display Camera on both T-OLED and P-OLED tracks.

Acknowledgement. This work is partially supported by the National Key *R&D* Program of China (NO. 2019YFB17050003, NO. 2018YFB1308801, NO. 2017YFB0306401), the Consulting Research Project of the Chinese Academy of Engineering (Grant no. 2019-XZ-7).

References

1. Abdelhamed, A., Timofte, R., Brown, M.S., Yu, S., Cao, Z.: Ntire 2019 challenge on real image denoising: Methods and results. In: 2019 IEEE/CVF Conference on Computer Vision and Pattern Recognition Workshops (CVPRW) (2020)

2. Bulat, A., Yang, J., Tzimiropoulos, G.: To learn image super-resolution, use a GAN to learn how to do image degradation first. In: Ferrari, V., Hebert, M., Sminchisescu, C., Weiss, Y. (eds.) ECCV 2018, Part VI. LNCS, vol. 11210, pp. 187–202. Springer, Cham (2018). https://doi.org/10.1007/978-3-030-01231-1_12
3. Cai, J., Zeng, H., Yong, H., Cao, Z., Zhang, L.: Toward real-world single image super-resolution: a new benchmark and a new model. In: Proceedings of the IEEE International Conference on Computer Vision, pp. 3086–3095 (2019)
4. Chen, C., Xiong, Z., Tian, X., Zha, Z.J., Wu, F.: Camera lens super-resolution. In: 2019 IEEE/CVF Conference on Computer Vision and Pattern Recognition (CVPR) (2020)
5. Dong, Y., Liu, Y., Zhang, H., Chen, S., Qiao, Y.: FD-GAN: generative adversarial networks with fusion-discriminator for single image dehazing (2020)
6. Gong, D., Sun, W., Shi, Q., Hengel, A.V.D., Zhang, Y.: Learning to zoom-in via learning to zoom-out: Real-world super-resolution by generating and adapting degradation (2020)
7. Gross, S., Wilber, M.: Training and investigating residual nets. Facebook AI Res. **6**, (2016)
8. He, K., Zhang, X., Ren, S., Sun, J.: Deep residual learning for image recognition. In: Proceedings of the IEEE conference on Computer Vision and Pattern Recognition, pp. 770–778 (2016)
9. Kupyn, O., Budzan, V., Mykhailych, M., Mishkin, D., Matas, J.: Deblurgan: blind motion deblurring using conditional adversarial networks. In: Proceedings of the IEEE Conference on Computer Vision and Pattern Recognition, pp. 8183–8192 (2018)
10. Ledig, C., et al.: Photo-realistic single image super-resolution using a generative adversarial network (2016)
11. Li, S., Cai, Q., Li, H., Cao, J., Li, Z.: Frequency separation network for image super-resolution. IEEE Access **8**, 1 (2020)
12. Liu, J., et al.: Learning raw image denoising with Bayer pattern unification and Bayer preserving augmentation (2019)
13. Liu, P., Zhang, H., Zhang, K., Lin, L., Zuo, W.: Multi-level wavelet-CNN for image restoration. In: Proceedings of the IEEE Conference on Computer Vision and Pattern Recognition Workshops, pp. 773–782 (2018)
14. Nah, S., et al.: Ntire 2019 challenge on video deblurring: Methods and results. In: Proceedings of the IEEE Conference on Computer Vision and Pattern Recognition Workshops (2019)
15. Orieux, F., Giovannelli, J.F., Rodet, T.: Bayesian estimation of regularization and point spread function parameters for wiener-hunt deconvolution. JOSA A **27**(7), 1593–1607 (2010)
16. Ronneberger, O., Fischer, P., Brox, T.: U-Net: convolutional networks for biomedical image segmentation. In: Navab, N., Hornegger, J., Wells, W.M., Frangi, A.F. (eds.) MICCAI 2015, Part III. LNCS, vol. 9351, pp. 234–241. Springer, Cham (2015). https://doi.org/10.1007/978-3-319-24574-4_28
17. Shi, W., et al.: Real-time single image and video super-resolution using an efficient sub-pixel convolutional neural network. In: Proceedings of the IEEE Conference on Computer Vision and Pattern Recognition, pp. 1874–1883 (2016)
18. Wang, X., et al.: ESRGAN: enhanced super-resolution generative adversarial networks. In: Leal-Taixé, L., Roth, S. (eds.) ECCV 2018, Part V. LNCS, vol. 11133, pp. 63–79. Springer, Cham (2019). https://doi.org/10.1007/978-3-030-11021-5_5

19. Xu, X., Ma, Y., Sun, W.: Towards real scene super-resolution with raw images. In: Proceedings of the IEEE Conference on Computer Vision and Pattern Recognition, pp. 1723–1731 (2019)
20. Yang, C.-Y., Ma, C., Yang, M.-H.: Single-image super-resolution: a benchmark. In: Fleet, D., Pajdla, T., Schiele, B., Tuytelaars, T. (eds.) ECCV 2014, Part IV. LNCS, vol. 8692, pp. 372–386. Springer, Cham (2014). https://doi.org/10.1007/978-3-319-10593-2_25
21. Zhang, H., Sindagi, V., Patel, V.M.: Image de-raining using a conditional generative adversarial network. IEEE Trans. Circuits Syst. Video Technol. **30**(11), 3943–3956 (2017)
22. Zhang, K., Zuo, W., Chen, Y., Meng, D., Zhang, L.: Beyond a gaussian denoiser: residual learning of deep CNN for image denoising. IEEE Trans. Image Process. **26**(7), 3142–3155 (2016)
23. Zhang, K., Zuo, W., Zhang, L.: FFDNet: toward a fast and flexible solution for CNN based image denoising. IEEE Trans. Image Process. **27**, 4608–4622 (2017)
24. Zhang, X., Chen, Q., Ng, R., Koltun, V.: Zoom to learn, learn to zoom. In: Proceedings of the IEEE Conference on Computer Vision and Pattern Recognition, pp. 3762–3770 (2019)
25. Zhang, Y., Tian, Y., Kong, Y., Zhong, B., Fu, Y.: Residual dense network for image restoration. IEEE Trans. Pattern Anal. Mach. Intell. (99), 1 (2020)
26. Zhou, Y., et al.: When awgn-based denoiser meets real noises (2019)
27. Zhou, Y., Ren, D., Emerton, N., Lim, S., Large, T.: Image restoration for under-display camera. arXiv preprint arXiv:2003.04857 (2020)

W31 - The Bright and Dark Sides of Computer Vision: Challenges and Opportunities for Privacy and Security (CV-COPS 2020)

W31 - The Bright and Dark Sides of Computer Vision: Challenges and Opportunities for Privacy and Security (CV-COPS 2020)

Computer vision is finally working in the real world, allowing computers and other devices to "see" the world around them. But what will be the consequences for our privacy and security? Recent work has shown that algorithms can spy on smartphone keypresses from meters away, steal information from inside homes via hacked cameras, and exploit social media to de-anonymize blurred faces. But other work shows how computer vision could enhance privacy and security, for example through assistive devices to help people with disabilities, spam detection techniques that incorporate visual features, and forensic tools for flagging fake images. Some technologies present both challenges and opportunities: biometric techniques could enhance personal security but create potential for abuse. We need to understand the potential threats and opportunities of computer vision as it enters the mainstream.

To work towards this goal, we held the Fourth International Workshop on The Bright and Dark Sides of Computer Vision: Challenges and Opportunities for Privacy and Security at ECCV 2020. The workshop featured three keynote speakers:

- Don't Use Computer Vision For Web Security, Florian Tramèr (Stanford University)
- In Search of Lost Performance in Privacy-Preserving Deep Learning, Reza Shokri (National University of Singapore)
- Responsible Computing Research and Some Takeaways for Computer Vision, Asia J. Biega (Microsoft Research)

The workshop solicited full papers and extended abstracts. The program committee consisted of 24 experts at the intersection of computer vision and security. Each submitted paper was reviewed by at least two PC members, double-blind, and the five organizers made the final decisions.

August 2020

Mario Fritz
David Crandall
Jan-Michael Frahm
Apu Kapadia
Vitaly Shmatikov

Body Shape Privacy in Images: Understanding Privacy and Preventing Automatic Shape Extraction

Hosnieh Sattar[1(✉)], Katharina Krombholz[2], Gerard Pons-Moll[1], and Mario Fritz[2]

[1] Max Planck Institute for Informatics, Saarbrücken, Germany
{sattar,gpons}@mpi-inf.mpg.de

[2] CISPA Helmholtz Center for Information Security, Saarland Informatics Campus, Saarbrücken, Germany
{krombholz,fritz}@cispa.saarland

Abstract. Modern approaches to pose and body shape estimation have recently achieved strong performance even under challenging real-world conditions. Even from a single image of a clothed person, a realistic looking body shape can be inferred that captures a users' weight group and body shape type well. This opens up a whole spectrum of applications – in particular in fashion – where virtual try-on and recommendation systems can make use of these new and automatized cues. However, a realistic depiction of the undressed body is regarded highly private and therefore might not be consented by most people. Hence, we ask if the automatic extraction of such information can be effectively evaded. While adversarial perturbations have been shown to be effective for manipulating the output of machine learning models – in particular, end-to-end deep learning approaches – state of the art shape estimation methods are composed of multiple stages. We perform the first investigation of different strategies that can be used to effectively manipulate the automatic shape estimation while preserving the overall appearance of the original image.

1 Introduction

Since the early attempts to recognize human pose in images [19,65], we have seen a transition to real-world applications where methods operate on challenging real-world conditions in uncontrolled pose and lighting. We have seen more recently progress towards extracting richer representations beyond the pose. Most notably, a full body shape that is represented by a 3D representation or a low dimensional manifold (SMPL) [32]. It has been shown that such representations can be obtained from fully clothed persons – even in challenging conditions from a single image [7] as well as from web images of a person [50].

Electronic supplementary material The online version of this chapter (https://doi.org/10.1007/978-3-030-68238-5_31) contains supplementary material, which is available to authorized users.

A. Bartoli and A. Fusiello (Eds.): ECCV 2020 Workshops, LNCS 12539, pp. 411–428, 2020.
https://doi.org/10.1007/978-3-030-68238-5_31

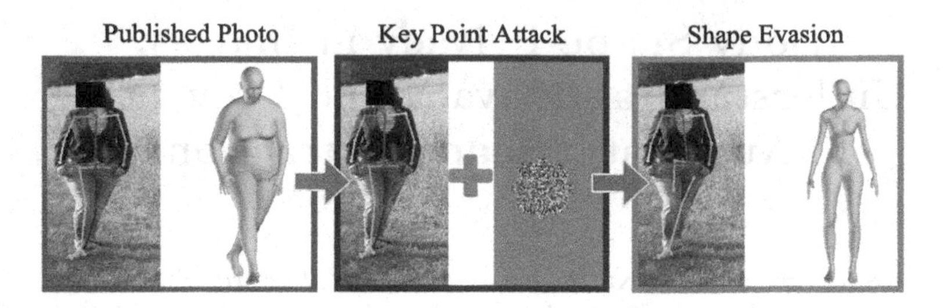

Fig. 1. A realistic depiction of the undressed body is considered highly private and therefore might not be consented by most people. We prevent automatic extraction of such information by small manipulations of the input image that keep the overall aesthetic of the image.

On the one hand, this gives rise to various applications – most importantly in the fashion domain. The more accurate judgment of fit could minimize clothing returns, and avatars and virtual try-on may enable new shopping experiences. Therefore, it is unsurprising that such technology already sees gradual adaption in businesses[1], as well as start-ups[2].

On the other hand, the automated extraction of such highly personal information from regular, readily available images might equally raise concerns about privacy. Images contain a rich source of implicit information that we are gradually learning to leverage with the advance of image processing techniques. Only recently, the first organized attempts were made to categorize private information in images [42] to raise awareness and to activate automatic protection mechanisms.

To control private information in images, a range of redaction and sanitization techniques have been introduced [41,55,59]. For example, evasion attacks have been used to disable classification routines to avoid extraction of information. Such techniques use adversarial perturbations to throw off a target classifier. It has been shown that such techniques can generalize to related classifiers [39], or can be designed under unknown/black-box models [8,12,27,36,54].

Unfortunately, such techniques are not directly applicable to state-of-the-art shape estimation techniques [2,3,7,21,31], as they are based on multi-stage processing. Typically, deep learning is used to extract person *keypoints*, and a model-fitting/optimization stage leads to the final keypoint estimation of pose and shape. As a consequence, there is no end-to-end architecture that would allow the computation of an image gradient needed for adversarial perturbations.

Today, we are missing successful evasion attacks on shape extraction methods. In this paper, we investigate to what extent shape extraction can be avoided by small manipulations of the input image (Fig. 1). We follow the literature on adversarial perturbations and require our changes in the input image to be of a

[1] https://www.cnet.com/news/amazon-buys-body-labs-a-3d-body-scanning-tech-startup/.
[2] https://bodylabs.io/en/.

small Euclidean norm. After analyzing a range of synthetic attack strategies that operate at the keypoints level, we experimentally evaluate their effectiveness to throw off multi-stage methods that include a model fitting stage. These attacks turn out to be highly effective while leaving the images visually unchanged. In summary, our contributions are:

- An orientative user study of concerns w.r.t. privacy and body shape estimation in different application contexts.
- Analysis of synthetic attacks on 2D keypoint detections.
- A new localized attack on keypoint feature maps that requires a smaller noise norm for the same effectiveness.
- Evaluation of overall effectiveness of different attacks strategies on shape estimation. We show the first successful attacks that offer an increase in privacy with negligible loss in visual quality.

2 Related Works

This work relates to 3D human shape estimation methods, privacy, and adversarial image perturbation techniques. We will here cover recent papers in these three domains and some of the key techniques directly relating to our approach.

Privacy and Computer Vision. Recent developments in computer vision techniques [16,24,30,37], increases concerns about extraction of private information from visual data such as age [6], social relationships [63], face detection [56,61], landmark detection [69], occupation recognition [52], and license plates [11,68,71]. Hence several studies on keeping the private content in visual data began only recently such as adversarial perturbations [33,43], automatic redaction of private information [41], predicting privacy risks in images [42], privacy-preserving video capture [1,35,46,49], avoiding face detection [22,64], full body re-identification [38] and privacy-sensitive life logging [25,29]. None of the previous work in this domain studied the users' shape privacy preferences. Hence, we present a new challenge in computer vision aimed at preventing automatic 3D shape extraction from images.

3D Body Shape Estimation. Recovery of 3D human shape from a 2D image is a very challenging task due to ambiguities such as depth and unknown camera data. This task has been facilitated by the availability of 3D generative body models learned from thousands of scans of people [4,32,48], which capture anthropometric constraints of the population and therefore reduce ambiguities. Several works [7,13,23,26,45,50,53,70,70] leverage these generative models to estimate 3D shape from single or multiple images, using shading cues, silhouettes and appearance. Recent model based approaches are using deep learning based 2D detections [10] – by either fitting a model to them at test time [2,3,7,21,50] or by using them to supervise bottom-up 3D shape predictors [2,28,40,44,58,60]. Hence, to evade recent shape estimators, we study different strategies to attack the 2D keypoint detections while preserving the overall appearance of the original image.

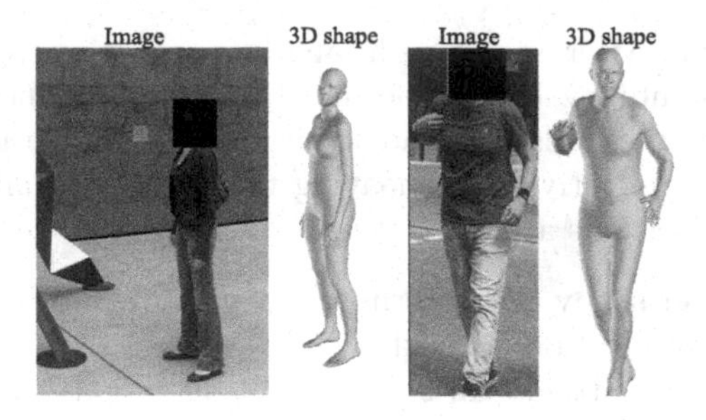

Fig. 2. Participants were asked to indicate their comfort level for sharing these images publicly, considering they are the subject in these images.

Adversarial Image Perturbation. Adversarial examples for deep neural networks were first reported in [20,57] demonstrating that deep neural networks are being vulnerable to small adversarial perturbations. This phenomenon was analyzed in several studies [5,17,18,51,66], and different approaches have been proposed to improve the robustness of neural networks [14,43]. Fast Gradient Sign Method (FGSM) and several variations of it were introduced in [20,34] for generating adversarial examples that are indistinguishable–to the human eye–from the original image, but can fool the networks. However, these techniques do not apply to state of the art body shape estimation as those are based on multi-stage processing. Typically, shape inference consists in fitting a body model to detected skeleton keypoints. Consequently, we perturb the 2D keypoints to produce an error in the shape fitting step. Cisse et al. [15], proposed a method to fool 2D pose estimation. None of these techniques propose a solution to evade model based shape estimation. In order to evade 3D shape estimation in a subtle manner, we attack by removing and flipping individual keypoints. Since these attacks simulate typical failure modes of detectors (mis-detections due to occlusion and keypoint flips), they are more difficult to identify by the defender.

3 Understanding Users Shape Privacy Preferences

Modern body shape methods [7,28,40,50] infer a realistic looking 3D body shape from a single photo of a person. The estimated 3D body captures user weight group and body shape type. However, such a realistic depiction of the undressed body is considered highly private and therefore might not be consented by most people. We performed a user study to explore the users' personal privacy preferences related to their body shape data. Our goal was to study the degree to which various users are sensitive to sharing their shape data such as height, different body part measurement, and their 3D body shape in different contexts. This study was approved by our university's ethical review board and is described next.

Fig. 3. In Question 2 participants were shown this image, and were asked to select the attributes from the list that could be extracted.

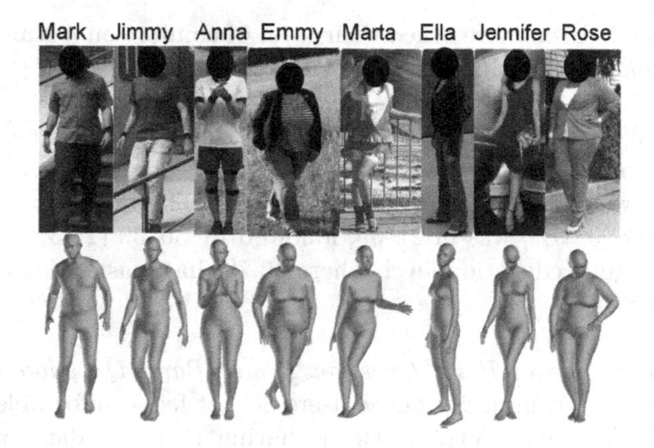

Fig. 4. Participants were asked to judge the closeness of the depicted 3D shape to the actual body of the person in the images.

User Study. We split the survey into three parts. In the first part of the survey, our goal was to understand users image sharing preferences and the users' knowledge of what type of information could be extracted from a single image.

Part1-Question 1: Users are shown Fig. 2 without the 3D shape data. Participants are asked how comfortable they are sharing such images publicly, considering they are the subject in these images. Responses are collected on a scale of 1 to 5, where: (1) Extremely comfortable, (2) Slightly comfortable, (3) Somewhat comfortable, (4) Not comfortable, and (5) Extremely uncomfortable.

Part1-Question 2: Participants were shown Fig. 3, and were asked which attributes could be extracted from this image.

In the second part, users were introduced to 3D shape models by showing them images of 8 people along with their 3D body shape, as shown in Fig. 4. The purpose of part 2 was to understand the user's perceived closeness of extracted 3D shapes to the original images, and their level of comfort with them.

Part2-Question 3: Participants were asked to rate how close the estimated 3D shape is to the person in the image. Responses are collected on a scale of 1 to 5, where: (1) Untrue of the person in the image, (2) somewhat untrue of the person in the image, (3) Neutral, (4) Somewhat true of the person in the image, and (5) True of the person in the image.

Part2-Question 4: Participants were shown Fig. 2 asked to indicate how comfortable they are sharing such a photograph along with 3D shape data publicly, considering they are the subject in these images. We collected responses on a scale of 1 to 5, similar to Question 1.

In the third part of this survey, we explore users preferences on what type of body shape information they would share for applications such as (a) Health insurance, (b) Body scanners at airport, (c) Online shopping platforms, (d) Dating platforms, and (e) Shape tracking applications (for sport, fitness, ...).

Part3-Question 5: Users were asked their level of comfort on a scale of 1 to 5 for the applications mentioned above.

Participants. We collect responses of 90 unique users in this survey. Participants were not paid to take part in this survey. Out of the 90 respondents, 43.3% were female, 55.6% were male, and 1.1% were queer. The dominant age range of our participants (63.3%) was in 21–39, followed by 30–39 (23.3%). Participants have a wide range of education level, where 46.7% has master degree, 21.1% has bachelor degree[3].

Analysis. The results of *Part1-Question 1* and *Part2-Question 4* are shown in Fig. 5a. We see that majority of the users do not feel comfortable or they feel extremely uncomfortable (36.0%, 30.0%) sharing their 3D data publicly compared to sharing only their images (29.0%, 14.0%).

In *Part1-Question 2*, the top three selected attributes were: gender (98.9%), pose (87.8%), and age (85.6%). Shape related attributes such as body mass index (BMI) (47.8%), weight (63.3%), and 3D body shape (66.7%) were not in the top selected attributes, indicating that many participants were unaware that such information could be extracted from an image using automatic techniques.

In *Part3-Question3*, users were asked to judge the quality of the presented 3D models. Around 43.0% of the participants believe the presented shape is Somewhat true of the person in the image, and 31.0% think the 3D mesh is true to the person in the picture. This indicates that recent approaches can infer perceptually faithful 3D body shapes under clothing from a single image.

Figure 5b presents the results from *Part3-Question 5*. Participants show a high level of discomfort in sharing their 3D shape data for multiple applications. In all investigated applications except for fitness, the majority of the users responded with "discomfort of some degree".

The user study demonstrates that users are concerned about the privacy of their body shape. Consequently, we present next a framework to prevent 3D shape extraction from images.

[3] Further details on participants demographic data are presented in the supplementary materials.

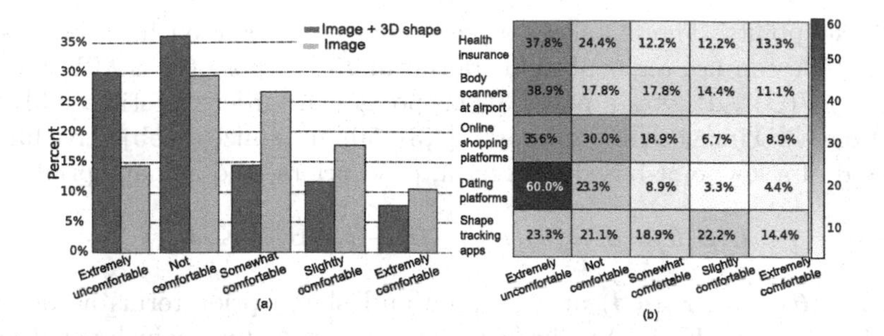

Fig. 5. (a) Comfort level of participants in sharing images with and without 3D mesh data, considering they are the subject in these images. (b) Comfort level of the participant for sharing their 3D mesh data with multiple applications. Results are shown as the percentage of times an answer is chosen.

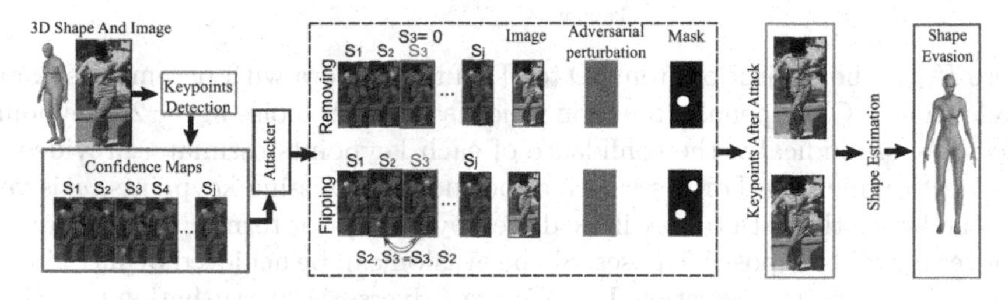

Fig. 6. The summary of our framework. We assume that we have full access to the parameters of the network. The attacker breaks the detections by removing or flipping of a keypoint. Hence the final estimated shape does not depict the person in the image.

4 Shape Evasion Framework

Model-based shape estimation methods from a 2D image are based on a two-stage approach. First, a neural network is used to detect a set of 2D body keypoints, then a 3D body model fits the detected keypoints. Since this approach is not end-to-end, it does not allows direct computation of the image gradient needed for adversarial perturbation. To this end, we approach the shape evasion by attacking the keypoints detection network. In Sect. 4.1, we give a brief introduction on model-based shape estimation method. In Sect. 4.2, we introduced a local attack that allows targeted attacks on keypoints. Figure 6 shows an overview of our approach.

4.1 Model Based Shape Estimation

The Skinned Multi-Person Linear Model (SMPL) [32] is a state of the art generative body model. The SMPL function $M(\boldsymbol{\beta}, \boldsymbol{\theta})$, uses shape $\boldsymbol{\beta}$ and poses $\boldsymbol{\theta}$ to parametrize the surface of the human body that is represented using $N = 6890$ vertices. The shape parameters $\boldsymbol{\beta} \in \mathbb{R}^{10}$ encode changes in height, weight and body proportions. The body pose $\boldsymbol{\theta} \in \mathbb{R}^{3P}$, is defined by a skeleton rig with

$P = 24$ keypoints. The 3D skeleton keypoints are predicted from body shape via $J(\boldsymbol{\beta})$. We can use a global rigid transform R_θ To pose the SMPL keypoint. Hence, $R_\theta(J(\boldsymbol{\beta})_i)$ denotes a posed 3D keypoint i. In order to estimate 3D body shape from a 2D image I, several works [7,31,50], minimize an objective function composed of a keypoint-based data term, pose priors, and a shape prior.

$$E(\boldsymbol{\beta}, \boldsymbol{\theta}) = E_{P_\theta}(\boldsymbol{\theta}) + E_{P_\beta}(\boldsymbol{\beta}) + E_J(\boldsymbol{\beta}, \boldsymbol{\theta}; \mathbf{K}, \mathbf{J}_{\text{est}}) \tag{1}$$

where $E_{P_\theta}(\boldsymbol{\theta})$, and $E_{P_\beta}(\boldsymbol{\beta})$ are the pose and shape prior terms as described in [7]. The $E_J(\boldsymbol{\beta}, \boldsymbol{\theta}; \mathbf{K}, \mathbf{J}_{\text{est}})$ is the keypoint-based data term which penalizes the weighted 2D distance between estimated 2D keypoints, $\mathbf{J}_{\text{est}}$, and the projected SMPL body keypoint $R_\theta(J(\boldsymbol{\beta}))$:

$$E_J(\boldsymbol{\beta}, \boldsymbol{\theta}; \mathbf{K}, \mathbf{J}_{\text{est}}) = \sum_{\text{keypoint} i} w_i \rho(\Pi_{\mathbf{K}}(R_\theta(J(\boldsymbol{\beta})_i)) - \mathbf{J}_{\text{est},i}) \tag{2}$$

where $\Pi_{\mathbf{K}}$ is the projection from 3D to 2D of the camera with parameters $\mathbf{K}$ and ρ a Geman-McClure penalty function which is robust to noise in the 2D keypoints detections. w_i indicates the confidence of each keypoints estimate, provided by 2D detection method. For cases such as occluded or missing keypoints, w is very low, and hence the data term will be driven by pose prior term. Furthermore, the prior term avoids impossible poses. Shape evasion can be achieved by introducing error in 2D keypoints detection $\mathbf{J}_{est}$. We use Adversarial perturbation to fool the pose detection method by either removing a keypoint or filliping two keypoints with each other.

4.2 Adversarial Image Generation

The State-of-the-art 2D pose detection methods such as [9] use a neural network f parametrized by ϕ, to predict a set of 2D locations of anatomical keypoints $\mathbf{J}_{est}$ for each person in the image. The network produces a set of 2D confidence maps $\mathbf{S} = \{\mathbf{S}_1, \mathbf{S}_2, \mathbf{S}_3, ..., \mathbf{S}_P\}$, where $\mathbf{S}_i \in \mathbb{R}^{w \times h}$, $i \in 1, 2, 3, ..., P$, is a confidence map for the keypoints i and P is total number of Keypoints. Assuming that a single person is in the image, then each confidence map contains a single peak if the corresponding part is visible. The final set of 2D keypoints $\mathbf{J}_{est}$ are achieved by performing non-maximum suppression per each confidence map. These confidence maps are shown in Fig. 6.

To attack a keypoint we used adversarial perturbation. Adding adversarial perturbation $\mathbf{a}$ to an image I will causes a neural network to change its prediction [57]. The adversarial perturbation $\mathbf{a}$ is defined as the solution to the optimization problem

$$\arg\min_{\mathbf{a}} \ \|\mathbf{a}\|_2 + L(f(I + \mathbf{a}; \phi), \mathbf{S}^*). \tag{3}$$

L is the loss function between the network output and desired confidence maps $\mathbf{S}^*$.

Removing and Flipping of Keypoints: The $\mathbf{S}^*$ is defined for removing and flipping of keypoints. To remove a keypoint, we put its confidence map to zero. For example if we are attacking the first keypoint we have: $\mathbf{S}^* = \{\mathbf{S}_1 = 0, \mathbf{S}_2, \mathbf{S}_3, ..., \mathbf{S}_P\}$. To flip two key points we exchanged the values of two confidence map as $\mathbf{S}^i, \mathbf{S}^j = \mathbf{S}^j, \mathbf{S}^i$. In case $i, j = 2, 3$ we have $\mathbf{S}^* = \{\mathbf{S}_1, \mathbf{S}_3, \mathbf{S}_2, ..., \mathbf{S}_P\}$. An example of removing and flipping of the keypoint is shown in Fig. 6.

Fast Gradient Sign Method (FGSM) [20]*:* FGSM is a first order optimization schemes used in practice for Eq. 3, which approximately minimizes the ℓ_∞ norm of perturbations bounded by the parameter ϵ. The adversarial examples are produced by increasing the loss of the network on the input I as

$$I^{adv} = I + \epsilon \ \text{sign}(\nabla_I L(f(I; \phi), \mathbf{S}^*)). \tag{4}$$

We call this type of attack global as the perturbation is applied to the whole image. This perturbation results in poses with several missing keypoints or poses outside of natural human pose manifold. While this will often make the subsequent shape optimization step fail (Eq. (2)), the approach has two limitations: i) this attack requires a large perturbation and ii) the attack is very easy to identify by the defender.

Masked Fast Gradient Sign Method (MFGSM): To overcome the limitations of the global approach, we introduced Masked FGSM. This allow for localized perturbation for more targeted attacks. This method will generate poses, which are close to ground truth pose, yet have a missing keypoint that will cause shape evasion–while requiring smaller perturbations as shown in the experiments. We will refer to this scheme as "local" in the rest of the paper. To attack a specific keypoint we solve the following optimization problem in a iterative manner as:

$$I_0^{adv} = I$$

$$I_{t+1}^{adv} = \text{clip}(I_t^{adv} - \alpha \cdot \text{sign}(\nabla_{I_t^{adv}} L(f(I_t^{adv}; \phi), \mathbf{S}^*) \odot M), \epsilon) \tag{5}$$

where mask $\mathbf{M} \in \mathbb{R}^{w \times h}$ is used to attack a keypoint $\mathbf{J}_{est,i} \in \mathbb{R}^2$ selectively. $\mathbf{M}$ is defined as:

$$\mathbf{M} = \begin{cases} 1 & \text{if } (x - \mathbf{J}_{est,i})^2 = r^2 \\ 0 & \end{cases}$$

r controls the spread of the attack and $\mathbf{x} \in \mathbb{R}^2$ are the pixel coordinates. To ensure the max norm constraint of perturbation **a** being no greater than ϵ is preserved, the *clip*(z, ϵ) is used, which keeps the values of z in the range$[z - \epsilon, z + \epsilon]$.

5 Experiments

The overall goal of the experimental section is to provide an understanding and the first practical approach to evade body shape estimation and hence protect the privacy of the user. We approach this by systematically studying the effect

Table 1. Shape estimation error on 3DPW with Procrustes analysis with respect to the ground truth shape. Error in cm. The minimum estimation error is 1.16 cm when we have no manipulation on input keypoints. The goal of each attack is to induce bigger error in the estimated shape. Hence, higher errors are indication of a successful attack.

Attack	Right ankle	Right knee	Right hip	Left hip	Left knee	Left ankle	Right wrist	Right elbow	Right shoulder	Left shoulder	Left elbow	Left wrist	Head top	Average
Adversarial	**1.32**	**1.4**	1.39	1.37	**1.38**	**1.32**	**1.36**	**1.41**	1.40	1.35	**1.28**	**1.37**	**1.35**	**1.37**
Synthetic	1.17	1.18	**1.79**	**1.94**	1.18	1.18	1.18	1.17	**1.43**	**1.49**	1.15	1.16	1.19	1.32

of attacking keypoint detections on the overall shape estimation pipeline. First, we study synthetic attacks based on direct manipulation of keypoints locations, where we can observe the effects on body shape estimation in an idealized scenario. This study is complemented by adversarial image-based attacks which make keypoint estimation fail. Together, we evaluate our approach that provides the first and effective defence against body shape estimation on real-world data.

Dataset. We used 3D Poses in the Wild Dataset (3DPW) [62], which includes 60 sequences with 7 actors. To achieve ground truth shape parameter $\boldsymbol{\beta}$, actors were scanned and SMPL was non-rigidly fit to them to obtain their 3D models similar to [47,67]. To the best of our knowledge, 3DPW is the only in wild image dataset which provides the ground truth shape data as well, which makes this dataset most suitable for our evaluation. For our evaluation, for each actor, we randomly selected multiple frames from different sequences. All reported results are averaged across subjects and sampled sequence frames.

Model. We used OpenPose [10] for keypoint detection as it is the most widely used. OpenPose consists of a two-branch multi-stage CNN, which process images at multi-scales. Each stage in the first branch predicts the confidence map S, and each stage in the second branch predicts the Part Affinity Fields (PAFs). For the shape estimation, we used the public code of Smplify [7], which infers a 3D mesh by fitting the SMPL body model to a set of 2D keypoints. To improve the 3D accuracy, we refined the estimations using silhouette as described in [50]. We used MFGSM (Eq. (5) with $\alpha = 1$) in an iterative manner. We evaluated attacks when setting the ℓ_∞ norm of the perturbations to $\epsilon = 0.035$ since we observed that higher values lead to noticeable artifacts in the image. We stop the iterations if we reach an Euclidean distance (between the original and perturbed images) of 0.02 in image space for local, and 0.04 for global attacks.

5.1 Synthetic Modification of the Keypoints

First, we studied the importance of each keypoint on the overall body shape estimation by removing one keypoint at a time–which simulates miss-detections. The error on shape estimation caused by this attack is reported in the second row of Table 1. We observe that removing "Hips", and "Shoulder" keypoints results in the highest increase of error of 60.78%, and 25.86% whereas "Elbows" and "Wrists" result in an increase of only 1%.

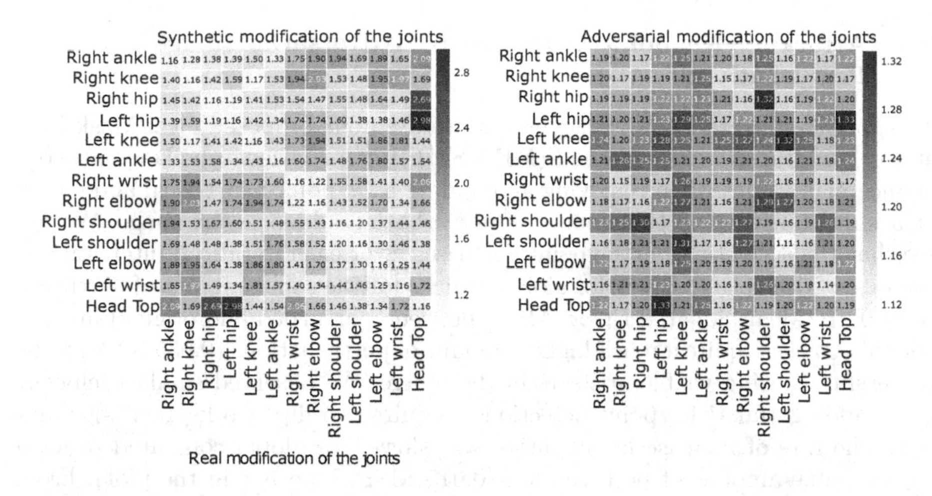

Fig. 7. Shape estimation error on 3DPW with Procrustes analysis. Error in cm for synthetic and adversarial flipping of the keypoints.

We also studied the effect of flipping keypoints. The results of this experiment are shown in Fig. 7. Flipping the "Head" with the left or right "Hip" caused an increase in error of 143.96%. Flipping the "Elbow" and "Knee" was the second most effective attack causing 67.0% increase of error in average. The least effective attack was by flipping the left and right knee (less than 0.1%). The average error introduced by removing or flipping of each keypoint is illustrated in Fig. 8 – higher error is larger in size and darker in colour. We can see that, overall "Hip", "Shoulder", and "Head" keypoints play a crucial role in the quality of the final estimated 3D mesh, and are the most powerful attacks.

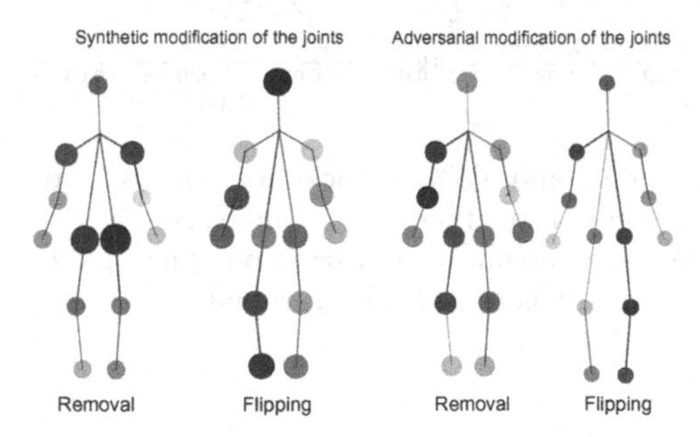

Fig. 8. The overall shape estimation error induced by synthetic and adversarial (local) attacks. The darker and bigger circles shows higher error.

5.2 Attacking Keypoint Detection by Adversarial Image Perturbation

To apply modifications to the keypoints, we used our proposed local Mask Iterative Fast Gradient Sign Method (MIFGSM). Figure 9 shows the keypoint confidence map values when removing and adding a keypoint using local and global attacks with respect to the amount of perturbation added to the image. We can see that the activation per keypoint decreases after each iteration. Interestingly, the rate of decrease is slower for global attacks for the same amount of perturbation (0.015 Mean Squared Error (MSE) between perturbed and original image). Global attacks require a much higher amount of perturbations (0.035 MSE) to be successful, causing visible artifacts in the image. We observed similar behavior when adding "fake" keypoint detections (required to flip two keypoints). Similarly, the rate of increase in activation was slower for global compared to local for the same amount of perturbation (0.015 MSE, blue bar in the plot). From Fig. 9 we can also see that shoulders and head are more resistant to the removal. Furthermore, the attack was the most successful in the creation of wrists. Since local attacks are more effective, we consider only the local attack method for further analysis.

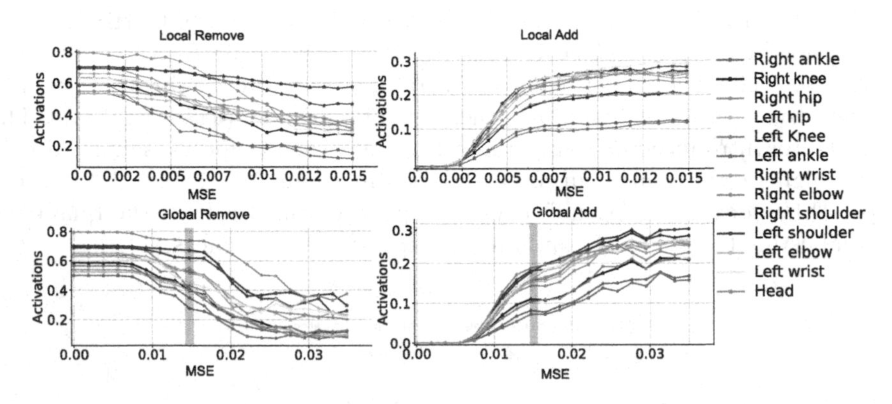

Fig. 9. Comparison of local and global attacks for removing and adding a keypoint. The local attack has a higher rate of decrease or increase of activation compared to the global method for the same amount of perturbation. The blue bar on the global plots shows the end of the local methods. (Color figure online)

5.3 Shape Evasion

In this section, we evaluate the effectiveness of the whole approach for evading shape estimation and therefore, protecting the users' privacy. We used our proposed local method to remove and flip keypoints instead of the synthetic modification of the keypoints as described in Sect. 5.1. Hence, we call this attack as a adversarial modification of keypoints.

The error on shape estimation caused by removing of the keypoints using our local method are reported in the first row of Table 1, we refer to it as adversarial. We see that attacks on “Right Elbow” and “Right knee” causes 21.55%, and 20.69% increase of error in shape estimation. The least amount of error 10.34% and 13.79% was produced by removing “Left Elbow” and “Ankles” respectively. However, “Hip” and “Shoulder” gained higher error in average for left and right keypoint by 19.0%. On average, the adversarial attack for removing keypoints caused an even higher error than the synthetic mode (18.1% to 13.79%), showing the effectiveness of this approach in shape evasion and hence protecting the users’ privacy.

The result for flipping the keypoints is shown in Fig. 7 (Adversarial modification of the keypoint). The highest increase in error was (14.66%) caused by flipping the “Head” with “Left Hip”, the second most effective attack was for flipping the “Right Shoulder” and “Left Knee” keypoints (12.93%). Overall the most effective attack was on “Knee” and the least effective attack was on “Wrist” with increase of 7.11%, and 1.72% error on average, respectively.

Adversarial flipping of keypoints achieves an error of 4.19% compared to adversarial removing attacks (18.1%), which shows they are less effective. In addition, similar to global attacks, filliping of keypoints causes more changes in the keypoints, making the detection of these attacks easier.

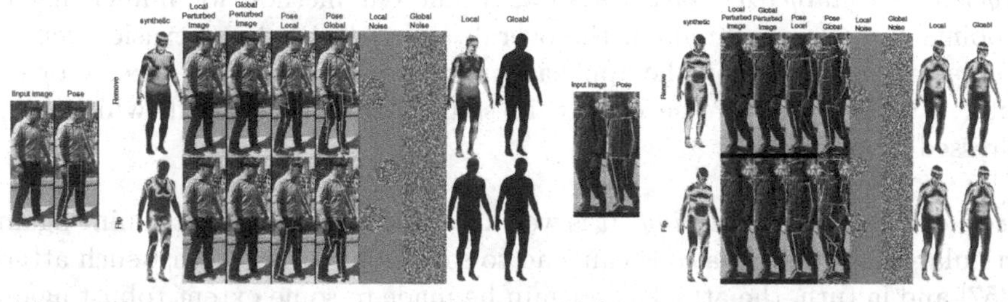

(a) Person with body shape close to SMPL template (0.04 cm)

(b) Person with a higher distance to SMPL template (2.0 cm)

Fig. 10. The left side shows the original image with the estimated pose, and the right the output when modified with local and global adversarial perturbations with corresponding error heatmaps with respect to ground truth shapes (red means > 2.0 cm). Here we applied local and global attack for removing the “Right Hip”, and flipping the “Right Hip” and “Head Top”. The global attack causes the pose estimation to hallucinate multiple people in the image, while our local attack only changes the selected keypoints. The predicted shape in case of a global attack is always close to the average template of SMPL causing a lower error for people with an average shape. (Color figure online)

5.4 Qualitative Results

In Fig. 10, we present example results obtained for each type of attack. The global attack causes pose estimation to hallucinate multiple people in the image, destroying the body signal of the person in the picture. As the predicted poses in the global attack are not in human body manifold, the optimization step in SMPL will fail to fit these keypoints resulting in average shape estimates. In the local attack, we were able to apply small changes in the keypoints. Hence, these small changes make the shape optimization stage predict shapes that are not average and also not close to the person in the image. Overall, shape evasion was most successful when removing the keypoints than flipping them, and when using the local attacks.

6 Discussion

As our study of privacy on automatically extracted body shapes and method for evading shape estimation is the first of its kind, it serves as a starting point – but naturally needs further investigations to extend on both lines of research that we have touched on. The following presents a selection of open research questions.

Targeted vs Untargeted Shape Evasion. While our method for influencing the keypoints detection is targeted, the overall approach to shape evasion remains untargeted. Depending on the application scenario, a consistent change or particular randomization of the change in shape might be desired, which is not addressed by our work.

Effects of Adversarial Training. It is well known that adversarial training against particular image perturbations can lead to some robustness against such attacks [39,57] and in turn, the attack can again be made to some extent robust against such defences. Preventing this cat-mouse-game is subject of on-going research and – while very important – we consider outside of the scope of our first demonstration of shape evasion methods.

Scope of the User Study. While our user study encompasses essential aspects of privacy of body shape information, clearly a more detailed understanding can be helpful to inform the design evasion techniques and privacy-preserving methodologies that comply with the users' expectations on handling personal data. As our study shows that such privacy preferences are personal as well as application domain specific, there seem ample opportunities to leverage the emerging methods of high-quality body shape estimation in compliance with user privacy.

7 Conclusion

Methods for body shape estimation from images of clothed people are getting more and more accurate. Hence we have asked the timely question to what extent this raises privacy concerns and if there are ways to evade shape estimation from images. To better understand the privacy concerns, we conduct a user study that sheds light on the privacy implication as well as the sensitivity of shape data in different application scenarios. Overall, we observe a high sensitivity, which is also dependent on the use case of the data. Based on this understanding, we follow up with a defence mechanism that can hamper or even prevent body shape estimation from real-world images. Today's state of the art body shape estimation approaches are frequently optimization based and therefore don't lend themselves to gradient-based adversarial perturbation. We tackle this problem by a two-stage approach that first analysis the effect of individual keypoints on the shape estimate and then proposes adversarial image perturbations to influence the keypoints. In particular, our novel localized perturbation techniques constitute an effective technique to evade body shape estimation at negligible changes to the original image.

References

1. Aditya, P., et al.: I-Pic: a platform for privacy-compliant image capture. In: ACM (2016)
2. Alldieck, T., Magnor, M., Bhatnagar, B.L., Theobalt, C., Pons-Moll, G.: Learning to reconstruct people in clothing from a single RGB camera. In: CVPR (2019)
3. Alldieck, T., Magnor, M., Xu, W., Theobalt, C., Pons-Moll, G.: Video based reconstruction of 3D people models. In: CVPR Spotlight (2018)
4. Anguelov, D., Srinivasan, P., Koller, D., Thrun, S., Rodgers, J., Davis, J.: SCAPE: shape completion and animation of people. ACM Trans. Graph. **24**, 408–416 (2005)
5. Arnab, A., Miksik, O., Torr, P.H.S.: On the robustness of semantic segmentation models to adversarial attacks. In: CVPR (2018)
6. Bauckhage, C., Jahanbekam, A., Thurau, C.: Age recognition in the wild. In: ICPR (2010)
7. Bogo, F., Kanazawa, A., Lassner, C., Gehler, P., Romero, J., Black, M.J.: Keep It SMPL: Automatic Estimation of 3D Human Pose and Shape from a Single Image. In: Leibe, B., Matas, J., Sebe, N., Welling, M. (eds.) ECCV 2016. LNCS, vol. 9909, pp. 561–578. Springer, Cham (2016). https://doi.org/10.1007/978-3-319-46454-1_34
8. Brendel, W., Bethge, M.: Comment on "biologically inspired protection of deep networks from adversarial attacks" arXiv:1704.01547 (2017)
9. Cao, Z., Hidalgo, G., Simon, T., Wei, S.-E., Sheikh, Y.: OpenPose: realtime multi-person 2D pose estimation using Part Affinity Fields. In arXiv preprint arXiv:1812.08008 (2018)
10. Cao, Z., Simon, T., Wei, S.-E., Sheikh, Y.: Realtime multi-person 2d pose estimation using part affinity fields. In: CVPR (2017)
11. Chang, S.-L., Chen, L.-S., Chung, Y.-C., Chen, S.-W.: Automatic license plate recognition. IEEE Trans. Intell. Transp. Syst. **5**, 42–53 (2004)

12. Chen, P.-Y., Zhang, H., Sharma, Y., Yi, J., Hsieh, C.-J.: Zoo: zeroth order optimization based black-box attacks to deep neural networks without training substitute models. In: Proceedings of the 10th ACM Workshop on Artificial Intelligence and Security. ACM (2017)
13. Chen, Yu., Kim, T.-K., Cipolla, R.: Inferring 3D Shapes and Deformations from Single Views. In: Daniilidis, K., Maragos, P., Paragios, N. (eds.) ECCV 2010. LNCS, vol. 6313, pp. 300–313. Springer, Heidelberg (2010). https://doi.org/10.1007/978-3-642-15558-1_22
14. Cissé, M., Bojanowski, P., Grave, E., Dauphin, Y., Usunier, N.: Parseval networks: improving robustness to adversarial examples. In: ICML (2017)
15. Cisse, M.M., Adi, Y., Neverova, N., Keshet, J.: Houdini: fooling deep structured visual and speech recognition models with adversarial examples. In: NIPS (2017)
16. Deng, J., Dong, W., Socher, R., Jia Li, L., Li, K., Fei-fei, L.: Imagenet: a large-scale hierarchical image database. In: CVPR (2009)
17. Fawzi, A., Fawzi, O., Frossard, P.: Analysis of classifiers' robustness to adversarial perturbations. Mach. Learn. **107**(3), 481–508 (2017). https://doi.org/10.1007/s10994-017-5663-3
18. Fawzi, A., Moosavi-Dezfooli, S.-M., Frossard, P.: Robustness of classifiers: from adversarial to random noise. In: NIPS (2016)
19. Gavrila, D.: The visual analysis of human movement: a survey. Comput. Vis. Image Underst. **73**, 82–98 (1999)
20. Goodfellow, I.J., Shlens, J., Szegedy, C.: Explaining and harnessing adversarial examples. arXiv:1412.6572 (2014)
21. Habermann, M., Xu, W., Zollhoefer, M., Pons-Moll, G., Theobalt, C.: Livecap: real-time human performance capture from monocular video. ACM Trans. Graph 38, 1–17 (2019). (Proc. SIGGRAPH
22. Harvey, A.: CV dazzle: Camouflage from computer vision. Technical report (2012)
23. Hasler, N., Ackermann, H., Rosenhahn, B., Thormahlen, T., Seidel, H.-P.: Multilinear pose and body shape estimation of dressed subjects from image sets. In: CVPR (2010)
24. He, K., Zhang, X., Ren, S., Sun, J.: Deep residual learning for image recognition. In: CVPR (2016)
25. Hoyle, R., Templeman, R., Anthony, D., Crandall, D., Kapadia, A.: Sensitive lifelogs: a privacy analysis of photos from wearable cameras. In: CHI (2015)
26. Huang, Y., et al.: Towards accurate marker-less human shape and pose estimation over time. In: 3DV (2017)
27. Ilyas, A., Engstrom, l., Madry, A.: Prior convictions: Black-box adversarial attacks with bandits and priors. arXiv preprint arXiv:1807.07978 (2018)
28. Kanazawa, A., Black, M.J., Jacobs, D.W., Malik, J.: End-to-end recovery of human shape and pose. In: CVPR (2018)
29. Korayem, M., Templeman, R., Chen, D., Crandall, D., Kapadia, A.: Enhancing lifelogging privacy by detecting screens. In: CHI (2016)
30. Krizhevsky, A., Sutskever, I., Hinton, G.E.: Imagenet classification with deep convolutional neural networks. In: NIPS (2012)
31. Lassner, C., Romero, J., Kiefel, M., Bogo, F., Black, M.J., Gehler, P.V.: Unite the people: closing the loop between 3D and 2D human representations. In: CVPR (2017)
32. Loper, M., Mahmood, N., Romero, J., Pons-Moll, G., Black, M.J.: SMPL: a skinned multi-person linear model. ACM Trans. Graph. 34, 1–16 (2015). Proc. SIGGRAPH Asia

33. Moosavi-Dezfooli, S.-M., Fawzi, A., Fawzi, O., Frossar, P.: Universal adversarial perturbations. In: CVPR (2007)
34. Moosavi-Dezfooli, S.-M., Fawzi, A., Frossard, P.: Deepfool: a simple and accurate method to fool deep neural networks. In: CVPR (2016)
35. Neustaedter, C., Greenberg, S., Boyle, M.: Blur filtration fails to preserve privacy for home-based video conferencing (2006)
36. Oh, S.J., Schiele, B., Fritz, M.: Towards Reverse-Engineering Black-Box Neural Networks. In: Samek, W., Montavon, G., Vedaldi, A., Hansen, L.K., Müller, K.-R. (eds.) Explainable AI: Interpreting, Explaining and Visualizing Deep Learning. LNCS (LNAI), vol. 11700, pp. 121–144. Springer, Cham (2019). https://doi.org/10.1007/978-3-030-28954-6_7
37. Oh, S.J., Benenson, R., Fritz, M., Schiele, B.: Person recognition in personal photo collections. In: ICCV (2015)
38. Oh, S.J., Benenson, R., Fritz, M., Schiele, B.: Faceless Person Recognition: Privacy Implications in Social Media. In: Leibe, B., Matas, J., Sebe, N., Welling, M. (eds.) ECCV 2016. LNCS, vol. 9907, pp. 19–35. Springer, Cham (2016). https://doi.org/10.1007/978-3-319-46487-9_2
39. Oh, S. J., Fritz, M., Schiele, B.: Adversarial image perturbation for privacy protection - a game theory perspective. In: ICCV (2017)
40. Omran, M., Lassner, C., Pons-Moll, G., Gehler, P., Schiele, B.: Neural body fitting: Unifying deep learning and model based human pose and shape estimation. In: 3DV (2018)
41. Orekondy, T., Fritz, M., Schiele, B.: Connecting pixels to privacy and utility: automatic redaction of private information in images. In: CVPR (2018)
42. Orekondy, T., Schiele, B., Fritz, M.: Towards a visual privacy advisor: Understanding and predicting privacy risks in images. In: ICCV (2017)
43. Papernot, N., McDaniel, P., Wu, X., Jha, S., Swami, A.: Distillation as a defense to adversarial perturbations against deep neural networks. In: 2016 IEEE Symposium on Security and Privacy (SP). IEEE (2016)
44. Pavlakos, G., Zhu, L., Zhou, X., Daniilidis, K.: Learning to estimate 3D human pose and shape from a single color image. In: CVPR (2018)
45. Guan, P., Weiss, A., Bãlan, A.O., Black, M.J.: Estimating human shape and pose from a single image. In: 2009 IEEE 12th International Conference on Computer Vision, pp. 1381–1388 (2009)
46. Pittaluga, F., Koppal, S.J.: Privacy preserving optics for miniature vision sensors. In: CVPR (2015)
47. Pons-Moll, G., Pujades, S., Hu, S., Black, M.: ClothCap: seamless 4D clothing capture and retargeting. ACM Trans. Graph. **36**(4), 1–15 (2017)
48. Pons-Moll, G., Romero, J., Mahmood, N., Black, M.J.: Dyna: a model of dynamic human shape in motion. ACM Trans. Graph. **34**, 120 (2015)
49. Raval, N., Srivastava, A., Lebeck, K., Cox, L., Machanavajjhala, A.: Markit: privacy markers for protecting visual secrets. In: UbiComp (2014)
50. Sattar, H., Pons-Moll, G., Fritz, M.: Fashion is taking shape: understanding clothing preference based on body shape from online sources. In: WACV (2019)
51. Shaham, U., Yamada, Y., Negahban, S.: Understanding adversarial training: Increasing local stability of neural nets through robust optimization. arXiv:1511.05432 (2015)
52. Shao, M., Li, L., Fu, Y.: What do you do? Occupation recognition in a photo via social context. In: CVPR (2013)
53. Sigal, L., Balan, A., Black, M.J.: Combined discriminative and generative articulated pose and non-rigid shape estimation. In: NIPS (2008)

54. Su, J., Vargas, D.V., Sakurai, K.: One pixel attack for fooling deep neural networks. arXiv:1710.08864 (2017)
55. Sun, Q., Tewari, A., Xu, W., Fritz, M., Theobalt, C., Schiele, B.: A Hybrid Model for Identity Obfuscation by Face Replacement. In: Ferrari, V., Hebert, M., Sminchisescu, C., Weiss, Y. (eds.) ECCV 2018. LNCS, vol. 11205, pp. 570–586. Springer, Cham (2018). https://doi.org/10.1007/978-3-030-01246-5_34
56. Sun, X., Wu, P., Hoi, S.C.H.: Face detection using deep learning: An improved faster RCNN approach. CoRR, abs/1701.08289 (2017)
57. Szegedy, C., et al.: Intriguing properties of neural networks. In: ICLR (2014)
58. Tan, V., Budvytis, I., Cipolla, R.: Indirect deep structured learning for 3d human body shape and pose prediction. In: BMVC (2017)
59. Tretschk, E., Oh, S. J., Fritz, M.: Sequential attacks on agents for long-term adversarial goals. In: ACM Computer Science in Cars Symposium - Future Challenges in Artificial Intelligence and Security for Autonomous Vehicles (CSCS) (2018)
60. Tung, H., Wei, H., Yumer, E., Fragkiadaki, K.: Self-supervised learning of motion capture. In: NIPS (2017)
61. Viola, P.A., Jones, M.J.: Robust real-time face detection. Int. J. Comput. Vis. **57**, 137–154 (2001). https://doi.org/10.1023/B:VISI.0000013087.49260.fb
62. von Marcard, T., Henschel, R., Black, M.J., Rosenhahn, B., Pons-Moll, G.: Recovering Accurate 3D Human Pose in the Wild Using IMUs and a Moving Camera. In: Ferrari, V., Hebert, M., Sminchisescu, C., Weiss, Y. (eds.) ECCV 2018. LNCS, vol. 11214, pp. 614–631. Springer, Cham (2018). https://doi.org/10.1007/978-3-030-01249-6_37
63. Wang, G., Gallagher, A., Luo, J., Forsyth, D.: Seeing People in Social Context: Recognizing People and Social Relationships. In: Daniilidis, K., Maragos, P., Paragios, N. (eds.) ECCV 2010. LNCS, vol. 6315, pp. 169–182. Springer, Heidelberg (2010). https://doi.org/10.1007/978-3-642-15555-0_13
64. Wilber, M.J., Shmatikov, V., Belongie, S.: Can we still avoid automatic face detection? In WACV (2016)
65. Wren, C.R., Azarbayejani, A., Darrell, T., Pentland, A.P.: Pfinder: real-time tracking of the human body. TPAMI **19**(7), 780–785 (1997)
66. Xu, X., Chen, X., Liu, C., Rohrbach, A., Darell, T., Song, D.: Can you fool ai with adversarial examples on a visual turing test? CoRR, abs/1709.08693 (2017)
67. Zhang, C., Pujades, S., Black, M., Pons-Moll, G.: Detailed, accurate, human shape estimation from clothed 3D scan sequences. In: CVPR (2017)
68. Zhang, H., Jia, W., He, X., Wu, Q.: Learning-based license plate detection using global and local features. In: ICPR (2006)
69. Zheng, Y., et al.: Tour the world: Building a web-scale landmark recognition engine. In: CVPR (2009)
70. Zhou, S., Fu, H., Liu, L., Cohen-Or, D., Han, X.: Parametric reshaping of human bodies in images. ACM Trans. Graph. **29**, 1–10 (2010)
71. Zhou, W., Li, H., Lu, Y., Tian, Q.: Principal visual word discovery for automatic license plate detection. IEEE Trans. Image Process. **21**, 4269–4279 (2012)

Adversarial Training Against Location-Optimized Adversarial Patches

Sukrut Rao(✉), David Stutz, and Bernt Schiele

Max Planck Institute for Informatics, Saarland Informatics Campus, Saarbrücken, Germany
{sukrut.rao,david.stutz,schiele}@mpi-inf.mpg.de

Abstract. Deep neural networks have been shown to be susceptible to adversarial examples – small, imperceptible changes constructed to cause mis-classification in otherwise highly accurate image classifiers. As a practical alternative, recent work proposed so-called adversarial patches: clearly visible, but adversarially crafted rectangular patches in images. These patches can easily be printed and applied in the physical world. While defenses against imperceptible adversarial examples have been studied extensively, robustness against adversarial patches is poorly understood. In this work, we first devise a practical approach to obtain adversarial patches while actively optimizing their location within the image. Then, we apply adversarial training on these location-optimized adversarial patches and demonstrate significantly improved robustness on CIFAR10 and GTSRB. Additionally, in contrast to adversarial training on imperceptible adversarial examples, our adversarial patch training does not reduce accuracy.

1 Introduction

While being successfully used for many tasks in computer vision, deep neural networks are susceptible to so-called adversarial examples [68]: *imperceptibly* perturbed images causing mis-classification. Unfortunately, achieving robustness against such "attacks" has been shown to be difficult. Many proposed "defenses" have been shown to be ineffective against newly developed attacks, e.g., see [5,6,17,26,70]. To date, adversarial training [50], i.e., training on adversarial examples generated on-the-fly, remains one of few approaches not rendered ineffective through advanced attacks. However, adversarial training regularly leads to reduced accuracy on clean examples [56,66,71,81]. While this has been addressed in recently proposed variants of adversarial training, e.g., [3,20,44,67], obtaining robust and accurate models remains challenging.

Besides imperceptible adversarial examples, recent work explored various attacks introducing *clearly visible* perturbations in images. Adversarial patches [13,42,47], for example, introduce round or rectangular patches that can be "pasted" on top of images, cf. Fig. 1 (left). Similarly, adversarial frames [80] add an adversarially-crafted framing around images, thereby only manipulating a small strip of pixels at the borders. While these approaches are limited

A. Bartoli and A. Fusiello (Eds.): ECCV 2020 Workshops, LNCS 12539, pp. 429–448, 2020.
https://doi.org/10.1007/978-3-030-68238-5_32

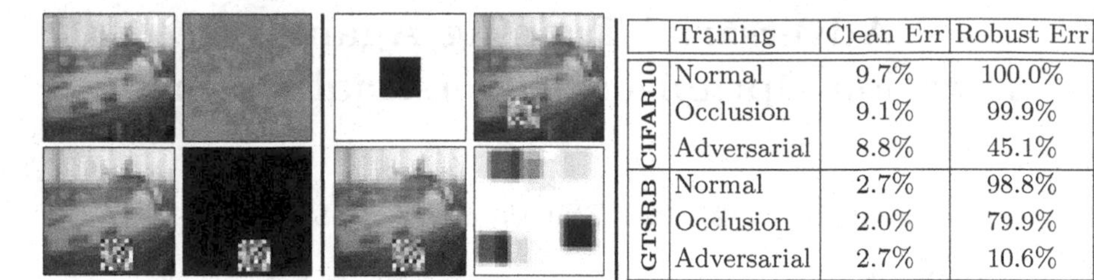

	Training	Clean Err	Robust Err
CIFAR10	Normal	9.7%	100.0%
	Occlusion	9.1%	99.9%
	Adversarial	8.8%	45.1%
GTSRB	Normal	2.7%	98.8%
	Occlusion	2.0%	79.9%
	Adversarial	2.7%	10.6%

Fig. 1. Adversarial patch training. *Left:* Comparison of imperceptible adversarial examples (top) and adversarial patches (bottom), showing an adversarial example and the corresponding perturbation. On top, the perturbation is within $[-0.03, 0.03]$ and gray corresponds to no change. *Middle:* Adversarial patches with location optimization. We constrain patches to the outer (white) border of images to ensure label constancy (top left) and optimize the initial location locally (top right and bottom left). Repeating our attack with varying initial location reveals adversarial locations of our adversarially trained model, AT-RandLO in Fig. 4. *Right:* Clean and robust test error for adversarial training on location-optimized patches in comparison to normal training and data augmentation with random patches. On both CIFAR10 and GTSRB, adversarial training improves robustness significantly, cf. Table 4.

in the number of pixels that can be manipulated, other works manipulate the whole image, e.g., by manipulating color [38,85] or directly generating images from scratch [12,58,64,84]. Such attacks can easily be printed and applied in the physical world [32,46] and are, thus, clearly more practical than imperceptible adversarial examples. As a result, such attacks pose a much more severe threat to applications such as autonomous driving [32,57,73] in practice.

While defenses against imperceptible adversarial examples has received considerable attention, robustness against adversarial patches is still poorly understood. Unfortunately, early approaches of localizing and in-painting adversarial patches [36,55] have been shown to be ineffective [24]. Recently, a certified defense based on interval bound propagation [34,52] has been proposed [24]. However, the reported certified robustness is not sufficient for many practical applications, even for small 2×2 or 5×5 patches. The sparse robust Fourier transform proposed in [7], targeted to both L_0-constrained adversarial examples and adversarial patches, reported promising results. However, the obtained robustness against L_0 adversarial examples was questioned in [70]. Overall, obtaining respectable robustness against adversarial patches is still an open problem.

Contributions: In this work, we address the problem of robustness against large adversarial patches by applying adversarial training on *location-optimized* adversarial patches. To this end, we introduce a simple heuristic procedure to optimize the location of the adversarial patch jointly with its content, cf. Fig. 1 (middle). Then, we conduct extensive experiments applying adversarial training against adversarial patches with various strategies for location optimization. On CIFAR10 [43] and GTSRB [65], we demonstrate that adversarial training is able

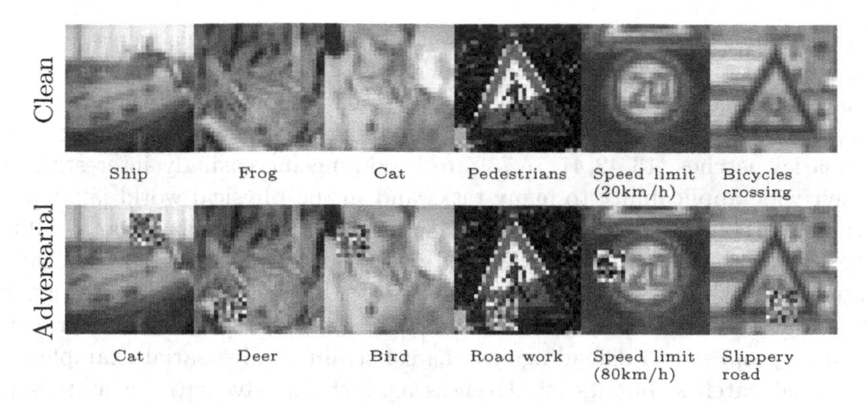

Fig. 2. Our adversarial patch attack on CIFAR10 and GTSRB. *Top:* correctly classified examples; *bottom:* incorrectly classified after adding adversarial patch. Adversarial patches obtained against a normally trained ResNet-20 [37].

to improve robustness against adversarial patches significantly while *not* reducing clean accuracy, cf. Fig. 1 (right), as often observed for adversarial training on imperceptible adversarial examples. We compare our adversarial patch training to [7], which is shown not to be effective against our adversarial patch attack. Our code is available at https://github.com/sukrutrao/adversarial-patch-training.

2 Related Work

Adversarial Examples: Originally proposed adversarial examples [68] were meant to be nearly *imperceptible*. In practice, L_p norms are used to enforce both visual similarity and class constancy, i.e., the *true* class cannot change. A common choice, $p = \infty$, results in limited change per feature. Examples include many popular white-box attacks, such as [15,23,28,33,49,50,78,82] with full access to the model including its weights and gradients, and black-box attacks, such as [4,9,11,14,21,22,25,35,40] without access to, e.g., model gradients. In the white-box setting, first-order gradient-based attacks such as [15,28,50] are the de-facto standard. Improving robustness against L_p-constrained adversarial examples, i.e., devising "defenses", has proved challenging: many defenses have been shown to be ineffective [5,6,16–19,26,30,48,54,62,70]. Adversarial training, i.e., training on adversarial examples generated on-the-fly, has been proposed in various variants [39,45,50,53,61,63,83] and has been shown to be effective. Recently, the formulation by Madry et al. [50] has been extended in various ways, tackling the computational complexity [59,72,74], the induced drop in accuracy [3,8,20,44] or the generalization to other L_p attacks [51,67,69]. Nevertheless, adversarial robustness remains a challenging task in computer vision. We refer to [1,10,77,79] for more comprehensive surveys.

Adversarial Patches: In contrast to (nearly) imperceptible adversarial examples, adversarial deformations/transformations [2,29,31,31,41,76], color change or image filters [38,85], as well as generative, so-called semantic adversarial examples [12,58,64,84] introduce clearly *visible* changes. Similarly, small but *visible* adversarial patches [13,42,47,57,73] are becoming increasingly interesting due to their wide applicability to many tasks and in the physical world [32,46]. For example, [13,42] use *universal* adversarial patches applicable to (nearly) all test images while the patch location is fixed or random. In [13,47,57], they can be printed and easily embedded in the real world. Unfortunately, defenses against adversarial patches are poorly studied. In [7], a L_0-robust sparse Fourier transformation is proposed to defend against L_0-constrained adversarial examples and adversarial patches, but its effectiveness against L_0 adversarial examples was questioned in [70]. In [24], the interval bound propagation approach of [34,52] is extended to adversarial patches to obtain certified bounds, but it is limited and not sufficient for most practical applications. Finally, in [36,55], an in-painting approach was used, but its effectiveness was already questioned in the very same work [36]. The recently proposed Defense against Occlusion Attacks (DOA) [75] is the closest to our work. However, unlike [75], we jointly optimize patch values and location. In addition, we evaluate against untargeted, image-specific patches, which have been shown to be stronger [60] than universal patches that were used for evaluating DOA against the adversarial patch attack.

3 Adversarial Training Against Location-Optimized Adversarial Patches

In the following, we first discuss our adversarial patch attack. Here, in contrast to related work, e.g., [13,42], we consider *image-specific* adversarial patches as a stronger alternative to the more commonly used universal adversarial patches. As a result, our adversarial patch attack is also *untargeted* and, thus, suitable for adversarial training following [50]. Then, we discuss our *location optimization* strategies, allowing to explicitly optimize patch location in contrast to considering random or fixed location only. Finally, we briefly introduce the idea of adversarial training on location-optimized adversarial patches in order to improve robustness, leading to our proposed adversarial patch training.

3.1 Adversarial Patches

Our adversarial patch attack is inspired by LaVAN [42]. However, following related work on adversarial training [50], we consider an image-specific, untargeted adversarial patch attack with an additional location optimization component:

- **Image-Specific Adversarial Patches:** The content and location of the adversarial patch is tailored specifically to each individual image. Thus, our adversarial patch attack can readily be used for adversarial training. As

Algorithm 1. Our location-optimized adversarial patch attack: Given image x with label y and trained classifier $f(\cdot; w)$, the algorithm finds an adversarial patch represented by the additive perturbation δ and the binary mask m such that $\tilde{x} = (1 - m) \odot x + m \odot \delta$ that maximizes the cross-entropy loss $L(f(\tilde{x}; w), y)$.

Input: image x of class y, trained classifier f, learning rate ϵ, number of iterations T, location optimization function NEXTLOCATION.
Output: adversarial patch given by $m^{(T)} \odot \delta^{(T)}$.

1: initialize perturbation $\delta^{(0)} \in [0,1]^{W \times H \times C}$ {e.g., uniformly}
2: initialize mask $m^{(0)} \in \{0,1\}^{W \times H \times C}$ {square, random or fixed location outside R}
3: **for** $t \leftarrow 0, \ldots, T-1$ **do**
4: $\quad \tilde{x}^{(t)} := (1 - m^{(t)}) \odot x + m^{(t)} \odot \delta^{(t)}$ {apply the patch}
5: $\quad l := L(f(\tilde{x}^{(t)}; w), y)$ {compute loss, i.e., forward pass}
6: $\quad \Delta^{(t)} := m^{(t)} \odot \text{sign}(\nabla_\delta l)$ {compute signed gradient, i.e., backward pass}
7: $\quad \delta^{(t+1)} := \delta^{(t)} + \epsilon \cdot \Delta^{(t)}$ {update patch values}
8: $\quad \delta^{(t+1)} := \text{CLIP}(\delta^{(t+1)}, 0, 1)$ {clip patch to image domain}
9: $\quad m^{(t+1)}, \delta^{(t+1)} := \text{NEXTLOCATION}(f, x, y, m^{(t)}, \delta^{(t+1)}, l)$ {update patch location}
10: **end for**
11: **return** $m^{(T)}, \delta^{(T)}$ {or return $m^{(t)}, \delta^{(t)}$ corresponding to highest cross-entropy loss}

experimentally shown in [60], training against image-specific attacks will also improve robustness to universal attacks. Thus, our adversarial patch training is also applicable against universal adversarial patches as considered in related work [13,42].

- **Untargeted Adversarial Patches:** Following common adversarial training practice, we consider untargeted adversarial patches. This means, we maximize the cross-entropy loss between the adversarial patch and the true label, as, e.g., in [50], and do not enforce a specific target label. This is also different from related work as universal adversarial patches usually target a pre-determined label.
- **Location-Optimized Adversarial Patches:** Most prior work consider adversarial patches to be applied randomly in the image [13,42] or consider a fixed location [42]. In contrast, we follow the idea of finding an optimal patch location for each image, i.e., the location where the attack can be most effective. This will improve the obtained robustness through adversarial training as the attack will focus on "vulnerable" locations during training.

Notation: We consider a classification task with K classes. Let $\{(x_i, y_i)\}_{i=1}^N$ be a training set of size N where $x_i \in [0,1]^{W \times H \times C}$ and $y_i \in \{0, 1, \ldots, K-1\}$ are images and labels with W, H, C denoting width, height and number of channels, respectively. Let f denote a trained classifier with weights w that outputs a probability distribution $f(x; w)$ for an input image x. Here, $f_i(x; w)$ denotes the predicted probability of class $i \in \{0, 1, \ldots, K-1\}$ for image x. The image is correctly classified when $y = \text{argmax}_i f_i(x; w)$ for the true label y. An adversarial patch can be represented by a perturbation $\delta \in [0,1]^{W \times H \times C}$ and a binary mask $m \in \{0,1\}^{W \times H \times C}$ representing the location of the patch, which we assume to be square. Then, an image x after applying the adversarial patch (δ, m) is given by $\tilde{x} = (1-m) \odot x + m \odot \delta$, where $\odot$ denotes the element-wise product. With $L(f(x;w), y)$ we denote the cross-entropy loss between the prediction $f(x;w)$ and the true label y.

Optimization Problem: For the optimization problem of generating an adversarial patch for an image x of class y, consisting of an additive perturbation δ and mask m, we follow [50] and intend to maximize the cross-entropy loss L. Thus, we use projected gradient ascent to solve:

$$\max_{\delta,m} L\left(f((1-m) \odot x + m \odot \delta; w), y\right) \tag{1}$$

where $\odot$ denotes the element-wise product and δ is constrained to be in $[0,1]$ through clipping, assuming all images lie in $[0,1]$ as well. The mask m represents a square patch and is ensured not to occlude essential features of the image by constraining it to the border of the image. For example, for CIFAR10 [43] and GTSRB [65], the patch is constrained not to overlap the center region R of size 10×10 pixels. As discussed below, the position of the patch, i.e., the mask, can be fixed or random, as in related work, or can be optimized. We also note that Eq. (1) is untargeted as we only seek to reduce the confidence in the true class y, and do not attempt to boost the probability of any other specific class.

Attack Algorithm: The attack algorithm is given in Algorithm 1. Here, Eq. (1) is maximized through projected gradient ascent. After randomly initializing $\delta^{(0)} \in [0,1]^{H\times W\times C}$ and initializing the mask $m^{(0)}$, e.g., as fixed or randomly placed square, T iterations are performed. In each iteration t, the signed gradient is used to update the perturbation $\delta^{(t)}$:

$$\delta^{(t+1)} = \delta^{(t)} + \epsilon \cdot \Delta^{(t)} \quad \text{with} \quad \Delta^{(t)} = m^{(t)} \odot \text{sign}\left(\nabla_\delta L(f(\tilde{x}^{(t)}; w), y)\right) \tag{2}$$

where ϵ denotes the learning rate, ∇_δ the gradient with respect to δ and $\tilde{x}^{(t)} = (1-m^{(t)}) \odot x + m^{(t)} \odot \delta^{(t)}$ is the adversarial patch of iteration t applied to image x. Note that the update is only performed on values of $\delta^{(t)}$ actually belonging to the patch as determined by the mask $m^{(t)}$. Afterwards, the Clip function clips the values in $\delta^{(t)}$ to $[0,1]$ and a location optimization step may takes place, cf. Line 9. The NextLocation function, described below and in Algorithm 2, performs a single location optimization step and returns the patch at the new location.

3.2 Location Optimization

The location of the patch in the image, given by the mask m, plays a key role in the effectiveness of the attack. While we ensure that the patch does not occlude essential features, which we assume to lie within the center R of the image, finding particularly "vulnerable" locations can improve the attack significantly. So far, related work mainly consider the following two ways to determine patch location:

Algorithm 2. NextLocation function for location optimization: To update the patch location in Algorithm 1, the patch is moved s pixels in each direction within the candidate set D to check whether cross-entropy loss increases. Then, the movement that maximizes cross-entropy loss is returned. If the cross-entropy loss cannot be increased, the location is left unchanged.

Input: image x of class y, trained classifier f, mask m, patch values δ, cross-entropy loss of current iteration l, stride s, center region R, candidate directions D
Output: new mask position m and correspondingly updated δ

```
 1: function NextLocation(f, x, y, m, δ, l)
 2:     l_max := l, d' := None
 3:     {full/random optimization: D={up, down, left, right}/|D|=1 random direction}
 4:     for d ∈ D do
 5:         m', δ' ← m, δ shifted in direction d by s pixels
 6:         x̃ := (1 − m') ⊙ x + m' ⊙ δ'
 7:         l' = L(f(x̃; w), y)
 8:         if l' > l_max then
 9:             l_max := l'
10:             d' := d
11:         end if
12:     end for
13:     if d' ≠ None then
14:         m, δ ← m, δ shifted in direction d' by s pixels if no intersection with R
15:     end if
16: return m, δ
17: end function
```

- **Fixed Location:** The patch is placed at a pre-defined location (e.g., the top left corner) of the image, outside of the center region.
- **Random Location:** The patch is placed randomly outside of the center region. In our case, this means that the patch location may differ from image to image as we consider image-specific adversarial patches.

Unfortunately, from an adversarial training perspective, both fixed and random locations are insufficient. Training against adversarial patches with fixed location is expected to generalize poorly to adversarial patches at different locations. Using random locations, in contrast, is expected to improve robustness to adversarial patches at various locations. However, the model is rarely confronted with particularly adversarial locations. Thus, we further allow the attack to actively optimize the patch location and consider a simple heuristic: in each iteration, the patch is moved by a fixed number of pixels, defined by the stride s, in a set of candidate directions $D \subseteq \{\text{up}, \text{down}, \text{left}, \text{right}\}$ in order to maximize Eq. (1). Thus, if the cross-entropy loss L increases in one of these directions, the patch is moved in the direction of greatest increase by s pixels, and not moved otherwise. We use two schemes to choose the set of candidate directions D:

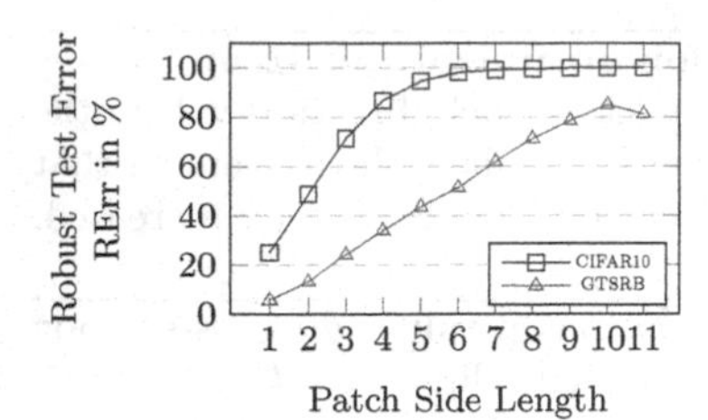

Fig. 3. Robust test error vs. patch size. Robust test error RErr in % and (square) patch size using AP-FullLO$_{(50,3)}$ against Normal, i.e., adversarial patches with full location optimation, 50 iterations and 3 random restarts. We use 8×8, where RErr on CIFAR10 stagnates.

Table 1. Clean test error. We report (clean) test error Err in % for our models on the full test sets of CIFAR10 and GTSRB. Our adversarial patch training does not increase test error compared to normal training, which is in stark contrast to adversarial training on imperceptible adversarial examples [66,71].

Model	CIFAR10	GTSRB
Normal	9.7	2.7
Occlusion	9.1	**2.0**
AT-Fixed	10.1	2.1
AT-Rand	9.1	2.1
AT-RandLO	**8.7**	2.4
AT-FullLO	8.8	2.7

- **Full Location Optimization:** Here, we consider all four directions, i.e., $D = \{\text{up}, \text{down}, \text{left}, \text{right}\}$, allowing the patch to explore all possible directions. However, this scheme requires a higher computation cost as it involves performing four extra forward passes on the network to compute the cross-entropy loss after moving in each direction.
- **Random Location Optimization:** This uses a direction chosen at random, i.e.$|D| = 1$, which has the advantage of being computationally more efficient since it requires only one extra forward pass. However, it may not be able to exploit all opportunities to improve the patch location.

The NEXTLOCATION function in Algorithm 1 is used to update the location in each iteration t, following the above description and Algorithm 2. It expects the stride s, the center region R to be avoided, and the candidate set of directions D as parameters. In addition to moving the pixels in the mask m, the pixels in the perturbation δ need to be moved as well.

3.3 Adversarial Patch Training

We now use the adversarial patch attack to perform adversarial training. The goal of adversarial training is to obtain a robust model by minimizing the loss over the model parameters on adversarial patches, which are in turn obtained by maximizing the loss over the attack parameters. As an adversarially trained model still needs to maintain high accuracy on clean images, we split each batch into 50% clean images and 50% adversarial patches. This effectively leads to the following optimization problem:

$$\min_{w} \left\{ \mathbb{E}\left[\max_{m,\delta} L(f((1-m) \odot x + m \odot \delta; w), y) \right] + \mathbb{E}\left[L(f(x; w), y) \right] \right\} \quad (3)$$

Table 2. Ablation study of AP-Rand **on** CIFAR10 . We report robust test error RErr in % for each model against AP-Rand with varying number of iterations T and random restarts r. More iterations or restarts generally lead to higher RErr.

Varying #iterations T and #restarts r on **CIFAR10**								
Model	T ($r=3$)						r ($T=100$)	
	10	**25**	**50**	**100**	**500**	**1000**	**3**	**30**
Normal	96.9	98.9	99.7	99.8	99.9	100.0	99.8	100.0
Occlusion	54.7	76.1	86.6	93.8	95.1	97.5	93.8	99.4
AT-Fixed	31.2	33.7	35.3	43.3	63.8	73.9	43.3	71.2
AT-Rand	16.4	16.7	16.8	18.1	37.9	57.2	18.1	33.0

Table 3. Results for $T = 50$ iterations and $r = 3$ restarts on CIFAR10. Using a limited attack cost ($T = 50, r = 3$) for each attack is still effective against Normal and Occlusion. However, RErr for adversarially trained models drops significantly.

$T=50, r=3$ on **CIFAR10**: Robust Test Error (RErr) in %				
Model	AP-Fixed	AP-Rand	AP-RandLO	AP-FullLO
Normal	99.0	99.7	99.7	99.6
Occlusion	77.3	86.6	87.4	88.9
AT-Fixed	12.7	35.3	45.5	48.4
AT-Rand	13.2	16.8	26.3	25.7
AT-RandLO	12.7	18.4	24.3	26.0
AT-FullLO	**11.1**	**14.2**	**22.0**	**24.4**

where $f(\cdot; w)$ denotes the classifier whose weights w are to be learned, and the perturbation δ and the mask m are constrained as discussed above. This balances cross-entropy loss on adversarial patches (left) with cross-entropy loss on clean images (right), following related work [68]. For imperceptible adversarial examples, 50%/50% adversarial training in Eq. (3) improves clean accuracy compared to training on 100% adversarial examples. However, it still exhibits reduced accuracy compared to normal training [66,67]. As we will demonstrate in our experiments, this accuracy-robustness trade-off is not a problem for our adversarial patch training.

4 Experiments

We evaluate our location-optimized adversarial patch attack and the corresponding adversarial patch training on CIFAR10 [43] and GTSRB [65]. We show that our adversarial patch attack with location-optimization is significantly more effective and allows to train robust models while not sacrificing accuracy.

Datasets: We use the 32×32 color images from CIFAR10 and the German Traffic Sign Recognition Benchmark (GTSRB) datasets. The CIFAR10 dataset consists of 50,000 training and 10,000 test images across 10 classes. We use the first 1,000 test images for adversarial evaluation. For GTSRB, we use a subset with 35,600 training images and 1,273 test images across 43 classes. The GTSRB dataset consists of signs that are commonly seen when driving, and hence represents a practical use case for autonomous driving.

Attack: Following the description of our attack in Section 3.1, we consider adversarial patches of size 8×8 (covering 6.25% of the image) constrained to a border region of 11 pixels along each side, i.e., the center region R of size 10×10

Table 4. Robust test error RErr on CIFAR10 and GTSRB: We report robust test error RErr in % for our adversarially trained models in comparison to the baselines. We tested each model against all four attacks, considering a fixed patch, a random patch and our strategies of location optimization. In all cases, results correspond to the per-example worst-case across 33 restarts with $T = 100$ or $T = 1000$ iterations. As can be seen, adversarial training with location-optimized adversarial patches improves robustness significantly and outperforms all baselines.

Model	Results on **CIFAR10**: RErr in %				Results on **GTSRB**: RErr in %			
	AP-Fixed	AP-Rand	AP-RandLO	AP-FullLO	AP-Fixed	AP-Rand	AP-RandLO	AP-FullLO
Normal	99.9	100.0	100.0	100.0	12.5	95.4	98.3	98.8
Occlusion	94.5	99.7	99.8	99.9	6.7	69.2	79.6	79.9
AT-Fixed	63.4	82.1	85.5	85.1	**3.0**	85.6	92.3	93.9
AT-Rand	51.0	60.9	61.5	63.3	3.4	11.3	15.6	16.4
AT-RandLO	40.4	54.2	60.6	62.8	3.1	7.6	**10.4**	**10.4**
AT-FullLO	**27.9**	**39.6**	**44.2**	**45.1**	3.3	**7.4**	10.6	10.6

is not changed to ensure label constancy. For location optimization, we consider a stride of $s = 2$ pixels. From all T iterations, we choose the patch corresponding to the worst (i.e., highest) cross-entropy error. To evaluate our location optimization strategies, we use four configurations: (1) AP-Fixed: Fixed patch location at coordinate $(3, 3)$ from the top left corner; (2) AP-Rand: Random patch location without optimizing location; (3) AP-RandLO: Random (initial) location with *random* location optimization; and (4) AP-FullLO: Random (initial) location with *full* location optimization. We use subscript (T, r) to denote an attack with T iterations and r attempts, i.e., random restarts. However, if not noted otherwise, as default attacks, we use $T = 100, r = 30$ and $T = 1000, r = 3$.

Adversarial Training: We train ResNet-20 [37] models from scratch using stochastic gradient descent with initial learning rate $\eta = 0.075$, decayed by factor 0.95 each epoch, and weight decay 0.001 for 200 epochs with batch size 100. The training data is augmented using random cropping, contrast normalization, and flips (flips only on CIFAR10). We train a model each per attack configuration: (1) AT-Fixed with $\text{AP-Fixed}_{(25,1)}$, (2) AT-Rand with $\text{AP-Rand}_{(25,1)}$, (3) AT-RandLO with $\text{AP-RandLO}_{(25,1)}$, and (4) AT-FullLO with $\text{AP-FullLO}_{(25,1)}$. By default, we use $T = 25$ iterations during training. However, as our location optimization based attacks, AP-RandLO and AP-FullLO, require additional forward passes, we later also consider experiments with equal computational cost, specifically 50 forward passes. This results in $T = 50$ iterations for AT-Fixed and AT-Rand, $T = 25$ for AP-RandLO, and $T = 10$ for AP-FullLO.

Baselines: We compare our adversarially trained models against three baselines: (1) Normal, a model trained without adversarial patches; (2) Occlusion, a model trained with randomly placed, random valued patches; and (3) SFT, the L_0-robust sparse Fourier transform defense from [7]. For the latter, we consider two configurations, roughly following [7,27]: SFT using hard thresholding with

$k = 500, t = 192, T = 10$, and $\mathsf{SFT_P}$ using patch-wise hard thresholding with $k = 50, t = 192, T = 10$ on 16×16 pixel blocks. Here, k denotes the sparsity of the image/block, t the sparsity of the (adversarial) noise and T the number of iterations of the hard thresholding algorithm. We refer to [7,27] for details on these hyper-parameters. Overall, SFT is applied at test time in order to remove the adversarial effect of the adversarial patch. As the transformation also affects image quality, the models are trained with images after applying the sparse Fourier transformation, but without adversarial patches.

Metrics: We use (regular) test error (Err), i.e., the fraction of incorrectly classified test examples, to report the performance on clean examples. For adversarial patches, we use the commonly reported robust test error (RErr) [50] which computes the fraction of test examples that are either incorrectly classified or successfully attacked. Following [67], we report robust test error considering the *per-example* worst-case across both our default attacks with a combined total of 33 random restarts.

4.1 Ablation

Patch Size: Figure 3 shows the robust test error RErr achieved by the $\mathsf{AP\text{-}FullLO}_{(50,3)}$ attack against Normal using various (square) patch sizes. For both datasets, RErr increases with increasing patch size, which is expected since a larger patch has more parameters and covers a larger fraction of the image. However, too large patches might restrict freedom of movement when optimizing location, explaining the slight drop on GTSRB for patches of size 11×11. In the following, we use a 8×8 patches, which is about where RErr saturates for CIFAR10. Note that for color images, a 8×8 patch has $8 \times 8 \times 3$ parameters. Fig. 2 shows examples of the 8×8 pixel adversarial patches obtained using full location optimization against Normal on CIFAR10 and GTSRB. We observed that the center region R is necessary to prevent a significant drop in accuracy due to occlusion (e.g., for Occlusion without R).

Number of Iterations and Attempts: In Table 2, we report robust test error RErr for various number of iterations T and random restarts r using AP-Rand. Across all models, RErr increases with increasing T, since it helps generating a better optimized patch. Also, increasing the number of restarts helps finding better local optima not reachable from all patch initializations. We use $T = 1000$ with $r = 3$ and $T = 100$ with $r = 30$ as our default attacks. Finally, considering that, e.g., AT-Rand, was trained with adversarial patches generated using $T = 25$, we see that it shows appreciable robustness against much stronger attacks.

4.2 Results

Adversarial Patch Training with Fixed and Random Patches: The main results can be found in Table 4, which shows the per-example worst-case robust

Table 5. Normalized cost results on CIFAR10 and GTSRB. We report robust test error RErr in % on models trained using attacks with exactly 50 forward passes, see text for details. As can be seen, training without location optimization might be beneficial when the cost budget is limited.

Model	Norm. Cost on **CIFAR10**: Robust Test Error (RErr) in %				Norm. Cost on **GTSRB**: Robust Test Error (RErr) in %			
	AP-Fixed	AP-Rand	AP-RandLO	AP-FullLO	AP-Fixed	AP-Rand	AP-RandLO	AP-FullLO
AT-Fixed$_{50}$	45.3	73.4	77.8	76.9	3.3	84.0	91.2	91.6
AT-Rand$_{50}$	**13.2**	**30.6**	**35.4**	**35.7**	3.6	12.8	18.7	20.0
AT-RandLO$_{50}$	40.4	54.2	60.6	62.8	**3.1**	**7.6**	**10.4**	**10.4**
AT-FullLO$_{50}$	40.8	50.0	56.9	56.5	4.6	17.7	23.6	23.2

test error RErr for each model and attack combination. Here, we focus on adversarial training with fixed and random patch location, i.e., AT-Fixed and AT-Rand, evaluated against the corresponding attacks, AP-Fixed and AP-Rand, and compare them against the baselines. The high RErr of the attacks against Occlusion shows that training with patches using random (not adversarial) content is not effective for improving robustness. Similarly, AT-Fixed performs poorly when attacked with randomly placed patches. However, AP-Rand shows that training with randomly placed patches also improves robustness against fixed patches. On CIFAR10, while using AP-Rand against AT-Rand results in an RErr of 60.9%, enabling location optimization in the attack increases RErr to 63.3%, indicating that training with location optimization might further improve robustness. On GTSRB, AT-Fixed even has higher RErr than Occlusion, which suggests that patch location might have a stronger impact than patch content on robustness.

Adversarial Patch Training with Location-Optimized Patches: Table 4 also includes results for adversarially trained models with location optimized adversarial patches, i.e., AT-FullLO and AT-RandLO. On CIFAR10, adversarial training with location-optimized patches has a much stronger impact in improving robustness as compared to the relatively minor 2.4% increase in RErr when attacking with AP-FullLO instead of AP-Rand on AT-Rand. Adversarial training with full location optimization in AT-FullLO leads to a RErr of 45.1% against AP-FullLO, thereby making it the most robust model and also outperforming training with random location optimization, AT-RandLO, significantly. On GTSRB, in contrast, AT-FullLO does not improve over AT-RandLO. This might be due to the generally lower RErr values, meaning GTSRB is more difficult to attack with adversarial patches. Nevertheless, training with random location optimization clearly outperforms training without, cf. AT-RandLO and AT-Rand, and leads to a drop of 88.4% in RErr as compared to Normal.

Table 3 additionally shows results for only $T = 50$ iterations with 3 random restarts on CIFAR10. Similar observations as above can be made, however, the RErr values are generally lower. This illustrates that the attacker is required to invest significant computational resources in order to increase RErr against our adversarially trained models. This can also be seen in our ablation, cf. Table 2.

Table 6. Results for robust sparse Fourier transformation (SFT). Robust test error RErr in % on CIFAR10 and GTSRB using the sparse Fourier transform [7] defense against our attacks. SFT does not improve robustness against our attack with location optimization and is outperformed by our adversarial patch training.

Model	Results on **CIFAR10**: RErr in %					Results on **GTSRB**: RErr in %				
	Clean	AP-Fixed	AP-Rand	AP-RandLO	AP-FullLO	Clean	AP-Fixed	AP-Rand	AP-RandLO	AP-FullLO
SFT	12.8	90.5	97.4	96.8	96.7	2.0	18.2	83.4	89.9	90.2
SFT_P	11.1	81.4	89.9	91.1	90.6	2.4	11.8	74.6	80.2	79.6

Preserved Accuracy: In contrast to adversarial training against imperceptible examples, Table 1 shows that adversarial patch training does not incur a drop in accuracy, i.e., increased test error Err . In fact, on CIFAR10, training with adversarial patches might actually have a beneficial effect. We expect that adversarial patches are sufficiently "far away" from clean examples in the input space, due to which adversarial patch training does not influence generalization on clean images. Instead, it might have a regularizing effect on the models.

Cost of Location Optimization: The benefits of location optimization come with an increased computational cost. Random location optimization and full location optimization introduce a factor of 2 and 5 in terms of the required forward passes, respectively. In order to take the increased cost into account, we compare the robustness of the models after normalizing by the number of forward passes. Specifically, we consider 50 forward passes for the attack, resulting in: (1) AT-Fixed_{50} with $\text{AP-Fixed}_{(50,1)}$, (2) AT-Rand_{50} with $\text{AP-Rand}_{(50,1)}$, (3) AT-RandLO_{50} with $\text{AP-RandLO}_{(25,1)}$ and (4) AT-FullLO_{50} with $\text{AP-FullLO}_{(10,1)}$, as also detailed in our experimental setup. Table 5 shows that for CIFAR10, AT-Rand_{50} has a much lower RErr than AT-RandLO_{50} and AT-FullLO_{50}. This suggests that with a limited computational budget, training with randomly placed patches without location optimization could be more effective than actively optimizing location. We also note that the obtained 35.7% RErr against AP-FullLO is lower than the 45.1% for AT-FullLO reported in Table 4. However, given that location optimization is done using greedy search, we expect more efficient location optimization approaches to scale better. On GTSRB, in contrast, AT-RandLO_{50} has a much lower RErr than AT-Rand_{50} and AT-FullLO_{50}.

Comparison to Related Work: We compare our adversarially trained models against models using the (patch-wise) robust sparse Fourier transformation, SFT, of [7]. We note that SFT is applied at test time to remove the adversarial patch. As SFT also affects image quality, we trained models on images after applying SFT. However, the models are not trained using adversarial patches. As shown in Table 6, our attacks are able to achieve high robust test errors RErr on CIFAR10 and GTSRB, indicating that SFT does not improve robustness. Furthermore, it is clearly outperformed by our adversarial patch training.

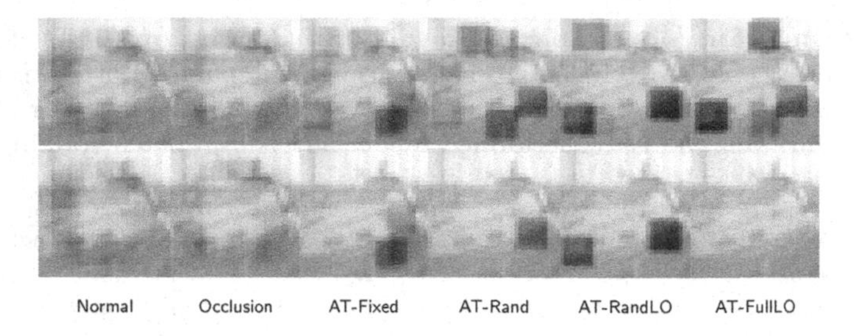

Fig. 4. Location heatmaps of our adversarial patch attacks. Heat maps corresponding to the final patch location using AP-FullLO$_{(10,1000)}$. *Top:* considering all $r = 1000$ restarts; *bottom:* considering only successful restarts. See text for details.

Universal Adversarial Patches: In a real-world setting, image-specific attacks might be less practical than universal adversarial patches. However, as also shown in [60], we found that our adversarial patch training also results in robust models against universal adversarial patches. To this end, we compute universal adversarial patches on the last 1000 test images of CIFAR10, with randomly selected initial patch locations that are then fixed across all images. On CIFAR10, computing universal adversarial patches for target class 0, for example, results in robust test error RErr reducing from 74.8% on Normal to 9.1% on AT-FullLO.

Visualizing Heatmaps: To further understand the proposed adversarial patch attack with location optimization, Fig. 4 shows heatmaps of vulnerable locations. We used our adversarial patch attack with full location optimization and $r = 1000$ restarts, AP-FullLO$_{(10,1000)}$. We visualize the frequency of a patch being at a specific location after $T = 10$ iterations; darker color means more frequent. The empty area in the center is the 10×10 region R where patches cannot be placed. The first row shows heatmaps of adversarial patches independent of whether they successfully flipped the label. The second row only considers those locations leading to mis-classification. For example, none of the 1000 restarts were successful against AT-FullLO. While nearly all locations can be adversarial for Normal or Occlusion, our adversarial patch training requires the patch to move to specific locations, as seen in dark red. Furthermore, many locations adversarial patches converged to do not necessarily cause mis-classification, as seen in the difference between both rows. Overall, Fig. 4 highlights the importance of considering patch location for obtaining robust models.

5 Conclusion

In this work, we addressed the problem of robustness against clearly visible, adversarially crafted patches. To this end, we first introduced a simple heuristic for explicitly optimizing location of adversarial patches to increase the attack's effectiveness. Subsequently, we used adversarial training on location-optimized adversarial patches to obtain robust models on CIFAR10 and GTSRB. We showed that our location optimization scheme generally improves robustness when used with adversarial training, as well as strengthens the adversarial patch attack. For example, visualizing patch locations after location optimization showed that adversarially trained models reduce the area of the image vulnerable to adversarial patches. Besides outperforming existing approaches [7], our adversarial patch training also preserves accuracy. This is in stark contrast to adversarial training on imperceptible adversarial examples, that usually cause a significant drop in accuracy. Finally, we observed that our adversarial patch training also improves robustness against universal adversarial patches, frequently considered an important practical use case [13,42].

References

1. Akhtar, N., Mian, A.: Threat of adversarial attacks on deep learning in computer vision: a survey. IEEE Access **6**, 14410–14430 (2018)
2. Alaifari, R., Alberti, G.S., Gauksson, T.: ADef: an iterative algorithm to construct adversarial deformations. In: International Conference on Learning Representations (2019). https://openreview.net/forum?id=Hk4dFjR5K7
3. Alayrac, J.B., Uesato, J., Huang, P.S., Fawzi, A., Stanforth, R., Kohli, P.: Are labels required for improving adversarial robustness? In: Wallach, H., Larochelle, H., Beygelzimer, A., d' Alché-Buc, F., Fox, E., Garnett, R. (eds.) Advances in Neural Information Processing Systems, vol. 32, pp. 12214–12223. Curran Associates, Inc. (2019). http://papers.nips.cc/paper/9388-are-labels-required-for-improving-adversarial-robustness.pdf
4. Andriushchenko, M., Croce, F., Flammarion, N., Hein, M.: Square attack: a query-efficient black-box adversarial attack via random search. arXiv: 1912.00049 (2019)
5. Athalye, A., Carlini, N.: On the robustness of the CVPR 2018 white-box adversarial example defenses. arXiv: 1804.03286 (2018)
6. Athalye, A., Carlini, N., Wagner, D.: Obfuscated gradients give a false sense of security: circumventing defenses to adversarial examples. In: Proceedings of Machine Learning Research, vol. 80, pp. 274–283. PMLR, Stockholmsmässan, Stockholm Sweden, 10–15 July 2018. http://proceedings.mlr.press/v80/athalye18a.html
7. Bafna, M., Murtagh, J., Vyas, N.: Thwarting adversarial examples: an L_0-robust sparse fourier transform. In: Bengio, S., Wallach, H., Larochelle, H., Grauman, K., Cesa-Bianchi, N., Garnett, R. (eds.) Advances in Neural Information Processing Systems, vol. 31, pp. 10075–10085. Curran Associates, Inc. (2018). http://papers.nips.cc/paper/8211-thwarting-adversarial-examples-an-l_0-robust-sparse-fourier-transform.pdf
8. Balaji, Y., Goldstein, T., Hoffman, J.: Instance adaptive adversarial training: improved accuracy tradeoffs in neural nets. arXiv:1910.08051 (2019)
9. Bhagoji, A.N., He, W., Li, B., Song, D.: Exploring the space of black-box attacks on deep neural networks. arXiv: 1712.09491 (2017)
10. Biggio, B., Roli, F.: Wild patterns: ten years after the rise of adversarial machine learning. Pattern Recogn. **84**, 317–331 (2018). https://doi.org/10.1016/j.patcog.2018.07.023, http://www.sciencedirect.com/science/article/pii/S0031320318302565
11. Brendel, W., Bethge, M.: Comment on "biologically inspired protection of deep networks from adversarial attacks". arXiv: 1704.01547 (2017)

12. Brown, T.B., Carlini, N., Zhang, C., Olsson, C., Christiano, P., Goodfellow, I.: Unrestricted adversarial examples. arXiv: 1809.08352 (2017)
13. Brown, T.B., Mané, D., Roy, A., Abadi, M., Gilmer, J.: Adversarial patch. arXiv: 1712.09665 (2017)
14. Brunner, T., Diehl, F., Knoll, A.: Copy and paste: a simple but effective initialization method for black-box adversarial attacks. arXiv: 1906.06086 (2019)
15. Carlini, N., Wagner, D.: Towards evaluating the robustness of neural networks. In: 2017 IEEE Symposium on Security and Privacy (SP), pp. 39–57 (2017)
16. Carlini, N.: Is ami (attacks meet interpretability) robust to adversarial examples? arXiv: 1902.02322 (2019)
17. Carlini, N., Wagner, D.: Adversarial examples are not easily detected: Bypassing ten detection methods. In: Proceedings of the 10th ACM Workshop on Artificial Intelligence and Security, AISec 2017, pp. 3–14. Association for Computing Machinery, New York (2017). https://doi.org/10.1145/3128572.3140444
18. Carlini, N., Wagner, D.A.: Defensive distillation is not robust to adversarial examples. arXiv: 1607.04311 (2016)
19. Carlini, N., Wagner, D.A.: Magnet and "efficient defenses against adversarial attacks" are not robust to adversarial examples. arXiv: 1711.08478 (2017)
20. Carmon, Y., Raghunathan, A., Schmidt, L., Duchi, J.C., Liang, P.S.: Unlabeled data improves adversarial robustness. In: Wallach, H., Larochelle, H., Beygelzimer, A., d' Alché-Buc, F., Fox, E., Garnett, R. (eds.) Advances in Neural Information Processing Systems, vol. 32, pp. 11192–11203. Curran Associates, Inc. (2019). http://papers.nips.cc/paper/9298-unlabeled-data-improves-adversarial-robustness.pdf
21. Chen, J., Jordan, M.I.: Boundary Attack++: Query-efficient decision-based adversarial attack. arXiv: 1904.02144 (2019)
22. Chen, P.Y., Zhang, H., Sharma, Y., Yi, J., Hsieh, C.J.: Zoo: zeroth order optimization based black-box attacks to deep neural networks without training substitute models. In: Proceedings of the 10th ACM Workshop on Artificial Intelligence and Security, AISec 2017 pp. 15–26. Association for Computing Machinery, New York (2017). https://doi.org/10.1145/3128572.3140448
23. Chiang, P., Geiping, J., Goldblum, M., Goldstein, T., Ni, R., Reich, S., Shafahi, A.: Witchcraft: efficient PGD attacks with random step size. In: ICASSP 2020–2020 IEEE International Conference on Acoustics, Speech and Signal Processing (ICASSP), pp. 3747–3751 (2020)
24. Chiang, P., Ni, R., Abdelkader, A., Zhu, C., Studor, C., Goldstein, T.: Certified defenses for adversarial patches. In: International Conference on Learning Representations (2020). https://openreview.net/forum?id=HyeaSkrYPH
25. Croce, F., Hein, M.: Sparse and imperceivable adversarial attacks. In: Proceedings of the IEEE/CVF International Conference on Computer Vision (ICCV), October 2019
26. Croce, F., Hein, M.: Reliable evaluation of adversarial robustness with an ensemble of diverse parameter-free attacks. In: Proceedings of the International Conference on Machine Learning, vol. 1, pp. 11571–11582 (2020). http://proceedings.mlr.press/v119/croce20b.html
27. Dhaliwal, J., Hambrook, K.: Recovery guarantees for compressible signals with adversarial noise. arXiv: 1907.06565 (2019)
28. Dong, Y., Liao, F., Pang, T., Su, H., Zhu, J., Hu, X., Li, J.: Boosting adversarial attacks with momentum. In: Proceedings of the IEEE Conference on Computer Vision and Pattern Recognition (CVPR), June 2018

29. Dumont, B., Maggio, S., Montalvo, P.: Robustness of rotation-equivariant networks to adversarial perturbations. arXiv: 1802.06627 (2018)
30. Engstrom, L., Ilyas, A., Athalye, A.: Evaluating and understanding the robustness of adversarial logit pairing. arXiv: 1807.10272 (2018)
31. Engstrom, L., Tsipras, D., Schmidt, L., Madry, A.: A rotation and a translation suffice: Fooling CNNs with simple transformations. arXiv: 1712.02779 (2017)
32. Eykholt, K., et al.: Robust physical-world attacks on deep learning visual classification. In: IEEE/CVF Conference on Computer Vision and Pattern Recognition, pp. 1625–1634 (2018)
33. Goodfellow, I.J., Shlens, J., Szegedy, C.: Explaining and harnessing adversarial examples. arXiv: 1412.6572 (2014)
34. Gowal, S., et al.: On the effectiveness of interval bound propagation for training verifiably robust models. arXiv: 1810.12715 (2018)
35. Guo, C., Gardner, J., You, Y., Wilson, A.G., Weinberger, K.: Simple black-box adversarial attacks. In: International Conference on Machine Learning, pp. 2484–2493 (2019)
36. Hayes, J.: On visible adversarial perturbations & digital watermarking. In: IEEE/CVF Conference on Computer Vision and Pattern Recognition Workshops (CVPRW), pp. 1597–1604 (2018)
37. He, K., Zhang, X., Ren, S., Sun, J.: Deep residual learning for image recognition. In: IEEE Conference on Computer Vision and Pattern Recognition (CVPR), pp. 770–778 (2016)
38. Hosseini, H., Poovendran, R.: Semantic adversarial examples. In: IEEE/CVF Conference on Computer Vision and Pattern Recognition Workshops (CVPRW), pp. 1614–1619 (2018)
39. Huang, R., Xu, B., Schuurmans, D., Szepesvári, C.: Learning with a strong adversary. arXiv: 1511.03034 (2015)
40. Ilyas, A., Engstrom, L., Athalye, A., Lin, J.: Black-box adversarial attacks with limited queries and information. In: Proceedings of the 35th International Conference on Machine Learning, ICML 2018, July 2018
41. Kanbak, C., Moosavi-Dezfooli, S.M., Frossard, P.: Geometric robustness of deep networks: Analysis and improvement. In: Proceedings of the IEEE Conference on Computer Vision and Pattern Recognition (CVPR), June 2018
42. Karmon, D., Zoran, D., Goldberg, Y.: LaVAN: localized and visible adversarial noise. In: Proceeding of the International Conference on Machine Learning (ICML), pp. 2512–2520 (2018)
43. Krizhevsky, A.: Learning multiple layers of features from tiny images. Technical Report (2009)
44. Lamb, A., Verma, V., Kannala, J., Bengio, Y.: Interpolated adversarial training: achieving robust neural networks without sacrificing too much accuracy. In: Proceedings of the ACM Workshop on Artificial Intelligence and Security, pp. 95–103 (2019)
45. Lee, H., Han, S., Lee, J.: Generative adversarial trainer: defense to adversarial perturbations with GAN. arXiv: 1705.03387 (2017)
46. Lee, M., Kolter, Z.: On physical adversarial patches for object detection. arXiv: 1906.11897 (2019)
47. Liu, X., Yang, H., Song, L., Li, H., Chen, Y.: DPatch: Attacking object detectors with adversarial patches. arXiv: 1806.02299 (2018)
48. Liu, Y., Zhang, W., Li, S., Yu, N.: Enhanced attacks on defensively distilled deep neural networks. arXiv: 1711.05934 (2017)

49. Luo, B., Liu, Y., Wei, L., Xu, Q.: Towards imperceptible and robust adversarial example attacks against neural networks. In: McIlraith, S.A., Weinberger, K.Q. (eds.) Proceedings of the Thirty-Second AAAI Conference on Artificial Intelligence, (AAAI-18), the 30th innovative Applications of Artificial Intelligence (IAAI-18), and the 8th AAAI Symposium on Educational Advances in Artificial Intelligence (EAAI-18), New Orleans, Louisiana, USA, 2–7 February 2018, pp. 1652–1659. AAAI Press (2018). https://www.aaai.org/ocs/index.php/AAAI/AAAI18/paper/view/16217
50. Madry, A., Makelov, A., Schmidt, L., Tsipras, D., Vladu, A.: Towards deep learning models resistant to adversarial attacks. In: International Conference on Learning Representations (2018). https://openreview.net/forum?id=rJzIBfZAb
51. Maini, P., Wong, E., Kolter, J.Z.: Adversarial robustness against the union of multiple perturbation models. In: Proceedings of the International Conference on Machine Learning (ICML) (2020)
52. Mirman, M., Gehr, T., Vechev, M.T.: Differentiable abstract interpretation for provably robust neural networks. In: Proceedings of the International Conference on Machine Learning (ICML), pp. 3575–3583 (2018)
53. Miyato, T., Maeda, S.i., Koyama, M., Nakae, K., Ishii, S.: Distributional smoothing with virtual adversarial training. arXiv: 1507.00677 (2015)
54. Mosbach, M., Andriushchenko, M., Trost, T.A., Hein, M., Klakow, D.: Logit pairing methods can fool gradient-based attacks. arXiv: 1810.12042 (2018)
55. Naseer, M., Khan, S., Porikli, F.: Local gradients smoothing: defense against localized adversarial attacks. In: Proceedings of the IEEE Winter Conference on Applications of Computer Vision (WACV), pp. 1300–1307 (2019)
56. Raghunathan, A., Xie, S.M., Yang, F., Duchi, J.C., Liang, P.: Adversarial training can hurt generalization. arXiv: 1906.06032 (2019)
57. Ranjan, A., Janai, J., Geiger, A., Black, M.J.: Attacking optical flow. In: Proceedings of the IEEE/CVF International Conference on Computer Vision (ICCV), October 2019
58. Schott, L., Rauber, J., Brendel, W., Bethge, M.: Robust perception through analysis by synthesis. arXiv: 1805.09190 (2018)
59. Shafahi, A., et al.: Adversarial training for free! In: Wallach, H.M., Larochelle, H., Beygelzimer, A., d'Alché-Buc, F., Fox, E.B., Garnett, R. (eds.) Advances in Neural Information Processing Systems (NIPS), pp. 3353–3364 (2019)
60. Shafahi, A., Najibi, M., Xu, Z., Dickerson, J.P., Davis, L.S., Goldstein, T.: Universal adversarial training. In: The Thirty-Fourth AAAI Conference on Artificial Intelligence, AAAI 2020, The Thirty-Second Innovative Applications of Artificial Intelligence Conference, IAAI 2020, The Tenth AAAI Symposium on Educational Advances in Artificial Intelligence, EAAI 2020, New York, NY, USA, 7–12 February 2020, pp. 5636–5643. AAAI Press (2020). https://aaai.org/ojs/index.php/AAAI/article/view/6017
61. Shaham, U., Yamada, Y., Negahban, S.: Understanding adversarial training: increasing local stability of neural nets through robust optimization. arXiv: 1511.05432 (2015)
62. Sharma, Y., Chen, P.Y.: Attacking the madry defense model with l1-based adversarial examples. arXiv: 1710.10733 (2017)
63. Sinha, A., Namkoong, H., Duchi, J.: Certifiable distributional robustness with principled adversarial training. In: International Conference on Learning Representations (2018). https://openreview.net/forum?id=Hk6kPgZA-
64. Song, Y., Shu, R., Kushman, N., Ermon, S.: Generative adversarial examples. arXiv: 1805.07894 (2018)

65. Stallkamp, J., Schlipsing, M., Salmen, J., Igel, C.: Man vs. computer: benchmarking machine learning algorithms for traffic sign recognition. Neural Netw. **32**, 323–332 (2012). https://doi.org/10.1016/j.neunet.2012.02.016, http://www.sciencedirect.com/science/article/pii/S0893608012000457
66. Stutz, D., Hein, M., Schiele, B.: Disentangling adversarial robustness and generalization. In: Proceedings of the IEEE/CVF Conference on Computer Vision and Pattern Recognition (CVPR), June 2019
67. Stutz, D., Hein, M., Schiele, B.: Confidence-calibrated adversarial training: generalizing to unseen attacks. In: Proceedings of the International Conference on Machine Learning ICML (2020)
68. Szegedy, C., et al.: Intriguing properties of neural networks. In: Proceedings of the International Conference on Learning Representations (ICLR) (2014)
69. Tramér, F., Boneh, D.: Adversarial training and robustness for multiple perturbations. In: Wallach, H., Larochelle, H., Beygelzimer, A., d' Alché-Buc, F., Fox, E., Garnett, R. (eds.) Advances in Neural Information Processing Systems, vol. 32, pp. 5866–5876. Curran Associates, Inc. (2019). http://papers.nips.cc/paper/8821-adversarial-training-and-robustness-for-multiple-perturbations.pdf
70. Tramèr, F., Carlini, N., Brendel, W., Madry, A.: On adaptive attacks to adversarial example defenses. arXiv: 2002.08347 (2020)
71. Tsipras, D., Santurkar, S., Engstrom, L., Turner, A., Madry, A.: Robustness may be at odds with accuracy. In: International Conference on Learning Representations (2019). https://openreview.net/forum?id=SyxAb30cY7
72. Wang, J., Zhang, H.: Bilateral adversarial training: towards fast training of more robust models against adversarial attacks. In: Proceedings of the IEEE/CVF International Conference on Computer Vision (ICCV), October 2019
73. Wiyatno, R., Xu, A.: Physical adversarial textures that fool visual object tracking. In: 2019 IEEE/CVF International Conference on Computer Vision (ICCV), pp. 4821–4830 (2019)
74. Wong, E., Rice, L., Kolter, J.Z.: Fast is better than free: revisiting adversarial training. In: International Conference on Learning Representations (2020). https://openreview.net/forum?id=BJx040EFvH
75. Wu, T., Tong, L., Vorobeychik, Y.: Defending against physically realizable attacks on image classification. In: International Conference on Learning Representations (2020). https://openreview.net/forum?id=H1xscnEKDr
76. Xiao, C., Zhu, J.Y., Li, B., He, W., Liu, M., Song, D.: Spatially transformed adversarial examples. In: International Conference on Learning Representations (2018). https://openreview.net/forum?id=HyydRMZC-
77. Xu, H., et al.: Adversarial attacks and defenses in images, graphs and text: a review. Int. J. Autom. Comput. **17**, 151–178 (2020)
78. Xu, K., et al.: Structured adversarial attack: towards general implementation and better interpretability. In: International Conference on Learning Representations (2019). https://openreview.net/forum?id=BkgzniCqY7
79. Yuan, X., He, P., Zhu, Q., Li, X.: Adversarial examples: attacks and defenses for deep learning. IEEE Trans. Neural Netw. Learn. Syst. **30**(9), 2805–2824 (2019)
80. Zajac, M., Zołna, K., Rostamzadeh, N., Pinheiro, P.O.: Adversarial framing for image and video classification. In: Proceedings of the AAAI Conference on Artificial Intelligence, vol. 33, pp. 10077–10078 (2019)
81. Zhang, H., Yu, Y., Jiao, J., Xing, E.P., Ghaoui, L.E., Jordan, M.I.: Theoretically principled trade-off between robustness and accuracy. In: Proceedings of the International Conference on Machine Learning (ICML), pp. 7472–7482 (2019)

82. Zhang, H., Chen, H., Song, Z., Boning, D.S., Dhillon, I.S., Hsieh, C.: The limitations of adversarial training and the blind-spot attack. In: 7th International Conference on Learning Representations, ICLR 2019, New Orleans, LA, USA, 6–9 May 2019. OpenReview.net (2019). https://openreview.net/forum?id=HylTBhA5tQ
83. Zhang, S., Huang, K., Zhu, J., Liu, Y.: Manifold adversarial learning. arXiv: 1807.05832v1 (2018)
84. Zhao, Z., Dua, D., Singh, S.: Generating natural adversarial examples. In: International Conference on Learning Representations (2018). https://openreview.net/forum?id=H1BLjgZCb
85. Zhao, Z., Liu, Z., Larson, M.A.: A differentiable color filter for generating unrestricted adversarial images. arXiv: 2002.01008 (2020)

Revisiting the Threat Space for Vision-Based Keystroke Inference Attacks

John Lim(✉), True Price(✉), Fabian Monrose(✉), and Jan-Michael Frahm(✉)

Department of Computer Science,
The University of North Carolina at Chapel Hill, Chapel Hill, NC 27514, USA
{jlim13,jtprice,fabian,jmf}@cs.unc.edu

Abstract. A vision-based keystroke inference attack is a side-channel attack in which an attacker uses an optical device to record users on their mobile devices and infer their keystrokes. The threat space for these attacks has been studied in the past, but we argue that the defining characteristics for this threat space, namely the strength of the attacker, are outdated. Previous works do not study adversaries with vision systems that have been trained with deep neural networks because these models require large amounts of training data and curating such a dataset is expensive. To address this, we create a large-scale synthetic dataset to simulate the attack scenario for a keystroke inference attack. We show that first pre-training on synthetic data, followed by adopting transfer learning techniques on real-life data, increases the performance of our deep learning models. This indicates that these models are able to learn rich, meaningful representations from our synthetic data and that training on the synthetic data can help overcome the issue of having small, real-life datasets for vision-based key stroke inference attacks. For this work, we focus on single keypress classification where the input is a frame of a keypress and the output is a predicted key. We are able to get an accuracy of 95.6% after pre-training a CNN on our synthetic data and training on a small set of real-life data in an adversarial domain adaptation framework.

Keywords: Side-channel attack · Domain adaptation · Synthetic data

1 Introduction

Mobile devices have become the main interface for many aspects of human life. People use their phones to connect with friends, send work emails, manage personal finances, and capture photos. Not only does the amount of information we channel through our devices increase as our dependence on our devices increases, but so does the level of sensitivity. It is not uncommon for users to enter social security numbers, credit card numbers, birth dates, addresses, or other private information onto mobile devices. As a result, it remains important to study

A. Bartoli and A. Fusiello (Eds.): ECCV 2020 Workshops, LNCS 12539, pp. 449–461, 2020.
https://doi.org/10.1007/978-3-030-68238-5_33

attacks that threaten mobile privacy and security. It is only by carefully studying the threat landscape can more robust defenses can be devised.

In this paper, we analyze Vision-based keystroke inference attacks wherein an attacker uses an optical device to record users on their mobile devices and extract user input. In past work, researchers have explored the ability of adversaries to extract information via direct surveillance or reflective surfaces [1–3,16,24], eye gaze [5], finger motion [25], and device perturbations [21]. Unfortunately, these works do not examine adversarial settings where the attacker applies deep learning methods—that have revolutionized computer vision in recent years—and they only consider limited capture scenarios. Consequently, a broad understanding of the threat space for vision-based keystroke inference attacks is missing.

To understand the threat posed by deep learning models, we consider the methods by which an attacker might train such a model for general and reliable use. One of the key factors for the overall success of deep learning is the availability of large, annotated datasets such as ImageNet [9] and MS COCO [13]. Given a large corpus of annotated real-world data (Fig. 1), it is reasonable to assume that an attacker could train a powerful model to predict user input from video data. However, collecting annotated data for vision-based keystroke inference attacks is a *prohibitively* expensive and time-consuming endeavor. Indeed, acquiring a large-enough real-world dataset with sufficient variability for this task would pose a daunting task. That said, an alternative strategy may be possible: leveraging a simulation engine that offers flexibility to generate a wide array of synthetic, yet realistic, data.

Fig. 1. Left: Example of real-life capture scenarios for vision-based keystroke inference attacks. Right: Warped images of the phone to a known template image to consolidate for the various viewpoints.

In what follows, we reexamine the threat space for vision-based direct surveillance attacks. Specifically, we examine ability of an adversary equipped with a deep learning systems that feeds off training data created in a systematic way that does not constrain the parameters of the attacker or capture scenario. To do so, we provide a framework for creating a simulation engine that models the capture pipeline of an attacker. Using this framework, we can model different capture scenarios of direct surveillance by permuting the parameters of the simulator: distance, brightness, user's skin tone, angle, capture device, user's device, screen contrast, and typing style. Armed with this framework, we can readily explore the power of adversaries with deep learning capabilities because of the abundance of data we can generate. While there are differences between the synthetic data and real-life data, notably the texture and finger kinematics, we show that our simulator produces data which allows us deep learning models to learn rich, meaningful representations that can be leveraged in the presence of a small set of real-life training data.

Our specific contributions include:

- The first analysis of vision-based keystroke inference attacks with adversaries employing deep learning algorithms and capturing with mobile devices.
- A systematic approach for evaluating inference attacks that can simulate various scenarios by permuting the capture parameters.

2 Related Work

Vision-Based Keystroke Inference Attacks: Some of the earliest works on keystroke inference attacks focused on direct line of sight and reflective surfaces *(i.e., teapots, sunglasses, eyes)* [1,2,16,24–26] to infer sensitive data. Under those threat models, an attacker trains a keypress classifier that accounts for various viewing angles and distances by aligning the user's mobile phone to a template keyboard. The success of these attacks rests on the ability to recover graphical pins, words, and full sentences by detecting the individual keypresses.

More recent work considers threat models where an attacker can not see the screen directly. For example, Sun et al. [21] study an attacker who is able to infer keystrokes on an iPad by only observing the back side of the tablet, focusing on the perturbations of the iPad as the user presses a key. They use steerable pyramid decomposition to detect and measure this motion of select areas of interest—the Apple logo, for example—in order infer keystrokes. Shukla et al. [20] infer keystrokes by exploiting the spatio-temporal dynamics of a user's hand while typing. No information about the user's screen activity is required, yet it is possible to infer phone, ATM, and door pins. Chen et al. [5] also create an attack where the user's device is not observed. They track a user's eye gaze to infer graphical pins, iOS pins, and English words. The major drawback of these methods that do not look to exploit the user's on-screen information and finger activity is that attacks do not perform as well compared to the methods that do focus on the user's finger motion and on-screen activity. The adversary trades rate-of-success for discreetness.

Synthetic-to-Real Domain Adaptation: Synthetic-to-real domain adaptation addresses the dataset bias problem between the synthetic domain $\mathcal{X}^s = \{\mathbf{x}_i^s, \mathbf{y}_i^s\}$ and real-life domain $\mathcal{X}^t = \{\mathbf{x}_i^t, \mathbf{y}_i^t\}$ where $\mathbf{x}_i \subset \mathbb{R}^d$ is the feature representation and $\mathbf{y}_i$ is the label. $\mathbf{x}_i^s$ and $\mathbf{x}_i^t$ are sampled from two different distributions but share the same feature representation and label space. Computer vision and machine learning algorithms that are trained with supervision require a considerable amount of annotated data that well covers the diverse distribution of application scenarios. Due to the high costs of curating such datasets, many researchers have worked on creating realistic, high-quality synthetic sources. Researchers have developed simulation engines to aid in training algorithms for optical flow [10], eye gaze estimation [23], and semantic segmentation [8,17,18]. Numerous other approaches [4,6,11,15,19] adopt adversarial training to learn a function to produce features that are domain invariant or to transform the pixels in the synthetic data to match distribution of the real data.

3 Overview

The general workflow of our model is highlighted in Fig. 2. This vision-based keystroke inference attack framework seeks to apply deep learning to a real-world domain in which the attacker has very few labeled datapoints. First, simulated training data is generated to model the space of attacker parameters, including different viewpoints and recording devices, as well as the victim's texting behavior, for example, finger kinematics. Note that training annotations come for free with this simulation, as the content of the victim's message is specified by the operator of the simulation engine. During the generation of synthetic data, we also collect and annotate a small set of real-life training data.

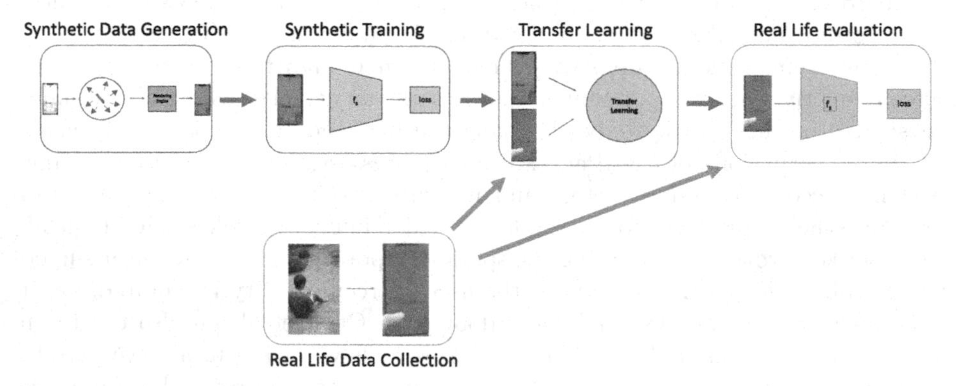

Fig. 2. Overview of our Approach. The blue indicates the flow of the synthetic data. The green indicates the flow of the real-life data. The orange indicates where the synthetic-to-real domain transfer learning happens. (Color figure online)

After simulation, we train a model, $\mathbf{f_s}$, on just the synthetic data. The representations learned from $\mathbf{f_s}$ are useful in the transfer learning step. $\mathbf{f_s}$ can be fine-tuned with real-life data, if available. Also, $\mathbf{f_s}$ can enforce task consistency when performing pixel-wise domain adaptation techniques [11] or the features learned while training the source classifier can be used in adversarial discriminative approaches [22]. Finally, after performing the transfer learning step, the model $\mathbf{f_s}$ can be applied to a real-life test set.

3.1 Synthetic Dataset Generation

We develop a simulation engine for keystroke inference attacks in which the attacker has a direct line of sight to the user's phone screen. The parameters that govern our simulations are: the attacker's capture device, user's mobile device, capture distance, screen brightness and contrast, the user's skin tone, and the typed message. Being able to permute these parameters allows us to systematically assess the threat space for this attack because we are not restricted to a fixed attack setting. The general pipeline is displayed in the "Synthetic Data Generation" module of Fig. 2.

Capture Stage. In the capture stage, the attacker uses an optical device such as a mobile phone's camera to record the user's behavior. In our scenario, the attacker focuses on recording the user's device and the associated finger movements that result from different keypresses. For our experiments, we set the attacker's capture device as the camera on the iPhone 6 and the user's device as an iPhone XR. For simulation, we utilize 3D models of the iPhone XR and the user's thumb. For a given keypress, we align the thumb model over the associated key and then render the thumb and iPhone models into a randomly selected attacker viewpoint, thereby simulating what a real-world attacker would observe.

Information Extraction and Alignment. This critical step allows us to consolidate all of the varying capturing positions of the attacker to one view. Given an image of the user's finger and phone (from the attacker's point of view), we need to extract meaningful information such as the type of phone or localization of the finger. This can be done via computer vision algorithms, for example, running a phone detection algorithm localize the phone, or by manually cropping out the phone in the image. Regardless of the approach, the goal of this step is to extract the most salient information from the given image. In our case, we assume that the attacker can manually crops out the phone the most salient information is the phone and user's fingers.

Next, we need to extract the four corners of the user's device in order to align the image of the phone to a reference template via a homography. A homography is a 2D projective transformation that relates two images of the same planar object. The phone is a planar object that is captured from varying viewpoints. The phone in the images that we capture are all related to each other by a homography transformation. Given any image of the phone, we can warp that image to a template image by the homography matrix, $\mathbf{H}$. We can calculate $\mathbf{H}$ using the four corners of the rendered image, which we know from simulation, and the four corners of the template image. In our explorations, we use the iPhone XR image from Apple Xcode's simulator as our template image and use the dimensions of the phone, which are available online, for the four corners. While the captured phone and template phone are both planar objects, the thumb is not. There is minimal distortion and our experiments show that this does not affect the learning process. Once the captured image and template image are aligned, we can train a classifier to predict the keypress. The input to the classifier is a single image of a single key press, and the output is a prediction of which key was pressed (Fig. 3).

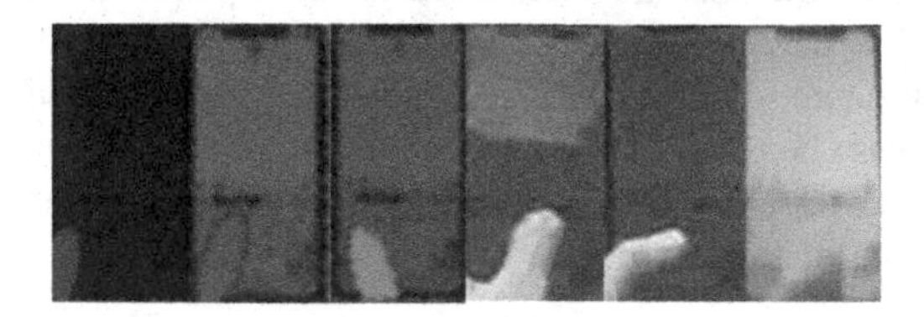

Fig. 3. Left: Examples of our synthetic data. Right: Examples of our real-life data.

3.2 Single Keypress Classification

For single keypress classification, the input is a single image of the user's thumb over a specific key. As previously mentioned, the input image is homography-aligned to the reference image to consolidate the different viewpoints from which the attacker can capture the user. We train a model to output a predicted key, p' on a QWERTY keyboard.

We train a Logistic Regression, Support Vector Machine (SVM) and a Convolutional Neural Network (CNN) for our evaluations, and show that the deep learning approach performs better than the shallow methods. This is a 26-way classification task that is trained by the cross entropy loss function:

$$L = E_{(x,y)\sim X} \sum_{n=1}^{N} \mathbb{1}_{[n=y]} \log(\sigma(f(x))) \tag{1}$$

Previous works have analyzed single keypress classification, but we differ in that we only focus on attackers with mobile devices. Xu et al. [24] and Raguram et al. [16] use high-end recording devices in their setup, and while those devices are considerably smaller and cheaper than telescopes used by Backes et al. [1,2] and Kuhn [12], the size and conspicuousness of such devices still restricts their use in discreet capturing scenarios. While mobile cameras have less capture capability than a high-end DSLR or telescope, they allow for more discreet capturing, making the attack less noticeable.

Transfer Learning. We adopt transfer learning techniques to bridge the gap between the synthetic and real-life data distributions. Recall that the majority of our data comes from a simulation engine and we do the majority of our training on this data to compensate for the difficulty in collecting real-life data samples. We adopt two approaches for transfer learning: fine-tuning and adversarial domain adaptation, similar to the technique introduced by Tzeng et al. [22].

CNNs are successful in vision tasks in which we have large amounts of training data because they are able to learn powerful representations with millions of parameters. Due to the high number of parameters in these CNNs, we are unable to learn meaningful representations on a small dataset. Oquab et al. [14] and Chu et al. [7] have shown that CNNs initially trained on a large dataset can transfer those representations to the target dataset by fine-tuning the CNN. Fine-tuning allows us to learn the key features for the general task of single key press classification using synthetic data and allows adjust the weights for a specific domain of single key press classification. Our results show that a CNN trained on our synthetic data learns a representation that can be transferred to real life data.

We follow the Adversarial Discriminative Domain Adaptation, ADDA, framework introduced by Tzeng et al. [22] where the purpose is to learn a domain invariant feature representation between the source and target domain. The source domain is denoted as $\mathcal{X}^s = \{\mathbf{x}_i^s, \mathbf{y}_i^s\}$ where $\mathbf{x}_i^s$ is the feature representation and $\mathbf{y}_i^s$ is the label. The target domain is denoted as $\mathcal{X}^t = \{\mathbf{x}_i^t, \mathbf{y}_i^t\}$. In the target domain we have a small set of labeled instances. A visual classifier, f, can be decomposed into two functions, $f = g \circ h$. g is the feature extractor that takes the input image into a d-dimensional feature space and h is the predictor that takes the feature representation and outputs a probability distribution over the label space. g_s and g_t represent the feature extractors for the source and target domains, respectively. h_s and h_t represent the predictors.

The training for ADDA is done in multiple stages; we do make some slight adjustments to the training process as we have access to a small set of labels in the target domain. First, we train g_s and h_s to minimize the loss function Eq. 1. Next, we train g_t in an adversarial fashion. We maximize the discriminator's ability to distinguish between the features outputted from g_t and g_s while also forcing g_t to extract features that are indistinguishable from those extracted from g_s. During this step, we also train h_t. Finally, we test using g_t and h_t. The optimization procedure is formally denoted below:

$$\min_{g_s, h_s} \mathcal{L}_{cls} = E_{(x_s, y_s) \sim X_s} \sum_{n=1}^{N} \mathbf{1}_{[n=y_s]} \log(\sigma(h_s(g_s(x_s)))) \quad (2)$$

$$\min_{D} \mathcal{L}_{adv}(X_s, X_t, g_s, g_t) = E_{x_s \sim X_s} [\log g_s(x_s)] + E_{x_t \sim X_t} [\log (1 - g_t(x_t))] \quad (3)$$

$$\min_{g_s, g_t} \mathcal{L}_g(X_s, X_t, D) = E_{x_t \sim X_t} [\log D(g_t(x_t))] \quad (4)$$

$$\min_{g_t, h_t} \mathcal{L}_{cls} = E_{(x_t, y_t) \sim X_t} \sum_{n=1}^{N} 1_{[n=y_t]} \log(\sigma(h_t(g_t(x_t)))) \quad (5)$$

Equation 2 is trained by itself in the first stage. Equations 3 and 4 are trained together in the next step, and when labels are available, Eq. 5 is also used.

4 Experiments

Next, we share our implementation details and experimental setup for single keypress classification. We also utilize our simulation engine to simulate different types of defenses and evaluate their effectiveness.

4.1 Single Keypress Classification

Perhaps not surprisingly, we show that a deep learning based approach outperforms shallow machine learning methods for single keypress classification. The input to these models is a single frame of a keypress and the output is a prediction of the pressed key. Our synthetic data consists of 15,000 keypress images and were labeled as one of the 26 letters in the alphabet. We randomized the lighting, screen blur, screen contrast, camera angle, distance (1–7 m), and skin tone in order to simulate various capture conditions and to diversify our dataset. Our real life training data consisted of 540 images of single key press frames. The dataset is split to 390, 80, and 80 images for training, testing, and validation, respectively. These images were captured at distances of up to 5 m and were taken in both indoor and outdoor settings. We show that our accuracy on our real-life test set is similar to that of our synthetic data after adopting transfer learning techniques. We also conduct experiments to see how the minimum number of instances for each class affects transfer learning. The full real-life dataset has 15 instances for each class (Table 2).

Table 1. Single Key Classification. The scores under the Synthetic column are trained and evaluated on only synthetic data. Similarly, for the Real-Life column.

Method	Synthetic	Real-Life
Logistic Regression	81.8%	78.3%
SVM	80.4%	75.1%
CNN	96.3%	76.2%

Table 2. Single Key Classification on real-life data. We compare the performance of a CNN on a real-life test set. No Adaptation means that the CNN has a random weight initialization and is trained using the real-life training set only. Finetuning and ADDA use the CNN trained on synthetic as the initialization.

	Real Only	Finetuning	ADDA
CNN	76.2%	93.08%	95.6%

Experimental Setup. We use a linear regression and a SVM as our baseline methods for this task. A 3-layer CNN is used for our deep learning model. Each layer follows a Conv2d-BatchNorm2d-ReLU-MaxPool structure. Each layer has filters of size 5 $\times$ 5, stride 1, and padding 2. The channels are 16, 32, and 16 for each layer. After these convolution layers, there is a linear layer, followed by a 26-way softmax layer. We set our initial learning rate to 0.0002 and use the Adam optimizer. We also crop out the image so that we are only focusing on the keyboard and location of the finger.

For the finetuning experiments, we take the same CNN architecture trained just on synthetic data, replace the last linear layer with a new one, and freeze the early layers. We used a learning rate of 0.00002 using the Adam optimizer and trained it for 60 epochs. For the ADDA results, we use a learning rates of 0.0002 and 0.0004 for the classifiers and feature extractors, respectively.

Results. Table 1 shows that the CNN significantly outperforms the two shallow methods when trained and evaluated synthetic data. However, if we train and evaluate on just the real-life data, we see a decrease in performance because the model is overfitting to the training data. Training and evaluating on such a small dataset does not give us any insight into this attack because a dedicated attacker could curate his own dataset large enough to benefit from deep learning approaches. We adopt transfer learning approaches to compensate for our lack of real-life training data. Finetuning gives us a classification score that approaches the synthetic data performance, which indicates that our simulation engine is capable of generating data to evaluate single key press classification when we are constrained with limited real-life data. ADDA yields the highest results. In Fig. 4, we show our performance on the real-life test while decreasing the number of per-class examples seen during training.

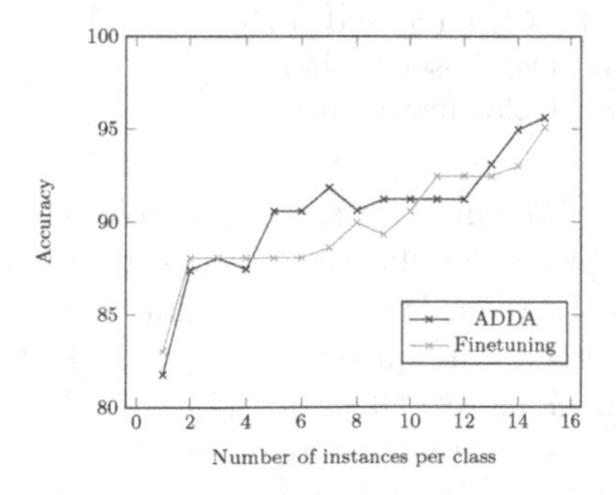

Fig. 4. The accuracy on the real-life test set is plotted against the number of per-class instances seen during training.

4.2 Defenses

Establishing defenses that generalize against multiple attacks is a very challenging problem. One of the main challenges for establishing defenses for vision-based keystroke inference attacks is that there are a few number of methods to prevent an attacker from capturing a user's behavior. The most effective method is abstaining from mobile phone usage, but that is not a practical solution for most people. Many of the previous suggest countermeasures to the attacks it presented, but a defense for attack A can be the threat scenario for attack B. For example, the defense against an attacker who exploits eye gaze would be to wear dark, protective eyewear such as sunglasses, but the user is at risk against an attacker who looks for compromising reflections. Some defenses that can generalize to multiple attacks is the user of randomized keyboards, typing fast, and dynamically moving the phone while typing. While we are not able to study all possible defenses, our simulation engine allows to study some subset of defenses in a systematic way (Table 3).

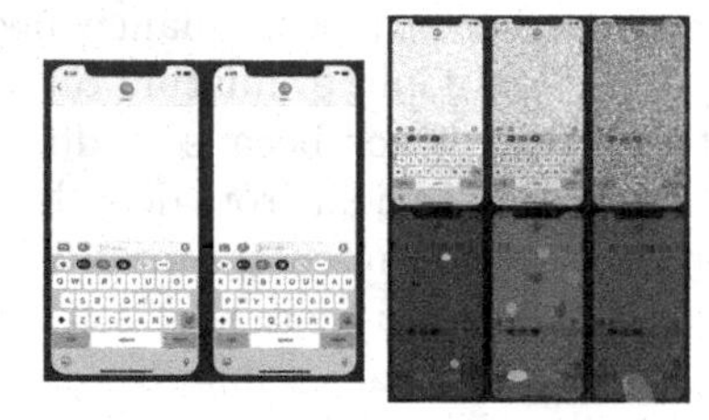

Fig. 5. Left: the standard QWERTY keyboard. Next to it is a randomly permuted keyboard. For our experiments, note that we only permute the 26 letters. Right: different on-screen perturbations. The top phone screens are corrupted with Gaussian noise with σs of 0.01, 0.05, and 0.75, respectively. The bottoms have on-screen blocks of varying colors and sizes (Color figure online)

Table 3. Evaluating On-Screen Perturbations as Defenses. We evaluate how a CNN trained without ever seeing any of these perturbations performs against them.

On-Screen Perturbation	Accuracy
Gaussian ($\sigma = 0.001$)	94.3%
Gaussian ($\sigma = 0.05$)	94.3%
Gaussian ($\sigma = 0.15$)	94.3%
Gaussian ($\sigma = 0.75$)	92%
Small Corruption	82.1%
Large Corruption	45.4%
Thumb Corruption	84.8%

Experimental Setup. We simulate a few possible on-screen perturbations as defenses for single key press classification. These perturbations can be emitted from the mobile phone's screen. We simulate different Gaussian noise patterns and different "phone screen corruptions" as types of on-screen perturbations. Small and Large Corruptions are those in which we have the phone randomly emit various shapes across the phone screen. A Thumb Corruption is when the phone emits a random shape around the user's thumb when he presses a key. We also study how our methods perform against randomized keyboards.

Randomized keyboards are one of the most effective ways to defend against keystroke inference attacks. Vision-based keystroke inference attacks learn a mapping between user behavior to a fixed keyboard layout. This mapping is broken when the keyboard layout changes. We generate a dataset of randomized keyboards and evaluate how a CNN trained on layout A performs on layout B.

Results. To evaluate the different on-screen perturbations, we first train a CNN on synthetic data without any of these perturbations. Then, we evaluate the model on a separate test set with these corruptions. The results are displayed in Fig. 5. The defense becomes more effective as the severity of the perturbations increase. Of course, doing so takes away from the user's usability. While some of the on-screen perturbations were effective in spoofing the CNN, if we were to train the CNN with these perturbations in the training set, the defenses do not hold. The CNN is unaffected as these perturbations become a form of data augmentation.

Randomized keyboards are an effective defense against the model used in our experiments. We first train a CNN on synthetic data on a QWERTY layout. Then, we generate a training set with a fixed layout, B, that is **not** a QWERTY. If we evaluate the CNN trained on the QWERTY, we do not get better than 0.04% accuracy, which means that the classifier is guessing. Similarly, if we generate a new training and testing set, all with instances of randomly permuted keyboards, we still do not do better than guessing. This indicates that the model can not recover any sort of information from the keyboard to indicate which key is pressed. The model effectively learns a mapping from the user's finger tip to an assumed keyboard layout. If that assumption is broken, then the model can not predict the key. Of course, the biggest sacrifice for using a randomized keyboard is the severe decrease in usability.

5 Conclusion

We explored a method to evaluate deep learning methods for vision based keystroke inference attacks; a domain in which curating a dataset large and diverse enough for deep learning methods is expensive. In doing so, we developed a simulation engine that generates data that allows us to systematically study these attacks by manipulating various parameters (e.g., capture distance, camera rotation, screen brightness, texting speed). Similarly, this capability allows us to study different defenses. We create synthetic data for the task of single key press classification, and show that deep learning models, when pre-trained on this data, are able to learn powerful representations that compensate for the lack of real-life training data. Our experiments not only show that deep learning approach outperforms shallow methods for single key press classification, but also show that an attacker does not need many real-life data points to train such a classifier. These experiments indicate that we need to rethink our beliefs of the threat space for vision-based keystroke inference attacks, as they are outdated.

References

1. Backes, M., Dürmuth, M., Unruh, D.: Compromising reflections-or-how to read LCD monitors around the corner. In: 2008 IEEE Symposium on Security and Privacy, SP 2008, pp. 158–169. IEEE (2008)

2. Backes, M., Chen, T., Duermuth, M., Lensch, H.P.A., Welk, M.: Tempest in a teapot: compromising reflections revisited. In: 2009 30th IEEE Symposium on Security and Privacy, pp. 315–327. IEEE (2009)
3. Balzarotti, D., Cova, M., Vigna, G.: ClearShot: eavesdropping on keyboard input from video. In: 2008 IEEE Symposium on Security and Privacy, SP 2008, pp. 170–183. IEEE (2008)
4. Bousmalis, K., Silberman, N., Dohan, D., Erhan, D., Krishnan, D.: Unsupervised pixel-level domain adaptation with generative adversarial networks. In: Proceedings of the IEEE Conference on Computer Vision and Pattern Recognition, pp. 3722–3731 (2017)
5. Chen, Y., Li, T., Zhang, R., Zhang, Y., Hedgpeth, T.: EyeTell: video-assisted touchscreen keystroke inference from eye movements. In: 2018 IEEE Symposium on Security and Privacy (SP), pp. 144–160. IEEE (2018)
6. Chen, Y., Li, W., Chen, X., Van Gool, L.: Learning semantic segmentation from synthetic data: a geometrically guided input-output adaptation approach. In: Proceedings of the IEEE Conference on Computer Vision and Pattern Recognition, pp. 1841–1850 (2019)
7. Chu, B., Madhavan, V., Beijbom, O., Hoffman, J., Darrell, T.: Best practices for fine-tuning visual classifiers to new domains. In: Hua, G., Jégou, H. (eds.) ECCV 2016. LNCS, vol. 9915, pp. 435–442. Springer, Cham (2016). https://doi.org/10.1007/978-3-319-49409-8_34
8. Cordts, M., et al.: The cityscapes dataset for semantic urban scene understanding. In: Proceedings of the IEEE Conference on Computer Vision and Pattern Recognition (CVPR) (2016)
9. Deng, J., Dong, W., Socher, R., Li, L.-J., Li, K., Fei-Fei, L.: ImageNet: a large-scale hierarchical image database (2009)
10. Dosovitskiy, A., et al.: FlowNet: learning optical flow with convolutional networks. In: IEEE International Conference on Computer Vision (ICCV) (2015). http://lmb.informatik.uni-freiburg.de/Publications/2015/DFIB15
11. Hoffman, J., et al.: CyCADA: cycle-consistent adversarial domain adaptation (2018). https://openreview.net/forum?id=SktLlGbRZ
12. Kuhn, M.G.: Compromising emanations: eavesdropping risks of computer displays. Ph.D. thesis, University of Cambridge (2002)
13. Lin, T.-Y., et al.: Microsoft COCO: common objects in context. In: Fleet, D., Pajdla, T., Schiele, B., Tuytelaars, T. (eds.) ECCV 2014. LNCS, vol. 8693, pp. 740–755. Springer, Cham (2014). https://doi.org/10.1007/978-3-319-10602-1_48
14. Oquab, M., Bottou, L., Laptev, I., Sivic, J.: Learning and transferring mid-level image representations using convolutional neural networks. In: Proceedings of the IEEE Conference on Computer Vision and Pattern Recognition, pp. 1717–1724 (2014)
15. Peng, X., Usman, B., Saito, K., Kaushik, N., Hoffman, J., Saenko, K.: Syn2Real: a new benchmark for synthetic-to-real visual domain adaptation. arXiv preprint arXiv:1806.09755 (2018)
16. Raguram, R., White, A.M., Goswami, D., Monrose, F.Frahm, J.-M.: iSpy: automatic reconstruction of typed input from compromising reflections. In: Proceedings of the 18th ACM Conference on Computer and Communications Security, pp. 527–536. ACM (2011)
17. Richter, S.R., Vineet, V., Roth, S., Koltun, V.: Playing for data: ground truth from computer games. In: Leibe, B., Matas, J., Sebe, N., Welling, M. (eds.) ECCV 2016. LNCS, vol. 9906, pp. 102–118. Springer, Cham (2016). https://doi.org/10.1007/978-3-319-46475-6_7

18. Ros, G., Sellart, L., Materzynska, J., Vazquez, D., Lopez, A.M.: The SYNTHIA dataset: a large collection of synthetic images for semantic segmentation of urban scenes. In: Proceedings of the IEEE Conference on Computer Vision and Pattern Recognition, pp. 3234–3243 (2016)
19. Shrivastava, A., Pfister, T., Tuzel, O., Susskind, J., Wang, W., Webb, R.: Learning from simulated and unsupervised images through adversarial training. In: Proceedings of the IEEE Conference on Computer Vision and Pattern Recognition, pp. 2107–2116 (2017)
20. Shukla, D., Kumar, R., Serwadda, A., Phoha, V.V.: Beware, your hands reveal your secrets! In: Proceedings of the 2014 ACM SIGSAC Conference on Computer and Communications Security, pp. 904–917. ACM (2014)
21. Sun, J., Jin, X., Chen, Y., Zhang, J., Zhang, Y., Zhang, R.: Visible: video-assisted keystroke inference from tablet backside motion. In: NDSS (2016)
22. Tzeng, E., Hoffman, J., Saenko, K., Darrell, T.: Adversarial discriminative domain adaptation. In: Proceedings of the IEEE Conference on Computer Vision and Pattern Recognition, pp. 7167–7176 (2017)
23. Wood, E., Baltrušaitis, T., Morency, L.-P., Robinson, P., Bulling, A.: Learning an appearance-based gaze estimator from one million synthesised images. In: Proceedings of the Ninth Biennial ACM Symposium on Eye Tracking Research & Applications, pp. 131–138. ACM (2016)
24. Xu, Y., Heinly, J., White, A.M., Monrose, F., Frahm, J.-M.: Seeing double: reconstructing obscured typed input from repeated compromising reflections. In: Proceedings of the 2013 ACM SIGSAC Conference on Computer & Communications Security, pp. 1063–1074. ACM (2013)
25. Ye, G., et al.: Cracking android pattern lock in five attempts (2017)
26. Yue, Q., Ling, Z., Fu, X., Liu, B., Ren, K., Zhao, W.: Blind recognition of touched keys on mobile devices. In: Proceedings of the 2014 ACM SIGSAC Conference on Computer and Communications Security, pp. 1403–1414. ACM (2014)

Black-Box Face Recovery from Identity Features

Anton Razzhigaev[1,2(✉)], Klim Kireev[1,2], Edgar Kaziakhmedov[1,2], Nurislam Tursynbek[1,2], and Aleksandr Petiushko[2,3]

[1] Skolkovo Institute of Science and Technology, Moscow, Russia
anton.razzhigaev@skoltech.ru
[2] Huawei Moscow Research Center, Moscow, Russia
[3] Lomonosov Moscow State University, Moscow, Russia

Abstract. In this work, we present a novel algorithm based on an iterative sampling of random Gaussian blobs for black-box face recovery, given only an output feature vector of deep face recognition systems. We attack the state-of-the-art face recognition system (ArcFace) to test our algorithm. Another network with different architecture (FaceNet) is used as an independent critic showing that the target person can be identified with the reconstructed image even with no access to the attacked model. Furthermore, our algorithm requires a significantly less number of queries compared to the state-of-the-art solution.

Keywords: Adversarial · Privacy · Black-box · Arcface · Face recognition

1 Introduction

The most common characteristic to identify a person from a still image is its face. Automatic face identification is an important computer vision task with real-world applications in smartphone cameras, video surveillance systems, human-computer interaction. Following rapid progress in image classification [11,16], object detection [25,26], semantic and instance segmentation [2,10], Deep Neural Networks demonstrated state-of-the-art performance in face identity recognition, even in extremely challenging scenarios with millions of identities [15].

Although end-to-end solutions exist, leading face recognition systems usually require a few-step procedure. First, the face is detected in the given image, and the alignment process is done. Then, the aligned face is fed to a face identification network, which converts it to descriptive feature vectors of the lower dimensionality. It is challenging to allocate those representations so different images of the same person are mapped to be closer to each other than to those of different.

Recent solutions incorporate different types of margins to the training loss to enhance the discriminative power. Current state-of-the-art publicly available model is ArcFace [4], a geometric method, that uses Additive Angular Margin Loss, to produce highly distinguishable features and stabilize training process.

A. Bartoli and A. Fusiello (Eds.): ECCV 2020 Workshops, LNCS 12539, pp. 462–475, 2020.
https://doi.org/10.1007/978-3-030-68238-5_34

Besides strong performance of face recognition models in the real world, it is crucial to study and overcome their vulnerabilities, since adversaries might harm security and privacy aspects of such systems. Face recognition systems might be maliciously attacked from different perspectives. Impersonation and dodging attacks aim to fool the network, by wearing specifically designed accessories such as glasses [28]. 2D and 3D spoofing attacks have been demonstrated in practical applications of face identification systems such as face unlock systems [17,24].

Another critical vulnerability of a face recognition system is the leakage of data, as face embeddings (face identity features) might be reconstructed into recognizable faces. In this paper, we consider black-box scenario (see Fig. 1), i.e. we only receive embedding produced by face identification model for our requesting image, without access to the model's architecture, since unknown embeddings might be exposed or hacked, and using corresponding target face recognition APIs we can request necessary output.

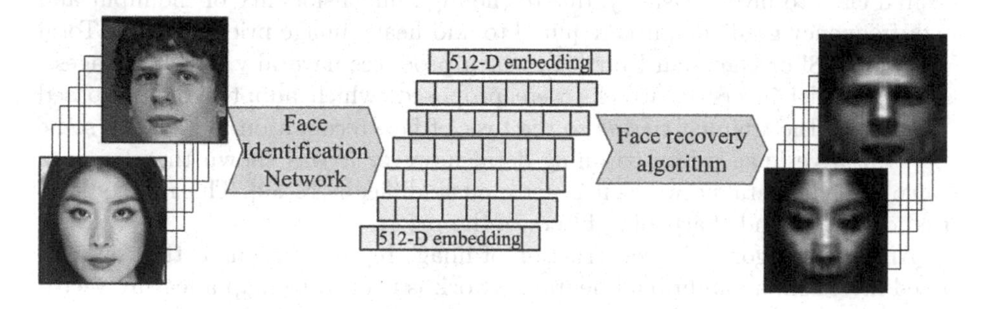

Fig. 1. The schematic of the face recovery procedure from the identity features.

Main contributions of this work are the following:

- We proposed the novel face recovery method in the black-box setup;
- We quantitatively and qualitatively demonstrated the superiority of our method compared to the previous one;
- The proposed method works without a prior knowledge such as a training dataset from the same domain;
- We evaluated our method and its competitor with an independent critic;
- We proved that the result is the same for the train and test datasets.

2 Related Work

The remarkable progress of deep neural networks in many areas attained significant attention in the scientific community to the nature of its internal representations. Researchers always questioned how to interpret the decisions of deep learning models. One way of interpretation of neural networks in the task of pattern recognition was found to be the inverting of class outputs or hidden feature

Table 1. Comparison table

Algorithm	Target model	Setting	Dataset-free
Ours	Arcface output	Black-Box	+
NBNet [19]	FaceNet output	Black-Box	–
Cole et al. [3]	FaceNet intermediate features	White-box	–
CNN[34]	FaceNet output features	White-box	–
Gradient wrt input [18]	Any classifier output	White-box	+

vectors. Using image pixel gradients, optimization-based inversion of image classification neural networks presented interesting results in [7,18,22,29,31]. The basic idea of this branch of works was to use white-box back-propagation to get input gradients and minimize the loss between the network output and the desired class to invert. Usually, due to the high dimensionality of the input and high-frequency gradients, it is required to add heavy image priors, such as Total Variation [18] or Gaussian Blur [31], which produces naturally looking images.

The model inversion attacks were proposed, which adopts gradient-based inversion of the training classes to the task of face recognition [8], leaking some representative images from training data, however it was shown that for deep convolutional neural networks it is notoriously difficult [12,30]. This method also used denoising and sharpening filters as the prior.

Another category of reconstruction of image representations is the training-based inversion: an additional neural network is trained to map a feature vector into an image. The resulting neural network is similar to a decoder part of an auto-encoder network with the face identification encoder. To train a network, usually, L1 or L2 loss is used between original and reconstructed images. The results of this method were shown in [5,6,23]. Compared to gradient-based inversion, training-based inversion is only costly during training the inversion model, which is a one-time effort. Reconstruction from a given prediction requires just one single forward pass through the network.

Training-based methods to recover faces from the facial embeddings were found to produce interesting results in [3,19–21,34]. In [20], it was proposed to use the radial basis function regression to reconstruct faces from its signatures. In [21], multidimensional scaling was used to construct a similarity matrix between faces and embeddings. It should be mentioned, both [20] and [21] were only tested for shallow neural networks. In [34], it was proposed to train a convolutional neural network that maps face embeddings to the photographs, however their method requires gradients of a face identification system. In [3], it was proposed to yield a reconstructed image from estimated face landmarks and textures, however high-quality face images are required for estimation. In [19], it was proposed to use the neighbourly de-convolution neural network to reconstruct recognizable faces, however this method requires input-output pairs for training process, and thus might be overfitted towards dataset or face identification model. To the best of our knowledge, no prior work on black-box zero-shot face reconstruction from identity features was presented before. To fill this gap, we propose our method. Since most of published results consider white-box setup, as direct competitor for our solution we see NBNet [19]. Brief comparison of various methods is collected in Table 1.

2.1 Black-Box Mode, Prior Knowledge and Number of Queries

In this paper, we consider a black-box attack procedure: we do not have access to the face recognition system and can only query it to get the output. In this setup, the number of queries is the main performance metric, along with the attack success rate. However, the number of queries is highly dependent on prior knowledge about the model. For adversarial examples this phenomenon is studied in [14]. Even models that are claimed to be fully black-box, such as NBNet [19], in fact exploit deep prior about the target model. They need to have a dataset from the same domain and the same alignment as it was for the target model, otherwise they cannot learn the function between a face image and an embedding since for the face-id network, this function is guaranteed to work well only for properly aligned images. In practice, we can have the model with a proprietary aligner and an unknown training dataset domain. One of the advantages of our method is that it is fully black-box and can work even in such a restrictive setup.

3 Gaussian Sampling Algorithm

We designed an iterative algorithm for reconstructing a face from its embedding. The algorithm is a zero-order optimization in the linear space of 2D Gaussian functions. One step of our algorithm is the following: we sample a batch of random Gaussian blobs and add them to the current state image. Then the batch is put into black-box feature extractor, and loss function is evaluated across embeddings. Based on the evaluation, one image is selected and set as the current image. Such a procedure is similar to the random descent in the linear space of 2D Gaussian functions.

3.1 Choosing Function-Family for Sampling

In our algorithm we sample Gaussian functions (Fig. 2):

$$G(x,y) = A \cdot \exp \frac{(x-x_0)^2}{2\sigma_1^2} \exp \frac{(y-y_0)^2}{2\sigma_2^2}$$

where,

x, y - pixel coordinates in the image,
x_0, y_0 - coordinates of a center of gaussian,
σ_1, σ_2 - vertical and horizontal standard deviations,
A - amplitude

Fig. 2. Gaussian blob with parameters: $x_0, y_0, \sigma_1, \sigma_2, A = 56, 72, 22, 42, 1$.

Hypothetically, any function representing a basis in a 2D space can be chosen as a function for the sampling. We tried sines/cosines, polynomial functions, random noise, but only Gaussians-based approach works well. We suppose that the reasons are the following:

1. Gaussian functions are semi-local, which means that the distortion of a picture is localized and hence more controllable. With even a small number of such functions, it is easier to fit many shapes.
2. Low frequencies are dominant in Gaussian functions (if we restrict the interval of possible σ). We suppose this prevents overfitting of an attacked network and prevents generating of non-semantic high-frequency adversarial patterns.

We found that the restriction of the vertical symmetry on the family of sampling functions improves the speed of convergence and the quality of the final result, which makes sense as human faces are mostly symmetrical, and bringing this constraint to our algorithm reduces search space. We symmetrize sampled Gaussians by adding a vertically flipped copy:

$$G_{sym} = G + \text{flip}(G)$$

To relax the problem further on, i.e. simplify the optimization process, we restore images in the grayscale colormap. In other words, the hypothesis is that embedding of deep face recognition systems is tolerant to color. To verify the assumption, we set up two experiments for the most popular publicly available face recognition systems: ArcFace [4] (model name "LResNet100E-IR, ArcFace@ms1m-refine-v2"[1], accessed March 21, 2020) and FaceNet [27] (model name "20180402-114759"[2], accessed March 21, 2020). We checked pairwise similarity of RGB image and its grayscale copy. We perform this experiment with images from LFW [13] and MS-Celeb-1M [9] (version named "MS1M-ArcFace"[3], accessed March 21, 2020) datasets. It can be clearly seen on Fig. 4 that for the majority of images moving to the grayscale domain did not affect much corresponding embeddings. Anyway, we tried to reconstruct faces in the RGB domain, but obtained colors turned out to be far from natural regardless of the shapes being correct (Fig. 3).

[1] https://github.com/deepinsight/insightface/wiki/Model-Zoo.
[2] https://github.com/davidsandberg/facenet.
[3] https://github.com/deepinsight/insightface/wiki/Dataset-Zoo.

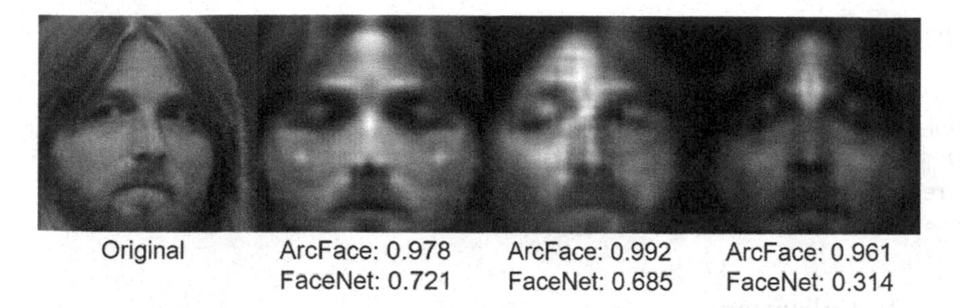

Fig. 3. From left to right: original image, reconstructed from embedding in grayscale setting with symmetric constraint, without symmetric constraint, RGB with symmetric constraint and corresponding cosine similarities by attacked model (ArcFace) and independent (FaceNet).

So, our finding is that face embeddings are mostly not sensitive to color; therefore it is not possible to recover properly the color information of initial picture. Most importantly, it relaxes the problem significantly, allowing us to sample only grayscale Gaussian blobs. But, despite the fact that we reconstruct faces in a grayscale color space it is still possible to colorize it naturally later on with the use of dedicated colorization models [33].

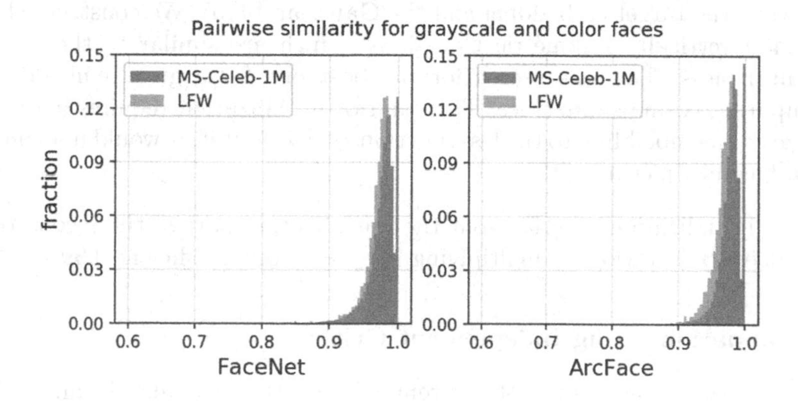

Fig. 4. Pairwise ArcFace cosine similarity of images and their grayscale analogs from LFW and MS-celeb-1M datasets.

3.2 Loss Function

A loss function is needed to choose the best sampled element from a batch. The suggested loss function depends on norms of embeddings (the target one and the embedding of a reconstructed image) and the cosine similarity between the target embedding and the embedding of a reconstructed image:

$$L(y, y') = \lambda \cdot (\|y\| - \|y'\|)^2 - s(y, y'),$$

where,

s - cosine similarity function,
$\|y\|$ - L_2 norm of the target embedding,
$\|y'\|$ - L_2 norm of the embedding of
a reconstructed image,
$\lambda = 0.0025$, empirically found hyperparameter.

3.3 Initialization

We found that proper initialization of the algorithm has crucial importance. Without an initialization, the algorithm most likely would not converge to a face. We tried two variants of initialization (Fig 5):

1. Initialize with a face. This kind of initialization uses a predefined image with a face. Such initialization works good and even let us not use norm of an embedding (works just with cosine similarity between embeddings as a loss function). But it has a strong disadvantage as the reconstructed face is "fitted" into initialization face: reconstructed face looks similar to a target person (facial traits), while it has the shape of initialization face. Thus, we did not use this method for further experiments.
2. Initialize with the optimal Gaussian blob (optimal in terms of cosine similarity between the target embedding and the Gaussian blob). We constructed a set of 4480 vertically symmetric Gaussians, which are similar to the shape of natural faces. Then we search for the best one for a given embedding by comparing cosine similarities. This kind of initialization requires adding the norm of an embedding to the loss function as, without it, it would not converge to a face-like picture.

For both initializations, we fade-out the initialization part of the reconstructed image at every iteration by multiplying it by fade-out coefficient 0.99.

3.4 Validation Using Independent Critic

While comparing the results of different variations of the algorithm, we faced a problem of the objective evaluation of the quality of a reconstructed face. In [19] and [3] the cosine similarity between embeddings of the target image and the reconstructed one was used as a criterion of quality, but they used the same network for evaluation as for reconstruction attack, that, we think, might cause some problems as the reconstructed face might have high similarity with the true image but does not look like the same person and even does not look face-like—so it is a kind of "adversarial face" which has high similarity but looks wrong. This happens as the network used for evaluation is the same as used in an algorithm, and it is a "dependent critic". What is more important, because of the specificity of our algorithm (minimizing loss), the reconstructed faces always have cosine similarities higher than 0.9 when attacking the same network as

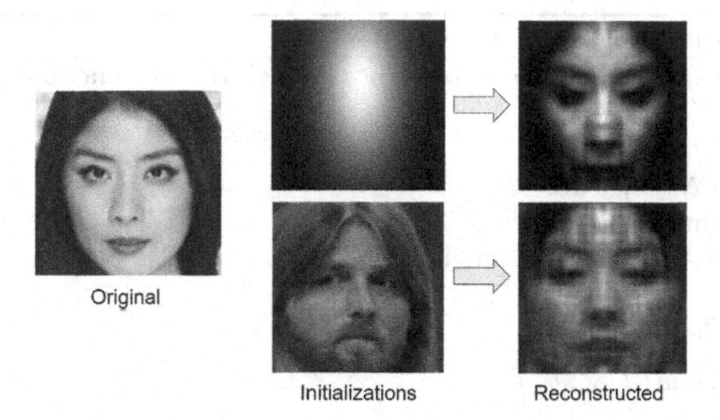

Fig. 5. From left to right: original image, two types initializations: an optimal Gaussian blob and a random face, corresponding reconstructed images.

for similarities computation. To quantify the quality of a reconstructed image in a robust way, we suggest using another network with different architecture compared to the attacked one as an "independent critic". Another solution is to use the human evaluation (like Mean Opinion Score), but as human opinion varies – some statistics are needed to quantify the quality of compared images.

We used FaceNet trained on VGGFace2 [1] as an independent critic. We provide results for both metrics—"dependent critic" and "independent critic".

3.5 Algorithm

The algorithm is a zero-order optimization in the space of 2D Gaussian functions. At every step the best one Gaussian function from a batch is chosen in terms of objective function and added to current reconstruction image. The formal description of an algorithm is presented using Algorithm 1. An example of the reconstruction process is presented in Fig. 6. The mean cosine similarity dynamics while doing queries is presented on Fig. 7.

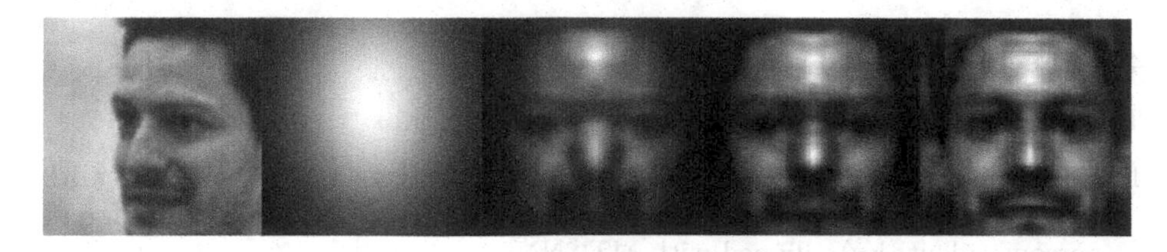

Fig. 6. Iterations of Gaussian sampling algorithm. From left to right: original image, initialization, 30k queries, 60k queries, 300k queries.

Algorithm 1. Face recovery algorithm

INPUT: target face embedding y, black-box model M, loss function L, $N_{queries}$

1: $X \leftarrow 0$
2: Initialize G_0
3: **for** $i \leftarrow 0$ to $N_{queries}$ **do**:
4: Allocate image batch $\mathbf{X}$
5: Sample batch $\mathbf{G}$ of random gaussians
6: $\mathbf{X}_j = X + G_0 + \mathbf{G}_j$
7: $\mathbf{y}' = M(\mathbf{X})$
8: $\text{ind} = \text{argmin}\Big(L(\mathbf{y}'_i, y)\Big)$
9: $X \leftarrow X + \mathbf{G}_{\text{ind}}$
10: $G_0 \leftarrow 0.99 \cdot G_0$
11: $i \leftarrow i + \text{batchsize}$
12: **end for**
13: $X \leftarrow X + G_0$

OUTPUT: reconstructed face X

4 Experiments

4.1 Baseline Reproduce

We use the original NBNet source code (author's git repository[4], accessed March 21, 2020) and trained it on MS1M-ArcFace dataset. Retrain is needed since the original model is trained with different alignment and worked poorly with photos aligned for ArcFace (by MTCNN [32]). In the original paper, it was trained on the DCGAN output, since there were no sufficient datasets at the time of publication. However, modern datasets are much larger than the number of queries needed for NBNet (MS1M-ArcFace contains 5.8M images). We followed the original paper training procedure as far as it was possible. The model was trained with MSE loss at the first stage, then the perceptual loss was added at the second stage. The MSE loss stage took 160K $\times$ 64 queries, then the loss stopped to decrease. The perceptual loss stage 100K iterations, as in the original paper.

4.2 Face Reconstruction

To evaluate our method we considered two main setups:

1. Reconstruction of faces from a MS-Celeb-1M dataset the attacked network (ArcFace) is trained on. We used a random subset of 1000 faces of different persons (identities), aligned with MTCNN;
2. Reconstruction of faces that are not presented in the training dataset. We selected a subset of 1000 unique faces of different persons (identities) from LFW that are not presented in MS1M-ArcFace: we checked each identity in LFW with all identities given in MS1M-ArcFace and left only ones for which cosine similarity was below 0.4. All images are aligned with MTCNN too.

[4] https://github.com/csgcmai/NBNet.

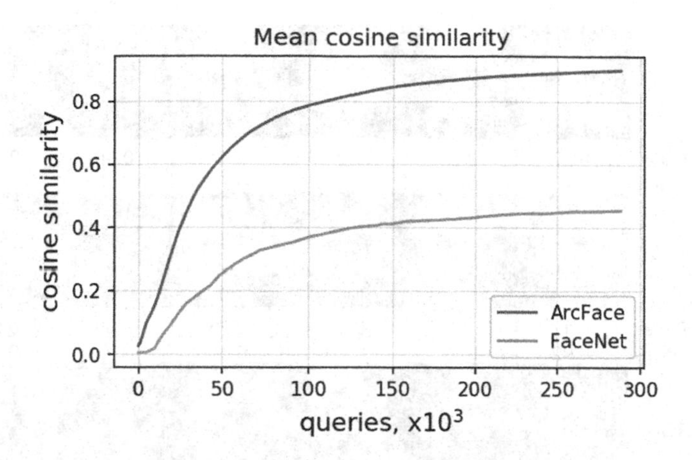

Fig. 7. Mean cosine similarity between target embedding and embedding of reconstructed image for filtered LFW subset for our algorithm.

Two sets of images are reconstructed: one with NBNet and another one with the proposed approach. The obtained faces are then fed to ArcFace and FaceNet to check the cosine similarity distribution. These setups allow the performance of given methods to be evaluated in two scenarios:

1. The network has already seen the photo, so it would ease the problem;
2. The network has never seen the photo to be restored.

In order to provide honest comparison, we trained NBNet with the same hyperparameters on grayscale version of dataset. Since our method restores a grayscale image, we thought that NBNet could also benefit from the color reduction. The results for the first setup are illustrated in Fig. 9. The first figure presents cosine similarity distribution for ArcFace network, where our method shows the superior performance. However, since ArcFace is the attacked model,

Table 2. Average cosine similarity by ArcFace and FaceNet (independent critic) between embedding of a reconstructed image and embedding of target image for subsets of 1000 images from LFW and MS1M-ArcFace and corresponding number of queries.

Method	ArcFace	FaceNet	# of queries
(Ours) Symmetric gauss, LFW (wb)	**0.92**	**0.46**	**300k**
(Ours) Asymmetric gauss, LFW (wb)	0.85	0.42	400k
NBNet, LFW (RGB)	0.25	0.34	3M
NBNet, LFW (wb)	0.19	0.27	3M
(Ours) Symmetric gauss, MS1M-ArcFace (wb)	**0.91**	**0.44**	**300k**
NBNet, MS1M-ArcFace (RGB)	0.26	0.38	3M
NBNet, MS1M-ArcFace (wb)	0.20	0.32	3M

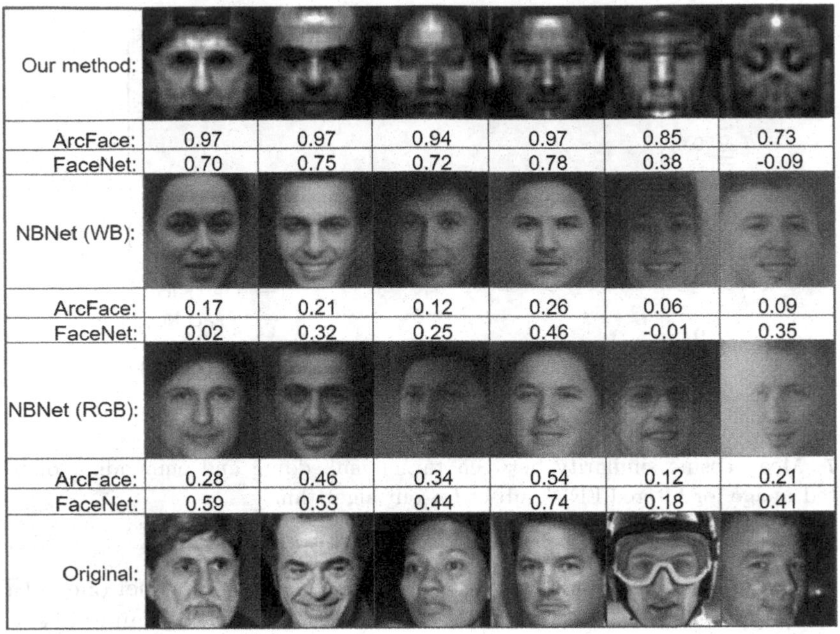

Our method:						
ArcFace:	0.97	0.97	0.94	0.97	0.85	0.73
FaceNet:	0.70	0.75	0.72	0.78	0.38	-0.09
NBNet (WB):						
ArcFace:	0.17	0.21	0.12	0.26	0.06	0.09
FaceNet:	0.02	0.32	0.25	0.46	-0.01	0.35
NBNet (RGB):						
ArcFace:	0.28	0.46	0.34	0.54	0.12	0.21
FaceNet:	0.59	0.53	0.44	0.74	0.18	0.41
Original:						

Fig. 8. Examples of recovered images from LFW dataset and the corresponding cosine similarities by ArcFace and FaceNet.

such results might be caused by overfitting. In order to avoid this, we also checked results with FaceNet. Facenet results also shows superior performance: the distribution of faces generated by the proposed method is shifted towards a higher similarity range compared to NBNet. Also it is shown that, in fact, grayscale train degrades NBNet performance.

We also checked reconstruction for the symmetric and asymmetric modes for LFW and MS1M-Arcface datasets. Since the symmetric mode reduces the complexity of an optimization process, it should show a superior performance taking less number of queries compared to the asymmetric mode: which is confirmed experimentally, and results are shown in Table 2 for both datasets.

The reconstruction process is shown in Fig. 6 with validation on FaceNet. The obtained faces are given on Fig. 8. We observed interesting behavior in the reconstruction process: faces with high validation similarity always have high-quality attributes while low similarity faces have a rather unnatural look (can be seen on the last column in Fig. 8). This is completely different from what happens with the face reconstruction by NBNet, where faces are always good to look at, and the quality does not correlate much with the cosine similarity. We attribute this problem to the NBNet training procedure, which optimizes MSE loss, which depends on insignificant features such as skin tone, while important features (eyebrows, nose form and so on) impact slightly.

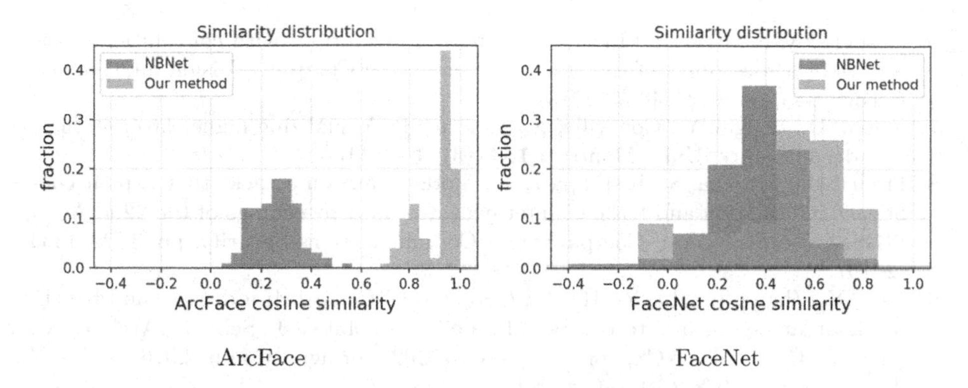

Fig. 9. Cosine similarity distribution for reconstructed faces and their true embeddings for subset of 1000 unique identities from LFW.

5 Conclusion and Future Work

We demonstrate that it is possible to recover recognizable faces from deep feature vectors of a face-recognition model in a black-box mode with no prior knowledge. The proposed method outperforms current solutions not only in terms of the average cosine similarity of embeddings produced by the attacked model but in terms of average cosine similarity given by an independent critic. Moreover, the proposed method requires a significantly smaller number of queries compared to previous solutions and does not need prior information such as proper training dataset, in other words – our algorithm works in a zero-shot mode and hence does not need to know how faces look like to recover them. As a future work, we see an investigation of poorly reconstructed faces and further minimization of the number of queries.

References

1. Cao, Q., Shen, L., Xie, W., Parkhi, O.M., Zisserman, A.: VGGFace2: a dataset for recognising faces across pose and age. In: 2018 13th IEEE International Conference on Automatic Face & Gesture Recognition (FG 2018), pp. 67–74. IEEE (2018)
2. Chen, L.-C., Zhu, Y., Papandreou, G., Schroff, F., Adam, H.: Encoder-decoder with atrous separable convolution for semantic image segmentation. In: Ferrari, V., Hebert, M., Sminchisescu, C., Weiss, Y. (eds.) ECCV 2018. LNCS, vol. 11211, pp. 833–851. Springer, Cham (2018). https://doi.org/10.1007/978-3-030-01234-2_49
3. Cole, F., Belanger, D., Krishnan, D., Sarna, A., Mosseri, I., Freeman, W.T.: Synthesizing normalized faces from facial identity features. In: Proceedings of the IEEE Conference on Computer Vision and Pattern Recognition, pp. 3703–3712 (2017)
4. Deng, J., Guo, J., Xue, N., Zafeiriou, S.: ArcFace: additive angular margin loss for deep face recognition. In: Proceedings of the IEEE Conference on Computer Vision and Pattern Recognition, pp. 4690–4699 (2019)
5. Dosovitskiy, A., Brox, T.: Generating images with perceptual similarity metrics based on deep networks. In: Advances in Neural Information Processing Systems, pp. 658–666 (2016)

6. Dosovitskiy, A., Brox, T.: Inverting visual representations with convolutional networks. In: Proceedings of the IEEE Conference on Computer Vision and Pattern Recognition, pp. 4829–4837 (2016)
7. Erhan, D., Bengio, Y., Courville, A., Vincent, P.: Visualizing higher-layer features of a deep network. Univ. Montreal **1341**(3), 1 (2009)
8. Fredrikson, M., Jha, S., Ristenpart, T.: Model inversion attacks that exploit confidence information and basic countermeasures. In: Proceedings of the 22nd ACM SIGSAC Conference on Computer and Communications Security, pp. 1322–1333 (2015)
9. Guo, Y., Zhang, L., Hu, Y., He, X., Gao, J.: MS-Celeb-1M: a dataset and benchmark for large-scale face recognition. In: Leibe, B., Matas, J., Sebe, N., Welling, M. (eds.) ECCV 2016. LNCS, vol. 9907, pp. 87–102. Springer, Cham (2016). https://doi.org/10.1007/978-3-319-46487-9_6
10. He, K., Gkioxari, G., Dollár, P., Girshick, R.: Mask R-CNN. In: Proceedings of the IEEE International Conference on Computer Vision, pp. 2961–2969 (2017)
11. He, K., Zhang, X., Ren, S., Sun, J.: Deep residual learning for image recognition. In: Proceedings of the IEEE Conference on Computer Vision and Pattern Recognition, pp. 770–778 (2016)
12. Hitaj, B., Ateniese, G., Perez-Cruz, F.: Deep models under the GAN: information leakage from collaborative deep learning. In: Proceedings of the 2017 ACM SIGSAC Conference on Computer and Communications Security, pp. 603–618 (2017)
13. Huang, G.B., Mattar, M., Berg, T., Learned-Miller, E.: Labeled faces in the wild: a database for studying face recognition in unconstrained environments (2008)
14. Ilyas, A., Engstrom, L., Madry, A.: Prior convictions: black-box adversarial attacks with bandits and priors. arXiv preprint arXiv:1807.07978 (2018)
15. Kemelmacher-Shlizerman, I., Seitz, S.M., Miller, D., Brossard, E.: The megaface benchmark: 1 million faces for recognition at scale. In: Proceedings of the IEEE Conference on Computer Vision and Pattern Recognition, pp. 4873–4882 (2016)
16. Krizhevsky, A., Sutskever, I., Hinton, G.E.: ImageNet classification with deep convolutional neural networks. In: Advances in Neural Information Processing Systems, pp. 1097–1105 (2012)
17. Liu, S., Yuen, P.C., Zhang, S., Zhao, G.: 3D mask face anti-spoofing with remote photoplethysmography. In: Leibe, B., Matas, J., Sebe, N., Welling, M. (eds.) ECCV 2016. LNCS, vol. 9911, pp. 85–100. Springer, Cham (2016). https://doi.org/10.1007/978-3-319-46478-7_6
18. Mahendran, A., Vedaldi, A.: Understanding deep image representations by inverting them. In: Proceedings of the IEEE Conference on Computer Vision and Pattern Recognition, pp. 5188–5196 (2015)
19. Mai, G., Cao, K., Yuen, P.C., Jain, A.K.: On the reconstruction of face images from deep face templates. IEEE Trans. Pattern Anal. Mach. Intell. **41**(5), 1188–1202 (2018)
20. Mignon, A., Jurie, F.: Reconstructing faces from their signatures using RBF regression (2013)
21. Mohanty, P., Sarkar, S., Kasturi, R.: From scores to face templates: a model-based approach. IEEE Trans. Pattern Anal. Mach. Intell. **29**(12), 2065–2078 (2007)
22. Mordvintsev, A., Olah, C., Tyka, M.: Inceptionism: going deeper into neural networks (2015)
23. Nash, C., Kushman, N., Williams, C.K.: Inverting supervised representations with autoregressive neural density models. arXiv preprint arXiv:1806.00400 (2018)
24. Patel, K., Han, H., Jain, A.K.: Secure face unlock: spoof detection on smartphones. IEEE Trans. Inf. Forensics Secur. **11**(10), 2268–2283 (2016)

25. Redmon, J., Farhadi, A.: YOLOv3: an incremental improvement. arXiv preprint arXiv:1804.02767 (2018)
26. Ren, S., He, K., Girshick, R., Sun, J.: Faster R-CNN: towards real-time object detection with region proposal networks. In: Advances in Neural Information Processing Systems, pp. 91–99 (2015)
27. Schroff, F., Kalenichenko, D., Philbin, J.: FaceNet: a unified embedding for face recognition and clustering. In: Proceedings of the IEEE Conference on Computer Vision and Pattern Recognition, pp. 815–823 (2015)
28. Sharif, M., Bhagavatula, S., Bauer, L., Reiter, M.K.: Accessorize to a crime: real and stealthy attacks on state-of-the-art face recognition. In: Proceedings of the 2016 ACM SIGSAC Conference on Computer and Communications Security, pp. 1528–1540 (2016)
29. Simonyan, K., Vedaldi, A., Zisserman, A.: Deep inside convolutional networks: visualising image classification models and saliency maps. arXiv preprint arXiv:1312.6034 (2013)
30. Yang, Z., Chang, E.C., Liang, Z.: Adversarial neural network inversion via auxiliary knowledge alignment. arXiv preprint arXiv:1902.08552 (2019)
31. Yosinski, J., Clune, J., Nguyen, A., Fuchs, T., Lipson, H.: Understanding neural networks through deep visualization. arXiv preprint arXiv:1506.06579 (2015)
32. Zhang, K., Zhang, Z., Li, Z., Qiao, Y.: Joint face detection and alignment using multitask cascaded convolutional networks. IEEE Signal Process. Lett. **23**(10), 1499–1503 (2016)
33. Zhang, R., Isola, P., Efros, A.A.: Colorful image colorization. In: Leibe, B., Matas, J., Sebe, N., Welling, M. (eds.) ECCV 2016. LNCS, vol. 9907, pp. 649–666. Springer, Cham (2016). https://doi.org/10.1007/978-3-319-46487-9_40
34. Zhmoginov, A., Sandler, M.: Inverting face embeddings with convolutional neural networks. arXiv preprint arXiv:1606.04189 (2016)

Privacy-Aware Face Recognition with Lensless Multi-pinhole Camera

Yasunori Ishii[1(✉)], Satoshi Sato[1], and Takayoshi Yamashita[2]

[1] Panasonic Corporation, 1006 Kadoma, Kadoma, Osaka, Japan
{ishii.yasunori,sato.satoshi}@jp.panasonic.com
[2] Chubu University, 1200 Matsumotocho, Kasugai, Aichi, Japan
takayoshi@isc.chubu.ac.jp

Abstract. Face recognition and privacy protection are closely related. A high-quality facial image is required to achieve a high accuracy in face recognition; however, this undermines the privacy of the person being photographed. From the perspective of confidentiality, storing facial images as raw data is a problem. If a low-quality facial image is used, to protect user privacy, the accuracy of recognition decreases. In this paper, we propose a method for face recognition that solves these problems. We train a neural network with an unblurred image at first, and then train the neural network with a blurred image, using the features of the neural network trained with the unblurred image, as an initial value. This makes it possible to train features that are similar to the features trained with the neural network using a high-quality image. This enables us to perform face recognition without compromising user privacy. Our method consists of a neural network for face feature extraction, which extracts suitable features for face recognition from a blurred facial image, and a face recognition neural network. After pretraining both networks, we fine-tune them in an end-to-end manner. In experiments, the proposed method achieved accuracy comparable to that of conventional face recognition methods, which take as input unblurred face images from simulations and from images captured by our camera system.

Keywords: Coded aperture · Lensless multi-pinhole camera · Face recognition · Image debluring

1 Introduction

Face recognition [12,25,46] is an important task in various applications. A facial image is personal information, and so we need to consider privacy when we include face recognition in these applications. The European Union has enforced the General Data Protection Regulation, which requires the protection of personal information. In addition, the regulation of privacy protection may expand worldwide in the future. We must therefore pay close attention to the protection of privacy when using facial images. However, it is difficult to successfully realize both privacy protection and face recognition.

A. Bartoli and A. Fusiello (Eds.): ECCV 2020 Workshops, LNCS 12539, pp. 476–493, 2020.
https://doi.org/10.1007/978-3-030-68238-5_35

Nodari et al. [29] proposed a method of decreasing the visibility of a facial region with a mosaic, in Google Street View. Padilla-Lopez et al. [30] proposed a method of replacing a facial image with a public image, to avoid privacy concerns. Fernande et al. [13] proposed a blurring method for a self-moving robot. Thorpe et al. [44] proposed a method using two types of blurred images, which differ depending on whether they are public or private images. These methods focus on image processing after an image is captured, and must overcome the serious problem of storing the facial image securely, because stored facial images can be leaked, and privacy can be violated through hacking.

To protect privacy, an image should be captured with enhanced security for personal information. As examples, an event camera records only the change in brightness of pixels for each frame [15], and thermal-image-based recognition also records information without detailed personal information [14]. Browarek et al. [4] proposed a method for human detection that uses an infrared sensor. Dai et al. [11] proposed a privacy protection method that uses a low-resolution camera. However, although these methods preserve privacy, they have difficulty in recognizing faces with high accuracy.

Computational photography [33] is another perspective on privacy protection that is worth considering. As an example of such technologies, Cossalter et al. [10] proposed a method that protects privacy by random projection based on compressive sensing. It captures an optically blurred image using a coded-aperture camera [6,17,19,23,24,26,41,45], in which a coded mask is arranged in front of the aperture. Pittaluga et al. [31,32] proposed a method of shooting an optically blurred image with a multi-aperture mask using three-dimensionally printed optics. However, its effectiveness in recognition technology is not clear because image recognition is not evaluated quantitatively. Wang et al. [48] proposed an action recognition method that protects privacy using a coded aperture camera. However, the recognition accuracy is low because it is insufficient as a feature extraction method for improving the accuracy of image recognition. Canh et al. [5] proposed a method in which it is possible to select whether to reconstruct the area excluding the face area or including the face area from the blurred image; however, they do not describe the application that uses the reconstructed image.

It is difficult to identify an individual from a blurred image captured by the coded aperture camera. However, it is possible to obtain a reconstructed image from the blurred image if we know a code pattern of the mask. Inspired by this technology, we propose a multi-pinhole camera (MPC) with a mask that has multiple pinholes. We also propose a face recognition method that achieves the accuracy of non-blurred images even when we use privacy-preserving blurred images. In this paper, an image captured with a normal camera is called an unblurred image, and an image captured with an MPC, in which privacy is protected, is called a blurred image. In general, the features of a blurred image are ineffective for face recognition, and reconstruction methods are employed in prepossessing, to obtain effective features. ConvNet-based methods reconstruct high-quality facial images [8,9,16,21,28,35,37,39,42,43,47,50,51]. However, these methods require substantial memory and have a high computational cost.

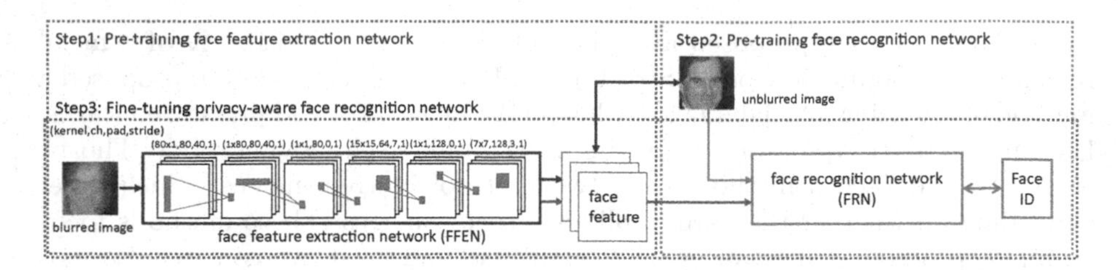

Fig. 1. Training steps of the proposed face recognition method. The training of the proposed method comprises three steps. First, we pretrain the face feature extraction network, using blurred images and unblurred images, for the extraction of face features. Second, we pretrain the face recognition network using unblurred images. Finally, we fine-tune the privacy-aware face recognition network

As an alternative approach, Xu et al. [49] and Ren et al. [36] proposed restoration methods for slight blur using shallow neural networks. These methods are based on the fact that a blurred image is constructed by convoluting an unblurred image with a point spread function (PSF). The methods can both reduce the calculation cost and suppress blurring in the blurred image, even when face reconstruction is difficult. Therefore, the methods are suitable for face feature extraction, while privacy is protected because face reconstruction is difficult, because of their use of a shallow neural network.

Face recognition and privacy protection are inseparable. It is difficult to protect privacy when effective features can be extracted from an unblurred image. However, face recognition accuracy decreases when the restored face features are protected by privacy because the blurred features remain. To solve this dilemma, we focus on the extraction of effective features, rather than the face image. Additionally, it is possible to recognize a face in an extremely blurred image.

We propose a method of improving the face recognition accuracy in a privacy-protected image to solve the above dilemma. Our method involves capturing an extremely blurred facial image using a lensless MPC, for privacy protection. The training of the proposed method comprises three steps, as shown in Figure 1. First, we pretrain the face feature extraction network (FFEN), which extracts face features from a blurred image. Second, we pretrain the face recognition network (FRN) using unblurred images. Finally, we train a privacy-aware FRN, which is the connected FFEN and FRN, as one network, in an end-to-end manner.

The contributions of the paper are as follows.

- New attempts are made to achieve both privacy protection and high recognition accuracy.
- Effective features are extracted from a restored facial image captured by a lensless MPC.
- High facial recognition accuracy is achieved even if the facial image is extremely blurred.

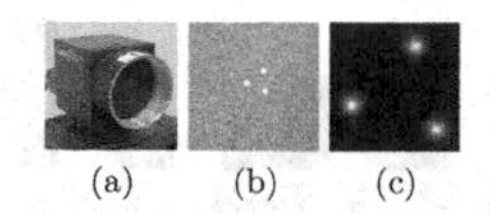

(a) (b) (c)

Fig. 2. Proposed MPC: (a) a lensless MPC, (b) an enlarged image of the coded mask captured with a microscope, and (c) a PSF of the coded mask (b)

Table 1. Specifications of the coded mask

PSFid	Diameter (μm)	Num. of pinholes	Distance (mm)	PSF size (width, height) (pixels)
3–025	120	3	0.25	(111,113)
3–050	120	3	0.50	(191,174)
9–025	120	9	0.25	(140,140)
9–050	120	9	0.50	(271,272)

2 Lensless Multi-pinhole Camera

The MPC captures a blurred image such that privacy is protected, even if a raw image stored in memory is leaked. Conventional coded apertures are intended to record light rays. Their image is therefore only slightly blurred, and privacy protection is not possible. In contrast, in the MPC, a coded mask with multiple pinholes is arranged in front of the aperture. Multiple blurs can be superimposed because light rays from the same object are incident on each pinhole. It is difficult to identify the individual in such a multiple-blurred facial image.

2.1 Design of the Proposed Lensless Multi-pinhole Camera

Various methods have been proposed to reconstruct deblurred images from images captured by a lensless camera [2,20,22,27,40]. These methods have a high cost because the imaging systems must be changed substantially. In contrast, the setup of the lensless MPC only requires a coded mask to be attached in front of the aperture. Therefore, the lensless MPC is more versatile and practical. Figure 2 shows our lensless MPC and coded mask. We employ an FLIR Blackfly BFLY-U3-23S6C-C for the body of the lensless MPC, as shown in Fig. 2(a). Figure 2(b) is an example of an enlarged image of the coded mask captured with a microscope. Figure 2(c) is the PSF corresponding to the coded mask in Fig. 2(b). The blurred image is a spatial convolution of an unblurred image and the PSF. Therefore, if the PSF is unknown, it is difficult to reconstruct the unblurred image from the blurred image. Conversely, if the PSF is known in advance, as it is for our system, the unblurred image can be reconstructed easily, by inversely convoluting the blurred image with the PSF. Because the PSF is different for each camera, a hacker would need to steal the camera to discover the PSF. The hacker would then need to install the camera in its original location after measuring the PSF. Therefore, the PSF leakage risk is lower than the image leakage risk in an actual scene. Even if a blurred image is leaked from the network or data storage, the risk of image restoration is small.

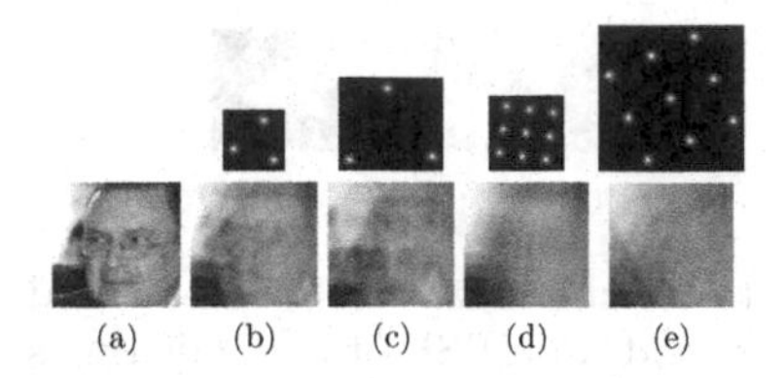

Fig. 3. Examples of simulated images convoluted by a PSF: (a) an unblurred image, and images convoluted by (b) PSFid:3–025, (c) PSFid:3–050, (d) PSFid:9–025, and (e) PSFid:9–050

2.2 PSF of the Proposed Camera

The PSF represents the response of an optical system to a point light source, in terms of the spread of spatial blur. If we know the PSF function as prior information, we can obtain the unblurred image by convolving the blurred image and the inverse PSF. We can measure the PSF before capturing an image.

We prepared four types of coded mask. The specifications of each coded mask are given in Table 1, including the number of pinholes and the distance from the center of the mask to the nearest pinhole. Figure 3 shows the measured PSF and an example of a blurred image, for each coded mask. The overlapping of objects increases with the number of pinholes. In the case where there are three pinholes and the distance from the center of the mask to each pinhole is 0.25 mm, the facial image is blurred, as shown in Fig. 3(b). It is difficult to recognize the individual in the image because parts of the face that exhibit individual characteristics, such as the eyes, nose, and mouth, are blurred. The facial image is extremely blurred when we combine nine pinholes with a distance from the center of the mask of 0.50 mm. It is even difficult to recognize the image as being that of a human face, with these settings.

3 Face Recognition from Images Captured by the Lensless Multi-pinhole Camera

We propose a method of recognizing a face in a blurred image captured by the lensless MPC. As shown in Fig. 1, the proposed method adopts the FFEN and FRN. In the FFEN, we extract effective face features from the blurred image. We can employ a state-of-the-art FRN if we obtain features similar to the features of the unblurred image. The important aspect of the proposed method is that, to preserve privacy, we do not reconstruct the deblurred facial image explicitly. In the proposed method, we train both networks in an end-to-end manner to obtain suitable face features for the FRN. The training of the proposed method comprises three steps. First, we pretrain the FFEN, which extracts face features from a blurred image. Second, we pretrain the FRN using unblurred images. Finally, we train a privacy-aware FRN, which is the connected FFEN and FRN, as one network, in an end-to-end manner. To protect privacy, we use these images only for training.

Many reconstruction methods have been proposed, but they focus on reconstruction of the entire face. This approach fails to reconstruct detail in the facial region. However, facial areas exhibiting individual characteristics are important features from the viewpoint of facial recognition. In contrast, in our approach, the FFEN focuses on the extraction of face features from a blurred image instead of a high-quality facial image.

The FRN extracts features that are effective in verifying the individual and can be easily used with state-of-the-art methods. We employ metric-learning-based methods that achieve high recognition accuracy, such as ArcFace [12], CosFace [46], and SphereFace [25]. After pretraining the FFEN and FRN, we fine-tune both networks using a blurred facial image to extract suitable features.

3.1 Pretraining of the FFEN

We measure the PSF of the lensless MPC before training the FFEN. We initialize the parameter of the network by calculating the inverse PSF, following [36,49]. The blurred image y is obtained by convolving the unblurred image x and PSF k, as expressed in Eq. (1).

$$y = k * x. \tag{1}$$

Here, $*$ is the convolution operation. Equation (1) is replaced by Eq. (2) in the frequency domain. The convolution operation is the product of each element in the frequency domain.

$$\mathcal{F}(y) = \mathcal{F}(k) \times \mathcal{F}(x). \tag{2}$$

Here, $\mathcal{F}(\cdot)$ is the discrete Fourier transform. After converting to the frequency domain, we convert the blurred image y to the unblurred image x by Eq. (3).

$$x = \mathcal{F}^{-1}(1/\mathcal{F}(k)) * y. \tag{3}$$

The function $\mathcal{F}^{-1}(\cdot)$ is the inverse Fourier transform. To prevent division by zero in the frequency domain, the Wiener filter, expressed in Eq. (4), is used.

$$\begin{aligned} x &= \mathcal{F}^{-1}(1/\mathcal{F}(k)\{\frac{|\mathcal{F}(k)|^2}{|\mathcal{F}(k)|^2 + \frac{1}{SNR}}\}) * y \\ &= k^{\dagger} * y. \end{aligned} \tag{4}$$

Here, $k^{\dagger}$ is the pseudo-inverse PSF and the SNR is the signal-to-noise ratio in the pseudo-inverse PSF. If the SNR is large, it is robust to noise.

The pseudo-inverse PSF can be resolved into $k^{\dagger} = USV^T$ through singular value decomposition (SVD). The elements of the jth rows of U and V are u_j and v_j, respectively, and the jth singular value is s_j. In Eq. (4), SVD replaces the convolution of the two-dimensional pseudo-inverse PSF with the product of the convolution of the one-dimensional vectors u_j and v_j and the scalar s_j, as in Eq. (5).

$$x = \sum_j s_j \cdot u_j * (v_j^T * y). \tag{5}$$

Conversion from the blurred image to the unblurred image using the pseudo-inverse PSF can be considered to be the adoption of a convolutional neural network taking s_j, u_j, and v_j^T as the convolutional kernels of three layers. These three layers have neither an activation function nor normalization, such as batch normalization. We use the outlier rejection subnetwork in addition to the last three layers, following [36,49].

The FFEN module in Fig. 1 shows the network architecture. The first and second layers have $K \times 1$ and $1 \times K$ kernels, respectively. Both layers have K channels. The third layer has a 1×1 kernel with K channels. The initial values of the kernels are the K eigenvectors and eigenvalues selected from the larger eigenvalue. The kernel sizes of the fourth, fifth, and sixth layers are 15×15, 1×1, and 7×7, respectively. The fourth, fifth, and sixth layers have 64, 128, and 128 channels, respectively. In optimization, we use the L_1 loss for FFEN.

$$loss_{FFEN} = \frac{1}{N} \sum_{n=1}^{N} |x_n - z_n| \\ z = DF(y) \tag{6}$$

Here, N is the number of pixels, x is the unblurred image, y is the blurred image, and $DF(y)$ is the face feature of y.

3.2 Pretraining of the FRN

The FFEN outputs features that are effective for face recognition. These features are then input to the FRN. We first perform pretraining of the FRN with unblurred facial images. The FRN network is based on ArcFace [12], CosFace [46], and SphereFace [25], which are state-of-the-art methods. ArcFace obtains effective features using cosine distance. The loss function for pretraining in ArcFace is given by Eq. (7).

$$loss_{arcface} = -\frac{1}{M} \sum_{i=1}^{M} log \frac{e^{s(cos(\theta_{y_i}+m)))}}{e^{s(cos(\theta_{y_i}+m))} + \sum_{i=1,j\neq y_i}^{n} e^{scos(\theta_j)}}. \tag{7}$$

Here, M is the number of data items, s is the scale parameter for cosine similarity, and m is the margin with other classes.

To recognize whether two facial images are of the same person, we extract features of the faces using the trained FRN. The two facial images are of the same person if the cosine distance between the extracted features is greater than or equal to a threshold.

3.3 Fine-Tuning of the Privacy-Aware FRN

The FFEN and FRN are trained independently. The output from the FFEN comprises face features, and the input to the FRN is the face features. We perform fine-tuning to adapt to the input and output of the two networks in an end-to-end manner, using the loss function given by Eq. (7). The proposed method is less affected by blur because of the combination of these networks.

For our networks, particularly the FFEN, the feature extraction accuracy of the entire face region is not essential. By fine-tuning both networks, it is possible to extract the feature only the region in which it is effective to extract features for recognizing the individual. Even if the subject wears eyeglasses, and there are few samples of faces wearing eyeglasses in the training data, the network can extract features in other important regions. When we do not pretrain the FFEN, it is necessary to extract features of the entire facial image that represent individual characteristics. However, it is difficult to extract them because the network cannot extract a feature of a small region that represents individual characteristics. The proposed method improves the accuracy of face recognition by training a FFEN that extracts feature maps representing individual characteristics.

4 Experiments

4.1 Details of Implementation

The parameters of each layer of the FFEN are shown in Fig. 1. An activation function is not arranged in the first three layers. In the second three layers, Leaky ReLU, with a gradient of 0.02, is arranged as an activation function. The mini-batch size is 1, the learning rate is 0.0001, and the number of iterations is 50 epochs in the FFEN. We use 58,346 images, randomly sampled from MS1MV2 [12] and LFW [18]. To validate the performance of feature extraction, we use 58,346 images, randomly sampled from MS1MV2 and LFW images, that are not used in training.

We employ SphereFace [25], CosFace [46], and ArcFace [12] as the FRN. The backbone network is ResNet50. We use the MS1MV2 dataset for training. The number of images is 5,822,653 and the number of IDs is 85,741. We use LFW [18], CPLFW [3], and CALFW [7] for the evaluation data; there are 12,000 images in each dataset. Each image is normalized in terms of orientation and cropped to 112 × 112 pixels. The mini-batch size is 256 and there are four epochs. The initial learning rate for pretraining is 0.1, and the learning rate is multiplied by 0.1 in epoch 3. The learning rate, momentum, and weight decay in fine-tuning are determined by adopting Bayesian optimization [1].

Public face recognition datasets do not include both unblurred and blurred images. Therefore, we first simulate the blurred images using the PSF of this camera, as shown in the leftmost four columns of Fig. 4. Blurring of PSFid:3–025 can be seen for the eyes, nose, and mouth, and it is difficult to recognize the individual. In the case of PSFid:3–050, the distance of each pinhole from the center of the mask is large, and it is possible to identify the individual, but the positional deviation is large and feature extraction is therefore difficult. It is generally possible to identify the shape of the facial contours in the blurred image of PSFid:9–025, but it is difficult to identify facial parts, because of the blur. For the blurred image of PSFid:9–050, it is difficult to identify the contours of the face and face parts. In order to prevent personal identification, our method blurs

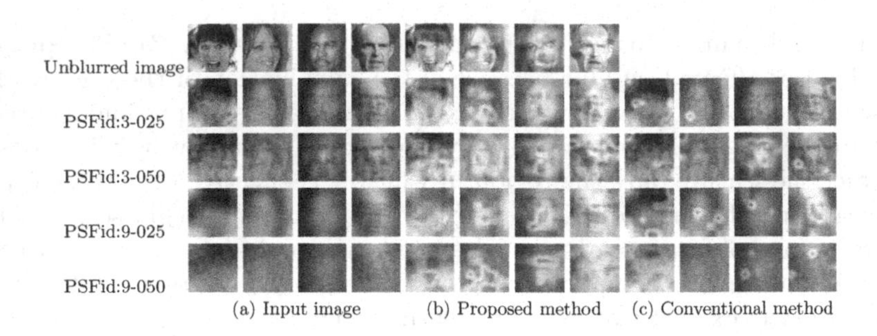

Fig. 4. Examples of unblurred and blurred images and attention maps of each image. (a) Left four columns: examples of unblurred images and blurred images for each PSFid. (b) Center four columns: attention maps of unblurred images and proposed method for each PSFid. (c) Right four columns: attention maps of conventional method for each PSFid

the face by overlaying the image. When the number of pinholes is small or the distance from the center of the mask is large, the area where the image overlaps decreases. In this case, it is difficult to protect privacy, so in order to make it difficult to identify individuals, it is desirable to make pinholes where face images overlap in many areas. We used blurred images captured by this apparatus as real images in the experiments reported below, in which we evaluated a simulated image against a real image captured by the lensless MPC.

4.2 Face Recognition Results of the Privacy-Aware FRN

In this section we present the results of the pretraining of the FFEN and FRN, and the results of the fine-tuning of the privacy-aware FRN for each PSF. Table 2 shows the face recognition results of each PSF for LFW, CPLFW, and CALFW. In Table 2, the first column shows the dataset, the second column shows the PSFid, and the third and subsequent columns show evaluation results using different FRN algorithms. (A), (B), (C), and (D) show the result of training with a blurred image, the result of training without pretraining the FRN, the result without fine-tuning, and the result of the proposed method, respectively. (A) is a conventional result. Each row shows the result for different coded masks, and the other rows show the results of SphereFace, CosFace, and ArcFace trained with unblurred images. The first value of each PSFid is the number of pinholes, and the second value is the distance of the pinholes from the mask center.

When there are three pinholes, the performance is similar to that when unblurred images are used in training. Even for nine pinholes, the performance of the proposed method is superior to that without pretraining or fine-tuning. This result shows that both pretraining and fine-tuning, which are the training steps of the proposed method, are effective. The recognition rate of CPLFW and CALFW is lower than that of LFW. This is not limited to this study, but it has been reported that this trend is similar in [12]. CPLFW performs face verification of pair images with different face pose, and CALFW performs face

Table 2. Comparison of face verification results (%)

		Basis network: SphereFace				Basis network: CosFace				Basis network: ArcFace			
Dataset	PSFid	(A)	(B)	(C)	(D)	(A)	(B)	(C)	(D)	(A)	(B)	(C)	(D)
LFW	3–025	98.6	97.2	96.5	**99.2**	99.1	98.8	98.2	**99.4**	98.4	99.1	98.2	**99.4**
	3–050	97.7	98.6	93.6	**99.4**	98.5	98.0	95.2	**99.2**	97.8	99.0	94.3	**99.4**
	9–025	97.4	98.4	90.8	**99.0**	97.8	97.1	92.6	**98.8**	97.0	95.4	92.0	**99.1**
	9–050	93.7	92.8	84.1	**97.7**	94.6	91.1	84.9	**97.3**	90.3	92.9	84.4	**97.7**
	SphereFace [25]	99.3				99.3				99.3			
	CosFace [46]	99.5				99.5				99.5			
	ArcFace [12]	99.5				99.5				99.5			
CPLFW	3–025	86.4	80.2	82.1	**87.1**	85.6	85.2	84.1	**88.0**	82.0	86.3	82.8	**89.0**
	3–050	82.4	85.0	77.4	**88.4**	83.8	77.9	79.2	**83.6**	80.1	85.4	77.8	**88.0**
	9–025	81.1	82.9	74.1	**86.1**	82.1	74.2	74.8	**82.4**	78.9	77.8	73.8	**85.3**
	9–050	74.2	72.8	65.9	**76.4**	73.3	71.2	66.3	**75.5**	67.4	73.4	66.1	**78.5**
	SphereFace [25]	87.7				87.7				87.7			
	CosFace [46]	87.6				87.6				87.6			
	ArcFace [12]	88.3				88.3				88.3			
CALFW	3–025	91.6	84.5	86.5	**93.4**	92.7	92.2	91.5	**94.5**	91.0	93.6	90.9	**95.0**
	3–050	90.1	92.6	82.8	**94.3**	91.3	90.6	87.0	**93.0**	90.4	93.1	85.9	**94.3**
	9–025	87.4	90.5	78.1	**92.7**	88.8	87.5	82.2	**91.7**	87.4	79.9	81.0	**93.3**
	9–050	81.9	75.4	71.6	**91.1**	83.2	72.7	74.3	**90.4**	76.3	75.5	73.1	**90.8**
	SphereFace [25]	94.3				94.3				94.3			
	CosFace [46]	94.9				94.9				94.9			
	ArcFace [12]	95.0				95.0				95.0			

verification of pair images of different ages. Therefore, CPLFW and CALFW are more difficult images than LFW.

4.3 Analysis Using the Area of Focus of Features and Extracted Features

We visualize whether a fine-tuning model extracts face features, using Grad-Cam [38]. ArcFace obtains similarity based on the cosine distance between feature vectors. The visualization is performed using a one-hot vector that has a value of 1 for the most similar person.

The leftmost four columns (a) of Fig. 4 show the unblurred image and the blurred image for each PSFid, the center four columns (b) show the attention maps of the proposed method for each PSFid, and the rightmost four columns (c) show the attention maps of the blurred image (conventional method) for each PSFid. When the face has little blur, such as in the case of PSFid:3–025, the attention maps of both the unblurred image and the proposed method are similar in position and strength within the area of the face. For other PSFs, the position of the attention map is slightly different, but parts of the face such as the eyes, nose, and mouth respond strongly. In the case of the conventional method, the attention maps are largely outside the face. It is therefore difficult to obtain effective features for face recognition from the blurred image.

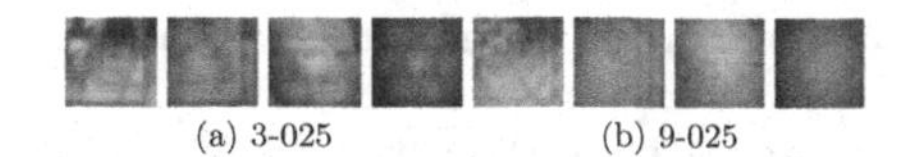

Fig. 5. Examples of captured images. Real captured images are more blurred than simulated images

Table 3. Face verification results for captured images (%)

		Basis network: SphereFace				Basis network: CosFace				Basis network: ArcFace			
Dataset	PSFid	(A)	(B)	(C)	(D)	(A)	(B)	(C)	(D)	(A)	(B)	(C)	(D)
LFW	3–025	92.4	88.5	80.7	**93.8**	92.9	88.5	84.1	**96.3**	92.7	73.0	82.4	**94.2**
	9–025	86.7	80.9	73.4	**90.4**	86.6	81.3	75.0	**89.8**	87.1	66.7	75.5	**88.9**
	SphereFace [25]	97.9				97.9				97.9			
	CosFace [46]	99.1				99.1				99.1			
	ArcFace [12]	96.7				96.7				96.7			
CPLFW	3–025	68.5	64.7	60.5	**71.1**	70.7	64.6	62.8	**75.0**	69.4	59.1	61.2	**72.0**
	9–025	64.0	60.5	56.9	**67.3**	63.6	60.4	59.2	**75.9**	64.9	55.6	58.0	**66.5**
	SphereFace [25]	75.7				75.7				75.7			
	CosFace [46]	81.9				81.9				81.9			
	ArcFace [12]	60.9				60.9				60.9			
CALFW	3-025	78.1	72.4	68.1	**93.7**	80.9	73.6	70.8	**86.6**	77.8	58.4	68.5	**79.9**
	9-025	71.0	64.7	61.0	**76.4**	71.8	65.3	62.7	**75.9**	68.6	55.5	62.0	**71.0**
	SphereFace [25]	89.4				89.4				89.4			
	CosFace [46]	92.9				92.9				92.9			
	ArcFace [12]	57.7				57.7				57.7			

4.4 Experiments Using Real Images

We compared the accuracy of face recognition for a real image, using blurred images captured by the lensless MPC for PSFid:3-025 and 9-025. The unblurred image was displayed on the monitor in a dark room and considered as a captured image with real blur. As a result, a pair, comprising an unblurred image and a real blurred image, was obtained. To train the FFEN, we used 53,143 images randomly sampled from MS1MV2 and LFW. The captured images are presented in Fig. 5. The real image used in the experiment was more blurred than the simulated image. To train the FRN, we used 147,464 images sampled from MS1MV2. The image size is 112×112 as well as simulation.

Comparison results are shown in Table 3. The proposed method achieved higher accuracy than the conventional method. Although the blurred image was extremely blurred and there was little training data, the setting for both PSFid:3–025 and PSFid:9–025 had higher accuracy than the other settings. In an experiment using real images, pretraining and fine-tuning were effective, as in an experiment using simulated data. Because the proposed method achieved high accuracy even with real images, we conclude that it achieves face recognition that can protect privacy.

4.5 Evaluation of Privacy Protection Performance of Proposed System

We evaluate the privacy protection performance of blurred images. As noted in Sect. 2.1, PSF does not leak. Therefore, we evaluated the privacy protection performance using CycleGAN [52], which is a generative model, and SelfDeblur [34], which is one of the state-of-the-art methods for blind deconvolution.

For training CycleGAN, we require unblurred and blurred images. Unblurred images were randomly selected from LFW. Two types of blurred images were used: The blurred images selected from LFW did not overlap with the unblurred images from LFW. The number of training images in each set (unblurred images from LFW and blurred images) was 5000. We used images that were not used for training, as the evaluation images. The number of training iteration is 10000. In each PSFid, losses of a generator are shown in Fig. 6. The vertical axis is the loss, and the horizontal axis is the number of the iteration. From this figure, losses converged in approximately 7000 iterations, so in this experiment, sufficient training has been done by 10000 iterations. SelfDeblur estimates the unblurred image and the PSF, given a single image.

Figure 7(a) shows an unblurred image, Fig. 7(b) shows the reconstruction result of the simulated image, and Fig. 7(c) shows the reconstruction result using the captured image. For each PSFid, the figure shows the blurred image, the results of CycleGAN, and the results of SelfDeblur. The image generated by CycleGAN is a sharp image. When the distance between pinholes is small, such as PSFid:3–025 and PSFid:9–025, the contour shape is similar to an unblurred image, but the face parts of the generated image are different from those of the unblurred image. In contrast, when the distance from the center of PSF is large, such as PSFid:3–050 and PSFid:9–050, the unblurred image and the generated image differ greatly in the shape of the contours, in addition to the face parts. Therefore, it is difficult to recognize the blurred image and the generated image as the same person. In general, the more training, the higher the accuracy. However, in GAN, the distribution of training data is trained. In this experiment,

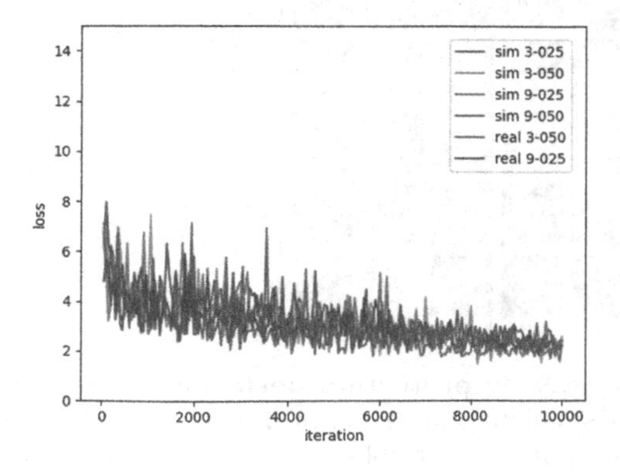

Fig. 6. Training losses of CycleGAN

Fig. 7. Example of privacy protection performance. (a) is an unblurred image, (b) is a result of the simulated image, and (c) is a result of the captured image. For each PSFid, the leftmost image is the blurred image, the center image is that generated by CycleGAN [52], and the rightmost image is that reconstructed by SelfDeblur [34]

the feature distribution for expressing the face is trained rather than the character which represents the individual. Therefore, it is possible to generate face images without blur, but since individuality is lost, the face recognition accuracy does not necessarily increase even if the number of images is increased in this experiment.

The deconvolution image created by SelfDeblur from the simulated image can approximately distinguish the face area from the background. However, the artifact is so large that the subject cannot be identified. This tendency is the same for all PSFids, but increasing the number of pinholes causes the face shape to collapse more, making deconvolution difficult. In the result of deconvolution using a captured image, it is difficult even to visually recognize the position of the face area. From these results, it was confirmed that it is difficult to identify the person in image generation and image reconstruction when the PSF is unknown, and the proposed system is effective for privacy protection.

5 Conclusion

We have proposed a privacy-aware face recognition method that solves the dilemma of simultaneously realizing good privacy protection and face recognition accuracy. To be successful at both, we constructed an acquisition system based on a lensless MPC that captures extremely blurred face images. The MPC has several pinholes and captures a blurred image. From this blurred image, we extract face features that are similar to those of an unblurred image using a FFEN. The FFEN is trained with initial parameters calculated using the inverse PSF. An FRN based on ArcFace recognizes a person using the face features. These networks are fine-tuned, in an end-to-end manner, after each is pretrained.

We are concerned that privacy may not be protected in the event that a hacker steals the captured image. If the PSF is unknown, it is difficult to reconstruct the image only from the blurred image; however, if the PSF is known, image reconstruction can be performed relatively easily. However, because the PSF is different for each camera, a hacker would need to measure the PSF, in addition to stealing the captured image. Therefore, in a real environment, it is unlikely that a hacker could recover a blurred image. By experiments using image reconstruction when the PSF is unknown, we showed that it is difficult to reconstruct a blurred image without PSF into an unblurred image.

We experimented with four types of coded masks, but these are not always optimal for privacy protection. In future studies, we intend to design a pattern that is optimal for both recognition and privacy protection, by treating the coded mask pattern as a training parameter. And, The loss of face recognition is backpropagated to FFEN by fine-tuning, but it does not specify explicitly whether to train the effective region for face recognition. We consider effective use of combining with attention and facial feature inspection.

References

1. Akiba, T., Sano, S., Yanase, T., Ohta, T., Koyama, M.: Optuna: A next-generation hyperparameter optimization framework. In: Proceedings of the 25th ACM SIGKDD International Conference on Knowledge Discovery & Data Mining, pp. 2623–2631 (2019)
2. Asif, M.S., Ayremlou, A., Sankaranarayanan, A., Veeraraghavan, A., Baraniuk, R.G.: Flatcam: thin, lensless cameras using coded aperture and computation. IEEE Trans. Comput. Imag. **3**(3), 384–397 (2016)
3. Best-Rowden, L., Bisht, S., Klontz, J.C., Jain, A.K.: Unconstrained face recognition: Establishing baseline human performance via crowdsourcing. In: IEEE International Joint Conference on Biometrics, pp. 1–8. IEEE (2014)
4. Browarek, S.: High resolution, Low cost, Privacy preserving Human motion tracking System via passive thermal sensing. Ph.D. thesis, Massachusetts Institute of Technology (2010)
5. Canh, T.N., Nagahara, H.: Deep compressive sensing for visual privacy protection in flatcam imaging. In: 2019 IEEE/CVF International Conference on Computer Vision Workshop (ICCVW), pp. 3978–3986. IEEE (2019)
6. Cannon, T., Fenimore, E.: Tomographical imaging using uniformly redundant arrays. Appl. Opt. **18**(7), 1052–1057 (1979)
7. Chen, B.C., Chen, C.S., Hsu, W.H.: Face recognition and retrieval using cross-age reference coding with cross-age celebrity dataset. IEEE Trans. Multimedia **17**(6), 804–815 (2015)
8. Chen, R., Mihaylova, L., Zhu, H., Bouaynaya, N.C.: A deep learning framework for joint image restoration and recognition. In: Circuits, Systems, and Signal Processing, pp. 1–20 (2019)
9. Chrysos, G.G., Zafeiriou, S.: Deep face deblurring. In: Proceedings of the IEEE Conference on Computer Vision and Pattern Recognition Workshops, pp. 69–78 (2017)
10. Cossalter, M., Tagliasacchi, M., Valenzise, G.: Privacy-enabled object tracking in video sequences using compressive sensing. In: 2009 Sixth IEEE International Conference on Advanced Video and Signal Based Surveillance, pp. 436–441. IEEE (2009)
11. Dai, J., Wu, J., Saghafi, B., Konrad, J., Ishwar, P.: Towards privacy-preserving activity recognition using extremely low temporal and spatial resolution cameras. In: Proceedings of the IEEE Conference on Computer Vision and Pattern Recognition Workshops, pp. 68–76 (2015)
12. Deng, J., Guo, J., Xue, N., Zafeiriou, S.: Arcface: additive angular margin loss for deep face recognition. In: Proceedings of the IEEE Conference on Computer Vision and Pattern Recognition, pp. 4690–4699 (2019)
13. Fernandes, F.E., Yang, G., Do, H.M., Sheng, W.: Detection of privacy-sensitive situations for social robots in smart homes. In: 2016 IEEE International Conference on Automation Science and Engineering (CASE), pp. 727–732. IEEE (2016)
14. Gade, R., Moeslund, T.B.: Thermal cameras and applications: a survey. Mach. Vis. Appl. **25**(1), 245–262 (2014)
15. Gallego, G., et al.: Event-based vision: A survey. arXiv preprint arXiv:1904.08405 (2019)
16. Gupta, K., Bhowmick, B., Majumdar, A.: Motion blur removal via coupled autoencoder. In: 2017 IEEE International Conference on Image Processing (ICIP), pp. 480–484. IEEE (2017)

17. Hiura, S., Matsuyama, T.: Depth measurement by the multi-focus camera. In: Proceedings. 1998 IEEE Computer Society Conference on Computer Vision and Pattern Recognition (Cat. No. 98CB36231), pp. 953–959. IEEE (1998)
18. Huang, G.B., Mattar, M., Berg, T., Learned-Miller, E.: Labeled faces in the wild: A database forstudying face recognition in unconstrained environments (2008)
19. Inagaki, Y., Kobayashi, Y., Takahashi, K., Fujii, T., Nagahara, H.: Learning to capture light fields through a coded aperture camera. In: Proceedings of the European Conference on Computer Vision (ECCV), pp. 418–434 (2018)
20. Jiao, S., Feng, J., Gao, Y., Lei, T., Yuan, X.: Visual cryptography in single-pixel imaging. arXiv preprint arXiv:1911.05033 (2019)
21. Jin, M., Hirsch, M., Favaro, P.: Learning face deblurring fast and wide. In: Proceedings of the IEEE Conference on Computer Vision and Pattern Recognition Workshops, pp. 745–753 (2018)
22. Khan, S.S., Adarsh, V., Boominathan, V., Tan, J., Veeraraghavan, A., Mitra, K.: Towards photorealistic reconstruction of highly multiplexed lensless images. In: Proceedings of the IEEE International Conference on Computer Vision, pp. 7860–7869 (2019)
23. Levin, A., Fergus, R., Durand, F., Freeman, W.T.: Image and depth from a conventional camera with a coded aperture. ACM Trans. Graph. (TOG) **26**(3), 70 (2007)
24. Liang, C.K., Lin, T.H., Wong, B.Y., Liu, C., Chen, H.H.: Programmable aperture photography: multiplexed light field acquisition. ACM Trans. Graph. (TOG) **27**, 55 (2008)
25. Liu, W., Wen, Y., Yu, Z., Li, M., Raj, B., Song, L.: Sphereface: Deep hypersphere embedding for face recognition. In: Proceedings of the IEEE Conference on Computer Vision and Pattern Recognition, pp. 212–220 (2017)
26. Nagahara, H., Zhou, C., Watanabe, T., Ishiguro, H., Nayar, S.K.: Programmable aperture camera Using LCoS. In: Daniilidis, K., Maragos, P., Paragios, N. (eds.) ECCV 2010. LNCS, vol. 6316, pp. 337–350. Springer, Heidelberg (2010). https://doi.org/10.1007/978-3-642-15567-3_25
27. Nguyen Canh, T., Nagahara, H.: Deep compressive sensing for visual privacy protection in flatcam imaging. In: Proceedings of the IEEE International Conference on Computer Vision Workshops (2019)
28. Nikonorov, A.V., Petrov, M., Bibikov, S.A., Kutikova, V.V., Morozov, A., Kazanskii, N.L.: Image restoration in diffractive optical systems using deep learning and deconvolution. Comput. Opt. **41**(6), 875–887 (2017)
29. Nodari, A., Vanetti, M., Gallo, I.: Digital privacy: replacing pedestrians from google street view images. In: Proceedings of the 21st International Conference on Pattern Recognition (ICPR2012), pp. 2889–2893. IEEE (2012)
30. Padilla-López, J.R., Chaaraoui, A.A., Flórez-Revuelta, F.: Visual privacy protection methods: a survey. Exp. Syst. Appl. **42**(9), 4177–4195 (2015)
31. Pittaluga, F., Koppal, S.J.: Privacy preserving optics for miniature vision sensors. In: Proceedings of the IEEE Conference on Computer Vision and Pattern Recognition, pp. 314–324 (2015)
32. Pittaluga, F., Koppal, S.J.: Pre-capture privacy for small vision sensors. IEEE Trans. Pattern Anal. Mach. Intell. **39**(11), 2215–2226 (2016)
33. Raskar, R.: Less is more: coded computational photography. In: Yagi, Y., Kang, S.B., Kweon, I.S., Zha, H. (eds.) ACCV 2007. LNCS, vol. 4843, pp. 1–12. Springer, Heidelberg (2007). https://doi.org/10.1007/978-3-540-76386-4_1

34. Ren, D., Zhang, K., Wang, Q., Hu, Q., Zuo, W.: Neural blind deconvolution using deep priors. In: IEEE/CVF Conference on Computer Vision and Pattern Recognition (CVPR) (2020)
35. Ren, D., Zuo, W., Zhang, D., Xu, J., Zhang, L.: Partial deconvolution with inaccurate blur kernel. IEEE Trans. Image Process. **27**(1), 511–524 (2017)
36. Ren, W., et al.: Deep non-blind deconvolution via generalized low-rank approximation. In: Bengio, S., Wallach, H., Larochelle, H., Grauman, K., Cesa-Bianchi, N., Garnett, R. (eds.) Advances in Neural Information Processing Systems, vol. 31, pp. 297–307. Curran Associates, Inc. (2018). http://papers.nips.cc/paper/7313-deep-non-blind-deconvolution-via-generalized-low-rank-approximation.pdf
37. Schuler, C.J., Christopher Burger, H., Harmeling, S., Scholkopf, B.: A machine learning approach for non-blind image deconvolution. In: Proceedings of the IEEE Conference on Computer Vision and Pattern Recognition, pp. 1067–1074 (2013)
38. Selvaraju, R.R., Cogswell, M., Das, A., Vedantam, R., Parikh, D., Batra, D.: Grad-cam: visual explanations from deep networks via gradient-based localization. In: Proceedings of the IEEE International Conference on Computer Vision, pp. 618–626 (2017)
39. Shen, Z., Lai, W.S., Xu, T., Kautz, J., Yang, M.H.: Deep semantic face deblurring. In: Proceedings of the IEEE Conference on Computer Vision and Pattern Recognition, pp. 8260–8269 (2018)
40. Sinha, A., Lee, J., Li, S., Barbastathis, G.: Lensless computational imaging through deep learning. Optica **4**(9), 1117–1125 (2017)
41. Sloane, N.J., Harwitt, M.: Hadamard transform optics (1979)
42. Son, H., Lee, S.: Fast non-blind deconvolution via regularized residual networks with long/short skip-connections. In: 2017 IEEE International Conference on Computational Photography (ICCP), pp. 1–10. IEEE (2017)
43. Tai, Y., Yang, J., Liu, X., Xu, C.: Memnet: a persistent memory network for image restoration. In: Proceedings of the IEEE International Conference on Computer Vision, pp. 4539–4547 (2017)
44. Thorpe, C., Li, F., Li, Z., Yu, Z., Saunders, D., Yu, J.: A coprime blur scheme for data security in video surveillance. IEEE Trans. Pattern Anal. Mach. Intell. **35**(12), 3066–3072 (2013)
45. Veeraraghavan, A., Raskar, R., Agrawal, A., Mohan, A., Tumblin, J.: Dappled photography: mask enhanced cameras for heterodyned light fields and coded aperture refocusing. ACM Trans. Graph. (TOG). **26**, 69 (2007)
46. Wang, H., et al.: Cosface: large margin cosine loss for deep face recognition. In: Proceedings of the IEEE Conference on Computer Vision and Pattern Recognition, pp. 5265–5274 (2018)
47. Wang, R., Tao, D.: Training very deep CNNs for general non-blind deconvolution. IEEE Trans. Image Process. **27**(6), 2897–2910 (2018)
48. Wang, Z.W., Vineet, V., Pittaluga, F., Sinha, S.N., Cossairt, O., Bing Kang, S.: Privacy-preserving action recognition using coded aperture videos. In: Proceedings of the IEEE Conference on Computer Vision and Pattern Recognition Workshops (2019)
49. Xu, L., Ren, J.S., Liu, C., Jia, J.: Deep convolutional neural network for image deconvolution. In: Advances in Neural Information Processing Systems, pp. 1790–1798 (2014)

50. Zhang, K., Xue, W., Zhang, L.: Non-blind image deconvolution using deep dual-pathway rectifier neural network. In: 2017 IEEE International Conference on Acoustics, Speech and Signal Processing (ICASSP), pp. 2602–2606. IEEE (2017)
51. Zhang, L., Zuo, W.: Image restoration: from sparse and low-rank priors to deep priors [lecture notes]. IEEE Signal Process. Mag. **34**(5), 172–179 (2017)
52. Zhu, J.Y., Park, T., Isola, P., Efros, A.A.: Unpaired image-to-image translation using cycle-consistent adversarial networks. In: Proceedings of the IEEE International Conference on Computer Vision, pp. 2223–2232 (2017)

Frequency-Tuned Universal Adversarial Perturbations

Yingpeng Deng[1(✉)] and Lina J. Karam[1,2]

[1] Image, Video and Usability Lab, School of ECEE, Arizona State University, Tempe, AZ, USA
{ypdeng,karam}@asu.edu

[2] Department of ECE, School of Engineering, Lebanese American University, Beirut, Lebanon

Abstract. The predictions of a convolutional neural network (CNN) for an image set can be severely altered by one single image-agnostic perturbation, or universal perturbation, even when the perturbation is small to restrict its perceptibility. Such universal perturbations are typically generated and added to an image in the spatial domain. However, it is well known that human perception is affected by local visual frequency characteristics. Based on this, we propose a frequency-tuned universal attack method to compute universal perturbations in the frequency domain. We show that our method can realize a good balance between perceptibility and effectiveness in terms of fooling rate by adapting the perturbations to the local frequency content. Compared with existing universal adversarial attack techniques, our frequency-tuned attack method can achieve cutting-edge quantitative results. We demonstrate that our approach can significantly improve the performance of the baseline on both white-box and black-box attacks.

Keywords: Universal adverarial attack · Frequency tuning · Just-noticeable difference (JND) · Discrete Cosine Transform (DCT)

1 Introduction

Convolutional neural networks (CNNs) have shown great potential in various computer vision problems; however, these CNNs were shown to be vulnerable to perturbations in their input, even when such perturbations are small [34]. By adding such a small perturbation to the input image, a CNN-based classifier can be easily fooled and can alter its predictions, while a human can still correctly classify the perturbed input image. Furthermore, researchers conducted various studies on the vulnerabilities of CNNs to adversarial attacks in the image understanding area, inlcuding image classification [6,12,20,21] and semantic segmentation [26,38,39]. Generally, the proposed adversarial attack algorithms can be categorized as black-box attacks and white-box attacks according to the accessibility to the attacked model. Usually, a black-box attack has little or no access to

A. Bartoli and A. Fusiello (Eds.): ECCV 2020 Workshops, LNCS 12539, pp. 494–510, 2020.
https://doi.org/10.1007/978-3-030-68238-5_36

the targeted model, with its predictions as the only possible feedback [3]. Thus, one can perform such an attack by making use of gradient estimation [2,5,15], transferability [20,25], local search [8,13] or combinatorics [19].

On the contrary, for a white-box attack, we have full knowledge about the targeted model, such as its architecture, weight and gradient information, during the computation of the perturbation. In this case, the perturbation can be optimized effectively with the model information and an objective function by backpropagation. Additionally, there are thousands of publicly available CNN models on which we can carry out white-box adversarial attacks. Thus, to find a small but effective adversarial perturbation, we tend to utilize the white-box attack. One interesting direction of the white-box attack is the universal adversarial attack. In this latter case, a single perturbation, which is computed using a training set with a relatively small number of images, can be applied to input images to fool a targeted model [20]. Given the good cross-model generalization of white-box attacks [20,27], we can also extend the universal perturbations that are pretrained on one or multiple popular CNN models to black-box attacks on other unseen models.

However, most of the universal adversarial attack methods attempt to decrease the perceptibility of the computed perturbation δ by limiting the l_p norm, $p \in [1, \infty)$, to not exceed a fixed threshold (for instance, $\|\delta\|_\infty \leq 10$ for an 8-bit image). This type of thresholding method treats different image/texture regions identically and does not take human perception sensitivity to different frequencies into consideration. Generally, a human is more sensitive to intensity changes that occur in the image regions that are dominated by low to middle frequency content than those with mid-high and high frequency content.

Therefore, instead of computing the perturbations directly in the spatial image as commonly done by many universal attack algorithms, we propose to conduct our attacks in the frequency domain by tuning the computed perturbation based on frequency content. For this purpose, we adopt a perceptually motivated just-noticeable difference (JND) model based on frequency sensitivity in various frequency bands. Moreover, we examine the effects of different frequency bands on training perturbations as well as find a balance between the effectiveness of universal attacks and the imperceptibility to human eyes. Some adversarial examples that are produced with our proposed frequency-tuned universal adversarial perturbation (FTUAP) method are shown in Fig. 1.

Our main contributions are summarized as follows:

1) To the best of our knowledge, we are the first ones to propose the generation of universal adversarial perturbations in the frequency domain by adaptively tuning the perturbation based on local frequency masking characteristics to compute imperceptible or quasi-imperceptible universal perturbations. We propose to adopt a JND threshold to mimic the perception sensitivity of human vision and guide the computation of universal perturbations. The JND threshold for each frequency band is calculated specifically based on the parametric model which approximates luminance-based contrast sensitivity [1,10].

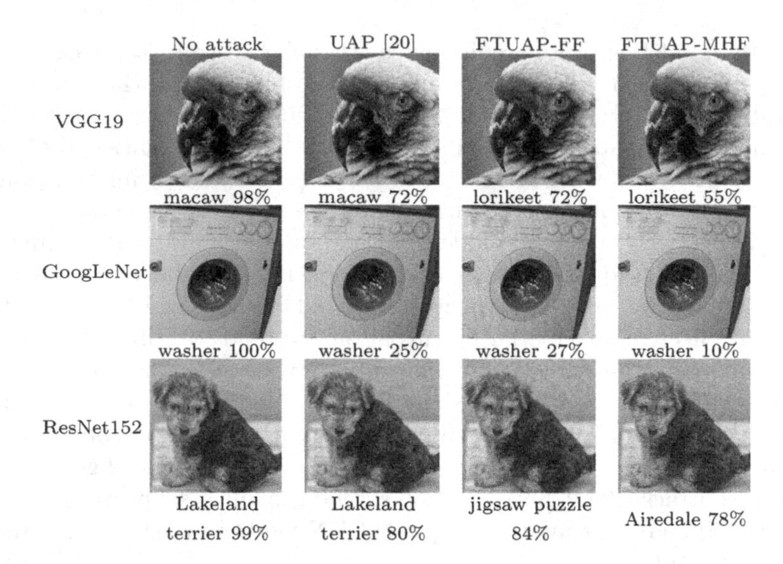

Fig. 1. Examples of perturbed images. From left to right: targeted model name, unperturbed image, perturbed images by universal adversarial perturbation (UAP) in the spatial domain [20], and our frequency-tuned UAP (FTUAP) in all frequency bands (FF) and middle/high frequency bands (MHF). The original (green) and perturbed predictions (red) with their corresponding confidence scores are listed under each image.

2) We conduct a series of ablation experiments on the universal perturbations that were generated based on frequency-domain training. We test and compare the results with respect to attack performance and visibility and show that the universal attacks on middle and high frequency bands can be both effective and nearly imperceptible.
3) We show that universal perturbations that are generated using our proposed method can significantly improve the performance in terms of fooling rate/top-1 accuracy, perceptibility and attack strength even when considering defended models.

2 Related Works

In this paper, we mainly focus on white-box universal adversarial attacks. First we will introduce some image-dependent attack algorithms because they can be adopted into the pipeline of some universal attack methods, and then the related works on universal attacks will be summarized, followed by some frequency-based and imperceptible attack algorithms.

2.1 Image-Dependent Attacks

The attack power can be maximized by computing an image-dependent perturbation for each input image separately. Goodfellow *et al.* [6] found that only a single step of gradient ascent based on the original input image and loss function, referred to as fast gradient sign method (FGSM), can generate an adversarial example with a large fooling probability, within an almost imperceptible scale compared to the image scale. Moosavi-Dezfooli *et al.* [21] proposed DeepFool to compute the minimum perturbation that reaches the decision boundary of the classifier. Later, Kurakin *et al.* [12] introduced an iterative gradient sign method (IGSM) algorithm, which consists of a multistep FGSM. An alternative for gradient calculation is the least likely class method [12]. Instead of maximizing the loss on the true label, it minimizes the loss on the label with the lowest prediction probability for the clean image. Modas *et al.* [18] showed that strong quasi-imperceptible attacks can be obtained by perturbing only few pixels without limiting the perturbation's norm.

2.2 Universal Attacks

While the aforementioned methods deal with image-dependent adversarial perturbations, the generation of a universal, image-agnostic perturbation is a more challenging topic because it aims to find one universal perturbation that can drastically reduce the prediction accuracy when applied to any input image. Based on the DeepFool algorithm [21], Moosavi-Dezfooli *et al.* [20] generated universal adversarial perturbations (UAP) by accumulating the updates iteratively for all the training images. Some universal attack methods, such as Fast Feature Fool [23], GD-UAP [22] and PD-UA [14], did not make use of training data but rather aimed to maximize the mean activations of different hidden layers or the model uncertainty. These data-independent methods are unsupervised and not as strong as the aforementioned supervised ones. Thus, in this paper, we will mainly focus on supervised universal perturbations. Recently, generative adversarial networks were adopted to generate universal perturbations in [26,27], where the adversarial networks were set as the targeted models. Poursaeed *et al.* [26] trained the generative network using a fixed random pattern as input to produce the generative adversarial perturbation (GAP), while Mopuri *et al.* [27] introduced both fooling and diversity objectives as the loss function of their network for adversary generation (NAG) to learn the perturbation distribution through random patterns sampled from the latent space. The authors of [24] and [29] use mini-batch based stochastic PGD (sPGD) during training by maximizing the average loss over each mini-batch.

2.3 Frequency Attacks

Prior to our work, Guo *et al.* [7] presented a black-box attack by randomly searching in low frequency bands in the DCT domain. Sharma *et al.* [30] showed that both undefended and defended models are vulnerable to low frequency attacks. However, the authors of [7,30] set the limit on the perturbation norm in the spatial domain instead of the frequency domain, and these existing methods generate significantly perceivable perturbations by focusing on low frequency components and suppressing the high frequency components. The resulting low-frequency perturbations can be easily perceived by humans even with a small perturbation norm. Wang *et al.* [36] improved the robustness of CNNs against blur-type adversarial attacks by preserving the low frequency components of input images. The attacks that are considered in [36] consist of blurring the input images by removing high frequency components, which is equivalent to low pass filtering the input images with the degree of the "blur" attack controlled by how many high frequency components get removed or filtered out. To test the robustness of CNNs in image classification problems, Yin *et al.* [40] constructed image-dependent corruptions by randomly sampling up to two non-zero elements in the Fourier domain. Tsuzuku *et al.* [35] came up with a black-box universal attack by characterizing the computed UAP in the Fourier domain and exploring the effective perturbations that are generated on different Fourier basis functions. Both of these attacks in the frequency domain are black-box and limit the attacks to only one or two frequency bands.

2.4 Imperceptible Adversarial Attacks

The concepts of JND and imperceptibility have been explored in previous image-dependent adversarial attack methods. Wang *et al.* [37] introduced JND into a penalization term in the loss function. Their adopted JND model is based on the pixel value in the image/spatial domain. Zhang *et al.* [41] adopted JND to compute the spatial threshold for each pixel. Both of these methods [37,41] use the pixel-wise JND model to constrain the perturbation specifically on each single image in the spatial domain. Luo *et al.* [17] selected m pixels by perturbation priority and tested if the human perceptual distance computed by local standard deviation is smaller than a chosen threshold, while Zhao *et al.* [42] adopted a perceptual color distance as the perceptual loss. Similarly, the imposed constraints are pixel-wise and non-universal. Overall, all these methods aimed to produce imperceptible image-dependent attacks and generate the adversarial examples in the spatial domain. In contrast, our proposed method generates universal image-independent attacks by adaptively tuning the perturbation characteristics in the frequency domain. Reducing the perceptibility of universal attacks is more challenging given that universal attacks are to fool the model on a whole validation set whose images are inaccessible during training, while [17,37,41,42] compute and evaluate the perturbation on the same single image. Even without a perceptibility constraint, image-dependent attacks can be easily made invisible by finding the minimal effective perturbation [21].

3 Proposed Frequency-Tuned Universal Adversarial Perturbations (FTUAP)

Let δ be the desired universal perturbation in the spatial domain. Given an input image x, a perturbed image $\hat{x}$ can be generated by adding δ to x, i.e., $\hat{x} = x + \delta$. Let δ_X, X and $\hat{X}$ be the DCTs of the perturbation δ, the input image x, and

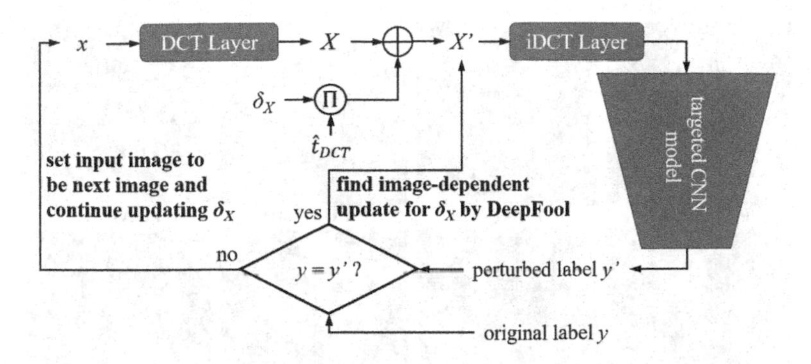

Fig. 2. Block diagram of the proposed frequency-tuned universal adversarial perturbation (FTUAP). $\prod$ denotes the projection operation, i.e., thresholding based on the l_∞ norm. X and δ_X represent the DCTs of the input image x and the spatial-domain perturbation δ, respectively. X' represents the perturbed image in the DCT domain. Initially, the input image x is set to be the first image in a randomly shuffled training set.

the perturbed image $\hat{x}$, respectively, then $\hat{X} = X + \delta_X$. The proposed FTUAP method computes the frequency-domain universal perturbation, δ_X, with the objective of maximizing the fooling rate while reducing the perceptibility of the spatial-domain perturbation δ, by adaptively tuning the perturbation in the various DCT frequency bands. Once δ_X is determined in the DCT frequency domain, the spatial-domain universal perturbation δ is obtained by simply computing the inverse DCT of δ_X. A block diagram of the proposed FTUAP is shown in Fig. 2. Given an input image x, its DCT X is first calculated. We then use the DeepFool algorithm [21] to compute the initial image-dependent δ_X in the DCT domain until δ_X can just alter the model prediction on the current image. Afterwards, we proceed with the next input image, compute its DCT X and add δ_X to X, followed by an inverse DCT to recover the perturbed image into the spatial domain. Then if the perturbed image cannot successfully fool the targeted CNN model, that is when the unperturbed label is equal to the perturbed label, the DeepFool algorithm is applied to update δ_X in the DCT domain until δ_X can just alter the model prediction. This updating procedure is applied iteratively on each subsequent input image in the training set, and it results in the final δ_X once the last input image is processed. The final spatial-domain universal perturbation is obtained as $\delta = DCT^{-1}\{\delta_X\}$. Figure 3 shows sample perturbations (after inverse DCT) as generated using our proposed FTUAP on different frequency bands. We also display a sample perurbation that was obtained using UAP for comparison. For FTUAP, the low frequency perturbation generally causes only color artifacts, and middle and high frequency perturbations produce more imperceptible but denser texture patterns. The perturbation in middle frequency bands contains more spiral patterns while high frequency patterns are visually more sparse yet locally intense. More details about the DCT-based perturbation and JND thresholds computation are given in the subsections below.

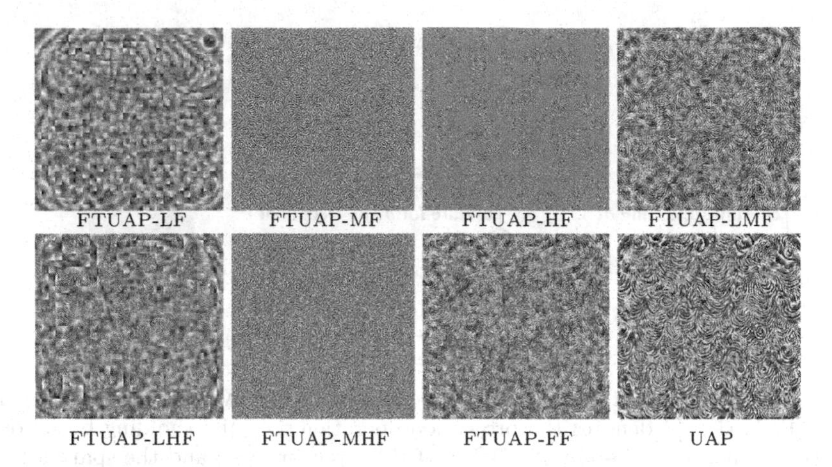

Fig. 3. Examples of different perturbations in the spatial domain. From left to right: (top) results by FTUAP in low (LF), middle (MF), high (HF) and low&middle frequency (LMF); (bottom) results by FTUAP in low&high (LHF), mid&high (MHF), full frequency (FF) and by UAP. The perturbations are scaled for visualization, and all the perturbations by FTUAP have been transformed back to the spatial domain before visualization.

3.1 Discrete Cosine Transform Based Perturbation Generation

Formula Representation. The Discrete Cosine Transform (DCT) is a fundamental tool that is used in signal, image and video processing, especially for data compression, given its properties of energy compaction in the frequency domain and real-number transform. In our paper, we adopt the orthogonal Type-II DCT, whose formula is identical to its inverse DCT. The DCT formula can be expressed as:

$$X(k_1,k_2)=\sum_{n_1=0}^{N_1-1}\sum_{n_2=0}^{N_2-1} x(n_1,n_2)c_1(n_1,k_1)c_2(n_2,k_2) \tag{1}$$

$$c_i(n_i,k_i)=\tilde{c}_i(k_i)\cos\left(\frac{\pi(2n_i+1)k_i}{2N_i}\right),\ 0\le n_i,k_i\le N_i-1,\ i=1,2 \tag{2}$$

$$\tilde{c}_i(k_i)=\begin{cases}\sqrt{\frac{1}{N_i}}, & k_i=0\\ \sqrt{\frac{2}{N_i}}, & k_i\ne 0\end{cases},\ i=1,2. \tag{3}$$

In Eq. 1, $x(n_1,n_2)$ is the image pixel value at location (n_1,n_2) and $X(k_1,k_2)$ is the DCT of the $N_1\times N_2$ image block. Usually, we set $N_1=N_2=8$. Using the DCT, an 8×8 spatial block $x(n_1,n_2)$ can be converted to an 8×8 frequency response block $X(k_1,k_2)$, with a total of 64 frequency bands. Suppose we have a 224×224 gray image x. Given that the DCT is block-wise with $N_1=N_2=8$, we can divide the image into $28\times 28=784$ non-overlapping 8×8 blocks. Then we can obtain its DCT $X(\hat{n}_1, \hat{n}_2, k_1, k_2)$, where k_1, k_2 denote 2-D frequency indices and the 2-D block indices $(\hat{n}_1, \hat{n}_2)$ satisfy $0 \leq \hat{n}_1, \hat{n}_2 \leq 27$.

Matrix Representation. For better computation efficiency, we can represent the DCT as a matrix operation. Here, let $c(n_1, n_2, k_1, k_2) = c_1(n_1, k_1)c_2(n_2, k_2)$, and further perform a flattening operation by representing the spatial indices n_1, n_2 as s and frequency indices k_1, k_2 as b, where $s = 8n_1 + n_2$ and $b = 8k_1 + k_2$, $0 \leq s, b \leq 63$. We define $C = \{c(s, b)\}$ as the 64×64 DCT coefficient matrix, where s corresponds to the flattened spatial index of the block and b indicates the flattened frequency index. If we consider an 8×8 image block x and its vectorized version x_v, the DCT can be written as:

$$X_v = C^T x_v, \tag{4}$$

which is a simple matrix dot product operation and X_v is the vectorized frequency response in the DCT domain. To compute the DCT for a 224×224 image, one can construct x_v as a stack of multiple vectorized image blocks with proper reshaping. In this case, x_v becomes a 64×784 image stack with each column corresponding to the vectorized 8×8 image block, and 784 indicates the number of image blocks. X_v is also a 64×784 stack with vectorized DCT blocks in each column. This can be easily extended to color images or image batches by adding channels to the stack. With this representation, we can directly plug the DCT in the backpropagation process to efficiently compute the perturbation within the frequency domain. In fact, by only flattening the frequency indices as b, we can directly add a convolutional layer with a fixed DCT matrix $C = \{c(n_1, n_2, b)\}$ and a stride of 8 to implement the block-wise DCT followed by updating and adding the perturbation in the DCT domain. This is further followed by the inverse DCT before the targeted CNN model.

3.2 Perception-Based JND Thresholds

Inspired by human perception sensitivity, we compute the JND thresholds in the DCT domain by adopting the luminance-model-based JND thresholds [1,10]. The computed JND thresholds are used to adaptively constrain the l_∞ norm of perturbations in different frequency bands as explained later in this subsection. The JND thresholds for different frequency bands can be computed as

$$t_{DCT}(k_1, k_2) = \frac{MT(k_1, k_2)}{2\tilde{c_1}(k_1)\tilde{c_2}(k_2)(L_{max} - L_{min})}, \tag{5}$$

where L_{min} and L_{max} are the minimum and maximum display luminance, and $M = 255$ for 8-bit image. To compute the background luminance-adjusted contrast sensitivity $T(k_1, k_2)$, Ahumada *et al.* proposed an approximating parametric model[1] [1]:

$$\log_{10}(T(k_1,k_2)) = \log_{10} \frac{T_{min}}{r + (1-r)\cos^2\theta(k_1,k_2)} + K(\log_{10} f(k_1,k_2) - \log_{10} f_{min})^2. \quad (6)$$

The frequency $f(k_1, k_2)$ and its orientation $\theta(k_1, k_2)$ are given as follows:

$$f(k_1,k_2) = \frac{1}{2N_{DCT}}\sqrt{\frac{k_1^2}{w_x^2} + \frac{k_2^2}{w_y^2}}, \quad \theta(k_1,k_2) = \arcsin\frac{2f(k_1,0)f(0,k_2)}{f^2(k_1,k_2)}, \quad (7)$$

and the luminance-dependent parameters are generated by the following equations:

$$T_{min} = \begin{cases} \left(\frac{L}{L_T}\right)^{\alpha_T}\frac{L_T}{S_0}, & L \leq L_T \\ \frac{L}{S_0}, & L > L_T \end{cases}, \quad f_{min} = \begin{cases} f_0\left(\frac{L}{L_f}\right)^{\alpha_f}, & L \leq L_f \\ f_0, & L > L_f \end{cases}, \quad K = \begin{cases} K_0\left(\frac{L}{L_K}\right)^{\alpha_K}, & L \leq L_K \\ K_0, & L > L_K \end{cases}. \quad (8)$$

The values of constants in Eqs. 6–8 are $r = 0.7$, $N_{DCT} = 8$, $L_T = 13.45$ cd/m^2, $S_0 = 94.7$, $\alpha_T = 0.649$, $f_0 = 6.78$ cycles/degree, $alpha_f = 0.182$, $L_f = 300$ cd/m^2, $K_0 = 3.125$, $\alpha_K = 0.0706$, and $L_K = 300$ cd/m^2. Given a viewing distance 60 cm and a 31.5 pixels-per-cm (80 pixels-per-inch), the horizontal width/vertical height of a pixel (w_x/w_y) is 0.0303 degree of visual angle [16]. In practice, for a measured luminance of $L_{min} = 0$ cd/m^2 and $L_{max} = 175$ cd/m^2, we use the luminance L corresponding to the median value of image during computation as following [11]:

$$L = L_{min} + 128\frac{L_{max} - L_{min}}{M} \quad (9)$$

Instead of setting the visibility threshold to be at the JND level, one can adjust/control the visibility tolerance by setting the threshold $\hat{t}_{DCT}(k_1, k_2) = \lambda t_{DCT}(k_1, k_2)$. In our experimental results, we use $\lambda = 2$, which was found to provide a good performance tradeoff in terms of fooling rate and perceived visibility. Let $\delta_X(k_1, k_2)$ be the perturbation that is being generated in the frequency domain for perturbing frequency band (k_1, k_2). $\delta_X(k_1, k_2)$ is projected to th unit ball with radius $\hat{t}_{DCT}(k_1, k_2)$ to constrain its l_∞ norm such that $\|\delta_X(k_1, k_2)\|_\infty \leq \hat{t}_{DCT}(k_1, k_2)$. This would also allow us to train the perturbations on the select frequency bands by setting $\hat{t}_{DCT}(k_1, k_2) = 0$ for bands that we want to keep unperturbed.

Using Eq. (5), the resulting DCT-based thresholds $t_{DCT}(k_1, k_2)$ are given below for 64 bands, $0 \leq k_1, k_2 \leq 7$:

[1] $T(k_1, k_2)$ can be computed for any k_1, k_2 which satisfy $k_1 k_2 \neq 0$ by this model, while $T(0, 0)$ is estimated as $\min(T(0, 1), T(1, 0))$.

$$t_{DCT}(k_1,k_2) = \begin{bmatrix} 17.31 & 12.24 & 4.20 & 3.91 & 4.76 & 6.39 & 8.91 & 12.60 \\ 12.24 & 6.23 & 3.46 & 3.16 & 3.72 & 4.88 & 6.70 & 9.35 \\ 4.20 & 3.46 & 3.89 & 4.11 & 4.75 & 5.98 & 7.90 & 10.72 \\ 3.91 & 3.16 & 4.11 & 5.13 & 6.25 & 7.76 & 9.96 & 13.10 \\ 4.76 & 3.72 & 4.75 & 6.25 & 8.01 & 10.14 & 12.88 & 16.58 \\ 6.39 & 4.88 & 5.98 & 7.76 & 10.14 & 13.05 & 16.64 & 21.22 \\ 8.91 & 6.70 & 7.90 & 9.96 & 12.88 & 16.64 & 21.30 & 27.10 \\ 12.60 & 9.35 & 10.72 & 13.10 & 16.58 & 21.22 & 27.10 & 34.39 \end{bmatrix}. \quad (10)$$

4 Experimental Results

We compute the universal perturbations on 10000 randomly sampled images from the ILSVRC 2012 ImageNet [28] training set, and all of our implemented results (fooling rate/top-1 accuracy) are reported through evaluation on the full validation set with 50000 images of ImageNet. To restrict the perturbations during training, we set $\|\delta\|_\infty \leq 10$ for UAP and compute the l_∞ norm thresholds as Eq. (10) for our FTUAP, one for each of the 64 DCT frequency bands. Given the 8×8 DCT frequency matrix, we recombine them as 15 nonoverlapping slanting frequency bands along the diagonal direction, i.e., the i-th slanting bands consists of all the previous frequency bands in (k_1, k_2) such that $k_1 + k_2 = i, 0 \leq i \leq 14$. Further, we divide the whole frequency bands as low $(0 \leq i \leq 3)$, middle $(4 \leq i \leq 10)$ and high $(11 \leq i \leq 14)$ frequency regions without overlap. We stop the training process of the universal perturbations after 5 epochs on VGG and ResNet models, and after 10 epochs on Inception models. For quantative results, we adopt commonly used top-1 accuracy and fooling rate as the metrics. Given an image set I and a universal perturbation δ in the spatial domain (obtained for our proposed FTUAP by applying the inverse DCT to the frequency-domain perturbation δ_X), the fooling rate can be expressed as $FR = \frac{1}{n}\sum_{i=1}^{n} [\![\hat{k}(x_i+\delta) \neq \hat{k}(x_i)]\!]$, $x_i \in I$, where $\hat{k}(x_i)$ is the predicted label by the classifier and $[\![\cdot]\!]$ denotes the indicator function.

4.1 Perceptibility Vs. Effectiveness

To show perceivability and effectiveness on different frequency bands, we use our FTUAP algorithm to train the perturbations against the pretrained ResNet50 model on low, middle, high frequency bands and all the combined frequency bands (full frequency) separately. Figure 4 shows samples of resulting perturbed visual examples and the fooling rates evaluated on the whole validation set, and provides comparisons with UAP (spatial) and unperturbed images. For the FTUAP method, all the visual results are perturbed in the frequency domain and then inversely transformed to the spatial domain. From Fig. 4, it can be seen that our proposed FTUAP achieves a significantly higher fooling rate than the spatial-domain UAP with similar image quality.

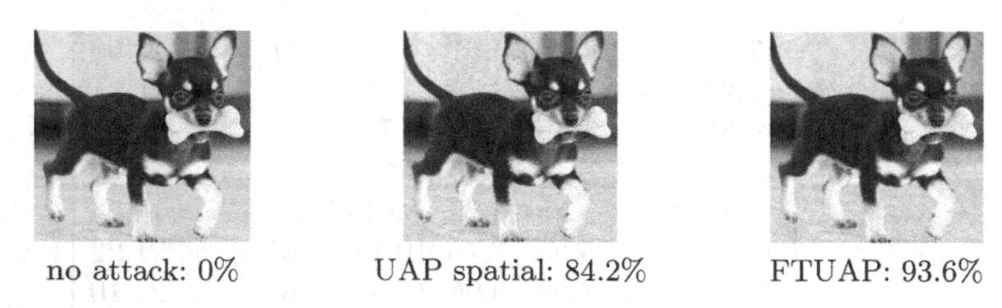

Fig. 4. Visual examples of UAP in the spatial domain and our FTUAP. The attack method and fooling rate on the validation set are listed below each image.

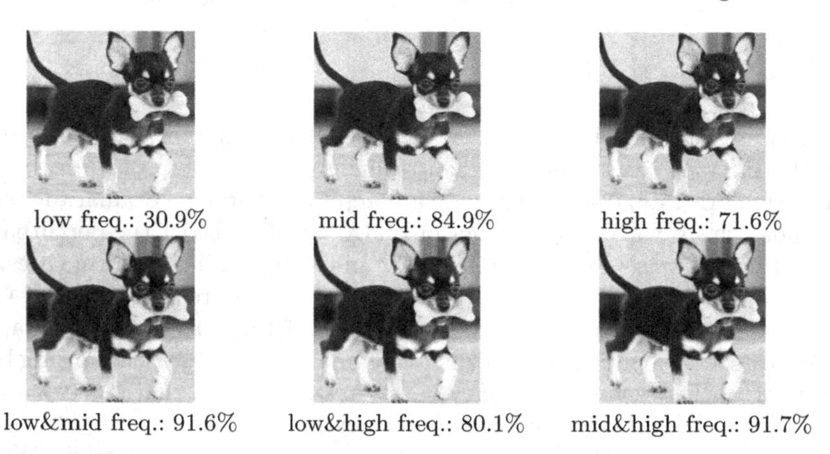

Fig. 5. Visual examples of FTUAP in various combinations of frequency bands. The corresponding attack bands and fooling rate on the whole validation set are listed below each image.

Figure 5 shows the performance that is obtained by applying our proposed FTUAP on select combinations of frequency bands. It can be seen that, compared to FTUAP with low frequency, FTUAP on middle and/or high frequency bands are more imperceptible. In terms of fooling rates, FTUAP on middle frequency bands can perform better than that on low or high frequency bands, even exceeding UAP by 0.7%. Further, FTUAP on middle and high frequencies can achieve a 91.7% fooling rate with the perturbation being nearly imperceptible, while more artifacts can be observed in the example by FTUAP on low and middle frequency bands even with a similar fooling rate result.

4.2 Fooling Capability

To demonstrate the attacking ability of our FTUAP method, we implement UAP and our FTUAP in middle frequency (FTUAP-MF), middle&high frequency (FTUAP-MHF) and full frequency (FTUAP-FF) separately, on several modern

Table 1. Comparison of fooling rates for different universal perturbations on pretrained CNN classifiers. Bold number indicates the best performance and underlined number indicates the second best performance.

	VGG16	VGG19	ResNet50	ResNet152	GoogLeNet	Inception3
UAP [20]	84.8	86.6	84.2	77.0	69.4	55.1
GAP [26]	83.7	80.1	–	–	–	**82.7**
NAG [27]	77.6	83.8	86.6	87.2	**90.4**	–
FTUAP-MF	86.4	86.0	84.9	83.1	72.8	60.9
FTUAP-MHF	<u>90.1</u>	<u>90.3</u>	<u>91.7</u>	<u>90.2</u>	82.3	<u>71.3</u>
FTUAP-FF	**93.5**	**94.5**	**93.6**	**92.7**	<u>85.8</u>	**82.7**

Table 2. Comparison of top-1 accuracy between sPGD [29] and our FTUAP-FF on 5000 training samples. Bold values correspond to best performance.

	VGG16	ResNet152	GoogLeNet	Inceptionv3	Mean
sPGD [29]	22.5	16.4	**19.8**	**20.1**	19.7
FTUAP-FF	**9.3**	**12.0**	20.9	23.8	**16.5**

CNN classifiers, including VGG [31], Inception[2] [32,33] and ResNet [9] classifiers pretrained on the ImageNet in the Pytorch library. Then we also compare our results with the published results of state-of-the-art universal adversarial attack algorithms - GAP [26], NAG [27] and sPGD [29], as listed in Tables 1 and 2.

In Table 1, our FTUAP-FF achieves the best performance in terms of fooling rate on most targeted models, especially on VGG and ResNet models, and our FTUAP-MHF provides less-perceivable perturbations (Figs. 1 and 5) while also performing well. It is worth noting that by only attacking the middle frequency bands, FTUAP-MF is able to achieve similar or higher fooling rates on all the listed models as compared to UAP. FTUAP-FF, which produces the best quantitative results among all the FTUAP variants, significantly outperforms UAP in terms of fooling rate by about 8% on VGG16, VGG19 and ResNet50 models, more than 15% on ResNet152 and GoogLeNet models, and over 20% on the Inception3 model. From Table 2, it can be seen that our FTUAP-FF exhibits a better overall attack power on the four models as compared to sPGD [29], with significantly lower top-1 accuracies on the VGG16 and ResNet152 networks and marginally higher ones on the GoogLeNet and Inception3 architectures. The obtained results show that the proposed FTUAP outperforms existing methods in terms of resulting in a higher fooling rate and reduced top-1 accuracy yet with similar image quality. Alternatively, the proposed FTUAP can reduce the perceptibility of the perturbation while maintaining similar or better performance in terms of fooling rate (MF & MHF).

[2] To implement block-wise DCT, the image size must be an integral multiple of 8, thus the 299×299 input image is resized to 296×296 for Inception3 model.

Table 3. Cross-model fooling rates. The first row displays the attacked models, and the first column indicates the targeted models for which adversarial perturbations were computed.

		VGG16	VGG19	ResNet50	ResNet152	GoogLeNet	Mean
VGG16	UAP	84.8	72.4	43.1	33.5	40.2	54.8
	FTUAP-MF	86.4	76.4	50.4	39.5	47.4	60.0
	FTUAP-MHF	90.1	79.9	54.0	40.5	46.3	62.2
	FTUAP-FF	93.5	84.7	50.3	38.3	42.4	61.8
VGG19	UAP	76.8	86.6	41.7	32.9	39.7	55.5
	FTUAP-MF	79.6	86.0	47.1	37.4	48.4	59.7
	FTUAP-MHF	82.1	90.3	50.8	39.6	49.5	62.5
	FTUAP-FF	87.0	94.5	47.7	36.8	42.3	61.7
ResNet50	UAP	64.9	61.0	84.2	45.3	45.0	60.1
	FTUAP-MF	73.2	64.6	84.9	50.3	47.8	64.2
	FTUAP-MHF	75.6	68.0	91.7	53.8	52.1	68.2
	FTUAP-FF	74.9	70.0	93.6	57.8	52.0	69.7
ResNet152	UAP	58.9	55.8	55.1	77.0	39.4	57.2
	FTUAP-MF	66.1	58.8	60.8	83.1	46.2	63.0
	FTUAP-MHF	79.3	71.4	75.3	90.2	59.0	75.0
	FTUAP-FF	74.5	69.7	77.1	92.7	56.0	74.0
GoogLeNet	UAP	57.9	56.8	43.7	35.0	69.4	52.6
	FTUAP-MF	66.2	64.3	49.1	42.0	72.8	58.9
	FTUAP-MHF	68.6	67.3	54.4	42.6	82.3	63.0
	FTUAP-FF	64.0	63.7	52.9	42.7	85.8	61.8

In another experiment, we use perturbations which are computed for the specific targeted model to attack other untargeted models, to test if transferability can also be improved by our frequency-tuned algorithm, which serves as a very useful property for black-box attacks. Table 3 shows that our method in addition to causing stronger white-box attacks, can result in a better cross-model generalization of the perturbation, i.e., stronger black-box attacks. From Table 3, UAP hardly produces a mean fooling rate of over 60%, while all perturbations that are generated by our FTUAP-MHF and FTUAP-FF result in a significantly higher mean fooling rate. In particular, FTUAP-MHF and FTUAP-FF reach 75% and 74%, respectively, in terms of mean fooling rate on ResNet152, whereas the highest one for UAP is only 60.1% on ResNet50. According to the mean fooling rates, FTUAP-MHF generally shows better generalization on various models than FTUAP-FF.

4.3 Attacks on Defended Models

Given that a number of advanced defense algorithms have been published against adversarial attacks, we also consider examining the potential of our FTUAP algorithm against the defended models. We adopt the latest defense algorithm [4].

Table 4. Comparison of top-1 accuracy between UAP and FTUAP-FF by attacking the defended models in [4].

	VGG16	ResNet152	GoogLeNet
No attack	66.3	79.0	67.8
UAP	62.5	76.7	65.5
FTUAP-FF	51.1	74.2	58.4

According to [4], the defended model can outperform existing state-of-the art defense strategies and even effectively withstand unseen attacks via resilient feature regeneration. From Table 4, it can be seen that our FTUAP-FF outperforms UAP on all three tested models. While the spatial-domain UAP can only results in a top-1 accuracy decrease of less than 4% when attacking the defended models, the proposed FTUAP-FF undermines the defense by reducing the top-1 accuracy by approximately 15% and 9% on VGG16 and GoogLeNet, respectively.

5 Conclusion

Motivated by the fact that the human contrast sensitivity varies in functions of frequency, in our proposed frequency-tuned attack method, we integrate contrast sensitivity JND thresholds for generating frequency-domain perturbations that are tuned to the two-dimensional DCT frequency bands. In comparison with the baseline UAP, our FTUAP is able to achieve much higher fooling rates in a similar or even more imperceptible manner. Additionally, according to the conducted experiments, the proposed FTUAP significantly improves the universal attack performance as compared to existing universal attacks on various fronts including an increased fooling rates of white-box attacks towards the targeted models, cross-model transferability for black-box attacks and attack power against defended models.

Acknowledgement. We would like to thank Tejas Borkar for evaluating the perturbations on his defended models and providing the results in Table 4.

References

1. Ahumada Jr., A.J., Peterson, H.A.: Luminance-model-based DCT quantization for color image compression. In: Human Vision, Visual Processing, and Digital Display III, vol. 1666, pp. 365–374. International Society for Optics and Photonics (1992)
2. Bhagoji, A.N., He, W., Li, B., Song, D.: Practical black-box attacks on deep neural networks using efficient query mechanisms. In: Ferrari, V., Hebert, M., Sminchisescu, C., Weiss, Y. (eds.) ECCV 2018. LNCS, vol. 11216, pp. 158–174. Springer, Cham (2018). https://doi.org/10.1007/978-3-030-01258-8_10
3. Bhambri, S., Muku, S., Tulasi, A., Buduru, A.B.: A survey of black-box adversarial attacks on computer vision models. arXiv preprint arXiv:1912.01667 (2019)

4. Borkar, T., Heide, F., Karam, L.: Defending against universal attacks through selective feature regeneration. In: IEEE Conference on Computer Vision and Pattern Recognition (2020)
5. Chen, P.Y., Zhang, H., Sharma, Y., Yi, J., Hsieh, C.J.: ZOO: zeroth order optimization based black-box attacks to deep neural networks without training substitute models. In: The 10th ACM Workshop on Artificial Intelligence and Security, pp. 15–26 (2017)
6. Goodfellow, I.J., Shlens, J., Szegedy, C.: Explaining and harnessing adversarial examples. arXiv preprint arXiv:1412.6572 (2014)
7. Guo, C., Frank, J.S., Weinberger, K.Q.: Low frequency adversarial perturbation. arXiv preprint arXiv:1809.08758 (2018)
8. Guo, C., Gardner, J.R., You, Y., Wilson, A.G., Weinberger, K.Q.: Simple black-box adversarial attacks. arXiv preprint arXiv:1905.07121 (2019)
9. He, K., Zhang, X., Ren, S., Sun, J.: Deep residual learning for image recognition. In: IEEE Conference on Computer Vision and Pattern Recognition, pp. 770–778 (2016)
10. Hontsch, I., Karam, L.J.: Adaptive image coding with perceptual distortion control. IEEE Trans. Image Process. **11**(3), 213–222 (2002)
11. Karam, L.J., Sadaka, N.G., Ferzli, R., Ivanovski, Z.A.: An efficient selective perceptual-based super-resolution estimator. IEEE Trans. Image Process. **20**(12), 3470–3482 (2011)
12. Kurakin, A., Goodfellow, I., Bengio, S.: Adversarial examples in the physical world. arXiv preprint arXiv:1607.02533 (2016)
13. Li, Y., Li, L., Wang, L., Zhang, T., Gong, B.: Nattack: learning the distributions of adversarial examples for an improved black-box attack on deep neural networks. In: International Conference on Machine Learning, pp. 3866–3876 (2019)
14. Liu, H., Ji, R., Li, J., Zhang, B., Gao, Y., Wu, Y., Huang, F.: Universal adversarial perturbation via prior driven uncertainty approximation. In: IEEE International Conference on Computer Vision, pp. 2941–2949 (2019)
15. Liu, Y., Moosavi-Dezfooli, S.M., Frossard, P.: A geometry-inspired decision-based attack. In: IEEE International Conference on Computer Vision, pp. 4890–4898 (2019)
16. Liu, Z., Karam, L.J., Watson, A.B.: JPEG2000 encoding with perceptual distortion control. IEEE Trans. Image Process. **15**(7), 1763–1778 (2006)
17. Luo, B., Liu, Y., Wei, L., Xu, Q.: Towards imperceptible and robust adversarial example attacks against neural networks. In: Thirty-Second AAAI Conference on Artificial Intelligence (2018)
18. Modas, A., Moosavi-Dezfooli, S.M., Frossard, P.: SparseFool: a few pixels make a big difference. In: IEEE Conference on Computer Vision and Pattern Recognition, pp. 9087–9096 (2019)
19. Moon, S., An, G., Song, H.O.: Parsimonious black-box adversarial attacks via efficient combinatorial optimization. In: International Conference on Machine Learning (2019)
20. Moosavi-Dezfooli, S.M., Fawzi, A., Fawzi, O., Frossard, P.: Universal adversarial perturbations. In: IEEE Conference on Computer Vision and Pattern Recognition, pp. 1765–1773 (2017)
21. Moosavi-Dezfooli, S.M., Fawzi, A., Frossard, P.: DeepFool: a simple and accurate method to fool deep neural networks. In: IEEE Conference on Computer Vision and Pattern Recognition, pp. 2574–2582 (2016)

22. Mopuri, K.R., Ganeshan, A., Babu, R.V.: Generalizable data-free objective for crafting universal adversarial perturbations. IEEE Trans. Pattern Anal. Mach. Intell. **41**(10), 2452–2465 (2018)
23. Mopuri, K.R., Garg, U., Babu, R.V.: Fast feature fool: a data independent approach to universal adversarial perturbations. In: British Machine Vision Conference (2017)
24. Mummadi, C.K., Brox, T., Metzen, J.H.: Defending against universal perturbations with shared adversarial training. In: IEEE International Conference on Computer Vision, pp. 4928–4937 (2019)
25. Papernot, N., McDaniel, P., Goodfellow, I., Jha, S., Celik, Z.B., Swami, A.: Practical black-box attacks against machine learning. In: 2017 ACM on Asia Conference on Computer and Communications Security, pp. 506–519 (2017)
26. Poursaeed, O., Katsman, I., Gao, B., Belongie, S.: Generative adversarial perturbations. In: IEEE Conference on Computer Vision and Pattern Recognition, pp. 4422–4431 (2018)
27. Reddy Mopuri, K., Ojha, U., Garg, U., Venkatesh Babu, R.: NAG: network for adversary generation. In: IEEE Conference on Computer Vision and Pattern Recognition, pp. 742–751 (2018)
28. Russakovsky, O., et al.: ImageNet large scale visual recognition challenge. Int. J. Comput. Vis. **115**(3), 211–252 (2015)
29. Shafahi, A., Najibi, M., Xu, Z., Dickerson, J., Davis, L.S., Goldstein, T.: Universal adversarial training. arXiv preprint arXiv:1811.11304 (2018)
30. Sharma, Y., Ding, G.W., Brubaker, M.A.: On the effectiveness of low frequency perturbations. In: The 28th International Joint Conference on Artificial Intelligence, pp. 3389–3396 (2019)
31. Simonyan, K., Zisserman, A.: Very deep convolutional networks for large-scale image recognition. In: International Conference on Learning Representations (2015)
32. Szegedy, C., et al.: Going deeper with convolutions. In: IEEE Conference on Computer Vision and Pattern Recognition, pp. 1–9 (2015)
33. Szegedy, C., Vanhoucke, V., Ioffe, S., Shlens, J., Wojna, Z.: Rethinking the inception architecture for computer vision. In: IEEE Conference on Computer Vision and Pattern Recognition, pp. 2818–2826 (2016)
34. Szegedy, C., et al.: Intriguing properties of neural networks. arXiv preprint arXiv:1312.6199 (2013)
35. Tsuzuku, Y., Sato, I.: On the structural sensitivity of deep convolutional networks to the directions of fourier basis functions. In: IEEE Conference on Computer Vision and Pattern Recognition, pp. 51–60 (2019)
36. Wang, H., Wu, X., Yin, P., Xing, E.P.: High frequency component helps explain the generalization of convolutional neural networks. arXiv preprint arXiv:1905.13545 (2019)
37. Wang, Z., Song, M., Zheng, S., Zhang, Z., Song, Y., Wang, Q.: Invisible adversarial attack against deep neural networks: an adaptive penalization approach. IEEE Trans. Depend. Secure Comput. (2019)
38. Xiao, C., Deng, R., Li, B., Yu, F., Liu, M., Song, D.: Characterizing adversarial examples based on spatial consistency information for semantic segmentation. In: European Conference on Computer Vision, pp. 217–234 (2018)
39. Xie, C., Wang, J., Zhang, Z., Zhou, Y., Xie, L., Yuille, A.: Adversarial examples for semantic segmentation and object detection. In: IEEE International Conference on Computer Vision, pp. 1369–1378 (2017)

40. Yin, D., Lopes, R.G., Shlens, J., Cubuk, E.D., Gilmer, J.: A fourier perspective on model robustness in computer vision. In: Advances in Neural Information Processing Systems, pp. 13276–13286 (2019)
41. Zhang, Z., Qiao, K., Jiang, L., Wang, L., Yan, B.: Advjnd: Generating adversarial examples with just noticeable difference. arXiv preprint arXiv:2002.00179 (2020)
42. Zhao, Z., Liu, Z., Larson, M.: Towards large yet imperceptible adversarial image perturbations with perceptual color distance. In: IEEE Conference on Computer Vision and Pattern Recognition, pp. 1039–1048 (2020)

Face-Image Source Generator Identification

Mohammad Salama(✉) and Hagit Hel-Or(✉)

University of Haifa, Mt Carmel, Haifa, Israel
mohamad.f1992@gmail.com, hagit@cs.haifa.ac.il

Abstract. Recent advances in deep networks and specifically, Generative Adversarial Networks, have introduced new ways of manipulating and synthesizing "fake" images. Concerns have been raised as to the sinister use of these images, and accordingly challenges have been raised to detect "fake" from "real" images. In this study we address a slightly different problem in image forensics. Rather than discriminating real from fake, we attempt to perform "Source Generator Identification", i.e. determine the source generator of the synthesized image. In this study we focus on face images. We exploit the specific characteristics associated with each fake face image generator and introduce a face generator representation space (the profile space) which allows a study of the distribution of face generators, their distinctions as well as allows estimating probability of images arising from the same generator.

Keywords: Image source identification · Image forensics · Fake vs Real · Auto-encoder · Generative adversarial networks · Deep learning

1 Introduction

In recent years, the increasing interest in image synthesis has led to the creation of numerous methods for generating new images. Recently, deep neural networks have been used to generate images and specifically for synthesizing "fake" faces. Many face generators have been introduced that produce examples that are so realistic that they are highly difficult to distinguish from real images of faces. Concerns have been raised about these algorithms and their ability to produce faces that do not exist and which may be used for sinister purposes such as producing fake incriminating photographs and generating fake realistic social media profiles [31]. These concerns have given rise to studies that attempt to distinguish real from fake images [47].

The study of forged images and forgery in images has long been a major player in the area of image forensics extending to various visual media (see surveys on 2D images [6,12,38,52], videos [1,42], depth images [39] and more). Although classically, forgery detection focused on finding evidence of tampering

This research was supported by grant no 1455/16 from the Israeli Science Foundation.

A. Bartoli and A. Fusiello (Eds.): ECCV 2020 Workshops, LNCS 12539, pp. 511–527, 2020.
https://doi.org/10.1007/978-3-030-68238-5_37

following the image's acquisition by the camera, current technology eliminates the camera from the forensic loop and forgery detection now attempts to distinguish complete synthetic images from those acquired by a camera sensor. A more appropriate definition of forgery detection would now be to determine whether an image faithfully renders a true existing scene in the world.

Regardless of the media, studies on forensics and forgery have typically been aimed at one of the following:

- **Authentication** - verifying that no modification has been applied on the data following its acquisition, or, verifying that it *has* been acquired by a camera rather than synthesized. Authentication algorithms typically output a measure of authenticity, often a binary output - authentic or not.
- **Forgery detection** - determining the type of forgery applied to the data (such as copy-paste, in 2D images, time manipulation in videos and depth manipulation in 3D images). These methods typically determine both the type of manipulation as well as the region affected within the data.
- **Source identification** - determining the source of the data, namely, the specific camera used to acquire the image.

With the introduction of high quality image generators, challenges have been called to develop algorithms to determine "Real" vs "Fake" (generated) images. This falls under the Authentication category of forgery detection, as described above. In this paper, we study forgery detection in deep networks generated fake images, from a different aspect. Rather than discriminating real from fake (authentication), we consider the problem of source identification or rather, "Source Generator Identification". Thus we attempt to determine the specific generator[1] that produced a given fake image. Specifically we focus on determining source generators of fake face images.

To do so, we first show that a multi-class classification model successfully determines the source image generator of input images, from amongst different GAN based face generators. The successful performance implies that, indeed, there are specific characteristics associated with each image generator. However, the multi-class classification approach is restrictive in that that it assumes knowledge on the possible source generators. Thus, we extend to a modular system of one-class classifier models that allows additional face image generators to be considered as source generators without the need for retraining. Finally, we map the face generators to a representation space (the profile space) based on the one-class classifier models. This space allows analyzing the distinction between face generators and allows evaluation of unknown images as arising from the same or different generators.

[1] We intentionally term the image source as "generator" since it does not necessarily require images to be sourced from GAN networks, and the approach may apply to any fake image generator, including manual fabrication.

2 Related Work

Since the pioneering work introducing the Generative Adversarial Network (GAN) [9], numerous studies presented models that generate images. Several algorithms and applications focus on synthesizing new images with high perceptual quality [2,3,15,17,23,30,34,40,41,57], super resolution [24], image synthesis and completion [14,27,49], image to image translation [15,53,58], domain transfer [21,45], and facial attribute and expressions [4,21,28,37].

One of the most interesting domains is human face generation using various deep networks such as Deep Convolutional Generative Adversarial Networks (DCGAN) [40], Energy-Based Generative Adversarial Networks (EBGAN) [57], Boundary Equilibrium Generative Adversarial Networks (BEGAN) [3], Least Squares Generative Adversarial Networks (LSGAN) [30], DCGAN with gradient penalty scheme (DRAGAN) [23], and the recent Style-Based Generator Architecture for Generative Adversarial Networks (StyleGAN) [18,19]. The quality of generated images have increased to the point where it is difficult to distinguish by the naked eye.

Following the concerns about the reliability of face images and the difficulty in distinguishing real from fake, several methods have been proposed to determine real from GAN generated images. Several approaches rely on classic forensic approaches such as pixel and image statistics [48], color cues [26,33] texture cues [36] and high frequency content (PRNU) [32]. Other approaches use machine learning and deep networks to extract features and binary classify real vs fake [16,25,31,35,46]. In a recent study [13] real-fake image pairs were used to train a network to learn the discriminative feature between fake images and real images across a set of GANs, allowing generalization to new GANs.

These studies attempt to authenticate images as real or fake. In this study, rather than determining real vs fake, we aim to detect the source generator which produced a given fake image. Although this work focuses on NN based generators, the approach is valid on any generative process.

Image Source Identification (image attribution) has been rigorously studied in the context of image forensics. In which, the source camera or sensor used to acquire the data is determined. Invasive approaches, using watermarking techniques, can uniquely identify the source camera[44] however, passive methods are more challenging. These assume that the acquisition device leaves a subtle trace or fingerprint in the data. Approaches have been proposed that rely on image statistics, and sensor noise, or on the camera's specific processing pipeline. See [43] for a review. Recently, machine learning and deep networks have been exploited for detecting source cameras (see [52] for a summary of these approaches). Source identification studies have also been presented for videos [42], 3D data [39], and other devices such as printers and RF devices [20].

In this paper, rather than source camera, we deal with determining the source generator of synthetically generated images. The approach assumes that generated images (specifically GAN generated images) leave some unique fingerprint that allows identifying the generator source. In [46], a shallow convolutional Network (ShallowNet) is used to determine whether a fake image was geberated by a GAN generated vs Human generated fake images. In [32], sensor noise type analysis (PRNU) was used to show that GAN generated images contain unique fingerprints that can be used to determine source generator. In a recent study [54], GAN fingerprints were determined using deep learning. These detected fingerprints were shown to perform source GAN identification at high success rate. The fingerprints were shown to be sensitive to GAN architectures as well as their training parameters and training samples. The fingerprints were shown to be stable under image attacks, including blur, noise and crop. The study in [54] assumes a given set of GAN images on which the network is trained. In the study presented in this paper, we propose a method for source generator identification that generalizes to new unknown fake image generators. We map the face generators to a representation space (the profile space) that allows analyzing the distinction between different face generators and allows evaluation of new image sets to determine whether they were sourced from the same fake image generator.

3 Fake Image Generator Source Identification

In this study, we propose a method that determines the source generator of an image. In this work, we restrict our study to face images. This increases the difficulty of the problem and removes biases due to image content. The basis of our study is the assumption that the space of generated face images associated with every generator, contains some unique characteristics, or fingerprint that will allow distinction from other generated images (as shown in [32,54]). We further assume, as will be discussed in Sect. 5, that these generators all perform very well on the facial component of the image and that the unique characteristic of each generator is expressed in the noise, high frequency content and background of these images that are often difficult to discern visually. This is supported by the findings in [32,54].

The paper consists of three parts. We first show that a multi-class classifier is able to successfully determine the source GAN generator of fake images (Sect. 4) from a set of GANs. This validates the fact that indeed generated face image spaces differ across different generators. We then proceed to determine source generators using one-class classifier models (Sect. 5). This has the advantage of modularity, in that new generators can be incorporated into the source identification scheme without the need for retraining over the whole database. We show that we are able to determine whether images with unknown source are likely to have originated from the same source generator or not. Finally, we combine the one-class classifiers with a multi-class classifier to form a network that maps images into a Generator Profile Space (Sect. 6). The space reflects the similarity of fake images to known generators. We show that, using the Generator Profile Space, we can evaluate novel synthetic images as originating from the same or from different source generators. That generalizes to new unknown fake image generators.

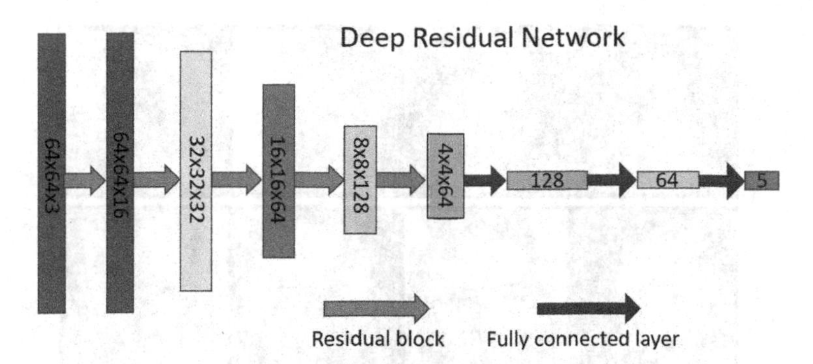

Fig. 1. Multi-class deep network for source generator classification

4 Multi-Class Classification Model

Multi-class classification is a standard supervised training model. We show that a Multi-class classification network successfully performs source generator identification on a set of GAN generators.

For our study, we consider five different face image generators, all based on the GAN architecture: BEGAN [3], DCGAN [40], EBGAN [57], LSGAN [30] and DRAGAN [23]. These GANs were chosen in the hope that they are on par in terms of quality, and we chose not to include GANS that created very high quality images [17,18] together with the low quality GANs. We felt this would present a biased base for testing. To form a baseline for all the GAN architectures, we trained each of the five GANS on the CelebA data set [29] consisting of over two hundred thousand face images. Training all GANs on the same dataset removes biases in source identification that might be due to the training data. Following training, each GAN model produced a data set of generated face images. The input to all the GAN models used in producing the face images, were noise vectors of size 64 with randomly selected values from a standard normal distribution. The datasets created by each of the five GAN architectures consisted of 160k images each. Examples of the produced images by each of the GAN architectures can be seen in Fig. 2.

The multi-class classifier for determining source generator was built using the Residual-Block [10] as the main block of the network. The network model consists of five residual blocks as the first five layers, followed by three fully connected layers, and a Softmax layer for the final classification. The model receives as input, an rgb image of size 64×64×3, and returns a vector of five entries as output, representing the probability of the input image originating from each of the five possible source generators. The model architecture is shown in Fig. 1. The size of the tensor is given for each layer of the network.

The classifier model was trained on the data sets of images, $\{I_{FG}\}$, created by the five face image generators: $FG \in$ {BEGAN, DCGAN, EBGAN, LSGAN, DRAGAN}. Every image was labeled with the generator that created it. Training data consisted of 120 K images per generator. 5-fold cross validation was used

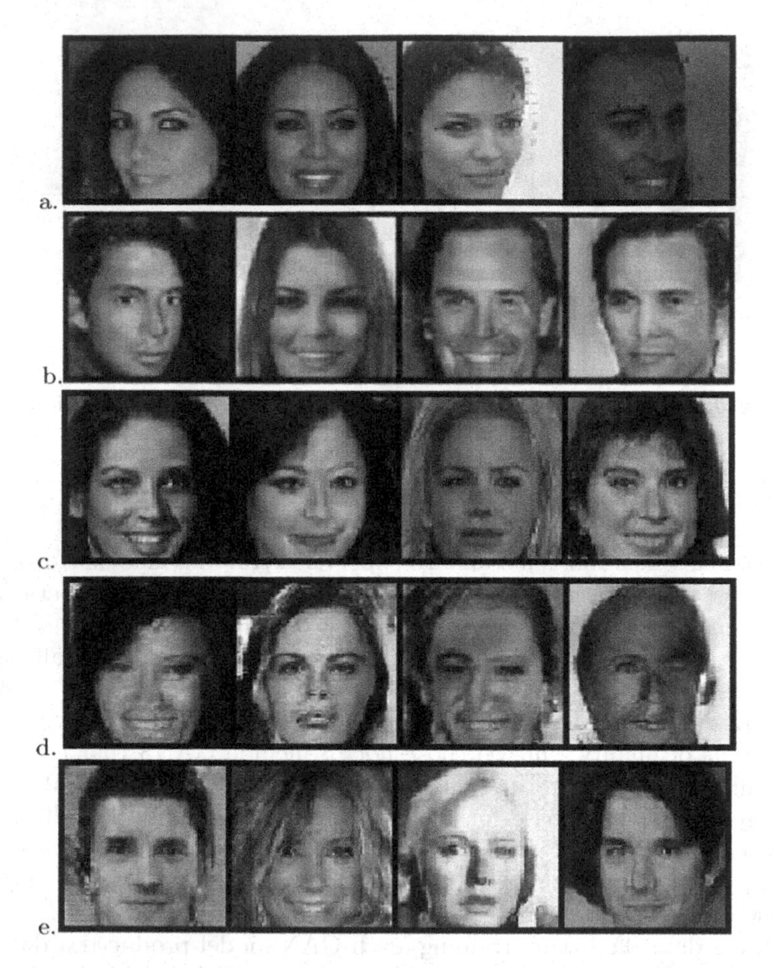

Fig. 2. Face images produced by five GAN architectures. a) BEGAN [3], b) DCGAN [40], c) DRAGAN. [23] d) EBGAN [57], e) LSGAN [30].

to evaluate performance. Training of the network was run for five epochs with batch size of 64. Adam [22] was used as optimizer. The trained classifier resulted in high accuracy averaging 99.6% over the testing sets. This is similar to the accuracy obtained in [54]. Figure 3 displays the confusion matrix. It can be seen that only LSGAN and DRAGAN have a slight error due to the large noise in a few of the sythesized images.

5 One-Class Classification Model

The successful source generator identification results using the multi-class classifier, imply that indeed there is a unique characteristic or signature that distinguishes the different generated face image classes. This is indicative of successfully applying one-class classification for source generator identification.

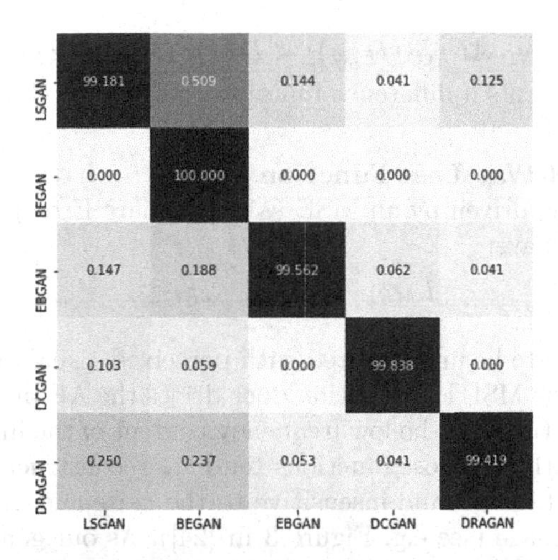

Fig. 3. Multi-Class source generator classification confusion matrix.

Using a set of One-class classifier models [11] rather than a single multi-class classifier, is advantageous in that new image generators which are constantly being developed and improved, can be incorporated into the source identification scheme without the need for retraining as in the multi-class models. Additionally, the one-class classification models train on positive samples only which in our case are well defined by the face image generator, and do not require any negative samples which, in our case, are difficult to obtain, again, due to the continuous introduction of new image generators.

We base our one-class classifiers and source generator identification, on training Convolution Auto-Encoders [7]. For a given face image generator FG we train an Auto-Encoder (AE) [7] to determine whether a given input image was generated by FG. The AE is trained on a large set of images generated by FG (positive training samples).

Typically, input to an AE is encoded into a compressed representation termed *latent representation* through an encoder. The compressed representation is then decoded back into an output. The AE is trained to produce an output that is as similar as possible to the input image. For source generator identification, we train each AE to output a reproduction $\hat{I}_{FG}$ of the input image I_{FG} which was generated by FG. Our goal in training the AE is to capture the unique characteristics of the face generator FG as expressed in the FG generated images. We show that after training, the AE produces a reconstructed image that strongly differs from the original when the latter was generated by a different face generator.

Formally, let AE_{FG} be the AE trained on the face images generated by FG. Define the output image of AE_{FG} as:

$$\hat{I} = AE_{FG}(I)$$

Following training, we show that if I_{FG1}, I_{FG2} are images generated by different generators $FG1$ and $FG2$, respectively, then:

$$DIFF(I_{FG1}, AE_{FG1}(I_{FG1})) < DIFF(I_{FG2}, AE_{FG1}(I_{FG2})) \quad (1)$$

where $DIFF$ represents a difference function that will be described below.

AE Model and 3-Way Loss Function

AE training is often driven by an MSE (Mean Square Error) loss function such that for AE_{FG} we have:

$$L_{MSE} = \|I_{FG} - \hat{I}_{FG}\|_2^2 \quad (2)$$

Although shown to be uncorrelated with perceived visual differences between images [8,55,55], the MSE loss function does drives the AE in the reconstruction of the general structure and the low frequency content of the image. We find that training the AE with MSE loss functions tends to reconstruct a slightly blurred version of the input image, and insensitive to the noise and high frequency content of the input image (see e.g. Figure 3 in [24]). As our goal is to specifically detect the latter, as it is assumed to contain the distinguishing factors between different FGs, we augment the loss function with a perceptual loss [56] in the form of a structural similarity based loss component.

The Structural Similarity Index (SSIM) [50,51] is a full reference measure of image quality. It is used in evaluating the difference between a corrupted image and a reference (original) image thus often used to evaluate effects of image compression, transmission, denoising etc. The SSIM as a component of our AE loss function, drives the network to reproduce the noise and high frequency patterns in the image, thus capturing the unique characteristics of the specific FG associated with AE_{FG}.

SSIM is defined between two images x,y, and is a function of the luminance, contrast and correlation values between corresponding regions of x and y. SSIM is given by:

$$SSIM(x,y) = \frac{(2\mu_x\mu_y + C1)(2\sigma_{xy} + C2)}{(\mu_x^2 + \mu_y^2 + C1)(\sigma_x^2 + \sigma_y^2 + C2)}$$

where μ_x, σ_x and μ_y, σ_y are the mean and std of x and y respectively, and σ_{xy} is the covariance of x and y. $C1$ and $C2$ are constants set to avoid instabilities. SSIM values are in $(0, 1)$ with larger values indicating similarity. For further details on SSIM see [50,51]. Thus, we incorporate the SSIM as a component of the AE_{FG} loss function:

$$L_{SSIM} = 1 - SSIM(I_{FG}, \hat{I}_{FG}) \quad (3)$$

Finally, in order to drive the convergence of the AE network we incorporate an adversarial loss into the AE loss function. This approach has been taken in various networks [5,24] The discriminator network D is optimized in an alternating manner along with the AE to solve the adversarial min-max problem. Thus forAE_{FG} we want to solve:

$$\min_{AE_{FG}} \max_{D} E_{I_{FG}}[log(D(I_{FG}))] + E_{\hat{I}_{FG}}[log(1 - D(\hat{I}_{FG}))] \quad (4)$$

where as defined above $\hat{I}_{FG}$ is the AE_{FG} reconstructed image.

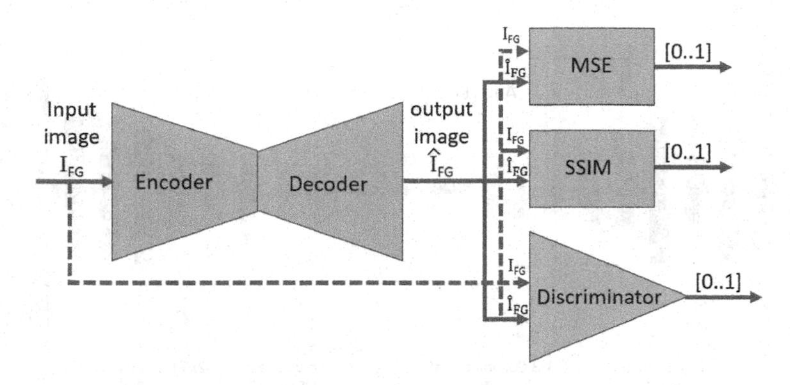

Fig. 4. One-class classification model: The Auto-Encoder is trained to reconstruct the generated face image given as input. Training is based on a 3-way loss function: an MSE loss, an SSIM loss and an adversarial loss given by a discriminator network.

The adversarial loss for training the AE is:

$$L_{ADV} = -log(1 - D(\hat{I}_{FG}))$$

Thus for training we use a loss function composed of 3 weighted components:

$$L_{AE} = w_1 L_{MSE} + w_2 L_{SSIM} + w_3 L_{ADV}$$

Following [24] we set $w_1 = w_2 = 1$ and $w_3 = 10^{-3}$ for avoiding mode collapse of the AE.

A schematic diagram of the one-class classifier network is shown in Fig. 4, the architecture of the auto-encoder network is detailed in Fig. 5. We used a symmetric convolutional auto encoder, where the encoder architecture is similar to the decoder architecture. As shown in Fig. 5, the AE has 22 layers with filters of size 2×2 and 3×3 throughout with Relu nonlinear functions. The discrimnator architecture is given in Fig. 6.

AE Training
We trained five AEs for the five face image generators: BEGAN [3], DCGAN [40], EBGAN [57], LSGAN [30] and DRAGAN [23]. Each AE was trained on images generated by the specific GAN generator. Training was run in batches of 64 for 20 epochs. Adam [22] was used as optimizer, with 0.0001 learning rate. The Discriminator trained in parallel with the AE, with learning rate of 0.0001. 5-fold cross validation was used to evaluate performance.

Results
Following training, each of the five AE_{FG_i} served as a classifier that evaluates whether any given generated face image I_{FG_j} was generated by FG. To do so, every input image I_{FG_j} was compared with its reproduction $\hat{I}_{FG_j}$ produced by

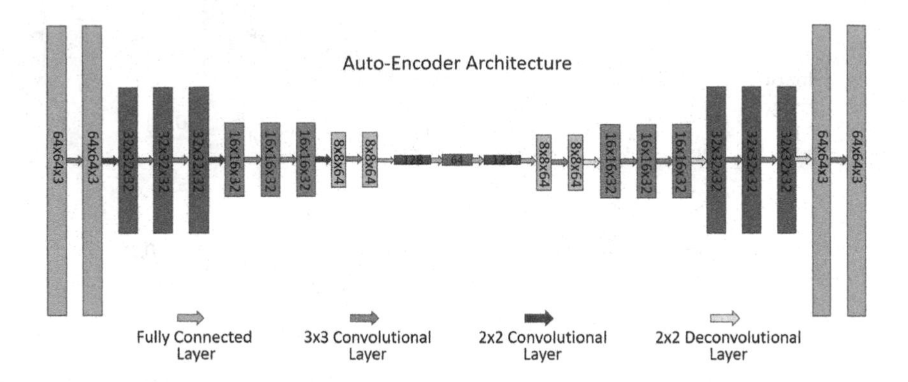

Fig. 5. One-class classification model architecture.

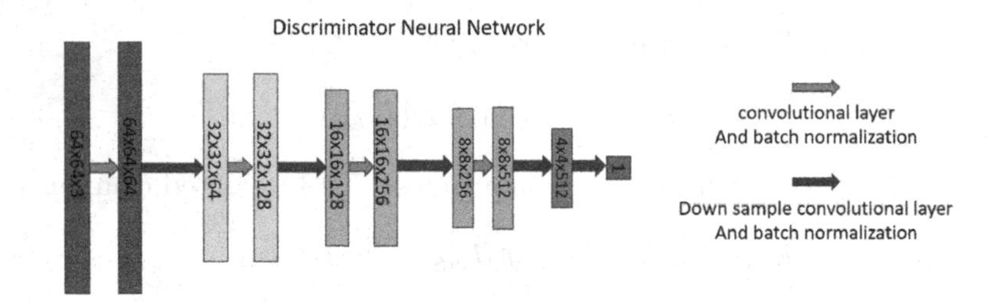

Fig. 6. Discriminator architecture.

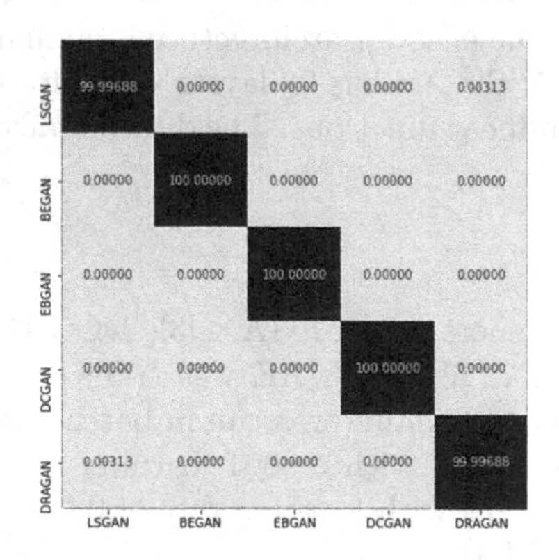

Fig. 7. One-Class classifier results - Confusion matrix

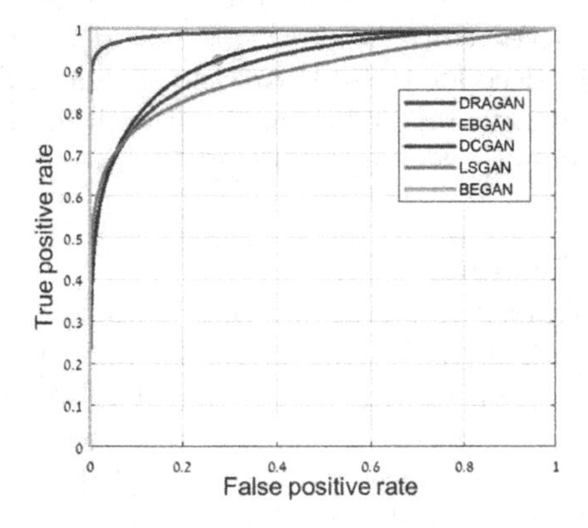

Fig. 8. ROC curves for one class classification, per each classifier. Positive test samples are from one image generator, negative test samples are from all other generators.

AE_{FG_i}. The reproduced output should be very similar for images generated by FG_i and dissimilar for images generated by any other generator FG_j as was expressed in Eq. 1. L_{SSIM} was used as the measure of difference since, as described above, the distinction between generated face images lies in the noise or high frequency content of the image and is captured by SSIM. Using argmax over the five SSIM values, the image is classified to its source generator. Using this measure, the five one-class classifiers show almost 100% accuracy on the testing set of 160K. Figure 7 shows the confusion matrix. Thus, the results confirm Eq. 1:

$$L_{SSIM}(I_{FG_i}, AE_{FG_i}(I_{FG_i})) < L_{SSIM}(I_{FG_j}, AE_{FG_i}(I_{FG_j})) \quad \text{for } FG_i \neq FG_j$$

We further consider each AE_{FG} as an independent source classifier. Thus we define a threshold value on the L_{SSIM} value that will allow a binary classification of an input image as originating from the source generator FG. To do so we consider the ROC curve associated with each classifier as created using only the positive samples. The ROC curves are shown in Fig. 8. Selecting a threshold that achieves an 80% true positive rate, will produce between 0% and 20% false positives. This was validated using 5-fold cross-validation.

6 Image Generator Profile Space

As described above, one-class networks can be trained on any new face image generator, if sufficient examples of generated images are provided. In this section we consider cases when such a large set is not available and AE training can not be performed successfully. We show that some information can still be deduced, e.g. determining whether a set of face images were all created by the same source generator.

To do so, we consider each one-class classifier AE_{FG}, described above as performing projection of the input image into the subspace of FG generated images $\{I_{FG}\}$. Given any image I, its distance to $\{I_{FG}\}$ can be an indication of similarity of the artifacts in I to those produced by the generator FG. Thus given N such classifiers AE_{FG_i}, we have an N-dimensional source generator space in which we can represent any image I, using an N-vector of distances:

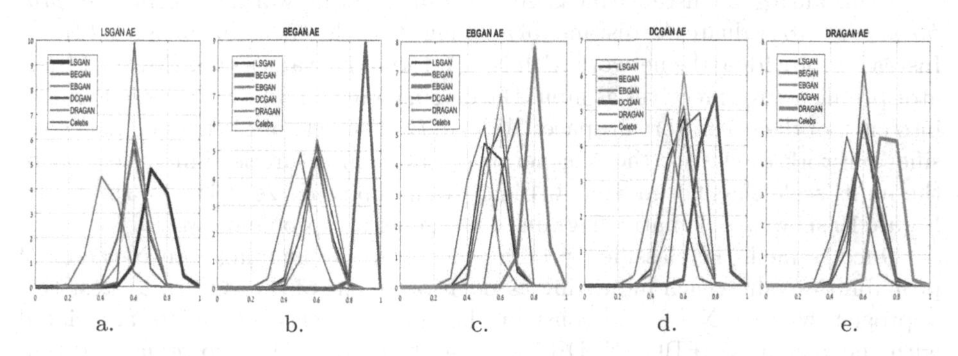

Fig. 9. Distribution of SSIM values produced with the AE_{FG} trained for: .a. LSGAN, b. BEGAN, c. EBGAN, d. DCGAN, e. DARGAN.

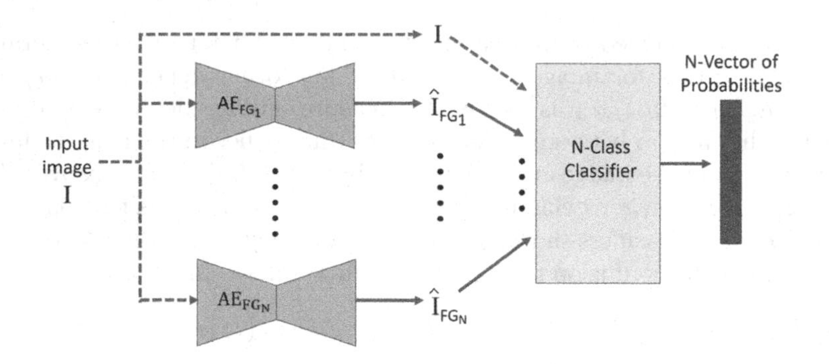

Fig. 10. Generator profile network.

$$DIST(I, AE_{FG_i}(I)) \quad \text{for } i = 1 \ldots N \tag{5}$$

This vector forms a *generator profile* for the generated image I within an N-dimensional generator profile space. We can exploit this profile vector to provide insight into the generator of I.

A natural choice for the distance function DIST would be the SSIM value produced by AE_{FG} on the input I. However this is not effective as a distance feature since the one-class classifier, being trained on positive samples only, distinguishes the positive from negative samples, but does *not* distinguish between different negative samples and thus assigns similar SSIM values to all negative samples. This can be seen in Fig. 9 which displays the distribution of SSIM values outputted for each of the AE_{FG} classifiers. The plots show 5 (color coded) distributions, one for each of the 5 datasets of generated images $\{I_{FG}\}$. It is clearly seen that for every AE_{FG_i}, the SSIM distribution value for the associated dataset I_{FG_i} (marked in bold line) is distinct of and of higher value than the 4 other datasets. This explains the successful performance of the one-class classifiers as shown in Fig. 7. However, Fig. 9 also shows that the SSIM value is not a distinguishing feature for the negative samples of each AE_{FG} and thus does not form a good measure for distances in the generator profile space (Eq. 5).

To obtain a good distance measure, we train a new network, the *generator profile network* to evaluate the distance of an image to each generator subspace $\{I_{FG}\}$. Instead of providing the network with N uninformative values of SSIM, the generator profile network receives as input the original image I and N projected images into each of the generator subspaces $AE_{FG_i}(I)$. The network is trained as a classifier network to classify the N generators, its architecture is similar to that of the multi-scale classifier in Sect. 4 (Fig. 1) with input of size (64×64×3)x(N+1). Figure 10 shows a schematic diagram of the generator profile network.

Once trained, the classifier, produces for every input image, an N-vector of probabilities values each indicating association to one of the AE_{FG}. To test this approach, we take N = 3 and consider the three one-class classifiers associated with the generators: EBGAN, DRAGAN and LSGAN. The two remaining generators: BEGAN and DCGAN serve for testing. 1000 samples were selected from each dataset $\{I_{FG}\}$ and their generator profile was produced by the generator profile network. Figure 11 displays the 3D generator profile space in which the resulting profile vectors are plotted as color coded points. It can be seen that the images from EBGAN, DRAGAN and LSGAN (blue, red yellow respectively) are clustered at high values on their corresponding axis (implying the network correctly assigned high probability to their correct generator class). We find that for the two datasets of BEGAN and DCGAN, the points are clustered in distinct regions of the profile space. This indicates that there is a basis for evaluating these two sets of data (and pairs of images within) as arising from different source generators. This is further quantified by using an SVM classifier to distinguish between the two sets, with success rate of 94.5%.

We further study the characteristics of the GAN generators by projecting images generated by STYLEGAN [18,19] as well as real images from the CelebA data set [29] into the profile space. There is a large overlap between the clusters, thus for visualization, Fig. 12 shows the center of mass of the clusters in profile space. It can be seen that in the profile space defined by EBGAN, DRAGAN and LSGAN, the center of the STYLEGAN cluster is closest to the cluster of real images. To further test this insight we created additional profile spaces based on different subsets of GANS. The distance between GAN cluster centers and real cluster center averaged across all profile spaces is given in the following table:

STYLEGAN	BEGAN	DCGAN	LSGAN	DRGAN	EBGAN
2.2384	4.0607	16.7950	10.2197	19.7449	19.0501

STYLEGAN is consistently closer to the cluster of real images, than other GAN clusters, conforming to the high quality of STYLEGAN images compared to the other tested GANs.

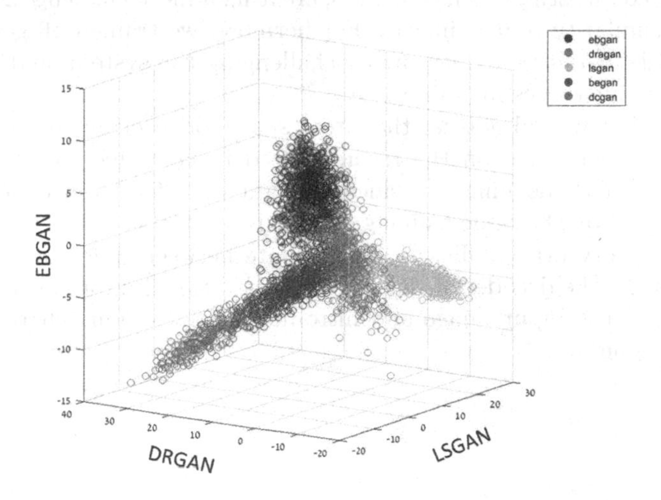

Fig. 11. GAN image clusters in generator profile space.

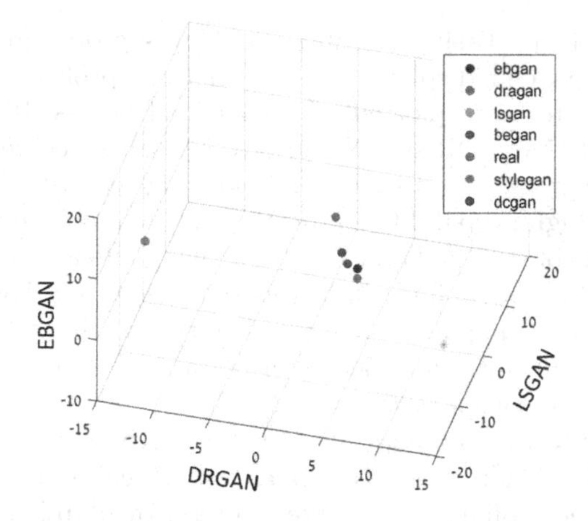

Fig. 12. GAN image cluster means in generator profile space.

7 Conclusion

In this study we dealt with source identification of "fake" images, ie determining the source generator that synthesized the image. Using multi-class classification we showed that indeed there is sufficient information, characteristic of each class of generated images, to achieve successful source identification. We then used one-class classifiers to determine per generator whether an input image originated from that generator. Finally we built a multi-class classifier on the one-class classifier outputs to create an Image Generator Profile Space in which generated images can be evaluated. We showed that the projected space of images of different GANS can be distinguished and showed that images generated by StoA generators are closer to real images in the Generator Profile Space.

We focused our study on face images, creating a more challenging set of data due to the similarity of face images. Furthermore, we trained all generators on the same CelebA dataset, thus further challenging the system, and eliminating biases due to differences in training sets.

We note that we did not use the latent space for discrimination source generators but rather relied on the reconstructed images, due to two factors: 1) The data was a challenging set, since all data were face images and all were generated based on the same training set. Thus our tests showed that the latent space showed very little ability to discriminate between different source generated images. 2) The decoder of the AE stores additional information in the way it reconstructs the input image and introduces those unique characteristics of each source generator.

References

1. Al-Sanjary, O.I., Sulong, G.: Detection of video forgery: a review of literature. J. Theor. Appl. Inf. Technol. **74**(2) (2015)
2. Arjovsky, M., Chintala, S., Bottou, L.: Wasserstein Gan. arXiv:1701.07875 (2017)
3. Berthelot, D., Schumm, T., Metz, L.: BEGAN: boundary equilibrium generative adversarial networks. arXiv:1703.10717 (2017)
4. Choi, Y., Choi, M., Kim, M., Ha, J.W., Kim, S., Choo, J.: Stargan: unified generative adversarial networks for multi-domain image-to-image translation. In: IEEE Conference on Computer Vision and Pattern Recognition, pp. 8789–8797 (2018)
5. Dou, Q., Ouyang, C., Chen, C., Chen, H., Heng, P.: Unsupervised cross-modality domain adaptation of convnets for biomedical image segmentations with adversarial loss. CoRR abs/1804.10916 (2018). http://arxiv.org/abs/1804.10916
6. Farid, H.: A survey of image forgery detection. IEEE Signal Process. Mag. **26**(2), 16–25 (2009)
7. Ghasedi Dizaji, K., Herandi, A., Deng, C., Cai, W., Huang, H.: Deep clustering via joint convolutional autoencoder embedding and relative entropy minimization. In: Proceedings of the IEEE International Conference on Computer Vision, pp. 5736–5745 (2017)
8. Girod, B.: What's wrong with mean-squared error? In: Digital Images and Human Vision, pp. 207–220 (1993)
9. Goodfellow, I., et al.: Generative adversarial nets. In: Advances in Neural Information Processing Systems (NIPS), pp. 2672–2680 (2014)
10. He, K., Zhang, X., Ren, S., Sun, J.: Deep residual learning for image recognition. In: IEEE Conference on Computer Vision and Pattern Recognition, pp. 770–778 (2016)
11. Hempstalk, K., Frank, E.: Discriminating against new classes: one-class versus multi-class classification. In: Australasian Joint Conference on Artificial Intelligence, pp. 325–336 (2008)
12. Ho, A.T.S., Li, S.: Handbook of Digital Forensics of Multimedia Data and Devices. Wiley-IEEE Press (2015)
13. Hsu, C.C., Zhuang, Y.X., Lee, C.Y.: Deep fake image detection based on pairwise learning. Appl. Sci. **10**(1), 370 (2020)
14. Iizuka, S., Simo-Serra, E., Ishikawa, H.: Globally and locally consistent image completion. ACM Trans. Graph. (ToG) **36**(4), 1–14 (2017)
15. Isola, P., Zhu, J.Y., Zhou, T., Efros, A.A.: Image-to-image translation with conditional adversarial networks. In: IEEE Conference on Computer Vision and Pattern Recognition, pp. 1125–1134 (2017)
16. Jain, A., Singh, R., Vatsa, M.: On detecting GANs and retouching based synthetic alterations. In: IEEE International Conference on Biometrics Theory, Applications and Systems (BTAS), pp. 1–7 (2018)
17. Karras, T., Aila, T., Laine, S., Lehtinen, J.: Progressive growing of gans for improved quality, stability, and variation. arXiv:1710.10196 (2017)
18. Karras, T., Laine, S., Aila, T.: A style-based generator architecture for generative adversarial networks. In: IEEE Conference on Computer Vision and Pattern Recognition, pp. 4401–4410 (2019)
19. Karras, T., Laine, S., Aittala, M., Hellsten, J., Lehtinen, J., Aila, T.: Analyzing and improving the image quality of style gan. In: Proceedings of the IEEE/CVF Conference on Computer Vision and Pattern Recognition, pp. 8110–8119 (2020)

20. Khanna, N., et al.: A survey of forensic characterization methods for physical devices. Dig. Invest. **3**, 17–28 (2006)
21. Kim, T., Cha, M., Kim, H., Lee, J.K., Kim, J.: Learning to discover cross-domain relations with generative adversarial networks. In: International Conference on Machine Learning, vol. 70, pp. 1857–1865 (2017)
22. Kingma, D.P., Ba, J.: Adam: A method for stochastic optimization. arXiv preprint arXiv:1412.6980 (2014)
23. Kodali, N., Abernethy, J., Hays, J., Kira, Z.: On convergence and stability of gans. arXiv:1705.07215 (2017)
24. Ledig, C., et al.: Photo-realistic single image super-resolution using a generative adversarial network. In: Proceedings of the IEEE Conference on Computer Vision and Pattern Recognition, pp. 4681–4690 (2017)
25. Li, H., Chen, H., Li, B., Tan, S.: Can forensic detectors identify GAN generated images? In: Asia-Pacific Signal and Information Processing Association Annual Summit and Conference (APSIPA ASC), pp. 722–727 (2018)
26. Li, H., Li, B., Tan, S., Huang, J.: Detection of deep network generated images using disparities in color components. arXiv:1808.07276 (2018)
27. Li, Y., Liu, S., Yang, J., Yang, M.H.: Generative face completion. In: IEEE Conference on Computer Vision and Pattern Recognition, pp. 3911–3919 (2017)
28. Liu, M.Y., Tuzel, O.: Coupled generative adversarial networks. In: Advances in Neural Information Processing Systems, pp. 469–477 (2016)
29. Liu, Z., Luo, P., Wang, X., Tang, X.: Deep learning face attributes in the wild. In: Proceedings of International Conference on Computer Vision (ICCV) (2015)
30. Mao, X., Li, Q., Xie, H., Lau, R.Y., Wang, Z., Paul Smolley, S.: Least squares generative adversarial networks. In: IEEE International Conference on Computer Vision, pp. 2794–2802 (2017)
31. Marra, F., Gragnaniello, D., Cozzolino, D., Verdoliva, L.: Detection of gan-generated fake images over social networks. In: IEEE Conference on Multimedia Information Processing and Retrieval (MIPR), pp. 384–389 (2018)
32. Marra, F., Gragnaniello, D., Verdoliva, L., Poggi, G.: Do gans leave artificial fingerprints? In: IEEE Conference on Multimedia Information Processing and Retrieval (MIPR), pp. 506–511 (2019)
33. McCloskey, S., Albright, M.: Detecting GAN-generated imagery using color cues. arXiv:1812.08247 (2018)
34. Mirza, M., Osindero, S.: Conditional generative adversarial nets. arXiv:1411.1784 (2014)
35. Mo, H., Chen, B., Luo, W.: Fake faces identification via convolutional neural network. In: ACM Workshop on Information Hiding and Multimedia Security, pp. 43–47 (2018)
36. Nataraj, L., et al.: Detecting GAN generated fake images using co-occurrence matrices. Electron. Imag. **2019**(5), 1–532 (2019)
37. Perarnau, G., Van De Weijer, J., Raducanu, B., Álvarez, J.M.: Invertible conditional GANs for image editing. arXiv:1611.06355 (2016)
38. Piva, A.: An overview on image forensics. ISRN Signal Process. **2013**, 496–701 (2013)
39. Privman-Horesh, N., Haider, A., Hel-Or, H.: Forgery detection in 3D-sensor images. In: IEEE Conference on Computer Vision and Pattern Recognition Workshops, pp. 1561–1569 (2018)
40. Radford, A., Metz, L., Chintala, S.: Unsupervised representation learning with deep convolutional generative adversarial networks (2015)

41. Salimans, T., Goodfellow, I., Zaremba, W., Cheung, V., Radford, A., Chen, X.: Improved techniques for training gans. In: Advances in Neural Information Processing Systems, pp. 2234–2242 (2016)
42. Sawant, R., Sabnis, M.: A review of video forgery and its detection. J. Comput. Eng. (IOSR-JCE) **20**, 1–4 (2018)
43. Sencar, H.T., Memon, N.: Digital Image Forensics - There is More to a Picture than Meets the Eye. Springer, New York (2012). https://doi.org/10.1007/978-1-4614-0757-7
44. Shih, F.Y.: Digital Watermarking and Steganography: Fundamentals and Techniques. CRC Press, Boca Raton (2017)
45. Taigman, Y., Polyak, A., Wolf, L.: Unsupervised cross-domain image generation. arXiv:1611.02200 (2016)
46. Tariq, S., Lee, S., Kim, H., Shin, Y., Woo, S.S.: Detecting both machine and human created fake face images in the wild. In: Proceedings of the 2nd International Workshop on Multimedia Privacy and Security, pp. 81–87 (2018)
47. Tolosana, R., Vera-Rodriguez, R., Fierrez, J., Morales, A., Ortega-Garcia, J.: Deepfakes and beyond: A survey of face manipulation and fake detection. arXiv preprint arXiv:2001.00179 (2020)
48. Valle, R., Cai, W., Doshi, A.: Tequilagan: How to easily identify GAN samples. arXiv:1807.04919 (2018)
49. Wang, T.C., Liu, M.Y., Zhu, J.Y., Tao, A., Kautz, J., Catanzaro, B.: High-resolution image synthesis and semantic manipulation with conditional GANs. In: IEEE Conference on Computer Vision and Pattern Recognition (2018)
50. Wang, Z., Bovik, A.C.: A universal image quality index. IEEE Signal Process. Lett. **9**(3), 81–84 (2002)
51. Wang, Z., Bovik, A.C., Sheikh, H.R., Simoncelli, E.P.: Image quality assessment: from error visibility to structural similarity. IEEE Trans. Image Process. **13**(4), 600–612 (2004)
52. Yang, P., Baracchi, D., Ni, R., Zhao, Y., Argenti, F., Piva, A.: A survey of deep learning-based source image forensics. J. Imag. **6**(3), 9 (2020)
53. Yi, Z., Zhang, H., Tan, P., Gong, M.: Dualgan: unsupervised dual learning for image-to-image translation. In: IEEE International Conference on Computer Vision, pp. 2849–2857 (2017)
54. Yu, N., Davis, L.S., Fritz, M.: Attributing fake images to gans: learning and analyzing gan fingerprints. In: Proceedings of the IEEE International Conference on Computer Vision, pp. 7556–7566 (2019)
55. Zhang, L., Zhang, L., Mou, X., Zhang, D.: A comprehensive evaluation of full reference image quality assessment algorithms. In: IEEE International Conference on Image Processing, pp. 1477–1480 (2012)
56. Zhao, H., Gallo, O., Frosio, I., Kautz, J.: Loss functions for image restoration with neural networks. IEEE Trans. Comput. Imag. **3**(1), 47–57 (2016)
57. Zhao, J., Mathieu, M., LeCun, Y.: Energy-based generative adversarial network. arXiv:1609.03126 (2016)
58. Zhu, J.Y., Park, T., Isola, P., Efros, A.A.: Unpaired image-to-image translation using cycle-consistent adversarial networks. In: Proceedings of the IEEE International Conference on Computer Vision, pp. 2223–2232 (2017)

Spatio-Temporal Handwriting Imitation

Martin Mayr[(✉)], Martin Stumpf, Anguelos Nicolaou, Mathias Seuret, Andreas Maier, and Vincent Christlein

Pattern Recognition Lab, Friedrich–Alexander-Universität Erlangen–Nürnberg, Erlangen, Germany
{martin.mayr,martin.stumpf,anguelos.nicolaou, mathias.seuret,andreas.maier,vincent.christlein}@fau.de
https://lme.tf.fau.de/

Abstract. Most people think that their handwriting is unique and cannot be imitated by machines, especially not using completely new content. Current cursive handwriting synthesis is visually limited or needs user interaction. We show that subdividing the process into smaller subtasks makes it possible to imitate someone's handwriting with a high chance to be visually indistinguishable for humans. Therefore, a given handwritten sample will be used as the target style. This sample is transferred to an online sequence. Then, a method for online handwriting synthesis is used to produce a new realistic-looking text primed with the online input sequence. This new text is rendered and style-adapted to the input pen. We show the effectiveness of the pipeline by generating in- and out-of-vocabulary handwritten samples that are validated in a comprehensive user study. Additionally, we show that also a typical writer identification system can partially be fooled by the created fake handwritings.

Keywords: Offline handwriting generation · Style transfer · Forgery · Handwriting synthesis

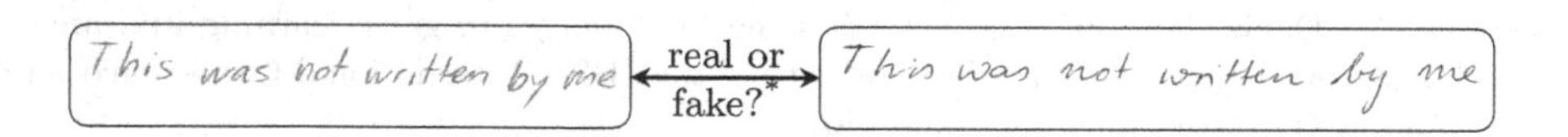

1 Introduction

Handwriting is still a substantial part of communication, note making, and authentication. Generating text in your handwriting without the need to actually take a pen in your hand can be beneficial, not only because we live in the age of digitization but also when the act of writing is physically impaired due to

*Both samples are synthesized.

Electronic supplementary material The online version of this chapter (https://doi.org/10.1007/978-3-030-68238-5_38) contains supplementary material, which is available to authorized users.

A. Bartoli and A. Fusiello (Eds.): ECCV 2020 Workshops, LNCS 12539, pp. 528–543, 2020.
https://doi.org/10.1007/978-3-030-68238-5_38

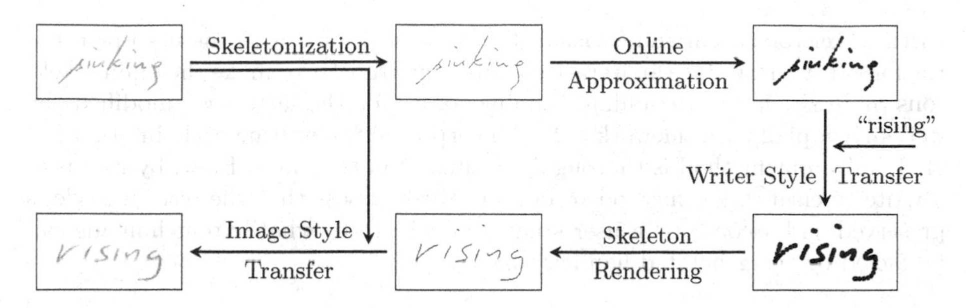

Fig. 1. Offline-to-Offline handwriting style transfer pipeline. (Color figure online)

injuries or diseases. Handwriting synthesis could also enable to send handwritten messages in a much more personal way than using standard handwriting fonts, e.g., for gift messages when sending presents. Personal handwriting could also be useful in virtual reality games, where parts could be written in a famous handwriting or in the player's own handwriting in order to identify more strongly with the avatar. Similarly, it could be used for augmented reality or in movies. Why not adapting someone's handwriting style when translating a note from one language into another one? Since the most deep learning methods need large datasets for training, another use case could be the improvement of automatic Handwritten Text Recognition (HTR). In fact, the simulation of single handwritten words is already used for data augmentation during the training of HTR methods [3].

With the help of the method we propose, it is possible to generate handwritten text in a personal handwriting style from only one *single* handwritten line of text of the person to be imitated. This method would also enable the creation of realistic-looking forgeries. To detect these forgeries, new and robust writer identification/verification methods may need to be developed. To train such systems, our method allows forensic researchers to generate and imitate handwriting without the need of skillful forgers.

In contrast to *offline* data, i.e., scanned-in handwriting, *online* handwriting stores for each data point not only its position but also the temporal information, representing the actual movement of a pen. This can for example be recorded with special pens or pads. In our pipeline, the offline data is transferred to the temporal domain, see Fig. 1 top branch. Then, we make use of an existing online handwriting synthesis method [10], which takes the temporal information and the target text as input. Finally, the resulting online data is rendered again in the style of the writer to be imitated, cf. Fig. 1 bottom branch.

The main contributions of this work are as follows: (1) creation of a full pipeline for the synthesis of artificial Western cursive handwriting, recreating both the visual and the writer-specific style; (2) development of a novel conversion method to approximate online handwriting from offline data to be able to utilize an existing online-based style transfer method; (3) the adaptation of conditional Generative Adversarial Networks (GANs) to compute a robust handwriting skeletonization and a visual style transfer to adapt to the used pen. For the former, we introduce an iterative knowledge transfer to make the offline skeletons more similar to the online training data. For the latter, we modified the well-known pix2pix framework [12] to incorporate the writing style information. (4) Finally, our method is thoroughly evaluated in two ways. First, by means of a writer identification method to quantitatively assess that the writing style is preserved and second, by a user study to evaluate to what extent humans can be fooled by the generated handwriting.

2 Related Work

Online Handwriting Synthesis. To produce convincing handwriting, it needs to reproduce the given content exactly, while keeping the style consistent, but not constant. Real human handwriting will repeat the same content almost identically, but still with some variance. This requires a solid long term memory, combined with some guidance from the required content. In the seminal work by Alex Graves [10] this is achieved by the use of a Recurrent Neural Network (RNN) with Long Short-Term Memory (LSTM) cells, which enables the network to make predictions in context of the previous pen positions. Each hidden layer uses skip connections to give the network control over the learning flow. Additionally, an early form of attention mechanism was developed that decides which part of the content the network focuses on. A problem of Graves' method [10] is that it tends to change the writing during the sequence generation. This can be overcome by the use of Variational RNNs (VRNNs) to generate consistent handwriting [6]. Another method [2] builds on the idea of predicting single pen positions. Instead of relying on the network's internal memory to store style, the goal was to explicitly extract style and content from the input. This is achieved by utilizing a Conditional VRNN (CVRNN) to split the input into two separate latent variables, representing style and content. A drawback of this method is the need to split the input strokes into words and characters in form of begin- and end-of-character (EOC) tokens, which typically cannot be automatically determined from offline handwriting. In contrast to Graves' attention mechanism, it uses the generated EOC tokens to switch letters. This could remove some of the predictive capability of the network, since it can only foresee which letter follows when the next letter is already about to be written. In a subsequent work [1], the CVRNN was replaced by Stochastic Temporal CNNs (STCNNs) showing more consistent handwriting generation.

Offline Handwriting Synthesis. There are approaches that use printed text style synthesis for augmentation [9] or text stylization for artistic typography creation [24]. Similarly, cursive handwriting synthesis [3] can be generated using a GAN combined with an auxiliary text recognition network. While the augmentation of HTR improves the recognition accuracy, the generated words cannot imitate a specific handwriting. Another approach [5] synthesis handwriting from public fonts by finding the best character matches in public handwritten fonts. The results are convincing but still far away from the actual user's style.

The closest work to ours is the work by Haines *et al.* [11]. To the best of our knowledge, this is the only other method using offline cursive handwritings as input. They produce convincing output by creating glyph models from segmented labeled ligatures and individual glyphs. The method has two main short-comings. First, the selection of the glyphs and ligatures involves human assistance. Second, only letters present in the handwriting to be imitated can be reproduced.

Another work [16] converted offline data to another modality and then rendered and adapted the style. Therefore, glyphs need to be segmented and matched to characters. From these matched glyphs, strokes are extracted by registering them to a trajectory database, and sampled with regular points. In contrast, we propose a method based on maximum acceleration that uses more points for curved strokes. The user's style in [16] is learned by a feed forward neural network, which is added during the rendering process. While no human interaction is needed to generate handwritten Chinese fonts with a large amount of characters, the "characters should be written separately without touching each other in a given order and consistently in size and style." [16]. This issue is targeted by Nakamuar *et al.* [18] who generate Chinese characters for samples with an incomplete character set by choosing the closest learned character distribution.

In contrast, our method works fully automatic for handwriting without any user interaction given a single line for priming. It is able to render consistent full lines instead of single glyphs/words as in other works. The method is able to produce out-of-vocabulary (OOV) letters and words as long as the trained model has seen some instances during training, but not necessarily from the writer to be imitated.

3 Offline-to-Offline Handwriting Style Transfer Pipeline

We decided to split the offline to offline handwriting style transfer system into several subtasks. While background, pen and writer style are static problems that could be solved with Convolutional Neural Networks (CNNs), the text content has a structural component and therefore makes some temporal generation, e.g., in form of an RNN preferable. Each of the subtasks is trainable on its own, allowing human prior knowledge to guide the process, and to evaluate single steps separately. Furthermore, not all these tasks require a neural network if ther is an algorithmic solution. For a writer imitation, we need to apply a style transfer on two levels: (1) the arrangement of writing strokes and (2) the pen style, i.e., thickness and color distribution. Figure 1 gives an overview over our pipeline.[1] First, a skeleton is computed from the input sequence, which is used as the writing style to imitate (commonly several words long). The skeleton is converted to online data with a novel sampling process that puts emphasis on the curved structure of handwriting. Afterwards, we make use of the writing generation method of Graves [10], which creates new online text in a given

[1] Code and models available below https://github.com/M4rt1nM4yr/spatio-temporal_handwriting_imitation.

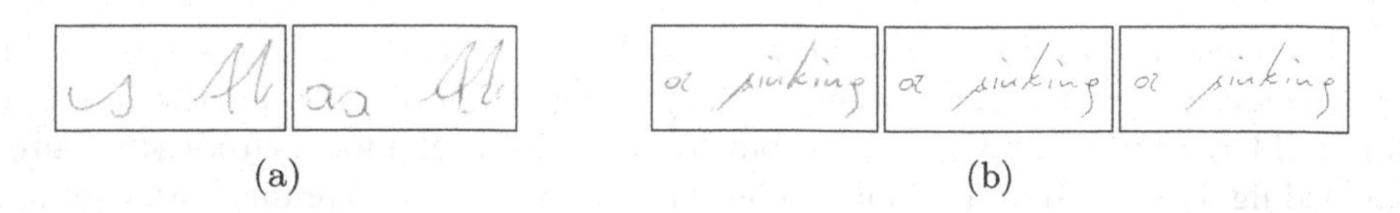

Fig. 2. (a) Shows a skeleton rendered from the IAM-Online dataset (left) and the estimated offline handwriting by the CycleGAN (right). (b) Shows an offline example excerpt (left), skeleton output of a basic method [25] (middle), and through our skeletonization network (right). (Color figure online)

writing style. This handwriting sequence is rendered as a skeleton and transferred to the visual appearance of the priming sequence using additional data created through the initial skeletonization process. Each step of the pipeline is described in the following subsections.

3.1 Skeletonization

In this stage, we convert images of real handwriting to their corresponding skeletons which are then subsequently mapped to the temporal domain and fed into the generative network. There are sophisticated learning-free [4] and Deep Learning (DL) [23]-based methods, which learn a mapping between natural objects and skeletons. We face the challenge that individual datasets exist of both real handwriting images and skeletons (online data). However, to the best of our knowledge, there is no dataset that annotates a mapping between those two.

Given an offline handwritten sample, the challenge is to produce a skeleton similar enough to the online data used in the generative network. Specifically, we use the CVL dataset [14] as the source for real offline data and the IAM-Online dataset [17] for real skeleton data, where the latter is also used for training the generative network. A natural choice would be the use of CycleGAN [26], which enforces cycle consistency, i.e., the output of a transfer from source to target and back is similar to the source again (and vice versa). A shortcoming of CycleGAN is that it has to guess the transfer function. This could lead to several problems because it is not guaranteed that the resulting mapping will still satisfy spatial consistency. It could freely add/remove strokes, as long as the result is GAN-consistent and contains enough information to perform an inverse mapping, see Fig. 2a for an example. Conversely, there are basic skeletonization algorithms [15,25]. These could be used to guide the training to incorporate prior knowledge about the mapping, cf. Fig. 2b. Therefore, we propose *iterative knowledge transfer* to extract the knowledge of one of those algorithms and transfer it to a neural network. The proposed method is not limited to our specific use case. It is rather a general method to transfer knowledge from an existing mapping function to a neural network while improving and generalizing it along the way. It requires a naive mapping function and two non-paired datasets for which we would like to achieve a mapping and consists of the following steps, as illustrated in Fig. 3.

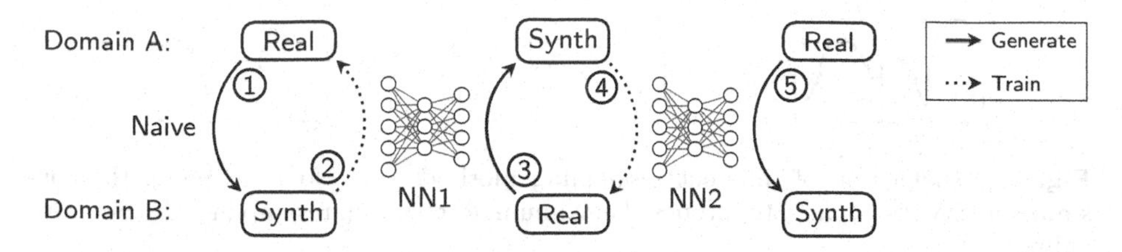

Fig. 3. Iterative knowledge transfer from naive algorithms. (1) A naive algorithm transfers real data from domain A to fake data of domain B. (2) Using this fake data, a neural network (NN1) is trained, (3) which is then used to convert real data from domain B to domain A. (4) With this new data the final NN2 is trained, which (5) can then be used for producing synthetic data. (Color figure online)

(1) Generation of synthetic skeletons from the real source dataset (CVL) using the naive mapping. This dataset is expected to be erroneous.
(2) Training of an inverse mapping (NN1) based on the synthetic dataset, enforcing generalization. The network capacity needs to be sized correctly, so that generalization happens without picking up on the errors of the naive implementation.
(3) Generation of synthetic CVL data from the real skeletons (IAM Online) using the trained mapping.
(4) This new dataset is used to train the desired mapping from offline handwriting to their corresponding skeletons (NN2).
(5) The final model is expected to produce realistic output data, because the target of the training procedures was the generation of real target data.

It is assumed that in steps 2 and 4 the synthetic data is almost similar to the real data, otherwise the training procedure would not produce reasonable results.

3.2 Online Sequence Approximation

The purpose of this stage of the pipeline is to take skeleton images as input and convert them to an approximate online representation. As skeletons do not contain temporal annotations, this step will require a heuristic approach to synthesize realistic online data analytically. This stage consists of two steps: (1) conversion of the bitmap representation to strokes, and (2) temporal resampling and ordering.

Conversion to Strokes. First, the bitmap data is converted to a graph. This is done by connecting the neighboring pixels because we assume that strokes are connected lines in the skeleton. Our target is to create strokes without cycles, but the generated graph contains pixel clusters. These pixel clusters are defined as triangles having at least one common edge. Graph cycles that do not consist of triangles are assumed to be style characteristics of the writer, like drawing

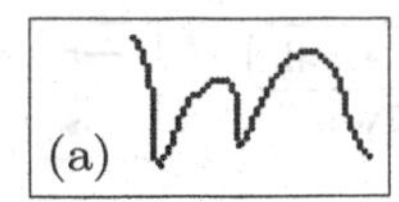

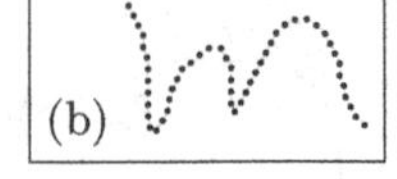

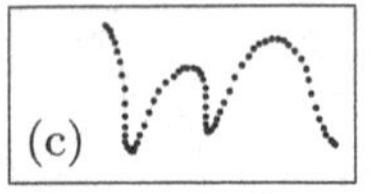

Fig. 4. Visualisation of different resampling methods. (a) No resampling, (b) constant velocity resampling, (c) proposed maximum acceleration resampling. (Color figure online)

the dot on the 'i' as a tiny circle. The cluster artifacts are solved by replacing all cluster nodes that are not connected to outside nodes with a single node at the mean position of the group. The final step for creating strokes is to remove intersections and cycles. This implies that the nodes of every line segment have either one or two neighbors. To produce a consistent behaviour the cycles are split at the upmost position, similarly to human writing.

Resampling. To obtain online handwriting, we still need to incorporate temporal information. The simplest way would be to sample at constant time steps. However, the training of the subsequent online network becomes very difficult because very small time steps are needed. We were unable to generate convincing results, presumably because the sequence is too far away from a real handwriting sequence. Therefore, we propose *maximum acceleration resampling* on the computed lines to imitate the dynamics of human writing. We constrain the velocity to be zero at both extremities of a line, increase it on straight parts, and decrease it on curved parts. This has the advantage that the network focuses on important parts of the handwriting. The difference between constant resampling and the proposed maximum acceleration resampling can be seen in Fig. 4. In detail, the algorithm consists of the following steps: (1) Resampling of the curve to sufficiently small intervals. (2) Creating a reachability graph between nodes to prevent cutting corners. (3) Analyzing acceleratability, i.e., the acceleration required between two nodes. This will push the problem into a 4D space: (x, y, v_x, v_y). (4) Searching shortest path using directed Dijkstra.

Constant Pre-sampling. The Dijkstra search [7] does not include the possibility to cut lines into pieces. It can only select a set of optimal nodes from existing nodes. Consequently, we need to make sure that the existing nodes are spaced appropriately by resampling them into small constant intervals. We empirically set the distance to 1/3 of the maximum acceleration value.

Reachability Graph. We have to avoid cutting corners, such that during curved sections the sampling rate is higher. Therefore, we create a graph, encoded in a boolean matrix of size $N \times N$ (where N is the number of nodes in the pre-sampled graph), which stores the pairwise reachability between all nodes. We define points $\boldsymbol{p}_i$ and $\boldsymbol{p}_{i+n}$ to be reachable, when $\max_{j=i..i+n} d(\boldsymbol{p}_j) < t$ with $d(\boldsymbol{p}_j)$ being the distance of the point $\boldsymbol{p}_j$ to the line between $\boldsymbol{p}_i$ and $\boldsymbol{p}_{i+n}$ and t

being a given threshold. We found the threshold parameter to be optimal at 3 times the node distance of the pre-sampling step.

Acceleratability. So far, our nodes are two dimensional: $\boldsymbol{p} = (x, y)$. We now enhance them with two more dimensions: the incoming velocities $\boldsymbol{v} = (v_x, v_y)$. The incoming velocity $\boldsymbol{v}_i$ of node $\boldsymbol{p}_i$ from node $\boldsymbol{p}_{i-n}$ is computed by $\boldsymbol{v}_i = \boldsymbol{p}_i - \boldsymbol{p}_{i-n}$. Thus, only one 4D node exists for each preceding 2D node, with multiple possible edges from different velocities of that preceding node. We now create a 4D acceleratability graph based on the 2D reachability graph that connects all 4D nodes $(\boldsymbol{p}_i, \boldsymbol{v}_i)$ and $(\boldsymbol{p}_j, \boldsymbol{v}_j)$ that fulfill the following 'acceleratability' criterion:

$$\boldsymbol{v}_j = \boldsymbol{p}_j - \boldsymbol{p}_i \ \wedge \ ||\boldsymbol{v}_j - \boldsymbol{v}_i|| < a \tag{1}$$

with a being the maximum acceleration hyperparameter. As we never go back, we have a directed graph, and thus it contains all possible pen trajectories that create the given curve.

Shortest Path Search. The set of possible paths is quite large, therefore we use a Dijkstra shortest path search with some optimizations specific to the 4D case. Since we have a directed graph, we will always start at one end of the stroke and move towards the other one. So, we can step through the curve, 2D node by 2D node, and compute all optimal paths to that node for all possible velocities at that node. The number of possible velocities is limited and is equivalent to the number of incoming edges to that node. Computing an optimal path to a given position $\boldsymbol{p}$ and velocity $\boldsymbol{v}$ can be done as follows. First, we get the position of the previous node: $\boldsymbol{p}_{prev} = \boldsymbol{p} - \boldsymbol{v}$. We now take all the possible velocities $\boldsymbol{v}_{prev}$ at position $\boldsymbol{p}_{prev}$ that can reach $\boldsymbol{p}$, based on the acceleratability criterion, cf. Eq. (1). The shortest path l to $(\boldsymbol{p}, \boldsymbol{v})$ is then:

$$l(\boldsymbol{p}, \boldsymbol{v}) = \min_{\boldsymbol{v}_{prev}} l(\boldsymbol{p}_{prev}, \boldsymbol{v}_{prev}) + 1 \tag{2}$$

We start the entire algorithm at one end of the curve, which we define as the starting point. Further, we set $\boldsymbol{p} = \boldsymbol{p}_{start}$, for which the only valid velocity is $\boldsymbol{v}_{start} = 0$ and $l(\boldsymbol{p}_{start}, \boldsymbol{v}_{start}) = 0$. We then iterate through the entire curve until we reach $\boldsymbol{p}_{end}$. As we defined both start and end velocities to be zero, we then look at $l(\boldsymbol{p}_{end}, \boldsymbol{v}_{end})$ with $\boldsymbol{v}_{end} = 0$ and backtrack to get the optimal path.

Ordering. Finally, the points are ordered by computing the mean of every stroke and ordering them from left to right to sort the list of strokes. In real human writing there are cases where this is not true, but in this way we produce a consistent behaviour, which is necessary for the further stages in the pipeline.

3.3 Writer Style Transfer

The produced online sequence is used to prime an online writing synthesis algorithm. We employ Graves' algorithm [10] who showed that LSTM cells are capable of generating complex structures with long-term contextual dependencies. To

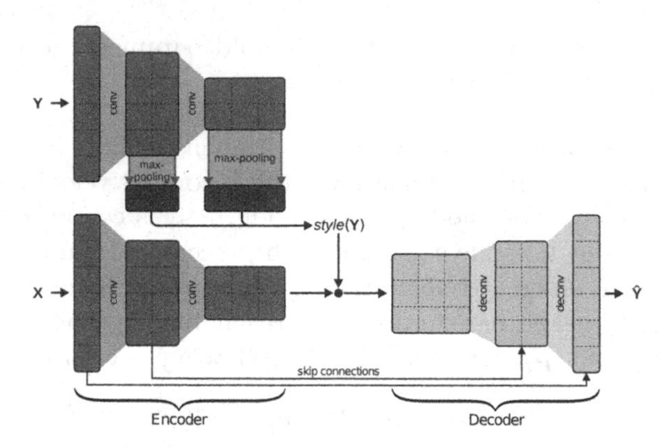

Fig. 5. Modified pix2pix generator network for conditional style transfer. The green network depicts the encoder of the pix2pix generator network. The max-pooling extracts the style information of all activation maps of style image Y. The style information gets concatenated with the output of the encoder for newly rendered image X used in the decoder to generate the output $\hat{Y}$. (Color figure online)

be able to predict both the content and the style at the same time, the content of the text is added to the network as a side input to one of the intermediate layers. The network does not see the entire content sequence at once, instead, a form of attention is used. To achieve this, the mixture density output from intermediate layers decide which part of the content gets delivered to the network. Note that we are not forced to use Gaves' method but can use any online handwriting generation approach.

3.4 Image Style Transfer

The produced new sequence of online data is transferred back to offline handwriting by means of drawing lines between the points of the online data. The last stage of the pipeline is to produce realistic offline handwriting by imitating the ink and style of the input image given the new skeleton of the produced online handwriting. Therefore, we modify the pix2pix [12] architecture to output the correct style.

The network consists of an encoder and a decoder network. The first step in creating a style transfer network is to extract the style information from our input image Y. We use the encoder part of the pix2pix network with the style image as input. It is important that the style only contains global information, otherwise the discriminator D could start to discriminate by content instead. Hence, we take the max-pooled outputs of all activation maps as style information. Then, we feed the style information into the pix2pix generator network by concatenating it with the innermost layer of the network, as seen in Fig. 5. To keep the size-agnostic property of the network, we repeat the style information along the two spatial axes to match the size of the innermost layer.

To include the style extraction network in the training process, we add the output of the style extraction network to the input of D. Incorporating this into the objective of pix2pix [12] results in

$$\begin{aligned}\mathcal{L}_{\mathrm{cGAN}(G,D)} &= \mathbb{E}_{X,Y}[\log D(X, Y, \mathrm{style}(Y))] \\ &+ \mathbb{E}_{X,Y}[\log(1 - D(X, G(X, \mathrm{style}(Y)), \mathrm{style}(Y)))] \\ &+ \lambda \mathbb{E}_{X,Y}[||Y - G(X, \mathrm{style}(Y))||_1].\end{aligned} \quad (3)$$

Feeding the style information to D has the effect that D can now differentiate between real and generated images by comparing their style output. This forces the generator G to generate images that include the style taken from the extraction network to produce a style consistent output. We encounter that the model sometimes produces wrong pen colors while the stroke width and all other characteristics are transferred correctly. Therefore, we find the most important principal component of the style image and transfer it to the output image.

4 Evaluation

We conduct a large-scale user study to assess how well a human could be fooled by our method. Furthermore, we utilize writer identification methods to have a non-biased measurement of how well the results are imitating the original writer.

4.1 Datasets

The following datasets are used to train the models, perform the experiments and user studies.

IAM On-Line Handwriting Dataset. The IAM On-Line Handwriting Database [17] consists of handwritten samples recorded on a whiteboard. The dataset consists of 221 writers with 13 049 isolated and labeled text lines in total containing 86 272 word instances of a 11 059 large word dictionary.

CVL Dataset. The CVL Database [14] consists of offline handwriting data providing annotated text line and word images. We used the official training set consisting of 27 writers contributing 7 texts each.

Out-of-vocabulary Dataset. For the OOV words, we used the 2016 English and German Wikipedia corpora [8], containing both 100 000 words. Words containing less than four characters and words already part of the CVL training set are removed from these dictionaries. For the user study, the number of German words is reduced such that the ratio of English and German words is equal to the ones of the CVL dataset.

4.2 Implementation Details

For the skeletonization, we make use of the proposed iterative knowledge transfer algorithm due to the nonexistent dataset for our problem description, see

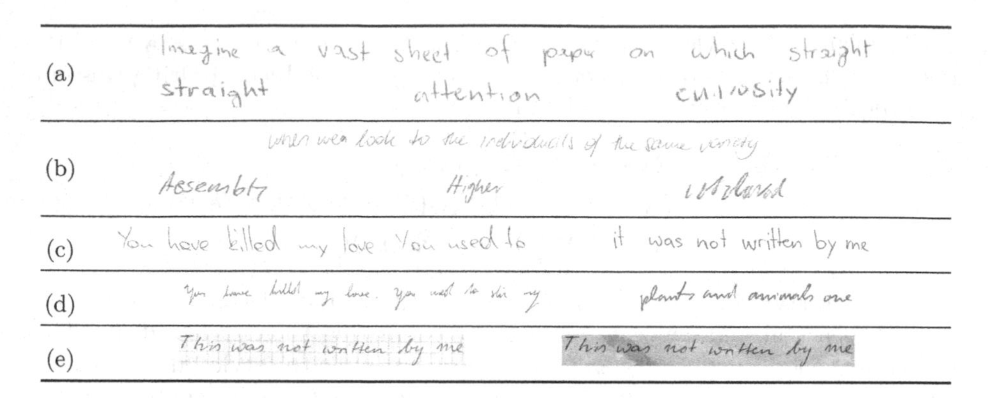

Fig. 6. (a) and (b) show three generated words each - two good results and one bad one with the given style line on top. (c) and (d) show two generated lines (right) with two style lines (left). (e) shows synthetic outputs with added backgrounds. (Color figure online)

Sect. 3.1. We use the pix2pix framework [12] as the mapping network, trained with augmentations, such as added noise, resizing, cropping and color jitter.

For the pen style transfer, we use another pix2pix model for which we employ an asymmetric version of the U-Net [21]. It is shortened to 4 layers (original 8) and reduced number of filters: [16, 32, 64, 64] for the encoder, [32, 64, 128, 128] for the style extractor, and [192, 256, 128, 64] for the decoder. The goal of this network is to synthesize a real-looking handwriting image given a skeleton, which was produced earlier through the pix2pix skeletonization, and a random sample of the CVL dataset as style input, cf. Sect. 3.4. Both models are trained with standard pix2pix [12] training parameters. We use ADAM with a learning rate of 0.0002, momentum parameters $\beta_1 = 0.5$, $\beta_2 = 0.999$, and $\lambda_{L_1} = 100$ for weighting the L_1 distance between Y and $\hat{Y}$.

The RNN model of the online handwriting style transfer is trained using the IAM On-Line Handwriting Database. The model consists of 400 LSTM cells. It is trained with RMSProp with $\beta_1 = 0.9$, a batch size of 32 and a learning rate of 0.0001 and gradient clipping in $[-10, 10]$.

4.3 Qualitative Results

Figure 6 shows four distinctive outputs. Figure 6a and 6b show three synthesized words, depicting two good results and one containing typical pipeline failures with the genuine anchor (=style) line at top. Figure 6c and 6d show two common outputs of the network with the anchor line on the left. For example, the addition of superfluous lines, incorrect splitting and positioning of lines, detail removal, such as missing t-, f- lines or i-dots, or skipped letters at the beginning of the message. We notice that the less legible a writer's handwriting is, the harder it is to imitate the writing style, see for example writer Figs. 6b and 6d. We also showcase generated lines with added backgrounds in Fig. 6e.

	Predicted		
	Real	Fake	
Real	33.9	16.1	R: 67.8
Synth	25.2	24.8	SP: 49.6
	P: 57.4	NPV: 60.6	ACC: 58.7

(a) Confusion matrix

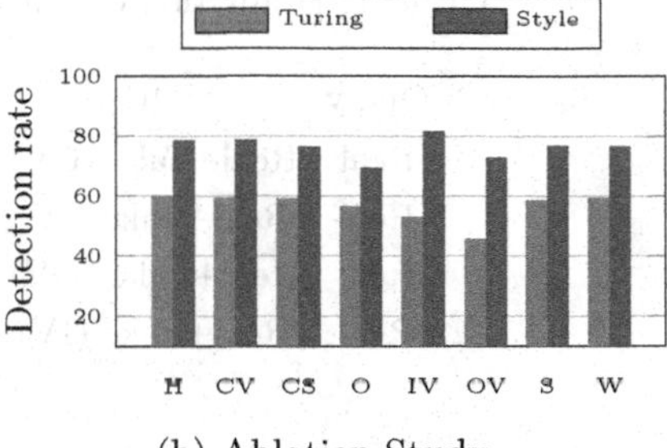

(b) Ablation Study

Fig. 7. (a) Displays the results of the Turing task. (b) shows the results of different groups (H: humanities scholars, CV: computer vision/image analysis background, CS: computer scientist (general), O: other) and different ablation studies (OV: out-of-vocabulary words, IV: in-vocabulary words, S: synthetic background, W: white background. (Color figure online)

4.4 User Study

We conducted a user study[2] to estimate the degree to which humans can distinguish real from synthesized data. Random samples from the CVL dataset serve as real data, and the synthesized data is generated with words from the CVL and OOV corpora while line images from CVL are randomly chosen as style inputs. Two sub-studies were performed. In the first one ( ), humans are shown a sample and asked if it was "written by a human" or "created by a machine". This is a kind of Turing test asking to differentiate between human and synthesized handwriting. In the second one ( ), we wanted to evaluate the ability to generate a person-specific handwriting. Therefore, we made an experimental setup consisting of a line written by a specific writer and two words. The user had to choose the word which was written by the same writer as the given line. Every subject had to answer 96 questions without time limit: 32 for the *Turing* task and 64 for the *Style* task. For each task, a background representing notebook paper was artificially introduced [20,22] in exactly 50% of the queries in order to investigate the impact of such a distraction. In total, 59 people participated in our user study with different knowledge background: *Humanities* (H), i.e., paleographers, book scientists, etc.; people working in *Computer Vision* (CV) or image processing; general *Computer Science* (CS); and *Other* (O). Most users needed between 8 and 24 min to answer the 96 questions.

The accuracy of the Turing task is 58.7%, shown in Fig. 7a. The accuracy of the Style task is 76.8%. Further, in the Turing task recall (R) is higher than specificity (SP), i.e. given a real sample the decisions of the users are better. Precision (P) and negative predictive value (NPV) are similar. Figure 7b shows that for the Turing task, the average results per user category range from 56.6% to 60.2% correctly recognized samples. Experts in the humanities are slightly

[2] https://forms.gle/MGCPk5UkxnR23FqT9.

Table 1. Evaluation of generated vs. nearest human writer.

Query	DB	Skeleton	mAP %	Acc. %
Real	Real+Fake (OV)	Naive	31.94	18.52
Real	Real+Fake (IV)	Naive	26.18	7.41
Real	Real+Fake (OV)	pix2pix	29.66	14.82
Real	Real+Fake (IV)	pix2pix	37.13	25.92

more accurate than others. There is very little difference between working in the field of computer vision or image analysis and other computer scientists (59.4% vs. 59.1%). For the Style task, the average results per user category range from 69.5% to 78.6%. This time, however, people with computer vision background reach an accuracy similar to the one of scholars in the humanities (78.8% and 78.6%, respectively). The gap to computer scientists and others is larger with a difference of 2.4%, and 9.3%, respectively. It seems that samples generated from the training set (IV) are easier to distinguish than words coming from pages of an unknown corpus (OV). This might come from the usage of a different, more complicated lexical field. There is no accuracy difference when using samples with synthetic background (S) in comparison to real, white background (W).

4.5 Writer Identification with Synthesized Offline Handwriting

We tested samples of our proposed method against an automatic learning-free writer identification approach [19]. We made a joint retrieval database having our synthetic paragraphs and a paragraph of the CVL training set from all 27 writers of the dataset, i.e., a total of 54 samples. For every query sample, we removed the real sample of the writer in question and retrieved the most similar ones from the 53 database samples. This procedure effectively benchmarks our forged samples vs. the most similar writer to the real one, which could also be considered a forgery system. We used the specific experimental pipeline to obtain a strong baseline for forgeries and in order to asses the importance our skeletonization method trained with the proposed iterative knowledge transfer vs. a naive skeletonization.

Table 1 (last row) shows that our method outperforms the nearest forgery on 25.9% of the queries when our forgery is created with text from the same corpus. When the samples are generated with out-of-corpus words, performance drops significantly (row three). It can also be seen that substituting our proposed skeletonization for the naive skeletonization (row 1 and 2) has also a large impact on system performance. As a sanity check, we tested the retrieval of real writers from real writers, naive synthetic from naive synthetic, and pix2pix synthetic from pix2pix synthetic. In all cases, we achieved 100% performance. It should be pointed out that automatic methods operate on the page/paragraph level while human users were challenged at the word level. Any comparison between

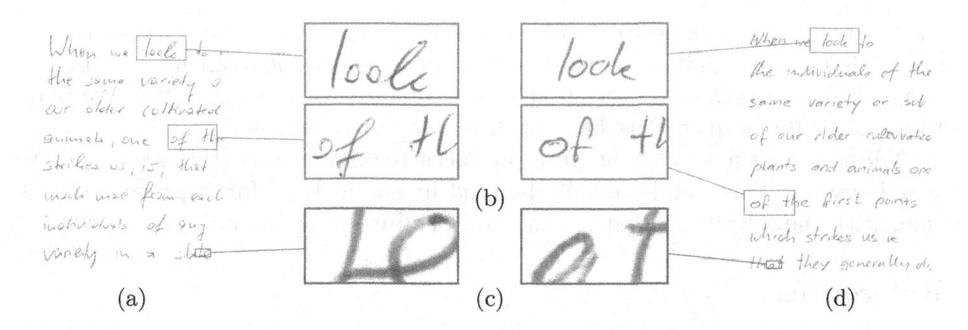

Fig. 8. Qualitative result of one whole paragraph: (a) original, (b) stroke level details, (c) pen level details, (d) synthesized. (Color figure online)

humans and automatic writer identification methods must not be implied from this experiment.

5 Discussion and Limitations

At general sight, the synthesized lines of Fig. 8d look very similar to the original writing shown in Fig. 8a. But a closer look reveals flaws in the writing style. The orange boxes in Fig. 8b show the stroke level difference. While there are only few minor differences in the word *look* (first row), there are some major style differences in the letter *f* (second row). On the micro level, i.e., the pen style, the imitation shows in general convincing results, cf. Fig. 8c.

At the current stage, there are a few limitations. First of all, every step of the pipeline is dependant of the previous actions. However, we encountered that the last pipeline step is capable of fixing errors of earlier stages. Graves model [10] needs a minimum amount of input composed of images and transcriptions to generate strokes with a new text sequence robustly. This part has also difficulties to imitate writers with bad handwriting because the writing style model has problems with matching the letters of the transcription with the corresponding letters in the input image. Our training data does not include punctuation marks, therefore the trained models are only capable of using standard characters. In practice, scaling is also an issue, because our pipeline was mainly trained with CVL data. This can be easily solved with augmentation in upcoming versions.

6 Conclusion

In this work, we proposed a fully automatic method to imitate full lines of offline handwriting using spatial-temporal style transfer. The pipeline is capable of producing letters and words that are not in the vocabulary of the writer's samples and therefore could be applied to many use cases. We show that generated results just from a single line show auspicious results, often indistinguishable from real handwriting. In future work, we will evaluate different alternatives to the chosen algorithmic steps. We believe that especially an alternative online synthesis method can improve the system's outcome. In the moment the pipeline is capable of changing the background in a very realistic manner [22]. We plan to additionally adapt the background from the input sample.

We note that a concurrent work produces realistic words [13] using a GAN-based system. This could partially be used in our method for improved stylization, with the advantage that our method produces full line images.

References

1. Aksan, E., Hilliges, O.: STCN: stochastic temporal convolutional networks. arXiv e-prints arXiv:1902.06568, February 2019
2. Aksan, E., Pece, F., Hilliges, O.: DeepWriting: making digital ink editable via deep generative modeling. In: Proceedings of the 2018 CHI Conference on Human Factors in Computing Systems, CHI 2018. Association for Computing Machinery, New York (2018)
3. Alonso, E., Moysset, B., Messina, R.: Adversarial generation of handwritten text images conditioned on sequences. In: 2019 International Conference on Document Analysis and Recognition (ICDAR), pp. 481–486, September 2019
4. Bai, X., Ye, L., Zhu, J., Zhu, L., Komura, T.: Skeleton filter: a self-symmetric filter for skeletonization in noisy text images. IEEE Trans. Image Process. **29**, 1815–1826 (2020)
5. Balreira, D.G., Walter, M.: Handwriting synthesis from public fonts. In: 2017 30th SIBGRAPI Conference on Graphics, Patterns and Images (SIBGRAPI), pp. 246–253, October 2017
6. Chung, J., Kastner, K., Dinh, L., Goel, K., Courville, A.C., Bengio, Y.: A recurrent latent variable model for sequential data. In: Cortes, C., Lawrence, N.D., Lee, D.D., Sugiyama, M., Garnett, R. (eds.) Advances in Neural Information Processing Systems 28, pp. 2980–2988. Curran Associates, Inc. (2015)
7. Dijkstra, E.W.: A note on two problems in connexion with graphs. Numer. Math. **1**(1), 269–271 (1959)
8. Goldhahn, D., Eckart, T., Quasthoff, U.: Building large monolingual dictionaries at the Leipzig corpora collection: from 100 to 200 languages, pp. 759–765 (2012)
9. Gomez, R., Furkan Biten, A., Gomez, L., Gibert, J., Karatzas, D., Rusiñol, M.: Selective style transfer for text. In: 2019 International Conference on Document Analysis and Recognition (ICDAR), pp. 805–812, September 2019
10. Graves, A.: Generating sequences with recurrent neural networks. arXiv e-prints arXiv:1308.0850, August 2013
11. Haines, T.S.F., Mac Aodha, O., Brostow, G.J.: My text in your handwriting. ACM Trans. Graph. **35**(3) (2016)
12. Isola, P., Zhu, J., Zhou, T., Efros, A.A.: Image-to-image translation with conditional adversarial networks. In: 2017 IEEE Conference on Computer Vision and Pattern Recognition (CVPR), pp. 5967–5976, July 2017
13. Kang, L., Riba, P., Wang, Y., Rusiñol, M., Fornés, A., Villegas, M.: GANwriting: content-conditioned generation of styled handwritten word images. arXiv e-prints arXiv:2003.02567, March 2020
14. Kleber, F., Fiel, S., Diem, M., Sablatnig, R.: CVL-DataBase: an off-line database for writer retrieval, writer identification and word spotting. In: Proceedings of the 2013 12th International Conference on Document Analysis and Recognition, ICDAR 2013, pp. 560–564. IEEE Computer Society, Washington, DC (2013)

15. Lee, T., Kashyap, R., Chu, C.: Building skeleton models via 3-D medial surface axis thinning algorithms. CVGIP: Graph. Models Image Process. **56**(6), 462–478 (1994)
16. Lian, Z., Zhao, B., Chen, X., Xiao, J.: EasyFont: a style learning-based system to easily build your large-scale handwriting fonts. ACM Trans. Graph. **38**(1) (2018)
17. Liwicki, M., Bunke, H.: IAM-OnDB - an on-line English sentence database acquired from handwritten text on a whiteboard. In: 8th International Conference on Document Analysis and Recognition, vol. 2, pp. 956–961 (2005)
18. Nakamura, K., Miyazaki, E., Nitta, N., Babaguchi, N.: Generating handwritten character clones from an incomplete seed character set using collaborative filtering. In: 2018 16th International Conference on Frontiers in Handwriting Recognition (ICFHR), pp. 68–73, August 2018
19. Nicolaou, A., Bagdanov, A.D., Liwicki, M., Karatzas, D.: Sparse radial sampling LBP for writer identification. In: 2015 13th International Conference on Document Analysis and Recognition (ICDAR), pp. 716–720. IEEE (2015)
20. Pérez, P., Gangnet, M., Blake, A.: Poisson image editing. In: ACM SIGGRAPH 2003 Papers, pp. 313–318 (2003)
21. Ronneberger, O., Fischer, P., Brox, T.: U-Net: convolutional networks for biomedical image segmentation. In: Navab, N., Hornegger, J., Wells, W.M., Frangi, A.F. (eds.) MICCAI 2015. LNCS, vol. 9351, pp. 234–241. Springer, Cham (2015). https://doi.org/10.1007/978-3-319-24574-4_28
22. Seuret, M., Chen, K., Eichenbergery, N., Liwicki, M., Ingold, R.: Gradient-domain degradations for improving historical documents images layout analysis. In: 2015 13th International Conference on Document Analysis and Recognition (ICDAR), pp. 1006–1010. IEEE (2015)
23. Shen, W., Zhao, K., Jiang, Y., Wang, Y., Bai, X., Yuille, A.: DeepSkeleton: learning multi-task scale-associated deep side outputs for object skeleton extraction in natural images. IEEE Trans. Image Process. **26**(11), 5298–5311 (2017)
24. Yang, S., Liu, J., Yang, W., Guo, Z.: Context-aware text-based binary image stylization and synthesis. IEEE Trans. Image Process. **28**(2), 952–964 (2019)
25. Zhang, T.Y., Suen, C.Y.: A fast parallel algorithm for thinning digital patterns. Commun. ACM **27**(3), 236–239 (1984)
26. Zhu, J., Park, T., Isola, P., Efros, A.A.: Unpaired image-to-image translation using cycle-consistent adversarial networks. In: 2017 IEEE International Conference on Computer Vision (ICCV), pp. 2242–2251, October 2017

W32 - The Visual Object Tracking Challenge Workshop VOT2020

W32 - The Visual Object Tracking Challenge Workshop VOT2020

This VOT2020 workshop is the eighth in a series of annual workshops on visual single-object tracking of a priori unknown objects. The five challenges hosted by the workshop are the VOT-ST on RGB short-term tracking, the VOT-RT on RGB short-term real-time tracking, the VOT-LT on RGB long-term tracking, the VOT-RGBT on RGB and thermal tracking and the VOT-RGBD on RGB and depth long-term tracking. A significant change has been made to VOT-ST by replacing the bounding box by a segmentation mask as label. Our aim is to promote trackers capable of per-pixel object localization and to close the gap between the areas of single-object tracking and video segmentation. The evaluation methodology of VOT-ST has been adapted for modern trackers by introducing a new failure recovery. The adapted toolkit was re-implemented in Python and makes the MATLAB version obsolete. The first paper covers the challenges' results, comprising work of 111 coauthors from 46 institutions. The paper (with results omitted) was shared with all co-authors who provided feedback and improved the quality. The VOT2020 Organizing Committee (OC) received five regular paper submissions which were reviewed in a double-blind process; each paper by three independent Program Committee (PC) members. We received additionally a rejected ECCV paper which included all reviews, rebuttal reports, and information about improvements to the original work. This paper was reviewed by the OC. All six papers were finally accepted to the workshop.

We thank all PC members for their efforts, which are essential for maintaining the scientific quality of the workshop, all the authors for their contributions, all the workshop participants for their valuable comments on future challenges during the panel discussion, and our sponsor – University of Ljubljana, Faculty of Computer and Information Science – for their financial support. We also express our gratitude to the ECCV workshop organization who offered us the conditions to make VOT2020 successful as a virtual event. More information including past challenges is available on the VOT web page: http://www.votchallenge.net/vot2020/.

August 2020

Matej Kristan
Aleš Leonardis
Jiři Matas
Michael Felsberg
Roman Pflugfelder
Joni-Kristian Kämäräinen
Martin Danelljan
Luka Čehovin Zajc
Alan Lukežič
Gustavo Fernández

The Eighth Visual Object Tracking VOT2020 Challenge Results

Matej Kristan[1(✉)], Aleš Leonardis[2], Jiří Matas[3], Michael Felsberg[4], Roman Pflugfelder[5,6], Joni-Kristian Kämäräinen[7], Martin Danelljan[8], Luka Čehovin Zajc[1], Alan Lukežič[1], Ondrej Drbohlav[3], Linbo He[4], Yushan Zhang[4,9], Song Yan[7], Jinyu Yang[2], Gustavo Fernández[5], Alexander Hauptmann[10], Alireza Memarmoghadam[39], Álvaro García-Martín[36], Andreas Robinson[4], Anton Varfolomieiev[25], Awet Haileslassie Gebrehiwot[36], Bedirhan Uzun[12], Bin Yan[11], Bing Li[18], Chen Qian[29], Chi-Yi Tsai[35], Christian Micheloni[43], Dong Wang[11], Fei Wang[29], Fei Xie[33], Felix Jaremo Lawin[4], Fredrik Gustafsson[44], Gian Luca Foresti[43], Goutam Bhat[8], Guangqi Chen[29], Haibin Ling[34], Haitao Zhang[46], Hakan Cevikalp[12], Haojie Zhao[11], Haoran Bai[32], Hari Chandana Kuchibhotla[17], Hasan Saribas[13], Heng Fan[34], Hossein Ghanei-Yakhdan[45], Houqiang Li[41], Houwen Peng[23], Huchuan Lu[11], Hui Li[19], Javad Khaghani[37], Jesus Bescos[36], Jianhua Li[11], Jianlong Fu[23], Jiaqian Yu[28], Jingtao Xu[28], Josef Kittler[42], Jun Yin[46], Junhyun Lee[21], Kaicheng Yu[16], Kaiwen Liu[18], Kang Yang[24], Kenan Dai[11], Li Cheng[37], Li Zhang[40], Lijun Wang[11], Linyuan Wang[46], Luc Van Gool[8], Luca Bertinetto[14], Matteo Dunnhofer[43], Miao Cheng[46], Mohana Murali Dasari[17], Ning Wang[24], Ning Wang[41], Pengyu Zhang[11], Philip H. S. Torr[40], Qiang Wang[26], Radu Timofte[8], Rama Krishna Sai Gorthi[17], Seokeon Choi[20], Seyed Mojtaba Marvasti-Zadeh[37], Shaochuan Zhao[19], Shohreh Kasaei[31], Shoumeng Qiu[30], Shuhao Chen[11], Thomas B. Schön[44], Tianyang Xu[42], Wei Lu[46], Weiming Hu[18,26], Wengang Zhou[41], Xi Qiu[22], Xiao Ke[15], Xiao-Jun Wu[19], Xiaolin Zhang[30], Xiaoyun Yang[27], Xuefeng Zhu[19], Yingjie Jiang[19], Yingming Wang[11], Yiwei Chen[28], Yu Ye[15], Yuezhou Li[15], Yuncon Yao[33], Yunsung Lee[21], Yuzhang Gu[30], Zezhou Wang[11], Zhangyong Tang[19], Zhen-Hua Feng[42], Zhijun Mai[38], Zhipeng Zhang[18], Zhirong Wu[23], and Ziang Ma[46]

[1] University of Ljubljana, Ljubljana, Slovenia
matej.kristan@fri.uni-lj.si
[2] University of Birmingham, Birmingham, UK
[3] Czech Technical University, Prague, Czech Republic
[4] Linköping University, Linköping, Sweden
[5] Austrian Institute of Technology, Seibersdorf, Austria
[6] TU Wien, Vienna, Austria
[7] Tampere University, Tampere, Finland
[8] ETH Zürich, Zürich, Switzerland
[9] Beijing Institute of Technology, Beijing, China
[10] Carnegie Mellon University, Pittsburgh, USA
[11] Dalian University of Technology, Dalian, China

A. Bartoli and A. Fusiello (Eds.): ECCV 2020 Workshops, LNCS 12539, pp. 547–601, 2020.
https://doi.org/10.1007/978-3-030-68238-5_39

[12] Eskisehir Osmangazi University, Eskişehir, Turkey
[13] Eskisehir Technical University, Eskişehir, Turkey
[14] Five AI, London, UK
[15] Fuzhou University, Fuzhou, China
[16] High School Affiliated to Renmin University of China, Beijing, China
[17] Indian Institute of Technology, Tirupati, Tirupati, India
[18] Institute of Automation, Chinese Academy of Sciences, Beijing, China
[19] Jiangnan University, Wuxi, China
[20] KAIST, Daejeon, Korea
[21] Korea University, Seoul, Korea
[22] Megvii, Beijing, China
[23] Microsoft Research, Redmond, USA
[24] Nanjing University of Information Science and Technology, Nanjing, China
[25] National Technical University of Ukraine, Kiev, Ukraine
[26] NLP, Beijing, China
[27] Remark Holdings, London, UK
[28] Samsung Research China-Beijing (SRC-B), Beijing, China
[29] Sensetime, Taiwan, Hong Kong
[30] Shanghai Institute of Microsystem and Information Technology, Chinese Academy of Sciences, Shanghai, China
[31] Sharif University of Technology, Tehran, Iran
[32] Sichuan University, Chengdu, China
[33] Southeast University, Nanjing, China
[34] Stony Brook University, Stony Brook, USA
[35] Tamkang University, New Taipei City, Taiwan
[36] Universidad Autónoma de Madrid, Madrid, Spain
[37] University of Alberta, Edmonton, Canada
[38] University of Electronic Science and Technology of China, Chengdu, China
[39] University of Isfahan, Isfahan, Iran
[40] University of Oxford, Oxford, UK
[41] University of Science and Technology of China, Hefei, China
[42] University of Surrey, Guildford, UK
[43] University of Udine, Udine, Italy
[44] Uppsala University, Uppsala, Sweden
[45] Yazd University, Yazd, Iran
[46] Zhejiang Dahua Technology, Binjiang, China

Abstract. The Visual Object Tracking challenge VOT2020 is the eighth annual tracker benchmarking activity organized by the VOT initiative. Results of 58 trackers are presented; many are state-of-the-art trackers published at major computer vision conferences or in journals in the recent years. The VOT2020 challenge was composed of five sub-challenges focusing on different tracking domains: (i) VOT-ST2020 challenge focused on short-term tracking in RGB, (ii) VOT-RT2020 challenge focused on "real-time" short-term tracking in RGB, (iii) VOT-LT2020 focused on long-term tracking namely coping with target disappearance and reappearance, (iv) VOT-RGBT2020 challenge focused on short-term tracking in RGB and thermal imagery and (v) VOT-RGBD2020 challenge focused on long-term tracking in RGB and depth imagery. Only the VOT-ST2020 datasets were refreshed. A significant novelty is introduction of a new VOT short-term tracking evaluation methodology, and introduction of segmentation ground truth in the VOT-ST2020 challenge – bounding boxes will no longer be used in the VOT-ST challenges. A new VOT Python toolkit that implements all these novelites was introduced. Performance of the tested trackers typically by far exceeds standard baselines. The source code for most of the trackers is publicly available from the VOT page. The dataset, the evaluation kit and the results are publicly available at the challenge website (http://votchallenge.net).

Keywords: Visual object tracking · Performance evaluation protocol · State-of-the-art benchmark · RGB · RGBD · Depth · RGBT · Thermal imagery · Short-term trackers · Long-term trackers

1 Introduction

Visual object tracking remains a core computer vision problem and a popular research area with many open challenges, which has been promoted over the last decade by several tracking initiatives like PETS [86], CAVIAR[1], i-LIDS[2], ETISEO[3], CDC [20], CVBASE[4], FERET [58], LTDT[5], MOTC [39,66] and Videonet[6]. However, prior to 2013, a consensus on performance evaluation was missing, which made objective comparison of tracking results across papers impossible. In response, the VOT[7] initiative has been formed in 2013. The primary goal of VOT was establishing datasets, evaluation measures and toolkits as well as creating a platform for discussing evaluation-related issues through organization of tracking challenges. This lead to organization of seven challenges, which have taken place in conjunction with ICCV2013 (VOT2013 [36]), ECCV2014 (VOT2014 [37]), ICCV2015 (VOT2015 [35]), ECCV2016 (VOT2016 [34]), ICCV2017 (VOT2017 [33]), ECCV2018 (VOT2018 [32]) and ICCV2019 (VOT2019 [31]).

Initially the VOT considered single-camera, single-target, model-free, causal trackers, applied to short-term tracking. The *model-free* property means that the only training information provided is the bounding box in the first frame. The *short-term* tracking means that trackers are assumed not to be capable of performing successful re-detection after the target is lost. *Causality* requires that the tracker does not use any future frames, or frames prior to re-initialization, to infer the object position in the current frame. In 2018, the VOT tracker categories were extended by an additional one: single-camera, single-target, model-free long-term

[1] http://homepages.inf.ed.ac.uk/rbf/CAVIARDATA1.
[2] http://www.homeoffice.gov.uk/science-research/hosdb/i-lids.
[3] http://www-sop.inria.fr/orion/ETISEO.
[4] http://vision.fe.uni-lj.si/cvbase06/.
[5] http://www.micc.unifi.it/LTDT2014/.
[6] http://videonet.team.
[7] http://votchallenge.net.

trackers. *Long-term* tracking means that the trackers are *required* to perform re-detection after the target has been lost and are therefore *not* reset after such an event.

This paper presents the VOT2020 challenge, organized in conjunction with the ECCV2020 Visual Object Tracking Workshop, and the results obtained. Several novelties are introduced in VOT2020 with respect to VOT2019, which consider encoding of the ground truth target positions, performance measures and a complete re-implementation of the VOT toolkit in a widely-used programming language for easier tracker integration. In the following we overview the most closely related works, discuss issues with exisiting performance measures and point out the contributions of VOT2020.

1.1 Short-Term Tracker Evaluation

Over the last eight years, the Visual Object Tracking initiative (VOT) has been gradually developing performance evaluation methodology with an overall guideline to develop interpretable measures that probe various tracking properties. At VOT inception in 2013, a simple evaluation protocol was popularized by OTB [77]. This methodology applies a no-reset experiment in which the tracker is initialized in the first frame and it runs unsupervised until the end of the sequence. The overall performance is summarized by area-under-the-curve principle, which has been showed in [70,72] to be an average overlap (AO) computed over the entire sequence of frames. A downside of the AO is that all frames after the first failure receive a zero overlap, which increases bias and variance of the estimator [38].

Alternatively, based on the analysis later published in [38,72], VOT proposed two basic performance measures: accuracy and robustness. The goal was to promote trackers that well approximate the target position, and even more importantly, do not fail often. The first methodology introduced in VOT2013 [36] was based on ranking trackers along each measure and averaging the ranks. Due to a reduced interpretation power and dependency of ranks on the tested trackers, this approach was replaced in VOT2015 [35] by the expected average overlap measure (EAO), which principally combines the individual basic measures.

To provide an incentive for community-wide exploration of a wide spectrum of well-performing trackers and to reduce the pressure for fine-tuning to benchmarks with the sole purpose of reaching the number one rank on particular test data, VOT introduced the so-called state-of-the-art bound (*SotA* bound). Any tracker exceeding this bound is considered state-of-the-art by the VOT standard.

While most of the tracking datasets [18,19,25,40,44,53,54,65,77,85,93] have partially followed the trend in computer vision of increasing the number of sequences, the VOT [31–38] datasets have been constructed with diversity in mind and were kept sufficiently small to allow fast tracker development-and-testing cycles. Several recent datasets [18,25] have adopted elements of the VOT dataset construction principles for rigorous tracker evaluation. In VOT2017 [33] a sequestered dataset was introduced to reduce the influence of tracker over-fitting without requiring to increase the public dataset size. Despite significant activity in dataset construction, the VOT dataset remains unique for its carefully chosen and curated sequences guaranteeing relatively unbiased assessment of performance with respect to attributes.

In 2015, the VOT introduced a new short-term tracking challenge dedicated to tracking in thermal imagery. The VOT short-term performance evaluation methodology was used with the LTIR [2] dataset. The challenge gradually evolved into an RGB+Thermal short-term tracking and constructed a new dataset based on [43]. The targets were re-annotated by rotated bounding boxes using a semi-automatic protocol [3].

The VOT evaluation protocols have promoted development of robust short-term trackers. But with increased robustness of modern trackers, a drawback of the reset-based evaluation protocol has emerged. In the VOT performance evaluation protocol a tracker is initialized in the first frame and whenever the overlap between the reported and the ground truth target location (i.e., bounding box) falls to zero, a failure is detected and the tracker is reset a fixed number of frames later. The robustness is measured as the number of times the tracker is reset and the accuracy is the average overlap between the periods of successful tracking. This setup reflects the tracker performance in a practical application, where the task is to track the target throughout the sequence, either automatically, or by user intervention, i.e., a tracker reset. Furthermore, this approach enables utilization of all sequence frames in the evaluation.

However, a point of tracking failure will affect the point of reset (tracker re-initialization) and initialization points profoundly affect the tracking performance. With recent development of very robust trackers, the initialization points started to play a significant role in the final tracker ranking. In particular, we have noticed that initialization at some frame might result in another failure later on in the sequence, while initializing a few frames later might not. This allows a possibility (although not trivially) for fine-tuning the tracker to fail on more *favorable* frames and by that reducing the failure rate and increase the overall apparent robustness as measured by the reset-based protocol.

Another potential issue of the existing VOT reset-based protocol is the definition of a tracking failure. A failure is detected whenever the overlap between the prediction and ground truth falls to zero. Since resets directly affect the performance, a possible way to reduce the resets is to increase the predicted bounding box size, so to avoid the zero overlap. While we have not observed such *gaming* often, there were a few cases in the last seven challenges where the trackers attempted this and one of the trackers has been disqualified upon identifying the use of the bounding box inflation strategy. But some trackers did resort to reporting a slightly larger bounding box due to the strictness of the failure protocol – the tracker will be reset if the zero overlap is detected in a single frame, even if the tracker would have jumped right back on the target in the next frame. We call this a short-term failure and the current protocol does not distinguish between trackers robust to short-term failures and trackers that fail completely.

1.2 Long-Term Tracker Evaluation

A major difference between short-term (ST) and long-term (LT) trackers is that LT trackers are required to handle situations in which the target may leave the field of view for a longer duration. This means that a natural evaluation protocol for LT trackers is a no-reset protocol. Early work [30,57] directly adapted existing object-detection measures precision, recall and F-measure based on 0.5 IoU (overlap) threshold and several authors [52,68] proposed a modification of the short-term average overlap measure. Valmadre et al. [28] introduced a measure that directly addresses the evaluation of the re-detection ability and most recently Lukežič et. al. [49] proposed *tracking* precision, *tracking* recall and *tracking* F-measure that do not depend on specifying the IoU threshold. Their primary measure, the tracking F-measure, reduces to a standard short-term measure (average overlap) when computed in a short-term setup, thus closing the gap between short- and log-term tracking measures. The measure is shown to be extremely robust and allows using a very sparse temporal target annotation, thus enabling using very long evaluation sequences at reduced annotation effort. For these reasons, the measures and the evaluation protocol from [49] were selected in 2018 as the main methodology for all VOT sub-challenges dealing with long-term trackers.

Several datasest have been proposed for RGB long-term tracking evaluation, starting with LTDT (See footnote 5) and followed by [28,48,49,52,53]. The authors of [48] argue that long-term tracking does not just refer to the sequence length, but more importantly to the sequence properties, like the number and the length of target disappearances, and the type of tracking output expected. Their dataset construction approach follows these guidelines and was selected in VOT2018 for the first VOT long-term tracking challenge and later replaced by the updated dataset from [49].

In 2019 VOT introduced another long-term tracking challenge to promote tracker operating with RGB and depth (RGBD). At the time, only two public datasets were available, namely the PTB [67] and STC [78], with relatively short sequences and limited range of scenes due to acquisition hardware restrictions. Recently, a more elaborate dataset called CDTB [46] was proposed, which contains many long sequences with many target disappearances, captured with a range of RGBD sensors both indoor and outdoor under various lighting conditions. This dataset was used in VOT2019 in the VOT-RGBD challenge.

1.3 The VOT2020 Challenge

Since VOT2020 considers short-term as well as long-term trackers in separate challenges, we adopt the definitions from [49] to position the trackers on the short-term/long-term spectrum:

- **Short-term tracker** (ST_0). The target position is reported at each frame. The tracker does not implement target re-detection and does not explicitly detect occlusion. Such trackers are likely to fail at the first occlusion as their representation is affected by any occluder.

- **Short-term tracker with conservative updating** (ST_1). The target position is reported at each frame. Target re-detection is not implemented, but tracking robustness is increased by selectively updating the visual model depending on a tracking confidence estimation mechanism.
- **Pseudo long-term tracker** (LT_0). The target position is not reported in frames when the target is not visible. The tracker does not implement explicit target re-detection but uses an internal mechanism to identify and report tracking failure.
- **Re-detecting long-term tracker** (LT_1). The target position is not reported in frames when the target is not visible. The tracker detects tracking failure and implements explicit target re-detection.

The evaluation toolkit and the datasets are provided by the VOT2020 organizers. The participants were required to use the new Python VOT toolkit that implements the new evaluation protocols and the new ground truth encoding. A toolkit beta testing period opened in early March 2020, and the challenge officially opened on March 20th 2020 with approximately a month available for results submission. Due to Covid-19 crisis, the VOT-RGBT team could not complete all the preparations in time and has decided to postpone the opening of the VOT-RGBT2020 sub-challenge to May 10th. The results submission deadline for all sub-challenges was May 3rd. The VOT2020 challenge thus contained five challenges:

1. **VOT-ST2020 challenge:** This challenge was addressing short-term tracking in RGB images and has been running since VOT2013 with annual updates and modifications. A significant novelty compared to 2019 was that the target position was encoded by a segmentation mask.
2. **VOT-RT2020 challenge:** This challenge addressed the same class of trackers as VOT-ST2020, except that the trackers had to process the sequences in real-time. The challenge was introduced in VOT2017. A significant novelty compared to 2019 was that the target position was encoded by a segmentation mask.
3. **VOT-LT2020 challenge:** This challenge was addressing long-term tracking in RGB images. The challenge was introduced in VOT2018. The target positions were encoded by bounding boxes.
4. **VOT-RGBT2020 challenge:** This challenge was addressing short-term tracking in RGB+thermal imagery. This challenge was introduced in VOT2019 and can be viewed as evolution of the VOT-TIR challenge introduced in VOT2015. The target positions were encoded by bounding boxes.
5. **VOT-RGBD2020 challenge:** This challenge was addressing long-term tracking in RGB+depth (RGBD) imagery. This challenge was introduced in VOT2019. The target positions were encoded by bounding boxes.

The authors participating in the challenge were required to integrate their tracker into the new VOT2020 evaluation kit, which automatically performed a set of standardized experiments. The results were analyzed according to the

VOT2020 evaluation methodology. Upon submission of the results, the participants were required to classify their tracker along the short-term/long-term spectrum.

Participants were encouraged to submit their own new or previously published trackers as well as modified versions of third-party trackers. In the latter case, modifications had to be significant enough for acceptance. Participants were expected to submit a single set of results per tracker. Changes in the parameters did not constitute a different tracker. The tracker was required to run with fixed parameters in all experiments. The tracking method itself was allowed to internally change specific parameters, but these had to be set automatically by the tracker, e.g., from the image size and the initial size of the bounding box, and were not to be set by detecting a specific test sequence and then selecting the parameters that were hand-tuned for this sequence.

Each submission was accompanied by a short abstract describing the tracker, which was used for the short tracker descriptions in Appendix A. In addition, participants filled out a questionnaire on the VOT submission page to categorize their tracker along various design properties. Authors had to agree to help the VOT technical committee to reproduce their results in case their tracker was selected for further validation. Participants with sufficiently well-performing submissions, who contributed with the text for this paper and agreed to make their tracker code publicly available from the VOT page were offered co-authorship of this results paper. The committee reserved the right to disqualify any tracker that, by their judgement, attempted to cheat the evaluation protocols.

To compete for the winner of VOT2020 challenge, learning on specific datasets (OTB, VOT, ALOV, UAV123, NUSPRO, TempleColor and RGBT234) was prohibited. In the case of GOT10k, a list of 1k prohibited sequences was created in VOT2019, while the remaining 9k+ sequences were allowed for learning. The reason was that part of the GOT10k was used for VOT-ST2020 dataset update.

The use of class labels specific to VOT was not allowed (i.e., identifying a target class in each sequence and applying pre-trained class-specific trackers was not allowed). An agreement to publish the code online on VOT webpage was required. The organizers of VOT2020 were allowed to participate in the challenge, but did not compete for the winner titles. Further details are available from the challenge homepage[8].

VOT2020 goes beyond previous challenges by updating the datasets in VOT-ST, VOT-RT, challenges, as well as introduction of the segmentation ground truth. New performance evaluation protocol and measures were used in the short-term tracking challenges and the new VOT2020 Python toolkit was developed that implements all the novelties and ensures seamless use of challenge-specific modalities and protocols.

[8] http://www.votchallenge.net/vot2020/participation.html.

2 Performance Evaluation Protocols

Since 2018 VOT considers two classes of trackers: short-term (ST) and long-term (LT) trackers. These two classes primarily differ on the target presence assumptions, which affects the evaluation protocol as well as performance measures. These are outlined in following two subsections. Section 2.1 introduces the new short-term performance evaluation protocol and measures, while the standard VOT long-term tracking evaluation protocol is overviewed in Sect. 2.2.

2.1 The New Anchor-Based Short-Term Tracking Evaluation Protocol

The main drawback of the existing VOT short-term performance evaluation protocol are the tracker-dependent resets, which induce a causal correlation between the first reset and the later ones. To avoid this, the notion of reset is replaced by *initialization points* (called *anchors* for short), which are made equal for all trackers in the new protocol. In particular, anchors are placed on each sequence Δ_{anc} frames apart, with the first and last anchor on the first and the last frame, respectively. A tracker is run from *each* anchor forward or backward in the sequences, whichever direction yields the longest sub-sequence. For example, if the anchor is placed before the middle of the sequence, the tracker is run forward, otherwise backward in the sequence. Each anchor is manually checked and potentially moved by a few frames to avoid placing the initialization point on an occluded target. Figure 1 shows example of the anchor placement and the tracking direction.

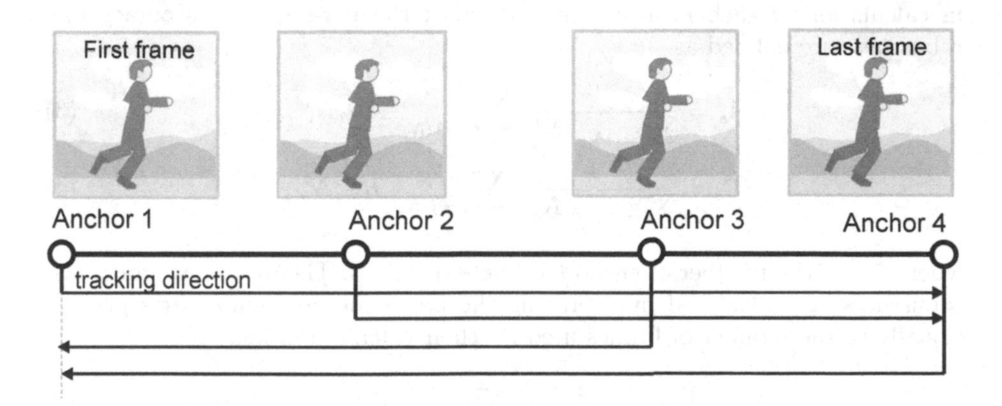

Fig. 1. Anchors are placed 50 frames apart. At each anchor the tracker is initialized and tracks in the direction that yields the longest subsequence.

The distance between the anchors was set to $\Delta_{\mathrm{anc}} = 50$. At approximately 25 frames per second, this amounts to 2 s distances. We have experimentally tested that this value delivers stable results for the measures described in the next section computed on typical-length short-term sequences, while keeping the computational complexity of the evaluation at a moderate level.

Like in previous VOT challenges, we use the accuracy and robustness as the basic measures to probe tracking performance and the overall performance is summarized by the expected average overlap (EAO). In the following we re-define these in the context of the new anchor-based evaluation protocol.

The New Accuracy and Robustness Measures. On a subsequence starting from an anchor a of sequence s, the accuracy $A_{s,a}$ is defined as the average overlap between the target predictions and the ground truth calculated from the frames before the tracker fails on that subsequence, i.e.,

$$A_{s,a} = \frac{1}{N^F_{s,a}} \sum\nolimits_{i=1:N^F_{s,a}} \Omega_{s,a}(i), \tag{1}$$

where $N^F_{s,a}$ is the number of frames before the tracker failed in the subsequence starting at anchor a in the sequence s (see Sect. 2.1 for the failure definition) and $\Omega_{s,a}(i)$ is the overlap between the prediction and the ground truth at frame i. The new robustness measure $R_{s,a}$ is defined as the *extent* of the sub-sequence before the tracking failure, i.e.,

$$R_{s,a} = N^F_{s,a}/N_{s,a}, \tag{2}$$

where $N_{s,a}$ is the number of frames of the subsequence.

The results from the sub-sequences are averaged in a weighted fashion such that each sub-sequence contributes proportionally to the number frames used in calculation of each measure. In particular, the per-sequence accuracy and robustness are defined as

$$A_s = \frac{1}{\sum_{a=1:N^A_s} N^F_{s,a}} \sum\nolimits_{a=1:N^A_s} A_{s,a} N^F_{s,a}, \tag{3}$$

$$R_s = \frac{1}{\sum_{a=1:N^A_s} N_{s,a}} \sum\nolimits_{a=1:N^A_s} R_{s,a} N_{s,a}, \tag{4}$$

where N^A_s is the number of anchors in the sequence s. The overall accuracy and robustness are calculated by averaging the per-sequence counterparts proportionally to the number of frames used for their calculation, i.e.,

$$A = \frac{1}{\sum_{s=1:N} N^F_s} \sum\nolimits_{s=1:N} A_s N^F_s, \tag{5}$$

$$R = \frac{1}{\sum_{s=1:N} N_s} \sum\nolimits_{s=1:N} R_s N_s, \tag{6}$$

where N is the number of sequences in the dataset, N_s is the number of frames in sequence s and $N^F_s = \sum_{a=1:N^A_s} N^F_{s,a}$ is the number of frames used to calculate the accuracy in that sequence.

The New EAO Measure. As in previous VOT challenges, the accuracy and robustness are principally combined into a single performance score called the expected average overlap (EAO). We use the same approach as in the previous VOT challenges, i.e., the expected average overlap curve is calculated and averaged over an interval of typical short-term sequence lengths into the EAO measure.

Note that the computation considers virtual sequences of overlaps generated from the sub-sequence results. In particular, if a tracker failed on a sub-sequence (s, a), the overlap falls to zero at the failure frame, and the overlaps can be extended to i-th frame by zeros, even if i exceeds the sub-sequence length. But if the tracker did not fail, the overlaps cannot be extrapolated beyond the original sub-sequence length.

The value of the EAO curve $\hat{\Phi}_i$ at sequence length i is thus defined as

$$\hat{\Phi}_i = \frac{1}{|\mathcal{S}(i)|} \sum_{s,a \in \mathcal{S}(i)} \Phi_{s,a}(i), \tag{7}$$

where $\Phi_{s,a}(i)$ is the average overlap calculated between the first and i-th frame of the extended sub-sequence starting at anchor a of sequence s, $\mathcal{S}(i)$ is the set of the extended sub-sequences with length greater or equal to i and $|\mathcal{S}(i)|$ is the number of these sub-sequences.

The EAO measure is then calculated by averaging the EAO curve from N_{lo} to N_{hi}, i.e.,

$$EAO = \frac{1}{N_{\mathrm{hi}} - N_{\mathrm{lo}}} \sum\nolimits_{i=N_{\mathrm{lo}}:N_{\mathrm{hi}}} \hat{\Phi}_i. \tag{8}$$

Similarly to VOT2015 [35], the interval bounds $[N_{\mathrm{lo}}, N_{\mathrm{hi}}]$ were determined from the mean $\pm$ one standard deviation of the anchor-generated sub-sequences.

Failure Definition. The tracking failure event is also redefined to (i) reduce the potential for the *gaming*, i.e., outputting the entire image as the prediction to prevent failure detection during an uncertain tracking phase, and (ii) allow for recovery from short-term tracking failures. A *tentative* failure is detected when the overlap falls below a non-zero threshold θ_{Φ}. The non-zero threshold punishes an actual drift from the target as well as speculation by outputting a very large bounding box to prevent failure detection. If a tracker does not recover within the next θ_N frames, i.e., the overlap does increase to over θ_{Φ}, a failure is detected. Note that in some rare situations a frame might contain the target fully occluded. Since short-term trackers are not required to report target disappearance, these frames are ignored in tracking failure detection.

By using several well-known trackers from different tracker design classes we experimentally determined that the threshold values $\theta_{\Phi} = 0.1$ and $\theta_N = 10$ reduce the *gaming* potential, allow recoveries from short-term failures, while still penalizing the trackers that fail more often.

Per-Attribute Analysis. Per-attribute accuracy and robustness are computed by accounting for the fact that the attributes are not equally distributed among the sequences and that the attribute at a frame may affect the tracker performance a few frames later in the sequence. Thus the two per-attribute measures $(A_{\mathrm{atr}}, R_{\mathrm{atr}})$ are defined as weighted per-sequence measures with weights proportional to the amount of attribute in the sequence, i.e.,

$$A_{\mathrm{atr}} = \frac{1}{\sum_{s=1:N} N_s^{\mathrm{atr}}} \sum_{s=1:N} N_s^{\mathrm{atr}} A_s, \tag{9}$$

$$R_{\mathrm{atr}} = \frac{1}{\sum_{s=1:N} N_s^{\mathrm{atr}}} \sum_{s=1:N} N_s^{\mathrm{atr}} R_s, \tag{10}$$

where N_s^{atr} is the number of frames with attribute "atr" in a sequence s.

2.2 The VOT Long-Term Performance Evaluation Protocol

In a long-term (LT) tracking setup, the target may leave the camera field of view for longer duration before re-entering it, or may undergo long-lasting complete occlusions. The tracker is thus required to report the target position only for frames in which the target is visible and is required to recover from tracking failures. Long-term sequences are thus much longer than short-term sequences to test the re-detection capability. LT measures should therefore measure the target localization accuracy as well as target re-detection capability.

In contrast to the ST tracking setup, the tracker is not reset upon drifting off the target. To account for the most general case, the tracker is required to report the target position at every frame and provide a confidence score of target presence. The evaluation protocol [49] first used in the VOT2018 is adapted.

Three long-term tracking performance measures proposed in [48] are adopted: tracking precision (Pr), tracking recall (Re) and tracking F-score. These are briefly described in the following.

The Pr and Re are derived in [48] from the counterparts in detection literature with important differences that draw on advancements of tracking-specific performance measures. In particular, the bounding box overlap is integrated out, leaving both measures $Pr(\tau_\theta)$ and $Re(\tau_\theta)$ depend directly on the tracker prediction certainty threshold τ_θ, i.e., the value of tracking certainty below which the tracker output is ignored. Precision and accuracy are combined into a single score by computing the tracking F-measure

$$F(\tau_\theta) = 2Pr(\tau_\theta)Re(\tau_\theta)/(Pr(\tau_\theta) + Re(\tau_\theta)). \tag{11}$$

Long-term tracking performance can thus be visualized by tracking precision, tracking accuracy and tracking F-measure plots by computing these scores for all thresholds τ_θ [48]. The final values of Pr, Re and F-measure are obtained by selecting τ_θ that maximizes tracker-specific F-measure. This avoids all manually-set thresholds in the primary performance measures.

Evaluation Protocol. A tracker is evaluated on a dataset of several sequences by initializing on the first frame of a sequence and run until the end of the sequence without re-sets. A precision-recall graph is calculated on each sequence and averaged into a single plot. This guarantees that the result is not dominated by extremely long sequences. The F-measure plot is computed according to (11) from the average precision-recall plot. The maximal score on the F-measure plot (tracking F-score) is taken as the long-term tracking primary performance measure.

3 Description of Individual Challenges

In the following we provide descriptions of all five challenges running in the VOT2020 challenge.

3.1 VOT-ST2020 Challenge Outline

This challenge addressed RGB tracking in a short-term tracking setup. The performance evaluation protocol and measures outlined in Sect. 2.1 were applied. In the following, the details of the dataset and the winner identification protocols are provided.

The Dataset. Results of the VOT2019 showed that the dataset was not saturated [31], and the public dataset has been refreshed by replacing one sequence (see Fig. 2). A single sequence in the sequestered dataset has been replaced as well to calibrate the attribute distribution between the two datasets. Following the protocols from VOT2019, the list of 1000 diverse sequences[9] from the GOT-10k [25] training set was used. The sequence selection and replacement procedure followed that of VOT2019. In addition, object category and motion diversity was ensured by manual review.

Fig. 2. The `pedestrian1` sequence of the VOT2019 public dataset has been replaced by a more challenging `hand02` sequence for VOT2020.

The bounding boxes are no longer used in the VOT-ST/RT tracking sub-challenges. The target position is now encoded by the segmentation masks. Since

[9] http://www.votchallenge.net/vot2019/res/list0_prohibited_1000.txt.

Fig. 3. Images from the VOT-ST2020 sub-challenge with segmentation masks superimposed (in cyan). (Color figure online)

2016, VOT has already been using segmentation masks for fitting the rotated bounding box ground truth in the previous years. However, a closer inspection revealed that while these masks were valid for fitting rectangles, their accuracy was insufficient for segmentation ground truth. Thus *the entire dataset* (public and sequestered) was re-annotated. The initial masks were obtained by a semi-automatic method and then all sequences were frame-by-frame manually corrected. Examples of segmentation masks are shown in Fig. 3.

Per-frame visual attributes were semi-automatically assigned to the new sequences following the VOT attribute annotation protocol. In particular, each frame was annotated by the following visual attributes: (i) occlusion, (ii) illumination change, (iii) motion change, (iv) size change, (v) camera motion.

The EAO interval bounds in (8) were estimated to be $[N_{\text{lo}}, N_{\text{hi}}] = [115, 755]$ on the public VOT-ST2020 dataset.

Winner Identification Protocol. The VOT-ST2020 winner was identified as follows. Trackers were ranked according to the EAO measure on the public dataset. Top five ranked trackers were then re-run by the VOT2020 committee on the sequestered dataset. The top ranked tracker on the sequestered dataset not submitted by the VOT2020 committee members was the winner of the VOT-ST2020 challenge.

3.2 VOT-RT2020 Challenge Outline

This challenge addressed *real-time* RGB tracking in a short-term tracking setup. The dataset was the same as in the VOT-ST2020 challenge, but the evaluation protocol was modified to emphasize the real-time component in tracking performance. In particular, the VOT-RT2020 challenge requires predicting bounding boxes faster or equal to the video frame-rate. The toolkit sends images to the tracker via the Trax protocol [71] at 20fps. If the tracker does not respond in time, the last reported bounding box is assumed as the reported tracker output at the available frame (zero-order hold dynamic model). The same performance evaluation protocol as in VOT-ST2020 is then applied.

Winner Identification Protocol. All trackers are ranked on the public RGB short-term tracking dataset with respect to the EAO measure. The winner was identified as the top ranked tracker not submitted by the VOT2020 committee members.

3.3 VOT-LT2020 Challenge Outline

This challenge addressed RGB tracking in a long-term tracking setup and is a continuation of the VOT-LT2019 challenge. We adopt the definitions from [48], which are used to position the trackers on the short-term/long-term spectrum. A long-term performance evaluation protocol and measures from Sect. 2.2 were used to evaluate tracking performance on VOT-LT2020.

Trackers were evaluated on the LTB50 [49], the same dataset as used in VOT-LT2019. The LTB50 dataset contains 50 challenging sequences of diverse objects (persons, car, motorcycles, bicycles, boat, animals, etc.) with the total length of 215294 frames. Sequence resolutions range between 1280×720 and 290×217. Each sequence contains on average 10 long-term target disappearances, each lasting on average 52 frames.

The targets are annotated by axis-aligned bounding boxes. Sequences are annotated by the following visual attributes: (i) Full occlusion, (ii) Out-of-view, (iii) Partial occlusion, (iv) Camera motion, (v) Fast motion, (vi) Scale change, (vii) Aspect ratio change, (viii) Viewpoint change, (ix) Similar objects. Note this is per-sequence, not per-frame annotation and a sequence can be annotated by several attributes. Please see [49] for more details.

Winner Identification Protocol. The VOT-LT2020 winner was identified as follows. Trackers were ranked according to the tracking F-score on the LTB50 dataset (no sequestered dataset available). The top ranked tracker on the dataset not submitted by the VOT2020 committee members was the winner of the VOT-LT2020 challenge.

3.4 VOT-RGBT2020 Challenge Outline

This challenge addressed short-term trackers using RGB and a thermal channel. The performance evaluation protocol and measures outlined in Sect. 2.1 were applied.

Trackers were evaluated on the VOT-RGBT2019 dataset, derived from [43], but extended with anchor frames for the new re-initialization approach. The VOT-RGBT2019 dataset contains 60 public and 60 sequestered sequences containing partially aligned RGB and thermal images. The longest three sequences in the sequestered dataset are among the simpler ones and were sub-sampled by factor five to avoid a positive bias in the EAO measure. Due to acquisition equipment, the RGB and thermal channels are slightly temporally de-synchronized, which adds to the challenge in RGBT tracking. All frames are annotated with the attributes (i) occlusion, (ii) dynamics change, (iii) motion change, (iv) size

Fig. 4. Example images from the VOT-RGBT2020 dataset. The left two frames illustrate the synchronization issue in the RGBT234 dataset [43] and the right two frames the small object issue.

change, and (v) camera motion. Due to the slight temporal de-synchronization and the partially very small objects, the consistency and accuracy of segmentation masks was not sufficient for unbiased tracker evaluation. A decision was thus made to use rotated bounding boxes already created for the VOT-RGBT2019 dataset instead. Examples of images from the dataset are shown in Fig. 4.

Winner Identification Protocol. The VOT-RGBT2020 winner has been identified as follows. Trackers were ranked according to the EAO measure on the public VOT-RGBT2020 dataset. The top five trackers have then been re-run by the VOT2020 committee on the sequestered VOT-RGBT dataset. The top ranked tracker on the sequestered dataset not submitted by the VOT2020 committee members was the winner of the VOT-RGBT2020 challenge.

3.5 VOT-RGBD2020 Challenge Outline

This challenge addressed long-term trackers using the RGB and depth channels (RGBD). The long-term performance evaluation protocol from Sect. 2.2 was used. The VOT-RGBD2020 trackers were evaluated on the CDTB dataset described in detail in [46]. The dataset contains 80 sequences acquired with three different sensor configurations: 1) a single Kinect v2 RGBD sensor, 2) a combination of the Time-of-Flight (Basler tof640) and RGB camera (Basler acA1920), and 3) a stereo-pair (Basler acA1920). Kinect was used in 12 indoor sequences, RGB-ToF pair in 58 indoor sequences and the stereo-pair in 10 outdoor sequences. The dataset contains tracking of various household and office objects (Fig. 5). The sequences contain target in-depth rotations, occlusions and disappearance that are challenging for only RGB and depth-only trackers. The total number of frames is 101,956 in various resolutions. For more details, see [46].

Winner Identification Protocol. The VOT-RGBD2020 winner was identified as follows. Trackers were ranked according to the F-score on the public VOT-RGBD2020 dataset (no sequestered dataset available). The reported numbers were computed using the submitted results, but the numbers were verified by re-running the submitted trackers multiple times. The top ranked tracker not submitted by the VOT2020 committee members was the winner of the VOT-RGBD2020 challenge.

Fig. 5. RGB and depth (D) frames from the VOT-RGBD dataset.

4 The VOT2020 Challenge Results

This section summarizes the trackers submitted, results analysis and winner identification for each of the five VOT2020 challenges.

4.1 The VOT-ST2020 Challenge Results

Trackers Submitted. In all, 28 valid entries were submitted to the VOT-ST2020 challenge. Each submission included the binaries or source code that allowed verification of the results if required. The VOT2020 committee and associates additionally contributed 9 baseline trackers. For these, the default parameters were selected, or, when not available, were set to reasonable values. Thus in total 37 trackers were tested on VOT-ST2020. In the following we briefly overview the entries and provide the references to original papers in the Appendix A where available.

Of all participating trackers, 17 trackers (46%) were categorized as ST_0, 18 trackers (49%) as ST_1 and 2 as LT_0. 92% applied discriminative and 8% applied generative models. Most trackers (95%) used holistic model, while 5% of the participating trackers used part-based models. Most trackers applied a locally uniform dynamic model[10] or a random walk dynamic model (95%) and only (5%) applied a nearly-constant-velocity dynamic model. 38% of trackers localized the target in a single stage, while the rest applied several stages, typically involving approximate target localization and position refinement. Most of the trackers (86%) use deep features, which shows that the field has moved away from using hand-crafted features, which were still widely used on their own or in combination with the deep features even a few years ago. 54% of these trackers re-trained their backbone on tracking or segmentation/detection datasets.

A particular novelty of the VOT-ST2020 is that target location ground truth is encoded as a segmentation mask. We observe a strong response in the VOT community to this: 57% of trackers reported target position as a segmentation mask, while the rest (43%) reported a bounding box. Among the segmentation trackers, 5 apply a deep grab-cut-like segmentation [79], 3 apply a nearest-

[10] The target was sought in a window centered at its estimated position in the previous frame. This is the simplest dynamic model that assumes all positions within a search region contain the target have equal prior probability.

neighbor segmentation akin to [50] and 13 apply patch-based Siamese segmentation akin to [74].

The trackers were based on various tracking principles. The two dominant tracking methodologies are discriminative correlation filters[11] (used in 68% of all submissions) and Siamese correlation networks, e.g. [4,41,74], (used in 46% of all submissions). 15 trackers were based only on DCFs (DET50 A.6, TRASTmask A.9, TRASFUSTm A.11, DPMT A.13, TRAT A.19, DiMP A.21, SuperDiMP A.22, LWTL A.23, TCLCF A.25, AFOD A.26, FSC2F A.27, KCF A.33, CSRpp A.31, ATOM A.30, UPDT A.37). 10 trackers were based only on Siamese correlation networks (DCDA A.3, igs A.4, SiamMaskS A.5, VPU-SiamM A.7, Ocean A.10, Siammask A.14, SiamMargin A.17, SiamEM A.18, AFAT A.24, SiamFc A.35). 6 trackers applied a combination of DCFs and Siamese networks (A3CTDmask A.2, RPT A.8, AlphaRef A.12, OceanPlus A.15, fastOcean A.16, DESTINE A.28), one tracker combined a DCF with nearest-neighbor deep feature segmentation D3S A.1. One tracker was based on generative adversarial networks InfoVital A.20, one entry was a state-of-the-art video segmentation method STM A.36, one entry was a subspace tracker (IVT A.32), one used multiple-instance learning (MIL A.34) and one entry was a scale-adaptive mean-shift tracker (ASMS A.29).

Results. The results are summarized in the AR-raw plots and EAO plots in Fig. 6, and in Table 7. The top ten trackers according to the primary EAO measure (Fig. 6) are RPT A.8, OceanPlus A.15, AlphaRef A.12, AFOD A.26, LWTL A.23, fastOcean A.16, TRASTmask A.9, DET50 A.6, D3S A.1 and Ocean A.10. All of these trackers apply CNN features for target localization. Nine (RPT, OceanPlus, AlphaRef, AFOD, LWTL, fastOcean, DET50, D3S, TRASTmask) apply a deep DCF akin to [13] (many of these in combination with a Siamese correlation net – RPT, OceanPlus, AlphaRef), while Ocean applies a Siamese correlation network without a DCF. All trackers provide the target location in form of a segmentation mask. Most trackers localize the target in multiple stages, except for AFOD, LWTL and D3S, which produce the target mask in a single stage. Several methods apply deep DCFs such as ATOM/DiMP [5,13] for target localization, bounding box estimation by region proposals [41] or fully-convolutional [69,84], while the final segmentation draws on approaches from [50,64,74]. A ResNet50 backbone pre-trained on general datasets is used in all top 10 trackers.

The top performer on the public dataset is RPT A.8. This is a two-stage tracker. The first stage combines the response of a fully-convolutional region proposal RepPoints [84] with a deep DCF DimP [5] response for initial bounding box estimation. In the second stage, a single-shot segmentation tracker D3S [50] is applied on the estimated target bounding box to provide the final segmentation mask. D3S appears to be modified by replacing the target presence map in geometrically constrained model by the more robust output from the approximate

[11] This includes standard FFT-based as well as more recent deep learning based DCFs (e.g., [5,13]).

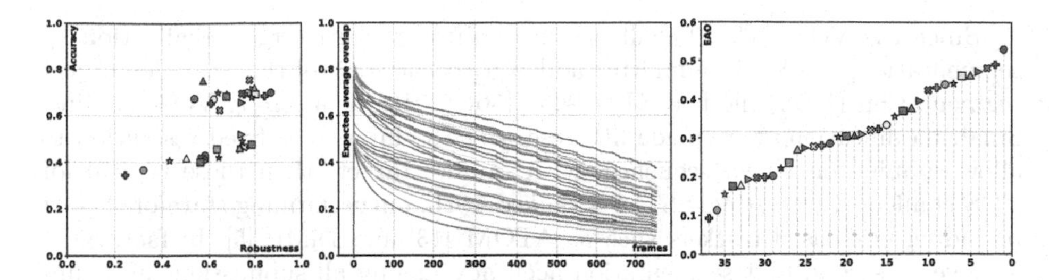

Fig. 6. The VOT-ST2020 AR-raw plots generated by sequence pooling (left) and EAO curves (center) and the VOT-ST2020 expected average overlap graph with trackers ranked from right to left. The right-most tracker is the top-performing according to the VOT-ST2020 expected average overlap values. The dashed horizontal line denotes the average performance of ten state-of-the-art trackers published in 2018 and 2019 at major computer vision venues. These trackers are denoted by gray circle in the bottom part of the graph.

localization stage. This tracker significantly stands out from the rest according to the EAO measure.

The second-best ranked tracker is OceanPlus A.15. This is a multi-stage tracker based on Siamese region proposal nets SiamDW [91] (a top-performer of several VOT2019 sub-challenges) that matches template features in three parallel branches with various filter dilation levels. Fused outputs are used to predict the target bonding box akin to Fcos [69] and DiMP [5] is applied for increased robustness. Attention maps akin to TVOS [89] are computed and a UNet-like architecture [61] is then applied to fuse it with the correlation features into the final segmentation. The tracker shows a comparable robustness to the top performer.

The third top-performing tracker is AlphaRef A.12. This is a two-stage tracker that applies DiMP [5] to localize the target region. The region is then passed to a refine network akin to the one used in [60] to merge pixel-wise correlation and non-local layer outputs into the final segmentation mask.

The top-three trackers stand out from the rest in different performance measures. RPT and OceanPlus are much more robust than the other trackers, meaning that they remain on the target for longer periods. Combined with their very accurate target segmentation mask estimation, they achieve a top EAO. The third tracker, AlphaRef, also obtains a very high EAO, but not due to robustness – its robustness is actually lower than a lower-ranked fastOcean. The very high EAO can be attributed to the high accuracy. This tracker achieves a remarkable 0.753 localization accuracy, meaning that the segmentation masks are of much higher quality than the competing trackers whenever the target is well localized. From the submitted tracker descriptions, we can speculate that all three trackers with top accuracy (AlphaRef, AFOD and LWTL) apply similar approaches for segmentation. This comes at a cost of a reduced robustness of several percentage points compared to the two top EAO performers.

Since the VOT-ST2020 challenge has shifted toward target localization by segmentation, the VOT committee added a recent state-of-the-art video object segmentation (VOS) method STM A.36 [56] (2019) as a strong VOS baseline. Small modifications were made like rescaling the input to a fixed resolution to allow running on longer sequences with smaller targets than those typical for VOS challenge. Interstingly STM is ranked 19th, outperforming state-of-the-art bounding-box-based trackers such as ATOM [13] and DiMP [5]. In fact, STM achieves a second-best segmentation accuracy among all submissions and runs with decent robustness – for example, it outperforms an improved bounding-box tracker SuperDiMP. These results show a great tracking potential for the video object segmentation methods.

The trackers which have been considered as baselines or state-of-the-art a few years ago (e.g., SiamFc, KCF, IVT, CSRpp, ASMS) are positioned at the lower part of the AR-plots and at the tail of the EAO rank list, and even some of the recent state-of-the-art like ATOM [13] and DiMP [5] are ranked in the lower third of the submissions. This is a strong indicator of the advancements made in the field. Note that 6 of the tested trackers have been published in major computer vision conferences and journals in the last two years (2019/2020). These trackers are indicated in Fig. 6, along with their average performance (EAO = 0.3173), which constitutes the VOT2020 state-of-the-art bound. Approximately 46% of submitted trackers exceed this bound, which speaks of significant pace of advancements made in tracking within a span of only a few years.

Table 1. VOT-ST2020 tracking difficulty with respect to the following visual attributes: camera motion (CM), illumination change (IC), motion change (MC), occlusion (OC) and size change (SC).

	CM	IC	OC	SC	MC
Accuracy	0.53③	0.54	0.45①	0.54	0.51②
Robustness	0.70	0.77	0.60①	0.69	0.63②

The per-attribute robustness analysis is shown in Fig. 7 for individual trackers. The overall top performers remain at the top of per-attribute ranks as well. None of the trackers consistently outperforms all others on all attributes, but RPT is consistently among the top two trackers. According to the median failure over each attribute (Table 1) the most challenging attributes remain occlusion and motion change as in VOT2019. The drop on these two attributes is consistent for all trackers (Fig. 7). Illumination change, motion change and scale change are challenging, but comparatively much better addressed by the submitted trackers.

The VOT-ST2020 Challenge Winner. Top five trackers from the baseline experiment (Table 7) were re-run on the sequestered dataset. Their scores

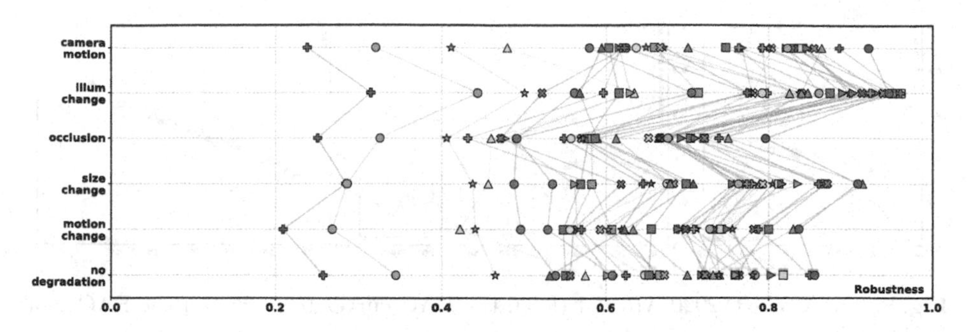

Fig. 7. Robustness with respect to the visual attributes.

obtained on sequestered dataset are shown in Table 2. The top tracker according to the EAO is RPT A.8 and is thus the VOT-ST2020 challenge winner.

Table 2. The top five trackers from Table 7 re-ranked on the VOT-ST2020 sequestered dataset.

	Tracker	EAO	A	R
1.	RPT	0.547①	0.766	0.850
2.	AFOD	0.536②	0.795	0.816
3.	LWTL	0.526③	0.781	0.822
4.	OceanPlus	0.513	0.760	0.818
5.	AlphaRef	0.510	0.823	0.762

4.2 The VOT-RT2020 Challenge Results

Trackers Submitted. The trackers that entered the VOT-ST2020 challenge were also run on the VOT-RT2019 challenge. Thus the statistics of submitted trackers was the same as in VOT-ST2020. For details please see Sect. 4.1 and Appendix A.

Results. The EAO scores and AR-raw plots for the real-time experiments are shown in Fig. 8 and Table 7. The top ten real-time trackers are AlphaRef A.12, OceanPlus A.15, AFOD A.26, fastOcean A.16, Ocean A.10, D3S A.1, AFAT A.24, SiamMargin A.17, LWTL A.23 and TRASTmask A.9.

The top three trackers, AlphaRef, OceanPlus and AFOD are ranked 3rd, 2nd and 4th on the VOT-ST2020 challenge, respectively. These, in addition to fastOcean stand out from the rest in EAO, owing this to an excellent robustness. AlphaRef has slightly lower robustness than OceanPlus, but a much better accuracy, which results in a higher EAO.

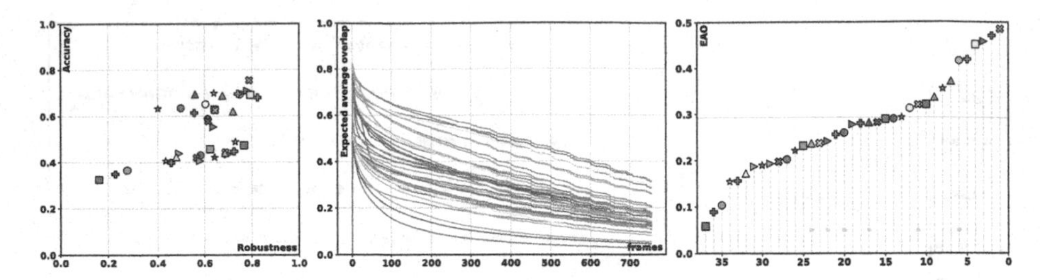

Fig. 8. The VOT-RT2020 AR plot (left), the EAO curves (center) and the EAO plot (right).

Astonishingly, 8 out of 10 top real-time trackers are among the top ten performers on VOT-ST challenge. This is in stark contrast to the previous years, where most of the top performers from VOT-ST challenge substantially dropped in ranks under the realtime constraint. The two additional trackers among to 10 are SiamMargin and TRASTmask. SiamMargin is the VOT-RT2019 challenge winner based on SiamRPN++ [41], while TRASTmask is a teacher-student network that uses DiMP [5] for the teacher for bounding box prediction. Both trackers apply SiamMask [74] for final segmentation.

Seven trackers (AlphaRef, OceanPlus, AFOD, fastOcean, Ocean, D3S and AFAT) outperform the VOT-RT2019 winner SiamMargin, which shows that the real-time performance bar has been substantially pushed forward this year. The tracking speed obviously depends on the hardware used, but overall, we see emergence of deep tracking architectures that no longer sacrifice speed and computational efficiency for performance (or vice versa).

Like in VOT-ST2020 challenge, 6 of the tested trackers have been published in major computer vision conferences and journals in the last two years (2019/2020). These trackers are indicated in Fig. 8, along with their average performance (EAO = 0.2932), which constitutes the VOT2020 realtime state-of-the-art bound. Approximately 32% of submitted trackers exceed this bound, which is slightly lower than in the VOT-ST2020 challenge.

The VOT-RT2020 Challenge Winner. According to the EAO results in Table 7, the top performer and the winner of the real-time tracking challenge VOT-RT2020 is AlphaRef (A.12).

4.3 The VOT-LT2020 Challenge Results

Trackers Submitted. The VOT-LT2020 challenge received 5 valid entries. The VOT2020 committee contributed additional three top performers from VOT-LT2019 as baselines, thus 8 trackers were considered in the challenge. In the following we briefly overview the entries and provide the references to original papers in Appendix B where available.

All participating trackers were categorized as LT_1 according to the ST-LT taxonomy from Sect. 1.3 in that they implemented explicit target re-detection. All methods are based on convolutional neural networks. Several methods are based on region proposal networks akin to [41] for approximate target localization at detection stage (Megtrack B.1, SPLT B.2, LTDSE B.7) and several approaches apply the MDNet classifier [55] for target presence verification (LTMUB B.3, ltMDNet B.5, CLGS B.6). One tracker is based purely on a deep DCF and applies the DCF for localization as well as for the detection module (RLTDiMP B.4) and one tracker applies an ensamble for improved robustness and accuracy (SiamDWLT B.8). Four trackers update their short-term and long-term visual models only when confident (Megtrack, LTMUB, RLTDiMP, LTDSE), while SPLT never updates the visual models, LTMDNet updates the short-term visual model at fixed intervals, but keeps the long-term model fixed, CLGS never updates the short-term model and updates the long-term model at fixed intervals, and SiamDWLT applies PN learning to update both visual models.

Table 3. List of trackers that participated in the VOT-LT2020 challenge along with their performance scores (Pr, Re, F-score) and ST/LT categorization.

Tracker	Pr	Re	F-Score
●LT_DSE	0.715②	0.677③	0.695①
✚LTMU_B	0.701	0.681②	0.691②
✖Megtrack	0.703③	0.671	0.687③
▶CLGS	0.739①	0.619	0.674
▲RLT_DiMP	0.657	0.684①	0.670
□SiamDW_LT	0.678	0.635	0.656
★ltMDNet	0.649	0.514	0.574
●SPLT	0.587	0.544	0.565

Results. The overall performance is summarized in Fig. 9 and Table 3. The top-three performers are LTDSE B.7, LTMUB B.3 and Megtrack B.1. LTDSE is the winner of the VOT-LT2019 challenge as was included by the VOT2020 committee as a strong baseline. This tracker applies a DCF [13] short-term tracker on top of extended ResNet18 features for initial target localization. The target position is refined by a SiamMask [74] run on the target initial position. The target presence is then verified by RT-MDNet [29]. If the target is deemed absent, an image-wide re-detection using a region proposal network akin to MBMD [90] is applied. The region proposals are verified by the online trained verifier.

LTMUB architecture is composed of a local tracker, verifier, global detector and meta-updater. Similarly to LTSDE, the short-term tracker is a combination

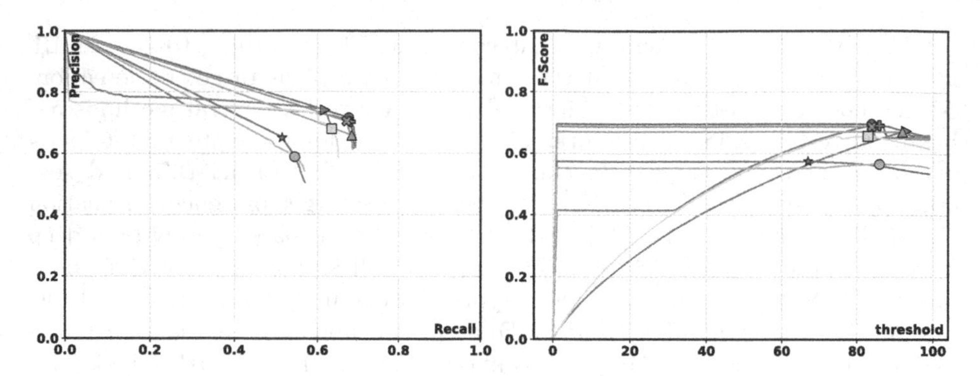

Fig. 9. VOT-LT2020 challenge average tracking precision-recall curves (left), the corresponding F-score curves (right). Tracker labels are sorted according to maximum of the F-score.

of DiMP [5] and SiamMask [74]. Adaptation of the MDNet [55] is used for a verifier, an LSTM-based meta updater from [11] to decide whether to update and [26] is used for image-wide re-detection. According to the authors, the method can be thought of as a simplified version of LTDSE.

Megtrack architecture applies a short-term tracker composed of ATOM [13] and SiamMask [74] for inter-frame target localization and presence verification. Target re-detection is performed by GlobalTrack [26] within a gradually increasing search region and verified using a combination of online-learned real-time MDNet [29] and offline-learned one-shot matching module. Short-term tracker is re-initialized on the re-detected target.

LTDSE achieves an overall best F-measure and slightly surpasses LTMUB (by 0.6%). LTDSE has a better precision, meaning that its target detections more reliably contain the target. On the other hand, LTMUB recovers more target positions, but at a cost of a reduced precision.

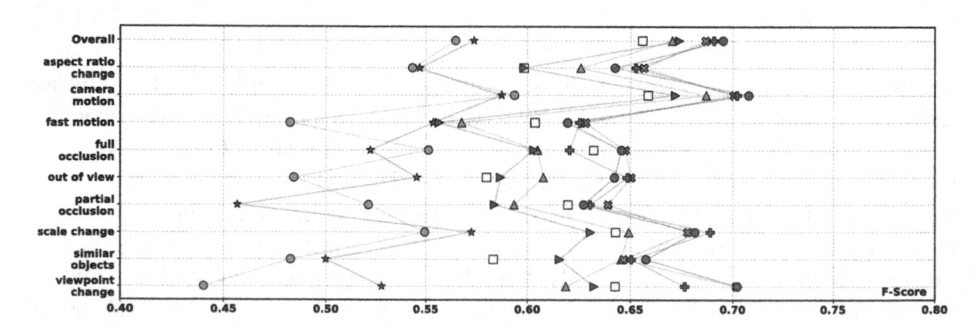

Fig. 10. VOT-LT2020 challenge maximum F-score averaged over overlap thresholds for the visual attributes. The most challenging attributes are partial and full occlusion and out-of-view.

Figure 10 shows tracking performance with respect to nine visual attributes from Sect. 3.3. The most challenging attributes are fast motion, partial and full occlusion and target leaving the field of view (out-of-view attribute).

The VOT-LT2020 Challenge Winner. According to the F-score, the top-performing tracker is LTDSE, closely followed by LTMUB. LTDSE was provided by the VOT committee as a baseline tracker and as such does not compete for the winner of the VOT-LT2020. Thus the winner of the VOT-LT2020 challenge is LTMUB B.3.

4.4 The VOT-RGBT2020 Challenge Results

Trackers Submitted. In all, 5 entries were submitted to the VOT-RGBT2020 challenge. All submissions included the source code that allowed verification of the results if required. Two additional trackers were contributed by the VOT committee: mfDiMP C.6 [87] and SiamDW-T C.7. Thus in total 7 trackers were compared on VOT-RGBT2020. In what follows we briefly overview the entries and provide the references to original papers in the Appendix C where available.

All five submitted trackers use discriminative models with a holistic representation. 2 trackers (40%) were categorized as ST_1 and 3 trackers (60%) as ST_0. All 5 trackers applied a locally uniform dynamic model.

The trackers were based on various tracking principles: 4 trackers (80%) are single-stage trackers based on discriminative correlation filters (M2C2Frgbt C.1, JMMAC C.2, AMF C.3, and SNDCFT C.4) and 1 tracker (20%) is a multi-stage tracker based on a Siamese network (DFAT C.5). Respectively 1 tracker (20%) makes use of subspace methods (M2C2Frgbt C.1) and RANSAC (JMMAC C.2). Most of the trackers (80%) use deep features, only M2C2Frgbt C.1 uses hand-crafted features. Except for JMMAC C.2 and SNDCFT C.4, all deep-feature-based trackers train their backbones.

Results. The results are summarized in the AR-raw plots, EAO curves, and the expected average overlap plots in Fig. 11. The values are also reported in Table 4.

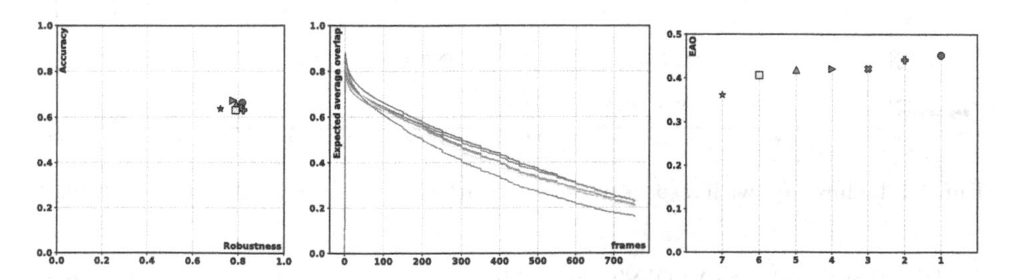

Fig. 11. The VOT-RGBT2020 AR plot (left), the EAO curves (center), and the EAO plot (right). The legend is given in Table 4.

Table 4. The ranking of the five submitted trackers and the two top-ranked trackers from VOT-RGBT2019 on the VOT-RGBT2020 public dataset.

Tracker	EAO	A	R
●JMMAC	0.420①	0.662②	0.818②
✚AMF	0.412②	0.630	0.822①
✖DFAT	0.390③	0.672①	0.779
▶SiamDW-T	0.389	0.654③	0.791
▲mfDiMP	0.380	0.638	0.793③
□SNDCFT	0.378	0.630	0.789
★M2C2Frgbt	0.332	0.636	0.722

The top performer on the public dataset is JMMAC C.2 with an EAO score of 0.420. This tracker thus repeats its top rank on the public dataset from 2019, even though it does not perform backbone training.

The second-best ranked tracker is AMF C.3 with an EAO score of 0.412. It follows the standard recipe of singe-stage discriminative correlation filter applied to deep features with backbone training.

The third top-performing position is taken by DFAT C.5, the only Siamese-network-based tacker among the submissions, with an EAO score of 0.390.

Since this has been the second RGBT-challenge within VOT, we can compare the newly submitted trackers to top-performing trackers from 2019: JMMAC C.2 (also submitted 2020), SiamDW-T C.7, and mfDiMP C.6. In comparison to these three trackers, only AMF C.3 and DFAT C.5 beat previous top-performers. Note that EAO scores from 2019 and 2020 differ due to the different restart policies. Note further that the number of trackers from 2019 is too small to introduce a state-of-the-art bound as for the VOT-ST challenge.

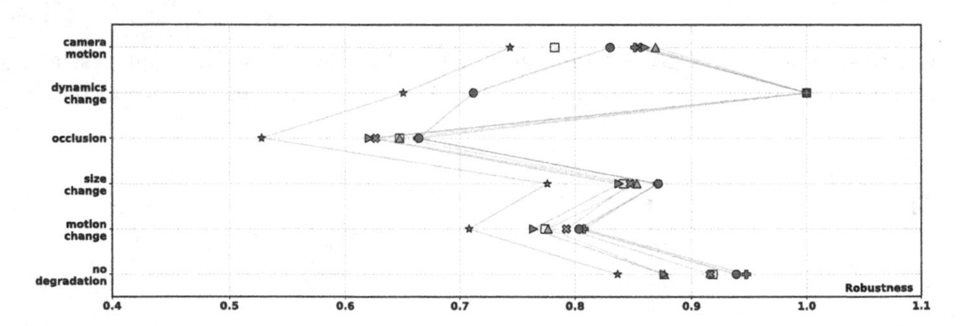

Fig. 12. Failure rate with respect to the visual attributes. The legend is given in Table 4.

However, similar to VOT-ST, we analyzed the number of failures with respect to the visual attributes (replacing illumination change with dyunamics change), see Fig. 12. The overall top performers remain at the top of per-attribute ranks as well, with the only exception that JMMAC shows degraded performance for dynamics changes. SiamDW-T and mfDiMP perform comparably weak if no degradation is present. The most challenging attributes in terms of failures are occlusion and motion change. Dynamics change is the least challenging attribute.

The VOT-RGBT2020 Challenge Winner. The top five trackers (besides 2019 top-performers AMF C.3 and DFAT C.5) were re-run on the sequestered dataset and their scores are shown in Table 5. Interestingly, the order of trackers has changed significantly. Looking at the individual scores, it becomes evident that the trackers submitted by the committee, SiamDW-T C.7 and mfDiMP C.6, were the only trackers with better EAO-score on the sequestered dataset than on the public dataset. The top tracker according to the EAO that has not been submitted by the committee is DFAT C.5, which achieved basically the same EAO-score on both datasets, and is thus the VOT-RGBT2020 challenge winner.

Table 5. The top five trackers from Table 4 re-ranked on the VOT-RGBT2020 sequestered dataset.

	Tracker	EAO	A	R
1.	SiamDW-T	0.403①	0.664①	0.702③
2.	mfDiMP	0.402②	0.623③	0.734①
3.	DFAT	0.385③	0.654②	0.674
4.	AMF	0.373	0.590	0.705②
5.	JMMAC	0.158	0.576	0.287

4.5 The VOT-RGBD2020 Challenge Results

Trackers Submitted. The VOT-RGBD2020 challenge received 4 valid entries: ATCAIS (D.1), DDiMP (D.2), CLGS_D (D.3) and Siam_LTD (D.4). We also included the best and the third best tracker from the previous year (VOT-RGBD2019): SiamDW_D and LTDSEd. The previous version of ATCAIS was submitted in 2019 as well and obtained the second best F-score (0.676). In addition, to study the performance gap between the best RGB and RGBD trackers the best performing RGB trackers from the VOT-LT2020 and VOT-ST2020 challenges were included: LTMU_B, Megtrack, RLT_DiMP, RPT, OceanPlus and AlphaRef. In total, 12 trackers were considered for the challenge. In the following we briefly overview the entries.

ATCAIS is based on the ATOM tracker [13] and HTC instance segmentation [9]. In ATCAIS the depth channel is used to detect occlusion and disappearance and in target re-detection. DDiMP is an extension of the original DiMP RGB tracker. DDiMP uses better features from ResNet50 and depth information is used to robustify scale changes during tracking. CLGS_D tracker utilizes a set of deep architectures (SiamMask, FlowNetV2, CenterNet and MDNet) and uses the optical flow (FlowNet) and depth maps to filter the region proposals for target re-detection. Similar to the other RGBD trackers Siam_LTD is based on deep architectures, but it is unclear how the depth information is integrated to the processing pipeline.

The SoTA RGB trackers are explained in more details in Sect. 4.3 and the two RGB-D trackers from the previous year can be found in the VOT2019 report [31].

Table 6. List of trackers that participated in the VOT-RGBD2020 challenge along with their performance scores (Pr, Re, F-score) and categorizations (ST/LT, RGB/RGBD). 2020 submissions are ATCAIS, DDiMP, CLGS_D and Siam_LTD. SiamDW_D and LTDSEd are 2019 submissions (SiamDW_D was the winner). RGB trackers are the three top performers of VOT-ST2020 and VOT-LT2020.

Tracker	Pr	Re	F-Score	ST/LT	RGB/RGBD
ATCAIS	0.709②	0.696①	0.702①	LT	RGBD
DDiMP	0.703③	0.689②	0.696②	ST	RGBD
CLGS_D	0.725①	0.664	0.693③	LT	RGBD
SiamDW_D	0.677	0.685③	0.681	LT	RGBD
LTDSEd	0.674	0.643	0.658	LT	RGBD
RLT_DiMP	0.625	0.632	0.629	LT	RGB
LTMU_B	0.680	0.581	0.626	LT	RGB
Megtrack	0.694	0.551	0.614	LT	RGB
RPT	0.601	0.546	0.572	ST	RGB
Siam_LTD	0.626	0.489	0.549	LT	RGBD
OceanPlus	0.577	0.502	0.537	ST	RGB
AlphaRef	0.491	0.547	0.518	ST	RGB

Results. The overall performances are summarized in Fig. 13 and Table 6. *ATCAIS* obtains the highest F-score in 2020 while it obtained the second best in 2019. The improvement on the same data is from 0.676 (F-score) to 0.702 while the last year winner (SiamDW_D) obtains 0.681. All the results are based on the submitted numbers, but these were verified by running the codes multiple times.

The three best RGBD trackers, ATCAIS, DDiMP and CLGS_D, provide better results than the last year winner, SiamDW_D, but the improvement of the best (ATCAIS) is only 3%. Moreover, the Precision, Recall and F-score values of the three best trackers are within 1.2% (F-score) to 4.5% (Recall) and the numbers are similar to VOT-LT2020 challenge which indicate that the results are saturating and a new dataset is needed to make the RGBD data more challenging.

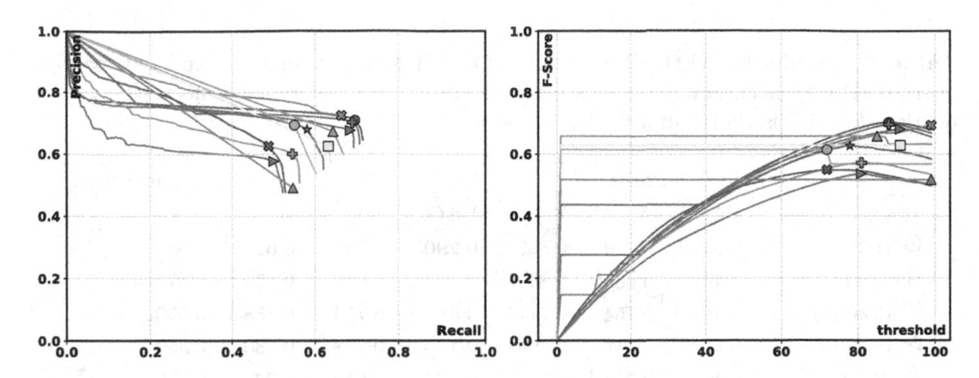

Fig. 13. VOT-RGBD2020 challenge average tracking precision-recall curves (left), the corresponding F-score curves (right). The tracker labels are sorted according to the maximum of the F-score.

ATCAIS is the best performer in 8 out of the 14 attribute categories (Fig. 14). It is noteworthy that in 2019 competition ATCAIS performance was particularly poor on *full occlusion* and *out-of-frame* categories, on which ATCAIS substantially improved this year. The second best RGBD tracker DDiMP has very similar per category performance to ATCAIS and DDiMP obtains better performance on *deformable*, *full occlusion. occlusion*, *out-of-frame*, and *similar objects* categories. On the other hand, the performance of the RLT_DiMP RGB tracker is moderately good across all attribute categories despite of not using the depth channel at all.

The VOT-RGBD2020 Challenge Winner. It should be noted that there are only minor differences among the four best RGBD trackers and the last year winner, SiamDW_D, is among them. They all achieve the maximum F-measure

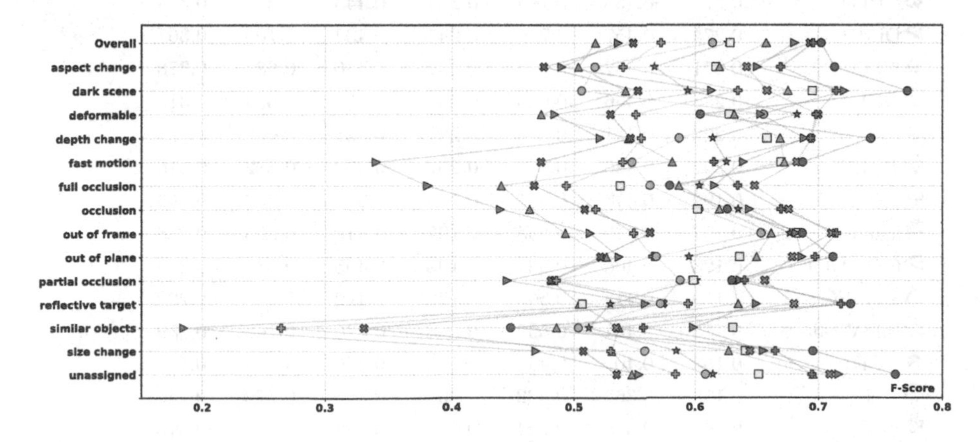

Fig. 14. VOT-RGBD2020 challenge: tracking performance w.r.t. visual attributes.

Table 7. Results for VOT-ST2020 and VOT-RT2020 challenges. Expected average overlap (EAO), accuracy and robustness are shown. For reference, a no-reset average overlap AO [76] is shown under *Unsupervised*.

	VOT-ST2020			VOT-RT2020			Unsupervised
Tracker	EAO	A	R	EAO	A	R	AO
RPT	0.530①	0.700	0.869①	0.290	0.587	0.614	0.632①
OceanPlus	0.491②	0.685	0.842②	0.471②	0.679	0.824①	0.575③
AlphaRef	0.482③	0.754①	0.777	0.486①	0.754①	0.788③	0.590②
AFOD	0.472	0.713	0.795	0.458③	0.708②	0.780	0.539
LWTL	0.463	0.719③	0.798	0.337	0.619	0.720	0.570
fastOcean	0.461	0.693	0.803③	0.452	0.691	0.792②	0.566
DET50	0.441	0.679	0.787	0.189	0.633	0.401	0.524
D3S	0.439	0.699	0.769	0.416	0.693	0.748	0.508
Ocean	0.430	0.693	0.754	0.419	0.695	0.741	0.533
TRASFUSTm	0.424	0.696	0.745	0.282	0.576	0.616	0.524
DESTINE	0.396	0.657	0.745	0.278	0.552	0.638	0.463
AFAT	0.378	0.693	0.678	0.372	0.687	0.676	0.502
TRASTmask	0.370	0.684	0.677	0.321	0.628	0.643	0.494
SiamMargin	0.356	0.698	0.640	0.355	0.698③	0.640	0.465
SiamMask_S	0.334	0.671	0.621	0.312	0.651	0.604	0.449
VPU_SiamM	0.323	0.652	0.609	0.280	0.613	0.555	0.405
siammask	0.321	0.624	0.648	0.320	0.624	0.645	0.405
SiamEM	0.310	0.520	0.743	0.187	0.438	0.491	0.418
STM	0.308	0.751②	0.574	0.282	0.694	0.559	0.445
SuperDiMP	0.305	0.477	0.786	0.289	0.472	0.767	0.417
DPMT	0.303	0.492	0.745	0.293	0.487	0.730	0.383
A3CTDmask	0.286	0.673	0.537	0.260	0.634	0.498	0.371
TRAT	0.280	0.464	0.744	0.256	0.445	0.724	0.367
UPDT	0.278	0.465	0.755	0.237	0.443	0.688	0.374
DiMP	0.274	0.457	0.740	0.241	0.434	0.700	0.367
ATOM	0.271	0.462	0.734	0.237	0.440	0.687	0.378
DCDA	0.236	0.456	0.635	0.232	0.456	0.624	0.315
igs	0.222	0.421	0.643	0.221	0.421	0.643	0.286
TCLCF	0.202	0.430	0.582	0.202	0.430	0.582	0.216
FSC2F	0.199	0.416	0.581	0.156	0.397	0.456	0.269
asms	0.197	0.419	0.565	0.197	0.419	0.565	0.256
CSR-DCF	0.193	0.406	0.582	0.193	0.405	0.580	0.242
SiamFC	0.179	0.418	0.502	0.172	0.422	0.479	0.229
InfoVital	0.175	0.401	0.562	0.058	0.326	0.163	0.239
KCF	0.154	0.407	0.432	0.154	0.406	0.434	0.178
MIL	0.113	0.367	0.322	0.104	0.366	0.276	0.146
IVT	0.092	0.345	0.244	0.089	0.349	0.229	0.096

near the same Precision-Recall region (Fig. 13) and differences between their Recall, Precision and F-score values are from 1% to 4%. The five best trackers are RGBD trackers which indicates the importance of the depth cue.

The winner is selected based on the best F-score and is ATCAIS (F-score 0.702). For the winning F-score ATCAIS also obtains the best recall (0.696) and the second best precision (0.709). ATCAIS also obtains the best performance on 7 out of 13 assigned attributes for all sequences (inc. "unassigned"). According to the VOT winner rules, the VOT-RGBD2020 challenge winner is therefore ATCAIS (D.1).

5 Conclusion

Results of the VOT2020 challenge were presented. The challenge is composed of the following five challenges focusing on various tracking aspects and domains: (i) the VOT2020 short-term RGB tracking challenge (VOT-ST2020), (ii) the VOT2020 short-term real-time RGB tracking challenge (VOT-RT2020), (iii) the VOT2020 long-term RGB tracking challenge (VOT-LT2020), (iv) the VOT2020 short-term RGB and thermal tracking challenge (VOT-RGBT2020) and (v) the VOT2020 long-term RGB and depth (D) tracking challenge (VOT-RGBD2020).

Several novelties were introduced in the VOT2020 challenge. A new VOT short-term performance evaluation methodology was introduced. The new methodology is an extension of the VOT2019 methodology that avoids tracker-dependent re-starts and addresses short-term failures. Another important novelty is transition from bounding boxes to segmentation masks in the VOT-ST challenge. Both, the VOT-ST public and sequestered dataset were refreshed and the targets were manually segmented in each frame.

A major technical novelty is transition to a new Python toolkit that implements the VOT2020 challenges and the new performance evaluation protocols. This toolkit will be extended to support VOT2019 challenges in near future, after which the previous Matlab-based toolkits will be made obsolete and only the VOT Python toolkit will be maintained.

The overall results of the VOT-ST2020 challenges show that the majority of tested trackers apply either deep discriminative correlation filters or Siamese networks. Majority of trackers report a segmentation mask, including the top performers. Interestingly, results show that video-object-segmentation state-of-the-art method STM [56] obtained competitive performance and outperformed several state-of-the-art bounding box trackers. Another observation is that, 8 out of 10 top real-time trackers (VOT-RT2020 challenge) are among the top ten performers on the VOT-ST2020, which shows emergence of deep learning architectures that no longer sacrifice the speed for tracking accuracy (assuming a sufficiently powerful GPU is available). As in VOT2019 short-term challenges, the most difficult attributes remain occlusion and motion change.

The VOT-LT2020 challenge top performers apply short-term localization and long-term re-detection tracker structure. Similarly to VOT2019, the dominant methodologies are deep DCFs [5,13] and Siamese correlation [4], region proposals and online trained CNN classifiers [55].

The participating trackers of the VOT-RGBT2020 challenge did not go significantly beyond the top-performers from VOT-RGBT2019. It was even observed that some of the participating trackers over-fitted to the public dataset such that the two top-ranked trackers from the VOT-RGBT2019 sequestered dataset are still top ranked in 2020. Notably, Siamese network approaches seem to have bypassed DCF-based trackers now, even though with a small margin.

All trackers submitted to the VOT-RGBD2020 challenge are based on the SoTA deep RGB trackers. Depth information is used to improve occlusion detection and target re-detection. The five best RGBD trackers (three of them submitted 2020 and two 2019) are better than the best RGB-only tracker with a clear margin. The results indicate that the depth provides complementary information for visual object tracking and therefore more research and new datasets are expected for RGBD tracking.

The top performer on the VOT-ST2020 *public dataset* is RPT A.8. This is a two-stage tracker that integrates a state-of-the-art region proposal network [84] with a state-of-the-art segmentation tracker [50]. On the public set, this tracker obtains a significantly better performance than the second-best tracker. RPT is also the top tracker on the sequestered dataset, on which the performance difference to the second-best is still large, albeit reduced compared to the difference observed on the public set. RPT A.8 is thus a clear winner of the VOT-ST2020 challenge.

The top performer and the winner of the VOT-RT2020 challenge is AlphaRef A.12. This is a two-stage tracker that applies DiMP [5] to localize the target region and refines it with a network akin to [60] to merge pixel-wise correlation and non-local layer outputs into the final segmentation mask. This tracker is also ranked quite high (3rd) on the public VOT-ST2020 challenge.

The top performer of the VOT-LT2020 challenge is LTDSE B.7, wich combines a CNN-based DCF [13] with a Siamese segmentation tracker [74] and an fast version of an online trained CNN classifier [55]. This tracker is also the VOT-LT2019 challenge winner, which was included by the VOT2020 committee as a strong baseline. The top submitted tracker and the winner of the VOT-LT2020 challenge, is LTMUB, which, accroding to the authors, can be considered as a simplified version of the LTDSE.

The top performer on the VOT-RGBT2020 *public dataset* is JMMAC (C.2), an approach that combines DCF-based tracking with RANSAC. The top-ranked participating tracker on the sequestered dataset and the VOT-RGBT2020 challenge winner is DFAT (C.5), the only Siamese-network-based participating tracker.

The top performer and the winner of the VOT-RGBD2020 challenge is ATCAIS (D.1) that improved its rank from the last year second place to this year first place. ATCAIS is based on the ATOM tracker [13] and HTC instance segmentation [9]. The depth value is used to detect the target occlusion or disappearance and re-find the target. For the winning F-score (0.702), ATCAIS obtains the best performance on 8 out of 14 attributes.

The VOT primary objective is to establish a platform for discussion of tracking performance evaluation and contributing to the tracking community with verified annotated datasets, performance measures and evaluation toolkits. The VOT2020 was the eighth effort toward this, following the very successful VOT2013, VOT2014, VOT2015, VOT2016, VOT2017, VOT2018 and VOT2019.

This VOT edition started a transition to a fully segmented ground truth. We believe this will boost the research in tracking, which will result in a class of robust trackers with per-pixel target localization. In future editions we expect more sub-challenges to follow this direction, depending on man-power, as producing high-quality segmentation ground truth requires substantial efforts. A new Python toolkit that implements the new evaluation protocols follows the trend of majority of trackers transiting to Python as the main programming language. Our future work will follow these lines of advancements.

Acknowledgements. This work was supported in part by the following research programs and projects: Slovenian research agency research programs P2-0214, Z2-1866, P2-0094, Slovenian research agency project J2-8175. Jiři Matas and Ondrej Drbohlav were supported by the Czech Science Foundation Project GACR P103/12/G084. Aleš Leonardis was supported by MURI project financed by MoD/Dstl and EPSRC through EP/N019415/1 grant. Michael Felsberg and Linbo He were supported by WASP, VR (ELLIIT and NCNN), and SSF (SymbiCloud). Roman Pflugfelder and Gustavo Fernández were supported by the AIT Strategic Research Programme 2020 Visual Surveillance and Insight. The challenge was sponsored by the Faculty of Computer Science, University of Ljubljana, Slovenia.

A VOT-ST2020 and VOT-RT2020 Submissions

This appendix provides a short summary of trackers considered in the VOT-ST2020 and VOT-RT2020 challenges.

A.1 Discriminative Sing-Shot Segmentation Tracker (D3S)

A. Lukezic
alan.lukezic@fri.uni-lj.si

Template-based discriminative trackers are currently the dominant tracking paradigm due to their robustness, but are restricted to bounding box tracking and a limited range of transformation models, which reduces their localization accuracy. We propose a discriminative single-shot segmentation tracker named D3S [50], which narrows the gap between visual object tracking and video object segmentation. A single-shot network applies two target models with complementary geometric properties, one invariant to a broad range of transformations, including non-rigid deformations, the other assuming a rigid object to simultaneously achieve high robustness and online target segmentation.

A.2 Visual Tracking by Means of Deep Reinforcement Learning and an Expert Demonstrator (A3CTDmask)

M. Dunnhofer, G. Foresti, C. Micheloni
{*matteo.dunnhofer, gianluca.foresti, christian.micheloni*}*@uniud.it*

A3CTDmask is the combination of the A3CTD tracker [16] with a one-shot segmentation method for target object mask generation. A3CTD is a real-time tracker built on a deep recurrent regression network architecture trained offline using a reinforcement learning based framework. After training, the proposed tracker is capable of producing bounding box estimates through the learned policy or by exploiting the demonstrator. A3CTDmask exploits SiamMask [74] by reinterpreting it as a one-shot segmentation module. The target object mask is generated inside a frame patch obtained through the bounding box estimates given by A3CTD.

A.3 Deep Convolutional Descriptor Aggregation for Visual Tracking (DCDA)

Y. Li, X. Ke
liyuezhou.cm@gmail.com, kex@fzu.edu.cn

This work aims to mine the target representation capability of pre-trained VGG16 model for visual tracking. Based on spatial and semantic priors, a central attention mask is designed for robust-aware feature aggregation, and an edge attention mask is used for accuracy aware feature aggregation. To make full use of the scene context, a regression loss is developed to learn a discriminative feature for complex scenes. DCDA tracker is implemented based on the Siamese network, with a feature fusion and template enhancement strategies.

A.4 IOU Guided Siamese Networks for Visual Object Tracking (IGS)

M. Dasari, R. Gorthi
{*ee18d001, rkg*}*@iittp.ac.in*

In the proposed IOU-SiamTrack framework, a new block called 'IOU module' is introduced. This module accepts the above feature domain response maps, convert them into image domain with the help of anchor boxes, as is done in the inference stage in [41,42]. Using the classification response map, top-K 'probable' bounding boxes, having top-K responses are selected. IOU module then calculates the IOU of probable bounding boxes w.r.t. estimated bounding box and produce the one with maximum IOU score as predicted output bounding box. Through training progress, predicted box is more aligned with ground truth, as network is guided to minimise the IOU loss.

A.5 SiamMask_SOLO (SiamMask_S)

Y. Jiang, Z. Feng, T. Xu, X. Song
yj.jiang@stu.jiangnan.edu.cn, {z.feng, tianyang.xu}@surrey.ac.uk,
x.song@jiangnan.edu.cn

The SiamMask_SOLO tracker is based on the SiamMask algorithm. It utilizes a multi-layer aggregation module to make full use of different levels of deep CNN features. Besides, to balance all the three branches, the mask branch is replaced by a SOLO [75] head that uses CoordConv and FCN, which improves the performance of the proposed SiamMask_SOLO tracker in terms of both accuracy and robustness. The original refined module is kept for a further performance boost.

A.6 Diverse Ensemble Tracker (DET50)

N. Wang, W. Zhou, H. Li
wn6149@mail.ustc.edu.cn, {zhwg, lihq}@ustc.edu.cn

In this work, we leverage an ensemble of diverse models to learn manifold representations for robust object tracking. Based on the DiMP method, a shared backbone network (ResNet-50) is applied for feature extraction and multiple head networks for independent predictions. To shrink the representational overlaps among multiple models, both model diversity and response diversity regularization terms are used during training. This ensemble framework is end-to-end trained in a data-driven manner. After box-level prediction, we use SiamMask for mask generation.

A.7 VPU_SiamM: Robust Template Update Strategy for Efficient Object Tracking (VPU_SiamM)

A. Gebrehiwot, J. Bescos, Á. García-Martín
awet.gebrehiwot@estudiante.uam.es, {j.bescos, alvaro.garcia}@uam.es

The VPU_SiamM tracker is an improved version of the SiamMask [74]. The SiamMask tracks without any target update strategy. In order to enable more discriminant features and to enhance robustness, the VPU_SiamM applies a target template update strategy, which leverages both the initial ground truth template and a supplementary updatable template. The initial template provides highly reliable information and increase robustness against model drift and the updatable template integrates the new target information from the predicted target location given by the current frame. During online tracking, VPU_SiamM applies both forward and backward tracking strategies by updating the updatable target template with the predicted target. The tracking decision on the next frame is determined where both templates yield a high response map (score) in the search region. Data augmentation strategy has been implemented during the training process of the refinement branch to become robust in handling motion-blurred and low-resolution datasets during inference.

A.8 RPT: Learning Point Set Representation for Siamese Visual Tracking (RPT)

H. Zhang, L. Wang, Z. Ma, W. Lu, J. Yin, M. Cheng
1067166127@qq.com, {wanglinyuan, kobebean, lwhfh01}@zju.edu.cn, {yin_jun, cheng_miao}@dahuatech.com

RPT tracker is formulated with a two-stage structure. The first stage is composed with two parallel subnets, one for target estimation with RepPoints [84] in an offline-trained embedding space, the other trained online to provide high robustness against distractors [13]. The online classification subnet is set to a lightweight 2-layer convolutional neural network. The target estimation head is constructed with Siamese-based feature extraction and matching. For the second stage, the set of RepPoints with highest confidence (i.e. online classification score) is fed into a modified D3S [50] to obtain the segmentation mask. A segmentation map is obtained by combining enhanced target location channel with target and background similarity channels. The backbone is ResNet50 pretrained on ImageNet, while the target estimation head is trained using pairs of frames from YouTube-Bounding Box [59], COCO [45] and ImageNet VID [63] datasets.

A.9 Tracking Student and Teacher (TRASTmask)

M. Dunnhofer, G. Foresti, C. Micheloni
{matteo.dunnhofer, gianluca.foresti, christian.micheloni}@uniud.it

TRASTmask is the combination of the TRAST tracker [17] with a one-shot segmentation method for target object mask generation. TRAST tracker consists of two components: (i) a fast processing CNN-based tracker, i.e. the Student; and (ii) an off-the-shelf tracker, i.e. the Teacher. The Student is trained offline based on knowledge distillation and reinforcement learning, where multiple tracking teachers are exploited. Tracker TRASTmask uses DiMP [5] as the Teacher. The target object mask is generated inside a frame patch obtained through the bounding box estimates given by TRAST tracker.

A.10 Ocean: Object-aware Anchor-free Tracking (Ocean)

Z. Zhang, H. Peng
zhangzhipeng2017@ia.ac.cn, houwen.peng@microsoft.com

We extend our object-aware anchor-free tracking framework [92] with novel transduction and segmentation networks, enabling it to predict accurate target mask. The transduction network is introduced to infuse the knowledge of the given mask in the first frame. Inspired by recent work TVOS [89], it compares the pixel-wise feature similarities between the template and search features, and then transfers the mask of the template to an attention map based on the similarities. We add the attention map to backbone features to learn target-background aware representations. Finally, a U-net shape segmentation pathway is designed to progressively refine the enhanced backbone features to target mask. The code will be completely released at https://github.com/researchmm/TracKit.

A.11 Tracking by Student FUSing Teachers (TRASFUSTm)

M. Dunnhofer, G. Foresti, C. Micheloni
{matteo.dunnhofer, gianluca.foresti, christian.micheloni}@uniud.it

The tracker TRASFUSTm is the combination of the TRASFUST tracker [17] with a one-shot segmentation method for target object mask generation. TRASFUSTm tracker consists of two components: (i) a fast processing CNN-based tracker, i.e. the Student; (ii) a pool of off-the-shelf trackers, i.e. Teachers. The Student is trained offline based on knowledge distillation and reinforcement learning, where multiple tracking teachers are exploited. After learning, through the learned evaluation method, the Student is capable to select the prediction of the best Teacher of the pool, thus performing robust fusion. Both trackers DiMP [5] and ECO [12] were chosen as Teachers. The target object mask is generated inside a frame patch obtained through the bounding box estimates given by TRASFUSTm tracker.

A.12 Alpha-Refine (AlphaRef)

B. Yan, D. Wang, H. Lu, X. Yang
yan_bin@mail.dlut.edu.cn, {wdice, lhchuan}@dlut.edu.cn,
xyang@remarkholdings.com

We propose a simple yet powerful two-stage tracker, which consists of a robust base tracker (super-dimp) and an accurate refinement module named Alpha-Refine [82]. In the first stage, super-dimp robustly locates the target, generating an initial bounding box for the target. Then in the second stage, based on this result, Alpha-Refine crops a small search region to predict a high-quality mask for the tracked target. Alpha-Refine exploits pixel-wise correlation for fine feature aggregation, and uses non-local layer to capture global context information. Besides, Alpha-Refine also deploys a delicate mask prediction head [60] to generate high-quality masks. The complete code and trained models of Alpha-Refine will be released at github.com/MasterBin-IIAU/AlphaRefine.

A.13 Hierarchical Representations with Discriminative Meta-Filters in Dual Path Network for Tracking (DPMT)

F. Xie, N. Wang, K. Yang, Y. Yao
220191672@seu.edu.cn, 20181222016@nuist.edu.cn,
yangkang779@163.con, 220191672@seu.edu.cn

We propose a novel dual path network with discriminative meta-filters and hierarchical representations to solve these issues. DPMT tracker consists of two pathways: (i) Geographical Sensitivity Pathway (GASP) and (ii) Geometrically Sensitivity Pathway (GESP). The modules in Geographical Sensitivity Pathway (GASP) are more sensitive to the spatial location of targets and distractors. Subnetworks in Geometrically Sensitivity Pathway (GESP) are designed to refine the bounding box to fit the target. According to this dual path network design, Geographical Sensitivity Pathway (GASP) should be trained to own more discriminative power between foreground and background while Geographical Sensitivity Pathway (GASP) should focus more on the appearance model of the object.

A.14 SiamMask (siammask)

Q. Wang, L. Zhang, L. Bertinetto, P.H.S. Torr, W. Hu
qiang.wang@nlpr.ia.ac.cn, {lz, luca}@robots.ox.ac.uk, philip.torr@eng.ox.ac.uk, wmhu@nlpr.ia.ac.cn

Our method, dubbed SiamMask, improves the offline training procedure of popular fully-convolutional Siamese approaches for object tracking by augmenting their loss with a binary segmentation task. In this way, our tracker gains a better instance-level understanding towards the object to track by exploiting the rich object mask representations offline. Once trained, SiamMask solely relies on a single bounding box initialisation and operates online, producing class-agnostic object segmentation masks and rotated bounding boxes. Code is publicly available at https://github.com/foolwood/SiamMask.

A.15 OceanPlus: Online Object-Aware Anchor-Free Tracking (OceanPlus)

Z. Zhang, H. Peng, Z. Wu, K. Liu, J. Fu, B. Li, W. Hu
zhangzhipeng2017@ia.ac.cn, houwen.peng@microsoft.com, Wu.Zhirong@microsoft.com, liukaiwen2019@ia.ac.cn, jianf@microsoft.com, bli@nlpr.ia.ac.cn, wmhu@nlpr.ia.ac.cn

This model is the extension of the Ocean tracker A.10. Inspired by recent online models, we introduce an online branch to accommodate to the changes of object scale and position. Specifically, the online branch inherits the structure and parameters from the first three stages of the Siamese backbone network. The fourth stage keeps the same structure as the original ResNet50, but its initial parameters are obtained through the pre-training strategy proposed in [5]. The segmentation refinement pathway is the same as Ocean. We refer the readers to Ocean tracker A.10 and https://github.com/researchmm/TracKit for more details.

A.16 fastOcean: Fast Object-Aware Anchor-Free Tracking (fastOcean)

Z. Zhang, H. Peng
zhangzhipeng2017@ia.ac.cn, houwen.peng@microsoft.com

To speed up the inference of our submitted tracker OceanPlus, we use TensorRT[12] to re-implement the model. All structure and model parameters are the same as OceanPlus. Please refer to OceanPlus A.15 and Ocean A.10 for more details.

[12] https://github.com/NVIDIA/TensorRT.

A.17 Siamese Tracker with Discriminative Feature Embedding and Mask Prediction (SiamMargin)

G. Chen, F. Wang, C. Qian
{chenguangqi, wangfei, qianchen}@sensetime.com

SiamMargin is based on the SiamRPN++ algorithm [41]. In the training stage, a discrimination loss is added to the embedding layer. In the training phase the discriminative embedding is offline learned. In the inference stage the template feature of the object in current frame is obtained by ROIAlign from features of the current search region and it is updated via a moving average strategy. The discriminative embedding features are leveraged to accommodate the appearance change with properly online updating. Lastly, the SiamMask [74] model is appended to obtain the pixel-level mask prediction.

A.18 Siamese Tracker with Enhanced Template and Generalized Mask Generator (SiamEM)

Y. Li, Y. Ye, X. Ke
liyuezhou.cm@gmail.com, yyfzu@foxmail.com, kex@fzu.edu.cn

SiamEM is a Siamese tracker with enhanced template and generalized mask generator. SiamEM improves SiamFC++ [81] by obtaining feature results of the template and flip template in the network header while making decisions based on quality scores to predict bounding boxes. The segmentation network presented in [10] is used as a mask generation network.

A.19 TRacker by Using ATtention (TRAT)

H. Saribas, H. Cevikalp, B. Uzun
{hasansaribas48, hakan.cevikalp, eee.bedirhan}@gmail.com

The tracker 'TRacker by using ATtention' uses a two-stream network which consists of a 2D-CNN and a 3D-CNN, to use both spatial and temporal information in video streams. To obtain temporal (motion) information, 3D-CNN is fed by stacking the previous 4 frames with one stride. To extract spatial information, the 2D-CNN is used. Then, we fuse the two-stream network outputs by using an attention module. We use ATOM [13] tracker and ResNet backbone as a baseline. Code is available at https://github.com/Hasan4825/TRAT.

A.20 InfoGAN Based Tracker: InfoVITAL (InfoVital)

H. Kuchibhotla, M. Dasari, R. Gorthi
{ee18m009, ee18d001, rkg}@iittp.ac.in

Architecture of InfoGAN (Generator, Discriminator and a Q-Network) is incorporated in the Tracking-By-Detection Framework using the Mutual Information concept to bind two distributions (latent code) to the target and the background samples. Additional Q Network helps in proper estimation of the assigned distributions and the network is trained offline in an adversarial fashion. During online testing, the additional information from the Q-Network is used to obtain the target location in the subsequent frames. This greatly helps to assess the drift from the exact target location from frame-to-frame and also during occlusion.

A.21 Learning Discriminative Model Prediction for Tracking (DiMP)

G. Bhat, M. Danelljan, L. Van Gool, R. Timofte
{goutam.bhat, martin.danelljan, vangool, timofter}@vision.ee.ethz.ch

DiMP is an end-to-end tracking architecture, capable of fully exploiting both target and background appearance information for target model prediction. The target model here constitutes the weights of a convolution layer which performs the target-background classification. The weights of this convolution layer are predicted by the target model prediction network, which is derived from a discriminative learning loss by applying an iterative optimization procedure. The model prediction network employs a steepest descent based methodology that computes an optimal step length in each iteration to provide fast convergence. The online learned target model is applied in each frame to perform target-background classification. The final bounding box is then estimated using the overlap maximization approach as in [13]. See [5] for more details about the tracker.

A.22 SuperDiMP (SuperDiMP)

G. Bhat, M. Danelljan, F. Gustafsson, T. B. Schön, L. Van Gool, R. Timofte
{goutam.bhat, martin.danelljan}@vision.ee.ethz.ch, {fredrik.gustafsson, thomas.schon}@it.uu.se, {vangool, timofter}@vision.ee.ethz.ch

SuperDiMP [23] combines the standard DiMP classifier from [5] with the EBM-based bounding-box regressor from [14,22]. Instead of training the bounding box regression network to predict the IoU with an L_2 loss [5], it is trained using the NCE+ approach [23] to minimize the negative-log likelihood. Further, the tracker uses better training and inference settings.

A.23 Learning What to Learn for Video Object Segmentation (LWTL)

G. Bhat, F. Jaremo Lawin, M. Danelljan, A. Robinson, M. Felsberg, L. Van Gool, R. Timofte
goutam.bhat@vision.ee.ethz.ch, felix.jaremo-lawin@liu.se, martin.danelljan@vision.ee.ethz.ch, {andreas.robinson, michael.felsberg}@liu.se, {vangool, timofter}@vision.ee.ethz.ch

LWTL is an end-to-end trainable video object segmentation VOS architecture which captures the current target object information in a compact parametric model. It integrates a differentiable few-shot learner module, which predicts the target model parameters using the first frame annotation. The learner is designed to explicitly optimize an error between target model prediction and a ground truth label, which ensures a powerful model of the target object. Given a new frame, the target model predicts an intermediate representation of the target mask, which is input to the offline trained segmentation decoder to generate the final segmentation mask. LWTL learns the ground-truth labels used by the few-shot learner to train the target model. Furthermore, a network module is trained to predict spatial importance weights for different elements in the few-shot learning loss. All modules in the architecture are trained end-to-end by maximizing segmentation accuracy on annotated VOS videos. See [7] for more details.

A.24 Adaptive Failure-Aware Tracker (AFAT)

T. Xu, S. Zhao, Z. Feng, X. Wu, J. Kittler
tianyang.xu@surrey.ac.uk, zsc960813@163.com, z.feng@surrey.ac.uk, wu_xiaojun@jiangnan.edu.cn, j.kittler@surrey.ac.uk

Adaptive Failure-Aware Tracker [80] is based on Siamese structure. First, multi-RPN module is employed to predict the central location with Resnet-50. Second, a 2-cell LSTM is established to perform quality prediction with an additional motion model. Third, fused mask branch is exploited for segmentation.

A.25 Ensemble Correlation Filter Tracking Based on Temporal Confidence Learning (TCLCF)

C. Tsai
chiyi_tsai@gms.tku.edu.tw

TCLCF is a real-time ensemble correlation filter tracker based on the temporal confidence learning method. In the current implementation, we use four different correlation filters to collaboratively track the same target. The TCLCF tracker is a fast and robust generic object tracker without GPU acceleration. Therefore, it can be implemented on the embedded platform with limited computing resources.

A.26 AFOD: Adaptive Focused Discriminative Segmentation Tracker (AFOD)

Y. Chen, J. Xu, J. Yu
{yiwei.chen, jingtao.xu, jiaqian.yu}@samsung.com

The proposed tracker is based on D3S and DiMP [5], employing ResNet-50 as backbone. AFOD calculates the feature similarity to foreground and background of the template as proposed in D3S. For discriminative features, AFOD updates the target model online. AFOD adaptively utilizes different strategies during tracking to update the scale of search region and to adjust the prediction. Moreover, the Lovasz hinge loss metric is used to learn the IoU score in offline training.

The segmentation module is trained using both databases YoutubeVOS2019 and DAVIS2016. The offline training process includes two stages: (i) BCE loss is used for optimization and (ii) the Lovasz hinge is applied for further fine tuning. For inference, two ResNet-50 models are used; one for the segmentation and another for the target.

A.27 Fast Saliency-Guided Continuous Correlation Filter-Based Tracker (FSC2F)

A. Memarmoghadam
a.memarmoghadam@yahoo.com

The tracker FSC2F is based on the ECOhc approach [12]. A fast spatio temporal saliency map is added using the PQFT approach [21]. The PQFT model utilizes intensity, colour, and motion features for quaternion representation of the search image context around the previously pose of the tracked object. Therefore, attentional regions in the coarse saliency map can constrain target confidence peaks. Moreover, a faster scale estimation algorithm is utilised by enhancing the fast fDSST method [15] via jointly learning of the sparsely-sampled scale spaces.

A.28 Adaptive Visual Tracking and Instance Segmentation (DESTINE)

S.M. Marvasti-Zadeh, J. Khaghani, L. Cheng, H. Ghanei-Yakhdan, S. Kasaei
mojtaba.marvasti@ualberta.ca, khaghani@ualberta.ca, lcheng5@ualberta.ca, hghaneiy@yazd.ac.ir, kasaei@sharif.edu

DESTINE is a two-stage method consisting of an axis-aligned bounding box estimation and mask prediction, respectively. First, DiMP50 [5] is used as the baseline tracker switching to ATOM [13] when IoU and normalized L1-distance between the results meet predefined thresholds. Then, to segment the estimated bounding box, the segmentation network of FRTM-VOS [60] uses the predicted mask by SiamMask [74] as its scores. Finally, DESTINE selects the best target mask according to the ratio of foreground pixels for two predictions. The codes are publicly released at https://github.com/MMarvasti/DESTINE.

A.29 Scale Adaptive Mean-Shift Tracker (ASMS)

Submitted by VOT Committee

The mean-shift tracker optimizes the Hellinger distance between template histogram and target candidate in the image. This optimization is done by a gradient descend. ASMS [73] addresses the problem of scale adaptation and presents a novel theoretically justified scale estimation mechanism which relies solely on the mean-shift procedure for the Hellinger distance. ASMS also introduces two improvements of the mean-shift tracker that make the scale estimation more robust in the presence of background clutter – a novel histogram colour weighting and a forward-backward consistency check. Code available at https://github.com/vojirt/asms.

A.30 ATOM: Accurate Tracking by Overlap Maximization (ATOM)

Submitted by VOT Committee

ATOM separates the tracking problem into two sub-tasks: i) target classification, where the aim is to robustly distinguish the target from the background; and ii) target estimation, where an accurate bounding box for the target is determined. Target classification is performed by training a discriminative classifier online. Target estimation is performed by an overlap maximization approach where a network module is trained offline to predict the overlap between the target object and a bounding box estimate, conditioned on the target appearance in first frame. See [13] for more details.

A.31 Discriminative Correlation Filter with Channel and Spatial Reliability - C++ (CSRpp)

Submitted by VOT Committee

The CSRpp tracker is the C++ implementation of the Discriminative Correlation Filter with Channel and Spatial Reliability (CSR-DCF) tracker [47].

A.32 Incremental Learning for Robust Visual Tracking (IVT)

Submitted by VOT Committee

The idea of the IVT tracker [62] is to incrementally learn a low-dimensional sub-space representation, adapting on-line to changes in the appearance of the target. The model update, based on incremental algorithms for principal component analysis, includes two features: a method for correctly updating the sample mean, and a forgetting factor to ensure less modelling power is expended fitting older observations.

A.33 Kernelized Correlation Filter (KCF)

Submitted by VOT Committee

This tracker is a C++ implementation of Kernelized Correlation Filter [24] operating on simple HOG features and Colour Names. The KCF tracker is equivalent to a Kernel Ridge Regression trained with thousands of sample patches around the object at different translations. It implements multi-thread multi-scale support, sub-cell peak estimation and replacing the model update by linear interpolation with a more robust update scheme. Code available at https://github.com/vojirt/kcf.

A.34 Multiple Instance Learning tracker (MIL)

Submitted by VOT Committee

MIL tracker [1] uses a tracking-by-detection approach, more specifically Multiple Instance Learning instead of traditional supervised learning methods and shows improved robustness to inaccuracies of the tracker and to incorrectly labelled training samples.

A.35 Robust Siamese Fully Convolutional Tracker (RSiamFC)

Submitted by VOT Committee

RSiamFC tracker is an extended SiamFC tracker [4] with a robust training method which puts a transformation on training sample to generate a pair of samples for feature extraction.

A.36 VOS SOTA Method (STM)

Submitted by VOT Committee

Please see the original paper for details [56].

A.37 (UPDT)

Submitted by VOT Committee

Please see the original paper for details [6].

B VOT-LT2020 Submissions

This appendix provides a short summary of trackers considered in the VOT-LT2020 challenge.

B.1 Long-Term Visual Tracking with Assistant Global Instance Search (Megtrack)

Z. Mai, H. Bai, K. Yu, X. QIu
marchihjun@gmail.com, 522184271@qq.com, valjean1832@outlook.com, qiuxi@megvii.com

Megtrack tracker applies a 2-stage method that consists of local tracking and multi-level search. The local tracker is based on ATOM [13] algorithm improved by initializing online correlation filters with backbone feature maps and by inserting a bounding box calibration branch in the target estimation module. SiamMask [74] is cascaded to further refining the bounding box after locating the centre of the target. The multi-level search uses RPN-based regression network to generate candidate proposals before applying GlobalTrack [26]. Appearance scores are calculated using both the online-learned RTMDNet [29] and the offline-learned one-shot matching module and linearly combine them to leverage the former's high robustness and the latter's discriminative power. Using a pre-defined threshold, the highest-scored proposal is considered as the current tracker state and used to re-initialize the local tracker for consecutive tracking.

B.2 Skimming-Perusal Long-Term Tracker (SPLT)

B. Yan, H. Zhao, D. Wang, H. Lu, X. Yang
{yan_bin, haojie_zhao}@mail.dlut.edu.cn, {wdice, lhchuan}@dlut.edu.cn, xyang@remarkholdings.com

This is the original SPLT tracker [83] without modification. SPLT consists of a perusal module and a skimming module. The perusal module aims at obtaining precise bounding boxes and determining the target's state in a local search region. The skimming module is designed to quickly filter out most unreliable search windows, speeding up the whole pipeline.

B.3 A Baseline Long-Term Tracker with Meta-Updater (LTMU_B)

K. Dai, D. Wang, J. Li, H. Lu, X. Yang
dkn2014@mail.dlut.edu.cn, {wdice, jianhual}@dlut.edu.cn, lhchuan@dlut.edu.cn, xyang@remarkholdings.com

The tracker LTMU_B is a simplified version of LTMU [11] and LTDSE with comparable performance adding a RPN-based regression network, a sliding-window based re-detection module and a complex mechanism for updating models and target re-localization. The short-term tracker LTMU_B contains two components. One is for target localization and based on DiMP algorithm [5] using ResNet50 as the backbone network. The update of DiMP is controlled by meta-updater which is proposed by LTMU[13]. The second component is the SiamMask network [74] used for refining the bounding box after locating the centre of the target. It also takes the local search region as the input and outputs the tight bounding boxes of candidate proposals. For the verifier, we adopts MDNet network [5] which uses VGGM as the backbone and is pre-trained on ILSVRC VID dataset. The classification score is finally obtained by sending the tracking result's feature to three fully connected layers. GlobalTrack [26] is utilised as the global detector.

B.4 Robust Long-Term Object Tracking via Improved Discriminative Model Prediction (RLTDiMP)

S. Choi, J. Lee, Y. Lee, A. Hauptmann
seokeon@kaist.ac.kr, {ljhyun33, swack9751}@korea.ac.kr, alex@cs.cmu.edu

We propose an improved Discriminative Model Prediction method for robust long-term tracking based on a pre-trained short-term tracker. The baseline tracker is SuperDiMP which combines the bounding-box regressor of PrDiMP [14] with the standard DiMP [5] classifier. To make our model more discriminative and robust, we introduce uncertainty reduction using random erasing, background augmentation for more discriminative feature learning, and random search with spatio-temporal constraints. Code available at https://github.com/bismex/RLT-DIMP.

[13] https://github.com/Daikenan/LTMU.

B.5 Long-Term MDNet (ltMDNet)

H. Fan, H. Ling
{hefan, hling}@cs.stonybrook.edu

We designate a long-term tracker by adapting MDNet [55]. In specific, we utilize an instance-aware detector [26] to generate target proposals. Then, these proposals are forwarded to MDNet for classification. Since the detector performs detection on the full image, the final tracker can locate the target in the whole image which can robustly deal with full occlusion and out-of-view. The instance-aware detector is implemented by on Faster R-CNN using ResNet-50. The MDNet is implemented as in the original paper.

B.6 (CLGS)

Submitted by VOT Committee

In this work, we develop a complementary local-global search (CLGS) framework to conduct robust long-term tracking, which is a local robust tracker based on SiamMask [74], a global detection based on cascade R-CNN [8], and an online verifier based on Real-time MDNet [29]. During online tracking, the SiamMask model locates the target in local region and estimates the size of the target according to the predicted mask. The online verifier is used to judge whether the target is found or lost. Once the target is lost, a global R-CNN detector (without class prediction) is used to generate region proposals on the whole image. Then, the online verifier will find the target from region proposals again. Besides, we design an effective online update strategy to improve the discrimination of the verifier.

B.7 (LT_DSE)

Submitted by VOT Committee

This algorithm divides each long-term sequence into several short episodes and tracks the target in each episode using short-term tracking techniques. Whether the target is visible or not is judged by the outputs from the short-term local tracker and the classification-based verifier updated online. If the target disappears, the image-wide re-detection will be conducted and output the possible location and size of the target. Based on these, the tracker crops the local search region that may include the target and sends it to the RPN based regression network. Then, the candidate proposals from the regression network will be scored by the online learned verifier. If the candidate with the maximum score is above the pre-defined threshold, the tracker will regard it as the target and re-initialize the short-term components. Finally, the tracker conducts short-term tracking until the target disappears again.

B.8 (SiamDW_LT)

Submitted by VOT Committee

SiamDW_LT is a long-term tracker that utilizes deeper and wider backbone networks with fast online model updates. The basic tracking module is a short-term Siamese tracker, which returns confidence scores to indicate the tracking reliability. When the Siamese tracker is uncertain on its tracking accuracy, an online correction module is triggered to refine the results. When the Siamese tracker is failed, a global re-detection module is activated to search the target in the images. Moreover, object disappearance and occlusion are also detected by the tracking confidence. In addition, we introduce model ensemble to further improve the tracking accuracy and robustness.

C VOT-RGBT2020 Submissions

This appendix provides a short summary of trackers considered in the VOT-RGBT2020 challenge.

C.1 Multi-model Continuous Correlation Filter for RGBT Visual Object Tracking (M2C2Frgbt)

A. Memarmoghadam
a.memarmoghadam@yahoo.com

Inspired by ECO tracker [12], we propose a robust yet efficient tracker namely as M2C2Frgbt that utilizes multiple models of the tracked object and estimates its position every frame by weighted cumulative fusion of their respective regressors in a ridge regression optimization problem [51]. Moreover, to accelerate tracking performance, we propose a faster scale estimation method in which the target scale filter is jointly learned via sparsely sampled scale spaces constructed by just the thermal infrared data. Our scale estimation approach enhances the running speed of fDSST [15] as the baseline algorithm better than 20% while maintaining the tracking performance as well. To suppress unwanted samples mostly belong to the occlusion or other non-object data, we conservatively update every model on-the-fly in a non-uniform sparse manner.

C.2 Jointly Modelling Motion and Appearance Cues for Robust RGB-T Tracking (JMMAC)

P. Zhang, S. Chen, D. Wang, H. Lu, X. Yang
pyzhang@mail.dlut.edu.cn, shuhaochn@mail.dlut.edu.cn, wdice@dlut.edu.cn, lhchuan@dlut.edu.cn, xyang@remarkholdings.com

Our tracker is based on [88], consisting of two components, i.e. multimodal fusion for appearance trackers and camera motion estimation. In multimodal fusion, we develop a late fusion method to infer the fusion weight maps of both RGB and thermal (T) modalities. The fusion weights are determined by using offline-trained global and local Multimodal Fusion Networks (MFNet), and then adopted to linearly combine the response maps of RGB and T modalities obtained from ECOs. In MFNet, the truncated VGG-M networks is used as backbone to extract deep feature. In camera motion estimation, when the drastic camera motion is detected, we compensate movement to correct the search region by key-point-based image registration technique. Finally, we employ YOLOv2 to refine the bounding box. The scale estimation and model updating methods are borrowed from ECO in default.

C.3 Accurate Multimodal Fusion for RGB-T Object Tracking (AMF)

P. Zhang, S. Chen, B. Yan, D. Wang, H. Lu, X. Yang
{pyzhang, shuhaochn, yan_bin}@mail.dlut.edu.cn, {wdice, lhchuan}@dlut.edu.cn, xyang@remarkholdings.com

We achieve multimodal fusion for RGB-T tracking by linear combining the response maps obtained from two monomodality base trackers, i.e., DiMP. The fusion weight is obtained by the Multimodal Fusion Network proposed in [88]. To achieve high accuracy, the bounding box obtained from fused DiMP is then refined by a refinement module in visible modality. The refinement module, namely Alpha-Refine, aggregates features via a pixel-level correlation layer and a non-local layer and adaptively selects the most adequate results from three branches, namely bounding box, corner and mask heads, which can predict more accurate bounding boxes. Note that the target scale estimated by IoUNet in DiMP is also applied in visible modality which is followed by Alpha-Refine and the model updating method is borrowed from DiMP in default.

C.4 SqueezeNet Based Discriminative Correlation Filter Tracker (SNDCFT)

A. Varfolomieiev
a.varfolomieiev@kpi.ua

The tracker uses FHOG and convolutional features extracted from both video and infrared modalities. As the convolutional features, the output of the 'fire2/concat' layer of the original SqueezeNet network [27] is used (no additional pre-training for the network is performed). The core of the tracker is the spatially regularized discriminative correlation filter, which is calculated using the ADMM optimizer. The calculation of the DCF filter is performed independently over different feature modalities. The filter is updated in each frame using simple exponential forgetting.

C.5 Decision Fusion Adaptive Tracker (DFAT)

H. Li, Z. Tang, T. Xu, X. Zhu, X. Wu, J. Kittler
hui_li_jnu@163.com, 1030415519@vip.jiangnan.edu.cn, tianyang.xu@surrey.ac.uk, xuefeng_zhu95@163.com, wu_xiaojun@jiangnan.edu.cn, j.kittler@surrey.ac.uk

Decision Fusion Adaptive Tracker is based on Siamese structure. Firstly, the multi-layer deep features are extracted by Resnet-50. Then, multi-RPN module is employed to predict the central location with multi-layer deep features. Finally, an adaptive weight strategy for decision level fusion is utilized to generate the final result. In addition, the template features are updated by a linear template update strategy.

C.6 Multi-modal Fusion for End-to-End RGB-T Tracking (mfDiMP)

Submitted by VOT Committee

The mfDiMP tracker contains an end-to-end tracking framework for fusing the RGB and TIR modalities in RGB-T tracking [87]. The mfDiMP tracker fuses modalities at the feature level on both the IoU predictor and the model predictor of DiMP [87] and won the VOT-RGBT2019 challenge.

C.7 Online Deeper and Wider Siamese Networks for RGBT Visual Tracking (SiamDW-T)

Submitted by VOT Committee

SiamDW-T is based on previous work by Zhang and Peng [91], and extends it with two fusion strategies for RGBT tracking. A simple fully connected layer is appended to classify each fused feature to background or foreground. SiamDW-T achieved the second rank in VOT-RGBT2019 and its code is available at https://github.com/researchmm/VOT2019.

D VOT-RGBD2020 Submissions

This appendix provides a short summary of trackers considered in the VOT-RGBD2020 challenge.

D.1 Accurate Tracking by Category-Agnostic Instance Segmentation for RGBD Image (ATCAIS)

Y. Wang, L. Wang, D. Wang, H. Lu, X. Yang
{wym097,wlj,wdice,lhchuan}@dlut.edu.cn, xyang@remarkholdings.com

The proposed tracker combines both instance segmentation and the depth information for accurate tracking. ATCAIS is based on the ATOM tracker and the HTC instance segmentation method which is re-trained in a category-agnostic manner. The instance segmentation results are used to detect background distractors and to re-fine the target bounding boxes to prevent drifting. The depth value is used to detect the target occlusion or disappearance and re-finding the target.

D.2 Depth Enhanced DiMP for RGBD Tracking (DDiMP)

S. Qiu, Y. Gu, X. Zhang
{shoumeng, gyz, xlzhang}@mail.sim.ac.cn

DDiMP is based on SuperDiMP which combines the standard DiMP classifier from [5] with the bounding box regressor from [5]. The update strategy of the model during the tracking process is enhanced by using the model's confidence for the current tracking results. Output of IoU-Net is used to determine whether to fine-tune the shape, size, and position of the target. To handle scale variations, the target is searched over five scales $1.025^{\{-2,-1,0,1,2\}}$, and depth information is utilized to prevent scale from changing too quickly. Finally, two trackers with different model update confidence thresholds run in parallel, and the output with higher confidence is selected as the tracking result of the current frame.

D.3 Complementary Local-Global Search for RGBD Visual Tracking (CLGS-D)

H. Zhao, Z. Wang, B. Yan. D. Wang, H. Lu, X. Yang
{haojie_zhao,zzwang,yan_bin,wdice,lhchuan@dlut.edu.cn}@mail.dlut.edu.cn, xyang@remarkholdings.com

CLGS-D tracker is based on SiamMask, FlowNetv2, CenterNet, Real-time MDNet and a novel box refine module. The SiamMask model is used as the base tracker. MDNet is used to judge whether the target is found or lost. Once the target is lost, CenterNet is used to generate region proposals on the whole image. FlowNetv2 is used to estimate the motion of the target by generating a flow map. Then, the region proposals are filtered with aid of the flow and depth maps. Finally, an online "verifier" will find the target from the remaining region proposals again. A novel module is also used in this work to refine the bounding box.

D.4 Siamese Network for Long-term RGB-D Tracking (Siam_LTD)

X.-F. Zhu, H. Li, S. Zhao, T. Xu, X.-J. Wu
{xuefeng_zhu95,hui_li_jnu,zsc960813,wu_xiaojun}@163.com, tianyang.xu@surrey.ac.uk

Siam_LTD employes ResNet-50 to extract backbone features and RPN branch to locate the centre. In addition, a re-detection mechanism is introduced.

References

1. Babenko, B., Yang, M.H., Belongie, S.: Robust object tracking with online multiple instance learning. IEEE Trans. Pattern Anal. Mach. Intell. **33**(8), 1619–1632 (2011)
2. Berg, A., Ahlberg, J., Felsberg, M.: A thermal object tracking benchmark. In: 12th IEEE International Conference on Advanced Video- and Signal-based Surveillance, Karlsruhe, Germany, 25–28 August 2015. IEEE (2015)

3. Berg, A., Johnander, J., de Gevigney, F.D., Ahlberg, J., Felsberg, M.: Semi-automatic annotation of objects in visual-thermal video. In: IEEE International Conference on Computer Vision, ICCV Workshops (2019)
4. Bertinetto, L., Valmadre, J., Henriques, J.F., Vedaldi, A., Torr, P.H.S.: Fully-convolutional Siamese networks for object tracking. In: Hua, G., Jégou, H. (eds.) ECCV 2016. LNCS, vol. 9914, pp. 850–865. Springer, Cham (2016). https://doi.org/10.1007/978-3-319-48881-3_56
5. Bhat, G., Danelljan, M., Gool, L.V., Timofte, R.: Learning discriminative model prediction for tracking. In: IEEE International Conference on Computer Vision, ICCV (2019)
6. Bhat, G., Johnander, J., Danelljan, M., Khan, F.S., Felsberg, M.: Unveiling the power of deep tracking. In: ECCV, pp. 483–498 (2018)
7. Bhat, G., et al.: Learning what to learn for video object segmentation. In: Vedaldi, A., Bischof, H., Brox, T., Frahm, J.-M. (eds.) ECCV 2020. LNCS, vol. 12347, pp. 777–794. Springer, Cham (2020). https://doi.org/10.1007/978-3-030-58536-5_46
8. Cai, Z., Vasconcelos, N.: Cascade R-CNN: delving into high quality object detection. In: CVPR (2018)
9. Chen, K., et al.: Hybrid task cascade for instance segmentation. In: IEEE Conference on Computer Vision and Pattern Recognition (2019)
10. Chen, L.C., Zhu, Y., Papandreou, G., Schroff, F., Adam, H.: Encoder-decoder with atrous separable convolution for semantic image segmentation. In: Proceedings of the European Conference on Computer Vision (ECCV), pp. 801–818 (2018)
11. Dai, K., Zhang, Y., Wang, D., Li, J., Lu, H., Yang, X.: High-performance long-term tracking with meta-updater. In: CVPR (2020)
12. Danelljan, M., Bhat, G., Khan, F.S., Felsberg, M.: ECO: efficient convolution operators for tracking. In: CVPR, pp. 6638–6646 (2017)
13. Danelljan, M., Bhat, G., Khan, F.S., Felsberg, M.: ATOM: accurate tracking by overlap maximization. In: CVPR, pp. 4660–4669 (2019)
14. Danelljan, M., Gool, L.V., Timofte, R.: Probabilistic regression for visual tracking. In: CVPR (2020)
15. Danelljan, M., Häger, G., Khan, F.S., Felsberg, M.: Discriminative scale space tracking. IEEE Trans. Pattern Anal. Mach. Intell. **39**(8), 1561–1575 (2016)
16. Dunnhofer, M., Martinel, N., Luca Foresti, G., Micheloni, C.: Visual tracking by means of deep reinforcement learning and an expert demonstrator. In: The IEEE International Conference on Computer Vision (ICCV) Workshops, October 2019
17. Dunnhofer, M., Martinel, N., Micheloni, C.: A distilled model for tracking and tracker fusion (2020)
18. Fan, H., et al.: Lasot: a high-quality benchmark for large-scale single object tracking. In: Computer Vision Pattern Recognition (2019)
19. Galoogahi, H.K., Fagg, A., Huang, C., Ramanan, D., Lucey, S.: Need for speed: a benchmark for higher frame rate object tracking. CoRR abs/1703.05884 (2017). http://arxiv.org/abs/1703.05884
20. Goyette, N., Jodoin, P.M., Porikli, F., Konrad, J., Ishwar, P.: Changedetection.net: a new change detection benchmark dataset. In: CVPR Workshops, pp. 1–8. IEEE (2012)
21. Guo, C., Zhang, L.: A novel multiresolution spatiotemporal saliency detection model and its applications in image and video compression. IEEE Trans. Image Process. **19**(1), 185–198 (2009)

22. Gustafsson, F.K., Danelljan, M., Bhat, G., Schön, T.B.: Energy-based models for deep probabilistic regression. In: Vedaldi, A., Bischof, H., Brox, T., Frahm, J.-M. (eds.) ECCV 2020. LNCS, vol. 12365, pp. 325–343. Springer, Cham (2020). https://doi.org/10.1007/978-3-030-58565-5_20
23. Gustafsson, F.K., Danelljan, M., Timofte, R., Schön, T.B.: How to train your energy-based model for regression. CoRR abs/2005.01698 (2020). https://arxiv.org/abs/2005.01698
24. Henriques, J., Caseiro, R., Martins, P., Batista, J.: High-speed tracking with kernelized correlation filters. PAMI **37**(3), 583–596 (2015)
25. Huang, L., Zhao, X., Huang, K.: Got-10k: a large high-diversity benchmark for generic object tracking in the wild. arXiv:1810.11981 (2018)
26. Huang, L., Zhao, X., Huang, K.: GlobalTrack: a simple and strong baseline for long-term tracking. In: AAAI (2020)
27. Iandola, F.N., Han, S., Moskewicz, M.W., Ashraf, K., Dally, W.J., Keutzer, K.: SqueezeNet: alexnet-level accuracy with 50x fewer parameters and <0.5mb model size. arXiv:1602.07360 (2016)
28. Jack, V., et al.: Long-term tracking in the wild: A benchmark. arXiv:1803.09502 (2018)
29. Jung, I., Son, J., Baek, M., Han, B.: Real-time MDNet. In: ECCV, pp. 83–98 (2018)
30. Kalal, Z., Mikolajczyk, K., Matas, J.: Tracking-learning-detection. IEEE Trans. Pattern Anal. Mach. Intell. (TPAMI) **34**(7), 1409–1422 (2012). https://doi.org/10.1109/TPAMI.2011.239
31. Kristan, M., et al.: The seventh visual object tracking vot2019 challenge results. In: ICCV2019 Workshops, Workshop on Visual Object Tracking Challenge (2019)
32. Kristan, M., et al.: The visual object tracking vot2018 challenge results. In: ECCV2018 Workshops, Workshop on Visual Object Tracking Challenge (2018)
33. Kristan, M., et al.: The visual object tracking vot2017 challenge results. In: ICCV2017 Workshops, Workshop on Visual Object Tracking Challenge (2017)
34. Kristan, M., et al.: The visual object tracking vot2016 challenge results. In: ECCV2016 Workshops, Workshop on Visual Object Tracking Challenge (2016)
35. Kristan, M., et al.: The visual object tracking vot2015 challenge results. In: ICCV2015 Workshops, Workshop on Visual Object Tracking Challenge (2015)
36. Kristan, M., et al.: The visual object tracking vot2013 challenge results. In: ICCV2013 Workshops, Workshop on Visual Object Tracking Challenge, pp. 98–111 (2013)
37. Kristan, M., et al.: The visual object tracking vot2014 challenge results. In: ECCV2014 Workshops, Workshop on Visual Object Tracking Challenge (2014)
38. Kristan, M., et al.: A novel performance evaluation methodology for single-target trackers. IEEE Trans. Pattern Anal. Mach. Intell. **38**(11), 2137–2155 (2016)
39. Leal-Taixé, L., Milan, A., Reid, I.D., Roth, S., Schindler, K.: Motchallenge 2015: towards a benchmark for multi-target tracking. CoRR abs/1504.01942 (2015). http://arxiv.org/abs/1504.01942
40. Li, A., Li, M., Wu, Y., Yang, M.H., Yan, S.: Nus-pro: a new visual tracking challenge. IEEE-PAMI (2015)
41. Li, B., Wu, W., Wang, Q., Zhang, F., Xing, J., Yan, J.: SiamRPN++: evolution of Siamese visual tracking with very deep networks. In: Proceedings of the IEEE Conference on Computer Vision and Pattern Recognition, pp. 4282–4291 (2019)
42. Li, B., Yan, J., Wu, W., Zhu, Z., Hu, X.: High performance visual tracking with Siamese region proposal network. In: The IEEE Conference on Computer Vision and Pattern Recognition (CVPR), pp. 8971–8980, June 2018

43. Li, C., Liang, X., Lu, Y., Zhao, N., Tang, J.: RGB-T object tracking: benchmark and baseline. Pattern Recogn. (2019, submitted)
44. Liang, P., Blasch, E., Ling, H.: Encoding color information for visual tracking: algorithms and benchmark. IEEE Trans. Image Process. **24**(12), 5630–5644 (2015)
45. Lin, T.-Y., et al.: Microsoft COCO: common objects in context. In: Fleet, D., Pajdla, T., Schiele, B., Tuytelaars, T. (eds.) ECCV 2014. LNCS, vol. 8693, pp. 740–755. Springer, Cham (2014). https://doi.org/10.1007/978-3-319-10602-1_48
46. Lukežič, A., Kart, U., Kämäräinen, J., Matas, J., Kristan, M.: CDTB: a color and depth visual object tracking dataset and benchmark. In: ICCV (2019)
47. Lukežič, A., Vojíř T., Čehovin Zajc, L., Matas, J., Kristan, M.: Discriminative correlation filter with channel and spatial reliability. In: The IEEE Conference on Computer Vision and Pattern Recognition (CVPR), pp. 6309–6318, July 2017
48. Lukežič, A., Čehovin Zajc, L., Vojíř T., Matas, J., Kristan, M.: Now you see me: evaluating performance in long-term visual tracking. CoRR abs/1804.07056 (2018). http://arxiv.org/abs/1804.07056
49. Lukezic, A., Cehovin Zajc, L., Vojir, T., Matas, J., Kristan, M.: Performance evaluation methodology for long-term single object tracking. IEEE Trans. Cybern. (2020)
50. Lukezic, A., Matas, J., Kristan, M.: D3S - a discriminative single shot segmentation tracker. In: CVPR (2020)
51. Memarmoghadam, A., Moallem, P.: Size-aware visual object tracking via dynamic fusion of correlation filter-based part regressors. Signal Process. **164**, 84–98 (2019). https://doi.org/10.1016/j.sigpro.2019.05.021. http://www.sciencedirect.com/science/article/pii/S0165168419301872
52. Moudgil, A., Gandhi, V.: Long-term visual object tracking benchmark. arXiv preprint arXiv:1712.01358 (2017)
53. Mueller, M., Smith, N., Ghanem, B.: A benchmark and simulator for UAV tracking. In: Leibe, B., Matas, J., Sebe, N., Welling, M. (eds.) ECCV 2016. LNCS, vol. 9905, pp. 445–461. Springer, Cham (2016). https://doi.org/10.1007/978-3-319-46448-0_27
54. Muller, M., Bibi, A., Giancola, S., Alsubaihi, S., Ghanem, B.: TrackingNet: a large-scale dataset and benchmark for object tracking in the wild. In: ECCV, pp. 300–317 (2018)
55. Nam, H., Han, B.: Learning multi-domain convolutional neural networks for visual tracking. In: CVPR, pp. 4293–4302 (2016)
56. Oh, S.W., Lee, J.Y., Xu, N., Kim, S.J.: Video object segmentation using space-time memory networks. In: ICCV (2019)
57. Pernici, F., del Bimbo, A.: Object tracking by oversampling local features. IEEE Trans. Pattern Anal. Mach. Intell. **36**(12), 2538–2551 (2013). https://doi.org/10.1109/TPAMI.2013.250
58. Phillips, P.J., Moon, H., Rizvi, S.A., Rauss, P.J.: The FERET evaluation methodology for face-recognition algorithms. IEEE Trans. Pattern Anal. Mach. Intell. **22**(10), 1090–1104 (2000)
59. Real, E., Shlens, J., Mazzocchi, S., Pan, X., Vanhoucke, V.: YouTube-BoundingBoxes: a large high-precision human-annotated data set for object detection in video. In: Computer Vision and Pattern Recognition, pp. 7464–7473 (2017)
60. Robinson, A., Lawin, F.J., Danelljan, M., Khan, F.S., Felsberg, M.: Learning fast and robust target models for video object segmentation. In: IEEE/CVF Conference on Computer Vision and Pattern Recognition (CVPR). Computer Vision Foundation, June 2020

61. Ronneberger, O., Fischer, P., Brox, T.: U-Net: convolutional networks for biomedical image segmentation. In: Navab, N., Hornegger, J., Wells, W.M., Frangi, A.F. (eds.) MICCAI 2015. LNCS, vol. 9351, pp. 234–241. Springer, Cham (2015). https://doi.org/10.1007/978-3-319-24574-4_28
62. Ross, D.A., Lim, J., Lin, R.S., Yang, M.H.: Incremental learning for robust visual tracking. Int. J. Comput. Vis. **77**(1–3), 125–141 (2008)
63. Russakovsky, O., et al.: ImageNet large scale visual recognition challenge. IJCV **115**(3), 211–252 (2015). https://doi.org/10.1007/s11263-015-0816-y
64. Seoung, W.O., Lee, J.Y., Kim, S.J.: Fast video object segmentation by reference-guided mask propagation. In: Computer Vision Pattern Recognition, pp. 7376–7385 (2018)
65. Smeulders, A.W.M., Chu, D.M., Cucchiara, R., Calderara, S., Dehghan, A., Shah, M.: Visual tracking: an experimental survey. TPAMI (2013). https://doi.org/10.1109/TPAMI.2013.230
66. Solera, F., Calderara, S., Cucchiara, R.: Towards the evaluation of reproducible robustness in tracking-by-detection. In: Advanced Video and Signal Based Surveillance, pp. 1–6 (2015)
67. Song, S., Xiao, J.: Tracking revisited using RGBD camera: unified benchmark and baselines. In: ICCV (2013)
68. Tao, R., Gavves, E., Smeulders, A.W.M.: Tracking for half an hour. CoRR abs/1711.10217 (2017). http://arxiv.org/abs/1711.10217
69. Tian, Z., Shen, C., Chen, H., He, T.: FCOS: fully convolutional one-stage object detection. arXiv preprint arXiv:1904.01355 (2019)
70. Čehovin, L., Kristan, M., Leonardis, A.: Is my new tracker really better than yours? Technical report 10, ViCoS Lab, University of Ljubljana, October 2013. http://prints.vicos.si/publications/302
71. Čehovin, L.: TraX: The visual Tracking eXchange Protocol and Library. Neurocomputing (2017). https://doi.org/10.1016/j.neucom.2017.02.036
72. Čehovin, L., Leonardis, A., Kristan, M.: Visual object tracking performance measures revisited. IEEE Trans. Image Process. **25**(3), 1261–1274 (2016)
73. Vojíř, T., Noskova, J., Matas, J.: Robust scale-adaptive mean-shift for tracking. Pattern Recogn. Lett. **49**, 250–258 (2014)
74. Wang, Q., Zhang, L., Bertinetto, L., Hu, W., Torr, P.H.: Fast online object tracking and segmentation: a unifying approach. In: CVPR, pp. 1328–1338 (2019)
75. Wang, X., Kong, T., Shen, C., Jiang, Y., Li, L.: SOLO: segmenting objects by locations. arXiv preprint arXiv:1912.04488 (2019)
76. Wu, Y., Lim, J., Yang, M.H.: Online object tracking: a benchmark. In: Computer Vision Pattern Recognition (2013)
77. Wu, Y., Lim, J., Yang, M.H.: Object tracking benchmark. PAMI **37**(9), 1834–1848 (2015)
78. Xiao, J., Stolkin, R., Gao, Y., Leonardis, A.: Robust fusion of color and depth data for RGB-D target tracking using adaptive range-invariant depth models and spatio-temporal consistency constraints. IEEE Trans. Cybern. **48**, 2485–2499 (2018)
79. Xu, N., Price, B., Yang, J., Huang, T.: Deep grabcut for object selection. In: Proceedings of British Machine Vision Conference (2017)
80. Xu, T., Feng, Z.H., Wu, X.J., Kittler, J.: AFAT: adaptive failure-aware tracker for robust visual object tracking. arXiv preprint arXiv:2005.13708 (2020)
81. Xu, Y., Wang, Z., Li, Z., Ye, Y., Yu, G.: SiamFC++: towards robust and accurate visual tracking with target estimation guidelines. arXiv preprint arXiv:1911.06188 (2019)

82. Yan, B., Wang, D., Lu, H., Yang, X.: Alpha-refine: boosting tracking performance by precise bounding box estimation. arXiv preprint arXiv:2007.02024 (2020)
83. Yan, B., Zhao, H., Wang, D., Lu, H., Yang, X.: Skimming-Perusal Tracking: a framework for real-time and robust long-term tracking. In: IEEE International Conference on Computer Vision (ICCV) (2019)
84. Yang, Z., Liu, S., Hu, H., Wang, L., Lin, S.: RepPoints: point set representation for object detection. In: The IEEE International Conference on Computer Vision (ICCV), pp. 9657–9666, October 2019
85. Yiming, L., Shen, J., Pantic, M.: Mobile face tracking: a survey and benchmark. arXiv:1805.09749v1 (2018)
86. Young, D.P., Ferryman, J.M.: PETS Metrics: on-line performance evaluation service. In: Proceedings of the 14th International Conference on Computer Communications and Networks, ICCCN 2005, pp. 317–324 (2005)
87. Zhang, L., Danelljan, M., Gonzalez-Garcia, A., van de Weijer, J., Khan, F.S.: Multimodal fusion for end-to-end RGB-T tracking. In: IEEE International Conference on Computer Vision, ICCV Workshops (2019)
88. Zhang, P., Zhao, J., Wang, D., Lu, H., Yang, X.: Jointly modeling motion and appearance cues for robust RGB-T tracking. CoRR abs/2007.02041 (2020)
89. Zhang, Y., Wu, Z., Peng, H., Lin, S.: A transductive approach for video object segmentation. In: IEEE/CVF Conference on Computer Vision and Pattern Recognition (CVPR), pp. 4000–4009, June 2020
90. Zhang, Y., Wang, D., Wang, L., Qi, J., Lu, H.: Learning regression and verification networks for long-term visual tracking. CoRR abs/1809.04320 (2018)
91. Zhang, Z., Peng, H.: Deeper and wider Siamese networks for real-time visual tracking. In: Proceedings of the IEEE Conference on Computer Vision and Pattern Recognition, pp. 4591–4600, June 2019
92. Zhang, Z., Peng, H., Fu, J., Li, B., Hu, W.: Ocean: object-aware anchor-free tracking. arXiv preprint arXiv:2006.10721 (2020)
93. Zhu, P., Wen, L., Bian, X., Haibin, L., Hu, Q.: Vision meets drones: a challenge. arXiv preprint arXiv:1804.07437 (2018)

Robust Long-Term Object Tracking via Improved Discriminative Model Prediction

Seokeon Choi[1], Junhyun Lee[2], Yunsung Lee[2], and Alexander Hauptmann[3](✉)

[1] KAIST, Daejeon, South Korea
seokeon@kaist.ac.kr
[2] Korea University, Seoul, South Korea
{ljhyun33,swack9751}@korea.ac.kr
[3] Carnegie Mellon University, Pittsburgh, USA
alex@cs.cmu.edu

Abstract. We propose an improved discriminative model prediction method for robust long-term tracking based on a pre-trained short-term tracker. The baseline pre-trained short-term tracker is SuperDiMP which combines the bounding-box regressor of PrDiMP with the standard DiMP classifier. Our tracker RLT-DiMP improves SuperDiMP in the following three aspects: (1) Uncertainty reduction using random erasing: To make our model robust, we exploit an agreement from multiple images after erasing random small rectangular areas as a certainty. And then, we correct the tracking state of our model accordingly. (2) Random search with spatio-temporal constraints: we propose a robust random search method with a score penalty applied to prevent the problem of sudden detection at a distance. (3) Background augmentation for more discriminative feature learning: We augment various backgrounds that are not included in the search area to train a more robust model in the background clutter. In experiments on the VOT-LT2020 benchmark dataset, the proposed method achieves comparable performance to the state-of-the-art long-term trackers. The source code is available at: https://github.com/bismex/RLT-DIMP.

Keywords: Long-term object tracking · Robust object tracking · Uncertainty reduction · Random erasing · Random search · Background augmentation · Discriminative feature learning

1 Introduction

Visual Object Tracking (VOT), the task of continuously locating an arbitrary target in the first frame of a video, has been drawn attention in both academic and industrial fields over the last decade [16,18,28]. It is because VOT

S. Choi, J. Lee and Y. Lee—This work was done while the authors were visiting researchers at CMU.

A. Bartoli and A. Fusiello (Eds.): ECCV 2020 Workshops, LNCS 12539, pp. 602–617, 2020.
https://doi.org/10.1007/978-3-030-68238-5_40

can be widely used in real-world applications such as autonomous vehicles [20], robotics [29], and video surveillance systems [1]. With the advance of deep learning techniques, trackers are not only getting better performance but also used at long-term tracking (minute-level) beyond short-term tracking (second-level).

Besides the length of input videos, the clear difference between short-term tracking and long-term tracking is whether the target exists in the field of view, as reflected in standard benchmark datasets [25,33]. In general, short-term trackers are designed on the assumption that the target always appears in every single frame, otherwise, the short-term tracker will drift and fail. Long-term trackers, on the other hand, need to keep track of the object even if it disappears from the field of view in the middle of the frames. Consequently, the re-detection module, which localizes the target with a confidence score of its absence, is the essential part of long-term trackers.

Because long-term trackers encounter unpredictable abrupt changes during relatively long sequences, the robustness is the most important property of long-term trackers. If the long-term tracker misestimates the location of a target because of visual deformation, there is a high risk of incorrect estimation in the following frame. Previous researches tried to construct robust modules and strategies in various ways such as P-N learning [15], memory model [21], and dynamic programming [30]. Those methods are focusing on the robustness against visual deformation.

In this work, we focus on the robustness against the re-detection module itself (i.e. reliability of the tracker's prediction) as well as against visual deformation. First, we propose a way to reduce the uncertainty of our model and correct the prediction accordingly. The estimated location of the target, which is robust against the background noise, would not be changed even if we remove a certain small area of the background. If the estimation is not robust then it would be changed and not reliable even though the confidence score is high. In view of these characteristics, our model estimates the location of the target from multiple images with randomly erased the small rectangular area. Secondly, we utilize spatio-temporal constraints to adjust the confidence score for robust re-detection. When the target re-appear after occlusion or disappear, the time-space gap should be related in physical. For instance, if the estimated location of the target that re-appears in a short time is far from the last observation location, we can say that the estimation is unreliable. To offset the distortion of both time and space and to make a robust estimation as a result, we explicitly adjust the confidence score by penalizing it. Finally, we perform background augmentation for more discriminative feature learning in the online stage.

Section 2 provides a brief description of existing short- and long-term trackers. In Sect. 3, we explain the details about how our approach can handle the robustness issues of long-term trackers. The experimental results with analysis and the conclusion are in Sect. 4 and 5, respectively.

2 Related Work

2.1 Short-Term Object Tracking

Visual object tracking (VOT) is a task to track an object in a video when the first frame bounding box of the target object is given. A number of deep convolutional neural network (DCNN) based studies have been conducted, such as MDNet [26] and FCNT [31], showing successful results in the VOT Challenge [17]. Among DCNN-based studies, Siamese architecture [5,22,23,30,32] satisfies end-to-end training capabilities while also showing high efficiency [16,18]. DiMP [6] and PrDiMP [11], motivated from the idea of ATOM [9] that solved the limited target estimation problem of the previous studies, are also Siamese architecture-based model that showed performance improvement in the VOT Challenge.

Most tracking models before ATOM [9] were only focused on the development of powerful classifiers. The problem of accurate target state estimation has been overlooked. To this end, ATOM architecture consists of dedicated target estimation and classification components. This target estimation component is trained to predict the overlap between the target object and an estimated bounding box. High-level knowledge is incorporated into the target estimation through extensive offline learning.

Siamese networks [5,22,23,30,32] have received much attention due to their end-to-end training capabilities and high efficiency. However, Siamese approaches are limited in their inability to incorporate information from the background region or previous tracked frames into the model prediction. To deal with this issue, DiMP [6] takes inspiration from the discriminative online learning procedures [9,10,26]. DiMP tracking architecture consists of two branches: a target classification branch for distinguishing the target from the background, and a bounding box estimation branch for predicting an accurate target box. The target classification branch is derived from two main principles: (i) a discriminative learning loss promoting robustness in the learned target model and (ii) a powerful optimization strategy ensuring rapid convergence. Bounding box estimation branch is utilized from the overlap maximization based architecture introduced in ATOM [9].

Most tacking models rely on estimating a state-dependent confidence score, but this value lacks a clear probabilistic interpretation, complicating its use. Therefore, in PrDiMP [11], a probabilistic regression formulation is proposed, and it is applied to track the target. PrDiMP network predicts the conditional probability density of the target state given an input image. Their formulation helps the model to be robust from inaccurate annotations and ambiguities in the task.

2.2 Long-Term Object Tracking

The target may disappear in the long-term tracking setting, so a re-detection module is essential. In addition, a robust online update method capable of accommodating changes in the visual appearance of the target dramatically affects performance. Therefore, we shortly introduce how long-term trackers overcome significant problems.

The most representative long-term object tracker is TLD [15], which is divided into tracking (T), learning (L), and detection (D). In the tracking part, the tracker predicts the position of the target object based on the median-flow tracker [14]. In the detection part, the detector judges the existence of the target in a cascade manner over the entire area of the image. Assuming that the tracker and detector can fail, the learning module estimates errors based on P-N learning and trains the trackers and detectors more robustly.

Due to its good performance in both accuracy and speed, various object trackers based on Siamese networks have been proposed. One of those methods, MMLT [21], is designed for long-term tracking to handle visual deformation and target disappearance. In the tracking step, to accommodate changes in the visual appearance, the target position is estimated by the Siamese features obtained from short-term and long-term memory stores inspired by Atkinson-Shiffrin Memory Model (ASMM) [2]. In the re-detection step, the target is detected in the entire image without the dependency of the previous position. In particular, the coarse-to-fine strategy is adopted for improving speed.

Tracking by re-detection paradigm [3,4,12,22] has a long history, but re-detection is challenging due to the existence of distractor objects that are very similar to the template object. Siam R-CNN [30], an adaptation of Faster R-CNN [27] with a Siamese architecture, has two key methods. First, they introduced a hard example mining procedure which trains the re-detector specifically for difficult distractors. Secondly, dynamic programming is used to select the best object in the current time step based on the complete history of all target objects and distractor object tracklets(short object tracks). While being resistant to tracker drift and being able to immediately re-detect object after the disappearance, Siam R-CNN is able to effectively perform long-term object tracking.

3 Proposed Method

3.1 Baseline Short-Term Tracker

We propose an improved Discriminative Model Prediction method for robust long-term tracking based on a pre-trained short-term tracker. Our baseline pre-trained short-term tracker is SuperDiMP[1] combining the bounding-box regressor of PrDiMP [11] with the standard DiMP classifier [6] for better training and inference.

3.2 Uncertainty Reduction Using Random Erasing

We focus on robustness, which is the consistent generalization (tracking) ability, particularly against the artifact of background features. The robust model can consistently track the target whatever occurs in the background, even whenever

[1] The pre-trained model is provided at https://github.com/visionml/pytracking.

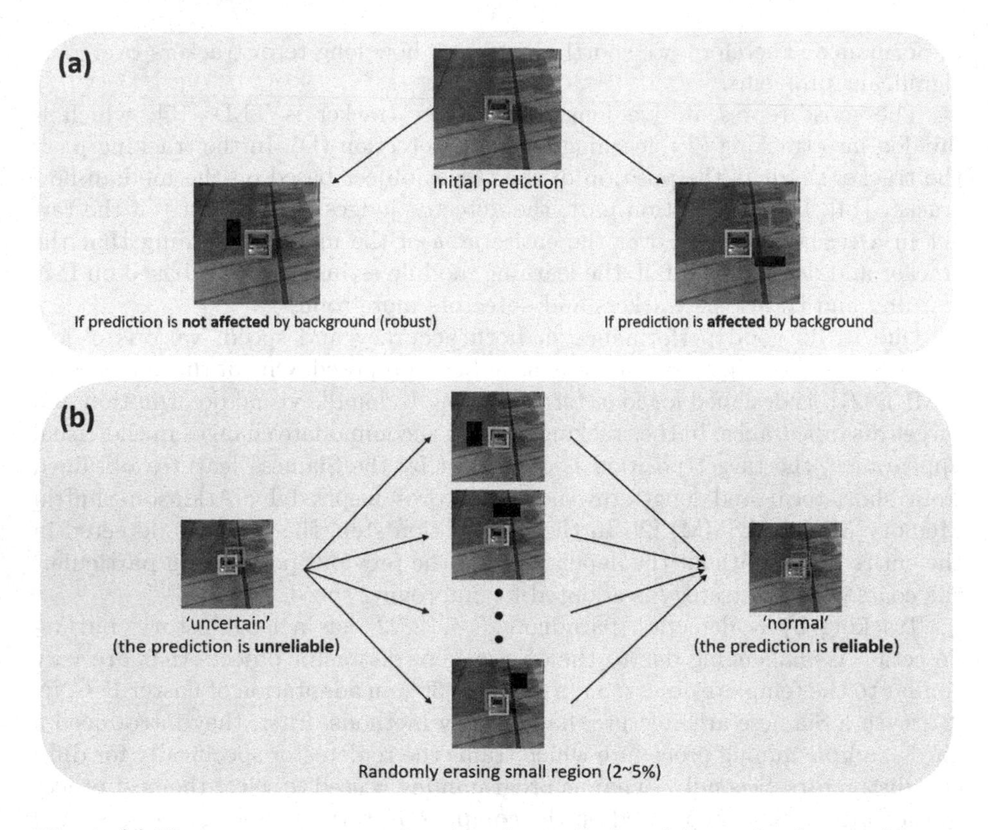

Fig. 1. (a) Illustrates our assumption that the prediction of the randomly erased image will be changed when the prediction is affected by background features or vice versa. (b) illustrates the concept of how to reduce the uncertainty and correct the prediction.

we remove some region of background. We consider uncertainty as an agreement (or consistency) and estimate the location of the target from multiple images after erasing random small rectangular areas. Because the target usually has a small portion of the whole image, even though we remove a small area, the target might be rarely removed. If the prediction of our model changes when the small region of background is randomly deleted, the prediction is affected by the background, so we can say that the certainty is low. Adapting this assumption, we correct the flag of our model according to the agreement (Fig. 1).

3.3 Random Search with Spatio-Temporal Constraints

General short-term trackers can only find the target if it is included in the searching area. Accordingly, global re-detection capacity is required for robust long-term tracking to deal with occlusions and target disappearances. Most long-term trackers adopt the global sliding window method, as shown in Fig. 2. However,

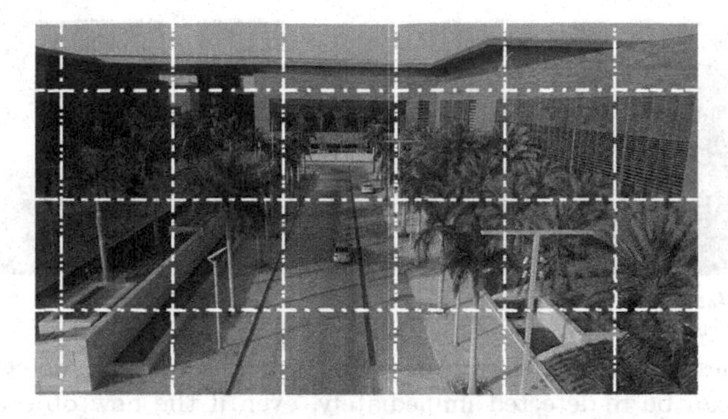

Fig. 2. A re-detection example using a general global search method. It takes a long time to find objects in all areas.

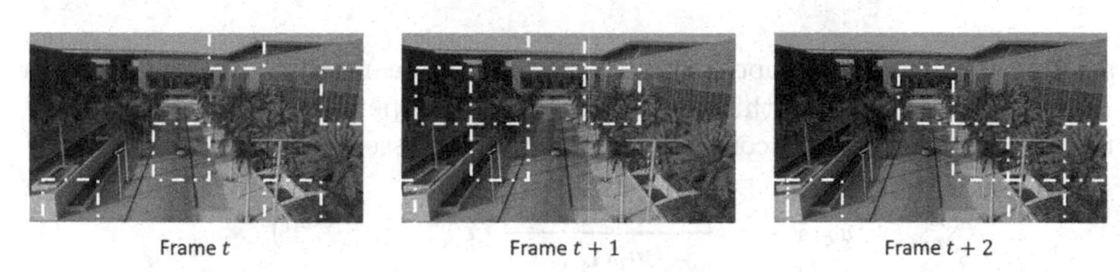

Fig. 3. Re-detection examples using the proposed random search method. From a stochastic point of view, it is possible to search the entire area within a few frames.

this approach not only requires high computation costs, but is not robust. To tackle this problem, we propose a robust random search method with spatio-temporal constraints.

Random Search. First, we create global searching templates with a predetermined interval. Next, we adaptively determine the number of searches according to the ratio of the image size to the target size. When the target size is relatively large, we set the fewer number to search. Otherwise, we assign more numbers to search when the target size is relatively small. Then, an object is detected within a randomly selected searching area. As visualized in Fig. 3, it is possible to search whole regions within a few frames from a stochastic point of view. This stochastic approach improves the re-detection speed compared to general sliding window methods, which is discussed in Sect. 4.3.

Score Penalty. When the confidence score s_{new} of the newly detected target is higher than a predetermined threshold, the frame in which the new object appeared is designated as the new first frame, and the object is again tracked. Here, we note that this naïve re-detection scheme is not robust. Figure 4 shows the examples of sudden detection of other similar objects or background distractors. Once the target is missing, a new object can be re-detected immediately, even if the new object is located far from the original target. However, the prob-

Fig. 4. Examples of the need for score penalty. In the naïve re-detection scheme, a similar object can be re-detected immediately, even if the new object is located far from the original target. In other words, this method is quite vulnerable to background clutter.

ability of an object disappearing and suddenly appearing at a distant location is very low. To prevent this sudden detection, we penalize a confidence score through spatio-temporal constraints, which is expressed as follows:

$$s'_{new} = w_b(1 - w_d \frac{||\mathbf{p}_{new} - \mathbf{p}_{old}||_2}{d_{max}} \cdot e^{-w_t|t_{new}-t_{old}|}) \cdot s_{new}, \quad (1)$$

where w_b, w_d, and w_t are hyper-parameters for basic re-detection, distance, and time, respectively. d_{max}, $\mathbf{p}$, and t indicate a diagonal length of an image (*i.e.* the maximum distance), a position vector, and a frame number, respectively. In terms of a spatial constraint, the score is more penalized when the distance between new and old positions is large. This distance penalty term prevents the problem of abnormal detection at a long distance. Meanwhile, this distance penalty is compensated by the time penalty (*i.e.* a temporal constraint), as time passes by not finding the target. This is because, if the period during which the object cannot be found becomes longer, the object may newly appear at a longer distance. This temporal constraint allows objects to be detected at relatively far locations. The revised re-detection module makes the tracker more robust.

3.4 Background Augmentation for More Discriminative Feature Learning

In the short-term baseline tracker, we extract features and trains the model by applying some transforms (blur, rotate, horizontal flip) in the searching area of the target. To train a robust model against the background clutter, we try to augment various backgrounds that are not included in the search area by combining the target image with another background. This augmentation skill makes the model capable of fully exploiting various background appearance information. Figure 5 shows our data augmentation methods for enhanced discriminative feature learning.

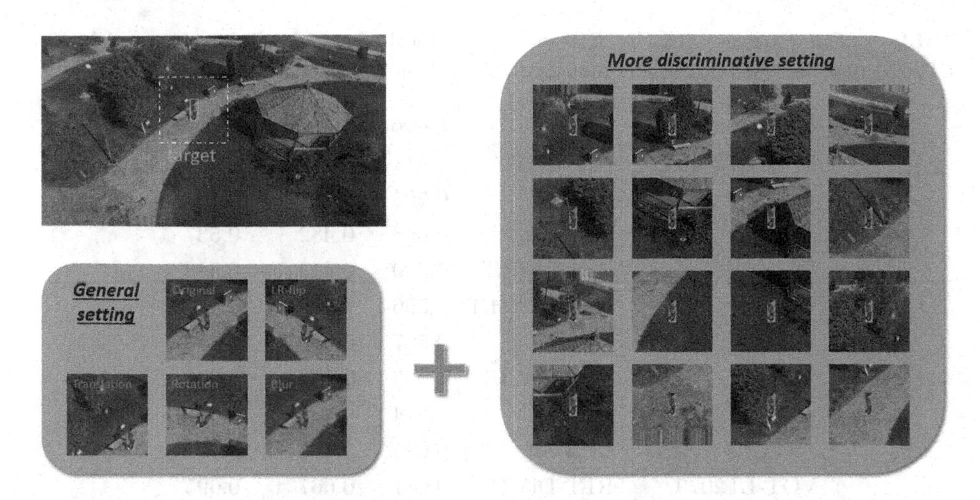

Fig. 5. General data augmentation and additional samples for more discriminative learning.

Learning at the First Frame. The bounding box is given in the first frame, which means that the target image is completely confident. Therefore, we train the short-term tracker using new data augmentation and additionally store the newly generated images in the memory system. This approach allows for improved discriminative learning in the first frame.

Online Learning. We also perform background augmentation during tracking in the online stage as same as in the first frame. However, this process is only used when the reliability of the bounding box is high since the estimated bounding box is not always confident. The background augmentation process is the same as the first frame one, and the filter is trained if more demanding conditions are met. Unlike in the first frame settings, images with background augmentation are not stored in memory. This prevents the problem of including negative features that cause drifting to other objects.

3.5 Confidence Score Assignment

The baseline short-term tracker classifies the state of object tracking into four types according to the score and various conditions: *normal*, *hard negative*, *uncertain*, and *not found*. In the case of *normal* and *hard negative*, the object tracking result is reliable, so the confidence score is given as 1. In the case of *uncertain*, it is difficult to determine whether it is an object or not, so the confidence score is given as 0.5. Lastly, in the case of *not found*, it is predicted that there is no object, so the confidence score is given as 0.

Table 1. Comparison with the state-of-the-art methods on VOT-LT2019 and VOT-LT2020 benchmarks. Both benchmarks are based on the LTB50 dataset.

Benchmark	Tracker	F-score	Precision	Recall
VOT-LT2019 [18]	FuCoLoT	0.411	0.507	0.346
	ASINT	0.505	0.517	0.494
	CooSiam	0.508	0.482	0.537
	Siamfcos-LT	0.520	0.493	0.549
	SiamRPNsLT	0.556	0.749	0.443
	mbdet	0.567	0.609	0.530
	SiamDW_LT	0.665	0.697	0.636
	CLGS	0.674	0.739	0.619
	LT_DSE	0.695	0.715	0.677
VOT-LT2020	RLT-DiMP	0.681	0.667	0.695

4 Experiments

4.1 Dataset and Settings

Dataset. We experiment with a long-term object tracking dataset, LTB50 [24]. This dataset is an extension of the LTB35 [24] used in the VOT-LT2018 challenge [16], and it is officially used in the VOT-LT2019 [18] challenge. In the VOT-LT2020 challenge, the LTB50 is used unchanged from last year. The LTB50 dataset contains 50 sequences of various objects with a total length of 215,294 frames for single-target object tracking. In each sequence, the target disappears an average of 10 times, and the disappeared target lasts an average of 52 frames. The resolution of video sequences is between 1280×720 and 290×217. All targets are marked with an axis-aligned bounding box.

Evaluation Protocol. The proposed RLT-DiMP is evaluated by the evaluation protocol of the VOT-LT2020 benchmark. An evaluation protocol for long-term trackers follows a no-reset protocol, which means that the object tracker will not restart even if the object tracking fails. Three evaluation metrics are adopted for the long-term tracking benchmark: tracking precision, tracking recall, and tracking F-measure. For additional information, please see [24]. This evaluation is automatically performed by the VOT toolkit [8,19]. All experiments are performed on a system with Intel(R) core(TM) i7-4770 3.40 GHz processor and a single NVIDIA GTX 1080 Ti with 11GB RAM.

4.2 Quantitative Evaluation

Overall Comparison with Long-Term Trackers. We compare our proposed method with the state-of-the-art methods in the VOT-LT2019 benchmark [18]. The VOT-LT2019 and VOT-LT2020 benchmarks are based on the LTB50

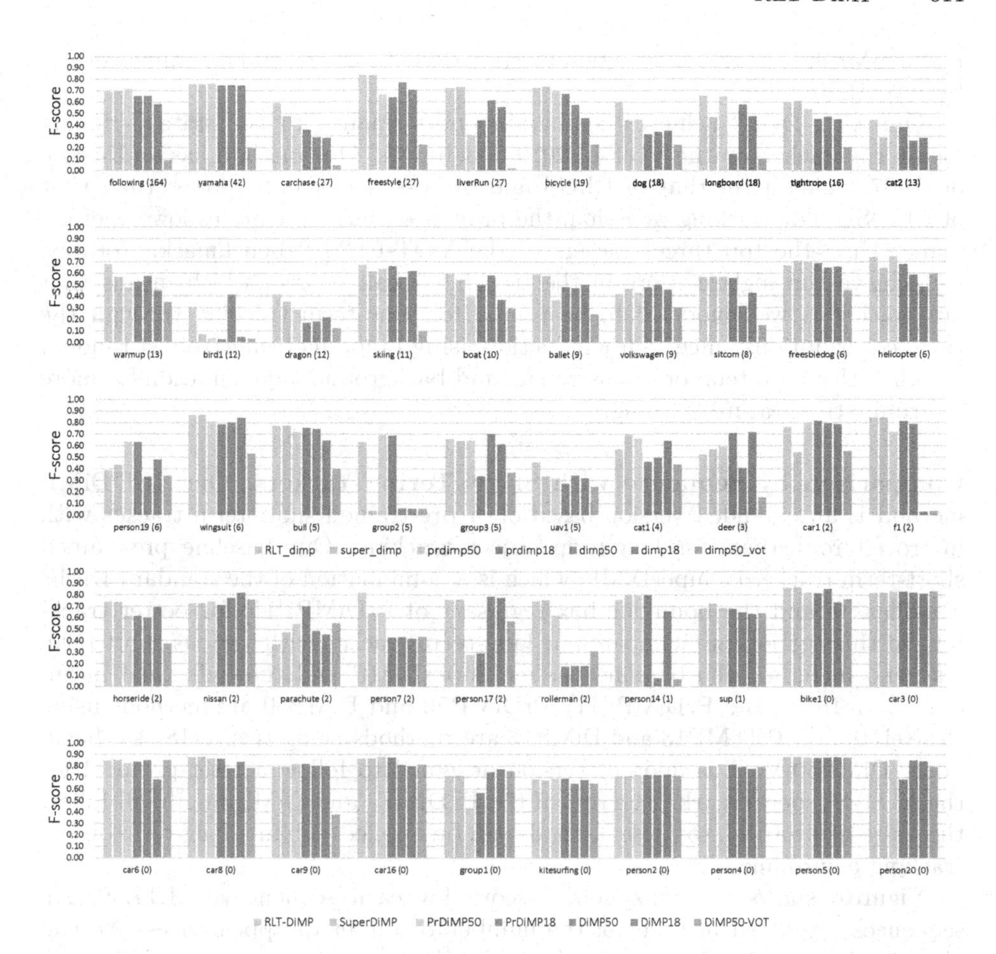

Fig. 6. The maximum F-score for each sequence on the LTB50 dataset. Sequences are sorted based on the number of target disappearances, which are indicated by the number in parentheses.

dataset [24]. Both of competing methods and our method are re-detecting long-term trackers (LT1). This term means that all of the trackers detect tracking failure, and an explicit re-detection technique is implemented, unlike the pseudo long-term tracker (LT0). The following taxonomy has been introduced explicitly in [24].

Table 1 shows the quantitative evaluation of long-term trackers on the LTB50 dataset. In the VOT-LT2019 benchmark, LT_DSE tracker has achieved the best F-score and the best tracking recall, and CLGS has achieved the best tracking precision. The tracker LT_DSE is designed based on target localization by ATOM [9], bounding box refinement by SiamMask [32], and a verifier network by RT-MDNet [13]. The tracker CLGS is designed based on target localization by SiamMask [32], global detection by cascade R-CNN [7], and an online verifier by RT-MDNet [13].

Our RLT-DiMP achieves comparable performance to the state-of-the-art long-term trackers in the VOT-LT2019 benchmark. Our method has an F-score of 0.007 higher than that of CLGS and an F-score of 0.014 lower than that of LT_DSE. For tracking precision, the proposed method achieves lower performance than the top three trackers of the VOT-LT2019 benchmark. Notably, we achieve the highest score in the tracking recall metric, which means that our tracker is well modeled to be robust for long-term tracking through the three contributions: uncertainty reduction using random erasing, robust random search with spatio-temporal constraints, and background augmentation for more discriminative feature learning.

Comparison by Sequence with Short-Term Trackers. Our RLT-DiMP method is an extended version based on a pre-trained short-term tracker with improved robustness for long-term object tracking. Our baseline pre-trained short-term tracker is SuperDiMP, which is a combination of the standard DiMP classifier [6] and the bounding-box regressor of PrDiMP [11] for better tracking. In this section, we compare our long-term tracker with various short-term trackers, including the baseline tracker SuperDiMP and the individual methods of DiMP [6] and PrDiMP [11]. PrDiMP50 and DiMP50 are methods using ResNet50, and PrDiMP18 and DiMP18 are methods using ResNet18 as a backbone. DiMP50-VOT is another version designed to follow a reset protocol for the VOT-ST benchmark. The rest of the PrDiMP- and DiMP-family all follow the no-reset protocol, so these methods can be applied well in a long-term object tracking environment.

Figure 6 shows the maximum F-score for each sequence on LTB50. All sequences are listed in order of the number of target disappearances. We can observe that the baseline tracker SuperDiMP has better reasoning skills than the individual methods of DiMP [6] and PrDiMP [11]. We emphasize that our RLT-DiMP method outperforms the baseline tracker in almost all sequences. This proves that our method is more suitable for long-term object tracking, and robust modeling via uncertainty reduction, robust random search, and background augmentation plays a significant role. This is analyzed in more detail through ablation studies in the next section.

4.3 Further Evaluations and Analysis

Visual Attributes on LTB50. In the LTB50 dataset, a total of 50 sequences are annotated by nine visual attributes as follows: 1) Full occlusion, 2) Out-of-view, 3) Partial occlusion, 4) Camera motion, 5) Fast motion, 6) Scale change, 7) Aspect ratio change, 8) Viewpoint change, 9) Similar objects. While a *longboard* sequence even has as many as 8 visual attributes, and a *person5* sequence has no visual attribute. As described above, each sequence includes several visual attributes, and we perform an ablation study by averaging performance for each visual characteristic.

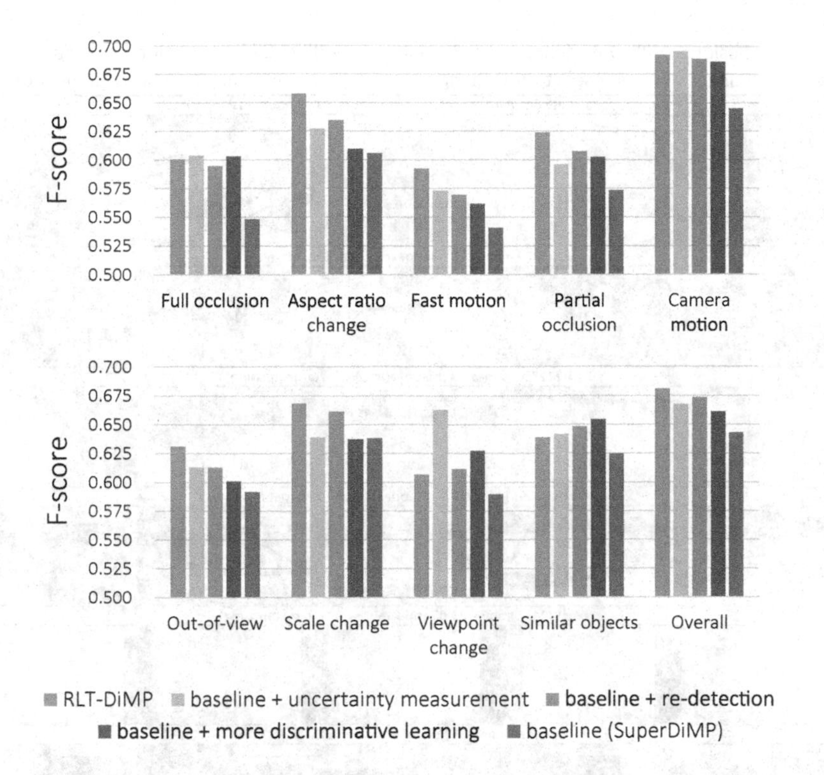

Fig. 7. The overall F-score and the average F-score for each attribute on the LTB50 dataset.

Ablation Studies. Figure 7 shows the overall F-score and the average F-score for each visual attribute with respect to various versions of the proposed method. RLT-DiMP and SuperDimp indicate our final version and baseline version, respectively. In this section, we compare performance by adding one proposed module each from the baseline.

As described in the overall performance in the bar graph, When applying uncertainty reduction, robust random search, background augmentation algorithms to baseline, the F-score is improved by 0.025, 0.031, and 0.018. Accordingly, these results prove that our sub-algorithms enable our tracker to estimate the target position more robustly. We note that our RLT-DiMP method outperforms the F-score by 0.038 compared to the baseline tracker.

In the visual attribute analysis, the proposed method surpasses the baseline tracker in all cases. Especially in situations with visual characteristics of full-occlusion, aspect ratio change, fast motion, and partial occlusion, our method has an F-score performance of 0.05 or higher than the baseline.

Processing Time Analysis. In the LTB50 dataset, our method records 14.17 fps, which is somewhat lower than the object tracking speed of PrDiMP [11]

Fig. 8. Qualitative results of the proposed RLT-DiMP, SuperDiMP, PrDiMP [11], DiMP [6]. Best viewed in color.

at 21.83 fps and DiMP [6] at 30.22 fps. All three proposed modules inevitably reduce the speed of object tracking. However, in the case of re-detection, we note that the random search method can improve the speed of about 3 fps compared to the global sliding window method.

4.4 Qualitative Evaluation

In this section, we perform qualitative evaluation by selecting several sequences, which are visualized in Fig. 8. The total of 10 selected sequences is sorted based on the number of times the target disappeared. Besides, not only the number of times disappeared for each sequence, but also the number of frames, the duration of the target disappearance, and the visual attributes are organized in the figure.

We show through various examples that our method RLT-DiMP estimates a target position more accurately than SuperDiMP, PrDiMP [11], and DiMP [6], without depending on the above situations. In particular, we emphasize that our method has a robust re-detection ability with the aid of spatio-temporal contraints in *dog*, *bird1*, *boat*, and *person7* sequences where the target object suddenly re-appears in a different location. In addition, we note that the target is robustly tracked in *carchase*, *warmup*, *uav1*, and *group2* sequences with similar objects or background clutter through uncertainty reduction using random erasing and enhanced discriminative feature learning with background augmentation.

5 Conclusion

We have proposed a robust long-term object tracker via an improved discriminative model prediction method. Conventional object trackers easily follow other objects when the target object disappears from view or is partially obscured due to the presence of background distractors or similar objects. Long-term object trackers need to be robust to these challenging issues because they have to track objects without restarting for a long period. To this end, our approach improves robustness for long-term tracking through uncertainty reduction using random erasing, robust random search with spatio-temporal constraints, and background augmentation for more discriminative feature learning. Quantitative and qualitative evaluation on the VOT-LT2020 benchmark dataset demonstrates the superiority of our method over the state-of-the-art long-term trackers.

Acknowledgment. This work was supported in part through NSF grant IIS-1650994, the financial assistance award 60NANB17D156 from U.S. Department of Commerce, National Institute of Standards and Technology (NIST) and by the Intelligence Advanced Research Projects Activity (IARPA) via Department of Interior/Interior Business Center (DOI/IBC) contract number D17PC0034. The U.S. Government is authorized to reproduce and distribute reprints for Governmental purposes notwithstanding any copy-right annotation/herein. Disclaimer: The views and conclusions contained herein are those of the authors and should not be interpreted as representing the official policies or endorsements, either expressed or implied, of NIST, IARPA, NSF, DOI/IBC, or the U.S. Government.

References

1. Ali, A., et al.: Visual object tracking–classical and contemporary approaches. Front. Comput. Sci. **10**(1), 167–188 (2016)
2. Atkinson, R.C., Shiffrin, R.M.: Human memory: a proposed system and its control processes (1968)
3. Avidan, S.: Support vector tracking. IEEE Trans. Pattern Anal. Mach. Intell. **26**(8), 1064–1072 (2004)
4. Babenko, B., Yang, M.H., Belongie, S.: Robust object tracking with online multiple instance learning. IEEE Trans. Pattern Anal. Mach. Intell. **33**(8), 1619–1632 (2010)
5. Bertinetto, L., Valmadre, J., Henriques, J.F., Vedaldi, A., Torr, P.H.S.: Fully-convolutional Siamese networks for object tracking. In: Hua, G., Jégou, H. (eds.) ECCV 2016. LNCS, vol. 9914, pp. 850–865. Springer, Cham (2016). https://doi.org/10.1007/978-3-319-48881-3_56
6. Bhat, G., Danelljan, M., Gool, L.V., Timofte, R.: Learning discriminative model prediction for tracking. In: Proceedings of the IEEE International Conference on Computer Vision, pp. 6182–6191 (2019)
7. Cai, Z., Vasconcelos, N.: Cascade R-CNN: delving into high quality object detection. In: Proceedings of the IEEE Conference on Computer Vision and Pattern Recognition, pp. 6154–6162 (2018)
8. Čehovin, L.: TraX: the visual tracking exchange protocol and library. Neurocomputing **260**, 5–8 (2017)
9. Danelljan, M., Bhat, G., Khan, F.S., Felsberg, M.: Atom: accurate tracking by overlap maximization. In: Proceedings of the IEEE Conference on Computer Vision and Pattern Recognition, pp. 4660–4669 (2019)
10. Danelljan, M., Robinson, A., Shahbaz Khan, F., Felsberg, M.: Beyond correlation filters: learning continuous convolution operators for visual tracking. In: Leibe, B., Matas, J., Sebe, N., Welling, M. (eds.) ECCV 2016. LNCS, vol. 9909, pp. 472–488. Springer, Cham (2016). https://doi.org/10.1007/978-3-319-46454-1_29
11. Danelljan, M., Van Gool, L., Timofte, R.: Probabilistic regression for visual tracking. arXiv preprint arXiv:2003.12565 (2020)
12. Grabner, H., Grabner, M., Bischof, H.: Real-time tracking via on-line boosting. In: BMVC, vol. 1, p. 6 (2006)
13. Jung, I., Son, J., Baek, M., Han, B.: Real-time MDNet. In: Proceedings of the European Conference on Computer Vision (ECCV), pp. 83–98 (2018)
14. Kalal, Z., Mikolajczyk, K., Matas, J.: Forward-backward error: automatic detection of tracking failures. In: 2010 20th International Conference on Pattern Recognition, pp. 2756–2759. IEEE (2010)
15. Kalal, Z., Mikolajczyk, K., Matas, J.: Tracking-learning-detection. IEEE Trans. Pattern Anal. Mach. Intell. **34**(7), 1409–1422 (2011)
16. Kristan, M., et al.: The sixth visual object tracking VOT2018 challenge results. In: Leal-Taixé, L., Roth, S. (eds.) ECCV 2018. LNCS, vol. 11129, pp. 3–53. Springer, Cham (2019). https://doi.org/10.1007/978-3-030-11009-3_1
17. Kristan, M., et al.: The visual object tracking VOT2015 challenge results. In: Proceedings of the IEEE International Conference on Computer Vision Workshops, pp. 1–23 (2015)
18. Kristan, M., et al.: The seventh visual object tracking VOT2019 challenge results. In: Proceedings of the IEEE International Conference on Computer Vision Workshops (2019)

19. Kristan, M., et al.: A novel performance evaluation methodology for single-target trackers. IEEE Trans. Pattern Anal. Mach. Intell. **38**(11), 2137–2155 (2016)
20. Laurense, V.A., Goh, J.Y., Gerdes, J.C.: Path-tracking for autonomous vehicles at the limit of friction. In: 2017 American Control Conference (ACC), pp. 5586–5591. IEEE (2017)
21. Lee, H., Choi, S., Kim, C.: A memory model based on the Siamese network for long-term tracking. In: Leal-Taixé, L., Roth, S. (eds.) ECCV 2018. LNCS, vol. 11129, pp. 100–115. Springer, Cham (2019). https://doi.org/10.1007/978-3-030-11009-3_5
22. Li, B., Wu, W., Wang, Q., Zhang, F., Xing, J., Yan, J.: SiamRPN++: evolution of Siamese visual tracking with very deep networks. In: Proceedings of the IEEE Conference on Computer Vision and Pattern Recognition, pp. 4282–4291 (2019)
23. Li, B., Yan, J., Wu, W., Zhu, Z., Hu, X.: High performance visual tracking with Siamese region proposal network. In: Proceedings of the IEEE Conference on Computer Vision and Pattern Recognition, pp. 8971–8980 (2018)
24. Lukezic, A., Zajc, L.C., Vojır, T., Matas, J., Kristan, M.: Now you see me: evaluating performance in long-term visual tracking. arXiv preprint arXiv:1804.07056 4 (2018)
25. Moudgil, A., Gandhi, V.: Long-term visual object tracking benchmark. In: Jawahar, C.V., Li, H., Mori, G., Schindler, K. (eds.) ACCV 2018. LNCS, vol. 11362, pp. 629–645. Springer, Cham (2019). https://doi.org/10.1007/978-3-030-20890-5_40
26. Nam, H., Han, B.: Learning multi-domain convolutional neural networks for visual tracking. In: Proceedings of the IEEE Conference on Computer Vision and Pattern Recognition, pp. 4293–4302 (2016)
27. Ren, S., He, K., Girshick, R., Sun, J.: Faster R-CNN: towards real-time object detection with region proposal networks. In: Advances in Neural Information Processing Systems, pp. 91–99 (2015)
28. Smeulders, A.W., Chu, D.M., Cucchiara, R., Calderara, S., Dehghan, A., Shah, M.: Visual tracking: an experimental survey. IEEE Trans. Pattern Anal. Mach. Intell. **36**(7), 1442–1468 (2013)
29. Šuligoj, F., Šekoranja, B., Švaco, M., Jerbić, B.: Object tracking with a multiagent robot system and a stereo vision camera. Procedia Eng. **69**, 968–973 (2014)
30. Voigtlaender, P., Luiten, J., Torr, P.H., Leibe, B.: Siam R-CNN: visual tracking by re-detection. arXiv preprint arXiv:1911.12836 (2019)
31. Wang, L., Ouyang, W., Wang, X., Lu, H.: Visual tracking with fully convolutional networks. In: Proceedings of the IEEE International Conference on Computer Vision, pp. 3119–3127 (2015)
32. Wang, Q., Zhang, L., Bertinetto, L., Hu, W., Torr, P.H.: Fast online object tracking and segmentation: a unifying approach. In: Proceedings of the IEEE Conference on Computer Vision and Pattern Recognition, pp. 1328–1338 (2019)
33. Wu, Y., Lim, J., Yang, M.H.: Object tracking benchmark. IEEE Trans. Pattern Anal. Mach. Intell. **37**(9), 1834–1848 (2015)

An Exploration of Target-Conditioned Segmentation Methods for Visual Object Trackers

Matteo Dunnhofer(✉), Niki Martinel, and Christian Micheloni

Machine Learning and Perception Lab, University of Udine, Udine, Italy
matteo.dunnhofer@uniud.it

Abstract. Visual object tracking is the problem of predicting a target object's state in a video. Generally, bounding-boxes have been used to represent states, and a surge of effort has been spent by the community to produce efficient causal algorithms capable of locating targets with such representations. As the field is moving towards binary segmentation masks to define objects more precisely, in this paper we propose to extensively explore target-conditioned segmentation methods available in the computer vision community, in order to transform any bounding-box tracker into a segmentation tracker. Our analysis shows that such methods allow trackers to compete with recently proposed segmentation trackers, while performing quasi real-time.

Keywords: Visual object tracking · Video object segmentation · Target-conditioned segmentation · Deep learning

1 Introduction

In its simplest definition, visual object tracking corresponds to the prediction of the state of a target object in a stream of images. It is considered one of the most difficult problems in computer vision. Object occlusion and fast motion, light changes, and motion blur are some of the challenges that algorithms have to deal with. Additionally, constraints of real-time operation are often demanded by the many applications, such as video surveillance, behavior understanding, autonomous driving, and robotics.

Despite the many state representation one can use to model the target's state, the bounding-box has been the most used until now. This is a rectangle that encloses the object of interest, and it is defined by the coordinates of its top-left corner or of its center, and by its width and height. Based on this representation, many model-free tracking algorithms have been studied. Early solutions were based on mean shift algorithms [9], part-based methods [5,41], and SVM learning [21]. Later, the correlation filter approach gained interest thanks to its fast processing time [1,4,12,24,36]. More recently, the features of convolutional neural networks (CNNs) have been exploited to develop more efficient trackers.

A. Bartoli and A. Fusiello (Eds.): ECCV 2020 Workshops, LNCS 12539, pp. 618–636, 2020.
https://doi.org/10.1007/978-3-030-68238-5_41

Trackers based on this image representation include deep regression networks [17,20,23], online tracking-by-detection methods [26,40], solutions that use reinforcement learning [6,16,17,48,64], CNN-based discriminative correlation filters [3,10,11,13] and siamese CNNs [2,30,31,59,65,67]. The last two methods raised the state-of-the-art year-by-year, showing a remarkable performance across all the available tracking benchmarks [19,25,28,29,38,39,60] that almost reaches the 70% of bounding-box overlap accuracy. Such a high and generalized performance poses the question about if the bounding-box based measures have been now saturated. Moreover, it was proven in the object detection community that humans hardly distinguish a bounding-box prediction that has a 30% overlap from one with 50% [51]. Hence, it is fair to ask ourselves if a 100% overlap accuracy tracker is really necessary for applications. From such considerations, we can wonder if the time is done for bounding-box representations.

Furthermore, starting from 2020, tracking communities (VOT2020[1] and MOT[2]) raised the bar in their annual challenges by requesting trackers binary segmentation maps – precise location and shape definitions through pixel-wise background-target classifications – as target state representation. Segmentation representations are not new in the visual tracking panorama. In many applications, model-based algorithms used contours [32,62] or masks [15,27,37] for tracking particular objects. From a more general point of view, the recent video object segmentation (VOS) problem requires to produce the segmentation masks of generic target objects in a video, given the mask of each in the first frame. The currently available solutions propose highly accurate methods in terms of segmentation ability [34,42,55–57,63], but with the drawback of poor speed performance. This is due to characteristics of the available benchmarks [44,46,61], that do not include challenging situations from a tracking point of view. In fact, these datasets provide temporally short sequences where the target covers a large fraction of frames, its appearance does not suffer major changes, or low background distractors are present. The performance of such methods on standard VOT benchmarks was proven very poor [28] and to mitigate such behavior, the SiamMask [59] and D3S [35] algorithms have been proposed recently. These solutions adapted, respectively, the siamese correlation approach and discriminative correlation filters to segmentation outputs, and showed promising results while performing in real-time.

We believe that the huge effort spent by the tracking community in developing bounding-box based trackers can be still exploited in the segmentation tracking domain. With such an idea in mind, in this paper, we propose to explore what is currently available in the computer vision literature that can be adapted to make any bounding-box tracker output segmentation masks. In particular, we propose to extensively evaluate three methods: Box2Seg [57], SiamMask [59], and AMP [53]. Two were already proposed for this task [53,57], but their capabilities were little explored. The other is a recent segmentation tracker [59] that we reinterpret as a segmentation module. Our evaluations are based on a framework that

[1] https://votchallenge.net/vot2020/.
[2] https://motchallenge.net.

requires a bounding-box tracker to provide a coarse localization of the object, and then a segmentation module conditioned on the target object is employed to provide its precise localization. Along with practical considerations, we will show that this combination can produce trackers able to compete with the recently proposed methods [35,59] on the VOT2020 and DAVIS benchmarks.

1.1 Related Work

Combining segmentation methods and trackers has been increasingly tackled in the last two years. SiamMask [59] and D3S [35] employed a CNN decoder module [45,50] to refine a latent representation constructed by a cross-correlation operation and discriminative filter, respectively. Zhang *et al.* [66] proposed to use ECO tracker's [10] bounding-box predictions to improve the segmentation performance of the OSVOS [55] VOS method. Similarly, [57] adapted a deep CNN for semantic segmentation to generate a segmentation mask after the bounding-box proposal of a tracking-by-detection approach. The combination of these methods achieved promising results, but they were mainly focused on the VOS task. Additionally, they did not provide any extensive evaluation by considering different trackers and segmentation methods. In this paper, we aim to provide a deep analysis of such combination on both visual tracking and VOS benchmarks.

2 Methodology

In this paper, we study how state-of-the-art off-the-shelf bounding-box trackers can be augmented to track an object with the requirement of a segmentation representation. Our idea is based on the belief that the much effort spent in developing algorithms to predict the motion of a target is relevant even if a segmentation is required. To implement our analysis, we design a framework where a bounding-box tracker is first used to get a coarse localization of the target object, and then a target-conditioned segmentation method is executed to generate a pixel-wise map. Under this setup, any bounding-box based tracker can be transformed into a segmentation tracker. Considering separately tracking and mask generation carries practical advantages: (i) the performance of a segmentation tracker can be analyzed more consistently, by separating the error committed in the localization from the error in shape definition; (ii) flexibility of easily switch tracking and segmentation modules to adapt to application needs; (iii) availability of two different forms of output (bounding-box and mask) that are obtained with independent modules.

In the following of this section, we first introduce the framework employed for the analysis. Then, an abstract description of each of the selected segmentation methods will be given.

2.1 Segmentation Tracking Framework

We first define the key elements of the framework. A video

$$\mathcal{V} = \{F_t\}, t \in \{0, \cdots, T\}, T \in \mathbb{N} \tag{1}$$

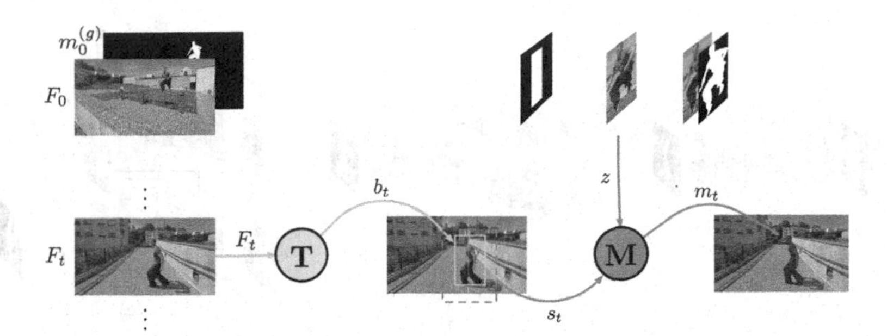

Fig. 1. Graphical representation of the framework used for the evaluation. $m_0^{(g)}$ outlines the target to be tracked in the first frame F_0 of a video. At every step t, the frame F_t is first given in input the the tracker $\mathbf{T}$ which outputs a bounding-box estimate b_t. This, together with a factor k, is employed to crop a searching area s_t in F_t. s_t is inputted to the mask generation algorithm $\mathbf{M}$ which is conditioned on the target template z computed in different form depending on $\mathbf{M}$'s input requirements. $\mathbf{M}$ returns the segmentation of the target inside s_t. The output mask m_t is finally built by placing $\mathbf{M}$'s output inside a zero-matrix at the location of s_t.

is considered as a T long sequence of frames $F_t \in \mathcal{I}$, where $\mathcal{I} = \{0, \cdots, 255\}^{W \times H \times 3}$ is the space of RGB images. We treat a bounding-box tracking algorithm as a function

$$\mathbf{T} : \mathcal{I} \rightarrow \mathbb{R}^4 \tag{2}$$

that is inputted with frame F_t and produces a bounding-box estimate $b_t = [x_t, y_t, w_t, h_t]$ as a real-valued vector containing the center coordinates x_t, y_t, and the width and height w_t, h_t (in the image coordinate system).[3] In a similar fashion, we consider a target-based segmentation algorithm as the function

$$\mathbf{M} : \mathcal{P} \times \mathcal{Z} \rightarrow \{0, 1\}^{W' \times H'} \tag{3}$$

which is given an image patch $s_t \in \mathcal{P} = \{0, \cdots, 255\}^{W' \times H' \times 3} \subseteq \mathcal{I}$ extracted from F_t and a template image $z \in \mathcal{Z}$ of the target object, and outputs a binary segmentation mask with zero-elements belonging to the background and one-elements defining the pixels of the target.

Given these concepts, the segmentation tracking procedure works as follows. At every time step t of a video $\mathcal{V}$, F_t is first given to the tracker $\mathbf{T}$ to produce b_t. Then, F_t and b_t are used to extract a searching area

$$s_t = F_t[x_t, y_t, k \cdot w_t, k \cdot h_t] \tag{4}$$

which is the area of F_t localized by the coordinates of b_t and which width and height are scaled by the factor $k \in \mathbb{R}$. s_t and z are given to the segmentation

[3] At $t = 0$, $\mathbf{T}$ is initialized with F_0 and the ground-truth bounding-box $b_0^{(g)}$.

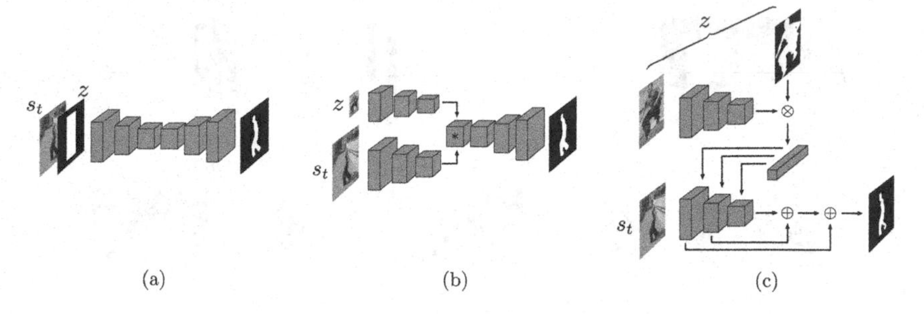

Fig. 2. Visual representation of the methodologies employed to generate target segmentations. (a) shows SemSeg, a deep segmentation model adapted to take in input a 4-channel tensor composed of s_t and z and produce the mask of the object. (b) presents SiamSeg, the siamese framework where a cross correlation operation between z and s_t features is employed to first locate the object, and then to produce its segmentation mask. (c) shows FewShotSeg, a few-shot segmentation algorithm that is adapted for visual segmentation tracking, by considering z as the support set and s_t as the query image.

algorithm $\mathbf{M}$ to produce the pixel-wise mask of the target inside s_t.[4] The output mask m_t is finally built by placing $\mathbf{M}$'s output at the s_t location of a zero-matrix with size $W \times H$.

A graphical representation of the described framework is shown in Fig. 1.

2.2 Target-Conditioned Segmentation Methods

In this subsection we describe the target-conditioned segmentation methodologies we analyzed. Three conceptually different approaches were chosen:

- an adapted semantic segmentation network [7,57], that we name SemSeg;
- a module based on the siamese correlation framework [59], referred as SiamSeg;
- a few-shot segmentation algorithm [53], called FewShotSeg.

SemSeg. The first target-conditioned segmentation method we analyzed was proposed as Box2Seg in [34,57]. The idea is to adapt a state-of-the-art fully convolutional deep neural network for image segmentation to target segmentation. Given an RGB image and an additional input channel containing coarse information regarding the position of the target, this module produces a detailed segmentation of the latter. In the context of our framework, the RGB channels of the searching area s_t are concatenated with the template channel

$$z = \{0, 1\}^{k \cdot w_t, k \cdot h_t} \tag{5}$$

[4] The details to obtain z are described in each subsection describing the segmentation methods.

which is a binary mask of the same size of s_t, and which positive elements are located inside the area defined by b_t. z is computed at every time step t, and the 4-channel input resulting from the concatenation is given to the network which produces the segmentation of the target inside the searching area. A visualization of this approach is proposed in Fig. 2(a). The network is trained offline by exploiting object segmentation, instance segmentation, and/or VOS datasets. The training pairs are formed as batches of inputs-targets, where the first are composed by searching area and template (built using the bounding-box that encloses the ground-truth segmentation), and the second are the actual object masks. Optimization is done by solving a two class segmentation problem (foreground-background) defined as the minimization of a pixel-wise classification loss (cross-entropy, Dice loss, etc.). This approach has the advantage of requiring just the bounding-box as first-frame target definition.

SiamSeg. As second mask generation method, we reinterpreted the siamese correlation framework for segmentation tracking [59]. The general view of this scheme is to first locate the target template in the higher-level feature space of template and searching area, and then project the localization into the segmentation space. These steps are jointly implemented with an encoder-decoder CNN architecture, which capabilities are acquired through an end-to-end offline procedure in which the whole model is optimized by minimizing a foreground-background pixel-wise classification loss. The training examples are pairs of searching area-template inputs, and ground-truth target masks, where searching area and template are sampled without temporal correlation.

Following this intuition, we adapted such method in our framework as follows. The target template is the image crop

$$z = F_0[x_0, y_0, w_0, h_0] \tag{6}$$

extracted from the first frame F_0 of the video, using the ground-truth bounding-box $b_0^{(g)} = [x_0, y_0, w_0, h_0]$. $b_0^{(g)}$ is obtained as the box that encloses the ground-truth mask $m_0^{(g)}$. The features $\widehat{z}$ of z are computed with a forward pass through the encoder module just at $t = 0$. At every other t, $\widehat{z}$ is cross-correlated to the encoded representation $\widehat{s_t}$, and the resulting activation map is then refined by the decoder module, and ultimately placed into m_t. The procedure is depicted in Fig. 2(b) and, as for SemSeg, it just requires the target definition as a bounding-box.

FewShotSeg. The last analyzed methodology treats target-conditioned segmentation as a few-shot segmentation problem. In such a setting, the goal is to provide a pixel-wise segmentation of a target object inside a query image, given a so-called support-set, i.e. one or more (few-shot) image and mask examples of the target. Algorithms for this problem are generally designed as fully CNNs, where the segmentation ability is guided by other convolutional branches or by model parameters that are made dependent on the support-set.

This view of few-shot segmentation can be reframed for the purpose of segmentation tracking. In our setting, the support-set is considered as the target template

$$z = (F_0[x_0, y_0, w_0, h_0], m_0^{(g)}[x_0, y_0, w_0, h_0]) \quad (7)$$

that is the pair of the image crop that contains the visual appearance of the target in F_0, and the relative cropped ground-truth mask. The crops are constructed considering $b_0^{(g)} = [x_0, y_0, w_0, h_0]$. The searching area s_t is extracted after every b_t of $\mathbf{T}$ and it is considered as the query image. Together with the template (the support-set), they are given to the few-shot segmentation model to produce the target segmentation. A graphical example of this methodology is proposed in Fig. 2(c). With respect to the previous methods, employing FewShotSeg requires the definition of the target object through a mask.

3 Experimental Setup

In this section, we report the experimental procedures we performed to implement and analyze the previously presented methodologies. All experiments were run on a machine with an Intel Xeon E5-2690 v4 @ 2.60 GHz CPU, 320 GB of RAM, and an NVIDIA TITAN V GPU. Code for tracker and segmentation methods was implemented in Python.

3.1 Trackers

The trackers selected for the analysis were KCF [24], DCFNet [58], MDNet [40], MetaCrest [43], SiamFC [2], SiamRPN [31], ECO [10], ATOM [11], and DiMP [3]. Such algorithms were chosen because they tackle visual tracking by different approaches and so can provide performance of various quality. For each of them, we used the public code made available by the authors. We tried the best to respect default parameters and settings.

3.2 Segmentation Modules

SemSeg. To implement this methodology, we followed the details of the Box2Seg refinement module provided in [34,57]. The DeepLab-v3 architecture [7] for image segmentation was translated for the task of interest. ImageNet [14] pre-trained ResNet-50 [22] was employed as backbone network and adapted to receive the 4-channel tensor. Before being inputted, the concatenated RGB and template channels were resized to 385 $\times$ 385 pixels. During training, the searching area was enlarged by the factor k, chosen uniformly in $\{1, 1.25, 1.5, 1.75, 2\}$. Batches of 12 input-target mask pairs were sampled from a training set composed of the training sets of COCO [33], YouTube-VOS [61], and DAVIS 2016 and 2017 [44,46]. Learning rate was set to 10^{-5} for the backbone layers, and to 10^{-4} for all the others. Training was carried on until the mIoU [18], computed over the foreground and background classes, stopped improving on a custom validation set composed of the validation sets of the aforementioned datasets.

SiamSeg. The second approach introduced in Subsect. 2.2 was implemented through the segmentation tracker SiamMask [59]. We used the code provided by the authors along with the pre-trained models. For completion, we present to the reader some information about the training procedures performed by the authors. The SiamMask architecture model was trained in two-stages: first, the encoder module based on ResNet-50 [22] was trained for target localization by optimizing a multi-task loss for similarity maximization and RPN [49] detection. After that, the decoder module designed as [45] was attached to the intermediate cross-correlation map and trained by minimizing a foreground-background pixel-wise cross-entropy loss. The training set used was a combination of ImageNet-VID [14], COCO [33] and YouTube-VOS [61]. Before being inputted to the model, z and s_t were resized to 127×127 and 255×255 pixels respectively.

FewShotSeg. As a few-shot segmentation module, we employed the strategy proposed in [53], which is a recently introduced state-of-the-art method that has been shown to perform well also in VOS tasks. The authors proposed a sample efficient method to segment an unseen class object via a multi-resolution imprinting procedure of adaptive masked proxies (AMP). AMPs are constructed by a Normalized Masked Average Pooling (NMAP) operation between the CNN embeddings of the support set's RGB sample and its relative binary mask. The AMP representations are used to imprint [47], at multiple resolutions, the CNN embeddings computed on the query image. The VGG16 [54] architecture is employed as a backbone feature extractor, and skip connections are also exploited as done similarly in FCN8s [52]. Data extracted from the PASCAL-VOC dataset [18] was used to compose training samples as query image, support-set image, support-set mask, and target mask. Optimization was performed by minimizing the pixel-wise cross-entropy loss between predicted and ground-truth masks. Code and pre-trained model provided by the authors were adapted to our implementation needs.

3.3 Benchmarks and Performance Measures

We performed analysis on the VOT2020 benchmark, and the validation sets of the DAVIS 2016 [44] and DAVIS 2017 [46] VOS benchmarks. All provide segmentations as target representations.

For VOT2020 we employed the newly introduced protocol.[5] The novel baseline protocol requires running a tracker on shorter sequences determined by predefined points (anchors). From such starting points, the tracker is initialized with the ground-truth mask and run either forward or backward, depending on the longest sub-sequence yielded by the two directions. The new accuracy ($A_{\uparrow}$) measures the average pixel-wise intersection-over-union between predicted and ground-truth masks, for frames where the tracker did not fail (i.e. the accuracy did not decrease after a certain threshold). The new robustness ($R_{\uparrow}$) expresses the normalized average number of frames where the algorithm

[5] https://data.votchallenge.net/vot2020/vot-2020-protocol.pdf.

successfully tracked the target before drifting. The two measures are joined in a refreshed single performance score known as expected average overlap ($\text{EAO}_{\uparrow}$). Version 0.4.2 of the Python toolkit was used to obtain the results.

The protocol used for DAVIS datasets is similar to the One-Pass evaluation (OPE) employed in OTB [60] benchmarks: the tracker is initialized with the mask of the target object in the first frame, and then it is run until the end of the sequence. Performance is measured in terms of the Jaccard index $\mathcal{J}$ which measures the pixel-wise intersection-over-union between the predicted and ground-truth masks. Along with this index, the F-measure $\mathcal{F}$ is employed to evaluate contour accuracy. For both measures, mean ($\mathcal{J}_{\mathcal{M}\uparrow}, \mathcal{F}_{\mathcal{M}\uparrow}$), recall ($\mathcal{J}_{\mathcal{R}\uparrow}, \mathcal{F}_{\mathcal{R}\uparrow}$), and decay ($\mathcal{J}_{\mathcal{D}\downarrow}, \mathcal{F}_{\mathcal{D}\downarrow}$) values are reported. For DAVIS 2017, where multiple objects must be tracked and segmented, we run the trackers independently for each object and then fuse the prediction masks by assigning each pixel to the object that received higher confidence in that location.

Table 1. Results of the baseline experiment on VOT2020. Best segmentation method results, per tracker, are highlighted in red (Rectangular Mask results are excluded).

Tracker	SemSeg			SiamSeg			FewShotSeg			Rectangular Mask		
	$\text{EAO}_{\uparrow}$	$\text{A}_{\uparrow}$	$\text{R}_{\uparrow}$	$\text{EAO}_{\uparrow}$	$\text{A}_{\uparrow}$	$\text{R}_{\uparrow}$	$\text{EAO}_{\uparrow}$	$\text{A}_{\uparrow}$	$\text{R}_{\uparrow}$	$\text{EAO}_{\uparrow}$	$\text{A}_{\uparrow}$	$\text{R}_{\uparrow}$
DCFNet	0.203	0.616	0.426	0.310	0.676	0.558	0.230	0.491	0.567	0.184	0.441	0.523
KCF	0.199	0.648	0.371	0.285	0.659	0.501	0.200	0.459	0.499	0.155	0.402	0.432
SiamFC	0.218	0.602	0.446	0.309	0.682	0.571	0.228	0.491	0.563	0.183	0.418	0.537
MetaCrest	0.240	0.602	0.513	0.336	0.657	0.624	0.250	0.479	0.647	0.189	0.390	0.587
SiamRPN	0.356	0.692	0.639	0.369	0.701	0.651	0.311	0.551	0.677	0.247	0.452	0.663
MDNet	0.295	0.638	0.609	0.371	0.662	0.689	0.308	0.546	0.723	0.234	0.440	0.687
ATOM	0.402	0.678	0.735	0.406	0.691	0.723	0.337	0.560	0.731	0.277	0.467	0.738
DiMP	0.410	0.675	0.744	0.410	0.691	0.730	0.347	0.556	0.749	0.278	0.464	0.733
ECO	0.322	0.632	0.735	0.414	0.694	0.729	0.349	0.561	0.759	0.275	0.459	0.746
b-oracle	0.806	0.809	0.996	0.697	0.744	0.970	0.541	0.623	0.941	0.516	0.519	1.0

4 Results

General Performance. Results on VOT2020 benchmark are presented in Table 1. Trackers combined with SiamSeg achieve the best overall performance in $\text{EAO}_{\uparrow}$ and $\text{A}_{\uparrow}$. This is explained by the fact the VOT benchmarks include difficult tracking scenarios for trackers, resulting in lower quality bounding-boxes that affect SemSeg and FewShotSeg. Thanks to its more robust segmentation method, SiamSeg allows to recover (to some extent) from inaccurate b_t estimates and so produce more accurate target segmentations. Interestingly, FewShotSeg is the approach that achieves the highest $\text{R}_{\uparrow}$, showing to be the method less susceptible to failure. For all the methods, employing a better tracker is fundamental to improve the overall performance.

Table 2. $\mathcal{J}$ results on DAVIS 2016 validation set. Best segmentation method results, per tracker, are highlighted in red (Rectangular Mask results are excluded).

Tracker	SemSeg			SiamSeg			FewShotSeg			Rectangular Mask		
	$\mathcal{J}_{\mathcal{M}\uparrow}$	$\mathcal{J}_{\mathcal{R}\uparrow}$	$\mathcal{J}_{\mathcal{D}\downarrow}$	$\mathcal{J}_{\mathcal{M}\uparrow}$	$\mathcal{J}_{\mathcal{R}\uparrow}$	$\mathcal{J}_{\mathcal{D}\downarrow}$	$\mathcal{J}_{\mathcal{M}\uparrow}$	$\mathcal{J}_{\mathcal{R}\uparrow}$	$\mathcal{J}_{\mathcal{D}\downarrow}$	$\mathcal{J}_{\mathcal{M}\uparrow}$	$\mathcal{J}_{\mathcal{R}\uparrow}$	$\mathcal{J}_{\mathcal{D}\downarrow}$
KCF	0.527	0.570	0.174	0.557	0.616	0.199	0.580	0.688	0.162	0.302	0.200	0.153
DCFNet	0.531	0.574	0.209	0.551	0.627	0.229	0.564	0.674	0.178	0.313	0.183	0.130
MetaCrest	0.574	0.624	0.169	0.595	0.672	0.145	0.598	0.712	0.136	0.323	0.151	0.108
MDNet	0.582	0.635	0.177	0.593	0.656	0.196	0.610	0.717	0.143	0.342	0.198	0.149
SiamFC	0.607	0.661	0.159	0.611	0.694	0.177	0.621	0.738	0.163	0.356	0.234	0.140
ECO	0.615	0.679	0.099	0.623	0.744	0.108	0.626	0.748	0.113	0.375	0.243	0.070
SiamRPN	0.689	0.772	0.089	0.663	0.782	0.111	0.681	0.859	0.089	0.417	0.340	0.066
ATOM	0.723	0.846	0.074	0.658	0.785	0.105	0.669	0.845	0.081	0.415	0.345	0.053
DiMP	0.723	0.827	0.086	0.704	0.844	0.100	0.699	0.886	0.095	0.443	0.379	0.027
b-oracle	0.812	0.920	0.020	0.732	0.896	0.044	0.739	0.946	0.052	0.455	0.418	0.008

Table 3. $\mathcal{F}$ results on DAVIS 2016 validation set. Best segmentation method results, per tracker, are highlighted in red (Rectangular Mask results are excluded).

Tracker	SemSeg			SiamSeg			FewShotSeg			Rectangular Mask		
	$\mathcal{F}_{\mathcal{M}\uparrow}$	$\mathcal{F}_{\mathcal{R}\uparrow}$	$\mathcal{F}_{\mathcal{D}\downarrow}$	$\mathcal{F}_{\mathcal{M}\uparrow}$	$\mathcal{F}_{\mathcal{R}\uparrow}$	$\mathcal{F}_{\mathcal{D}\downarrow}$	$\mathcal{F}_{\mathcal{M}\uparrow}$	$\mathcal{F}_{\mathcal{R}\uparrow}$	$\mathcal{F}_{\mathcal{D}\downarrow}$	$\mathcal{F}_{\mathcal{M}\uparrow}$	$\mathcal{F}_{\mathcal{R}\uparrow}$	$\mathcal{F}_{\mathcal{D}\downarrow}$
DCFNet	0.553	0.587	0.210	0.536	0.596	0.187	0.530	0.599	0.158	0.155	0.017	0.068
KCF	0.559	0.577	0.180	0.525	0.572	0.190	0.542	0.598	0.149	0.136	0.018	0.119
MetaCrest	0.599	0.632	0.193	0.561	0.637	0.136	0.572	0.634	0.137	0.139	0.019	0.063
MDNet	0.603	0.623	0.170	0.570	0.616	0.197	0.582	0.635	0.170	0.163	0.050	0.112
SiamFC	0.633	0.665	0.152	0.592	0.663	0.159	0.597	0.675	0.157	0.156	0.037	0.126
ECO	0.637	0.696	0.097	0.590	0.692	0.102	0.592	0.673	0.117	0.170	0.020	0.066
SiamRPN	0.713	0.783	0.105	0.629	0.707	0.127	0.642	0.752	0.105	0.186	0.059	0.081
ATOM	0.739	0.856	0.098	0.628	0.697	0.111	0.626	0.751	0.090	0.178	0.025	0.060
DiMP	0.744	0.821	0.108	0.658	0.754	0.130	0.657	0.767	0.118	0.191	0.071	0.015
b-oracle	0.843	0.918	0.033	0.693	0.805	0.064	0.717	0.873	0.056	0.219	0.073	0.015

Results on the DAVIS 2016 benchmark are reported in Tables 2 and 3. More weak trackers like DCFNet, KCF, MDNet, MetaCrest, and SiamFC, benefit of FewShotSeg for pixel-wise accuracy. When more precise bounding-box estimates are provided, through ECO, SiamRPN, ATOM, DiMP, SemSeg allows the best $\mathcal{J}_{\mathcal{M}\uparrow}$ performance. For $\mathcal{J}_{\mathcal{R}\uparrow}$ and $\mathcal{J}_{\mathcal{D}\downarrow}$, FewShotSeg is almost always the best approach. For contour accuracy, SemSeg is generally the best method at $\mathcal{F}_{\mathcal{M}\uparrow}$. Better trackers also benefit the same for $\mathcal{F}_{\mathcal{R}\uparrow}$ and $\mathcal{F}_{\mathcal{D}\downarrow}$. For the others, FewShotSeg gets the best results. SiamSeg is the weakest method on this benchmark, justified by the presence of easy tracking situations that put the focus on providing more accurate target segmentations.

On DAVIS 2017, which results are presented in Tables 4 and 5, SemSeg is still the best approach to use with stronger bounding-box trackers for $\mathcal{J}_{\mathcal{M}\uparrow}$ and $\mathcal{J}_{\mathcal{R}\uparrow}$. FewShotSeg is the method that achieves the most consistent masks across time. For low-performance tracking algorithms, SiamSeg results to be better than the

Table 4. $\mathcal{J}$ results on DAVIS 2017 validation set. Best segmentation method results, per tracker, are highlighted in red (Rectangular Mask results are excluded).

Tracker	SemSeg			SiamSeg			FewShotSeg			Rectangular Mask		
	$\mathcal{J}_{\mathcal{M}\uparrow}$	$\mathcal{J}_{\mathcal{R}\uparrow}$	$\mathcal{J}_{\mathcal{D}\downarrow}$	$\mathcal{J}_{\mathcal{M}\uparrow}$	$\mathcal{J}_{\mathcal{R}\uparrow}$	$\mathcal{J}_{\mathcal{D}\downarrow}$	$\mathcal{J}_{\mathcal{M}\uparrow}$	$\mathcal{J}_{\mathcal{R}\uparrow}$	$\mathcal{J}_{\mathcal{D}\downarrow}$	$\mathcal{J}_{\mathcal{M}\uparrow}$	$\mathcal{J}_{\mathcal{R}\uparrow}$	$\mathcal{J}_{\mathcal{D}\downarrow}$
DCFNet	0.443	0.474	0.299	0.455	0.497	0.281	0.424	0.434	0.214	0.283	0.166	0.176
KCF	0.433	0.464	0.277	0.461	0.517	0.272	0.425	0.451	0.209	0.268	0.167	0.198
MDNet	0.444	0.478	0.284	0.465	0.515	0.260	0.444	0.493	0.216	0.284	0.156	0.168
MetaCrest	0.447	0.468	0.276	0.468	0.518	0.262	0.426	0.443	0.178	0.273	0.145	0.155
SiamFC	0.466	0.499	0.260	0.468	0.523	0.277	0.431	0.454	0.225	0.280	0.176	0.196
ECO	0.498	0.556	0.244	0.503	0.567	0.222	0.458	0.501	0.178	0.310	0.220	0.132
SiamRPN	0.536	0.600	0.233	0.506	0.578	0.237	0.470	0.518	0.180	0.321	0.248	0.141
ATOM	0.566	0.659	0.148	0.544	0.626	0.188	0.488	0.547	0.168	0.321	0.251	0.103
DiMP	0.583	0.671	0.148	0.553	0.639	0.170	0.498	0.555	0.162	0.323	0.251	0.093
b-oracle	0.762	0.891	0.0	0.618	0.738	0.073	0.578	0.694	0.059	0.408	0.340	0.0

Table 5. $\mathcal{F}$ results on DAVIS 2017 validation set. Best segmentation method results, per tracker, are highlighted in red (Rectangular Mask results are excluded).

Tracker	SemSeg			SiamSeg			FewShotSeg			Rectangular Mask		
	$\mathcal{F}_{\mathcal{M}\uparrow}$	$\mathcal{F}_{\mathcal{R}\uparrow}$	$\mathcal{F}_{\mathcal{D}\downarrow}$	$\mathcal{F}_{\mathcal{M}\uparrow}$	$\mathcal{F}_{\mathcal{R}\uparrow}$	$\mathcal{F}_{\mathcal{D}\downarrow}$	$\mathcal{F}_{\mathcal{M}\uparrow}$	$\mathcal{F}_{\mathcal{R}\uparrow}$	$\mathcal{F}_{\mathcal{D}\downarrow}$	$\mathcal{F}_{\mathcal{M}\uparrow}$	$\mathcal{F}_{\mathcal{R}\uparrow}$	$\mathcal{F}_{\mathcal{D}\downarrow}$
KCF	0.517	0.542	0.307	0.500	0.544	0.287	0.506	0.565	0.266	0.172	0.035	0.178
DCFNet	0.532	0.567	0.322	0.512	0.561	0.284	0.511	0.572	0.237	0.194	0.049	0.140
MDNet	0.525	0.563	0.288	0.513	0.565	0.270	0.526	0.598	0.255	0.184	0.059	0.170
MetaCrest	0.545	0.593	0.305	0.520	0.567	0.278	0.521	0.583	0.241	0.176	0.042	0.146
SiamFC	0.556	0.611	0.299	0.523	0.583	0.294	0.524	0.601	0.267	0.184	0.064	0.184
ECO	0.592	0.663	0.255	0.553	0.620	0.236	0.553	0.637	0.226	0.214	0.055	0.128
SiamRPN	0.626	0.713	0.259	0.552	0.628	0.257	0.567	0.670	0.219	0.210	0.070	0.150
DiMP	0.663	0.765	0.181	0.591	0.675	0.215	0.584	0.685	0.195	0.206	0.059	0.093
ATOM	0.640	0.751	0.195	0.584	0.664	0.218	0.574	0.666	0.201	0.203	0.053	0.104
b-oracle	0.829	0.945	0.017	0.654	0.779	0.097	0.685	0.847	0.078	0.280	0.116	0.028

others in $\mathcal{J}_{\mathcal{M}\uparrow}$ and $\mathcal{J}_{\mathcal{R}\uparrow}$, mitigating the lower tracking performance with its target search strategy and showing the increased difficulty of this benchmark than its previous version. In terms of contour performance, SemSeg is the most appropriate method for $\mathcal{F}_{\mathcal{M}\uparrow}$ and $\mathcal{F}_{\mathcal{R}\uparrow}$ performance. For $\mathcal{F}_{\mathcal{D}\downarrow}$, FewShotSeg results in the best solution. Overall, as for VOT2020, in both DAVIS 2016 and 2017 employing better trackers lets achieve the best performances.

Comparison with a Rectangular Segmentation Tracker. In the last block of columns of Tables 1, 2, 3, 4, 5, we report the performance of the trackers considering their b_t predictions as segmentation masks, i.e. binary mask where the rectangular area defined by b_t is filled with 1. Overall, all the considered segmentation methods improve those baseline results on all the benchmarks and across all measures. This proves that employing the approaches presented in this paper lets bounding-box trackers improve their accuracy in terms of

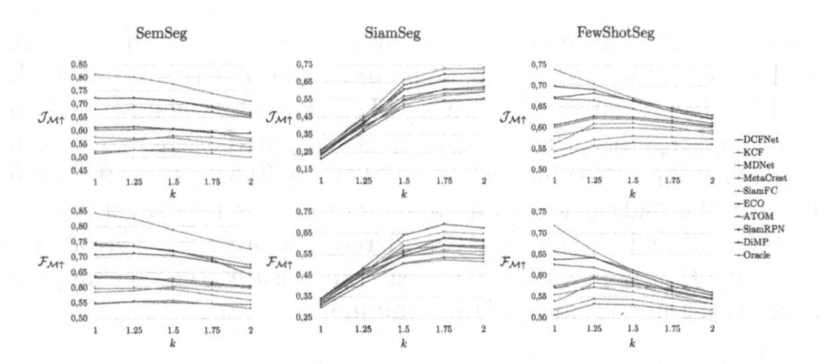

Fig. 3. Results on the DAVIS 2016 validation set of the sensibility of the target segmentation methods to the size of the searching area. Performance is evaluated in terms of $\mathcal{J}_{\mathcal{M}\uparrow}$ and $\mathcal{F}_{\mathcal{M}\uparrow}$.

precise target definition. SemSeg is the method that achieves generally the best improvement, followed by SiamSeg and FewShotSeg.

Comparison with a Bounding-Box Oracle Tracker. In the last row of Tables 1, 2, 3, 4, 5, the performance of a bounding-box oracle based tracker, *b*-oracle (i.e. the tracker that returns the ground-truth bounding-box $b_t^{(g)}$ at every t), is presented. Given this ground-truth information, SemSeg is the approach that best segments the target object, on every considered benchmark and performance measure. On VOT2020, accuracy and robustness performances reach almost 80% and 100%, meaning that its segmentation capabilities are effective for the objects contained in this dataset. SiamSeg follows with a decrease of 9% and 2.6%, while FewShotSeg shows a much bigger performance loss in $\mathrm{A}_{\uparrow}$ (−25% than SemSeg) than in $\mathrm{R}_{\uparrow}$ (−5.5%). FewShotSeg comes after SemSeg in terms of $\mathcal{J}$ and $\mathcal{F}$ on DAVIS 2016, and in terms of $\mathcal{F}$ on DAVIS 2017. SiamSeg gets the weakest performance on DAVIS 2016 but surpasses FewShotSeg in $\mathcal{J}$ on DAVIS 2017.

SemSeg is also the method that suffers the major gap between the *b*-oracle performance and the best tracker DiMP ($\mathrm{EAO}_{\uparrow}$ loss -48%, average $\mathcal{J}_{\mathcal{M}\uparrow}$ loss −17.2%, average $\mathcal{F}_{\mathcal{M}\uparrow}$ loss −15.9%). This shows the susceptibility to misaligned bounding-box predictions (we hypothesize this can be mitigated introducing some noise to the input bounding-boxes in the training procedure). The performance decrease happens also for the other methods, although with less magnitude.

Separating Localization and Segmentation Error. The results obtained with *b*-oracle and the rectangular mask output allow us to determine the tracking and segmentation error committed by $\mathbf{T}$ and $\mathbf{M}$ respectively. The error $e_{\mathbf{T}}$ committed by the tracker is just the performance difference between *b*-oracle and $\mathbf{T}$, both considered with rectangular mask output. The error $e_{\mathbf{M}}$ of $\mathbf{M}$ can be computed as the performance difference between *b*-oracle with $\mathbf{M}$ and $\mathbf{T}$ with $\mathbf{M}$

which tracking performance is corrected by summing $e_{\mathbf{T}}$. In this setting, $e_{\mathbf{M}}$ is considered as the distance from $\mathbf{M}$'s maximum achievable performance, that happens when b-oracle is employed as tracker. For example, when MDNet and SemSeg are executed together, the $A_{\uparrow}$ error $e_{\mathbf{T}}$ is computed as $e_{\mathbf{T}} = 0.519 - 0.440 = 0.079$, while the $e_{\mathbf{M}}$ is obtained as $e_{\mathbf{M}} = 0.809 - (0.638 + 0.079) = 0.092$. So, it results that the highest loss in accuracy is due to the segmentation than to tracking. If DiMP and SiamSeg are considered, we have an $A_{\uparrow}$ error $e_{\mathbf{T}} = 0.055$ and $e_{\mathbf{T}} = -0.002$ meaning that SiamSeg compensates the tracking error and even improves the performance of the combination.

Impact of the Searching Area Size. We analyzed how sensible the three segmentation methods are to different sizes of the searching area. In particular, the factor k was studied across the values $\{1, 1.25, 1.5, 1.75, 2\}$ on the DAVIS 2016 benchmark, and the results are shown in Fig. 3. With SemSeg, all the trackers show a slow decrease in $\mathcal{J}_{\mathcal{M}\uparrow}$ and $\mathcal{F}_{\mathcal{M}\uparrow}$ performance by enlarging the searching area. Best performance are obtained with $k = 1$ or $k = 1.25$ (proven also by the b-oracle based tracker). Similar conclusions can be made for FewShotSeg. The highest $\mathcal{J}_{\mathcal{M}\uparrow}$ is achieved with $k = 1.25$. For larger k, the performance of more weak trackers remains constant, while the performance of stronger trackers slightly decreases. $\mathcal{F}_{\mathcal{M}\uparrow}$ tends to decrease for all the trackers. SiamSeg shows the opposite trend. Better results are obtained with larger searching areas. Specifically, best $\mathcal{J}_{\mathcal{M}\uparrow}$ and $\mathcal{F}_{\mathcal{M}\uparrow}$ performance are obtained with $k \geq 1.75$. Weaker trackers have a smaller performance decrease between 1.75 and 1.5 than stronger ones, while for $k < 1.5$ the performance of all the trackers quickly drops. This can be explained by SiamSeg's training methodology, where the objective is set as target localization and segmentation in large image patches.

Speed Analysis. In Table 6 an analysis of the speed of the algorithms is presented. The fastest method to produce a segmentation is SiamSeg which runs at 43 FPS. With this method, DCFNet and SiamFC run in real-time (39 and 34 FPS respectively). Stronger trackers like ECO, ATOM, and DiMP, achieve a speed of 12, 13, and 15 FPS respectively. SemSeg runs independently at 16 FPS, and combined with SiamRPN and DiMP allows a speed of 12 and 9 FPS respectively. FewShotSeg is the slowest method and takes around 7 FPS. In this setup, the speed performance is almost completely taken by the segmentation method and best trackers reach a speed of 5–6 FPS.

State-of-the-art Comparison. Comparison with the state-of-the-art is presented in Table 7. The VOS methods outperform every studied $\mathbf{T} - \mathbf{M}$ combination on DAVIS 2016 and 2017, but they show poor speed results. DiMP and ATOM with SemSeg perform better than SiamMask in $\mathcal{J}$ on both DAVIS 2016 and 2017. In terms of $\mathcal{F}$ they outperform also D3S. On DAVIS 2017, D3S is improved by DiMP-SemSeg in every measure. On VOT2020, SiamMask is largely beaten by all the best trackers, combined both with SemSeg and SiamSeg. All

Table 6. Results on the speed analysis (in seconds and FPS) of the combined tracker-segmentation methods. The original tracker speeds are reported in the last two columns (times of the employed implementations). The last row shows the average speed of running just the segmentation methods.

Tracker	SemSeg		SiamSeg		FewShotSeg		Tracker speed	
	s	FPS	s	FPS	s	FPS	s	FPS
MDNet	0.628	1.6	0.580	1.7	0.704	1.4	0.550	1.8
MetaCrest	0.181	5.5	0.134	7.5	0.258	3.9	0.109	9.1
ECO	0.126	8	0.081	12.4	0.220	4.6	0.059	17.0
ATOM	0.123	8.1	0.079	12.7	0.198	5.1	0.050	20.0
DiMP	0.109	9.1	0.068	14.7	0.189	5.3	0.038	26.3
SiamRPN	0.026	12.1	0.047	21.4	0.172	5.8	0.026	38.4
KCF	0.070	14.3	0.034	29.5	0.167	6.0	0.013	78.8
SiamFC	0.064	15.6	0.030	33.5	0.157	6.4	0.008	125.3
DCFNet	0.062	16.2	0.026	39.0	0.143	7.0	0.004	227.8
No tracker	0.062	16.3	0.024	42.8	0.151	6.7	–	–

Table 7. State-of-the-art comparison for the best combinations. Best results are highlighted in red, second-best in blue.

Method	DAVIS 2016				DAVIS 2017				VOT2020			FPS
	$\mathcal{J}_{\mathcal{M}\uparrow}$	$\mathcal{J}_{\mathcal{R}\uparrow}$	$\mathcal{F}_{\mathcal{M}\uparrow}$	$\mathcal{F}_{\mathcal{R}\uparrow}$	$\mathcal{J}_{\mathcal{M}\uparrow}$	$\mathcal{J}_{\mathcal{R}\uparrow}$	$\mathcal{F}_{\mathcal{M}\uparrow}$	$\mathcal{F}_{\mathcal{R}\uparrow}$	$\text{EAO}_{\uparrow}$	$\text{A}_{\uparrow}$	$\text{R}_{\uparrow}$	
OSMN [63]	0.740	0.876	0.729	0.840	0.525	0.609	0.571	0.661	–	–	–	7
BoLTVOS [57]	0.781	–	0.812	–	**0.684**	–	**0.754**	–	–	–	–	1
OSVOS [55]	0.798	0.936	0.806	**0.926**	0.566	**0.638**	0.639	**0.738**	–	–	–	0.1
FAVOS [8]	0.824	0.965	0.795	0.894	0.546	0.611	0.618	0.723	–	–	–	0.8
RGMP [42]	0.815	0.917	0.820	0.908	0.648	–	0.686	–	–	–	–	8
OnAVOS [56]	0.857	–	**0.842**	–	0.610	–	0.661	–	–	–	–	0.1
PReMVOS [34]	**0.849**	**0.961**	0.886	0.947	0.739	0.831	0.817	0.889	–	–	–	0.03
SiamMask[a]	0.692	0.848	0.639	0.743	0.522	0.597	0.559	0.645	0.321	0.686	0.569	43
D3S [35]	0.754	–	0.726	–	0.578	–	0.638	–	0.439	**0.699**	0.769	25
SiamRPN-SiamSeg	0.663	0.782	0.629	0.707	0.506	0.578	0.552	0.628	0.369	0.701	0.651	21
ECO-SiamSeg	0.623	0.744	0.590	0.692	0.503	0.567	0.553	0.620	**0.414**	0.694	0.729	12
ATOM-SiamSeg	0.658	0.785	0.628	0.697	0.544	0.626	0.584	0.664	0.406	0.691	0.723	13
ATOM-SemSeg	0.723	0.846	0.739	0.856	0.566	0.659	0.640	0.751	0.402	0.678	0.735	8
DiMP-SemSeg	0.723	0.827	0.744	0.821	0.583	0.671	0.663	0.765	0.410	0.675	**0.744**	9
DiMP-SiamSeg	0.704	0.844	0.658	0.754	0.553	0.639	0.591	0.675	0.410	0.691	0.730	15

[a] Since we used SiamMask to implement SiamSeg, for fair comparison we report the results of the same implementation used for segmentation tracking, which has slightly worse performance than presented in the original paper.

the trackers using the second method improves SiamMask, showing its limitations in target localization. ECO and SiamSeg reaches an $\text{EAO}_{\uparrow}$ of 0.414, slightly improving DiMP and ATOM. With the same segmentation method, SiamRPN outperforms D3S in $\text{A}_{\uparrow}$, achieving the best 0.701, while maintaining a quasi real-time speed of 21 FPS.

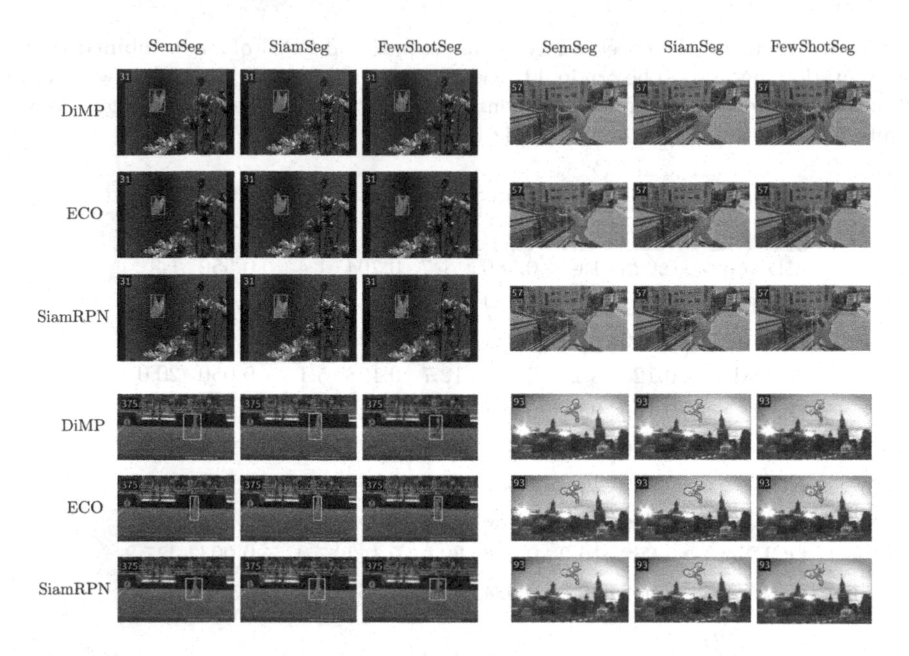

Fig. 4. Qualitative examples of the segmentation (red superimposed mask) proposed by the three target-conditioned segmentation methods, based on the bounding-box proposals (green rectangles) given by three different trackers. (Color figure online)

In Fig. 4 some qualitative examples of the segmentation methods are proposed.[6]

5 Conclusions

In this paper, three target-conditioned segmentation methods, SemSeg, SiamSeg, and FewShotSeg, were extensively analyzed to transform any bounding-box tracker into a segmentation tracker. SemSeg and SiamSeg resulted in the stronger methods, and their combination with trackers like SiamRPN, ECO, ATOM, and DiMP, allows to compete with the most recent segmentation trackers SiamMask and D3S on the DAVIS 2016 and 2017, and VOT2020 benchmarks.

Acknowledgements. This work was supported by the ACHIEVE-ITN H2020 project.

References

1. Bertinetto, L., Valmadre, J., Golodetz, S., Miksik, O., Torr, P.H.: Staple: complementary learners for real-time tracking. In: IEEE Conference on Computer Vision and Pattern Recognition (2016)

[6] For more, please see https://youtu.be/SODiKBD84_g.

2. Bertinetto, L., Valmadre, J., Henriques, J.F., Vedaldi, A., Torr, P.H.S.: Fully-convolutional siamese networks for object tracking. In: Hua, G., Jégou, H. (eds.) ECCV 2016. LNCS, vol. 9914, pp. 850–865. Springer, Cham (2016). https://doi.org/10.1007/978-3-319-48881-3_56
3. Bhat, G., Danelljan, M., Van Gool, L., Timofte, R.: Learning discriminative model prediction for tracking. In: Proceedings of the IEEE/CVF International Conference on Computer Vision (2019)
4. Bolme, D.S., Beveridge, J.R., Draper, B.A., Lui, Y.M.: Visual object tracking using adaptive correlation filters. In: IEEE Conference on Computer Vision and Pattern Recognition. IEEE (2010)
5. Čehovin, L., Kristan, M., Leonardis, A.: Robust visual tracking using anadaptive coupled-layer visual model. IEEE Trans. Pattern Anal. Mach. Intell. **35**(4), 941–953 (2013)
6. Chen, B., Wang, D., Li, P., Wang, S., Lu, H.: Real-time 'actor-critic' tracking. In: Ferrari, V., Hebert, M., Sminchisescu, C., Weiss, Y. (eds.) ECCV 2018. LNCS, vol. 11211, pp. 328–345. Springer, Cham (2018). https://doi.org/10.1007/978-3-030-01234-2_20
7. Chen, L.C., Papandreou, G., Schroff, F., Adam, H.: Rethinking Atrous Convolution for Semantic Image Segmentation, June 2017
8. Cheng, J., Tsai, Y.H., Hung, W.C., Wang, S., Yang, M.H.: Fast and accurate online video object segmentation via tracking parts. In: Proceedings of the IEEE Computer Society Conference on Computer Vision and Pattern Recognition. IEEE Computer Society, December 2018
9. Comaniciu, D., Ramesh, V., Meer, P.: Real-time tracking of non-rigid objects using mean shift. In: IEEE Conference on Computer Vision and Pattern Recognition (2000)
10. Danelljan, M., Bhat, G., Khan, F.S., Felsberg, M.: ECO: efficient convolution operators for tracking. In: IEEE Conference on Computer Vision and Pattern Recognition, November 2017
11. Danelljan, M., Bhat, G., Khan, F.S., Felsberg, M.: ATOM: accurate tracking by overlap maximization. In: IEEE Conference on Computer Vision and Pattern Recognition (2019)
12. Danelljan, M., Hager, G., Khan, F.S., Felsberg, M.: Discriminative Scale space tracking. IEEE Trans. Pattern Anal. Mach. Intell. **39**, 1561–1575 (2017)
13. Danelljan, M., Robinson, A., Shahbaz Khan, F., Felsberg, M.: Beyond correlation filters: learning continuous convolution operators for visual tracking. In: Leibe, B., Matas, J., Sebe, N., Welling, M. (eds.) ECCV 2016. LNCS, vol. 9909, pp. 472–488. Springer, Cham (2016). https://doi.org/10.1007/978-3-319-46454-1_29
14. Deng, J., Dong, W., Socher, R., Li, L.J., Kai Li, Li Fei-Fei: ImageNet: a large-scale hierarchical image database. In: IEEE Conference on Computer Vision and Pattern Recognition. IEEE, June 2009
15. Dunnhofer, M., et al.: Siam-U-Net: encoder-decoder siamese network for knee cartilage tracking in ultrasound images. Med. Image Anal. **60**, 101631 (2020)
16. Dunnhofer, M., Martinel, N., Foresti, G.L., Micheloni, C.: Visual tracking by means of deep reinforcement learning and an expert demonstrator. In: Proceedings of The IEEE/CVF International Conference on Computer Vision Workshops (2019)
17. Dunnhofer, M., Martinel, N., Micheloni, C.: A distilled model for tracking and tracker fusion (2020)
18. Everingham, M., Van Gool, L., Williams, C.K., Winn, J., Zisserman, A.: Thepascal visual object classes (VOC) challenge. Int. J. Comput. Vis. **88**, 303–338 (2010). https://doi.org/10.1007/s11263-009-0275-4

19. Fan, H., et al.: LaSOT: a high-quality benchmark for large-scale single object tracking. In: IEEE Conference on Computer Vision and Pattern Recognition, September 2019
20. Gordon, D., Farhadi, A., Fox, D.: Re 3: real-time recurrent regression networks for visual tracking of generic objects. IEEE Robot. Autom. Lett. **3**, 788–795 (2018)
21. Hare, S., et al.: Struck: structured output tracking with kernels. IEEE Trans. Pattern Anal. Mach. Intell. **38**, 2096–2109 (2016)
22. He, K., Zhang, X., Ren, S., Sun, J.: Deep residual learning for image recognition. In: IEEE Conference on Computer Vision and Pattern Recognition (2016)
23. Held, D., Thrun, S., Savarese, S.: Learning to track at 100 FPS with deep regression networks. In: Leibe, B., Matas, J., Sebe, N., Welling, M. (eds.) ECCV 2016. LNCS, vol. 9905, pp. 749–765. Springer, Cham (2016). https://doi.org/10.1007/978-3-319-46448-0_45
24. Henriques, J.F., Caseiro, R., Martins, P., Batista, J.: High-speed tracking with kernelized correlation filters. IEEE Trans. Pattern Anal. Mach. Intell. **37**, 583–596 (2015)
25. Huang, L., Zhao, X., Huang, K.: GOT-10k: a large high-diversity benchmark for generic object tracking in the wild. IEEE Trans. Pattern Anal. Mach. Intell. (2019)
26. Jung, I., Son, J., Baek, M., Han, B.: Real-time MDNet. In: Ferrari, V., Hebert, M., Sminchisescu, C., Weiss, Y. (eds.) ECCV 2018. LNCS, vol. 11208, pp. 89–104. Springer, Cham (2018). https://doi.org/10.1007/978-3-030-01225-0_6
27. Kim, C., Hwang, J.N.: Fast and automatic video object segmentation and tracking for content-based applications. IEEE Trans. Circ. Syst. Video Technol. **12**, 122–129 (2002)
28. Kristan, M., et al.: The sixth visual object tracking VOT2018 challenge results. In: Leal-Taixé, L., Roth, S. (eds.) ECCV 2018. LNCS, vol. 11129, pp. 3–53. Springer, Cham (2019). https://doi.org/10.1007/978-3-030-11009-3_1
29. Kristan, M., et al.: The seventh visual object tracking VOT2019 challenge results. In: Proceedings of the IEEE/CVF International Conference on Computer Vision Workshops (2019)
30. Li, B., Wu, W., Wang, Q., Zhang, F., Xing, J., Yan, J.: SIAMRPN++: evolution of siamese visual tracking with very deep networks. In: IEEE Conference on Computer Vision and Pattern Recognition (2019)
31. Li, B., Yan, J., Wu, W., Zhu, Z., Hu, X.: High performance visual tracking with siamese region proposal network. In: IEEE Conference on Computer Vision and Pattern Recognition. IEEE, June 2018
32. Li, M., Kambhamettu, C., Stone, M.: Automatic contour tracking in ultrasound images. Clin. Linguist. Phonet. **19**, 545–554 (2005)
33. Lin, T.-Y., et al.: Microsoft COCO: common objects in context. In: Fleet, D., Pajdla, T., Schiele, B., Tuytelaars, T. (eds.) ECCV 2014. LNCS, vol. 8693, pp. 740–755. Springer, Cham (2014). https://doi.org/10.1007/978-3-319-10602-1_48
34. Luiten, J., Voigtlaender, P., Leibe, B.: PReMVOS: proposal-generation, refinement and merging for video object segmentation. In: Jawahar, C.V., Li, H., Mori, G., Schindler, K. (eds.) ACCV 2018. LNCS, vol. 11364, pp. 565–580. Springer, Cham (2019). https://doi.org/10.1007/978-3-030-20870-7_35
35. Lukežič, A., Matas, J., Kristan, M.: D3S - a discriminative single shot segmentation tracker. In: IEEE/CVF Conference on Computer Vision and Pattern Recognition, November 2020
36. Lukežič, A., Vojíř, T., Čehovin Zajc, L., Matas, J., Kristan, M.: Discriminative correlation filter tracker with channel and spatial reliability. Int. J. Comput. Vis. **126**, 671–688 (2018)

37. McFarlane, N.J., Schofield, C.P.: Segmentation and tracking of piglets in images. Mach. Vis. Appl. **8**, 187–193 (1995)
38. Mueller, M., Smith, N., Ghanem, B.: A benchmark and simulator for UAV tracking. In: Leibe, B., Matas, J., Sebe, N., Welling, M. (eds.) ECCV 2016. LNCS, vol. 9905, pp. 445–461. Springer, Cham (2016). https://doi.org/10.1007/978-3-319-46448-0_27
39. Müller, M., Bibi, A., Giancola, S., Alsubaihi, S., Ghanem, B.: TrackingNet: a large-scale dataset and benchmark for object tracking in the wild. In: Ferrari, V., Hebert, M., Sminchisescu, C., Weiss, Y. (eds.) ECCV 2018. LNCS, vol. 11205, pp. 310–327. Springer, Cham (2018). https://doi.org/10.1007/978-3-030-01246-5_19
40. Nam, H., Han, B.: Learning multi-domain convolutional neural networks for visual tracking. In: IEEE Conference on Computer Vision and Pattern Recognition (2016)
41. Nam, H., Hong, S., Han, B.: Online graph-based tracking. In: Fleet, D., Pajdla, T., Schiele, B., Tuytelaars, T. (eds.) ECCV 2014. LNCS, vol. 8693, pp. 112–126. Springer, Cham (2014). https://doi.org/10.1007/978-3-319-10602-1_8
42. Oh, S.W., Lee, J.Y., Sunkavalli, K., Kim, S.J.: Fast video object segmentation by reference-guided mask propagation. In: 2018 IEEE/CVF Conference on Computer Vision and Pattern Recognition. IEEE, June 2018
43. Park, E., Berg, A.C.: Meta-tracker: fast and robust online adaptation for visual object trackers. In: Ferrari, V., Hebert, M., Sminchisescu, C., Weiss, Y. (eds.) ECCV 2018. LNCS, vol. 11207, pp. 587–604. Springer, Cham (2018). https://doi.org/10.1007/978-3-030-01219-9_35
44. Perazzi, F., Pont-Tuset, J., McWilliams, B., Gool, L.V., Gross, M., Sorkine-Hornung, A.: A benchmark dataset and evaluation methodology for video object segmentation. In: Proceedings of the IEEE Computer Society Conference on Computer Vision and Pattern Recognition (2016)
45. Pinheiro, P.O., Lin, T.-Y., Collobert, R., Dollár, P.: Learning to refine object segments. In: Leibe, B., Matas, J., Sebe, N., Welling, M. (eds.) ECCV 2016. LNCS, vol. 9905, pp. 75–91. Springer, Cham (2016). https://doi.org/10.1007/978-3-319-46448-0_5
46. Pont-Tuset, J., Perazzi, F., Caelles, S., Arbeláez, P., Sorkine-Hornung, A., Van Gool, L.: The 2017 DAVIS challenge on video object segmentation, April 2017
47. Qi, H., Brown, M., Lowe, D.G.: Low-shot learning with imprinted weights. In: Proceedings of the IEEE Computer Society Conference on Computer Vision and Pattern Recognition (2018)
48. Ren, L., Yuan, X., Lu, J., Yang, M., Zhou, J.: Deep reinforcement learning with iterative shift for visual tracking. In: Ferrari, V., Hebert, M., Sminchisescu, C., Weiss, Y. (eds.) ECCV 2018. LNCS, vol. 11213, pp. 697–713. Springer, Cham (2018). https://doi.org/10.1007/978-3-030-01240-3_42
49. Ren, S., He, K., Girshick, R., Sun, J.: Faster R-CNN: towards real-time object detection with region proposal networks. IEEE Trans. Pattern Anal. Mach. Intell. **39**, 1137–1149 (2017)
50. Ronneberger, O., Fischer, P., Brox, T.: U-net: convolutional networks for biomedical image segmentation. In: Navab, N., Hornegger, J., Wells, W.M., Frangi, A.F. (eds.) MICCAI 2015. LNCS, vol. 9351, pp. 234–241. Springer, Cham (2015). https://doi.org/10.1007/978-3-319-24574-4_28
51. Russakovsky, O., Li, L.J., Fei-Fei, L.: Best of both worlds: human-machine collaboration for object annotation. In: Proceedings of the IEEE Computer Society Conference on Computer Vision and Pattern Recognition (2015)
52. Shelhamer, E., Long, J., Darrell, T.: Fully convolutional networks for semantic segmentation. IEEE Trans. Pattern Anal. Mach. Intell. **39**, 640–651 (2017)

53. Siam, M., Oreshkin, B., Jagersand, M.: AMP: adaptive masked proxies for few-shot segmentation. In: Proceedings of the IEEE International Conference on Computer Vision, February 2019
54. Simonyan, K., Zisserman, A.: Very deep convolutional networks for large-scale image recognition. In: 3rd International Conference on Learning Representations (2015)
55. Voigtlaender, P., Leibe, B.: Online adaptation of convolutional neural networks for video object segmentation (2017)
56. Voigtlaender, P., Leibe, B.: Online adaptation of convolutional neural networks for video object segmentation. In: British Machine Vision Conference 2017. BMVA Press, June 2017
57. Voigtlaender, P., Luiten, J., Leibe, B.: BoLTVOS: box-level tracking for video object segmentation, April 2019
58. Wang, Q., Gao, J., Xing, J., Zhang, M., Hu, W.: DCFNet: discriminant correlation filters network for visual tracking, April 2017
59. Wang, Q., Zhang, L., Bertinetto, L., Hu, W., Torr, P.H.S.: Fast online object tracking and segmentation: a unifying approach. In: IEEE Conference on Computer Vision and Pattern Recognition (2019)
60. Wu, Y., Lim, J., Yang, M.H.: Online object tracking: a benchmark. In: IEEE Conference on Computer Vision and Pattern Recognition. IEEE Computer Society (2013)
61. Xu, N., et al.: YouTube-VOS: a large-scale video object segmentation benchmark, September 2018
62. Yang, F., Mackey, M.A., Ianzini, F., Gallardo, G., Sonka, M.: Cell segmentation, tracking, and mitosis detection using temporal context. In: Duncan, J.S., Gerig, G. (eds.) MICCAI 2005. LNCS, vol. 3749, pp. 302–309. Springer, Heidelberg (2005). https://doi.org/10.1007/11566465_38
63. Yang, L., Wang, Y., Xiong, X., Yang, J., Katsaggelos, A.K.: Efficient video object segmentation via network modulation. In: Proceedings of the IEEE Computer Society Conference on Computer Vision and Pattern Recognition. IEEE Computer Society, February 2018
64. Yun, S., Choi, J., Yoo, Y., Yun, K., Choi, J.Y.: Action-decision networks for visual tracking with deep reinforcement learning. In: IEEE Conference on Computer Vision and Pattern Recognition. IEEE, July 2017
65. Zhang, Z., Peng, H.: Deeper and wider siamese networks for real-time visual tracking. In: IEEE Conference on Computer Vision and Pattern Recognition, January 2019
66. Zhang, Z., et al.: Tracking-assisted weakly supervised online visual object segmentation in unconstrained videos. In: MM 2018 - Proceedings of the 2018 ACM Multimedia Conference. Association for Computing Machinery Inc., New York, October 2018
67. Zhu, Z., Wang, Q., Li, B., Wu, W., Yan, J., Hu, W.: Distractor-aware siamese networks for visual object tracking. In: Ferrari, V., Hebert, M., Sminchisescu, C., Weiss, Y. (eds.) ECCV 2018. LNCS, vol. 11213, pp. 103–119. Springer, Cham (2018). https://doi.org/10.1007/978-3-030-01240-3_7

AF2S: An Anchor-Free Two-Stage Tracker Based on a Strong SiamFC Baseline

Anfeng He[1(✉)], Guangting Wang[1], Chong Luo[2], Xinmei Tian[1], and Wenjun Zeng[2]

[1] CAS Key Laboratory of Technology in Geo-Spatial Information Processing and Application System, University of Science and Technology of China, Hefei, Anhui, China
heanfeng@mail.ustc.edu.cn, wgting96@gmail.com, xinmei@ustc.edu.cn
[2] Microsoft Research Asia, Beijing, China
{cluo,wezeng}@microsoft.com

Abstract. Siamese network based trackers have become a mainstream in visual object tracking. Recently, several high-performance multi-stage trackers have been proposed and some of them adopt SiamRPN for the first-stage region proposal. We argue that an anchor-based region proposal network is not necessary for the tracking task, as a tracker has a strong prior about the location and size of the target. In this paper, we propose a two-stage visual tracker which uses SiamFC for region proposal. SiamFC defines a bounding box by its center, which is a typical anchor-free (AF) network, so we dub our tracker AF2S. As the model size of SiamFC is only about 1/10 that of SiamRPN, AF2S results in a significantly lighter model than its SiamRPN-based counterparts. In the design of AF2S, we first build a strong AlexNet-based SiamFC baseline which improves the AUC on OTB-100 from 0.582 to 0.665. Further, we propose a position-sensitive convolutional layer which can be stacked after SiamFC backbone to increase the robustness of proposals without losing localization precision. Finally, a relation network is used for box refinement. Experimental results show that AF2S achieves the best performance on OTB-100 and VOT-18 among the state-of-the-art trackers which use AlexNet as backbone. On LaSOT-test, AF2S achieves an AUC of 0.480, which is among the first-tier performance even when trackers with more powerful backbone and much larger model size are considered.

Keywords: Visual object tracking · Siamese network · Two-stage tracker · Anchor free

1 Introduction

Visual object tracking (VOT) is a challenging computer vision problem and the foundation of a wide variety of artificial intelligence (AI) applications, including

A. He and G. Wang—This work is carried out while Anfeng He and Guangting Wang are interns in MSRA.

A. Bartoli and A. Fusiello (Eds.): ECCV 2020 Workshops, LNCS 12539, pp. 637–652, 2020.
https://doi.org/10.1007/978-3-030-68238-5_42

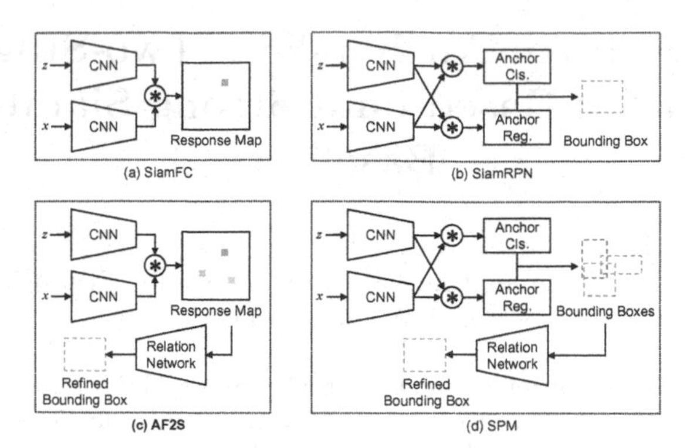

Fig. 1. Illustrating the architecture of SiamFC, SiamRPN, SPM, and the proposed AF2S. z is the template from the first frame and x is the search region image patch.

autonomous driving, video analytics, and human-computer interaction. In the deep learning era, great progress has been made in this research area since the groundbreaking work called SiamFC tracker [1] was proposed in 2016. In just three years or so, we are witnessing the paradigm shift from SiamFC to SiamRPN [16], and to SiamRPN-based multi-stage trackers [7,22]. The area under curve (AUC) metric on the tracking benchmark OTB-100 [24] has been greatly improved from 0.582 to 0.696. New benchmark datasets with a larger size and more diversified objects are proposed [5] while conventional benchmark datasets, such as VOT series [13], are continuously evolving.

Now we believe it is time to revisit the classical trackers and to see whether any hint can be obtained. Along the siamese network-based tracker design, we find that a lot of experience is drawn from the design of object detector. Interestingly, it is this obvious analogy that makes us take certain components for granted. As a result, unnecessary complexity is introduced and the growth in tracker performance does not match the increase in model size.

Figure 1 provides sketches of different tracking frameworks. The very original SiamFC employs siamese fully convolutional network to extract features from the target and the search region, and uses cross correlation operation to compute matching scores. SiamFC uses AlexNet [14] as the backbone and the model size is only 8.3MB. However, SiamFC matches the target object only with potential objects of the same size in the search region. It is not capable of adjusting the aspect ratio of the bounding box (bbox) and the scale adjustment is achieved by cumbersome multi-scale testing. In view of this, SiamRPN [16] is proposed. By introducing anchors and stacking anchor classification and bbox regression branches, SiamRPN avoids multi-scale testing and naturally adjusts bbox aspect ratios by region proposals. Although the performance of SiamRPN is improved with more precise object boundaries, the model size is dramatically increased to 87MB, which is over 10 times that of SiamFC.

If we compare the object tracking and detection frameworks, analogies can be drawn between SiamFC and sliding-window based R-CNN, and between

SiamRPN and the region proposal network (RPN). Naturally, one could design a two-stage or multi-stage tracker which is analogous to faster R-CNN object detector. Recently, SPM [22] and C-RPN (cascade RPN) [7] are proposed. They use SiamRPN for region proposal and use subsequent stages for bbox refinement. SPM achieves 0.671 AUC on OTB-100, but the model size is increased to 101.4 MB. Now a big question is: is a standard RPN really necessary for a multi-stage tracker? We ask this question simply because a tracker has a strong prior about the location and size of the target object. This is totally different from the situation in object detection, where objects of all sizes can appear at any location in an image. With this fundamental difference, it might not be necessary for a multi-stage tracker to use an anchor-based RPN which evaluates a large number of anchors with varying sizes.

Motivated by this discovery and inspired by the recent SPM, we propose a two-stage object tracker, named AF2S, which uses SiamFC for region proposal. In SiamFC response map, each pixel is uniquely associated with a bounding box centered at this location. Actually, defining a bounding box by its center is a typical treatment of anchor-free network. This is where AF2S gets its name. AF2S selects bounding boxes with the highest scores as proposals. Then, the second-stage relation network is dedicated to fine matching and bbox regression. In the design of AF2S, we find that using a pre-trained model as initialization and freezing the first a few layers result in a very strong SiamFC baseline. Without bells and whistles, this baseline achieves an AUC of 0.665 on OTB-100, outperforming a large number of complicated trackers with much larger model size. In order to further improve the robustness and precision of the first-stage, we propose a position sensitive convolutional layer (PS-Conv) which can be stacked after the SiamFC backbone. Results show that PS-Conv improves the tracker performance in both one-stage and two-stage trackers. This work makes four major contributions:

- We propose an anchor-free two-stage object tracker which uses SiamFC instead of SiamRPN for region proposal. This design significantly reduces the model size of the first stage by 8 times.
- We revisit the design principles and training procedure of SiamFC. A simple and strong SiamFC baseline is produced.
- We propose a PS-Conv layer that can be stacked after backbone of SiamFC to achieve an even higher tracking accuracy for both single-stage and two-stage trackers.
- We perform comprehensive ablation study to justify our design choices. System comparisons are made on OTB, VOT and LaSOT. Results show that AF2S achieves significant better performance than tracks with similar model size. Its performance is among the first-tier even when compared with models that are a few times larger.

The rest of the paper is organized as follows. Section 2 reviews related work. Section 3 introduces a strong AlexNet-based SiamFC baseline. Section 4 describes the proposed AF2S tracker and our novel PS-Conv layer. Section 5 presents the experimental results. Finally, Sect. 6 concludes the paper with some discussions.

2 Related Work

In this section, we mainly review Siamese network based object trackers. SiamFC [1] is a pioneer work that utilizes fully convolutional Siamese network for object tracking. The Siamese network encodes template patches and search region patches into a high-dimensional feature space, where the similarity matching can be achieved by the cross-correlation operation. Due to its simple yet effective architecture, SiamFC has attracted the attentions of many researchers and led to a large number of follow-up works. RASNet [23] introduces some attention modules before the cross-correlation operation. SA-Siam [9] proposes to utilize complementary features to represent the tracked object. SiamDW [25] intends to build a deeper and wider backbone for a better tracker. GCT [8] utilizes spatial-temporal graph convolution so that the tracker can benefit from more information. TADT [17] tries to enhance target-aware features to let the tracker focus on the object.

A major drawback of SiamFC is that it does not change the aspect ratio of bounding boxes. This prevents it from the precise bounding-box prediction. SiamRPN [16] and its extensions DaSiamRPN [26] and SiamRPN++ [15] address the issue by introducing two independent head branches, one for the classification and another for the bounding-box regression. Analogous to the RPN in object detection, the predefined boxes (called anchors) are trained to fit the tracking target. A side benefit of SiamRPN is that it does not need to do multi-scale testing any more, so it is over twice as fast as SiamFC. However, while the anchors with varying shapes can effectively cover different targets, the model size grows linearly with the number of anchors.

Multi-stage refinement has been proved useful in many vision tasks. PTAV [6] utilizes a parallel tracker to verify the tracking results. ACT [2] proposes that the tracker should first predict an action of location change and scale change. In the second stage, with more accurate location, the tracker classifies whether or not the target is in this region. ATOM [4] utilizes IoU-Net [10] to optimize the bounding box to maximization overlap after the target classification. Recent tracker SPM [22] and C-RPN [7] use SiamRPN as their base models. The candidate boxes proposed by the base models are refined stage by stage. Our work is inspired by these multi-stage trackers. The key observation is that anchor-based RPN might not be necessary if there is a second stage for box refinement.

3 A Strong SiamFC Baseline

The original SiamFC adopts AlexNet as the backbone. The entire network is trained from scratch in a similarity matching problem using ImageNet VID [21] dataset. In the original paper, an AUC of 0.582 is reported on OTB-100 and an EAO of 0.235 is reported on VOT-16. This section describes a strong baseline,

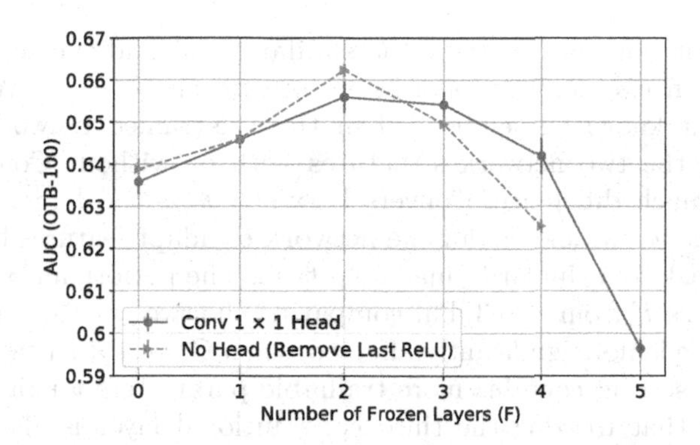

Fig. 2. Change of SiamFC tracking performance on OTB-100 with respect to the number of frozen layers (F). We show the mean and standard deviation of the top 10 epochs for each data point.

named SBT, which still uses AlexNet as the backbone. We leverage ImageNet pre-trained network for initialization and expand the offline training dataset. Surprisingly, SBT achieves an AUC of 0.665 on OTB-100 and an EAO of 0.345 on VOT-16.

Although it is a common practice in many computer vision tasks to use ImageNet pre-trained backbone for initialization, it is not adopted in training the original SiamFC tracker. One possible reason is that AlexNet is pretty small and the authors believe it can be trained sufficiently with the tracking training data. To the best of our knowledge, SA-Siam [9] is the first work that has tried to use a pre-trained model in SiamFC framework. Recently, quite a few SiamFC-based trackers have followed this practice and they confirmed that a pre-trained model is indeed helpful.

In the semantic branch of SA-Siam, all five convolutional layers of a pre-trained AlexNet are frozen and an additional 1×1 convolutional layer is appended. This Conv 1×1 layer is trainable and the purpose is to transform the classification features into a domain suitable for the cross-correlation operation. An AUC of 0.588 on OTB-100 is reported for the semantic branch in SA-Siam. In our experiment, we achieve a similar result under the same setting. However, in this work, we mainly explore how many convolutional layers should be frozen during training and what the implications are.

Figure 2 shows the AUC results on OTB-100 when different numbers of convolutional layers are frozen during training. There are two curves in the figure. The solid curve is achieved when a trainable Conv 1×1 layer is appended to AlexNet. The dotted curve is achieved when we remove ReLU from the last convolutional layer. In such a way, the last layer can be trained to transform features suitable for cross-correlation operation. But when this strategy is adopted, we cannot freeze all five layers as there will be no trainable parameters. This is why the dotted curve does not have data point when $F = 5$.

We find that both curves show the similar trend and the top performance is achieved when two convolutional layers are frozen ($F = 2$). When F is too small, the prior knowledge embedded in the pre-trained network is not fully exploited, and the two network structures with or without Conv 1×1 head do not make much difference. Conversely, when F is too large, there are not sufficient trainable parameters for the network to adapt features from an image classification task to a similarity matching task. The performance starts to drop when we increase F from 2 to 3, but comparing the two curves, we can find that the performance is not significantly affected when Conv 1×1 head is used. This is because this setting contains more trainable parameters for task adaptation. We conjecture that freezing the third convolutional layer is also beneficial as long as the subsequent network has sufficient trainable parameters. In the next section, we will present position sensitive convolutional layer that unlocks the potential of a SiamFC network with $F = 3$.

Another small change we make with this implementation is to expand the offline training datasets. According to recent common practice, we use YouTube-BoundingBoxes [20], COCO [19] and ImageNet Det [21] for training in addition to ImageNet VID. We tested the network with Conv 1×1 head and $F = 3$, the AUC will drop from 0.654 to 0.649 when the two image datasets are excluded, and it will further drop to 0.627 when the YouTube-BoundingBoxes is also excluded from training.

4 An Anchor-Free Two-Stage Object Tracker

4.1 AF2S Overview

Following the faster R-CNN framework for object detection, we intend to design a two-stage network for object tracking, where the first stage generates proposals and the second stage performs bbox refinement. We observe that there is a basic difference between the detection task in images and the tracking task in videos, and this difference makes the standard RPN not necessary for trackers. It is based on this observation that we design an anchor-free two-stage (AF2S) network for visual object tracking.

In particular, a tracker has a strong prior about the location and size of the target object because consecutive frames in a video have strong temporal correlations. The target object always appears near its position in the previous frame, and its size and shape (aspect ratio) will not change dramatically. This is different from object detection in images where objects of interest may appear anywhere in the image and their sizes may vary. It is necessary for object detectors to have several anchors of different aspect ratios. But for object tracker, we could simply use the object bounding box of the last frame as the default and the only anchor. In other words, a tracker can be designed anchor-free.

SiamFC happens to be a good candidate for region proposal in the first stage of AF2S. Essentially, SiamFC generates a response map comprised of matching scores between the target object and all the sliding windows with the same size in the search region. After non-maximum suppression (NMS), proposals with

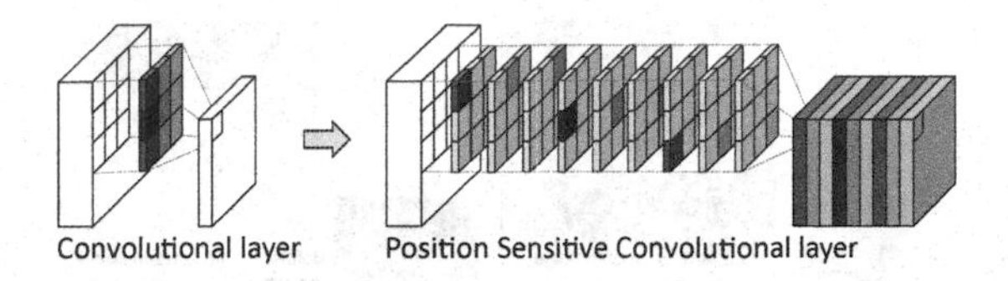

Fig. 3. Illustration of how PS-Conv is evolved from the conventional convolutional layer.

the top K highest scores are passed to the second stage for a fine matching and bbox regression. We follow the design of SPM [22] to use a relation network for the second stage. Figure 1(c) provides a sketch of the proposed AF2S network. The existence of the second stage not only makes the aspect ratios of proposals freely adjustable, but also renders multi-scale testing unnecessary. As such, the two-stage framework overcomes the two biggest drawbacks of SiamFC.

Now a key issue in AF2S design is how to improve the region proposal capability of SiamFC, since an anchor-free RPN is in general inferior to a full-fledged RPN. We have already obtained a strong base line SBT as described in the previous section. In the following subsection, we will present a position-sensitive convolutional layer which we design to further boost the SBT performance.

4.2 PS-Conv Layer

In the design of a SiamFC tracker, there is a well-known dilemma about how deep the convolutional neural network (CNN) should be. On the one hand, the deeper the network, the more robust the tracker is. This is because deep CNN features contain high-level semantics and are not sensitive to object appearance changes. On the other, the deeper the network, the less precise the tracker is. As the network goes deeper, the size of receptive field increases. A large receptive field will result in a diffused high response region in the response map and create difficulty in precise localization. PS-Conv layer is designed to alleviate this dilemma. The objective is to obtain higher-level semantics without losing localization precision. This layer is appended to SiamFC backbone to increase the robustness of proposals.

Figure 3 illustrates the difference between a conventional convolutional layer and PS-Conv. As the figure shows, PS-Conv decomposes a conventional $k \times k$ convolutional kernel into k^2 sub-kernels. Here we omit the input channel dimension for simplicity. Each sub-kernel still has a spatial resolution of $k \times k$, but only one position contains non-zero value. When we use such sub-kernels to compute convolution, the size of receptive field is not increased, but the features are refined by the convolution across channels. The number of output channels is increased by k^2 times.

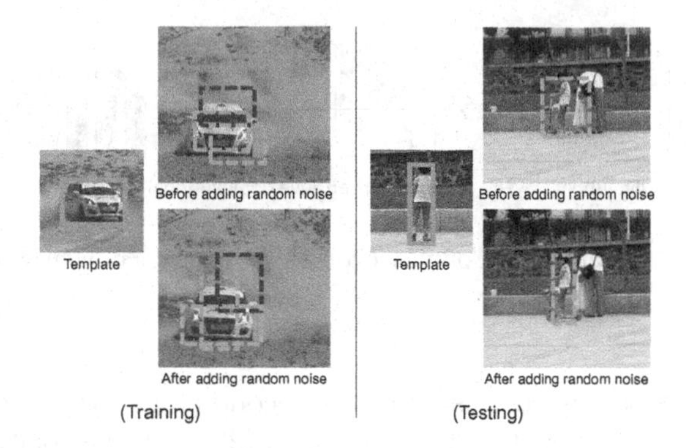

Fig. 4. Illustration of a critical data augmentation mechanism in training the second-stage network and how it affects the performance in testing.

Mathematically, the conventional convolutional output at a given location and channel can be written as:

$$v = \sum_{i,j} \sum_{m} u^{(i,j,m)} \cdot w^{(i,j,m)}, \tag{1}$$

where w is the convolution kernel and u is the corresponding window of input features. (i, j) is the spatial location at the kernel and m is the channel index. We can find that the sum pooling over i and j expands the receptive field and make v position insensitive. Conversely, if we want to make the next-layer feature position sensitive, we shall avoid the sum pooling operation over different positions. Our solution is to keep the product at each position and put them into different channels. The PS-Conv output at the same location can be written as:

$$\hat{v}^{(i,j)} = \sum_{m} u^{(i,j,m)} \cdot w^{(i,j,m)}. \tag{2}$$

In this formula, we only sum over different input channels, but do not compute summation over i and j. Actually, if we sum these $\hat{v}^{(i,j)}$, we will get exactly the same result as the conventional convolution:

$$v = \sum_{i,j} \hat{v}^{(i,j)} \tag{3}$$

Therefore, the amount of information contained in PS-Conv features is not less than conventional Conv features. It might be more powerful as it contains location information.

Before correlation operation, dimension of i, j is reshaped into the channel dimension. After this operation, the template matching progress is the same as conventional SiamFC. During template matching, the similarity scores across channels are summed up to generate the final response map. Note that only large values in exactly the same channel (defined by i and j) will produce a high similarity score. This way, PS-Conv achieves position sensitivity.

Our work is inspired by the position sensitive RoI pooling (PS-RoI-Pooing) proposed for object detection [3,18]. Although we share the same vision, PS-Conv is a completely different operation from PS-RoI-Pooing. PS-Conv operates on convolutional features and outputs a feature map with higher-level semantics. PS-RoI-Pooing operates on a regional score map and outputs a classification score map. Essentially, PS-Conv is a convolutional layer while PS-RoI-Pooing is a pooling operator.

4.3 Anchor-Free Bounding Box Refinement

The second-stage network in AF2S performs bbox refinement. We follow the design in SPM [22] and use a relation network for this purpose. Specifically, when we initialize the tracker, we use ROI-Align to extract regional features of the target from the first frame. During tracking, the first stage passes down N candidate positions with top scores after NMS. All the proposals share the same size which is drawn from the tracking result of the last frame. Similarly, we use ROI-Align to extract the regional features for each proposal. Then features of each proposal are concatenated with features of the target. Finally, the N concatenated features are fed into a relation network for classification and bbox regression.

However, training of the second stage in AF2S has a major difference from that in SPM. During AF2S inference, all proposals share the same size as the bbox result from the last frame. Then, if the bbox result is not accurate, it affects the input of the second stage. During training, we shall simulate such errors in proposal sizes. As shown in Fig. 4, we add random noise to both dimensions of bounding boxes during the training of bbox refinement. This data augmentation mechanism significantly improves the performance of the second stage. Details of data augmentation are listed below:

$$\begin{aligned} w &= \hat{w} + \hat{w} * d_w, d_w = e^{n_w}, n_w \sim N(0, \sigma_w^2) \\ h &= \hat{h} + \hat{h} * d_h, d_h = e^{n_h}, n_h \sim N(0, \sigma_h^2) \end{aligned} \tag{4}$$

where annotations with ^ are the shape of bounding boxes from the previous frame. Here we choose $\sigma_w = \sigma_h = 0.1$, d_w and d_h are truncated between [0.5, 2.0].

5 Experiments

In this section, we evaluate the performance of our trackers against state-of-the-art real-time trackers and carry out ablation studies to validate the contribution of PS-Conv layer and anchor-free two-stage refinement.

5.1 Benchmarks and Evaluation Metrics

OTB: The object tracking benchmark OTB-2015 [24] consists of two major datasets, namely TB-50 and TB-100. TB-50 consists of the more challenging

Table 1. Ablation study on OTB benchmarks. SBT denotes the strong baseline. PS denotes PS-Conv. AF2S is our final tracker.

Method	TB-50		TB-100	
	AUC	Prec.	AUC	Prec.
SBT	0.608	0.826	0.665	0.878
SBT+PS	0.611	0.820	0.671	0.884
AF2S w/o PS	0.623	0.841	0.673	0.887
AF2S	**0.636**	**0.877**	**0.679**	**0.907**

Table 2. AUC score comparison of adding different head to the backbone on OTB-100 benchmark. #O is the channel number of the output feature. #Params. is the number of parameters of the head.

	$F = 2$	$F = 3$	#O	#Params.
None	0.665	0.652	256	–
Conv 3×3	0.647	0.650	256	$256 \times 3 \times 3$
Conv 1×1	0.661	0.659	256	$256 \times 1 \times 1$
PS-Conv	**0.668**	**0.671**	288	$32 \times 3 \times 3$

sequences in TB-100. Success rate and precision are the two standard evaluation metrics on OTB. Conventionally, we report the area-under-curve (AUC) of the success plot and the precision at a threshold of 20 pixels. We use the standard OTB toolkit to obtain all the numbers.

VOT: We use two recent versions of the VOT benchmark, denoted by VOT2016 [11],and VOT2018 [12]. The VOT benchmarks evaluate a tracker by applying a reset-based methodology. Major evaluation metrics of VOT benchmarks are accuracy (A), robustness (R) and expected average overlap (EAO). A good tracker has high A and EAO scores but low R scores.

LaSOT: LaSOT stands for large-scale single object tracking benchmark [5]. It is one of the latest tracking benchmarks. There are 1400 sequences with more than 3.5M frames. All frames are manually annotated with bounding boxes of targets. There are 280 sequences in LaSOT-test, which is a subset of LaSOT-all.

5.2 Implementation Details

The strong baseline SBT is implemented as described in Sect. 3. For the analysis of PS-Conv, we select the top two settings, $F = 2$ and $F = 3$, as our baseline model. For *PS-Conv*, we set the channel number to 32 and the kernel size to 3×3 in the PS-Conv layer. Thus, the number of output channel is $32 \times 3 \times 3$, which is similar to the channel number of *Conv 1* $\times$ *1* and *Conv 3* $\times$ *3* when they use 256 output channels. Besides, *Conv 3* $\times$ *3* have 8 $\times$ parameter size of *PS-Conv* and *Conv 1* $\times$ *1*.

Table 3. Comparison with state-of-the-art Siamese trackers on OTB-100, VOT-18 and VOT-16 benchmarks. BB denotes backbone. AN denotes AlexNet. RN denotes ResNet or ResInception. For fairness, we take the AlexNet version of SiamRPN++ for comparison. The results in SPM* are achieved with the same four training datasets.

Tracker	BB	FPS	OTB-100 (AUC)	VOT-18 (EAO)	VOT-16 (EAO)
SiamFC	AN	86	0.582	0.188	0.235
RASNet	AN	83	0.642	0.281	–
SA-Siam	AN	50	0.657	0.236	0.291
GCT	AN	50	0.648	0.274	–
TADT	AN	34	0.660	–	0.299
SiamDW-FC	*RN*	*70*	*0.644*	*0.234*	*0.300*
SBT	AN	86	**0.665**	**0.294**	**0.345**
SiamRPN++	*RN*	*35*	*0.696*	*0.414*	*0.464*
SiamRPN++	AN	180	0.666	0.352	0.393
SiamDW-RPN	*RN*	*150*	*0.666*	*0.294*	*0.376*
SiamRPN	AN	200	0.637	0.244	0.344
DaSiamRPN	AN	160	0.658	0.326	0.411
SPM*	AN	120	0.671	0.347	**0.432**
C-RPN	AN	36	0.663	0.289	0.363
AF2S	AN	120	**0.679**	**0.353**	0.410

In AF2S implementation, we follow most of the settings in [22]. The batch size during training is 32, which is divided into 4 GPU devices and there are 8 samples in each GPU. We train the model for 50 epochs. The initial learning rate is 0.04. The learning rate is multiplied by 0.1 for every 20 epochs. There is no template updating in this work. We implement our model in PyTorch 1.0.0 framework in Python 3.6.7 environment. Our experiments are trained on a computer with a Xeon E5-2690 2.60GHz CPU and 4 × Tesla P100 GPU and evaluated with 1 × Tesla P40 GPU.

5.3 Ablation Analysis

We use the OTB benchmark for the ablation analysis.

PS-Conv and Two-Stage Architecture Bring Gains. Table 1 shows the performance of *SBT*, *SBT+PS*, *AF2S w/o PS*, and *AF2S* on OTB-100. Compare line 1 with line 2 and compare line 3 with line 4, we can conclude that PS-Conv can improve the performance of a SiamFC tracker and a two-stage tracker. Compare line 1 with line 3 and compare line 2 with line 4, it is clear that the second stage bbox refinement brings gains and the improvement on precision metric is prominent.

Detailed Analysis of PS-Conv: Table 2 provides a detailed analysis of PS-Conv. It shows that the performance gain achieved by PS-Conv does not simply

Table 4. Performance of state-of-the-art trackers on LaSOT.

Tracker	AUC of OPE		#Params.	FPS
	All	Test	(Million)	
MDNet	0.413	0.397	4.4	1
VITAL	0.412	0.390	4.4	1
CSRDCF	0.263	0.244	–	13
ECO	0.340	0.324	–	8
ECO-hc	0.311	0.304	–	60
SiamFC	0.358	0.336	3.8	86
CFNet	0.296	0.275	–	75
SiamRPN	0.440	0.433	22.6	160
SPM	**0.485**	0.471	26.4	120
C-RPN	0.459	0.455	89.4	36
SBT (ours)	0.415	0.402	2.4	86
AF2S (ours)	**0.483**	**0.480**	7.8	120

come from the deeper network. As the table shows, if we simply stack a conventional 3×3 convolutional layer to the backbone, performance drops. When we freeze three layers during training, adding a 1×1 convolution layer brings a small gain. This shows that even without a powerful convolution, only if the size of receptive field remains constant, performance can be improved. Our proposed PS-Conv layer is able to benefit from the pros of Conv 1×1 and Conv 3×3 but avoid their cons. Note that when only two layers are frozen, adding a Conv 3×3 incurs a large loss. It might due to too much trainable parameters. Despite of this, PS-Conv still gets some improvements.

In additon, we visualize the response maps of SiamFC when different convolutional head is added. Figure 5 shows that PS-Conv is able to generate more accurate responses than conventional convolutional layer.

Anchor-Based RPN is Not Necessary for Tracking. Figure 6 visualizes Δw and Δh of SiamRPN (which is also the first stage of SPM), and compares the two values in the second stage of SPM and AF2S. The first figure shows that the regression deltas tend to cluster into groups. It is because the shape of a specific target does not change too much, and it is usually different from the pre-defined anchors. The scattered (Δw, Δh) distribution suggests that the regression from pre-defined shapes is not efficient.

Bbox regression in the second stage of SPM and AF2S is used to refine the proposals provided by the first stage. The distribution of (Δw, Δh) in the second stage is actually an indication of how good the first-stage proposals are. Compare the second and the third subfigure in Fig. 6, we find that the predicted delta values of AS2F are more concentrated near $(0, 0)$. The mean length of (Δw, Δh) is 5.55 for SPM and 4.42 for AF2S. A smaller mean value not only

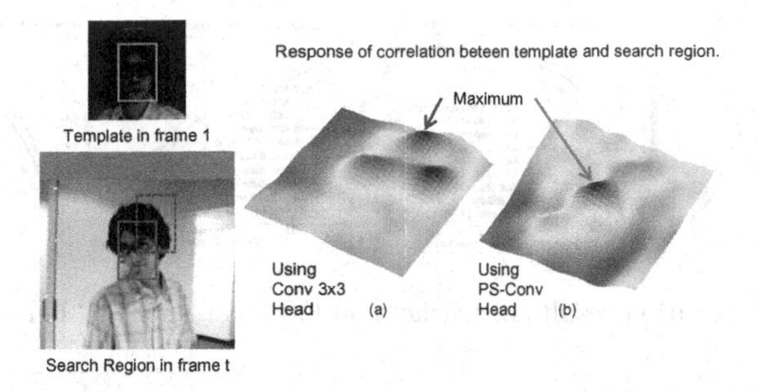

Fig. 5. Response maps of correlation between template and search region by adding different convolutional head after the backbone. Response (a) corresponds to using features of 3×3 Conv after backbone for correlation. The corresponding bounding box is red dotted line. High response at the wrong location causes tracking failure. Response (b) corresponds to using features of PS-Conv layer after backbone for correlation. The corresponding bounding box is green solid line. Best viewed in color. (Color figure online)

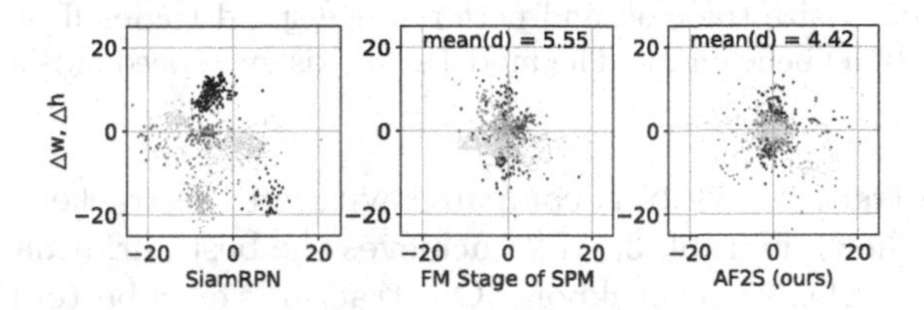

Fig. 6. Scatters of Δw and Δh of prediction steps. 5 colors in this figure denote 5 different video sequences. Each point is one of the tracking steps. mean(d) is the average length of every prediction.

confirms that anchor-based RPN is not necessary, but it simplifies the training of the second step.

Random Noise is Important for Training Refinement Stage. We have claimed that bounding boxes of refinement stage should be added with noise. In order to verify this statement, we remove the noise. The AUC on OTB-100 benchmark drops to 0.642. This is even worse than SBT without refinement stage.

5.4 Comparison with State-of-the-Arts

In this section we compare SBT and AF2S with state-of-the-art trackers on OTB, VOT and LaSOT benchmarks. We divide the trackers into three categories. The first one contains SiamFC-based trackers which are not able to change the aspect ratio of a target. As shown in Table 3, SBT achieves the best performance in this category although we do not add bells and whistles. The second category is SiamRPN-based trackers. These trackers are able to predict the change of aspect ratio with anchor classification and regression. The third category is two-stage

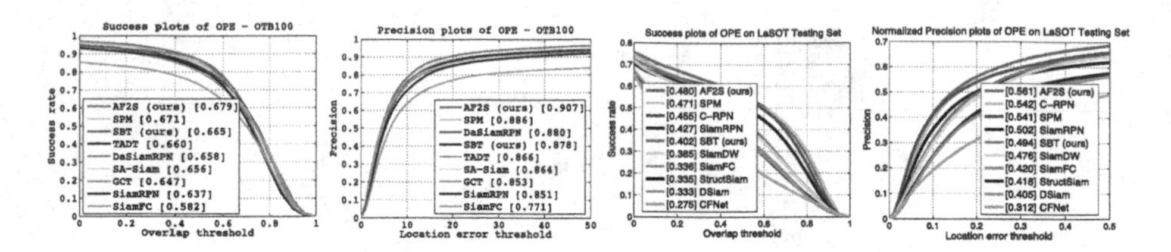

Fig. 7. Evaluation results of trackers on OTB and LaSOT benchmarks.

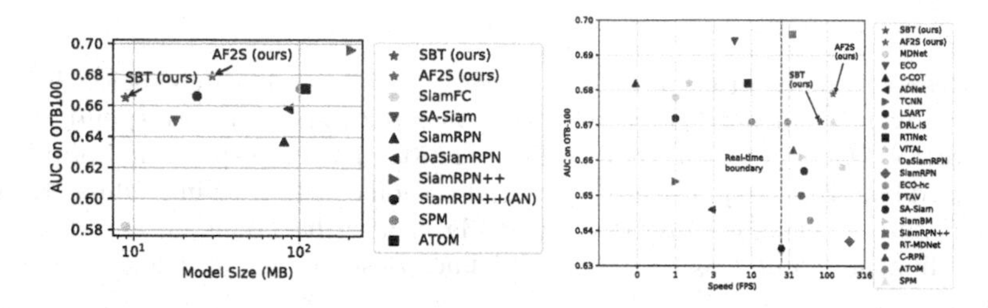

Fig. 8. Performance-size trade-off and performance-speed trade-off of top-performing trackers on OTB-100 benchmark. The model size axis and speed axis are logarithmic.

or multi-stage trackers. AF2S is compared with all the trackers in these three categories. As shown in Table 3, AF2S achieves the best performance among the trackers that use AlexNet as backbone. Our tracker is even better than SiamDW-RPN which uses a more powerful ResInception backbone. AF2S achieves 0.236 EAO score on VOT-20 dataset.

In addition, we compare AF2S with more state-of-the-art trackers on the LaSOT benchmark. As Table 4 shows, the performance of AF2S is among the top tier even when compared with some non-real-time trackers. Success plots and precision plots of AF2S and some highly related trackers on OTB and LaSOT benchmarks are shown in Fig. 7. As shown in Fig. 8, both SBT and AS2F are able to get a good performance with small model size and fast speed.

6 Conclusion

In this paper, we have proposed an anchor-free tracker named AF2S for visual object tracking. In the development of AF2S, we discover that just with a pre-trained model and to freeze a few layers during training, a vanilla SiamFC tracker can do surprisingly well. Motivated by the tradeoff between using deep features for robustness and using shallow features for precision, we have proposed PS-Conv layer which can get the benefit of both while avoiding their disadvantages. By adding a second stage to SiamFC, AF2S allows for bounding box refinement and avoids multi-scale testing, thus it becomes faster and more accurate. Compared with SiamRPN and SiamRPN-based multi-stage trackers, AF2S benefits from the anchor-free design and results in a much smaller model size. Extensive experiments show that the performance of AF2S is among the first tier on OTB,

VOT, and LaSOT datasets, although it only uses AlexNet as the backbone. In the future, we plan to explore more mechanisms to achieve position sensitive.

Acknowledgement. This work was supported by National Key Research and Development Program of China under Grant 2017YFB1002203 and the National Natural Science Foundation of China under Grant 61872329.

References

1. Bertinetto, L., Valmadre, J., Henriques, J.F., Vedaldi, A., Torr, P.H.S.: Fully-convolutional siamese networks for object tracking. In: Hua, G., Jégou, H. (eds.) ECCV 2016. LNCS, vol. 9914, pp. 850–865. Springer, Cham (2016). https://doi.org/10.1007/978-3-319-48881-3_56
2. Chen, B., Wang, D., Li, P., Wang, S., Lu, H.: Real-time'actor-critic'tracking. In: Proceedings of the European Conference on Computer Vision (ECCV), pp. 318–334 (2018)
3. Dai, J., Li, Y., He, K., Sun, J.: R-FCN: object detection via region-based fully convolutional networks. In: NIPS, pp. 379–387 (2016)
4. Danelljan, M., Bhat, G., Khan, F.S., Felsberg, M.: ATOM: accurate tracking by overlap maximization. In: CVPR, pp. 4660–4669 (2019)
5. Fan, H., Lin, L., Yang et al., F.: LaSOT: a high-quality benchmark for large-scale single object tracking. arXiv preprint arXiv:1809.07845 (2018)
6. Fan, H., Ling, H.: Parallel tracking and verifying: a framework for real-time and high accuracy visual tracking. In: ICCV, October 2017
7. Fan, H., Ling, H.: Siamese cascaded region proposal networks for real-time visual tracking. In: Proceedings of the IEEE Conference on Computer Vision and Pattern Recognition, pp. 7952–7961 (2019)
8. Gao, J., Zhang, T., Xu, C.: Graph convolutional tracking. In: ICCV, pp. 4649–4659 (2019)
9. He, A., Luo, C., Tian, X., Zeng, W.: A twofold siamese network for real-time object tracking. In: CVPR, June 2018
10. Jiang, B., Luo, R., Mao, J., Xiao, T., Jiang, Y.: Acquisition of localization confidence for accurate object detection. In: Proceedings of the European Conference on Computer Vision (ECCV), pp. 784–799 (2018)
11. Kristan, M., Leonardis, A., Matas, J., Felsberg, M.: The visual object tracking vot2016 challenge results. In: ECCV Workshop (2016)
12. Kristan, M., Leonardis, A., Matas, J., et al.: The sixth visual object tracking vot2018 challenge results. In: ECCV Workshop (2018)
13. Kristan, M., et al.: A novel performance evaluation methodology for single-target trackers. IEEE Trans. Pattern Anal. Mach. Intell. **38**(11), 2137–2155 (2016). https://doi.org/10.1109/TPAMI.2016.2516982
14. Krizhevsky, A., Sutskever, I., Hinton, G.E.: ImageNet classification with deep convolutional neural networks. In: NIPS, pp. 1097–1105 (2012)
15. Li, B., Wu, W., Wang, Q., Zhang, F., Xing, J., Yan, J.: SiamRPN++: evolution of siamese visual tracking with very deep networks. In: CVPR, pp. 4282–4291 (2019)
16. Li, B., Yan, J., Wu, W., Zhu, Z., Hu, X.: High performance visual tracking with siamese region proposal network. In: CVPR, June 2018
17. Li, X., Ma, C., Wu, B., He, Z., Yang, M.H.: Target-aware deep tracking. In: CVPR, pp. 1369–1378 (2019)

18. Li, Y., Qi, H., Dai, J., Ji, X., Wei, Y.: Fully convolutional instance-aware semantic segmentation. In: CVPR, pp. 2359–2367 (2017)
19. Lin, T.-Y., Maire, M., Belongie, S., Hays, J., Perona, P., Ramanan, D., Dollár, P., Zitnick, C.L.: Microsoft COCO: common objects in context. In: Fleet, D., Pajdla, T., Schiele, B., Tuytelaars, T. (eds.) ECCV 2014. LNCS, vol. 8693, pp. 740–755. Springer, Cham (2014). https://doi.org/10.1007/978-3-319-10602-1_48
20. Real, E., Shlens, J., Mazzocchi, S.: YouTube-BoundingBoxes: a large high-precision human-annotated data set for object detection in video. In: CVPR, pp. 7464–7473 (2017)
21. Russakovsky, O., Deng, J., Su, H., Krause, J.: Imagenet large scale visual recognition challenge. Int. J. Comput. Vis. **115**(3), 211–252 (2015)
22. Wang, G., Luo, C., Xiong, Z., Zeng, W.: SPM-tracker: series-parallel matching for real-time visual object tracking. In: CVPR, pp. 3643–3652 (2019)
23. Wang, Q., Teng, Z., Xing, J., Gao, J.: Learning attentions: residual attentional siamese network for high performance online visual tracking. In: CVPR, June 2018
24. Wu, Y., Lim, J., Yang, M.H.: Object tracking benchmark. IEEE Trans. Pattern Anal. Mach. Intell. **37**(9), 1834–1848 (2015)
25. Zhang, Z., Peng, H.: Deeper and wider siamese networks for real-time visual tracking. In: CVPR, pp. 4591–4600 (2019)
26. Zhu, Z., Wang, Q., Li, B., Wu, W.: Distractor-aware siamese networks for visual object tracking. In: ECCV, pp. 101–117 (2018)

RPT: Learning Point Set Representation for Siamese Visual Tracking

Ziang Ma(✉), Linyuan Wang, Haitao Zhang, Wei Lu, and Jun Yin

Zhejiang Dahua Technology, Hangzhou, China
{ma_ziang,wang_linyuan,zhang_haitao1,lu_wei,yin_jun}@dahuatech.com

Abstract. While remarkable progress has been made in robust visual tracking, accurate target state estimation still remains a highly challenging problem. In this paper, we argue that this issue is closely related to the prevalent bounding box representation, which provides only a coarse spatial extent of object. Thus an efficient visual tracking framework is proposed to accurately estimate the target state with a finer representation as a set of representative points. The point set is trained to indicate the semantically and geometrically significant positions of target region, enabling more fine-grained localization and modeling of object appearance. We further propose a multi-level aggregation strategy to obtain detailed structure information by fusing hierarchical convolution layers. Extensive experiments on several challenging benchmarks including OTB2015, VOT2018, VOT2019 and GOT-10k demonstrate that our method achieves new state-of-the-art performance while running at over 20 FPS.

Keywords: Visual tracking · Point set representation · Mutil-level aggregation

1 Introduction

Robust visual tracking is a fundamental task for various computer vision applications, including video monitoring analysis, man-machine interaction and intelligent driving. It aims at locating an arbitrary target of interest during a whole video sequence. Although substantial progress has been made in recent years, the design of a high-performance tracker is still highly challenging due to moving camera, occlusions and variations in structure.

Much attention has been invested for accurately estimating the target state in recent years. Its difficulty lies in frequently changed appearance caused by target or camera movement and varying postures. A region proposal network (RPN) was introduced to estimate the target bounding box with a pre-defined set of anchor boxes [15,16,38]. However, it hinders the generalization and efficiency of the Siamese-based tracking framework. The no-prior box design was subsequently presented [3,8,32], which is free of prior knowledge about target

A. Bartoli and A. Fusiello (Eds.): ECCV 2020 Workshops, LNCS 12539, pp. 653–665, 2020.
https://doi.org/10.1007/978-3-030-68238-5_43

Fig. 1. Qualitative comparisons with state-of-the-art trackers. Our approach RPT obtains more accurate bounding box predictions when handling variations in posture, scale and aspect ratio

scale/ratio disrtibution. It directly views locations as training samples and predicts the relative offsets to the corners or sides of bounding box. The location is assigned with a positive label if it falls within a preset center area of the ground-truth bounding box, ignoring the target appearance and geometric structure. Besides, the target state is commonly described with a bounding box for the conveniences of feature extraction and ground truth annotation. It provides only a coarse spatial extent of object, and lacks the modeling capability for geometric transformations, thereby severely restricting the localization accuracy [33,34] (Fig. 1).

In this paper, we propose a novel tracking method named Representative Points based Tracker (RPT) to address the issue of accurate target estimation. RPT models the target state with a new finer representation as a set of representative points, and learns to arrange them to indicate the target's spatial extent and geometrically significant positions. In contrast to the coarse encoding of bounding box, the point set representation facilitates more fine-grained localization and modeling of object appearance. The RPT framework is constructed with two parallel branches, one primarily accounting for target estimation with the point set representation, the other trained online to provide high robustness against distractors.

The main contributions of this paper are threefold.

- A novel point set representation facilitating more accurate target estimation is employed in the field of visual tracking.

Fig. 2. Comparison for methods used to estimate the target state. Multi-level aggregation provides detailed structure information of objects, facilitating more fine-grained localization

- We aggregate hierarchical convolution layers to provide detailed structural information of target and high discriminative power when handling distractors.
- Our work achieves new state-of-the-art performances and runs at 20 FPS on several benchmarks, including OTB2015 [30], VOT2018 [14], VOT2019 [13] and GOT-10k [12].

2 Related Work

Visual tracking is a fundamental computer vision problem, which can be divided into a foreground-background classification task and a target state estimation task. The former case is responsible for distinguishing the target appearance from distractors and the surrounding background. The latter one aims to accurately describe the target region against variations in posture, scale and aspect ratio.

Discriminative correlation filters [7,22,27] and online learning approaches [6, 37] have remarkably advanced the target classification task in recent years. On the other hand, accurately estimating the target state is still challenging and severely limited with a multi-scale search strategy [1,5,11,27]. The target state is usually represented with a bounding box, and further estimated with various methods. SAMF [17] adopts a multiple scales searching strategy to address the issue of scale variations. GOTURN [10] designs a box regression strategy with the Laplace distribution hypothesis, which is restricted to small motions and limited scale changes. SiamRPN and the succeeding works [15,16,38] introduce a region proposal network (RPN) to regress the target region from pre-defined anchors. It leads to an apparent degeneration in tracking efficiency and generalization. ATOM [6] iteratively refines the coarse initial location to obtain a higher overlap between the predicted bounding box and ground truth. Inspired from anchor-free detection pipeline, a no-prior box design is utilized to predict offsets from candidate locations to the corners or sides of desired bounding box [3,8,32].

In order to describe more spatial details of objects, rotated bounding box and segmentation mask are further utilized to represent the target state in visual tracking [18,21,29]. LDES simultaneously estimates the orientation and scale variation of target via polar coordinate transformation [18]. SiamMask produces

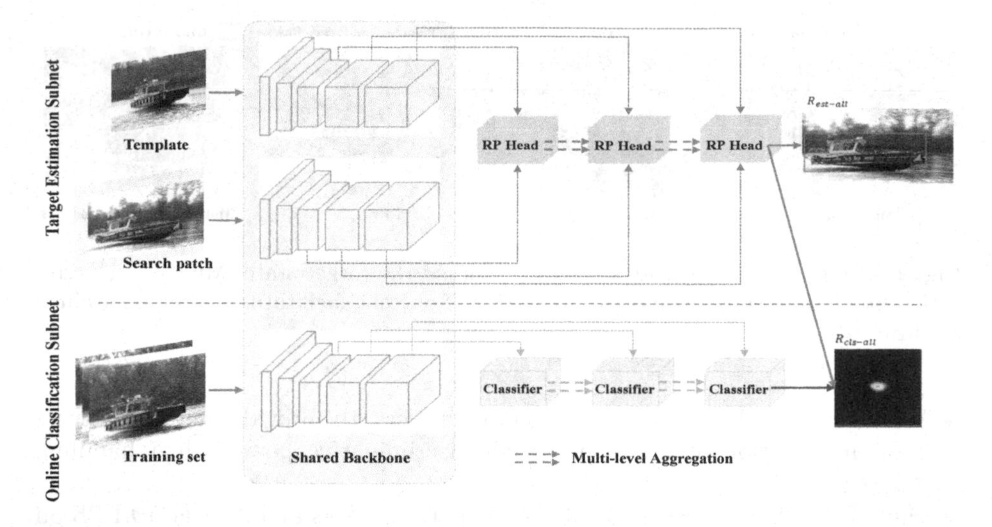

Fig. 3. The RPT framework with multi-level aggregation. The target estimation subnet predicts the spatial extent of object with a set of representative points in an offline-trained embedding space. The online classification subnet is responsible for distinguishing the target appearance from distractors and the surrounding background

a binary segmentation mask via an extra branch trained on YouTube-VOS [31], which is a large video dataset with pixel-wise annotations [29]. D3S obtains a noticeable advancement in segmentation accuracy by constructing a parallel structure with complementary geometric models [21].

3 RPT Framework

The RPT framework is constructed with a shared backbone network for feature extraction, and two parallel subnets accounting for target estimation and online classification respectively, as illustrated in Fig. 3. Following the architectural design guidelines in [36], we adopt ResNet-50 [9] as the backbone network, and extract hierarchical convolutional features from the last three residual blocks for multi-level prediction. The target estimation subnet is driven by a finer representation of object as a point set, which provides more fine-gained localization. The online classification subnet is trained exclusively online to enhance the discriminative capability in the presence of distractors.

3.1 Target Estimation with Point Set Representation

In contrast to object detection, tracking target can be an arbitrary object with unknown class. We therefore exploit Siamese-based feature extraction and matching for the requirement of target-specific estimation. Multi-level features from a target template (denoted as z) and a search region (denoted as x) are

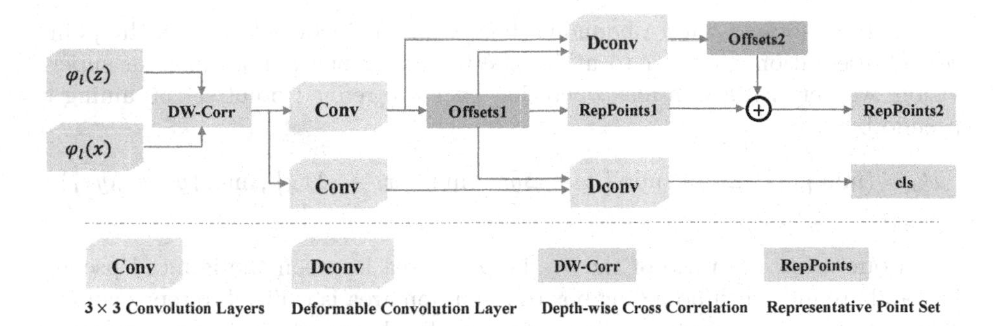

Fig. 4. The architecture of RP Head

extracted. For each feature level, a correlation map between the target patch and the search patch is obtained via depth-wise cross correlation [15] as:

$$g_l(z,x) = \varphi_l(z) * \varphi_l(x) \tag{1}$$

where $\varphi_l(z)$ and $\varphi_l(x)$ represent the corresponding feature outputs of the l-th level.

Inspired from [33], we design a classification head and a two-stage regression head over the correlation map. The head architecture is illustrated in Fig. 4. Each location of the correlation map is regarded as a target candidate. For the location (i,j), the target state is modeled with a set of sample points, which are uniformly initialized with the corresponding position as:

$$R = \{(x_k, y_k)\}_{k=1}^{n} \tag{2}$$

$$x_k = i, y_k = j, k = 1, 2, ..., n \tag{3}$$

where n represents the capacity of the point set. For each candidate, the regression head outputs a set of offsets to refine the distribution of sample points, while the classification head outputs two-channels for foreground-background classification. Specifically, the initial sample points are progressively refined with extra offsets in a point-wise manner:

$$R_r = \{(x_k + \Delta x_k, y_k + \Delta y_k)\}_{k=1}^{n} \tag{4}$$

where the predicted offsets $\{(\Delta x_k, \Delta y_k)\}_{k=1}^{n}$ are paired with the deformable convolution [4]. In contrast to its standard counterpart, deformable convolution augments the regular spatial sampling locations with additional offsets. It has the capability of modeling various geometric transformations for scale, aspect ratio and rotation. The kernel size of deformable convolutions is commonly set as 3×3. The number of representative points n is accordingly set to 9 in this work.

The refinement process is offline-trained with a multi-task loss for simultaneous localization and recognition. In the field of visual tracking, the target region is usually annotated with a bounding box, which is inconsistent with the point set representation. In order to utilize the bounding box annotations for supervision, we perform a min-max operation over the refined point set obtaining a pseudo box as:

$$B_p = (\min\{x_k + \Delta x_k\}, \min\{y_k + \Delta y_k\}, \max\{x_k + \Delta x_k\}, \max\{y_k + \Delta y_k\}) \tag{5}$$

In our work, the ratio of the overlapping area between the induced pseudo box and ground-truth bounding box to the union area is utilized to represent the regression loss. The focal loss [19] focusing on hard examples is further employed for foreground-background classification. Driven by both target regression and classification losses, a set of representative points are automatically learned that indicates the object boundaries and semantically prominent regions.

3.2 Discriminative Online Classification

The classification head discussed above is employed to distinguish foreground from background. However, it lacks the capacity to discriminate the target object from similar surrounding instances. In this section, we propose to complement the target estimation pipeline with an online-trained classifier to provide high robustness against distractors. Similar to [6,37], it is modeled as a light weight 2-layer fully convolutional neural network for efficiency.

The online classifier is trained with a certain amount of target regions obtained from the last few frames, which share the same size to the search patch. The training sample is commonly labeled with a two-dimensional Gaussion function centered at the target position, where the geometric structure of object is neglected. Here, we propose to label the confidence of target presence according to the representative point set. It is offline-learned to indicate the spatial extent of object and semantic key points. The desired classification score for each pixel of the training sample is therefore constructed by calculating the average position deviation to the point set.

After obtaining the online classification score, we combine it with the classification head output of the target estimation subnet as

$$f_l = \alpha f^l_{online} + (1 - \alpha) f^l_{offline} \tag{6}$$

where f^l_* represents the corresponding response maps for the l-th level, and α is the parameter controlling the impact of these two confidence scores.

3.3 Multi-level Aggregation

As pointed by [23,24], convolutional features extracted from earlier layers preserve finer spatial information benefiting for precise localization, while latter activations capture rich semantic information facilitating robustness against variation in appearance. Thus we propose to employ hierarchical convolutional layers contributing to the inference of both online classification and target estimation.

For online classification, the per-pixel confidences of target presence obtained from each classifier are combined via a weighted-fusion layer as

$$R_{cls-all} = \sum_{l=3}^{5} w_l * f_l \tag{7}$$

The outputs of the last three residual blocks share the same spatial resolution, weighted sum is therefore implemented in a pixel-wise manner. The set of weights w_l are end-to-end optimized together with the network.

For the target estimation subnet, each head outputs only a small set of representative points, which are insufficient when handling complicated object structures. Tracking accuracy is also limited as an inaccurate pseudo box is obtained from the sparse point set. Thus we propose to utilize a significantly larger set of points as the target state representation. The dense point set is simply constructed as a collection of representative points obtained by each head:

$$R_{est-all} = \bigcup_{l=3}^{5} R_l, \ R_l = \left\{\left(x_k^l, y_k^l\right)\right\}_{k=1}^{n} \tag{8}$$

As shown in Fig. 2, the fusion of sample points provides more elaborate structure of objects that is beneficial for accurate target estimation.

4 Experiments Results

4.1 Implementation Details

Following the architectural design guidelines in [36], ResNet-50 [9] pre-trained on ImageNet [26] is adopted as our backbone network. Multi-level features from the last three residual blocks are extracted for aggregation. The down-sampling operations are removed in the *conv4* and *conv5* blocks to preserve finer spatial information. Meanwhile, dilated convolutions with different atrous rates are exploited to improve the receptive fields.

The target estimation subnet is trained offline with image pairs selecting from YouTube-BoundingBox [25], COCO [20] and ImageNet VID [26]. For the first stage regression, location (i, j) on the correlation map is considered as a positive sample if its corresponding location on the search region is closest to the center of tracking target. The second stage regression and classification are then conducted on the first set of representative points. If the IOU (intersection-over-union) between the induced pseudo box and ground-truth is larger than 0.5, it is assigned with a positive label, and if the IOU is smaller than 0.4, it is assigned with a negative label, and otherwise ignored. For both stages, only the positive samples are exploited for the target state regression.

The training process is driven by SGD on 8 GPUs with 128 image pairs per mini-batch for 20 epochs. We employ a warm-up learning rate linearly ascended from 0.001 to 0.005 for the first 5 epochs, while the learning rate for the last

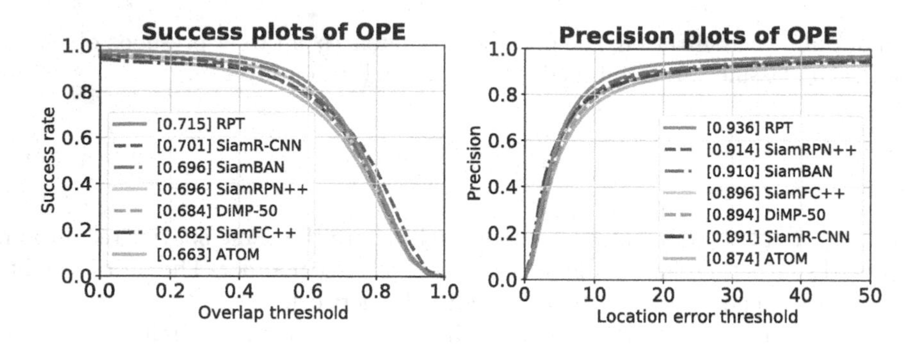

Fig. 5. Success and precision plots on OTB2015

15 epochs is exponentially decayed from 0.005 to 0.0005. In the first 10 epochs, only the head architecture of target estimation is trained. The whole network is fine-tuned in the last 10 epoch, where the learning rate of backbone is set to be 10 times smaller than the base value. The online classification subnet is trained with target regions obtained from the last few frames. Conjugate Gradient [6] is employed as the optimizer for efficient online learning. Our tracker is implemented on a PC with a GeForce GTX 1080Ti GPU using PyTorch.

4.2 Comparison with SOTA

The proposed RPT is compared with state-of-the-art trackers on several popular tracking Benchmarks, including OTB2015 [30], VOT2018 [14], VOT2019 [13] and GOT-10k [12].

OTB2015. OTB2015 [30] is one of the most widely used datasets for quantitative evaluation in visual tracking. We compare RPT with numerous recent and advanced visual trackers, including SiamRPN++ [15], ATOM [6], DiMP [2], SiamFC++ [32], SiamR-CNN [28] and SiamBAN [3], in terms of precision score and success rate. The precision score is formulated as the fraction of frames where the distance from the center of tracking result to the ground truth is under 20 pixels, while the success rate denoted as the proportion of frames whose intersection-over-union with the ground truth surpasses 0.5. As illustrated in Fig. 5, our tracker obtains an AUC (area under curve of success plot) score of 0.715 and precision score of 0.936, which are 1.4 and 2.2% points higher than the previous best result of SiamR-CNN [28] and SiamRPN++ [15].

GOT-10k. GOT-10k [12] is a large-scale dataset for tracking evaluation, which contains almost over 10 thousand sequences in training subset and 180 sequences in both validation and test subsets. Average overlap (AO) is employed for evaluations on this benchmark. For fair comparison, methods are required to be trained using the given dataset. The proposed RPT achieves a competitive result compared with the state-of-the-art methods with an AO score of 0.624, as shown in Table 1.

Table 1. Comparison with state-of-the-art trackers on GOT-10k. The best two results are marked in red and blue fonts

	ATOM [6]	SiamFC++ [32]	D3S [21]	DiMP-50 [2]	SiamR-CNN [28]	RPT
$SR_{0.5}$ (↑)	0.635	0.695	0.676	0.717	0.728	0.730
$SR_{0.75}$ (↑)	0.402	0.479	0.462	0.492	0.597	0.504
AO (↑)	0.556	0.595	0.597	0.611	0.649	0.624

VOT2018. VOT2018 [14] consists of 60 sequences with various challenging factors. In contrast to the aforementioned datasets, it's annotated with a rotated bounding box and restarted the tracker in case of failures. Methods are measured with three evaluation metrics, namely accuracy (A), robustness (R) and expected average overlap (EAO). Table 2 shows the comparison with the existing top-performing approaches on VOT2018 [14], in which RPT attains the best robustness and EAO among all methods. D3S [21] is superior in terms of accuracy as numerous segmentation sequences from Youtube-VOS [31] are utilized to produce a binary mask.

VOT2019. VOT2019 [13] is refreshed by replacing the 12 least difficult sequences from the previous version with several carefully selected sequences in GOT-10k dataset. The same measurements are exploited for performance evaluation. As illustrated in Table 2, we achieve an accuracy of 0.623, a robustness of 0.186 and an EAO of 0.417, outperforming DRNet [13] (leading performance on the public dataset of VOT2019 challenge) and other SOTA methods in terms of all metrics.

Table 2. Comparison with state-of-the-art trackers on VOT2018 and VOT2019. The best two results are marked in red and blue fonts

	VOT2018			VOT2019		
	EAO (↑)	R (↓)	A (↑)	EAO (↑)	R (↓)	A (↑)
ATOM [6]	0.401	0.204	0.590	0.292	0.411	0.603
SiamR-CNN [28]	0.408	0.220	0.609	–	–	–
SiamRPN++ [15]	0.414	0.234	0.600	0.285	0.482	0.599
SiamFC++ [32]	0.426	0.183	0.587	–	–	–
DiMP-50 [2]	0.440	0.153	0.597	0.379	0.278	0.594
SiamBAN [3]	0.452	0.178	0.597	0.327	0.396	0.602
SiamAttn [35]	0.470	0.160	0.630	–	–	–
DRNet [13]	–	–	–	0.395	0.261	0.605
D3S [21]	0.489	0.150	0.640	–	–	–
RPT	0.510	0.103	0.629	0.417	0.186	0.623

4.3 Ablation Study

In this section, an extensive ablation study is conducted to verify the impact of individual components on OTB2015 [30]. The modified version of SiamFC++ [32] using ResNet-50 [9] as backbone network is exploited as baseline. As illustrated in Table 3, the baseline approach obtains an AUC score of 0.675. By gradually adding the components of point set representation, online classification and multi-level aggregation, we achieve sustained performance gains of 2.5%, 0.9% and 0.6% in terms of AUC score. It is proven that the main contributions of our work facilitate more accurate target estimation. A similar conclusion can be obtained on other benchmarks.

Table 3. Ablation Study on OTB2015. Baseline denotes a modified SiamFC++ with ResNet-50 as backbone. PSR, OC and MLA represent the components of point set representation, online classification and multi-level aggregation, respectively

Baseline	PSR	OC	MLA	Precision (↑)	AUC (↑)	ΔAUC
√	×	×	×	0.898	0.675	–
√	√	×	×	0.925	0.700	+2.5%
√	√	√	×	0.928	0.709	+0.9%
√	√	√	√	0.936	0.715	+0.6%

4.4 Evaluation on VOT2020 Challenge

In contrast to previous versions, VOT2020 is annotated with segmentation masks using a new performance evaluation protocol. The novel protocol avoids tracker-dependent resets and re-defines the VOT basis measures. For evaluations on segmentation, the RPT framework is complemented with a modified D3S [21] to obtain class-agnostic object mask. In particular, the target location channel in D3S [21] is enhanced with the average position deviations from the representative point set. The augmented version of RPT achieves an EAO of 0.539 on VOT2020 challenge.

5 Conclusions

In this paper, we propose a novel tracking architecture named RPT to accurately estimate the target state with a finer point set representation. The RPT framework consists of two parallel branches, one primarily accounting for accurate target state estimation in an offline-trained embedding space, the other trained online to obtain high discriminative power in the presence of distractors. Besides, multiple convolution layers are aggregated to provide more fine-grained localization and detailed structural information of target. Comprehensive evaluations on various visual tracking datasets indicate that our method outperforms the recent and advanced trackers in terms of robustness and accuracy.

References

1. Bertinetto, L., Valmadre, J., Henriques, J.F., Vedaldi, A., Torr, P.H.S.: Fully-convolutional siamese networks for object tracking. In: Hua, G., Jégou, H. (eds.) ECCV 2016, Part II. LNCS, vol. 9914, pp. 850–865. Springer, Cham (2016). https://doi.org/10.1007/978-3-319-48881-3_56
2. Bhat, G., Danelljan, M., Gool, L.V., Timofte, R.: Learning discriminative model prediction for tracking. In: 2019 IEEE/CVF International Conference on Computer Vision, ICCV 2019, Seoul, Korea (South), 27 October - 2 November 2019, pp. 6181–6190. IEEE (2019)
3. Chen, Z., Zhong, B., Li, G., Zhang, S., Ji, R.: Siamese box adaptive network for visual tracking. CoRR abs/2003.06761 (2020). https://arxiv.org/abs/2003.06761
4. Dai, J., et al.: Deformable convolutional networks. In: IEEE International Conference on Computer Vision, ICCV 2017, Venice, Italy, 22–29 October 2017, pp. 764–773. IEEE Computer Society (2017)
5. Danelljan, M., Bhat, G., Khan, F.S., Felsberg, M.: ECO: efficient convolution operators for tracking. In: 2017 IEEE Conference on Computer Vision and Pattern Recognition, CVPR 2017, Honolulu, HI, USA, 21–26 July 2017, pp. 6931–6939. IEEE Computer Society (2017)
6. Danelljan, M., Bhat, G., Khan, F.S., Felsberg, M.: ATOM: accurate tracking by overlap maximization. In: IEEE Conference on Computer Vision and Pattern Recognition, CVPR 2019, Long Beach, CA, USA, 16–20 June 2019, pp. 4660–4669. Computer Vision Foundation/IEEE (2019)
7. Danelljan, M., Häger, G., Khan, F.S., Felsberg, M.: Learning spatially regularized correlation filters for visual tracking. In: 2015 IEEE International Conference on Computer Vision, ICCV 2015, Santiago, Chile, 7–13 December 2015, pp. 4310–4318. IEEE Computer Society (2015)
8. Guo, D., Wang, J., Cui, Y., Wang, Z., Chen, S.: SiamCAR: siamese fully convolutional classification and regression for visual tracking. CoRR abs/1911.07241 (2019). http://arxiv.org/abs/1911.07241
9. He, K., Zhang, X., Ren, S., Sun, J.: Deep residual learning for image recognition. In: 2016 IEEE Conference on Computer Vision and Pattern Recognition, CVPR 2016, Las Vegas, NV, USA, 27–30 June 2016, pp. 770–778. IEEE Computer Society (2016)
10. Held, D., Thrun, S., Savarese, S.: Learning to track at 100 FPS with deep regression networks. In: Leibe, B., Matas, J., Sebe, N., Welling, M. (eds.) ECCV 2016, Part I. LNCS, vol. 9905, pp. 749–765. Springer, Cham (2016). https://doi.org/10.1007/978-3-319-46448-0_45
11. Henriques, J.F., Caseiro, R., Martins, P., Batista, J.: High-speed tracking with kernelized correlation filters. IEEE Trans. Pattern Anal. Mach. Intell. **37**(3), 583–596 (2015)
12. Huang, L., Zhao, X., Huang, K.: Got-10k: a large high-diversity benchmark for generic object tracking in the wild. IEEE Trans. Pattern Anal. Mach. Intell. 1 (2019)
13. Kristan, M., et al.: The seventh visual object tracking VOT2019 challenge results. In: 2019 IEEE/CVF International Conference on Computer Vision Workshops, ICCV Workshops 2019, Seoul, Korea (South), 27–28 October 2019, pp. 2206–2241. IEEE (2019)
14. Kristan, M., et al.: The sixth visual object tracking VOT2018 challenge results. In: Leal-Taixé, L., Roth, S. (eds.) ECCV 2018, Part I. LNCS, vol. 11129, pp. 3–53. Springer, Cham (2019). https://doi.org/10.1007/978-3-030-11009-3_1

15. Li, B., Wu, W., Wang, Q., Zhang, F., Xing, J., Yan, J.: SiamRPN++: evolution of siamese visual tracking with very deep networks. In: IEEE Conference on Computer Vision and Pattern Recognition, CVPR 2019, Long Beach, CA, USA, 16–20 June 2019, pp. 4282–4291. Computer Vision Foundation/IEEE (2019)
16. Li, B., Yan, J., Wu, W., Zhu, Z., Hu, X.: High performance visual tracking with siamese region proposal network. In: 2018 IEEE Conference on Computer Vision and Pattern Recognition, CVPR 2018, Salt Lake City, UT, USA, 18–22 June 2018, pp. 8971–8980. IEEE Computer Society (2018)
17. Li, Y., Zhu, J.: A scale adaptive kernel correlation filter tracker with feature integration. In: Agapito, L., Bronstein, M.M., Rother, C. (eds.) ECCV 2014, Part II. LNCS, vol. 8926, pp. 254–265. Springer, Cham (2015). https://doi.org/10.1007/978-3-319-16181-5_18
18. Li, Y., Zhu, J., Hoi, S.C.H., Song, W., Wang, Z., Liu, H.: Robust estimation of similarity transformation for visual object tracking. In: The Thirty-Third AAAI Conference on Artificial Intelligence, AAAI 2019, The Thirty-First Innovative Applications of Artificial Intelligence Conference, IAAI 2019, The Ninth AAAI Symposium on Educational Advances in Artificial Intelligence, EAAI 2019, Honolulu, Hawaii, USA, 27 January - 1 February 2019, pp. 8666–8673. AAAI Press (2019)
19. Lin, T., Goyal, P., Girshick, R.B., He, K., Dollár, P.: Focal loss for dense object detection. In: IEEE International Conference on Computer Vision, ICCV 2017, Venice, Italy, 22–29 October 2017, pp. 2999–3007. IEEE Computer Society (2017)
20. Lin, T.-Y., et al.: Microsoft COCO: common objects in context. In: Fleet, D., Pajdla, T., Schiele, B., Tuytelaars, T. (eds.) ECCV 2014, Part V. LNCS, vol. 8693, pp. 740–755. Springer, Cham (2014). https://doi.org/10.1007/978-3-319-10602-1_48
21. Lukezic, A., Matas, J., Kristan, M.: D3S - a discriminative single shot segmentation tracker. CoRR abs/1911.08862 (2019). http://arxiv.org/abs/1911.08862
22. Lukezic, A., Vojír, T., Zajc, L.C., Matas, J., Kristan, M.: Discriminative correlation filter tracker with channel and spatial reliability. Int. J. Comput. Vis. **126**(7), 671–688 (2018)
23. Ma, C., Huang, J., Yang, X., Yang, M.: Hierarchical convolutional features for visual tracking. In: 2015 IEEE International Conference on Computer Vision, ICCV 2015, Santiago, Chile, 7–13 December 2015, pp. 3074–3082. IEEE Computer Society (2015)
24. Ma, Z., Lu, W., Yin, J., Zhang, X.: Robust visual tracking via hierarchical convolutional features-based sparse learning. In: 10th International Conference on Wireless Communications and Signal Processing, WCSP 2018, Hangzhou, China, 18–20 October 2018, pp. 1–7. IEEE (2018)
25. Real, E., Shlens, J., Mazzocchi, S., Pan, X., Vanhoucke, V.: YouTube-BoundingBoxes: a large high-precision human-annotated data set for object detection in video. In: 2017 IEEE Conference on Computer Vision and Pattern Recognition, CVPR 2017, Honolulu, HI, USA, 21–26 July 2017, pp. 7464–7473. IEEE Computer Society (2017)
26. Russakovsky, O., et al.: Imagenet large scale visual recognition challenge. Int. J. Comput. Vis. **115**(3), 211–252 (2015)
27. Sun, C., Wang, D., Lu, H., Yang, M.: Correlation tracking via joint discrimination and reliability learning. In: 2018 IEEE Conference on Computer Vision and Pattern Recognition, CVPR 2018, Salt Lake City, UT, USA, 18–22 June 2018, pp. 489–497. IEEE Computer Society (2018)
28. Voigtlaender, P., Luiten, J., Torr, P.H.S., Leibe, B.: Siam R-CNN: visual tracking by re-detection. CoRR abs/1911.12836 (2019). http://arxiv.org/abs/1911.12836

29. Wang, Q., Zhang, L., Bertinetto, L., Hu, W., Torr, P.H.S.: Fast online object tracking and segmentation: a unifying approach. In: IEEE Conference on Computer Vision and Pattern Recognition, CVPR 2019, Long Beach, CA, USA, 16–20 June 2019, pp. 1328–1338. Computer Vision Foundation/IEEE (2019)
30. Wu, Y., Lim, J., Yang, M.: Object tracking benchmark. IEEE Trans. Pattern Anal. Mach. Intell. **37**(9), 1834–1848 (2015)
31. Xu, N., et al.: YouTube-VOS: sequence-to-sequence video object segmentation. In: Ferrari, V., Hebert, M., Sminchisescu, C., Weiss, Y. (eds.) ECCV 2018, Part V. LNCS, vol. 11209, pp. 603–619. Springer, Cham (2018). https://doi.org/10.1007/978-3-030-01228-1_36
32. Xu, Y., Wang, Z., Li, Z., Ye, Y., Yu, G.: SiamFC++: towards robust and accurate visual tracking with target estimation guidelines. CoRR abs/1911.06188 (2019). http://arxiv.org/abs/1911.06188
33. Yang, Z., Liu, S., Hu, H., Wang, L., Lin, S.: RepPoints: point set representation for object detection. In: 2019 IEEE/CVF International Conference on Computer Vision, ICCV 2019, Seoul, Korea (South), 27 October - 2 November 2019, pp. 9656–9665. IEEE (2019)
34. Yang, Z., Xu, Y., Xue, H., Zhang, Z., Urtasun, R., Wang, L., Lin, S., Hu, H.: Dense reppoints: representing visual objects with dense point sets. CoRR abs/1912.11473 (2019)
35. Yu, Y., Xiong, Y., Huang, W., Scott, M.R.: Deformable siamese attention networks for visual object tracking. CoRR abs/2004.06711 (2020). https://arxiv.org/abs/2004.06711
36. Zhang, Z., Peng, H.: Deeper and wider siamese networks for real-time visual tracking. In: IEEE Conference on Computer Vision and Pattern Recognition, CVPR 2019, Long Beach, CA, USA, 16–20 June 2019, pp. 4591–4600. Computer Vision Foundation/IEEE (2019)
37. Zhou, J., Wang, P., Sun, H.: Discriminative and robust online learning for siamese visual tracking. CoRR abs/1909.02959 (2019). http://arxiv.org/abs/1909.02959
38. Zhu, Z., Wang, Q., Li, B., Wu, W., Yan, J., Hu, W.: Distractor-aware siamese networks for visual object tracking. In: Ferrari, V., Hebert, M., Sminchisescu, C., Weiss, Y. (eds.) ECCV 2018, Part IX. LNCS, vol. 11213, pp. 103–119. Springer, Cham (2018). https://doi.org/10.1007/978-3-030-01240-3_7

AFOD: Adaptive Focused Discriminative Segmentation Tracker

Yiwei Chen[1], Jingtao Xu[1](✉), Jiaqian Yu[1], Qiang Wang[1], ByungIn Yoo[2], and Jae-Joon Han[2]

[1] Samsung Research China - Beijing, Beijing, China
{yiwei.chen,jingtao.xu,jiaqian.yu,qiang.w}@samsung.com
[2] Samsung Advanced Institute of Technology, Suwon, South Korea
{byungin.yoo,jae-joon.han}@samsung.com

Abstract. Visual object tracking is a fundamental task in computer vision which could be integrated into numerous real-world applications. Traditional object tracking methods focus on providing the bounding box as object position, while some recent trackers start to consider the combination of segmentation module to generate the binary segmentation mask, pursuing more accurate localization. However, how to effectively integrate different information for accurate and robust tracking is an open question. In this paper, we propose a novel Adaptive FOcused Discriminative (AFOD) segmentation tracker with the following advanced components. For localization, a more discriminative light weight online target appearance model is employed to provide robust position estimation. For segmentation, leveraging the backbone semantic feature, the coarse segmentation feature, and the localization feature, an offline trained fine segmentation model with IoU optimization is utilized to generate the accurate high resolution masks. The boundary detection further enhances the segmentation quality. For combination, an adaptive prediction strategy is proposed to better integrate the information from two types of predictions, i.e. box and mask. AFOD achieves leading performance on two tracking benchmarks including the bounding box annotated VOT2018 and the segmentation mask annotated VOT2020, while running close to real-time.

Keywords: Visual object tracking · Adaptive segmentation tracker · Discriminative model

1 Introduction

Visual object tracking is one of the most fundamental tasks in computer vision which could be integrated into numerous applications, e.g. automatic surveillance, vehicle navigation, and activity recognition. The basic scenario is to track one or more specific objects in a video sequence, with the only annotation provided in the first frame. In traditional object tracking frameworks and databases [11,15,19,39], annotations are generally given in the form of a

A. Bartoli and A. Fusiello (Eds.): ECCV 2020 Workshops, LNCS 12539, pp. 666–682, 2020.
https://doi.org/10.1007/978-3-030-68238-5_44

simple axis-aligned or rotated bounding box. Modern template based trackers [4,9,21] adopt exhaustive search on the target region with bounding box regression to solve rotation, translation, scale and aspect ratio change issue. However, the bounding box representation has its own limitations resulting in compromised performance. When the tracking object is highly deformed, elongated, or rotated, the moderate variation of box estimation would lead to obvious prediction error [24].

Compared to the bounding box representation, the pixel-wise binary segmentation mask is more accurate and reliable to represent the object. It is natural that the segmentation mask provides more precise information (shown in Fig. 1) than a rough closing hull, i.e. a box including background area. Moreover, the mask representation has advantages for a variety of video applications with dense pixel-wise prediction, e.g. object boundary estimation for grasping [1], autonomous driving [33], video editing [40].

Fig. 1. The proposed method aims to develop a more accurate and robust segmentation tracker. With more advanced target appearance model, segmentation model, and adaptive prediction, AFOD continuously focuses on the tracking target (woman in iceskater2 sequence from VOT2020) and produces more accurate masks than the leading segmentation tracker D3S [24].

With the aforementioned segmentation mask output, tracking with masks is well aligned to the video object segmentation (VOS) task: a binary labeling problem aiming to separate the foreground object from the background region in a video. It is flourished by DAVIS [30] and YoutubeVOS [42] challenges in recent years. Different to tracking task, VOS task generally considers large main object on a relative simple background without obvious distractors. And the evaluation sequence is not as long as the tracking task. Recent works in VOS focus to address this issue via e.g. fine-tuning generic segmentation network [25,29,41], incorporating generative models of foreground and background appearance [14, 35], employing assistance from light-weight discriminative target model [31,36, 38]. However, VOS methods still have relatively limited capacity to distinguish the target from similar distractors in tracking sequences [19].

More recently, the segmentation mask has been taken into consideration for tracking. SiamMask [37] extends SiamRPN network [22] with an additional separate branch to predict the dense segmentation masks of each location. D3S [24] designs two parallel branches framework to encode geometrically invariant features and geometrical features. With a U-Net [32] liked refinement pathway, the finer mask is predicted. Although the segmentation mask is generated, the tracking evaluation of above approaches is performed on the fitted bounding box. For VOT2020 challenge [20], the dense pixel mask is chosen as the ground truth for demonstration. However, how to effectively integrate different types of information for accurate and robust tracking is a remaining open question.

To overcome the above-mentioned limitations, we propose a novel adaptive focused discriminative segmentation tracker, noted as AFOD. Our approach involves both accuracy and robustness improvement on D3S tracker from three aspects. For online position prediction, a powerful online target appearance model is integrated to provide more discriminative geometrical features. Operating on deep features, the one layer target model gradually updates according to the target appearance change and generates the robust target feature map.

For mask segmentation, except for the coarse segmentation model, an accurate fine segmentation network is incorporated to predict more precise representation. The semantic higher level backbone feature, the coarse segmentation feature and the online target appearance feature are integrated in a stage-wise way to maintain the prediction consistency. With a loss function optimizing on IoU, AFOD directly learns for the evaluation of segmentation masks. Furthermore, a boundary detection head is added to purify the mask edge area to be more distinguishable.

For information combination, an adaptive prediction strategy is employed to combine segmentation mask prediction and online position prediction according to the prediction quality, which further refines the update of online target model. Our tracker tends to be more discriminative for challenging tracking sequences, e.g. deformation, occlusion, low video quality.

For experiments, the proposed tracker is demonstrated on the bounding box annotated VOT2018 [19] and the segmentation mask annotated VOT2020 [20] dataset respectively. AFOD presents very competitive performance to state-of-the-art (SOTA) trackers and leading segmentation trackers, while running almost real-time. The remainder of this paper is organized as follows. In Sect. 2, we briefly review the related work. Section 3 describes the proposed AFOD tracker. In Sect. 4, we perform both quantitative and qualitative experiments to validate the proposed method. The conclusion is given in Sect. 5.

2 Related Work

In the past few years, numerous approaches have been proposed for both video object segmentation and visual object tracking, leading to the rapid performance improvement. However, they still suffer from the balance of accuracy and robustness on the challenging sequences.

Video Object Segmentation. Generally VOS approaches can be divided into following categories: first-frame fine-tuning [25,29,41], mask propagation [29,40], generative approaches [14,17] and tracking based methods [8,31]. Recent works focus on the last two types of methods to fully utilize the target appearance feature for segmentation. Unlike fine-tuning the network on the first frame, generative approaches first construct the target appearance model from features corresponding to the initial target ground truth. For the incoming frames, the prediction is produced by methods like classical clustering methods [7,17] or video matching [14,35,38]. Chen *et al.* [7] and Hu *et al.* [14] propose to segment by matching the current frame feature to the first frame feature. [27] further utilizes dynamic memory from multiple frames feature to accurately predict mask.

Visual Object Tracking. Two main categories of tracking methods are popularized in the most recent research. The first one is called discriminative correlation filters (DCF) [5] focusing on learning discriminative filters to classify target from background distractors. The problem is formulated as ridge regression and solved by circular correlation in Fourier domain [13]. With the progress in deep learning, deep feature has been considered into DCF as well. [10] uses pre-trained deep features for detection, while [34] develops DCF optimization on pretrained features. Recently [4,9] exploit efficient backpropagation method for DCF optimization. The second class, noted as Siamese trackers [3,21–23,43,46–48], is widely adopted to investigate the similarity between target and background through offline training. The template feature and backbone are fixed during inference, leading to an excellent real-time performance [3].

Segmentation Tracker. To further investigate the improvement of both tracking and segmentation, researchers consider to develop a type of hybrid approach. From the VOS perspective, [8] develops a VOS framework with part based tracker to simultaneously track local regions of target. In [31], an online discriminative target model is adopted to continuously predict coarse segmentation mask with variations. However, the proposed target model suffers from indistinguishable distractors in general tracking sequences. From the tracking perspective, SiamMask [37] is one pioneering work to bridge visual object tracking and video object segmentation. It extends the successful SiamRPN tracker [22] with an additional branch for mask prediction. The tracking and the segmentation are treated separately missing the opportunity to exchange beneficial information of two tasks. D3S [24] further narrows the gap of two missions by an elaborately designed mechanism to combine online and offline learned features. Our AFOD tracker is an improved approach to D3S. And there are several major differences. First, a more powerful online target appearance model is incorporated. Second, we adopt a more effective segmentation network with the assistance from IoU optimization and boundary prediction. Third, an adaptive prediction strategy is proposed to further combine the two types of predictions.

3 Method

In this section, we first present the details of our adaptive focused discriminative segmentation tracker in Sect. 3.1, 3.2 and 3.3 3.4. AFOD mainly includes four parts: a coarse segmentation model, a target appearance model which is updated online and provides box prediction, a fine segmentation model which is trained offline, and an adaptive prediction module. Then we describe the offline training in Sect. 3.5 and the inference in Sect. 3.6. The overview of our AFOD tracker is shown in Fig. 2.

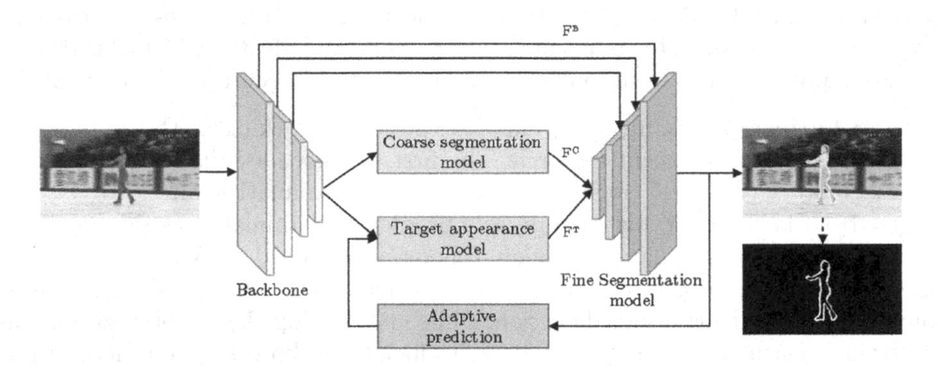

Fig. 2. Overview of AFOD tracker.

3.1 Coarse Segmentation Model

Feature matching has been utilized in previous video object segmentation and tracking [14,24] to extract geometrically invariant target features. In AFOD, we follow the process in these approaches to generate the coarse segmentation feature map. The target feature is first extracted by the backbone and split into the foreground and background part as two sets of templates, $\mathbf{Z}^F$ and $\mathbf{Z}^B$. For the incoming search image, the correlation matrix between the extracted search region feature $\mathbf{X}$ and two sets of templates are calculated via pixel-wise normalized inner product:

$$s^F_{i,j} = <\tilde{\mathbf{x}}_i, \tilde{\mathbf{z}}^F_j>, s^B_{i,j} = <\tilde{\mathbf{x}}_i, \tilde{\mathbf{z}}^B_j> \tag{1}$$

where $\tilde{()}$ is L2 normalization, i is the pixel location of search region feature and j is the index for foreground or background features. The average of the top-K correlation scores ($s^F_{i,j}$ and $s^B_{i,j}$) for each location i is calculated to generate two feature maps with the same size. The final concatenated feature is noted as $\mathbf{F}^C$. More details can be referred to [14,24].

3.2 Target Appearance Model

Although feature matching with an initial target mask generates the coarse segmentation feature, it cannot provide sufficient discriminative information to recognize similar instances and segment accurately. Conversely, the robust localization, which is well studied in discriminative correlation filters, produces complementary features for accurate tracking. Several efficient optimization techniques [4,9,34] have been exploited to operate on deep features. For the deep

DCF approach described in [9,24], a two fully connected layers model is optimized. All the parameters are learned through online learning instead of an end-to-end manner, which decreases the practical localization performance.

Here another recently proposed robust deep DCF tracker [4] is employed as our target appearance model for localization. The target appearance model $\mathbf{T}$ is generated from the target feature $\mathbf{Z}$:

$$\mathbf{T} = PrROIPool(conv(\mathbf{Z})), \tag{2}$$

where *conv* is an offline pretrained convolution layer for better initialization, and *PrROIPool* is the precise ROI pooling [16] to extract the target region feature. For prediction, the response map is calculated via $\mathbf{X} * \mathbf{T}$. Location with the maximum value of the response map is selected as the target location (box center). And a distance map is constructed as the target location feature $\mathbf{F}^T$ by calculating the Euclidean distance from each pixel to the center. Moreover, for online update, the final target position is calculated by adaptive prediction after generating one segmentation mask. The correlation filters $\mathbf{T}$ are then efficiently updated with steepest descent optimization [4,26].

3.3 Fine Segmentation Model

How to effectively integrate the coarse segmentation feature $\mathbf{F}^C$, the target location feature $\mathbf{F}^T$, and the backbone feature $\mathbf{F}^B$ will definitely affect tracking accuracy and robustness. AFOD adopts an offline pretrained fine segmentation model to combine these features and improve the accuracy from three aspects.

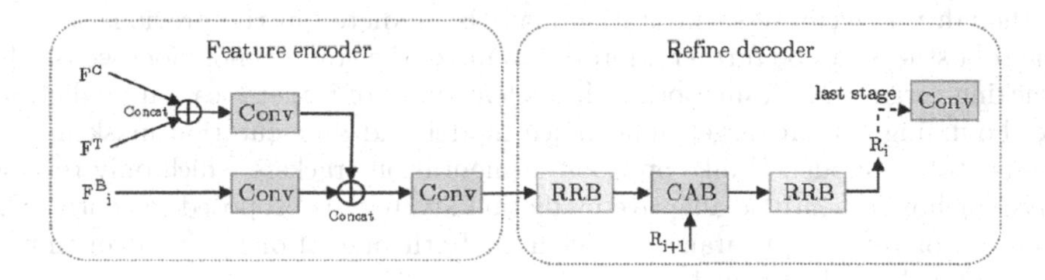

Fig. 3. Example of one stage structure in our fine segmentation model. Conv stands for the block including several convolution layers.

First, a more comprehensive encoder-decoder structure is developed to effectively integrate the feature from higher low resolution stages to lower high resolution stages while maintaining the content consistency. It includes a feature encoder (FE) to combine features from different models and a refine decoder (RD) to predict the segmentation mask in a stage-wise manner. FE takes the aforementioned three types of feature maps as input. The feature map $\mathbf{F}^C$ and $\mathbf{F}^T$ are concatenated first, and the channel number is increased. Then the reduced feature $\mathbf{F}_i^B$ (i stands for the ith stage in the backbone) is concatenated to the above feature. Lastly, RD outputs the refined feature with the attention from higher stage feature $\mathbf{R}_{i+1}$ by: $\mathbf{R}_i = RD_i(FE_i(\mathbf{F}_i^B, (\mathbf{F}^C, \mathbf{F}^T)), \mathbf{R}_{i+1})$. The RD module in each stage consists of two refinement residual blocks (RRB) and one channel attention block (CAB) as described in [31,44]. The refined feature $\mathbf{R}_{i+1}$ from previous stage constrains the prediction to be consistent with high level semantic information.

Second, generally the evaluation tool is an IoU based metric for dense pixel-wise or bounding box prediction. According to [2,45], the direct optimization on IoU for training could benefit corresponding IoU based testing. Therefore except for the typical binary cross-entropy loss in dense pixel-wise prediction task, the Lovász hinge loss [2,45] for binary segmentation is incorporated to optimize IoU as a surrogate function during offline training, while generating higher IoU score and accurate segmentation mask in inference.

Third, the clear edge is essential to distinguish the target from the complicated background and spatially adjacent similar distractors in segmentation. With this motivation, a light weight boundary detection module is added to the lowest stage of predicted mask to enhance its quality. Two 3×3 convolution layers are utilized to learn mask boundary B with explicit semantic boundary supervision. This further refines the mask to be more distinguishable. The ground truth for training is calculated by operating Canny operator [6] on the given mask.

The structure of our fine segmentation model (one stage) is illustrated in Fig. 3.

3.4 Adaptive Prediction

Although the accurate segmentation mask is predicted in the previous section, the robustness can be further improved to make the tracker more focused on the tracking target. Our framework includes two types of target location prediction, i.e. bounding box in target appearance model and segmentation mask in fine segmentation model. Unlike previous segmentation trackers which only rely on mask or box, a heuristic adaptive prediction strategy is proposed to control the learning of target appearance model from both predictions. The algorithm is summarized in Algorithm 1.

Two key changes are described here. The first item is to conditionally combine the two predictions according to the distance between the box center and the mask center. The mask center and scale are calculated by trivially fitting the maximum connected region to an axis-aligned bounding box. If they are close enough, the mask tends to be more reliable as location prediction. The second one is to smoothly update the scale with both box and mask scales. The weight α decides the contribution of two parts. Finally, the target appearance model in Sect. 3.2 is updated with the predicted center and scale in backpropagation.

Algorithm 1. Adaptive prediction

1: Input: the ith frame box prediction $B^i_{center}, B^i_{scale}, B^i_{uncertain}$ (probability score from the online model), the ith frame mask prediction $M^i_{center}, M^i_{scale}, M^i_{size}$, the average mask size Ave_{size}(previous P frames)
2: **if** $M^i_{size} > TH_1$ and $B^i_{uncertain} < TH_2$ **then**
3:    **if** $\|B^i_{center} - M^i_{center}\| < TH_3$ **then** # Check distance
4:       **if** $|M^i_{size} - Ave_{size}|/Ave_{size} < TH_4$ **then**
5:          $P^i_{scale} = (1-\alpha) * B^i_{scale} + \alpha * M^i_{scale}$ # Update scale
6:       **else**
7:          $P^i_{scale} = B^i_{scale}$
8:       **end if**
9:       **if** $|M^i_{size} - Ave_{size}|/Ave_{size} < TH_5$ **then**
10:          $P^i_{center} = M^i_{center}$
11:       **else**
12:          $P^i_{center} = B^i_{center}$
13:       **end if**
14:    **else**
15:       $P^i_{center} = B^i_{center}, P^i_{scale} = B^i_{scale}$
16:    **end if**
17: **else**
18:    $P^i_{center} = P^{i-1}_{center}, P^i_{scale} = P^{i-1}_{scale}$
19: **end if**
20: Update target appearance model with $P^i_{center}, P^i_{scale}$

3.5 Offline Training

For the coarse and fine segmentation network, the parameters are obtained through offline training on the standard VOS datasets with mask annotation. And the backbone is fixed during this process. The training data includes 3471 videos in YoutubeVOS-2018 training set [42] and 60 videos in DAVIS-2016 train/validation set [30]. Two frames are randomly selected from the same video as the target image and the search image to compose a pair of data. A mini-batch contains 128 pairs of images. The target appearance model is turned off and the corresponding feature is simulated by adding a small uniform perturbation on the ground truth mask center.

According to Sect. 3.3, additional loss function and boundary module are included, therefore the whole training process consists of three phases. In the first phase, only binary cross-entropy loss (ℓ_m) is back-propagated, resulting in a relatively good optimization start point. 5 epochs warm-up where the learning rate increases from $2e-4$ to $1e-3$ is also used. For the second phase, the Lovász hinge loss (ℓ_l) is utilized to assist the network to optimize with IoU information. In the last phase, a boundary head is placed on the segmentation mask and optimized by binary cross-entropy loss (ℓ_b). ℓ_m and ℓ_l loss are also back-propagated in this phase with a smaller weight $\lambda = 0.1$ in Eq. (6). For every phase (45 epochs), the start learning rate is $1e-3$ and reduced by 0.2 every 15 epochs. Weight decay is set to $5e-4$ and the optimizer is Adam [18]. All offline training can be finished on 4 Tesla V100 GPUs in one day.

$$\ell_m = BinaryCrossEntropyLoss(y_{mask}; w) \quad (3)$$

$$\ell_l = LovászLoss(y_{mask}; w) \quad (4)$$

$$\ell_b = BinaryCrossEntropyLoss(y_{boundary}; w) \quad (5)$$

$$L = \ell_m + \lambda\ell_l + \ell_b \quad (6)$$

3.6 Inference

During inference, AFOD is initialized by the first frame and the given mask. The axis-aligned bounding box for the target appearance model is obtained by fitting on the mask. If no mask is available, it is approximated by the method in [24]. The target appearance model, and the template feature in the coarse segmentation model are initialized separately on the first cropped target image with a region of four times the target size. Then for the incoming new frame, the target location feature and the coarse segmentation feature are extracted on the cropped search image according to the last frame location. The segmentation mask is predicted via sigmoid activation from the fine segmentation model. If the test data only has bounding box annotation, the location prediction in Algorithm 1 is adopted. Finally the target model is updated for the next frame.

3.7 Implementation Details

The proposed method is implemented on Pytorch [28] with ResNet-50 [12] pre-trained on ImageNet as backbone. Only the first four stages in the backbone are utilized. 384×384 input image is four times larger of the target.

For parameters in each module, K is 3 in the coarse segmentation model. The optimization parameter and update interval for the target appearance model follow [4]. For speed up, the channel number in the fine segmentation model is maintained as 64. Thus the channel number for dimension increase and reduction is all 64. For adaptive prediction strategy, the thresholds are $TH_1 = 50, TH_2 = 15, TH_3 = 10, TH_4 = 2, TH_5 = 0.1$. For comparison, the box and mask are projected back to the normalized 384×384 region. And previous $P = 25$ frames mask size is stored for average mask size calculation.

The speed is tested on an Nvidia 1080Ti GPU and AFOD runs at 20FPS.

4 Experiments

For experiments, the proposed AFOD tracker and other leading trackers are compared on two visual object tracking benchmarks, VOT2018 [19] and VOT2020 [20], with different bounding box and segmentation mask annotation respectively. Both quantitative and qualitative results are presented.

4.1 Experiment Setup

Both VOT2018 [19] and VOT2020 [20] contain 60 videos under highly challenging scenarios. The targets in VOT2018 are annotated by rotated rectangles while the segmentation masks are provided in VOT2020 for more accurate tracking evaluation. Both experiments are evaluated with the official tool[1]. For the evaluation criteria, VOT2018 is measured by accuracy (average overlap over successfully tracked frames), robustness (failure rate) and expected average overlap (EAO), which is a combination metric of accuracy and robustness. The protocol for VOT2020 is different, which avoids tracker-dependent resets and reduces the variance of the performance evaluation measures. Multiple anchors are set as initialization start points to generate various sequences for the baseline experiment. Accuracy is evaluated as the average overlap of mask before tracking failure and robustness is calculated as the average of the maximum tracking frame number before failure. Therefore, higher robustness is better for VOT2020. For unsupervised test, the anchor set is removed.

Table 1. Quantitative experiment results on VOT2020. A: accuracy, R: Robustness. ↑ means higher is better.

Method	Baseline (EAO/A/R)↑	Unsupervised (A)↑
DiMP [4]	0.282/0.454/0.748	0.364
FRTM-VOS [31]	0.198/0.543/0.518	0.320
SiamMask [37]	0.320/0.666/0.588	0.430
D3S [24]	0.425/0.674/0.765	0.495
AFOD	0.495/0.734/0.809	0.553

4.2 Comparison to State-of-the-art

Due to different annotations in two benchmarks, diverse SOTA trackers are selected for comparison. AFOD is compared to Siamese tracker SiamRPN++ [21], two top deep DCF trackers including ATOM [9] and DiMP [4], two leading segmentation trackers including SiamMask [37] and D3S [24]. Additionally, a recent VOS method FRTM-VOS [31] is chosen for comparison.

The first experiment is performed on VOT2020 with IoU evaluation on the segmentation mask. The results illustrated in Table 1 show AFOD outperforms existing top trackers in terms of all measures. For the SOTA discriminative tracker DiMP, no segmentation mask is generated which obviously decreases its accuracy and EAO. FRTM-VOS, one of the top-performing real-time VOS methods including a discriminative model, shows insufficient discrimination capability (36% relative lower robustness compared to AFOD) which further indicates the

[1] https://github.com/votchallenge/vot-toolkit.

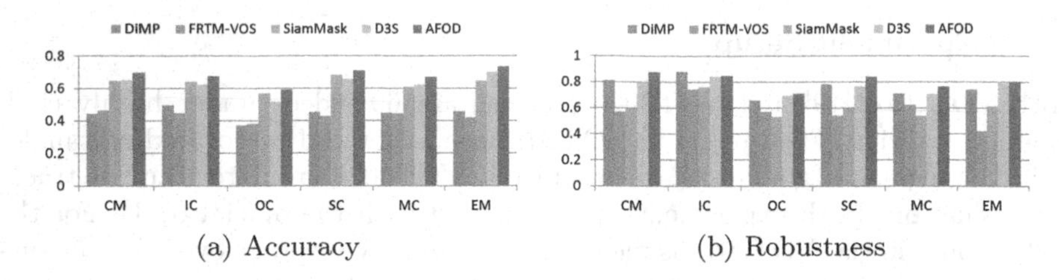

(a) Accuracy (b) Robustness

Fig. 4. Accuracy and robustness with respect to the visual attributes on VOT2020. CM: camera motion, IC: illumination change, OC: occlusion, SC: size change, MC: motion change, EM: empty.

essential difference between video object segmentation and visual object tracking. Compared to the leading segmentation tracker D3S, AFOD presents better results under all metrics, e.g. 0.495 of EAO, 16% relative improvement. This is mainly attributed to the leverage of both effective online localization and offline segmentation model. Better target appearance learning assists the tracker to be more concentrated on the tracking target with low failure probability, while high quality segmentation mask directly increases the accuracy. Moreover, the performance on different attributes is provided in Fig. 4.

Table 2. Quantitative experiment results on VOT2018. ↑ means higher is better while ↓ means lower is better.

Method	SiamRPN++ [21]	ATOM [9]	DiMP [4]	SiamMask [37]	D3S [24]	AFOD
EAO↑	0.414	0.401	0.440	0.380	0.489	0.491
Accuracy↑	0.600	0.590	0.597	0.609	0.640	0.634
Robustness↓	0.234	0.204	0.153	0.276	0.150	0.117

Experimental results on VOT2018 are shown in Table 2. Again, AFOD achieves more favorable performance compared to other competitive trackers, especially with a low robustness score. This demonstrates that the proposed adaptive prediction strategy successfully integrates two types of predictions, focusing to distinguish the challenging distractors. It is worth noting that the accuracy of AFOD is not better than D3S which can be attributed to the mask-to-box approach. Our mask is trivially fitted by an axis-aligned bounding box while D3S uses ellipse fitting and IoU optimization to increase accuracy.

4.3 Qualitative Result

Rather than the quantitative experiment, the visualization is performed in this section. We compare AFOD with the leading segmentation tracker D3S [24] and VOS method FRTM-VOS [31] in Fig. 5. The frames are sampled from the beginning stage to the end stage of sequences in VOT2020. Several challenging

Fig. 5. Qualitative results on VOT2020: basketball, dinosaur, butterfly.

scenarios are covered, e.g. fast motion, low video quality, deformation. As shown in the figure, D3S can perform tracking and segmentation well for the early frames. However, as time goes on, D3S struggles with similar distractors, e.g. basketball palyers. And the segmentation quality is not satisfying due to the simple network structure, e.g. the dinosaur head and butterfly. FRTM-VOS utilizes a more advanced sophisticated segmentation model, and performs better on generating high quality mask, e.g. butterfly with a nearly clean background. But its target model cannot perform well for difficult scenarios in tracking. In contrast, AFOD presents more promising results for both segmentation and localization which is in correspondence to the improvement of accuracy and robustness. From the beginning to the end, AFOD can continuously focus on the tracking target and distinguish it from distractors. Moreover, accompanied by the boundary refinement, the produced mask has more accurate and reliable edge.

4.4 Ablation Study

Since AFOD contains several components to increase the performance, we analyze the contribution of each model on VOT2020. The following experiments are performed separately.

Target Appearance Model. We replace the target appearance model with the two-layer model in [9,24]. As illustrated in Table 3, the performance gain from a powerful discriminative model is apparent. The robustness increases by more than 2% and the accuracy is slightly improved.

Table 3. Ablation study 1 on VOT2020: replace the target appearance model (#1); remove two key components in adaptive prediction strategy (#2).

	AFOD	#1	#2
EAO ↑	0.495	0.463	0.398
Accuracy ↑	0.734	0.715	0.711
Robustness ↑	0.809	0.786	0.682

Adaptive Prediction. We remove the proposed two key changes, distance condition and scale combination (choose either box scale or mask scale). In Table 3, we can see that robustness and EAO decrease severely without two key components, while the accuracy almost does not change. This further demonstrates the role of adaptive prediction in distinguishing distractors. The prediction from the discriminative model and the segmentation model should be adaptively integrated to achieve a more focused tracker.

Fine Segmentation Model. In order to test the performance of the segmentation model, we formulate an experiment with VOT2020 data. The first frame in each sequence is set as the target frame, while masks are predicted for the following frames. To alleviate the influence of the online target model, the distance map is calculated by using the ground truth mask center in every frame. Thus only the segmentation branch could affect the mask quality. We compare the mean IoU (mIoU) on all tested frames for models from different training phases and different segmentation networks (trained with the same setup) in Table 4. With the assistance from Lovász hinge loss and boundary detection, the mIoU increases 2.1%. Moreover, compared to U-Net in D3S, AFOD achieves a mIoU of 0.765, 31% relative better. A more accurate segmentation model guarantees the tracking performance to some extent.

Table 4. Ablation study 2 on VOT2020: 1st phase in offline training with ℓ_m (#1); 2nd phase with ℓ_l (#2); 3rd phase adding ℓ_b (#3); U-Net in [24] (#4).

	#1	#2	#3	#4
mIoU↑	0.744	0.761	0.765	0.584

To better illustrate our improvement, the qualitative test results on models from different training phases are shown in Fig. 6. Compared to the first phase using the binary cross-entropy loss only, optimizing IoU explicitly improves the segmentation quality, e.g. tail light and crab legs. In particular, the boundary detection module generates sufficient clear edge information to distinguish the target, leading to more accurate masks as well as correcting the mispredictions.

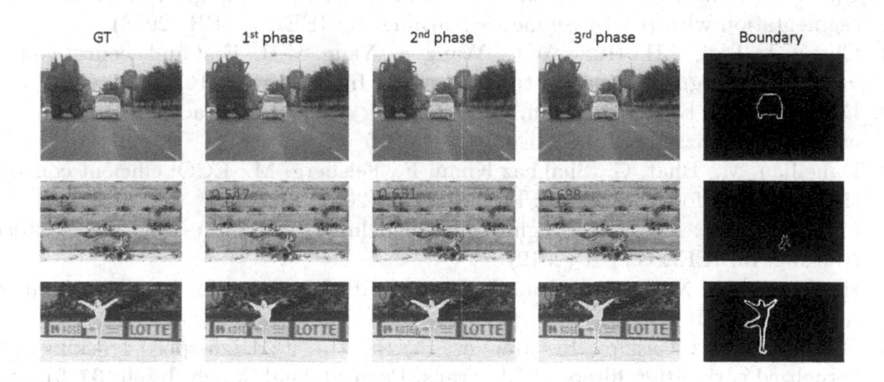

Fig. 6. Example results of AFOD in different training phases on VOT2020. The last column is the output from the boundary detection head. Red number is IoU score.

5 Conclusions

In this paper, we propose an adaptive focused discriminative segmentation tracker, referred to as AFOD, for accurate and robust tracking. Three major modules are developed to better integrate the predictions of bounding box and segmentation mask. With more advanced network structure, appropriate optimization, additional supervision, and adaptive prediction strategy, the proposed tracker achieves state-of-the-art performance on the bounding box annotated VOT2018 and the segmentation mask annotated VOT2020 respectively, showing the effectiveness and generalization ability.

References

1. Allen, P.K., Timcenko, A., Yoshimi, B., Michelman, P.: Automated tracking and grasping of a moving object with a robotic hand-eye system. IEEE Trans. Robot. Autom. **9**(2), 152–165 (1993)
2. Berman, M., Rannen Triki, A., Blaschko, M.B.: The lovász-softmax loss: a tractable surrogate for the optimization of the intersection-over-union measure in neural networks. In: IEEE CVPR (2018)
3. Bertinetto, L., Valmadre, J., Henriques, J.F., Vedaldi, A., Torr, P.H.S.: Fully-convolutional siamese networks for object tracking. In: Hua, G., Jégou, H. (eds.) ECCV 2016. LNCS, vol. 9914, pp. 850–865. Springer, Cham (2016). https://doi.org/10.1007/978-3-319-48881-3_56
4. Bhat, G., Danelljan, M., Gool, L.V., Timofte, R.: Learning discriminative model prediction for tracking. In: IEEE ICCV (2019)
5. Bolme, D.S., Beveridge, J.R., Draper, B.A., Lui, Y.M.: Visual object tracking using adaptive correlation filters. In: 2010 IEEE Computer Society Conference on Computer Vision and Pattern Recognition (2010)
6. Canny, J.: A computational approach to edge detection. IEEE Trans. Pattern Anal. Mach. Intell. **6**, 679–698 (1986)
7. Chen, Y., Pont-Tuset, J., Montes, A., Van Gool, L.: Blazingly fast video object segmentation with pixel-wise metric learning. In: IEEE CVPR (2018)
8. Cheng, J., Tsai, Y.H., Hung, W.C., Wang, S., Yang, M.H.: Fast and accurate online video object segmentation via tracking parts. In: IEEE CVPR (2018)
9. Danelljan, M., Bhat, G., Khan, F.S., Felsberg, M.: ATOM: accurate tracking by overlap maximization. In: IEEE CVPR (2019)
10. Danelljan, M., Bhat, G., Shahbaz Khan, F., Felsberg, M.: ECO: efficient convolution operators for tracking. In: IEEE CVPR (2017)
11. Fan, H., et al.: LaSOT: a high-quality benchmark for large-scale single object tracking. In: IEEE CVPR (2019)
12. He, K., Zhang, X., Ren, S., Sun, J.: Deep residual learning for image recognition. In: IEEE CVPR (2016)
13. Henriques, J.F., Caseiro, R., Martins, P., Batista, J.: High-speed tracking with kernelized correlation filters. IEEE Trans. Pattern Anal. Mach. Intell. **37**(3), 583–596 (2014)
14. Hu, Y.-T., Huang, J.-B., Schwing, A.G.: VideoMatch: matching based video object segmentation. In: Ferrari, V., Hebert, M., Sminchisescu, C., Weiss, Y. (eds.) ECCV 2018. LNCS, vol. 11212, pp. 56–73. Springer, Cham (2018). https://doi.org/10.1007/978-3-030-01237-3_4
15. Huang, L., Zhao, X., Huang, K.: GOT-10k: a large high-diversity benchmark for generic object tracking in the wild. IEEE Trans. Pattern Anal. Mach. Intell. (2019)

16. Jiang, B., Luo, R., Mao, J., Xiao, T., Jiang, Y.: Acquisition of localization confidence for accurate object detection. In: Ferrari, V., Hebert, M., Sminchisescu, C., Weiss, Y. (eds.) ECCV 2018. LNCS, vol. 11218, pp. 816–832. Springer, Cham (2018). https://doi.org/10.1007/978-3-030-01264-9_48
17. Johnander, J., Danelljan, M., Brissman, E., Khan, F.S., Felsberg, M.: A generative appearance model for end-to-end video object segmentation. In: IEEE CVPR (2019)
18. Kingma, D.P., Ba, J.: Adam: a method for stochastic optimization. arXiv preprint arXiv:1412.6980 (2014)
19. Kristan, M., et al.: The sixth visual object tracking VOT2018 challenge results. In: ECCV (2018)
20. Kristan, M., Leonardis, A., Matas, J., et al.: VOT 2020 challenge. https://www.votchallenge.net/vot2020/
21. Li, B., Wu, W., Wang, Q., Zhang, F., Xing, J., Yan, J.: SiamRPN++: evolution of siamese visual tracking with very deep networks. In: IEEE CVPR (2019)
22. Li, B., Yan, J., Wu, W., Zhu, Z., Hu, X.: High performance visual tracking with siamese region proposal network. In: IEEE CVPR (2018)
23. Li, P., Chen, B., Ouyang, W., Wang, D., Yang, X., Lu, H.: GradNet: gradient-guided network for visual object tracking. In: IEEE ICCV (2019)
24. Lukežič, A., Matas, J., Kristan, M.: D3S-a discriminative single shot segmentation tracker (2020)
25. Maninis, K.K., et al.: Video object segmentation without temporal information. IEEE Trans. Pattern Anal. Mach. Intell. **41**(6), 1515–1530 (2018)
26. Nocedal, J., Wright, S.: Numerical Optimization. Springer, Heidelberg (2006). https://doi.org/10.1007/978-0-387-40065-5
27. Oh, S.W., Lee, J.Y., Xu, N., Kim, S.J.: Video object segmentation using space-time memory networks. In: IEEE ICCV (2019)
28. Paszke, A., et al.: Automatic differentiation in PyTorch (2017)
29. Perazzi, F., Khoreva, A., Benenson, R., Schiele, B., Sorkine-Hornung, A.: Learning video object segmentation from static images. In: IEEE CVPR (2017)
30. Perazzi, F., Pont-Tuset, J., McWilliams, B., Van Gool, L., Gross, M., Sorkine-Hornung, A.: A benchmark dataset and evaluation methodology for video object segmentation. In: IEEE CVPR (2016)
31. Robinson, A., Lawin, F.J., Danelljan, M., Khan, F.S., Felsberg, M.: Learning fast and robust target models for video object segmentation. arXiv preprint arXiv:2003.00908 (2020)
32. Ronneberger, O., Fischer, P., Brox, T.: U-net: convolutional networks for biomedical image segmentation. In: Navab, N., Hornegger, J., Wells, W.M., Frangi, A.F. (eds.) MICCAI 2015. LNCS, vol. 9351, pp. 234–241. Springer, Cham (2015). https://doi.org/10.1007/978-3-319-24574-4_28
33. Ros, G., Ramos, S., Granados, M., Bakhtiary, A., Vázquez, D., Lopez, A.M.: Vision-based offline-online perception paradigm for autonomous driving. In: IEEE WACV (2015)
34. Valmadre, J., Bertinetto, L., Henriques, J., Vedaldi, A., Torr, P.H.: End-to-end representation learning for correlation filter based tracking. In: IEEE CVPR (2017)
35. Voigtlaender, P., Chai, Y., Schroff, F., Adam, H., Leibe, B., Chen, L.C.: FEELVOS: fast end-to-end embedding learning for video object segmentation. In: IEEE CVPR (2019)

36. Vondrick, C., Shrivastava, A., Fathi, A., Guadarrama, S., Murphy, K.: Tracking emerges by colorizing videos. In: Ferrari, V., Hebert, M., Sminchisescu, C., Weiss, Y. (eds.) ECCV 2018. LNCS, vol. 11217, pp. 402–419. Springer, Cham (2018). https://doi.org/10.1007/978-3-030-01261-8_24
37. Wang, Q., Zhang, L., Bertinetto, L., Hu, W., Torr, P.H.: Fast online object tracking and segmentation: a unifying approach. In: IEEE CVPR (2019)
38. Wang, Z., Xu, J., Liu, L., Zhu, F., Shao, L.: RANet: ranking attention network for fast video object segmentation. In: IEEE ICCV (2019)
39. Wu, Y., Lim, J., Yang, M.H.: Object tracking benchmark. IEEE Trans. Pattern Anal. Mach. Intell. **37**(9), 1834–1848 (2015)
40. Wug Oh, S., Lee, J.Y., Sunkavalli, K., Joo Kim, S.: Fast video object segmentation by reference-guided mask propagation. In: IEEE CVPR (2018)
41. Xu, N., et al.: YouTube-VOS: sequence-to-sequence video object segmentation. In: Ferrari, V., Hebert, M., Sminchisescu, C., Weiss, Y. (eds.) ECCV 2018. LNCS, vol. 11209, pp. 603–619. Springer, Cham (2018). https://doi.org/10.1007/978-3-030-01228-1_36
42. Xu, N., et al.: YouTube-VOS: a large-scale video object segmentation benchmark. arXiv preprint arXiv:1809.03327 (2018)
43. Xu, Y., Wang, Z., Li, Z., Ye, Y., Yu, G.: SiamFC++: towards robust and accurate visual tracking with target estimation guidelines. In: AAAI (2020)
44. Yu, C., Wang, J., Peng, C., Gao, C., Yu, G., Sang, N.: Learning a discriminative feature network for semantic segmentation. In: IEEE CVPR (2018)
45. Yu, J., Blaschko, M.B.: The lovász hinge: a novel convex surrogate for submodular losses. IEEE Trans. Pattern Anal. Mach. Intell. **42**(3), 735–748 (2020)
46. Zhang, L., Gonzalez-Garcia, A., Weijer, J.V.D., Danelljan, M., Khan, F.S.: Learning the model update for siamese trackers. In: IEEE ICCV (2019)
47. Zhang, Z., Peng, H.: Deeper and wider siamese networks for real-time visual tracking. In: IEEE CVPR (2019)
48. Zhu, Z., Wang, Q., Li, B., Wu, W., Yan, J., Hu, W.: Distractor-aware siamese networks for visual object tracking. In: Ferrari, V., Hebert, M., Sminchisescu, C., Weiss, Y. (eds.) ECCV 2018. LNCS, vol. 11213, pp. 103–119. Springer, Cham (2018). https://doi.org/10.1007/978-3-030-01240-3_7

Cascaded Tracking via Pyramid Dense Capsules

Ding Ma and Xiangqian Wu[✉]

School of Computer Science and Technology, Harbin Institute of Technology, Harbin 150001, China
{madingcs,xqwu}@hit.edu.cn

Abstract. The tracking-by-detection is a two-stage framework including, collecting the candidates around the target object and classifying each candidate as the target object or as background. Despite Convolutional Neural Networks (CNNs) based methods have been successful in tracking-by-detection framework, the own set of flaws of CNNs will still affect the performance. The underlying mechanism of CNNs that are based on the positional invariance (i.e., lose the spatial relationships between features) cannot capture the small affine transformations. This would ultimately result in drift. To solve this problem, we dig into spatial relationships endowed by the Capsule Networks (CapsNets) for tracking-by-detection framework. To strengthen the encoded power of convolutional capsules, we generate the convolutional capsules through a pyramid dense capsules (PDCaps) architecture. Our pyramid dense capsule representation is useful in producing comprehensive spatial relationships within the input. Besides, the critical challenges in the tracking-by-detection framework are how to avoid overfitting and mismatch during training and inference, where a reasonable intersection over union (IoU) threshold that defines the true/false positives is hard to set. To address the issue of the IoU threshold setting, a cascaded PDCaps model is proposed to improve the quality of candidates, and it consists of a sequential PDCaps model trained with increasing IoU thresholds to improve the quality of candidates sequentially. Extensive experiments demonstrate that our tracker performs favorably against state-of-the-art approaches.

Keywords: Visual tracking · Pyramid and dense capsules · Cascaded architecture

1 Introduction

Visual tracking plays a fundamental role in computer vision. Inspired by the great success of convolutional neural networks (CNNs), some tracking-by-detection trackers [14,16,17,23,26,30] achieve high performance with the robust feature representation in CNNs. Although the CNNs have been demonstrated to improve the expected outcome, there remain inherent flaws in CNNs, which will affect the accuracy of locating the target. 1) The predicted result of CNNs

A. Bartoli and A. Fusiello (Eds.): ECCV 2020 Workshops, LNCS 12539, pp. 683–696, 2020.
https://doi.org/10.1007/978-3-030-68238-5_45

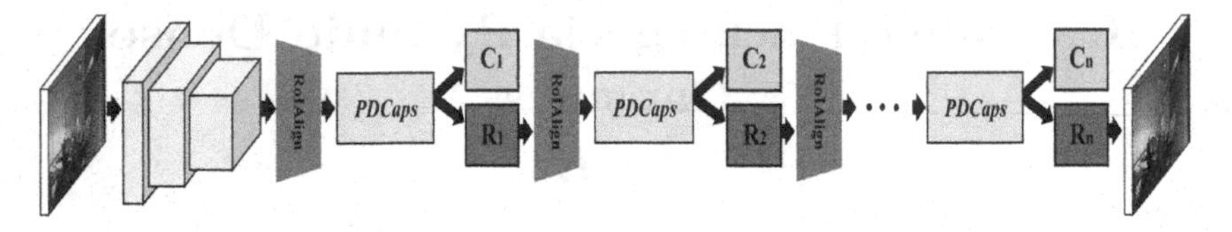

Fig. 1. An overview of the proposed *TCPDCaps*. The *TCPDCaps* contains multi-stage *PDCaps* which shares the same architecture. The optimal state of the target is determined by the classification and regression heads (i.e., $C*$ and $R*$).

is sensitive to a slight movement of the object. 2) Scalar neurons of CNNs pay more attention to the presence of features, rather than the spatial relationships within features. To solve these problems, Sabour et al. [29] proposed the idea of *CapsNets*, which models the spatial relationships with capsules and dynamic routing. Different from the scalar neurons in CNNs, the capsules utilize vector to store information (including magnitude, orientation, and so on) at the neuron level. After that, capsules in different layers are communicated with a dynamic routing algorithm that considers the agreement between these capsules, thus forming meaningful spatial relationships not found in original CNNs.

We believe that *CapsNets* is more suitable for the tracking-by-detection framework. 1) *CapsNets* is based on the principle of positional equivariance whereas, CNNs are based on the principal of positional invariance of the features. CNNs lose the relationships between features, especially if using fully-connected layers. Each capsule encodes the properties of the feature it represents, enables *CapsNets* more robust to small affine transformations, which is necessary to separate the target from distractors. 2) tracking-by-detection framework requires that the classifier learn the key features occurring in the target and use them to classify the target from the background. *CapsNets* have achieved state-of-the-art performance for classification, which indicates that *CapsNets* can extract discriminative features that are better than captured by CNNs.

To these end, we take advantage of the recent progress in *CapsNets* to generate multi-scaled spatial relationships to facilitate the tracking-by-detection framework. Although *CapsNets* achieves state-of-the-art accuracy on the MNIST dataset, their performance on more complex datasets such as CIFAR-10 is unsatisfactory. Indeed, towards accurate image classification, the classifier relies on an appropriate representation of capsules. We notice that the standard *CapsNets* may lack of a well-suited strategy to generate richer capsules. For details, the standard *CapsNets* consists of one convolution layer and two capsule layers. The convolution layer acts as a feature extractor. Then, the output of the feature extractor is fed into the first capsule layer (i.e. primary capsules), where 32×8D capsules are generated using 9×9 kernels with stride 2. In practice, the information for forming such capsules is limited in a fixed region, which is not enough to encode continuous spatial relationships within features. Ideally, the capsules should cover the attributes of the input at different scales. Motivated by the above observations, we propose a novel capsule architecture, referred to as *Pyramid Dense Capsules*

(*PDCaps*), to learn the pyramid dense capsule representation from multi-scaled features and achieve competitive results on CIFAR-10 (see ablation stdies). In our *PDCaps*, the primary capsules are generated in a pyramid dense way. By employing such a strategy, on the one hand, capsules generated from small-scaled features, which capture local spatial relationships; on the other hand, the global spatial relationships are extracted by large-scaled features. Furthermore, these capsules are inter-connected by dense-connection to complement each other.

Moreover, a crucial problem in the tracking-by-detection framework is that the quality of candidates depends on the appropriate IoU threshold. In [14,23,30], the positive candidates are drawn by $0.7 \leq IoU \leq 1$, and the negative candidates are collected by $0 \leq IoU \leq 0.5$. The candidates are selected by $IoU \geq 0.6$ during the inference time. Such settings will result in two problems. First, the example intensive neural networks will make the training stage prone to overfitting. Second, the quality of the detector and the candidates at the inference stage results in a mismatch. In fact, to produce a higher quality detector, it does not suffice to increase the IoU threshold during training, which will result in overfitting [5]. Precise detection requires candidates that match the detector quality. So we solve this problem by designing a multi-stage extension of the proposed *PDCaps*, where we gradually increase the IoU threshold for *PDCaps*. The cascaded *PDCaps* is trained sequentially, using the output (closer match candidates) of one stage to train the next. Here, each stage aims to find a good set of candidates for training the next stage (see Fig. 1).

We summarize the main contributions of this work as follows:

I. We propose a pyramid dense capsule model (*PDCaps*) to capture the multi-scaled spatial relationships within the target.
II. We propose a cascaded *PDCaps* to further improve the tracking performance with gradually improve the quality of candidates during training and inference time.
III. Quantitative and qualitative evaluations demonstrate the outstanding performance of our proposed tracker on OTB100 [35], TC128 [20], UAV123 [22], LaSOT [13] and VOT2018 [1] benchmarks.

The rest of the paper is organized as follows. We first review the related work in Sect. 2. The detailed configuration of the proposed algorithm is described in Sect. 3. We present the overall tracking algorithm in Sect. 4. Section 5 illustrates the experimental results on five large tracking benchmarks. Finally, conclusions and future work are drawn in Sect. 6.

2 Related Work

2.1 Tracking-by-Detection

The tracking-by-detection framework consists of a sampling stage and a classification stage. In practice, a large number of candidates are drawn in the sampling stage. The purpose of the classification stage is to compute the positive probability of each candidate. By the outstanding representation power of CNNs, many trackers have been proposed [17,23,26,30]. Nam et al. [23] propose to learn the feature and classifier jointly, which is referred to as MDNet. MDNet is composed of shared layers and domain-specific layers, where each domain is formulated as a binary classification to identify the target object from the background. To solve the problem of appearance variations and class imbalance in MDNet, Song et al. [30] use adversarial learning to obtain the most robust features of the objects over a long period and propose a *high-order cost sensitive loss* to decrease the effect of class imbalance. Inspire by Fast R-CNN, Jung et al. [17] introduce a tracker named RTMDNet to speed up the MDNet by improved RoIAlign technology. Nevertheless, RTMDNet takes little advantage of ROI pooling [15] as it cannot encode the difference between highly spatial overlapped candidates. Shi et al. [26] facilitate attention maps by selectively paying attention to robust temporal features. Despite the success of these methods, the sampling strategy can be challenging to solve accurately. This is partly because of overfitting/quality mismatch during training/inference. Since the model is easy to overfitting by a higher IoU training strategy. Besides, there is a mismatch between the quality of the detector and that of the candidates at inference time, when ground truth is unavailable. In this paper, we propose a cascaded architecture that overcomes the overfitting and quality mismatch to enable high-quality object tracking.

2.2 Some Variants of *CapsNets*

Researchers have introduced some variations of *CapsNets* for different vision tasks. Duarte et al. [11] propose a 3D capsule network that can jointly perform pixel-wise action segmentation along with action classification. Duarte et al. [12] design a novel attention-based EM routing algorithm to condition capsules for video segmentation. However, such methods ignore the basis for generating the capsules. Since the convolutional capsules are generated by convolution operation with a fixed kernel size, thus the single scaled spatial relationships are not enough to cope with the complex distributions.

3 *PDCaps* and Cascaded Learning

3.1 Network Architecture

Figure 1 shows the overall architecture of our network. The whole network consists of a feature extractor and multi-stage models. In each stage, given the RoI features, the *PDCaps* model is designed to generate comprehensive spatial relationships. Subsequently, we feed the features of *PDCaps* into two heads for classifying and regressing each proposal bounding box.

3.2 The Proposed *PDCaps*

Before introducing the designed architecture, we review the basics of *CapsNets*, and highlight their potential drawbacks, which leads to the proposed *PDCaps*.

The Original *CapsNets*. First, one convolutional layer is used to generate feature maps. Then, the feature maps are fed into a convolutional capsule layer, which transforms the feature maps to initial capsules (i.e., primary capsules). After that, another convolutional capsule layer is applied, which forms final capsules (i.e., digital capsules). The communication between these two layers is an iterative *Dynamic Rounting* [29] algorithm. Although *CapsNets* achieves competitive results on digital dataset, the accuracy on the more complex dataset is unsatisfactory. We analyze the observations on this occasion and propose *Pyramid Dense Capsules* (*PDCaps*).

Observation 1. The primary capsules are generated by the fixed receptive fields, and different objects may exceed the receptive field and cause discontinuous spatial relationships which increase the chance of misclassification for image classification. Besides, different objects may have different sizes in a complex scene, which is more suitable for analysis using the multi-scaled(pyramid) method. On such occasions, the convolutional capsule should pay more attention to different scaled regions enriching the continuous spatial relationships.

Observation 2. In [25], it has proven that the former representation is used as an input to the next capsule units, which can improve the accuracy. It implies that the convolutional capsule layer shares the same characteristics as the CNNs, which can capture richer representation as the architecture goes deeper. Along these lines, the connections from former capsule layers to subsequent capsule layers can further improve the information flow between capsules.

Observation 3. The convolutional capsules generated from differently scaled feature maps encode various spatial relationships, these spatial relationships are complementary. With this compensation, the capsules can gather much more comprehensive spatial relationships. So, there should be an dense-connection between any different scaled capsules to enrich the spatial relationships.

With the above observations, we introduce the idea of *PDCaps*, which form different scaled spatial relationships in a pyramid and dense way. The architecture of *PDCaps* is shown in Fig. 2. The number of channels and the size of feature maps are shown below the name of each layer. As illustrated in Fig. 2, given the features ($512 \times 3 \times 3$) of RoIAlign layer, we split the features into three groups (i.e., $128 \times 3 \times 3$, $128 \times 3 \times 3$ and $256 \times 3 \times 3$).

At the beginning of each *Level*, variously scaled primary capsules are generated (orange boxes). It aims to seed representations on all scales. The output of the three groups forms our first set of capsules, where we have three single capsule types with a grid of 1×1, 2×2, and 3×3 capsules, each of which contains a 64, 128 and 256-dimensional vector, respectively. Depending on the first set of capsules, multiple convolutional capsules are generated by different scaled capsules, which focus on different sub-regions of the feature maps. There are two-layer convolutional capsules in each *Caps_ * _**, which is denoted as (*in_num_caps*, *out_num_caps*, *channels*, *kernel*, *stride*). The parameters in the first layer are set as $(1, 4, channels, 1, 1)$, and the second layer is designed as $(4, 1, channels, 1, 1)$.

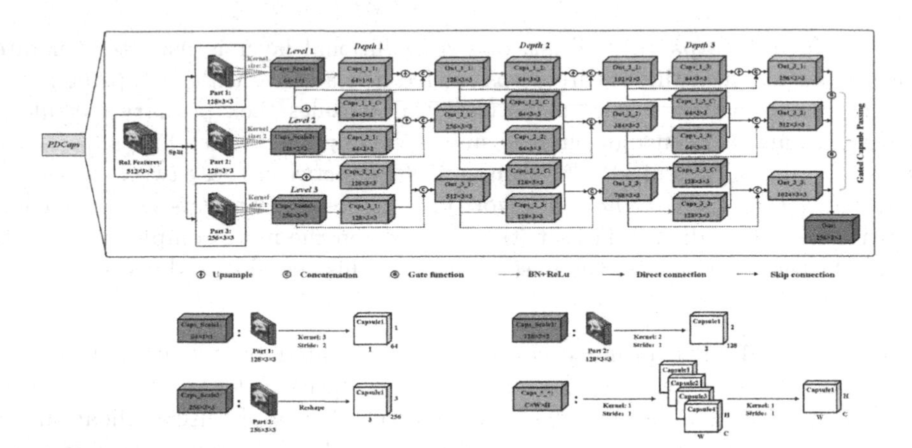

Fig. 2. The detailed framework of our proposed *PDCaps*. First, RoIAlign is used to extract RoI pooled features ($512 \times 3 \times 3$). After that, we generate multi-scaled primary capsules (orange boxes) within three pyramid level. The multi-scaled convolutional capsules are generated and expanded in a dense-connected way (gray boxes). Their integration (green boxes) is performed by a gated capsule passing module, where the gate function is employed to control the capsule passing rate. The integrated capsules (red boxes) are used for classification and bounding box regression. The detailed architecture of *Caps_Scale1*, *Caps_Scale2*, *Caps_Scale3*, and *Caps_*_** are shown below. (Color figure online)

Then, the output of capsules *Out_*_** produced at subsequent layers is transformed capsules from all previous capsules *Caps_*_*_C* with the concatenation operation. The dense-connection between each *Level* and *Depth* enriches the representation of spatial relationships. In this way, we create multiple sets of convolutional capsules at multi-*Depth*. Besides, we add the BN and ReLu operations between each *Out_*_** and *Caps_*_** for later *Depth* to improve nonlinearity.

For the final process of spatial relationships transfer, a decision should be made about whether the relationships of the current *Level* are useful for the next *Level* (the green boxes in Fig. 2). Inspired by [37], a gate function is designed as convolutional layers with sigmoid activation, which is used to control the relationships passing. The gated capsule passing module are expressed as:

$$Caps_{Level(i+1)} = G(Caps_{Level(i)}) \otimes Caps_{Level(i)} + Caps_{Level(i+1)} \tag{1}$$

where $\otimes$ is the element-wise product. $G(\cdot)$ is the gate function to control the relationships passing, which is defined as:

$$G(Caps) = Sig(Conv(Caps)) \tag{2}$$

where $Sig(\cdot)$ is the element-wise sigmoid function, $Conv(\cdot)$ is a 1×1 convolutional layer having the same number of channels with *Caps*. By adding the gate function into the *PDCaps*, only useful relationships are passed between different *Levels*,

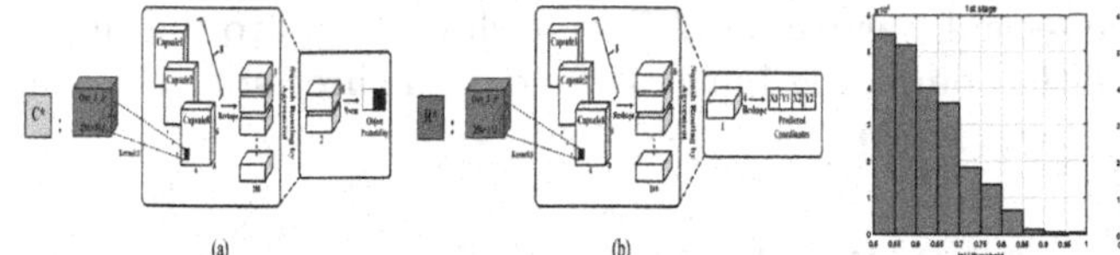

Fig. 3. (a) and (b) are the structures of the classification and regression head, respectively.

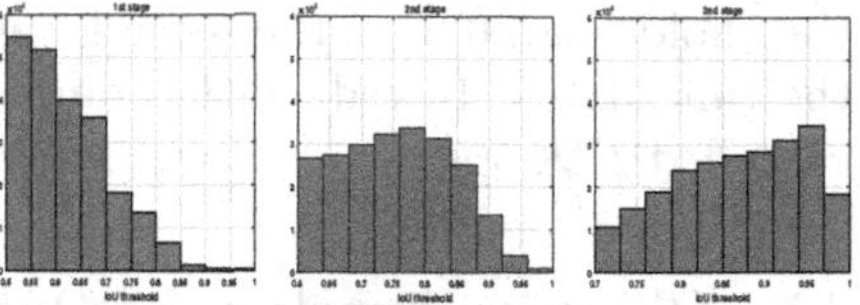

Fig. 4. The positive distributions for the corresponding IoU threshold within the three stages on ImageNet-Vid dataset.

and the useless relationships are prevented. Followed by the *PDCaps*, we design two heads to classify and regress each candidate (Fig. 4).

3.3 Classification and Regression Heads

Classification Head. Inspired by the multi-domain learning [23], which uses domain-specific layers to learn the representations distinguishing between target and background, we propose multi-branches with full capsules architecture to separate the target from the background. As shown in Fig. 3(a), each branch in the classification head has two-layer capsules which communicate with the *Dynamic Rounting* [29]. The output of the classification head is a 2D binary classification score, which is given to a loss function for determining whether a bounding box is a target or a background patch in the current domain.

Regression Head. We design a generic regression model to predict the coordinates of the optimal bounding box through all videos. The regression model shares the same architecture with the classification head (Fig. 3(b)). The output of the regression head model is a 4D coordinates (i.e., the top/down coordinates of the bounding box).

3.4 Cascaded Classification and Regression

To generate closer match candidates for higher performance detection, we propose a multi-stage extension of the proposed *PDCaps*. The cascade of *PDCaps* stages is trained sequentially, using the output of one stage to train the next. Thus, we decompose the classification and regression task into a sequence to improve the tracking performance. The cascaded *PDCaps* can be expressed as:

$$\mathcal{C}\&\mathcal{R}(x, \mathbf{B}) = \mathcal{C}\&\mathcal{R}_S \leftarrow \mathcal{C}\&\mathcal{R}_{S-1} \leftarrow \cdots \leftarrow \mathcal{C}\&\mathcal{R}_1(x, \mathbf{B}) \tag{3}$$

where x is the image patch. S indicates the number of cascaded stages. $\mathcal{C}\&\mathcal{R}_S(x, \mathbf{B})$ is the classifier and regressor at stage S with the corresponding candidate distribution $\mathbf{B}^s$ at the current stage. Here, the bounding box classifier and regressor are trained for a certain *IoU* threshold, which tends to produce a set of bounding boxes of higher *IoU*. So the classifier and regressor are starting from a lower *IoU* threshold (e.g., 0.5), and successively increases the *IoU* threshold (e.g., 0.6 for the second stage and 0.7 for the third stage) to re-sample the candidates. To reduce the computing burden, we design three stages in the cascaded *PDCaps*.

3.5 Objective Function for Optimizing

The cascaded *PDCaps* is trained end-to-end using an objective function which is the sum of two losses: classification losses and regression losses in different stages. The classification loss for each capsule is expressed as:

$$\mathcal{L}_{cls} = \frac{\exp(v_k)}{\sum_k \exp(v_k)} \tag{4}$$

where the output of capsule $*$ is denoted as v_*. This function indicates that the loss value is high when an entity of class k is absent.

The network predicts the coordinates of the bounding box for the object, and the *GIoU* loss $\mathcal{L}_{GIoU}$ [27] is used to compute the regression loss. Then, the cost function to be minimized by:

$$\mathcal{L}_{total} = \mathcal{L}_{cls_{stage1\sim3}} + \lambda_1 \mathcal{L}_{GIoU_{stage1\sim3}} \tag{5}$$

where λ is used to balance the regression loss and classification loss. In all experiments, we set $\lambda_1 = 0.5$.

4 Tracking with the Cascaded *PDCaps*

4.1 Main Process of Tracking

Given a test frame t, and a set of candidate samples $\{cs_t^i\}_{i=1,\ldots,300}$ generated from a Gaussian distribution centered at the target state of the previous frame, the optimal target state through three stages is defined as:

$$cs_t^* = \arg\max_{cs_t^i} f^+_{stage_{1\sim3}}(cs_t^i) \tag{6}$$

where $f^+_{stage_{1\sim3}}(cs_t^i)$ indicates the positive score of the ith sample. Subsequently, we choose 5 outputs of the regression head that corresponds to the top 5 indexes in the classification head. In the end, the coordinates of the optimal state are obtained by taking an average operation on the 5 outputs of the regression head.

4.2 Model Training and Optimization Setting

Offline Training. For offline pre-training during each iteration, a minibatch is constructed with samples collected from each video. We randomly select 4 frames in each video and draw 64 positive and 192 negative samples from each frame. During each stage, we only use the positive samples to train the regression head. Each branch in the classification head model corresponds to each video. We train our models on ImageNet-Vid [28] with 4500 videos.

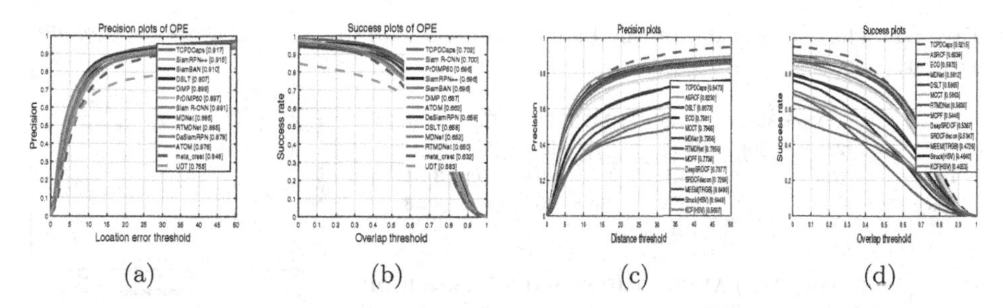

Fig. 5. (a) and (b) are the precision and success plots on OTB100, respectively. (c) and (d) are the precision and success plots on TC128, respectively.

Online Training. Given a test sequence, we fine-tune the pre-trained model with the first frame of each video. We draw 500 positive and 5000 negative samples based on the same IoU criteria with the offline training for each stage. From the second frame, the trackers gather 64 positive and 192 negative examples that have the same IoU criteria with the offline training. Here, we update our tracker online in an efficient way, which is almost identical to [23].

Optimization Setting. The input is resized to 107×107. For offline training, we train the network for 1100 epochs with a learning rate 0.0001. For online updates, the number of iterations for fine-tuning is 15, and the learning rate is set to 0.0003. The weight decay and momentum are fixed to 0.0005 and 0.9, respectively. The routing iteration is set to 3 by experiments. The proposed method is implemented in PyTorch, runs with NVIDIA Tesla K40c GPUs.

5 Experimental Results and Analysis

To validate our *TCPDCaps*, we first compare it with the state-of-the-art methods on 5 public benchmarks. In the OTB100 [35] dataset, we use the one-pass evaluation (OPE) with precision and success plots metrics. The precision and success criteria are also utilized in TC128 [20], UAV123 [22], and LaSOT [13] datasets. In the VOT2018 dataset [1], each tracker is measured by the metrics of Accuracy Ranks (A), Robustness Ranks (R), and Expected Average Overlap (EAO). Then, we explore the effectiveness of each part in the ablation studies.

5.1 Experiments on 5 Public Benchmarks

OTB100. We compare the proposed *TCPDCaps* tracker on OTB100 dataset [35] with the following recent published twelve trackers: SiamRPN++ [18], DSLT [21], DiMP [3], RTMDNet [17], MDNet [23], DaSiamRPN [38], ATOM [8], Siam R-CNN [33], PrDIMP50 [10], meta_crest [24], SiamBAN [6] and UDT [34]. The results are shown in Fig. 5. According to Fig. 5, the *TCPDCaps* tracker achieves competitive performance among the state-of-the-arts on this dataset. The values of the precision plot and the success plot are 0.917 and 0.702 on OTB100,

Table 1. The comparisons on the LaSOT and VOT2018 dataset.

	Ours	MDNet [23]	VITAL [30]	GradNet [19]	MLT [7]	SINT [32]	ECO [9]	SiamFC [2]
Prec	39.1	37.4	37.2	35.1	-	29.9	29.8	34.1
Succ	43.4	41.3	41.2	36.5	36.8	33.9	34.0	35.8

(a) LaSOT

	Ours	RCO [1]	UPDT [4]	LADCF [36]	DiMP [3]	SiamRPN [1]	SASiamR [1]	DRT [31]
EAO	0.437	0.376	0.378	0.389	0.440	0.383	0.337	0.356
A	0.61	0.51	0.54	0.50	0.59	0.59	0.57	0.52
R	0.15	0.16	0.18	0.16	0.15	0.28	0.26	0.20

(b) VOT2018

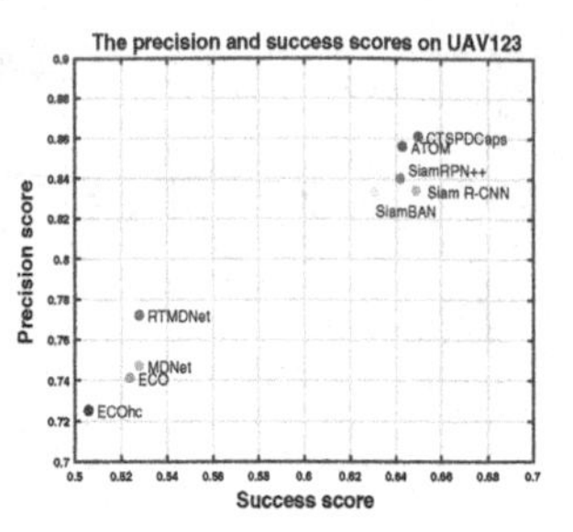

Fig. 6. The precision and success scores on UAV123.

respectively. Compared with the tracking-by-detection tracker, our *TCPDCaps* significantly outperforms the MDNet which is also pre-trained on ImageNet-Vid by reporting in [17].

TC128. For experiments on the TC128 [20] dataset containing 128 color videos, a comparison with 12 state-of-the-art trackers is shown in Fig. 5. Among the compared methods, our approach improves the precision score from 0.8473 of the state-of-the-art tracker ASRCF to 0.8230. The *TCPDCaps* tracker outperforms state-of-the-art approaches with an AUC score of 0.6215. The top rank verifies the robustness of the proposed *TCPDCaps*.

UAV123. In the UAV123 [22], *TCPDCaps* is compared with eight methods, including ECOhc, ECO, MDNet, RTMDNet, SiamBAN, SiamRPN++, ATOM, and Siam R-CNN. Figure 6 shows that our *TCPDCaps* performs better.

LaSOT. The LaSOT dataset [13] is a very large-scale dataset consisting of 1400 sequences with 70 categories. As shown in Table 1a, our tracker performs favorably against state-of-art trackers in both precision and success metrics.

VOT2018. We evaluate our *TCPDCaps* on VOT2018 [1] with 7 state-of-the-art trackers, including RCO, UPDT, LADCF, CPT, SiamRPN, SASiamR, and DRT. As illustrated in Table 1b, the *TCPDCaps* tracker achieves competitive results with higher ranking among all the compared trackers.

5.2 Ablation Study

To investigate the effectiveness of the components of *TCPDCaps*, we conducted various variants of *TCPDCaps* and evaluated them using OTB100 [35].

Ablation on the Pyramid and Dense Architecture. We explore the pyramid and dense architecture in *PDCaps*. We show the success rate and its corresponding parameters in Table 2. As shown in Table 2, the darker the gray color, the higher the success rate. *PDCaps* with S(i.e., Scale)3_D(i.e., Depth)3(28 MB)

Table 2. PDCaps.

	D1	D2	D3
S1	59.8 (5.5)	61.1 (9.2)	61.6 (13)
S2	61.7 (9.2)	62.9 (16.7)	63.4 (24.3)
S3	64.3 (13)	65.1 (19.3)	66.2 (28)

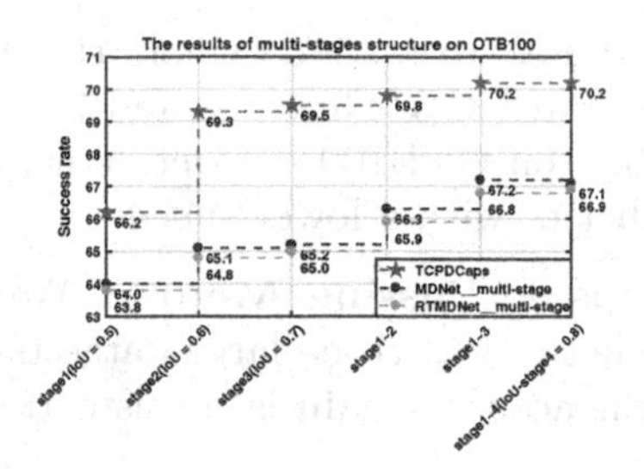

Fig. 7. Cascade architecture.

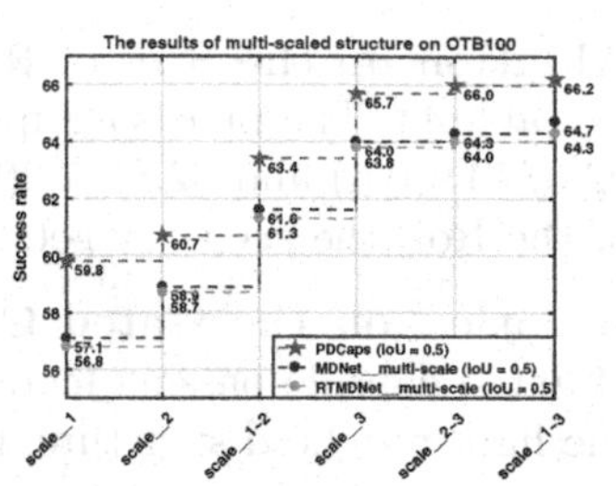

Fig. 8. Multi-scaled structure.

Table 3. Number of routing iterations.

Routing (Num)		1	2	3	4
OTB	Prec	90.0	91.2	91.7	91.6
	Succ	69.8	70.1	70.2	69.9
LaSOT	Prec	33.8	35.5	39.1	38.5
	Succ	39.7	40.8	43.4	42.9
VOT2018	A	0.51	0.55	0.61	0.59
	R	0.26	0.19	0.15	0.16

Table 4. Ablation study.

(a) Size of RoI features.

Feature Size	Prec	Succ	FPS
3 × 3	91.7	70.2	1
5 × 5	92.3	70.9	0.4
7 × 7	92.8	71.2	0.1

(b) Gated capsule passing module.

Gate	OTB100		LaSOT		VOT2018	
	Prec	Succ	Prec	Succ	A	R
w/o	91.5	70.0	38.8	43.1	61.1	15.1
w/	91.7	70.2	39.1	43.4	61.2	15.1

beats $S1\_D1$(5.5 MB) by more than 6%. It suggests that the pyramid dense capsules can gather more useful relationships to locate the target precisely. Besides, the accuracy of our pyramid dense module (constructed after the *Conv*1 [29]) with three *Level* and *Depth* is 94.83%, which outperforms the *CapsNets*(68.74%) and the 7 ensemble *CapsNet*(89.40%) [29] on CIFAR-10.

Ablation on the Performance in Different Stages. The red line of Fig. 7 shows the impact of the number of stages. $TCPDCaps_{stage_{1\sim *}}$ indicates the ensemble of * detectors with the $*^{th}$ stages. When $TCPDCaps$ equipped with $stage_{1\sim 3}$, the highest accuracy is achieved. We also notice that there is no more gain in accuracy when the *stage* goes to 4. An explanation is that when adding the 4th stage ($IoU = 0.8$), candidates are almost the same as the current detected target. Such false-positive candidates will make the model overfitting.

Ablation on the Number of Routing Iterations. Since the routing is one of the most distinctive features of the capsules, we conduct comprehensive experiments on the number of routing iterations in Table 3. The numbers of iterations is range from 1 to 4. The highest success rate is achieved when the iteration = 3. A possible explanation is that the overfitting has occurred as iteration = 4.

Ablation on the Size of RoI Features. We explore the size of RoI features in Table 4a. The prec/succ(fps) rates of 3×3, 5×5 and 7×7 are 91.7/70.2(1), 92.3/70.9(0.4) and 92.8/71.2(0.1) on OTB100 dataset. With the larger outputs of the RoI, the accuracy gets better with a lower speed.

Ablation on the Gated Capsule Passing Module. We explore the effect of gated capsule passing module through three large datasets in Table 4b. With the light-weighted structure, the accuracy gain is consistent.

Ablation on Capsule Architecture. We explore the necessity of capsule architecture from two aspects: the cascaded predictor and multi-scaled structure.

For the cascaded predictor, we construct a multi-stage MDNet and a multi-stage RTMDNet by multi-IoU thresholds, which is the same as *TCPDCaps*. For each IoU threshold, MDNet and RTMDNet are pre-trained independently. The MDNet and RTMDNet in each stage classify a candidate into a positive or a negative. The results of multi-stage MDNet, multi-stage RTMDNet and *TCPDCaps* are shown in Fig. 7. We notice that capsules perform better than the convolutional MDNet and RTMDNet. For the multi-scaled structure, we construct 3-scaled MDNet and RTMDNet depending on three groups of feature maps, which is the same as *PDCaps*. On each scale, it contains two fully-connected layers ($fc4-5$). The fusion of multi-scaled features (i.e., $fc5$) is implemented by concatenation operation. The results of multi-scaled MDNet, multi-scaled RTMDNet and *PDCaps* are shown in Fig. 8. Our *PDCaps* outperforms the 3-scaled MDNet and RTMDNet by a significant margin, which also proves the necessity of capsule architecture. So, it suggests that the capsules architecture is necessary for our method to achieve such a high performance.

6 Conclusions and Future Work

In this paper, we integrate the capsule into the tracking-by-detection framework. We first design a *PDCaps* model to gather comprehensive spatial relationships of the target. Then we introduce a gated capsule passing module, in which capsules from different levels adaptively pass useful relationships. The *PDCaps* model is trained in a cascaded way, which improves the quality of candidates during training and inference phases. The optimal state of the target is obtained by a classification head and a regression head. Experimental results on five benchmarks demonstrate that our tracker outperforms the state-of-the-art methods under different evaluation metrics.

Acknowledgments. This work was supported in part by the National Key R&D Program of China under Grant 2018YFC0832304, by the Natural Science Foundation of China under Grant 61672194, by the Distinguished Youth Science Foundation of Heilongjiang Province of China under Grant JC2018021, by the State Key Laboratory of Robotics and System (HIT) under Grant SKLRS-2019-KF-14, by the ZTE Industry-Academia-Research Cooperation Funds, by the Heilongjiang Touyan Innovation Team Program.

References

1. The sixth visual object tracking vot2018 challenge results (2018)
2. Bertinetto, L., Valmadre, J., Henriques, J.F., Vedaldi, A., Torr, P.H.S.: Fully-convolutional siamese networks for object tracking. In: Hua, G., Jégou, H. (eds.) ECCV 2016. LNCS, vol. 9914, pp. 850–865. Springer, Cham (2016). https://doi.org/10.1007/978-3-319-48881-3_56
3. Bhat, G., Danelljan, M., Van Gool, L., Timofte, R.: Learning discriminative model prediction for tracking (2019)
4. Bhat, G., Johnander, J., Danelljan, M., Khan, F.S., Felsberg, M.: Unveiling the power of deep tracking (2018)
5. Cai, Z., Vasconcelos, N.: Cascade R-CNN: delving into high quality object detection. arXiv preprint arXiv:1712.00726 (2017)
6. Chen, Z., Zhong, B., Li, G., Zhang, S., Ji, R.: Siamese box adaptive network for visual tracking. In: CVPR (2020)
7. Choi, J., Kwon, J., Lee, K.M.: Deep meta learning for real-time target-aware visual tracking. In: ICCV (2019)
8. Danelljan, M., Bhat, G., Khan, F.S., Felsberg, M.: ATOM: accurate tracking by overlap maximization. In: CVPR (2019)
9. Danelljan, M., Bhat, G., Shahbaz Khan, F., Felsberg, M.: ECO: efficient convolution operators for tracking. In: CVPR (2017)
10. Danelljan, M., Gool, L.V., Timofte, R.: Probabilistic regression for visual tracking. In: CVPR (2020)
11. Duarte, K., Rawat, Y., Shah, M.: VideoCapsuleNet: a simplified network for action detection. In: NIPS (2018)
12. Duarte, K., Rawat, Y.S., Shah, M.: CapsuleVOS: semi-supervised video object segmentation using capsule routing (2019)
13. Fan, H., et al.: LaSOT: a high-quality benchmark for large-scale single object tracking (2019)
14. Fan, H., Ling, H.: SANet: structure-aware network for visual tracking. In: CVPRW, pp. 2217–2224 (2017)
15. Girshick, R.: Fast R-CNN. In: ICCV (2015)
16. Hong, S., You, T., Kwak, S., Han, B.: Online tracking by learning discriminative saliency map with convolutional neural network. In: ICML (2015)
17. Jung, I., Son, J., Baek, M., Han, B.: Real-time MDNet (2018)
18. Li, B., Wu, W., Wang, Q., Zhang, F., Xing, J., Yan, J.: SiamRPN++: evolution of siamese visual tracking with very deep networks (2018)
19. Li, P., Chen, B., Ouyang, W., Wang, D., Yang, X., Lu, H.: GradNet: gradient-guided network for visual object tracking. In: ICCV (2019)
20. Liang, P., Blasch, E., Ling, H.: Encoding color information for visual tracking: algorithms and benchmark. TIP **24**, 5630–5644 (2015)
21. Lu, X., Ma, C., Ni, B., Yang, X., Reid, I., Yang, M.H.: Deep regression tracking with shrinkage loss. In: ECCV, pp. 353–369 (2018)
22. Mueller, M., Smith, N., Ghanem, B.: A benchmark and simulator for UAV tracking. In: Leibe, B., Matas, J., Sebe, N., Welling, M. (eds.) ECCV 2016. LNCS, vol. 9905, pp. 445–461. Springer, Cham (2016). https://doi.org/10.1007/978-3-319-46448-0_27
23. Nam, H., Han, B.: Learning multi-domain convolutional neural networks for visual tracking. In: CVPR, pp. 4293–4302 (2016)

24. Park, E., Berg, A.C.: Meta-tracker: fast and robust online adaptation for visual object trackers. arXiv (2018)
25. Phaye, S.S.R., Sikka, A., Dhall, A., Bathula, D.R.: Multi-level dense capsule networks. In: Jawahar, C.V., Li, H., Mori, G., Schindler, K. (eds.) ACCV 2018. LNCS, vol. 11365, pp. 577–592. Springer, Cham (2019). https://doi.org/10.1007/978-3-030-20873-8_37
26. Pu, S., Song, Y., Ma, C., Zhang, H., Yang, M.H.: Deep attentive tracking via reciprocative learning. In: NIPS (2018)
27. Rezatofighi, H., Tsoi, N., Gwak, J., Sadeghian, A., Reid, I., Savarese, S.: Generalized intersection over union: a metric and a loss for bounding box regression. In: CVPR (2019)
28. Russakovsky, O., et al.: Imagenet large scale visual recognition challenge. IJCV **115**(3), 211–252 (2015). https://doi.org/10.1007/s11263-015-0816-y
29. Sabour, S., Frosst, N., Hinton, G.E.: Dynamic routing between capsules (2017)
30. Song, Y., et al.: VITAL: visual tracking via adversarial learning. In: CVPR (2018)
31. Sun, C., Wang, D., Lu, H., Yang, M.H.: Correlation tracking via joint discrimination and reliability learning. In: CVPR (2018)
32. Tao, R., Gavves, E., Smeulders, A.W.M.: Siamese instance search for tracking, pp. 1420–1429 (2016)
33. Voigtlaender, P., Luiten, J., Torr, P.H.S., Leibe, B.: Siam R-CNN: visual tracking by re-detection (2019)
34. Wang, N., Song, Y., Ma, C., Zhou, W., Liu, W., Li, H.: Unsupervised deep tracking. In: CVPR (2019)
35. Wu, Y., Lim, J., Yang, M.H.: Object tracking benchmark. TPAMI **37**(9), 1834–1848 (2015)
36. Xu, T., Feng, Z.H., Wu, X.J., Kittler, J.: Learning adaptive discriminative correlation filters via temporal consistency preserving spatial feature selection for robust visual object tracking. IEEE Trans. Image Process. **28**, 5596–5609 (2019)
37. Zhang, L., Dai, J., Lu, H., He, Y., Wang, G.: A bi-directional message passing model for salient object detection. In: CVPR (2018)
38. Zhu, Z., Wang, Q., Li, B., Wu, W., Yan, J., Hu, W.: Distractor-aware siamese networks for visual object tracking. In: ECCV (2018)

W33 - Video Turing Test: Toward Human-Level Video Story Understanding

W33 - Video Turing Test: Toward Human-Level Video Story Understanding

The 2nd Workshop on Video Turing Test (VTT): Toward Human-Level Video Story Understanding was held fully virtually on 28th August 2020, in conjunction with the 16th ECCV. This workshop aims to encourage the research and development of human-like artificial intelligence by studying and discussing fundamental principles of how humans understand video stories. We invited six leading researchers to initiate a discussion on future challenges in data-driven video understanding. The subjects of the invited talks are as below:

- Towards Generating Stories about Video, Anna Rohrbach (UC Berkeley)
- Imagination Supervised Visiolinguistic Learning, Mohamed H. Elhoseiny (King Abdullah University of Science and Technology)
- Commonsense Intelligence: Cracking the Longstanding Challenge in AI, Yejin Choi (University of Washington)
- Reasoning about Complex Media from Weak Multi-modal Supervision, Adriana Kovashka (University of Pittsburgh)
- Machine Understanding of Social Situations, Makarand Tapaswi (Inria Paris)
- Ten Questions for a Theory of Vision, Marco Gori (University of Siena)

For paper submission, we provided both archival and non-archival tracks. Except for the non-archival long-paper track (published papers from previous conferences), all submissions were reviewed single-round and double-blind by the program committee. Among the total 13 submissions, 3 archival full papers, 4 non-archival short papers and 1 non-archival long paper were accepted at this workshop, and only the archival track papers are included in the proceedings.

We hosted the 2nd DramaQA challenge, which is regarded as a proxy task of the Turing Test for video story understanding. Among 9 teams, the top 3 teams were selected as winners and each received a $1,200 prize in cash.

We thank all of the invited speakers, all members of the program committee and all contributing authors for their work. Also, we thank all our sponsors, IITP, Ministry of Science and ICT in South Korea, and especially for the challenge, Naver and Kakao Brain. The 2nd VTT workshop would not have been such a great success without them.

August 2020

Yu-Jung Heo
Seongho Choi
Kyoung-Woon On
Minsu Lee
Vicente Ordóñez Román
Leonid Sigal
Chang Dong Yoo
Gunhee Kim
Marcello Pelill
Byoung-Tak Zhang

GCF-Net: Gated Clip Fusion Network for Video Action Recognition

Jenhao Hsiao(✉), Jiawei Chen, and Chiuman Ho

InnoPeak Technology, Palo Alto, CA, USA
{mark,jiawei.chen,chiuman}@innopeaktech.com

Abstract. In recent years, most of the accuracy gains for video action recognition have come from the newly designed CNN architectures (e.g., 3D-CNNs). These models are trained by applying a deep CNN on single clip of fixed temporal length. Since each video segment are processed by the 3D-CNN module separately, the corresponding clip descriptor is local and the inter-clip relationships are inherently implicit. Common method that directly averages the clip-level outputs as a video-level prediction is prone to fail due to the lack of mechanism that can extract and integrate relevant information to represent the video.

In this paper, we introduce the Gated Clip Fusion Network (GCF-Net) that can greatly boost the existing video action classifiers with the cost of a tiny computation overhead. The GCF-Net explicitly models the inter-dependencies between video clips to strengthen the receptive field of local clip descriptors. Furthermore, the importance of each clip to an action event is calculated and a relevant subset of clips is selected accordingly for a video-level analysis. On a large benchmark dataset (Kinetics-600), the proposed GCF-Net elevates the accuracy of existing action classifiers by 11.49% (based on central clip) and 3.67% (based on densely sampled clips) respectively.

Keywords: Video action recognition · 3D-CNNs · Dense slip sampling · Clip fusion

1 Introduction

The explosive growth in video and its applications has drawn considerable interest in the computer vision community and boost the need of high-level video understanding. Action recognition and localization in videos are the key research problems for high-level video understanding, where action recognition is to classify a video by assigning a pre-defined action classes, while action localization determines whether a video contains specific actions and also identifies temporal boundaries (e.g., start time and end time) of each action instance.

Video action recognition has witnessed much good progress in recent years. Most of the accuracy gains have come from the introduction of new powerful 3D-CNN architectures [1,2,13,22,24,29]. However, there are some drawbacks and

A. Bartoli and A. Fusiello (Eds.): ECCV 2020 Workshops, LNCS 12539, pp. 699–713, 2020.
https://doi.org/10.1007/978-3-030-68238-5_46

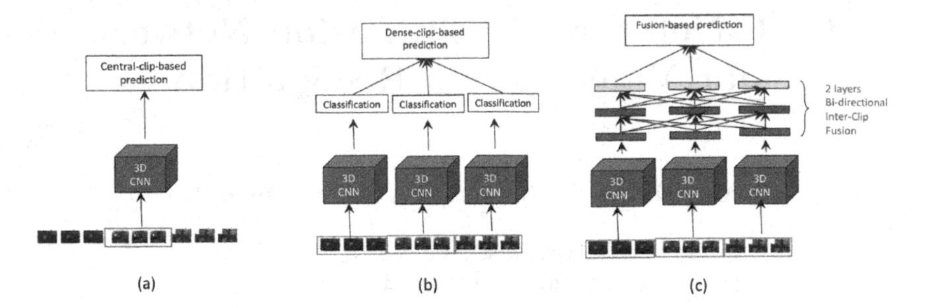

Fig. 1. Methods of video-level prediction: (a) central clip prediction; (b) dense clips prediction; (c) our bi-directional inter-clip fusion based prediction

limitations of these 3D-CNN based architectures. Firstly, most modern action recognition models are trained by applying a deep CNN on single clip of fixed temporal length (e.g., commonly 16 or 32 frames as a clip in existing works). This will become a performance bottleneck since the duration of different actions are variant and complex actions span multiple video segments (i.e., multiple clips can be involved). Single clip captures only very limited local temporal knowledge (e.g., may be either relevant or irrelevant to the target action), and it can hardly describe an action accurately from a global view. It is thus prone to fail due to the challenges in extracting global information.

Secondly, to achieve a video-level action prediction, most existing architectures adopt the following naive methods: central clip prediction or dense clip prediction. Central clip prediction (as shown in Fig. 1(a)) directly assume that the central clips (e.g., the central one or few clips) are the most related event, and average the action predictions of those clips to deliver a video-level action prediction. Dense clips prediction (as shown in Fig. 1(b)), on the other hand, aggregates the clip-level predictions over the entire video via a similar averaging scheme to model the temporal structure. Although these naive methods are easy to implement, they could somewhat hurt the action recognition accuracy. Videos in the real-world actually exhibit very different properties: they are often several minutes long, where brief relevant clips are often interleaved with segments of extended duration containing irrelevant information. Directly pooling information from all clips (or a few central clips) without consideration of their relevance will cause poor video-level prediction, as irrelevant clips may dominate the decision process.

Lastly, to detect an action's temporal location in a video, a strong supervision, in the form of training videos that have been manually collected, labeled and annotated, will be heavily relied on. However, collecting large amounts of accurately annotated action videos is already very expensive. Acquiring a large amount of labeled video data with temporal boundary information will be prohibitively difficult.

In this paper, we introduce the Gated Clip Fusion Network (GCF-Net) that uses a more sophisticated method to extract and integrate the information from each clip, and greatly boost the existing video action classifier with the cost of a tiny computation overhead. The proposed GCF-Net explicitly models the inter-dependencies between video clips and captures the clip-wise importance to increase the deep network's sensitivity to informative features across video segments, which turn out to deliver a more accurate recognition due to the use of a more comprehensive video-level feature. As a by-product, the enhanced prediction can be unsupervisedly back-propagated to generate an estimated spatio-temporal map that localizes possible actions in videos. An overview of our algorithm is shown in Fig. 2.

The contributions of this paper are summarized as below.

- We present a novel Bi-directional Inter-Clip Fusion method (Fig. 1(c)) that is able to utilize both short- and long-range video segments to model the inter-clip relationships and generate better clip representations. Comparing to traditional methods that mainly rely on 3D-CNN to separately generate a local feature with very limited local temporal knowledge, our method provides a better clip representation with broader receptive field.
- A Gated Clip-Wise Attention is proposed as the means to further suppress irrelevant clips for improving the video-level prediction accuracy. As a by-product, the attention weights generated by this module can be used to locate the time interval of an action event in a video (e.g., based on relevant clips).
- We demonstrate that the proposed GCF-Net, which models the inter-clip relationships and clip-wise importance in a much finer granularity, yields significant gain in video action recognition accuracy comparing to traditional methods that conduct analysis on all clips (e.g., dense sampling) or randomly/centrally selected clips. On a large benchmark video dataset (Kinetics-600), our method elevates the accuracy of an already state-of-the-art action classifier by by 11.49% (based on central clip) and 3.67% (based on densely clips sampling) respectively with the same amount of training data set and backbone network.

The rest of this paper is organized as follows. We discuss the related work in Sect. 2 and describe our action GCF-Net in Sect. 3. Section 4 presents the details of our experiment and Sect. 5 concludes this paper.

2 Related Work

The goal of video action recognition aims to identify a single or multiple actions per video, while action localization further attempts to determine the action intervals in a video. In recent years, most of the accuracy gains for video action recognition have come from the introduction of new powerful architectures.

Before the success of CNNs, hand-designed video features [3,12,15] was the mainstream approach and methods on improved dense trajectories [26] presented good performance. When it comes to the era of deep learning, convolutional neural networks have delivered a major paradigm shift and have been widely used to learn video features and classify video in an end-to-end manner. Two-stream method [20] is one of the popular frameworks that integrates spatial and temporal information via 2D network. 3D ConvNets [22,24] extend aforementioned 2D image models to the spatio-temporal domain, handling both spatial and temporal dimensions in a similar convolution way. A combination of two-stream networks and 3D convolutions, known as I3D [1], was proposed as a generic video representation learning method. There are also methods focusing on decomposing the convolutions into separate 2D spatial and 1D temporal filters [2,29], while methods in [5,6,27] model long-term filtering and pooling using different strategies (e.g., temporal strides, slow and fast network, self-attention, and etc.). In [13], the authors try to achieve the performance of 3D CNN but maintain 2D CNN's complexity.

For action localization, the objective is to localize the temporal start and end of each action within a given untrimmed video and to recognize the action class. Most of the existing methods [4,9,19,23] hugely rely on large-scale annotated video data, and achieve the detection goal through a two-step mechanism: an action proposal method firstly identifies candidate action segments, and then a classifier validates the class of each candidate and refines its temporal boundaries.

However, the networks proposed by the above methods mainly focused on the design of convolution network architecture, and are trained by single clip (e.g., gradients are updated based on one clip point of view), where irrelevant video segments could lead the gradient to the wrong direction and thus disrupt training performance. During inference, clips are sampled densely or at random. This naive averaging strategy can hardly model a complex action event that spans multiple video segments, and will hurt the recognition accuracy since, again, irrelevant clips could negatively affect the prediction decision.

Another branch of methods is the so-called learning-to-skip approach [8,18,30], where they skip segments by leveraging past observations to predict which future frames to consider next. In [11], the authors proposed a salient clip sampling strategy that prevents uninformative clips from joining the video-level prediction. However, the above methods merely skip or sample clips in videos, and didn't fully utilize the inter-clip relationships to model the temporal relationship of an action. The accuracy improvement can thus be limited.

3 Method

We claim that a successful action recognition network should consider both the inter-clip dependencies and clip-wise importance. Inter-clip dependencies expand the view of local clip descriptor to model a complex and multi-segment action event, while clip-wise importance identifies a set of key segments presenting important action component so that irrelevant clips can be prevented from joining the video-level prediction. To achieve this goal, our proposed GCF-Net, which is an end-to-end deep neural network, introduces two key modules: Bi-directional

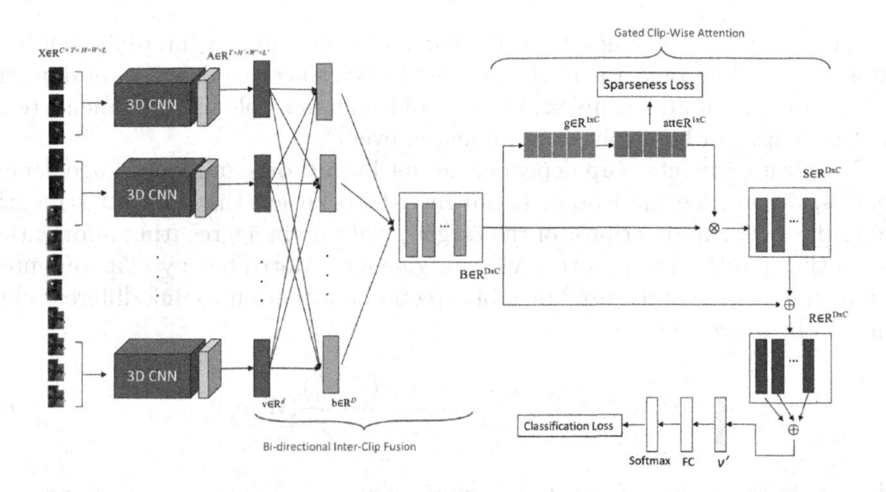

Fig. 2. The proposed GCF-Net.

Inter-Clip Fusion module and Gated Clip-Wise Attention module. We describe each step of the proposed GCF-Net in the rest of this section.

3.1 Gated Clip Fusion Network

Backbone Network. Figure 2 shows an overview of the proposed framework. For the proposed framework, clips $X = \{x_1, x_2, \ldots, x_C\}$ are set as the inputs and each clip x contains T stacked frames. Each input clip will be firstly processed by 3D-CNN, which contain a set of 3D convolutional layers, to extract corresponding clip features. The input shape for one batch data is $C \times T \times H \times W \times L$, where C denotes the number of clips, T frames are stacked together with height H and width W, and the channel number L is 3 for RGB images. The convolutional kernel for each 3D convolutional layer in 3D-CNN is in 3 dimensions. Then for each 3D convolutional layer, data will be computed among the three dimensions simultaneously.

To maintain the temporal accuracy of action localization, no strides will be applied along the T dimension during the 3D convolution. The feature map in the last convolution layer, denoted as A, is thus having dimension of $T \times H' \times W' \times L'$, where H' and W' are the height and width of feature maps, and L' is the number of convolution filters in the last convolution layer. Global pooling is then used to pool the content of feature maps and generate a summary vector of single clip. Outputs of 3D-CNN step are thus a set of raw clip descriptors $V = \{v_1, v_2, \ldots, v_C\}$, where $v \in R^d$ is the output of global pooling layer in 3D-CNN.

Bi-directional Inter-Clip Fusion. Since each clip descriptor is produced by the 3D-CNN module separately, the inter-clip relationships modeled by convolution are inherently implicit and local. That is, each clip descriptor can only observe an extremely limited local event and there are no inter-clip relationships. This will become a performance bottleneck since the duration of different actions are variant and complex actions could span multiple video segments (e.g., multiple clips can be involved for an action event).

To capture the inter-clip dependencies for both short- and long-range dependencies, we propose the Bi-directional Inter-Clip Fusion that aims at strengthening the local clip descriptor of the target position via aggregating information from other positions (e.g., other video segments). Motivated by [25], the inter-clip relationships can be fused by a bi-directional attention to link different clips and can be expressed as:

$$BA(v_i) = W_Z \sum_j \frac{(W_q v_i)(W_k v_j^T)}{N(v)}(W_v v_j), \tag{1}$$

where i is the index of the target positions, and j enumerates all possible other clip positions. W_q, W_k, W_v and W_z denote linear transform matrices. $(W_q v_i)(W_k v_j)$ denotes the relationship between video clip i and j, and $N(v)$ is the normalization factor. The resulting fused clip descriptor is named as bi-directional clip descriptor $b_i = BA(v_i)$, and $B = \{b_1, b_2, ..., b_C\}$, where $b \in R^D$ is a D-dimensional vector.

Gated Clip-Wise Attention. An action typically represents only a subset of objects and events which are most relevant to the context of a video. To suppress irrelevant clips, we further introduce the clip-wise attention to re-weight the above bi-directional clip descriptors and remove less related clips to join the final video-level action recognition decision. With a designated loss function (details described in next section), the resulting attention vector can also help locate temporal intervals of an action unsupervisely.

To be more specific, we first generate clip-wise statistics for each bi-directional clip descriptor by global average pooling

$$g = [mean(b_1), mean(b_2), ..., mean(b_C)] \tag{2}$$

The pooled output can be interpreted as a summary of the bi-directional clip descriptor whose statistics are expressive for the whole clip. To fully capture clip-wise dependencies, here we employ a gating mechanism with a sigmoid activation:

$$att = \sigma_{sigmoid}(W_2 \sigma_{ReLU}(W_1 g)), \tag{3}$$

where σ_{ReLU} refers to the ReLU function, W_1 and W_2 are the fully connected layer weights, and $\sigma_{sigmoid}$ is the sigmoid function. The *att* weight vector is defined in a clip-agnostic way, which is useful to identify video segments that are relevant to the action of interest and estimate the temporal intervals of the detected actions.

The final output of the block is obtained by re-scaling clip descriptors with the activation *att*:

$$S = \{s_1, s_2, \ldots, s_C\}, \tag{4}$$

$$s_i = att_i \times b_i \tag{5}$$

The *att* introduces dynamics conditioned on the input (i.e., bi-directional clip descriptor), which can be regarded as a gated function that re-weights the clips based on their significance to an action event.

To make the learning more robust and effective, a residual module is introduced

$$R = B + S \tag{6}$$

where $R = \{r_1, r_2, \ldots, r_c\}$ ($r \in R^D$) can be considered as the residual clip descriptors. Finally, a video-level representation, denoted by v', corresponds to the residual clip descriptors R, is given by

$$v' = \frac{\sum_i r_i}{C} \tag{7}$$

The action recognition can be performed based on v' and is defined as

$$y = \sigma_{softmax}(W_3 v'), \tag{8}$$

where W_3 is the fully connected layer weights, and $\sigma_{softmax}$ is the softmax function.

3.2 Loss Function

The loss function in the proposed network is composed of two terms, the action classification loss and the sparsity loss, which is given by

$$L = L_c + \lambda L_s \tag{9}$$

where L_c denotes the classification loss computed on the video-level action labels, L_s is the sparsity loss on the clip-wise attention weights, and λ is a constant to control the trade-off between the two terms. The classification loss is based on the standard cross-entropy loss between ground truth and the prediction y (generated by GCF-Net), while the sparsity loss is given by the L_1 norm on attention weights *att*:

$$L_s = \| att \|_1 \tag{10}$$

The output of attention weights will have the tendency towards 0 or 1 due to the use of L_1 loss and sigmoid function. In this case, action-related clips can be recognized with a sparse subset of key segments in a video, which will help locating the relevant clips for action detection.

3.3 Training Details

For the training of GCF-Net on Kinetics-600, Stochastic Gradient Descent (SGD) with standard categorical cross entropy loss is applied, and we use 128 videos as mini-batch size of SGD. The momentum, dampening and weight decay are set to 0.9, 0.9 and 1×10^{-3}, respectively. Learning rate is initialized with 0.1 and reduced 3 times with a factor of 10^{-1} when the validation loss converges. Note that the network was trained from scratch and no other dataset is used as prior. For the training of UCF-101 benchmark, we have used the pretrained models of Kinetics-600. We have frozen the network parameters and fine-tuned only the last layer. For fine-tuning, we start with a learning rate of 0.01 and reduce it with a factor of 10^{-1} when the validation loss converges. For spatial augmentation, we perform multi-scale cropping to augment video data. For Kinetics-600 and UCF-101, input clips are flipped with 50% probability.

For action recognition inference on Kinetics-600 and UCF-101, we select non-overlapping 16-frame from each video sample as one clip, and 10 clips are used as the network input. If the video contains smaller number of clips than the input size, loop padding is applied. After the padding, input to the network has the size of $10 \times 16 \times 112 \times 112 \times 3$ referring to number of clip numbers, frames, width, height, and input channels respectively.

4 Experiments

In this section, we first explain the experimented datasets. Then, we discuss about the achieved results for the experimented network architectures.

4.1 Datasets

- Kinetics-600 dataset [10] contains 600 human action classes, with at least 600 video clips for each action. Each clip is approximately 10 seconds long and is taken from a different YouTube video. There are in total 392,622 training videos. For each class, there are also 50 and 100 validation and test videos,
- UCF101 dataset [21] is an action recognition dataset of realistic action videos, collected from YouTube. It consists of 101 action classes, over 13k clips and 27 hours of video data. Compared to Kinetics-600, UCF-101 contains very little amount of training videos, hence prone to over-fitting. We thus conduct transfer learning (from network learned by Kinetics-600) to avoid this problem and show the effectiveness of the proposed method.

4.2 Action Recognition Accuracy

In this subsection, we study the effectiveness of the proposed model on learning video representations on different dataset.

Our GCF-Net can be used with any clip-based action classifiers and immediately boost the recognition accuracy. We demonstrate the general applicability

Table 1. Accuracy comparison of different methods

Method	Top1 Accuracy (%)	
	Kinetics-600	UCF101
3D-MobileNet+C-Clip	42.79	60.29
3D-ShuffleNet+C-Clip	45.61	61.27
R(2+1)D+C-Clip	54.69	76.92
3D-ResNeXt-101+C-Clip	58.58	80.12
3D-MobileNet+D-Clip	48.35	71.60
3D-ShuffleNet+D-Clip	53.70	73.32
R(2+1)D-101+D-Clips	62.18	87.02
3D-ResNeXt-101+D-Clips	66.40	89.08
GCF-Net(3D-MobileNet)	52.53	80.32
GCF-Net(3D-ShuffleNet)	57.10	81.23
GCF-Net(R(2+1)D)	68.01	95.12
GCF-Net(3D-ResNeXt-101)	70.07	96.82

of our approach by evaluating it with several popular 3D CNNs. For R(2+1)D network [2], we implement it by replacing the original 3D convolutional kernels (e.g.,$t \times k \times k$) with two (virtually) 2D blocks (e.g., $1 \times k \times k$ and $t \times 1 \times 1$ kernels) in 3D-ResNet-101 (i.e., 3D-ResNet with 101 layers). 3D-ResNext-101 [7,28] is an more efficient version of 3D-ResNet due to the use of cardinality block, and it also has 101 layers in our experiments. In addition to the above full networks, we also implement light weight networks for performance comparison, including 3D-MobileNet [16] and 3D-ShuffleNet [14]. 3D-MobileNet and 3D-ShuffleNet are the extentions of their original 2D versions (i.e., adding temporal dimension into kernel).

Table 1 shows the action recognition results of all methods. The baseline for our evaluation is the central-clip-based prediction (noted as C-Clip in the Table). For Kinetics-600 dataset, as can be seen that the baseline central-clip method has the poorest top-1 accuracy. Full networks, such as R(2+1)D and 3D-ResNeXt-101, can only achieves 54.69% and 58.58% top-1 accuracy, while light weight network, such as 3D-MobileNet and 3D-ShuffleNet, delivers even poorer 42.79% and 45.61% top-1 accuracy. The poor performance is mainly due to the lack of fully utilizing the information in the video (e.g., the rest relevant clips). Another popular method, dense-clips-based prediction (noted as D-Clips in the table) achieves better recognition accuracy (comparing to central-clip prediction). Among all dense-clips-based methods, 3D-ResNeXt-101 (i.e., 3D-ResNeXt-101+D-Clips) achieves the best 66.40% top-1 accuracy.

However, since an action is usually complex and spans video segments, uniformly averaging all clips is obviously not the best strategy and can only achieve limited accuracy. As can be seen in Table 1, our GCF-Net outperforms all the central-clip and dense-clips-based methods on the same backbone networks.

The best top-1 accuracy, delivered by GCF-Net based on 3D-ResNeXt-101, achieves 70.07% top-1 accuracy, which outperforms 3D-ResNeXt-101+D-Clips and 3D-ResNeXt-101+C-Clip by 3.67% and 11.49% respectively. The improvement on light weight networks are more significant (in terms of relative percentage of change), where GCF-Net(3D-ShuffleNet) outperforms 3D-ShuffleNet+D-Clip and 3D-ShuffleNet+C-Clip by 3.4% (i.e., 6.3% boosting) and 11.49% (i.e., 25.2% boosting) respectively. The proposed GCF-Net is thus proved to be able to build a better video-level feature representation that can capture short- and long-range statistics, which delivers a significant better recognition accuracy.

For UCF-101 dataset, we observed similar performance improvement. Comparing to traditional methods, such as 3D-ResNeXt-101+D-Clips and 3D-ResNeXt-101+C-Clip, that only have 80.12% and 89.08% accuracy, our GCF-Net(3D-ResNeXt-101) delivers a significant boosting in accuracy, which is 96.82%.

4.3 Ablation Studies

In this subsection we further investigate the accuracy improvement brought by the individual inter-clip (i.e., Bi-directional Inter-Clip Fusion) and clip-wise (Gated Clip-Wise Attention) strategies.

Table 2. Accuracy comparison of different GCF-Net modules on Kinetics dataset (all methods are based on 3D-ResNeXt-101)

Method	Accuracy (%)
C-clip prediction	58.58
D-clips prediction	66.40
Inter-clip fusion only	69.01
Clip-wise attention only	68.46
GCF-net full	70.07

To implement inter-clip-fusion-only strategy, the output of bi-directional clip descriptor B is directly used to generate a video-level representation (i.e., $v' = \frac{\sum_i b_i}{C}$), where the rest clip-wise attention module in the original GCF-Net is skipped. For clip-wise-attention-only strategy, the output of the raw 3D-CNN feature V is directly fed into clip-wise attention module (i.e., $g = [mean(v_1), mean(v_2), ..., mean(v_C)]$), and the same fusion process is proceeded to generate a video-level representation for later prediction.

Table 2 shows the accuracy comparison of different GCF-Net modules. As can be seen both strategies outperform traditional video-level prediction methods (e.g., C-Clip and D-Clips). Inter-clip-fusion-only strategy achieve slightly better top-1 accuracy of 69.07% than clip-wise-attention-only strategy that has 68.46% accuracy, which shows the importance of inter-clip dependencies. Since the raw

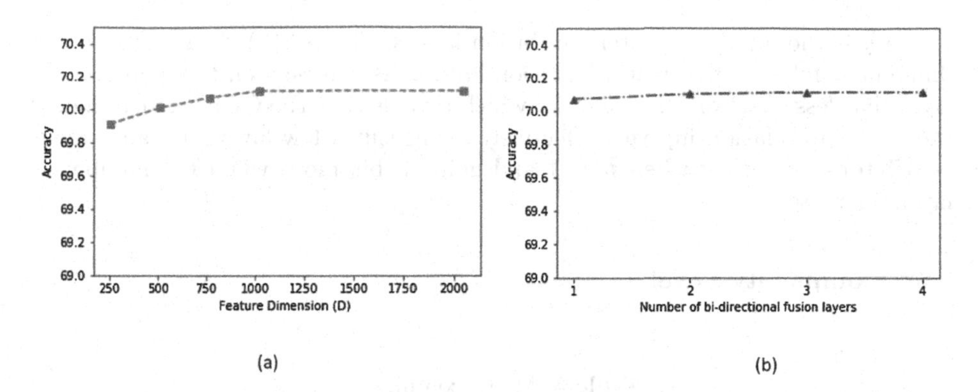

Fig. 3. Performance comparison of feature dimension and fusion layers

clip descriptor generated by 3D-CNN can only observe an extremely limited local information, using the clues from neighbors will largely extend the receptive field and strengthen the clip descriptor. The inter-clip fusion thus shows its excellence in fusing the inter-clip relationships to better describe a video segment.

Despite the fact that clip-wise-attention-only strategy's performance is a bit inferior to inter-clip-fusion-only strategy, it can be a complement to inter-clip strategy and can further refine the bi-directional clip feature (e.g., by filtering out irrelevant ones), which helps generating a better video-level representation. By concatenating clip-wise module with inter-clip, we can deliver the best video-level prediction accuracy, which proves the effectiveness of each module and the overall strategy.

4.4 Feature Dimension and Fusion Layers

In this subsection we explore the effect of network parameters in terms of prediction accuracy.

Figure 3(a) shows the result of different length of features, where D is the parameters to determine the feature dimension of a clip descriptor b, residual clip descriptor r, and video-level descriptor v'. As can be seen the GCF-Net has good performance in generating compact representation of video-level descriptor. Comparing to the raw 3D-CNN pooled feature (e.g., 2048-d in 3D-RestNeXt), GCF-Net only needs as small as a 256-d vector to represent a video-level, which can still achieve better top-1 accuracy than the central- and dense-clips-based methods, and is 8 times smaller than the raw 3D-CNN feature. Increasing the size of feature dimension in GCF-Net can slightly improve the accuracy, but it soon gets saturated after $D > 1024$.

For Bi-directional Inter-Clip Fusion, we can concatenate multiple bi-directional fusion layers to have a deeper fusion of inter-clip dependencies:

$$BA^n = BA(BA^{n-1}) \tag{11}$$

$$BA^1 = BA(v_i) \tag{12}$$

where n is the number of inter-clip fusion layers. Figure 3(b) show the result of different number of bi-directional fusion layers. As can be seen the increment of layers has less effect on the accuracy, which reveals that the GCF-Net can model the inter-clip relationship very effectively using jsut a few layers. It can achieve >70% top-1 accuracy when $n = 1$ and gains a bit more when the number of layers increases.

4.5 Complexity Level

Table 3. Model complexity

Method	MFLOPs	Params
3D-MobileNet	446	3.12M
3D-ShuffleNet	360	6.64M
R(2+1)D	6,831	50.91M
3D-ResNeXt-101	6,932	48.34M
GCF-Net (w/o backbone network)	15	0.86M

Table 3 shows the complexity level of each model. As expected that the light weight networks (e.g., 3D-MobileNet and 3D-ShuffleNet) have much smaller MFLOPs and parameters than full network (e.g., 3D-ResNeXt and R(2+1)D). We can observe that, comparing to the backbone networks, the overhead brought by GCF-Net (without the backbone network) is actually quite tiny. It has 15 MFOPs and 0.86M network parameters, which occupies only 3.2% and 0.2% computation effort of the overall prediction pipeline for 3D-MobileNet- and 3D-ResNeXt-based frameworks. Hence, GCF-Net successfully upgrades the video-level prediction framework with the cost of a tiny MFLOPs increment in exchange of a significant boost of accuracy.

4.6 Visualization

To verify the ability of GCF-Net on locating action event in a video, here we extend the the 2D Grad-CAM [17] to a 3D spatio-temporal action localization map. Similar to [17], we first compute the gradient of the predicted action class m (i.e., y^m) with respect to feature maps A_{tk} of the last convolutional layer at moment t (e.g., the output of k-th convolution kernel for t-th frame in video). These flowing-back gradients are global average-pooled to obtain the neuron importance weights $\alpha_{t,k}$:

$$\alpha_{t,k}^{m} = \frac{1}{Z} \sum_{i} \sum_{j} \frac{\partial y^m}{\partial A_{ij}^{t,k}} \tag{13}$$

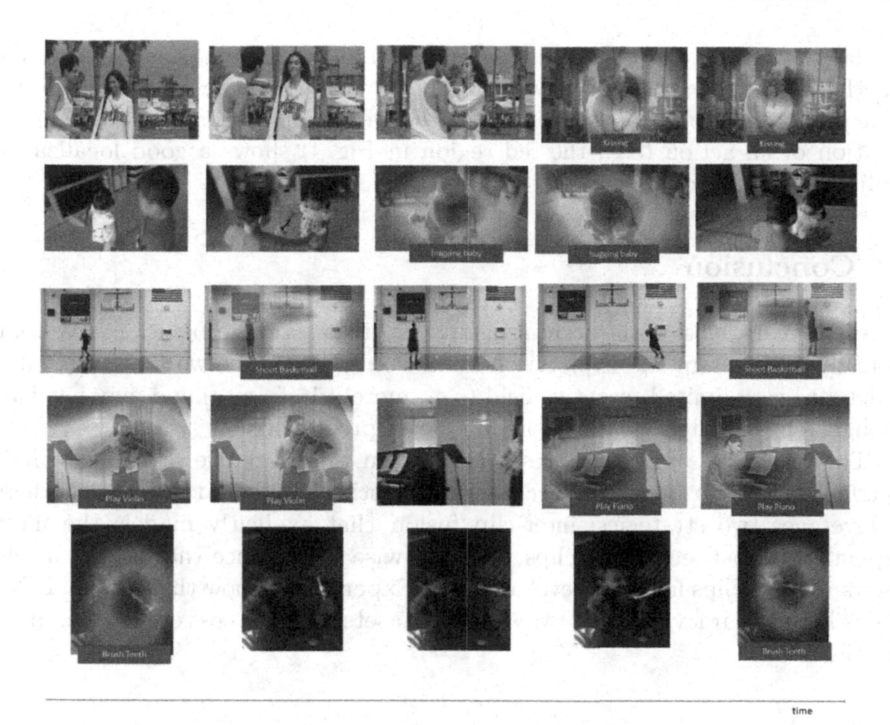

Fig. 4. Visualization of spatial-temporal action event. (Color figure online)

where Z is the normalization factor (i.e., the size of feature map). The weight $\alpha_{t,k}^{m}$ represents a partial linearization of the deep network downstream from $A_{t,k}^{m}$, and captures the 'importance' of feature map k for a target action m.

After having the importance estimation of each channel, we perform a weighted combination of forward activation maps, and follow it by a ReLU to obtain the localization map:

$$M_t = \sigma_{ReLU}(\sum_{k} \alpha_{t,k}^{m} A^{t,k}) \tag{14}$$

The output can then be resized to match the size of original video frames to showcase the spatial-temproal heatmap of an action event. Since only positive weights are what we interested (e.g., positive influence on the target action class), we can apply a ReLU to the linear combination of maps. In this case, we can get the intensity level of pixels whose intensity should be increased in order to increase y^m.

Figure 4 shows visualization examples for several videos. Here only clips with attention weights (i.e., *att*) that surpass the threshold (e.g., 0.5) will be marked as relevant segments with corresponding gradients being back-propagated to generate the spatio-temporal heatmap. As can be seen in Fig. 4, the GCF-Net nicely capture relevant clips in terms of temporal and spatial space. Despite the lack of temporal labels during training, the GCF-Network learns to judge whether it is a relevant or irrelevant clip unsupervisedly. The spatial location of an action is also nicely captured based on the back-propagation. The spatial pixel location of an action (e.g., the red region in Fig. 4) shows a good localization ability of the proposed network.

5 Conclusion

In recent years, most of the accuracy gains for video action recognition have come from the new design and exploration of 3D convolutional network architectures, which has very limited receptive field (e.g., merely 16 frames) and may not have enough representative power for a video level prediction.

This work aims to fill this research gap. In this work, we presented a light weight network to boost the accuracy of existing clip-based action classifiers. It leverages two strategies: inter-clip fusion that explicitly models the inter-dependencies between video clips, and clip-wise importance that selects a relevant subset of clips for video level analysis. Experiments show that our GCF-Net yields large accuracy gains on two action datasets with the cost of tiny increment in MFLOPs.

References

1. Carreira, J., Zisserman, A.: Quo vadis, action recognition? A new model and the kinetics dataset. In: CVPR, pp. 4724–4733 (2017)
2. Tran, D., Wang, H., Torresani, L., Ray, J., LeCun, Y., Paluri, M.: A closer look at spatiotemporal convolutions for action recognition. In: CVPR (2018)
3. Dollár, P., Rabaud, V., Cottrell, G., Belongie, S.: Behavior recognition via sparse spatio-temporal features. In: ICCV VS-PETS, pp. 65–72 (2005)
4. Caba Heilbron, F., Carlos Niebles, J., C.N., Ghanem, B.: Fast temporal activity proposals for efficient detection of human actions in untrimmed videos. In: CVPR (2016)
5. Feichtenhofer, C., Fan, H., Malik, J., He, K.: Slowfast networks for video recognition. In: ICCV (2019)
6. Varol, G., Laptev, I., Schmid, C.: Long-term temporal convolutions for action recognition. IEEE PAMI **40**, 1510–1517 (2018)
7. Hara, K., Kataoka, H., Satoh, Y.: Can spatiotemporal 3D CNNs retrace the history of 2D CNNs and ImageNet? In: CVPR (2018)
8. Fan, H., Xu, Z., Zhu, L., Yan, C., Ge, J., Yang, Y.: Watching a small portion could be as good as watching all: towards efficient video classification. In: IJCAI (2018)
9. Alwassel, H., Caba Heilbron, F., Ghanem, B.: Action search: spotting actions in videos and its application to temporal action localization. In: ECCV (2018)
10. Carreira, J., Noland, E., Banki-Horvath, A., Hillier, C., Zisserman, A.: A short note about kinetics-600. arXiv preprint arXiv:1212.0402 (2018)
11. Korbar, B., Tran, D., Torresani, L.: SCSampler: sampling salient clips from video for efficient action recognition. In: ICCV, October 2019
12. Laptev, I., Lindeberg, T.: Space-time interest points. In: ICCV, pp. 432–439 (2003)

13. Lin, J., Gan, C., Han, S.: Temporal shift module for efficient video understanding. CoRR abs/1811.08383 (2018). http://arxiv.org/abs/1811.08383
14. Ma, N., Zhang, X., Zheng, H.T., Sun, J.: ShuffleNet v2: practical guidelines for efficient CNN architecture design. In: The European Conference on Computer Vision (ECCV) (2018)
15. Sadan, S., Corso, J.J.: Action bank: a high-level representation of activity in video. In: CVPR (2012)
16. Sandler, M., Howard, A., Zhu, M., Zhmoginov, A., Chen, L.C.: MobileNeTv 2: inverted residuals and linear bottlenecks. In: CVPR (2018)
17. Selvaraju, R.R., Cogswell, M., Das, A., Vedantam, R., Parikh, D., Batra, D.: Grad-CAM: visual explanations from deep networks via gradient-based localization. In: ICCV, October 2017
18. Yeung, S., Russakovsky, O., Mori, G., Fei-Fei, L.: End-to-end learning of action detection from frame glimpses in videos. In: CVPR (2016)
19. Buch, S., Escorcia, V., Shen, C., Ghanem, B., Niebles, J.C.: SST: single-stream temporal action proposals. In: CVPR (2017)
20. Simonyan, K., Zisserman, A.: Two-stream convolutional networks for action recognition in videos. In: NIPS (2014)
21. Soomro, K., Zamir, A.R., Shah, M.: UCF101: a dataset of 101 human actions classes from videos in the wild. arXiv preprint arXiv:1212.0402 (2012)
22. Taylor, G.W., Fergus, R., Lecun, Y., Bregler, C.: Convolutional learning of spatio-temporal features. In: ECCV (2010)
23. Lin, T., Zhao, X., Su, H., Wang, C., Yang, M.: BSN: boundary sensitive network for temporal action proposal generation. In: ECCV (2018)
24. Tran, D., Bourdev, L., Fergus, R., Torresani, L., Paluri, M.: Learning spatiotemporal features with 3D convolutional networks. In: ICCV (2015)
25. Vaswani, A., et al.: Attention is all you need. In: Guyon, I., et al. (eds.) Advances in Neural Information Processing Systems, vol. 30, pp. 5998–6008. Curran Associates, Inc. (2017). http://papers.nips.cc/paper/7181-attention-is-all-you-need.pdf
26. Wang, H., Schmid, C., Wang, H., Schmid, C.: Action recognition with improved trajectories. In: ICCV (2013)
27. Wang, X., Girshick, R., Gupta, A., He, K.: Non-local neural networks. In: CVPR, June 2018
28. Xie, S., Girshick, R., Dollár, P., Tu, Z., He, K.:
29. Z. Qiu, Yao, T., Mei, T.: Learning spatio-temporal representation with pseudo-3D residual networks. In: ICCV (2017)
30. Wu, Z., Xiong, C., Ma, C.Y., Socher, R., Davis, L.S.: AdaFrame: adaptive frame selection for fast video recognition. In: CVPR (2019)

Watch Hours in Minutes: Summarizing Videos with User Intent

Saiteja Nalla[1], Mohit Agrawal[1], Vishal Kaushal[1(✉)], Ganesh Ramakrishnan[1], and Rishabh Iyer[2]

[1] Indian Institute of Technology Bombay, Mumbai, India
{saitejan,mohitagr,vkaushal,ganesh}@cse.iitb.ac.in
[2] The University of Texas at Dallas, Richardson, TX, USA
rishabh.iyer@utdallas.edu

Abstract. With the ever increasing growth of videos, automatic video summarization has become an important task which has attracted lot of interest in the research community. One of the challenges which makes it a hard problem to solve is presence of multiple 'correct answers'. Because of the highly subjective nature of the task, there can be different "ideal" summaries of a video. Modelling user intent in the form of queries has been posed in literature as a way to alleviate this problem. The query-focused summary is expected to contain shots which are relevant to the query in conjunction with other important shots. For practical deployments in which very long videos need to be summarized, this need to capture user's intent becomes all the more pronounced. In this work, we propose a simple two stage method which takes user query and video as input and generates a query-focused summary. Specifically, in the first stage, we employ attention within a segment and across all segments, combined with the query to learn the feature representation of each shot. In the second stage, such learned features are again fused with the query to learn the score of each shot by regressing through fully connected layers. We then assemble the summary by arranging the top scoring shots in chronological order. Extensive experiments on a benchmark query-focused video summarization dataset for long videos give better results as compared to the current state of the art, thereby demonstrating the effectiveness of our method even without employing computationally expensive architectures like LSTMs, variational autoencoders, GANs or reinforcement learning, as done by most past works.

Keywords: Query-focused · Video summarization · Attention · User-intent

1 Introduction

Videos have become an indispensable medium for capturing and conveying information. The increasing availability of cheaper and better video capturing and

S. Nalla and M. Agrawal—Equal contribution.

A. Bartoli and A. Fusiello (Eds.): ECCV 2020 Workshops, LNCS 12539, pp. 714–730, 2020.
https://doi.org/10.1007/978-3-030-68238-5_47

storage devices have led to the unprecedented growth in the amount of video data available today. Most of this data, however, comes with a lot of redundancy, partly because of the inherent nature of videos (as a set of *many* images) and partly due to the 'capture-now-process-later' mentality. Consequently this has given rise to the need of automatic video summarization techniques which essentially aim at producing shorter videos without significantly compromising the quality and quantity of information contained in them. A Video Summarization technique aims to select important, diverse (non-redundant) and representative frames (static video summarization) or shots (dynamic video summarization) from a video to enable quicker and easier consumption of information contained in the video. In this work we focus on producing summary as a sequence of shots (set of frames), i.e. dynamic video summarization. One of the characteristic challenges which make this problem hard to solve is the fact that there is no single correct answer (summary). Owing to the highly subjective nature of the task, summaries produced by different users tend to be different due to varying intents and perception. Researchers have looked at query-focused video summarization as a way to alleviate this problem. The user intent is taken as an additional input in the form of a query and the summary produced is influenced to contain more shots which are relevant to the query (Fig. 1). The summary thus produced is semantically relevant to the query, modelling user intent and preferences. This is especially welcome in the real-world setting where very long videos need to be summarized.

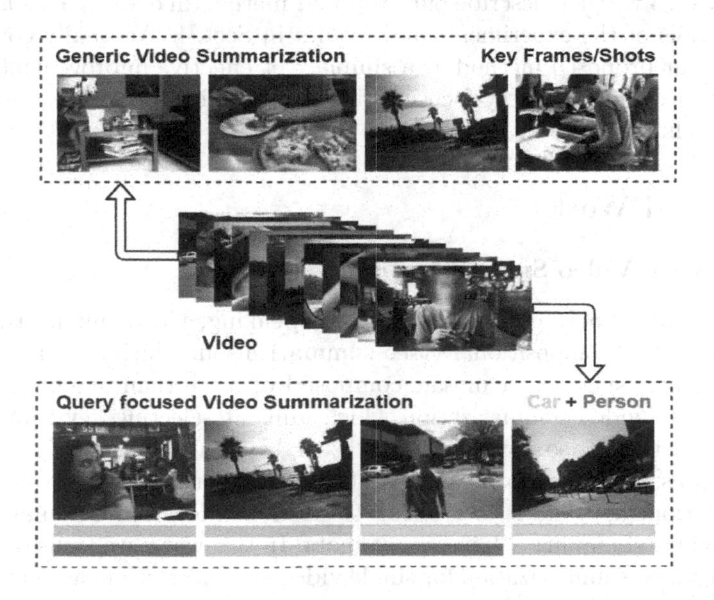

Fig. 1. Illustration of generic vs query focused video summarization for a given video

Query-focused Video Summarization has attracted a lot of attention in the recent past. To the best of our knowledge, the first work in Query-focused Video

Summarization was by Sharghi et. al. [34] which employed determinantal point processes [21]. This was followed by use of memory networks [35], submodular mixtures [27,36], adversarial networks [43] and attention [17,39,40] as different ways of computing the query relevance. Motivated by [15,17,40] we propose a simple attention based two-stage method to address query-focused video summarization enhanced by query fusion. Let us say we are provided a video that is divided into segments of fixed size shots. In the first stage, local attention is employed to model the query agnostic importance of the shots in the segments and local attention features are learnt. In addition, for each segment, query relevant shots within a segment are identified to represent the aggregate segment level semantic features relevant to the query. These segment representatives are then combined with the visual features to learn the global attention features of all the shots, considering the query. To enhance the effectiveness of query-relevance further, in the second stage, local attention features along with the global attention features are fused with the query representation vector. This is followed by a regression through fully connected layers to obtain shot scores, indicative of the rank of shots with respect to the query. We then assemble the summary by arranging the top scoring shots in chronological order. For a fair comparison with recent techniques, we test our method on the benchmark dataset for query-focused video summarization [35] and demonstrate the effectiveness of our method both quantitatively and qualitatively.

In the following sections, we begin by talking about the related work in this area. In Sect. 3 we then describe our proposed method in details. This is followed by the details of the experiments and results in Sect. 4. We finally conclude by reporting our proposed method as a simple, yet effective improvement over the current state of the art Query-focused Video Summarization techniques as tested on the benchmark dataset.

2 Related Work

2.1 Generic Video Summarization

In terms of the generated output, broadly speaking, video summarization can be categorized as compositional video summarization, which aims at producing spatio-temporal synopsis or mosaic composed of more than one frames [29–32] and extractive video summarization which aims at selecting key frames or key shots. Extractive video summarization with key frame selection is also often referred to as key frame extraction, static story board creation or static video summarization [5] while it is referred to as dynamic video summarization or dynamic video skimming [13] in case of shots. In this work, we focus on dynamic extractive video summarization for single video summarization, as against multi-video summarization [26]. Video summarization, at least from what appears at the surface, boils down to identify important, representative portions of a video while eliminating redundancy. Early approaches were mainly unsupervised and summarized a video using low level cues [24,38]. More advanced approaches looked into better indicators of 'important' portions of a video through presence of people or objects in egocentric videos and more recently, actionness [6,14,22]. First truly supervised approach, in terms of learning directly from a ground truth summary, was presented by [12] who adapted determinantal point processes (DPP) [21] to videos. Motivated by the fact that video is a form of sequence data where LSTMs have demonstrated superior performance [42] was the first work to use LSTMs for video summarization. They also proposed an additional DPP layer on top to ensure diversity. Another body of work looks at using external clues as an aide to summarization [4,19,20,46]. [3,23] explicitly focus on enhancing the diversity and representativeness of the generated summary. [23] for example employed sequential DPP to learn the time span of a video segment upon which the local diversity is imposed to guarantee the diversity of a long video. The absence of a large annotated dataset and the fact that there are multiple ground truth summaries possible for a video, has led to a recent rise of unsupervised techniques [1,18,25,41,45]. [45] was the first to apply reinforcement learning to unsupervised video summarization motivated by the fact that reward is available only at the end of the sequence. [25] used a generative adversarial framework, consisting of the summarizer and discriminator in an unsupervised setting to achieve comparable performance to supervised techniques. There is also a lot of recent work combining adversarial and attention based networks with an aim to produce better video summaries [1,7,8,16,25,41]. [16] was the first to use attentive encoder decoder based network to video summarization.

2.2 Query-Focused Video Summarization

SeqDPP introduced in [12] was used to model the problem of video summarization as diverse sequential subset selection. Based on this idea, in [34] Sharghi et al proposed sequential hierarchical DPP (SH-DPP) where the first layer modeled query relevance and the second layer modeled importance conditioned on first. Diversity was naturally modelled by DPP [21]. In [4], topic-based summary is generated by finding shots which co-occur mostly across videos collected using the given topic and a MBF (Maximal Biclique Finding) algorithm is optimized to find sparsely co-occuring pattern. In [35] Sharghi et al introduced QC-DPP (Query Conditioned DPP) where a memory network was used to model query importance as well as contextual importance of a shot. This is then fed into the seqDPP. They also, for the first time, introduced a dataset specifically prepared for the task of Query-focused Video Summarization. They also introduced a new evaluation metric which focuses on the semantic relationship between the shots in predicted and ground truth summary. This has emerged as a benchmark dataset for Query-focused Video Summarization with several recent techniques reporting their results on it. In [36] Vasudevan *et. al.*, model Query-focused Video Summarization as a subset selection problem where the best subset is found by maximizing a mixture of different submodular terms which capture (i) query similarity between frame and query in a common semantic embedding space, (i) quality score, (iii) diversity and (iv) representativeness. A similar approach is adopted by [27] where they demonstrate the importance of using joint vision-langauge embedding in addition to visual features. [43] use adversarial networks where the generator learns the joint representation of the user query and the video content, and the discriminator takes three pairs of query-conditioned summaries (generator, ground truth and random) as the input to discriminate the real summary from a generated and a random one, trained via a three-player loss. In [17] Jiang et al use a query-focused attention module to combine the semantic information of the query and a multilevel self-attention variational block to obtain context-important information and add user-oriented diversity and stochasticity. Reinforcement Learning is used in [44] to target this problem where a Mapping Network (MapNet) is used to map video shot and query in same space and after that a deep RL-based summarization network is used to provide query based summary by including parameters like relatedness, representativeness and diversity as rewards. Xiao et al in [40] employ a hierarchical attention network and demonstrate the effectiveness of local and global attention. In [39], Xiao et al extended their work [40] and used a pre-trained RL caption generator to generate captions for the video shots for textual information and along with semantic information generated from self-attentive module which helped to decide the important shots and then use a query-aware scoring module to generate query-focused summary. Huang et al. [15] addressed query-focused video summarization by learning the query relevance scores using a combination of visual features and a vector representation of input query.

3 Proposed Method

3.1 Problem Formulation

The objective of the Query-focused Video Summarization is to output a video summary which is a sequence of diverse, representative and query relevant video shots given a long video and a query. We extract visual and textual features from shots, compute shot scores using the proposed method and finally construct the summary based on the shot scores. We denote a video as a sequence of non-overlapping shots of fixed length. Let there be n shots in the video denoted by $\{s_1, s_2, \cdots, s_n\}$ These shots are further grouped together into fixed-sized non overlapping segments. This can easily be extended to using variable sized segments formed using Kernel Temporal Segmentation [28] or other alternate techniques for shot detection. We use the lexicon of concepts constructed by [35] and represent the textual query t_q as a collection of two concepts $\{c_1, c_2\}$. Each concept is a noun like 'SKY', 'LADY', 'FLOWER', 'COMPUTER' etc. More complex queries can easily be supported by using appropriate embedding for them as in some of the video localization works [9,10,33,37] or following the approach in [27]. We leave that to future work in this area. For a textual query t_q shot scores are calculated corresponding to visual features of each shot to construct a query-focused video summary.

3.2 Feature Embedding

Motivated by the success of I3D features [2] in better modelling of temporal resolution and in capturing long spatial and temporal dependencies, which is especially important for long videos, we extract p dimensional I3D features for every shot in the video. These features are reduced from p to d dimensions using a fully connected layer. The output of the fully connected layer corresponding to the extracted visual features $\{v_1^{'}, v_2^{'}, \cdots, v_n^{'}\}$ is $\{v_1, v_2, \cdots, v_n\}$, $v_i \in R^d$ is referred to as 'visual features' here after in this paper. For representing queries, we use one hot feature encoding of the concepts to form a query representation vector of 48 dimensions (corresponding to the 48 concepts introduced in [35]). The query representation vector is passed through another fully connected layer and the resultant d dimensional features are referred to as 'textual features' $f_q \in R^d$ here after in this paper. In order to generate the video summary of a long video, it is important to look at both local context as well as global context in determining query-relevance and importance of shots. To facilitate this, we define fixed-size windows, called segments, which are non-overlapping groups of shots. For local context, we consider shots within a segment and for global context we consider query-relevance and importance across segments. Visual features are used for the computation of local attention feature vectors within a segment in the Local Attention Module (LAM). Visual features along with textual features are used to compute query relevant segment representatives in the Query Relevant Segment Representation Module (QSRM). Query relevant features along with the local features are used to compute the global attention features across the segments in the Global Attention Module (GAM) which are further regressed to compute the shot scores. We illustrate this end to end pipeline in 2. In what follows, we give details of each of these modules.

3.3 Local Attention Module

The Local Attention Module (LAM) computes the attentive features of the shots within a segment for all segments of a video. It captures the semantic relations among the shots within a segment. We represent the visual features corresponding to shots $\{s_1, s_2, \cdots, s_n\}$ as $\{v_1, v_2, \cdots, v_n\}$ respectively. LAM takes these visual features as input and outputs attention vectors corresponding to each shot. Semantic similarity matrix is calculated from the visual features corresponding to each segment. Semantic similarity score of shots (v_i, v_j) i.e., $(i, j)^{th}$ element of semantic similarity matrix for a segment is computed as

$$\phi(v_i, v_j) = \mathbf{Z_l} \tanh(\mathbf{W_1^l} v_i + \mathbf{W_2^l} v_j + \mathbf{b}) \quad (1)$$

$W_1^l, W_2^l, Z_l \in R^{d \times d}$ and $b \in R^d$ are parameters to be learnt. The semantic score vector $\phi(v_i, v_j) \in R^d$ and the shape of semantic similarity score matrix Φ is $k \times k \times d$, where k is the segment length. Semantic similarity matrix captures the semantic relations between the shots within a segment (also representing the temporal features) and this is done for all the segments. The semantic relations of one segment are interacted with other segments temporally by sharing the

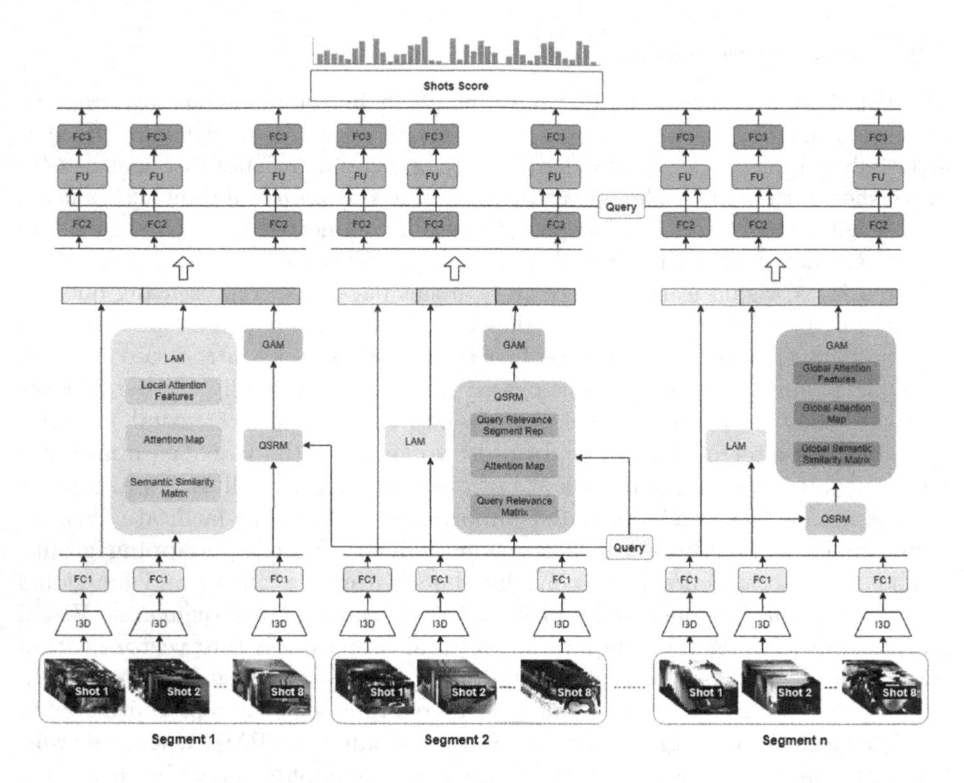

Fig. 2. Architecture of the proposed method. We split a video into non-overlapping shots and group them into non-overlapping segments. We extract the visual features using a pretrained model and compute the local attention features in LAM using visual features. We encode the textual query and compute the segment representatives in QSRM using textual and visual features. We pass the outputs of QSRM and visual features to GAM to compute the global attention vectors. Local and global attention vectors are concatenated with visual features, down-sampled and are passed to the Fusion Unit followed by FC layer to generate the shot scores which are then used to construct the summary

trainable parameters across the segments and thereby reducing the number of trainable parameters in the model. Softmax interactions within a segment are then calculated from the semantic similarity matrix and the values are computed corresponding to visual features (v_i, v_j) as,

$$\chi_{ij} = \frac{\exp(\phi(v_i, v_j))}{\sum_{t=0}^{k} \exp(\phi(v_i, v_t))} \tag{2}$$

The shape of the softmax interaction matrix is $k \times k \times d$ corresponding to every combination of the shots with the segment.

The local attentive features for i^{th} shot are then calculated from the softmax interactions and visual features as,

$$v_i^l = \sum_{j=0}^{k} \chi_{ij} v_j \tag{3}$$

The shape of the local attention feature matrix of a segment is $k \times d$ corresponding to every shot of the segment.

3.4 Query Relevant Segment Representation Module

Query relevant Segment Representation Module (QSRM) captures the semantic relations of the shots with the textual query and outputs the query-relevant representative features for each segment which are further used in the computation of global attention vectors in the GAM. Visual features $\{v_1, v_2, \cdots, v_n\}$ along with the textual features of a query are fed as input to the QSRM. Query relevant semantic scores are calculated as,

$$r_i = \mathbf{Z_g} \tanh(\mathbf{W_1^g} v_i + \mathbf{W_2^g} f_q + \mathbf{b})) \tag{4}$$

$W_1^g, W_2^g, Z_g \in R^{d \times d}$ and $b \in R^d$ are trainable parameters. Query relevant semantic matrix is of shape $k \times d$ for each segment. Query relevant softmax interactions are then computed from query relevant semantic scores as,

$$\chi_i = \frac{\exp(r_i)}{\sum_{t=0}^{k} \exp(r_t)} \tag{5}$$

Query relevant softmax interaction matrix is of shape $k \times d$ corresponding to every shot of a segment.

Query relevant segment representations for a segment are computed from the visual features $\{v_1, v_2, ..v_n\}$ of the shots and their corresponding query-relevant softmax interactions as,

$$v^{(s)} = \sum_{i=0}^{k} \chi_i v_i \tag{6}$$

Query relevant representation vectors for m segments are represented as $\{v_1^{(s)}, v_2^{(s)}, ..v_m^{(s)}\}$ and have a shape of $m \times d$ representing the aggregated attention representations for a query.

3.5 Global Attention Module

Global attention features are computed from the visual features and query relevant segment representations. These capture semantic interactions between the intra segment semantic features and query-relevant inter segment attention features. Global semantic similarity scores are computed from the visual features $\{v_1, v_2, \cdots, v_n\}$ and query relevant semantic representation features $\{v_1^{(s)}, v_2^{(s)}, \cdots, v_m^{(s)}\}$ as

$$r_j^g = \mathbf{Z_g} \tanh(\mathbf{W_1^g} v_i + \mathbf{W_2^g} v_j^{(s)} + \mathbf{b}) \tag{7}$$

Global semantic similarity score matrix has a shape of $n \times d$. $W_1^g, W_2^g, Z_g \in R^{d \times d}$ and $b \in R^d$ are trainable parameters. The softmax interaction scores are calculated from the semantic segment scores as,

$$\chi_j^g = \frac{\exp(r_j^g)}{\sum_{k=0}^{m} \exp(r_k^g)} \tag{8}$$

The shape of the softmax interaction matrix is $n \times d$. The global attentive features are computed from the softmax interaction scores and query relevant segment representatives as,

$$v_i^g = \sum_{j=0}^{m} \chi_j^g v_j^{(s)} \tag{9}$$

v^g is the global attention vector corresponding to each shot which captures the query relevant global semantic attention features of all the segments in a video. Shape of v^g is $n \times d$.

3.6 Fusion Unit

To better learn the shot scores, we make use of visual features v_i, local attention vectors v_i^l and global attention vectors v_i^g for each shot, these features are concatenated to form a single shot feature vector $v_i^{c'}$, $v_i^{c'} = [v_i, v_i^l, v_i^g], v_i^{c'} \in R^{3d}$. The concatenated feature vector $v_i^{c'}$ is reduced to d dimensions using a fully connected layer and the output of the fully connected layer is represented as v_i^c. To enhance the effectiveness of query-relevance with respect to the given query, the condensed features vector v_i^c along with the textual features $f_q \in R^d$ of the query t_q are fed as inputs to the Fusion Unit. The Fusion Unit (FU) aggregates the features by performing point wise additions, multiplications and concatenating the features and outputs a feature vector $v_i^f \in R^{4d}$. These features are used to finally predict the shot scores through a fully connected layer and the top ranked shots are used as predicted selections.

We use Adam optimizer to train the model end-to-end based on the Binary Cross Entropy (BCE) loss between the predicted shots and ground truth shots.

4 Experiments and Results

4.1 Dataset

For a fair comparison with other techniques, we evaluate our model's performance on the benchmarking dataset introduced by [35] which was built upon UTE dataset [22]. It contains four egocentric consumer grade videos captured in uncontrolled everyday scenarios and each video is 3 to 5 h long containing a diverse set of events. A set of 48 concepts is defined by [35] and every query is made up of two concepts. The queries are so defined by [35] as to cover four different scenarios: 1) all concepts in the query appear in the same video shots together 2) all concepts appear in video occur but never jointly in a shot 3) only one of the concepts in the query appears in some shots and 4) none of the concept in the query are present in the video. We follow the same convention in our work. The dataset provides four ground truth query-focused summaries for each video and query pair, 1 oracle summary and 3 user summaries.

4.2 End to End Pipeline

Preprocessing. The videos in the UTE dataset are divided into non-overlapping shots of 5 seconds each. We sample the frames at 3 fps and compute features of each shot (15 frames per shot) as follows: we use a pre trained I3D model as a feature extractor. For each shot, 15 frames each of size 224 × 224 is given as input to the I3D Model and it generates a 512 dimension feature vector (output from the temporal layer in the I3D Model) to be further used in our pipeline. We define segments as non-overlapping groups of 8 shots. As far as queries are concerned, we deal with bi-concept queries. There are a total of 48 concepts defined in the dataset. We represent the query by the addition of the one-hot vectors corresponding to each concept resulting in a 48-dimensional vector which is then used in our pipeline.

Training. We performed 4 experiments by using one video for test and rest for training and validation in turn. We train the model for 25 epochs. In each epoch, for each training video, I3D features are passed through a FC layer to create a 300-dimensional visual feature representation for each shot. These features are also used to create the local and global representation for each shot. The visual feature of each shot of a segment is given as input to the LAM (Local Attention Module). This module uses attention (refer Sect. 3.3 in the main paper) to generate a 300-dimensional local feature representation of each shot of the segment. Since these are non-overlapping segments, we improve the efficiency by doing this in parallel. The visual feature of each shot of a segment is also given as input to QSRM (Query relevant Segment Representative Module) in parallel along with the textual feature of the query. QSRM uses attention (refer Sect. 3.4 in the main paper) to generate a 300-dimensional segment representative vector. Again, since we have non overlapping segments, QSRM operates on all segments in parallel. For each segment, visual features of the shots are given as input to the GAM (Global Attention Module) which uses all the 300-dimensional segment representatives to generate a global representation of the shots using attention (refer Sect. 3.5 in the main paper). The visual features, local and global attention vector for each shot thus generated are concatenated together to generate a 900-dimensional representation which is passed through a fully connected layer to create a 300-dimensional embedding. This 300-dimensional embedding is fused with the 300-dimensional textual feature vector of the query using Fusion Unit (refer Sect. 3.6 in the main paper) which performs concatenation, pointwise addition and multiplication to generate a 1200-dimensional representation for each shot. This is regressed through a fully-connected layer followed by a sigmoid function to generate a score between 0 and 1 (inclusive) which is the predicted shot importance. We initialize all learnable parameters with Xavier Uniform initialization and learn the parameters end-to-end by Adam optimizer with a learning rate of 1e04 and weight decay of 1e01 using Binary Cross Entropy loss.

Inference. Inference involves taking a video and a query as input and generate a query-focussed summary for this video. We generate the visual features, local and global attention vector for each shot of the test video as described above along with the textual features of the input query. The trained model predicts the scores for each shot. The top ranking shots based on a threshold (empirically found using validation video) are assembled chronologically to construct the query-focussed summary.

4.3 Evaluation

Authors in [35] have also defined an evaluation metric which first finds a mapping between the ground truth shots and the generated summary shots by doing the maximum weight matching on a bipartite graph where weights are based on intersection-over-union (IoU) between the shots using the dense concept annotations of the shots provided in the dataset. This notion of distance or similarity takes the semantics into account and has been shown to be better than matching in visual domain or matching based on shot numbers [35]. Standard Precision, Recall and F1 scores are than calculated based on the number of matches.

Table 1. Quantitative results comparing our method against some existing query-focused video summarization techniques

	Video 1			Video 2			Video 3			Video 4			Average		
	Pre	Rec	F1	Pre	Rec	F1	Pre	Rec	F1	Pre	Rec	F1	Pre	Rec	F1
SeqDPP [12]	53.43	29.81	36.59	44.05	46.65	43.67	49.25	17.44	25.26	11.14	63.49	18.15	39.47	39.35	30.92
SH-DPP [34]	50.56	29.64	35.67	42.13	46.81	42.72	51.92	29.24	36.51	11.51	62.88	18.62	39.03	42.14	33.38
QC-DPP [35]	49.86	**53.38**	48.68	33.71	62.09	41.66	55.16	62.40	56.47	21.39	**63.12**	29.96	40.03	**60.25**	44.19
TPAN [43]	49.66	50.91	48.74	43.02	48.73	45.30	58.73	56.49	56.51	36.70	35.96	33.64	47.03	48.02	46.05
CHAN [40]	54.73	46.57	49.14	45.92	50.26	46.53	59.75	**64.53**	58.65	25.23	51.16	33.42	46.40	53.13	46.94
HVN [17]	52.55	52.91	**51.45**	38.66	**62.70**	47.49	**60.28**	62.58	**61.08**	26.79	54.21	35.47	44.57	58.10	48.87
QSAN [39]	48.41	52.34	48.52	46.51	51.36	46.64	56.78	61.14	56.93	30.54	46.90	34.25	45.56	52.94	46.59
Ours	**54.58**	52.51	50.96	**48.12**	52.15	**48.28**	58.48	61.66	58.41	**37.40**	43.90	**39.18**	**49.64**	52.55	**49.20**

4.4 Implementation Details

As defined by the UTE dataset used by [35] each shot is 5 seconds long. In this work we use fixed size segments, and empirically chose the size of each segment to be 8 shots. Since the dataset contains long ego-centric videos which do not contain fast changing events, this choice is neither too small (hence retaining sufficient local context) nor too big (hence not missing out on event changes). We leave out one video for testing and use remaining 3 videos for training and validation. We report the results when each video is used as a test video, retaining the remaining for train and validation and we also report the average performance over the four experiments. We use MLP with fully connected layers to increase and reduce the features dimensionality as required by the architecture. In each MLP, we use 3 fully connected (FC) layers and all the FC layers has 300 hidden units followed by ReLU activation functions. We initialized the weights using Xavier Uniform initialization [11], used Adam Optimizer with a learning rate $1e-04$ and weight decay $1e-01$. We use binary cross entropy loss to train the model. We compare the results of our method with some of the existing methods for Query-focused Video Summarization that have reported their results on the UTE benchmark dataset. Specifically, we chose SeqDPP [12], SH-DPP [34], QC-DPP [35], TPAN [43], CHAN [40], HVN [17] and QSAN [39] for comparison.

4.5 Results and Analysis

Table 1 shows the Precision, Recall and F1 score of our method as compared to other methods. On an average across all four videos and in three out of four videos, our method scores higher than all other methods on Precision without significant compromise in Recall. With regards to F1, which considers both precision and recall, we perform better than all techniques on an average. This is because of the use of better temporal and spatial representation using I3D features and enhancing the effect of query in selecting shots by fusing the textual features of query with the local and global attentive visual features.

In Fig. 3(a), (b), (c) and (d) we plot the shots selected by machine generated summary against the ground truth shot selections for queries (Face and Phone), (Book and Garden), (Cupglass and Desk) and (Car and Food) on video numbers 1, 2, 3, and 4 respectively (video index same as in the dataset). We observe following cases - i) for some shots, there are exact matches (as can be seen by the matching green and blue lines in each sub-figure and through sample frame visualizations), ii) there are some shots in ground truth which are not in our summary, and iii) there are some shots in our summary which are not in ground truth. With regards to ii) and iii), it is important to note, that since the evaluation is based on matching using semantic similarity and not using shot numbers, exact match based on shot numbers is not expected. As long as there is a semantic match between the generated summary shots and ground truth shots, the generated summary is still considered good. To validate this, we generated one more visualization. We plot the IoU values between the predicted shots and matching ground truth shot. We see that even those shots which are not in ground truth, have a considerably high value of IoU (>0.5) with a matching ground truth shot. With regards to IoU values, it may be noted that when the shot numbers exactly match, the IoU need not be 1. This is because of the maximum weight bipartite matching algorithm which is not greedy. For

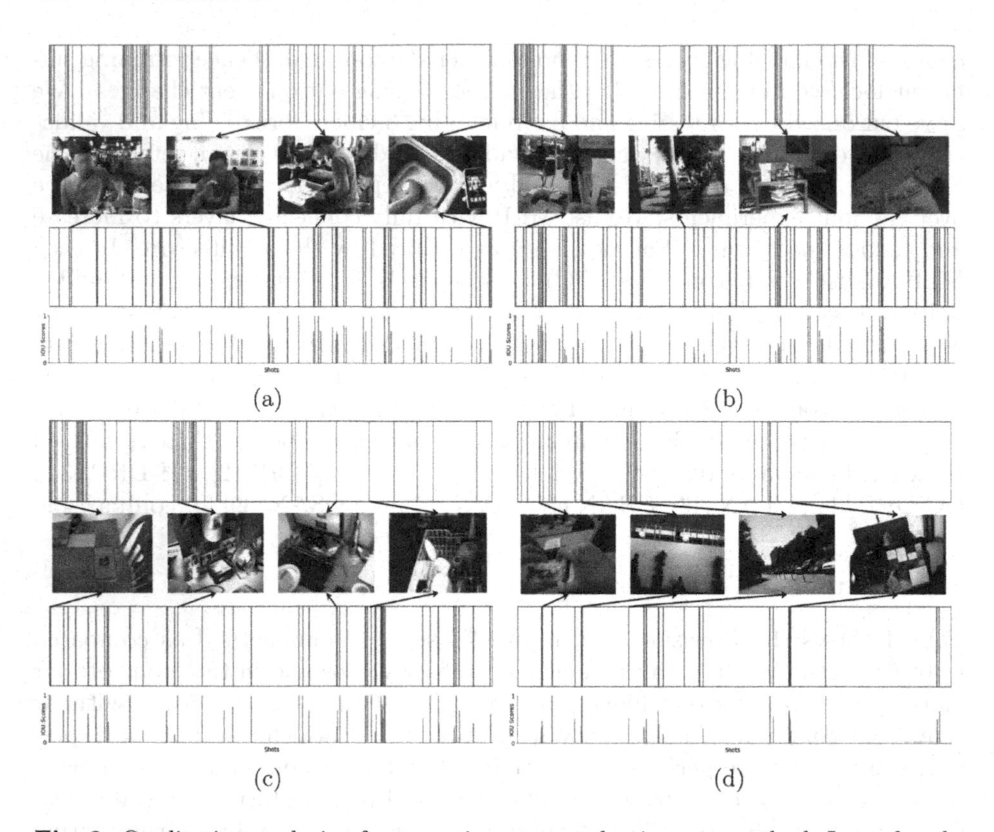

Fig. 3. Qualitative analysis of summaries generated using our method. In each sub-figure, the x-axis represents the video shot numbers, the green lines represents the ground truth shots, blue lines represents the predicted shots for a query and red bars shows the IoU score between predicted shot and the matching ground truth shot. Sub-figure (a), (b), (c) and (d) correspond to video 1, 2, 3 and 4 and queries (Face and Phone), (Book and Garden), (Cupglass and Desk) and (Car and Food) respectively (Color figure online)

example, in the case of the first visualized frame in Fig. 3(a) though it is in both ground truth as well as prediction, the IoU is seen to be less than 1.

4.6 Analysis of Model Complexity

Extensive experiments on a benchmark query-focused video summarization dataset for long videos give better results as compared to the current state of the art, thereby demonstrating the effectiveness of our method even without employing computationally expensive architectures like LSTMs, variational autoencoders, GANs or reinforcement learning, as done by most past works. To better understand the simplicity of our model as compared to some of the previous methods, we estimate a rough lower bound of the number of learnable parameters used in those methods based on the information published in the respective works. Wherever details are not mentioned, we have made assumptions, if required. Methods presented in [17,39,43,44] use LSTMs or BiLSTMs as a key component in their architecture with input dimensions ranging from 512 to 4096 and number of hidden units in these LSTMs/BiLSTMs ranging from 512 to 1024. This makes the number of learnable parameters in these models to be greater than ∼1e7. On the other hand the number of parameters in our proposed method is of the order of ∼1e5. In our method we have 3 fully connected layers with the maximum number of parameters for one of them being 900(input) × 300(output). The local and global attention weight matrices are 300 × 300 each. In addition, for mapping the features from one dimensionality to another, we have 4 weight matrices of max dimensionality for one of them being 512 × 300. Hence the total number of learnable parameters in our method is ∼1e6 is less than the lower bound estimate of other methods by an order of magnitude.

5 Conclusion

Query-focused video summarization is an important step forward in addressing the challenges associated with automatic video summarization. Past work has employed DPPs, memory networks, adversarial networks, submodular mixtures and attention networks in coming up with better techniques. In this work we proposed a simple architecture based on attention networks and query fusion and used I3D features to further the sate of the art. Extensive quantitative and qualitative evaluation of our method on the currently available benchmark data set of long videos especially made for this task establishes the effectiveness of our method.

Acknowledgements. This work is supported in part by the Ekal Fellowship (www.ekal.org) and National Center of Excellence in Technology for Internal Security, IIT Bombay (NCETIS, https://rnd.iitb.ac.in/node/101506).

References

1. Apostolidis, E., Adamantidou, E., Metsai, A.I., Mezaris, V., Patras, I.: Unsupervised video summarization via attention-driven adversarial learning. In: Ro, Y.M., et al. (eds.) MMM 2020. LNCS, vol. 11961, pp. 492–504. Springer, Cham (2020). https://doi.org/10.1007/978-3-030-37731-1_40
2. Carreira, J., Zisserman, A.: Quo vadis, action recognition? A new model and the kinetics dataset. In: proceedings of the IEEE Conference on Computer Vision and Pattern Recognition, pp. 6299–6308 (2017)
3. Chen, X., Li, X., Lu, X.: Representative and diverse video summarization. In: 2015 IEEE China Summit and International Conference on Signal and Information Processing (ChinaSIP), pp. 142–146. IEEE (2015)
4. Chu, W.S., Song, Y., Jaimes, A.: Video co-summarization: video summarization by visual co-occurrence. In: Proceedings of the IEEE Conference on Computer Vision and Pattern Recognition, pp. 3584–3592 (2015)

5. De Avila, S.E.F., Lopes, A.P.B., da Luz Jr, A., de Albuquerque Araújo, A.: VSUMM: a mechanism designed to produce static video summaries and a novel evaluation method. Pattern Recogn. Lett. **32**(1), 56–68 (2011)
6. Elfeki, M., Borji, A.: Video summarization via actionness ranking. In: 2019 IEEE Winter Conference on Applications of Computer Vision (WACV), pp. 754–763. IEEE (2019)
7. Fajtl, J., Sokeh, H.S., Argyriou, V., Monekosso, D., Remagnino, P.: Summarizing videos with attention. In: Carneiro, G., You, S. (eds.) ACCV 2018. LNCS, vol. 11367, pp. 39–54. Springer, Cham (2019). https://doi.org/10.1007/978-3-030-21074-8_4
8. Fu, T.J., Tai, S.H., Chen, H.T.: Attentive and adversarial learning for video summarization. In: 2019 IEEE Winter Conference on Applications of Computer Vision (WACV), pp. 1579–1587. IEEE (2019)
9. Gao, J., Sun, C., Yang, Z., Nevatia, R.: TALL: temporal activity localization via language query. In: Proceedings of the IEEE International Conference on Computer Vision, pp. 5267–5275 (2017)
10. Ge, R., Gao, J., Chen, K., Nevatia, R.: MAC: mining activity concepts for language-based temporal localization. In: 2019 IEEE Winter Conference on Applications of Computer Vision (WACV), pp. 245–253. IEEE (2019)
11. Glorot, X., Bengio, Y.: Understanding the difficulty of training deep feedforward neural networks. In: Proceedings of the Thirteenth International Conference on Artificial Intelligence and Statistics, pp. 249–256 (2010)
12. Gong, B., Chao, W.L., Grauman, K., Sha, F.: Diverse sequential subset selection for supervised video summarization. In: Advances in Neural Information Processing Systems, pp. 2069–2077 (2014)
13. Gygli, M., Grabner, H., Riemenschneider, H., Van Gool, L.: Creating summaries from user videos. In: Fleet, D., Pajdla, T., Schiele, B., Tuytelaars, T. (eds.) ECCV 2014. LNCS, vol. 8695, pp. 505–520. Springer, Cham (2014). https://doi.org/10.1007/978-3-319-10584-0_33
14. Gygli, M., Grabner, H., Riemenschneider, H., Van Gool, L.: Creating summaries from user videos. In: Fleet, D., Pajdla, T., Schiele, B., Tuytelaars, T. (eds.) ECCV 2014. LNCS, vol. 8695, pp. 505–520. Springer, Cham (2014). https://doi.org/10.1007/978-3-319-10584-0_33
15. Huang, J.H., Worring, M.: Query-controllable video summarization. arXiv preprint arXiv:2004.03661 (2020)
16. Ji, Z., Xiong, K., Pang, Y., Li, X.: Video summarization with attention-based encoder-decoder networks. IEEE Trans. Circ. Syst. Video Technol. (2019)
17. Jiang, P., Han, Y.: Hierarchical variational network for user-diversified & query-focused video summarization. In: Proceedings of the 2019 on International Conference on Multimedia Retrieval, pp. 202–206 (2019)
18. Jung, Y., Cho, D., Kim, D., Woo, S., Kweon, I.S.: Discriminative feature learning for unsupervised video summarization. In: Proceedings of the AAAI Conference on Artificial Intelligence, vol. 33, pp. 8537–8544 (2019)
19. Khosla, A., Hamid, R., Lin, C.J., Sundaresan, N.: Large-scale video summarization using web-image priors. In: Proceedings of the IEEE Conference on Computer Vision and Pattern Recognition, pp. 2698–2705 (2013)
20. Kim, G., Sigal, L., Xing, E.P.: Joint summarization of large-scale collections of web images and videos for storyline reconstruction. In: Proceedings of the IEEE Conference on Computer Vision and Pattern Recognition, pp. 4225–4232 (2014)
21. Kulesza, A., Taskar, B., et al.: Determinantal point processes for machine learning. Found. Trends® Mach. Learn. **5**(2–3), 123–286 (2012)

22. Lee, Y.J., Ghosh, J., Grauman, K.: Discovering important people and objects for egocentric video summarization. In: 2012 IEEE Conference on Computer Vision and Pattern Recognition, pp. 1346–1353. IEEE (2012)
23. Li, Y., Wang, L., Yang, T., Gong, B.: How local is the local diversity? Reinforcing sequential determinantal point processes with dynamic ground sets for supervised video summarization. In: Proceedings of the European Conference on Computer Vision (ECCV), pp. 151–167 (2018)
24. Ma, Y.F., Lu, L., Zhang, H.J., Li, M.: A user attention model for video summarization. In: Proceedings of the tenth ACM International Conference on Multimedia, pp. 533–542. ACM (2002)
25. Mahasseni, B., Lam, M., Todorovic, S.: Unsupervised video summarization with adversarial LSTM networks. In: Proceedings of the IEEE Conference on Computer Vision and Pattern Recognition (CVPR) (2017)
26. Panda, R., Mithun, N.C., Roy-Chowdhury, A.K.: Diversity-aware multi-video summarization. IEEE Trans. Image Process. **26**(10), 4712–4724 (2017)
27. Plummer, B.A., Brown, M., Lazebnik, S.: Enhancing video summarization via vision-language embedding. In: Proceedings of the IEEE Conference on Computer Vision and Pattern Recognition, pp. 5781–5789 (2017)
28. Potapov, D., Douze, M., Harchaoui, Z., Schmid, C.: Category-specific video summarization. In: Fleet, D., Pajdla, T., Schiele, B., Tuytelaars, T. (eds.) ECCV 2014. LNCS, vol. 8694, pp. 540–555. Springer, Cham (2014). https://doi.org/10.1007/978-3-319-10599-4_35
29. Pritch, Y., Ratovitch, S., Hendel, A., Peleg, S.: Clustered synopsis of surveillance video. In: 2009 Advanced Video and Signal Based Surveillance, pp. 195–200. IEEE (2009)
30. Pritch, Y., Rav-Acha, A., Gutman, A., Peleg, S.: Webcam synopsis: peeking around the world. In: IEEE 11th International Conference on Computer Vision, ICCV 2007, pp. 1–8. IEEE (2007)
31. Pritch, Y., Rav-Acha, A., Peleg, S.: Nonchronological video synopsis and indexing. IEEE Trans. Pattern Anal. Mach. Intell. **30**(11), 1971–1984 (2008)
32. Rav-Acha, A., Pritch, Y., Peleg, S.: Making a long video short: dynamic video synopsis. In: 2006 IEEE Computer Society Conference on Computer Vision and Pattern Recognition, vol. 1, pp. 435–441. IEEE (2006)
33. Shao, D., Xiong, Y., Zhao, Y., Huang, Q., Qiao, Y., Lin, D.: Find and focus: retrieve and localize video events with natural language queries. In: Proceedings of the European Conference on Computer Vision (ECCV), pp. 200–216 (2018)
34. Sharghi, A., Gong, B., Shah, M.: Query-focused extractive video summarization. In: Leibe, B., Matas, J., Sebe, N., Welling, M. (eds.) ECCV 2016. LNCS, vol. 9912, pp. 3–19. Springer, Cham (2016). https://doi.org/10.1007/978-3-319-46484-8_1
35. Sharghi, A., Laurel, J.S., Gong, B.: Query-focused video summarization: dataset, evaluation, and a memory network based approach. In: Proceedings of the IEEE Conference on Computer Vision and Pattern Recognition, pp. 4788–4797 (2017)
36. Vasudevan, A.B., Gygli, M., Volokitin, A., Van Gool, L.: Query-adaptive video summarization via quality-aware relevance estimation. In: Proceedings of the 25th ACM International Conference on Multimedia, pp. 582–590 (2017)
37. Wang, W., Huang, Y., Wang, L.: Language-driven temporal activity localization: a semantic matching reinforcement learning model. In: Proceedings of the IEEE Conference on Computer Vision and Pattern Recognition, pp. 334–343 (2019)
38. Wolf, W.: Key frame selection by motion analysis. In: 1996 IEEE International Conference on Acoustics, Speech, and Signal Processing, ICASSP-96. Conference Proceedings, vol. 2, pp. 1228–1231. IEEE (1996)

39. Xiao, S., Zhao, Z., Zhang, Z., Guan, Z., Cai, D.: Query-biased self-attentive network for query-focused video summarization. IEEE Trans. Image Process. **29**, 5889–5899 (2020)
40. Xiao, S., Zhao, Z., Zhang, Z., Yan, X., Yang, M.: Convolutional hierarchical attention network for query-focused video summarization. arXiv preprint arXiv:2002.03740 (2020)
41. Yuan, L., Tay, F.E., Li, P., Zhou, L., Feng, J.: Cycle-sum: cycle-consistent adversarial LSTM networks for unsupervised video summarization. arXiv preprint arXiv:1904.08265 (2019)
42. Zhang, K., Chao, W.-L., Sha, F., Grauman, K.: Video summarization with long short-term memory. In: Leibe, B., Matas, J., Sebe, N., Welling, M. (eds.) ECCV 2016. LNCS, vol. 9911, pp. 766–782. Springer, Cham (2016). https://doi.org/10.1007/978-3-319-46478-7_47
43. Zhang, Y., Kampffmeyer, M., Liang, X., Tan, M., Xing, E.P.: Query-conditioned three-player adversarial network for video summarization. arXiv preprint arXiv:1807.06677 (2018)
44. Zhang, Y., Kampffmeyer, M., Zhao, X., Tan, M.: Deep reinforcement learning for query-conditioned video summarization. Appl. Sci. **9**(4), 750 (2019)
45. Zhou, K., Qiao, Y., Xiang, T.: Deep reinforcement learning for unsupervised video summarization with diversity-representativeness reward. In: Thirty-Second AAAI Conference on Artificial Intelligence (2018)
46. Zhu, X., Loy, C.C., Gong, S.: Learning from multiple sources for video summarisation. Int. J. Comput. Vis. **117**(3), 247–268 (2016)

Late Temporal Modeling in 3D CNN Architectures with BERT for Action Recognition

M. Esat Kalfaoglu[1,3(✉)], Sinan Kalkan[2,3], and A. Aydin Alatan[1,3]

[1] Department of Electrical and Electronics Engineering, Middle East Technical University, Ankara, Turkey
[2] Department of Computer Engineering, Middle East Technical University, Ankara, Turkey
[3] Center for Image Analysis (OGAM), Middle East Technical University, Ankara, Turkey
{esat.kalfaoglu,skalkan,alatan}@metu.edu.tr

Abstract. In this work, we combine 3D convolution with late temporal modeling for action recognition. For this aim, we replace the conventional Temporal Global Average Pooling (TGAP) layer at the end of 3D convolutional architecture with the Bidirectional Encoder Representations from Transformers (BERT) layer in order to better utilize the temporal information with BERT's attention mechanism. We show that this replacement improves the performances of many popular 3D convolution architectures for action recognition, including ResNeXt, I3D, SlowFast and R(2+1)D. Moreover, we provide the-state-of-the-art results on both HMDB51 and UCF101 datasets with 85.10% and 98.69% top-1 accuracy, respectively. The code is publicly available github.com/artest08/LateTemporalModeling3DCNN.

Keywords: Action recognition · Temporal attention · BERT · Late temporal modeling · 3D convolution

1 Introduction

Action Recognition (AR) pertains to identifying the label of the action observed in a video clip. With cameras everywhere, AR has become essential in many domains, such as video retrieval, surveillance, human-computer interaction and robotics.

A video clip contains two critical pieces of information for AR: Spatial and temporal information. Spatial information represents the static information in the scene, such as objects, context, entities etc., which are visible in a single frame of the video, whereas temporal information, obtained by integrating the spatial information over frames, mostly captures the dynamic nature of the action.

In this work, the joint utilization of two temporal modeling concepts from the literature, which are 3D convolution and late temporal modeling, is proposed and

A. Bartoli and A. Fusiello (Eds.): ECCV 2020 Workshops, LNCS 12539, pp. 731–747, 2020.
https://doi.org/10.1007/978-3-030-68238-5_48

analyzed. Briefly, 3D convolution is a way of generating a temporal relationship hierarchically from the beginning to the end of CNN architectures. On the other hand, late temporal modeling is typically utilized with 2D CNN architectures, where the features extracted by 2D CNN architectures from the selected frames are usually modeled with recurrent architectures, such as LSTM, Conv LSTM.

Despite its advantages, temporal global average pooling (TGAP) layer which is used at the end of all 3D CNN architectures [1,2,7,12,22,27,28,35] hinders the richness of final temporal information. The features before TGAP can be considered as features of different temporal regions of a clip or video. Although, the receptive field might cover the whole clip, the effective receptive field has a Gaussian distribution [20], producing features focusing on different temporal regions of a clip. In order to discriminate an action, one part of the temporal feature might be more important than the others or the order of the temporal features might be more beneficial than simply averaging the temporal information. Therefore, TGAP ignores this ordering and fails to fully exploit the temporal information.

Therefore, we propose using attention mechanism of BERT for better temporal modeling than TGAP. BERT determines which temporal features are more important with its multi-head attention mechanism.

To the best of our knowledge, our work is the first to propose replacing TGAP in 3D CNN architectures with late temporal modeling. We also consider that this study is the first to utilize BERT as a temporal pooling strategy in AR. We show that BERT performs better temporal pooling than average pooling, concatenation pooling and standard LSTM. Moreover, we demonstrate that late temporal modeling with BERT improves the performances of various popular 3D CNN architectures for AR which are ResNeXt101, I3D, SlowFast, and R(2+1)D by using the split-1 of HMDB51 dataset. Using BERT R(2+1)D architecture, we obtain new state of the art results; 85.10% and 98.69% top-1 performances in HMDB51 and UCF101 datasets, respectively.

2 Related Work on Action Recognition

In this section, the AR literature is analyzed in two aspects: (i) temporal integration using pooling, fusion or recurrent architectures and (ii) 3D CNN architectures.

2.1 Temporal Integration Using Pooling, Fusion or Recurrent Architectures

Pooling is a well-known technique to combine various temporal features; concatenation, averaging, maximum, minimum, ROI, feature aggregation techniques and time-domain convolution are some of the possible pooling techniques [11,21].

Fusion frequently used for AR is very similar to pooling. Fusion is sometimes preferred instead of pooling in order to emphasize pooling location in the architecture or to differentiate information from different modalities. Late fusion, early fusion and slow fusion models on 2D CNN architectures can be performed by combining temporal information along the channel dimension at various points in CNN architectures [15]. As a method, the two-stream fusion architecture in [8] creates spatio-temporal relationship with extra 3D convolution layer inserted towards the end of the architecture and fuses information from RGB and optical flow streams.

Recurrent networks are also commonly used for temporal integration. LSTMs are utilized for temporal (sequential) modeling on 2D CNN features extracted from the frames of a video [5,21]. E.g., VideoLSTM [17] performs this kind of temporal modeling by using convolutional LSTM with spatial attention. RSTAN [6] implements both temporal and spatial attention concepts on LSTM and the attention weights of RGB and optical flow streams are fused.

2.2 3D CNN Architectures

3D CNNs are networks formed of 3D convolution throughout the whole architecture. In 3D convolution, filters are designed in 3D, and channels and temporal information are represented as different dimensions. Compared to the temporal fusion techniques, 3D CNNs process the temporal information hierarchically and throughout the whole network. Before 3D CNN architectures, temporal modeling was generally achieved by using an additional stream of optical flow or by using temporal pooling layers. However, these methods were restricted to 2D convolution and temporal information was put into the channel dimension. The downside of the 3D CNN architectures is that they require huge computational costs and memory demand compared to its 2D counterparts.

The first 3D CNN for AR is the C3D model [26]. Another successful implementation of 3D convolution is the Inception 3D model (I3D) [1], in which 3D convolution is modeled in a much deeper fashion compared to C3D. The ResNet version of 3D convolution is introduced in [12]. Then, R(2+1)D [28] and S3D [35] architectures are introduced in which 3D spatio-temporal convolutions are factorized into spatial and temporal convolutions, and shown to be more effective than traditional 3D convolution architectures. Another important 3D CNN architecture is Channel-Separated Convolutional Networks (CSN) [27] which separates the channel interactions and spatio-temporal interactions which can be thought as the 3D CNN version of depth-wise separable convolution [14].

Slow-fast networks [7] can be considered as a joint implementation of both fusion techniques and 3D CNN architectures. There are two streams, namely fast and slow paths. Slow stream operates at low frame and focuses on spatial information, as the RGB stream in traditional two stream architectures, while fast stream operates at high frame and focuses on temporal information as optical flow stream in traditional two-stream architectures. There is information flow from the fast stream to slow stream.

Although 3D CNNs are powerful, they still lack an effective temporal fusion strategy at the end of the architecture.

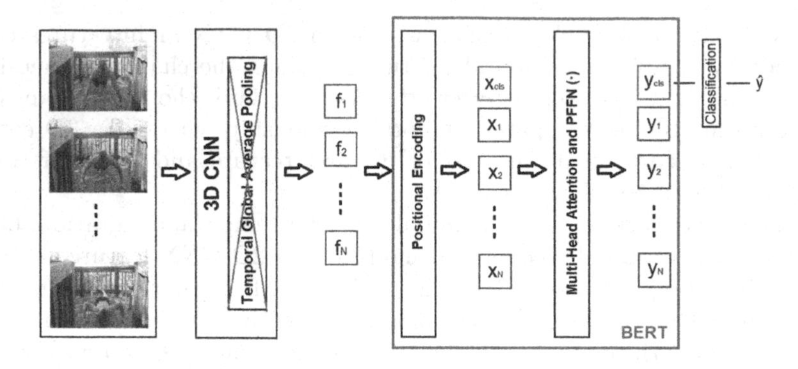

Fig. 1. BERT-based late temporal modeling

3 Proposed Method: BERT-Based Temporal Modeling with 3D CNN for Activity Recognition

Bi-directional Encoder Representations from Transformers (BERT) [4] is a bidirectional self-attention method, which has provided unprecedented success in many downstream Natural Language Processing (NLP) tasks. The bidirectional property enables BERT to fuse the contextual information from both directions, instead of relying upon only a single direction, as in former recurrent neural networks or other self-attention methods, such as Transformer [29]. Moreover, BERT introduces challenging unsupervised pre-training tasks which leads to useful representations for many tasks.

Our architecture utilizes BERT-based temporal pooling as shown in Fig. 1. In this architecture, the selected K frames from the input sequence is propagated through a 3D CNN architecture without temporal global average pooling at the end of architecture. Then, in order to preserve the positional information, a learned positional encoding is added to the extracted features. To perform classification with BERT, additional classification embedding ($\mathbf{x_{cls}}$) is appended as in [4] (represented as red box in Fig. 1). The classification of the architecture is implemented with the corresponding classification vector $\mathbf{y_{cls}}$ which is given to the fully connected layer, producing the predicted output label $\hat{y}$.

The general single head self-attention model of BERT is formulated as:

$$\mathbf{y_i} = PFFN\left(\frac{1}{N(x)}\sum_{\forall j} g(\mathbf{x_j})f(\mathbf{x_i},\mathbf{x_j})\right), \tag{1}$$

where $\mathbf{x_i}$ values are the embedding vectors that consists of extracted temporal visual information and its positional encoding; i indicates the index of the target output temporal position; j denotes all possible combinations; and $N(x)$ is the normalization term. Function $g(\cdot)$ is the linear projection inside the self-attention mechanism of BERT, whereas function $f(\cdot,\cdot)$ denotes the similarity between $\mathbf{x_i}$ and $\mathbf{x_j}$: $f(\mathbf{x_i},\mathbf{x_j}) = \text{softmax}_j(\theta(\mathbf{x_i})^T\phi(\mathbf{x_j}))$, where the functions $\theta(\cdot)$ and $\phi(\cdot)$

are also linear projections. The learnable functions $g(\cdot)$, $\theta(\cdot)$ and $\phi(\cdot)$ try to project the feature embedding vectors to a better space where the attention mechanism works more efficiently. The outputs of $g(\cdot)$, $\theta(\cdot)$ and $\phi(\cdot)$ functions are also defined as value, query and key, respectively [29]. $PFFN(\cdot)$ is Position-wise Feed-forward Network applied to all positions separately and identically: $PFFN(x) = \mathbf{W_2}GELU(\mathbf{W_1x} + \mathbf{b1}) + \mathbf{b2}$, where $GELU(\cdot)$ is the Gaussian Error Linear Unit (GELU) activation function [13].

The final decision of classification is performed with one more linear layer which takes $\mathbf{y_{cls}}$ as input. The explicit form of $\mathbf{y_{cls}}$ can be written as:

$$\mathbf{y_{cls}} = PFFN\left(\frac{1}{N(x)}\sum_{\forall j} g(\mathbf{x_j})f(\mathbf{x_{cls}}, \mathbf{x_j})\right). \quad (2)$$

Therefore, our use of temporal attention mechanism for BERT is not only to learn the convenient subspace where the attention mechanism work efficiently but also learn the classification embedding which learns how to attend the temporal features of the 3D CNN architecture properly.

A similar work for action recognition is implemented with non-local neural networks (NN) [34]. The main aim of Non-local block is to create global spatio-temporal relations because convolution operation is limited to local regions. Another similar study for action recognition is video action transformer network [10] where transformer is utilized in order to aggregate contextual information from other people and objects in the surrounding video. Video action transformer network deals with both action localization and action recognition, therefore its problem formulation is different from ours and its attention mechanism needs to be reformulated for the late temporal modeling for action recognition. Differently from video action transformer network, our proposed BERT-based late temporal modeling utilizes the learnt classification token, instead of using the feature of the backbone architecture.

4 Experiments

In this part, dataset, implementation details, ablation study, results on different architectures, and comparison with state-of-the-art sections are presented, respectively.

4.1 Dataset

Four datasets are relevant for our study: HMDB51 [16], UCF101 [25], Kinetics-400 [1] and IG65M [9] datasets. HMDB51 consists of ~7k clips with 51 classes whereas UCF101 includes ~13k clips with 101 classes. Both HMDB51 and UCF101 define three data splits and performances are calculated by averaging the results on these three splits. Kinetics-400 consists of about 240k clips with 400 classes. IG65M is a weakly supervised dataset which is collected by using the Kinetics-400 [1] class names as hashtags on Instagram. There are 65M clips from 400 classes. The dataset is not public for the time being but the pre-trained models are available.

For analyzing the improvements of BERT on individual architectures (Sect. 4.4), split 1 of the HMDB51 dataset is used whereas the comparisons with the state of the art (See Sect. 4.5) are performed using the three splits of the HMDB51 and UCF101 datasets. Additionally, the ablation study (See Sect. 4.3) is conducted using the three splits of HMDB51. Moreover, Kinetics-400 and IG65M are used for pre-trained weights of the architectures before fine-tuning on HMDB51 and UCF101. The pre-trained weights are obtained from the authors of architectures, which are ResNeXt, I3D, Slowfast and R(2+1)D. Among these architectures, R(2+1)D is pre-trained with IG65M but the rest of the architectures are pre-trained with Kinetics-400.

4.2 Implementation Details

For the standard architectures (with TGAP and without any modification to architectures), SGD with learning rate $1e{-}2$ is utilized, except I3D in which learning rate is set to $1e{-}1$ empirically. For architectures with BERT, the ADAMW optimizer [19] with learning rate $1e{-}5$ is utilized except I3D for which the learning rate is set to $1e{-}4$ empirically. For all training runs, the "reducing learning rate on plateau" scheduling is followed. The data normalization schemes are selected conforming with the data normalization schemes of the pre-training of the architectures in order to benefit fully the from pre-training weights. Multi-scale cropping scheme is applied for fine-tuning and testing of all architectures [32]. In the test time, the scores of non-overlapping clips are averaged. The optical flow of the frames are extracted with TV-L1 algorithm [36].

In the BERT architecture, there are eight attention heads and one transformer block. The dropout ratio in $PFFN(\cdot)$ is set to 0.9. Mask operation is applied with 0.2 probability. Instead of using a mask token, the attention weight of masked feature is set to zero. The classification token ($\mathbf{x_{cls}}$) and the learned positional embeddings are initialized as zero mean normal weight with 0.02 standard deviation.

4.3 Ablation Study

We will now analyze each step of our contribution and how our method compares with alternative pooling strategies – see Table 1. For this analysis, ResNeXt101 backbone is utilized with RGB modality, with 112×112 input image size and with 64-frame clips. In this table, temporal pool types, the existence of Feature Reduction with Modified Block (FRMB), the type of the optimizer, top1 performances, the number of parameters and the number of operations are presented as the columns of the analysis.

One important issue is the optimizer. For training BERT architectures in NLP tasks, the ADAM optimizer is chosen [4]. However, SGD is preferred for 3D CNN architectures [1,3,7,12,28]. Therefore, for training BERT, we choose

Table 1. Ablation Study of RGB ResNeXt101 architecture for temporal pooling analysis on HMDB51. FRMB: Feature Reduction with Modified Block.

Type of temporal pooling	FRMB?	Optimizer	Top1 (%)	# of Params	# of operations
Average pooling (baseline)		SGD	74.46	47.63 M	38.56 GFlops
Average pooling		ADAMW	75.99	47.63 M	38.56 GFlops
Average pooling	✓	ADAMW	74.97	44.22 M	38.36 GFlops
LSTM	✓	ADAMW	74.18	47.58 M	38.37 GFlops
Non-local + Concatenation + Fully connected layer	✓	ADAMW	76.36	47.35 M	38.43 GFlops
Concatenation	✓	ADAMW	76.49	44.30 M	38.36 GFlops
Concatenation + Fully Connected Layer	✓	ADAMW	76.84	47.45 M	38.36 GFlops
BERT pooling (ours)	✓	ADAMW	**77.49**	47.38 M	38.37 GFlops

ADAMW and not ADAM because ADAMW improves the generalization capability of ADAM [19]. In this ablation study, ResNeXt101 architecture (with Average Pooling in Table 1) is also trained with both ADAMW in Table 1 which shows 1.5% increase in performance compared to SGD.

In order to utilize BERT architecture in a more parameter efficient manner, the feature dimension of the output of the ResNeXt101 backbone is reduced from 2048 to 512. For this, two possible methods are considered. These are Feature Reduction with Modified Block (FRMB) and Feature Reduction with Additional Block (FRAB). In FRMB, the final bottleneck block of ResNeXt101 block is replaced with a newer bottleneck block for the feature dimension reduction. In FRAB, an additional bottleneck block is appended to the backbone to reduce dimensionality. The visualization for the implementation of FRMB and FRAB is presented in Fig. 2. For this ablation study, FRMB implementation is chosen for two reasons over FRAB. Firstly, FRMB yields about 0.5% better top1 performance than FRAB. Secondly, FRMB has a better computational complexity and parameter efficiency than FRAB because FRAB introduces an additional block to the whole architecture. Therefore, we choose FRMB owing to its lower computational complexity and better parameter efficiency at the cost of $\sim$1% decrease in top1 performance compared to the standard backbone (Table 1).

For a fair comparison, we set the hyper-parameters of the other pooling strategies (LSTM, Non-Local + concatenation + fully connected layer, and concatenation + fully connected layer) such that the number of parameters and the number of operations of different temporal pooling strategies are more or less the same with the proposed BERT pooling: LSTM is implemented in two stacks and with a hidden-layer size 450. The dimension of inter-channels of a Non-Local Attention block (the dimension size of attention mechanism) is set equal to the input size to the Non-Local block which is 512. The number of nodes of a fully connected layer is determined according to the need for equal parameter size with the proposed BERT temporal pooling for fair comparison.

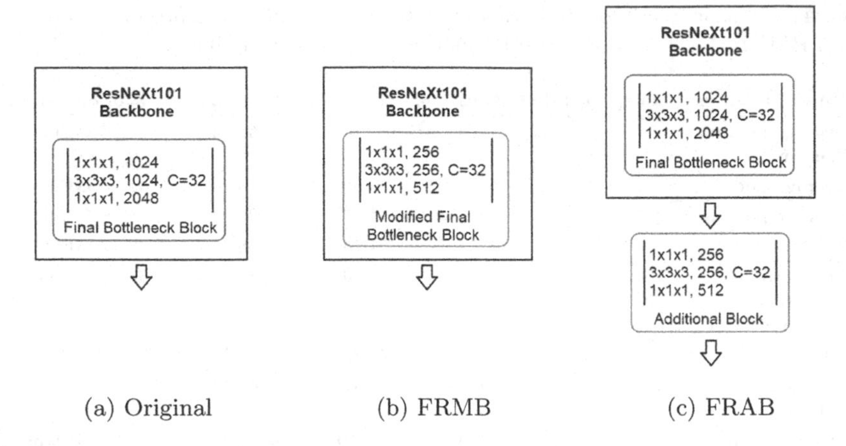

Fig. 2. The implementations of Feature Reduction with Modified Block (FRMB) and Feature Reduction with Additional Block (FRAB)

When we analyze Table 1, we observe that, among the 5 different alternatives (with FRMB), BERT is the best temporal pooling strategy. Additionally, our proposed FRMB-ResNeXt101-BERT provides 3% better top1 accuracy than the ResNeXt101-Average Pooling (Baseline) despite the fact that FRMB-ResNeXt101-BERT has a better computational complexity and parameter efficiency than the ResNeXt101-Average Pooling (Baseline) – See Table 1). The BERT layer itself has about 3M parameters and negligible computational complexity with respect to the ResNeXt101 backbone. For the other temporal pooling strategies, LSTM worsens the performance with respect to the temporal average pooling. Concatenation + fully connected layer is also another successful strategy in order to utilize the temporal features better than the average pooling. The addition of a Non-Local Attention block previously to the concatenation + fully connected layer also worsens the performance compared to only concatenation + fully connected layer pooling implementation. It should be highlighted that the original implementation of Non-Local study [34] also prefers not to utilize the Non-Local block at the end of final three bottleneck blocks, which is a consistent fact with the experimental result of this study related with Non-Local implementation.

In addition, the experiment of BERT pooling with two layers is also implemented. Top1 accuracy of it is 77.24% which is 0.25% lower than single layer BERT pooling implementation. Moreover, the memory trade-off of every layer of BERT is about 3M. The reason behind the deterioration might be the fact that late temporal modeling is not as much complex as capturing rich linguistic information and single layer might be enough to capture the temporal relationship between the output features of 3D CNN architectures. Moreover, the experiment of single head attention is performed in order to understand the impact of multi-head attention mechanism. Top1 accuracy of it is 76.97% which is 0.52% lower than multi-head implementation with eight attention heads, indicating

Table 2. Analysis of ResNeXt101 architecture with and without BERT for RGB, Flow, and two-stream modalities on HMDB51 split-1

BERT	Modality	Top1	# parameters	# operations
	RGB	74.38	47.63 M	38.56 GFlops
✓	RGB	**77.25**	47.38 M	38.37 GFlops
	Flow	79.48	47.60 M	34.16 GFlops
✓	Flow	**82.03**	47.36 M	33.97 GFlops
	Both	82.09	95.23 M	72.72 GFlops
✓	Both	**83.99**	94.74 M	72.34 GFlops

that learning multiple temporal relations is also beneficial. Additionally, the experiment of replacing learnt classification token with the average of extracted temporal features is implemented and top1 accuracy of it is 76.07% which is 1.42% lower than the implementation with the classification token.

4.4 Results on Different Architectures

In this part, the improvements brought by the replacement of TGAP with BERT pooling on popular 3D convolution architectures for action recognition is presented, including ResNeXt101 [12], I3D [1], SlowFast [7] and R(2+1)D [28].

ResNeXt Architecture. ResNeXt architecture is essentially ResNet with group convolutions [12]. For testing this architecture, the input size is selected as 112 × 112 as in the study of [3,12] and 64 frame length is utilized.

The results of the ResNeXt101 architecture is given in Table 2. The performance of the architectures are compared over RGB modality, (optical) Flow modality and Both (two-stream) in which both RGB and Flow-streams are utilized and the scores are summed from each stream. In this table, the number of parameters and operations of the architectures are also presented. The implementation of FRMB is chosen over FRAB for this analysis (See Sect. 4.3 for more details about FRAB and FRMB). Based on the results in Table 2, the most important conclusion is the improvement of the performance by using BERT over the standard architectures (without BERT) in *all* modalities.

I3D Architecture. I3D architecture is an Inception-type architecture. During I3D experiments, the input size is selected as 224 × 224 and 64 frame length is used conforming with the I3D study [1]. The result of BERT experiments on I3D architecture is given in Table 3. In the table, there are two BERT implementation which are with and without FRAB. For I3D-BERT architectures with FRAB, the final feature dimension of I3D backbone is reduced from 1024 to 512 in order

Table 3. The performance analysis of I3D architecture with and without BERT for RGB, Flow, and two-stream modalities on HMDB51 split-1

BERT	Modality	Feature reduction	Top1	# parameters	# operations
	RGB	X	75.42	12.34 M	111.33 GFlops
✓	RGB	FRAB	**75.75**	16.40 M	111.72 GFlops
✓	RGB	X	75.69	24.95 M	111.44 GFlops
	Flow	X	77.97	12.32 M	102.52 GFlops
✓	Flow	FRAB	77.25	16.37 M	102.91 GFlops
✓	Flow	X	**78.37**	24.92 M	102.63 GFlops
	Both	X	82.03	24.66 M	213.85 GFlops
✓	Both	FRAB	**82.68**	32.77 M	214.63 GFlops
✓	Both	X	**82.68**	49.87 M	214.07 GFlops

to utilize BERT in a more parameter efficient manner. However, contrary to the ResNeXt101-BERT architecture, FRAB is chosen instead of FRMB because FRAB obtains about 3.6% better top1 result for RGB-I3D-BERT architecture on split1 of HMDB51 (See Sect. 4.3 for more details about FRAB and FRMB). The reason behind the success of FRAB over FRMB might be that the final Inception block of I3D does not benefit from the pre-trained weights of the larger dataset because of the modification in FRMB.

The experimental results in Table 3 indicate that BERT increases the performance of I3D architectures in all modalities but the improvements are less and limited compared to ResNeXt. For the flow modality, although there is a performance improvement for BERT without FRAB, the implementation of BERT with FRAB performs worse than the standard I3D architecture, implying that preserving the feature size is more important for flow modality compared to RGB modality in I3D architecture. For two-stream setting, both of the proposed BERT architectures performs equally with each other and perform better than standard I3D with 0.65 % top1 performance increase.

SlowFast Architecture. SlowFast architecture [7] introduces a different perspective for the two-stream architectures. Instead of utilizing two different modalities as two identical streams, the overall architecture includes two different streams (namely fast and slow streams or paths) with different capabilities with only RGB modality. In SlowFast architecture, slow stream has a better spatial capability, while fast stream has a better temporal capability. Fast stream has better temporal resolution and less channel capacity compared to the slow stream. Although it might be possible to utilize SlowFast architecture with also optical flow modality, the authors of SlowFast did not consider this in their study. Therefore, in this study, the analysis of BERT is also implemented by only considering the RGB modality.

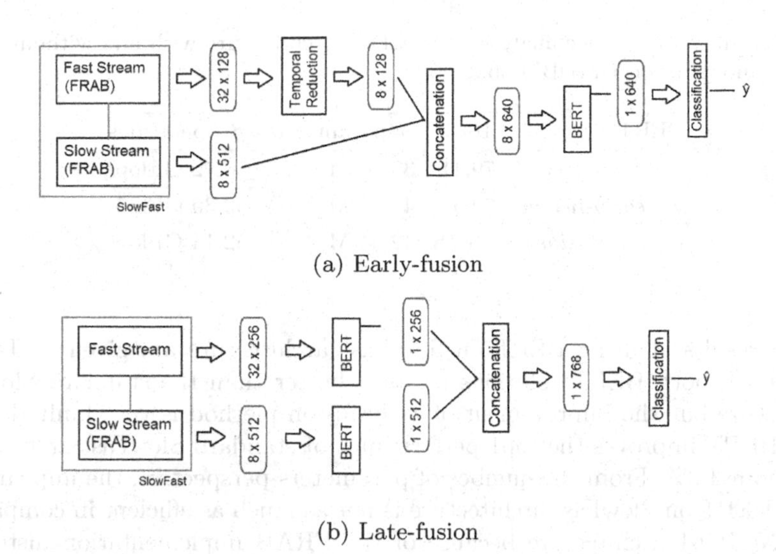

Fig. 3. Early-fusion and late-fusion implementations of BERT on SlowFast architecture.

The SlowFast architecture in our experiment is derived from a ResNet-50 architecture. The channel capacity of the fast streams is one eighth of the channel capacity of the slow stream. The temporal resolution of fast stream is four times the temporal resolution for the slow stream. The input size is selected as 224 × 224 and 64-frame length is utilized with the SlowFast architecture conforming with the SlowFast study [7].

For the implementation of BERT on SlowFast architecture, two alternative solutions are proposed: Early-fusion BERT and late-fusion BERT. In early-fusion BERT, the temporal features are concatenated before BERT layer and only a single BERT module is utilized. To make the concatenation feasible, the temporal resolution of the fast stream is decreased to the temporal resolution of the slow stream. In late-fusion BERT, two different BERT modules are utilized, one for each stream, and the outputs of two BERT modules from two streams are concatenated. The figure for early-fusion and late-fusion is shown in Fig. 3.

In order to utilize BERT architecture with less parameters, the final feature dimension of SlowFast backbone is reduced similar to the ResNeXt101-BERT and I3D-BERT architectures. Similar to the I3D-BERT architecture, FRAB is chosen instead of FRMB because FRAB obtains about 1.5% better top1 result for SlowFast-BERT architecture on the split1 of HMDB51 (See Sect. 4.3 for more details about FRAB and FRMB). For early-fusion BERT, the feature dimension of the slow stream is reduced from 2048 to 512 and the feature dimension of the fast stream is reduced from 256 to 128. For late-fusion BERT, only the feature dimension of the slow stream is reduced from 2048 to 512. The details about the size of the dimensions can also be seen in Fig. 3.

Table 4. The performance analysis of SlowFast architecture with and without BERT for RGB modality on HMDB51 split-1

BERT	Top1	# parameters	# operations
	79.41	33.76 M	50.72 GFlops
✓ (*early-fusion*)	79.54	43.17 M	52.39 GFlops
✓ (*late-fusion*)	**80.78**	42.04 M	52.14 GFlops

The results of using BERT on SlowFast architecture are given in Table 4. First of all, both BERT solutions perform better than the standard SlowFast architecture but the improvement of early-fusion method is very limited. Late-fusion BERT improves the top1 performance of standard SlowFast architecture with about 1.3%. From the number of parameters perspective, the implementation of BERT on SlowFast architecture is not as much as efficient in comparison to ResNeXt101 architecture because of the FRAB implementation instead of FRMB as in the case of I3D-BERT. Moreover, the parameter increase of RGB-SlowFast-BERT is even higher than RGB-I3D-BERT because of the two-stream implementation of SlowFast network for RGB input modality. The increase in the number of operations is also higher in the implementation of SlowFast-BERT than the I3D-BERT and ResNeXt101-BERT because of the higher temporal resolution in SlowFast architecture and two-stream implementation for RGB modality.

For the two alternative proposed BERT solution in Table 4, late-fusion BERT yields better performance with better computational complexity in contrast with early-fusion BERT. Although the attention mechanism is implemented jointly on the concatenated features, the destruction of the temporal richness of fast stream at some degree might be the reason for the worse performance of early-fusion BERT.

R(2+1)D Architecture. R(2+1)D [28] architecture is a ResNet-type architecture consisting of separable 3D convolutions in which temporal and spatial convolutions are implemented separately. For this architecture, 112×112 input dimensions are applied following the paper and 32-frame length is applied instead of 64-frame because of huge memory demand of this architecture and to be consistent with the paper [28]. The selected R(2+1)D architecture has 34 layers and implemented with basic block type instead of bottleneck block type (for further details about block types, see [12]). The most important difference of R(2+1)D experiments from the previous architectures is the utilization of the IG65M pre-trained weights, instead of Kinetics pre-trained weights (see Sect. 4.1 for details). Therefore, this detail should be considered while comparing this architecture with the aforementioned ones. The analysis of R(2+1)D BERT architecture is limited to RGB modality since the study [9] of the IG65M dataset where R(2+1)D architecture is preferred is limited to RGB modality.

Table 5. The performance analysis of R(2+1)D architecture with and without BERT for RGB modality on HMDB51 split-1

# BERT	Top1	# parameters	# operations
	82.81	63.67 M	152.95 GFlops
✓	**84.77**	66.67 M	152.97 GFlops

The experiments of BERT on R(2+1)D architecture are presented in Table 5. The feature dimension of R(2+1)D architecture is already 512 which is the same with the reduced feature dimension of ResNeXt101 and I3D backbones for BERT implementations. Therefore, we do not use FRMB or FRAB for R(2+1)D. There is an increase of about 3M parameters and the increase in the number of operations is still negligible. The performance increase of BERT on R(2+1)D architecture is about 2% which is a significant increase for RGB modality as in the case of ResNeXt101-BERT architecture.

4.5 Comparison with State-of-the-Art

In this section, the results of the best BERT architectures from the previous section are compared against the state-of-the-art methods. For this aim, two leading BERT architectures are selected: Two-Stream BERT ResNeXt101 and RGB BERT R(2+1)D (see Sect. 4.4). Note that these two architectures use different pre-training datasets, namely IG65 and Kinetics-400 for ResNext101 and R(2+1)D, respectively.

The results of the architectures on HMDB51 and UCF101 datasets are presented in Table 6. The table indicates if an architecture employs explicit optical flow. Moreover, the table lists the pre-training dataset used by the methods.

As shown in Table 6, BERT increases the Top-1 performance of the two-stream ResNeXt101 with 1.77% and 0.41% in HMDB51 and UCF101, respectively. Additionally, BERT improves the Top-1 performance of RGB R(2+1)D (32f) with 3.5 % and 0.48% in HMDB51 and UCF101, respectively, where 32f corresponds to 32-frame length. The results obtained by the R(2+1)D BERT (64f) architecture is the current state-of-the-art result in AR, to the best of our knowledge. Among the architectures pre-trained in Kinetics-400, the two-stream ResNeXt101 BERT is again the best in HMDB51 but the second best in the UCF101 dataset. This might be owing to the fact that HMDB51 involves some actions that can be resolved only using temporal reasoning and therefore benefits from BERT's capacity.

An important point to note from the table is the effect of pre-training with the IG65M dataset. RGB R(2+1)D (32f) (without Flow) pre-trained with IG65M obtains about 6% and 1.4% better Top-1 performance in HMDB51 and UCF101, respectively, than the one pre-trained with Kinetics-400, indicating the importance of the number of samples in pre-training dataset even if the samples are collected in a weakly-supervised manner.

Table 6. Comparison with the state-of-the-art.

Model	Uses flow?	Extra training data	HMDB51	UCF101
IDT [30]	✓		61.70	–
Two-Stream [24]	✓	ImageNet	59.40	88.00
Two-stream Fusion + IDT [8]	✓	ImageNet	69.20	93.50
ActionVlad + IDT [11]	✓	ImageNet	69.80	93.60
TSN [33]	✓	ImageNet	71.00	94.90
RSTAN + IDT [6]	✓	ImageNet	79.90	95.10
TSM [18]		Kinetics-400	73.50	95.90
R(2+1)D [28]		Kinetics-400	74.50	96.80
R(2+1)D [28]	✓	Kinetics-400	78.70	97.30
I3D [1]	✓	Kinetics-400	80.90	97.80
MARS + RGB + Flow [3]	✓	Kinetics-400	80.90	**98.10**
FcF [23]		Kinetics-400	81.10	–
ResNeXt101	✓	Kinetics-400	81.78	97.46
EvaNet [22]	✓	Kinetics-400	82.3	–
HAF+BoW/FV halluc [31]		Kinetics-400	82.48	–
ResNeXt101 BERT **(Ours)**	✓	Kinetics-400	**83.55**	97.87
R(2+1)D (32f)		IG65M	80.54	98.17
R(2+1)D BERT (32f) **(Ours)**		IG65M	83.99	98.65
R(2+1)D BERT (64f) **(Ours)**		IG65M	**85.10**	**98.69**

5 Conclusions

This study combines the two major components from AR literature, namely late temporal modeling and 3D convolution. Although there are many pooling, fusion and recurrent modeling strategies that are applied to the features from 2D CNN architectures, we firmly believe that this manuscript is the first study that removes temporal global average pooling (TGAP) and better employs temporal information at the output of 3D CNN architectures. To utilize these temporal features, an attention-based mechanism called BERT is selected. The effectiveness of this idea is proven on most of the popular 3D CNN architectures which are ResNeXt, I3D, SlowFast and R(2+1)D. In addition, significant improvements over the-state-of-the-art techniques are obtained in HMDB51 and UCF101 datasets.

Additionally as a future work, unsupervised concepts can still be proposed on BERT 3D CNN architectures, since the real benefits of BERT architecture rises to the surface with unsupervised techniques.

Acknowledgments. This work was supported by an Institutional Links grant under the Newton-Katip Celebi partnership, Grant No. 217M519 by the Scientific and Technological Research Council of Turkey (TUBITAK) and ID [352335596] by British Council, UK. The numerical calculations reported in this paper were partially performed at TUBITAK ULAKBIM, High Performance and Grid Computing Center (TRUBA resources).

References

1. Carreira, J., Zisserman, A.: Quo Vadis, action recognition? A new model and the kinetics dataset. In: Proceedings - 30th IEEE Conference on Computer Vision and Pattern Recognition, CVPR 2017, vol. 2017-Janua, pp. 4724–4733. Institute of Electrical and Electronics Engineers Inc. (2017). https://doi.org/10.1109/CVPR.2017.502
2. Chen, Y., Kalantidis, Y., Li, J., Yan, S., Feng, J.: Multi-fiber networks for video recognition. In: Ferrari, V., Hebert, M., Sminchisescu, C., Weiss, Y. (eds.) ECCV 2018. LNCS, vol. 11205, pp. 364–380. Springer, Cham (2018). https://doi.org/10.1007/978-3-030-01246-5_22
3. Crasto, N., Weinzaepfel, P., Alahari, K., Schmid, C.: MARS: motion-augmented RGB stream for action recognition. In: Proceedings of the IEEE Computer Society Conference on Computer Vision and Pattern Recognition, vol. 2019-June, pp. 7874–7883. IEEE Computer Society, June 2019. https://doi.org/10.1109/CVPR.2019.00807
4. Devlin, J., Chang, M.W., Lee, K., Toutanova, K.: BERT: pre-training of Deep Bidirectional Transformers for Language Understanding, October 2018. http://arxiv.org/abs/1810.04805
5. Donahue, J., et al.: Long-term recurrent convolutional networks for visual recognition and description. IEEE Trans. Pattern Anal. Mach. Intell. **39**(4), 677–691 (2017). https://doi.org/10.1109/TPAMI.2016.2599174
6. Du, W., Wang, Y., Qiao, Y.: Recurrent spatial-temporal attention network for action recognition in videos. IEEE Trans. Image Process. **27**(3), 1347–1360 (2018). https://doi.org/10.1109/TIP.2017.2778563
7. Feichtenhofer, C., Fan, H., Malik, J., He, K.: Slowfast networks for video recognition. In: Proceedings of the IEEE International Conference on Computer Vision. vol. 2019-October, pp. 6201–6210. Institute of Electrical and Electronics Engineers Inc. (2019). https://doi.org/10.1109/ICCV.2019.00630
8. Feichtenhofer, C., Pinz, A., Zisserman, A.: Convolutional Two-Stream Network Fusion for Video Action Recognition. In: Proceedings of the IEEE Computer Society Conference on Computer Vision and Pattern Recognition. vol. 2016-Decem, pp. 1933–1941. IEEE Computer Society (2016). https://doi.org/10.1109/CVPR.2016.213
9. Ghadiyaram, D., Feiszli, M., Tran, D., Yan, X., Wang, H., Mahajan, D.: Large-scale weakly-supervised pre-training for video action recognition. Proceedings of the IEEE Computer Society Conference on Computer Vision and Pattern Recognition 2019-June, pp. 12038–12047 (2019), http://arxiv.org/abs/1905.00561
10. Girdhar, R., Carreira, J., Doersch, C., Zisserman, A.: Video action transformer network. In: Proceedings of the IEEE Computer Society Conference on Computer Vision and Pattern Recognition 2019-June, pp. 244–253, December 2018. http://arxiv.org/abs/1812.02707
11. Girdhar, R., Ramanan, D., Gupta, A., Sivic, J., Russell, B.: ActionVLAD: learning spatio-temporal aggregation for action classification. In: Proceedings - 30th IEEE Conference on Computer Vision and Pattern Recognition, CVPR 2017 (2017). https://doi.org/10.1109/CVPR.2017.337
12. Hara, K., Kataoka, H., Satoh, Y.: Can spatiotemporal 3D CNNs retrace the history of 2D CNNs and ImageNet? In: Proceedings of the IEEE Computer Society Conference on Computer Vision and Pattern Recognition, pp. 6546–6555. IEEE Computer Society (2018). https://doi.org/10.1109/CVPR.2018.00685

13. Hendrycks, D., Gimpel, K.: Bridging nonlinearities and stochastic regularizers with Gaussian error linear units. CoRR abs/1606.08415 (2016). http://arxiv.org/abs/1606.08415
14. Howard, A.G., et al.: MobileNets: efficient convolutional neural networks for mobile vision applications (2017). http://arxiv.org/abs/1704.04861
15. Karpathy, A., Toderici, G., Shetty, S., Leung, T., Sukthankar, R., Li, F.F.: Large-scale video classification with convolutional neural networks. In: Proceedings of the IEEE Computer Society Conference on Computer Vision and Pattern Recognition (2014). https://doi.org/10.1109/CVPR.2014.223
16. Kuehne, H., Jhuang, H., Garrote, E., Poggio, T., Serre, T.: HMDB: a large video database for human motion recognition. In: Proceedings of the IEEE International Conference on Computer Vision, pp. 2556–2563 (2011).https://doi.org/10.1109/ICCV.2011.6126543
17. Li, Z., Gavrilyuk, K., Gavves, E., Jain, M., Snoek, C.G.: VideoLSTM convolves, attends and flows for action recognition. Comput. Vis. Image Underst. **166**, 41–50 (2018). https://doi.org/10.1016/j.cviu.2017.10.011
18. Lin, J., Gan, C., Han, S.: TSM: temporal shift module for efficient video understanding. In: Proceedings of the IEEE International Conference on Computer Vision 2019-October, pp. 7082–7092 (2018). http://arxiv.org/abs/1811.08383
19. Loshchilov, I., Hutter, F.: Decoupled weight decay regularization. In: 7th International Conference on Learning Representations, ICLR 2019 (2017). http://arxiv.org/abs/1711.05101
20. Luo, W., Li, Y., Urtasun, R., Zemel, R.: Understanding the effective receptive field in deep convolutional neural networks. In: Advances in Neural Information Processing Systems, pp. 4905–4913 (2017). http://arxiv.org/abs/1701.04128
21. Ng, J.Y.H., Hausknecht, M., Vijayanarasimhan, S., Vinyals, O., Monga, R., Toderici, G.: Beyond short snippets: deep networks for video classification. In: Proceedings of the IEEE Computer Society Conference on Computer Vision and Pattern Recognition, 07–12 June 2015, pp. 4694–4702. IEEE Computer Society (2015). https://doi.org/10.1109/CVPR.2015.7299101
22. Piergiovanni, A., Angelova, A., Toshev, A., Ryoo, M.S.: Evolving space-time neural architectures for videos. In: Proceedings of the IEEE International Conference on Computer Vision, October 2019, pp. 1793–1802 (2018). http://arxiv.org/abs/1811.10636
23. Piergiovanni, A., Ryoo, M.S.: Representation flow for action recognition. In: Proceedings of the IEEE Computer Society Conference on Computer Vision and Pattern Recognition, June 2019, pp. 9937–9945 (2018). http://arxiv.org/abs/1810.01455
24. Simonyan, K., Zisserman, A.: Two-stream convolutional networks for action recognition in videos. In: Advances in Neural Information Processing Systems, vol. 1, pp. 568–576. Neural information processing systems foundation (2014)
25. Soomro, K., Zamir, A.R., Shah, M.: UCF101: a dataset of 101 human actions classes from videos in the wild (2012). http://arxiv.org/abs/1212.0402
26. Tran, D., Bourdev, L., Fergus, R., Torresani, L., Paluri, M.: Learning spatiotemporal features with 3D convolutional networks. In: Proceedings of the IEEE International Conference on Computer Vision (2015). https://doi.org/10.1109/ICCV.2015.510
27. Tran, D., Wang, H., Feiszli, M., Torresani, L.: Video classification with channel-separated convolutional networks. In: Proceedings of the IEEE International Conference on Computer Vision, vol. 2019-Octob, pp. 5551–5560. Institute of Electrical

and Electronics Engineers Inc. (2019). https://doi.org/10.1109/ICCV.2019.00565. http://arxiv.org/abs/1904.02811
28. Tran, D., Wang, H., Torresani, L., Ray, J., Lecun, Y., Paluri, M.: A closer look at spatiotemporal convolutions for action recognition. In: Proceedings of the IEEE Computer Society Conference on Computer Vision and Pattern Recognition, pp. 6450–6459. IEEE Computer Society (2018). https://doi.org/10.1109/CVPR.2018.00675
29. Vaswani, A., et al.: Attention is all you need. In: Advances in Neural Information Processing Systems, vol. 2017-Decem, pp. 5999–6009. Neural Information Processing Systems Foundation (2017)
30. Wang, H., Schmid, C.: Action recognition with improved trajectories. In: Proceedings of the IEEE International Conference on Computer Vision (2013). https://doi.org/10.1109/ICCV.2013.441
31. Wang, L., Koniusz, P., Huynh, D.Q.: Hallucinating IDT descriptors and I3D optical flow features for action recognition with CNNs. Proceedings of the IEEE International Conference on Computer Vision, October 2019, pp. 8697–8707 (2019). http://arxiv.org/abs/1906.05910
32. Wang, L., Xiong, Y., Wang, Z., Qiao, Y.: Towards good practices for very deep two-stream ConvNets (2015). http://arxiv.org/abs/1507.02159
33. Wang, L., et al.: Temporal segment networks for action recognition in videos (2018). https://doi.org/10.1109/TPAMI.2018.2868668
34. Wang, X., Girshick, R., Gupta, A., He, K.: Non-local Neural Networks. In: Proceedings of the IEEE Computer Society Conference on Computer Vision and Pattern Recognition, pp. 7794–7803. IEEE Computer Society (2018). https://doi.org/10.1109/CVPR.2018.00813
35. Xie, S., Sun, C., Huang, J., Tu, Z., Murphy, K.: Rethinking spatiotemporal feature learning: speed-accuracy trade-offs in video classification. In: Ferrari, V., Hebert, M., Sminchisescu, C., Weiss, Y. (eds.) ECCV 2018. LNCS, vol. 11219, pp. 318–335. Springer, Cham (2018). https://doi.org/10.1007/978-3-030-01267-0_19
36. Zach, C., Pock, T., Bischof, H.: A duality based approach for realtime TV-L1 optical flow. In: Hamprecht, F.A., Schnörr, C., Jähne, B. (eds.) DAGM 2007. LNCS, vol. 4713, pp. 214–223. Springer, Heidelberg (2007). https://doi.org/10.1007/978-3-540-74936-3_22

Correction to: Real-Time Detection of Multiple Targets from a Moving 360° Panoramic Imager in the Wild

Boyan Yuan and Nabil Belbachir

Correction to:
Chapter "Real-Time Detection of Multiple Targets from a Moving 360° Panoramic Imager in the Wild" in: A. Bartoli and A. Fusiello (Eds.): *Computer Vision – ECCV 2020 Workshops*, LNCS 12539, https://doi.org/10.1007/978-3-030-68238-5_8

In the originally published version of chapter 8 an acknowledgement was missing. This has been corrected.

The updated version of this chapter can be found at https://doi.org/10.1007/978-3-030-68238-5_8

A. Bartoli and A. Fusiello (Eds.): ECCV 2020 Workshops, LNCS 12539, p. C1, 2021.
https://doi.org/10.1007/978-3-030-68238-5_49

Author Index

Printed in the United States
by Baker & Taylor Publisher Services